HUMAN
MICROANATOMY

HUMAN MICROANATOMY
Cell Tissue and Organ Histology with Celebrity Medical Histories

STEPHEN A. STRICKER

CRC Press
Taylor & Francis Group
Boca Raton London New York

CRC Press is an imprint of the
Taylor & Francis Group, an **informa** business

First edition published 2022
by CRC Press
2 Park Square, Milton Park, Abingdon, Oxon, OX14 4RN

and by CRC Press
6000 Broken Sound Parkway NW, Suite 300, Boca Raton, FL 33487-2742

CRC Press is an imprint of Informa UK Limited

British Library Cataloguing-in-Publication Data
A catalogue record for this book is available from the British Library

ISBN: 978-0-367-77184-3 (hbk)
ISBN: 978-0-367-36457-1 (pbk)
ISBN: 978-0-429-35330-7 (ebk)

DOI: 10.1201/9780429353307

Typeset in Utopia
by KnowledgeWorks Global Ltd.

With thanks and love to the two best people I know: S.M.D.S. and J.M.S.

Contents

Preface

Microanatomy, which is also termed histology, constitutes a key discipline that connects molecular and cellular fields at one end of the biological spectrum with anatomy and physiology at the other end. Over the years while teaching histology, it became clear that students were particularly interested in microanatomical topics that were related to ailments suffered by famous people. Thus, this book includes short medical histories of deceased celebrities to help provide context for accompanying descriptions of normally functioning cells, tissues, and organs. Such accounts of normal microanatomy routinely emphasize medically relevant information while also covering developmental and evolutionary correlates that are seldom addressed in other histology texts, and throughout the book, microanatomical features are illustrated in over 1400 drawings and microscopic images that are supplemented by a digital atlas of additional light and electron micrographs. To help synthesize a robust understanding of human histology, each chapter also includes a preview, pictorial summary, and self-study quiz. Alternatively, readers can obtain a quick overview of microanatomy by simply focusing on chapter previews, summaries, and figure captions. Collectively, the various novel features of this book aim to provide an engaging and comprehensive analysis of how the human body functions from the subcellular to organ level of organization.

Stephen A. Stricker
Friday Harbor, WA

Acknowledgments

Instead of completely re-inventing the wheel of histological illustrations, various drawings and micrographs already published in the literature are included here as supplements to the numerous author-generated images and newly commissioned line drawings that were produced for this project. In particular, depictions of anatomy greatly benefit from figures that are reproduced via creative commons licenses from the Openstax Anatomy and Physiology online textbook (OpenStax College [2013] *Anatomy and Physiology*. OpenStax. http://cnx.org/content/col11496/latest) and the Blausen Medical gallery of images (Blausen.com staff [2014] "Medical gallery of Blausen Medical 2014". WikiJournal of Medicine 1 (2): 10. doi:10.15347/wjm/2014.010). Both of these sources deserve to be commended for their highly informative and beautifully crafted illustrations.

In addition, many colleagues were extremely kind and generous to donate images to the cause. Such contributions as well as insightful and helpful comments regarding the text were provided by: Drs. F. Alessandrini, D. Bohorquez, V. Baena, V. Borelli, E. Bossen, A. Buckley, T. Chakraborty, A. Ceylan, R. Cripps, T. Finger, H. Florman, C. Franzini-Amstrong, A. Giessl, B. Giepmans, A. Hand, T. Iwanaga, H. Jastrow, K. Kadler, S. Kempf, W. Lamers, W. Landis, P. Lewis, D. Longnecker, T. Lovato, P. Mansfield, K. Meek, M. Morroni, M. Ochs, M. Raspanti, C. Recordati, F. Shapiro, G. Shinn, C. Slater, A. Suen, W. Sun, H. Takahashi-Iwanaga, M. Terasaki, J. Verlander, H. Wartenberg, S. Weiner, G. Wessel, T. White, C. Williams, D. Willows, and J. Wilting. Although these scientists have greatly improved the book, there are undoubtedly further corrections and enhancements that need to be made. Such defects are solely the fault of the author, who welcomes comments and suggestions to help upgrade possible future editions (StephenStrickerHM1e@gmail.com).

A project like this could not have been completed without all the professional and helpful assistance received from the Taylor & Francis Company and its subcontractors. In particular, the excellent work done by Mansi Agarwal, Chuck Crumly, Ana Eberhart, Patrick Lane, Kyle Myer, and Jordan Wearing is greatly appreciated. The author also thanks Drs. M. Cavey, R. Cloney, E. Kozloff, G. Schatten, and T. Schroeder for their expert instruction in histological and EM techniques during the author's graduate and postdoctoral training. Finally, two fellow histologists, Dr. G. Shinn and the late Dr. C. Reed, deserve a loud shout-out for all their enthusiasm during the many hours we shared over the years investigating the fascinating world of microanatomy.

Front Cover Figure Citations

Top row, left photograph: John Kennedy; downloaded from: https://commons.wikimedia.org/wiki/File:John_F._Kennedy,_White_House_color_photo_portrait.jpg; public domain image created by a federal employee as part of his/her duties.

Top row, middle color light micrograph: original image.

Top row, right TEM: adrenal gland; image courtesy of Drs. H. Jastrow and H. Wartenberg, reproduced with permission from Jastrow's Electron Microscopic Atlas (htttp://www.drjastrow.de).

Middle row, left light micrograph: cardiac myocyte T-tubules; from Guo, A et al. (2013) Emerging mechanisms of T-tubule remodeling in heart failure. Cardiovas Res 98: 204–215 reproduced with publisher permission.

Middle row, middle TEM: cardiac myocytes; image courtesy of Drs. H. Jastrow and H. Wartenberg reproduced with permission from Jastrow's Electron Microscopic Atlas (htttp://www.drjastrow.de).

Middle row, right photograph: Mother Teresa; image by Manfredo Ferrari and reproduced by a creative commons license; downloaded from https://commons.wikimedia.org/wiki/File:Mutter_Teresa_von_Kalkutta.jpg.

Bottom row, left photograph: Idi Amin; downloaded from: https://commons.wikimedia.org/wiki/File:Idi_Amin_-Archives_New_Zealand_AAWV_23583,_KIRK1,_5(B),_R23930288.jpg, reproduced under a Creative Commons 2.0 license.

Bottom row, middle color light micrograph: original image.

Bottom row, right SEM: renal corpuscle; from: Miyaki, T. et al. (2020) Three-dimensional imaging of podocyte ultrastructure using FE-SEM and FIB-SEM tomography. Cell Tissue Res 379: 245–254 https://doi.org/10.1007/s00441-019-03118-3, reproduced under a Creative Commons 4.0 license.

Online Digital Atlas and Instructor Resources

ONLINE DIGITAL ATLAS

A supplementary digital atlas of informative histological images has been compiled as a free resource available from the 'Support Material' section on this book's product page: https://www.routledge.com/Human-Microan-atomy-Cell-Tissue-and-Organ-Histology-with-Celebrity-Medical/Stricker/p/book/9780367364571.

The 300+ micrographs of the atlas include not only alternative views of structures already depicted in the text but also images of various microanatomical features not illustrated in the printed chapters. In combination with the 1400 figures available in the text, these supplemental digital images provide a comprehensive set of illustrations illuminating the microanatomical organization of cells, tissues, and organs. Students and instructors are encouraged to download this free resource, which will provide a deeper understanding of the fascinating world of anatomy invisible to the naked eye.

INSTRUCTOR RESOURCES

Qualified instructors can also register to download the text's illustrations in PDF or PowerPoint format. Such files are available in the 'Instructor Resources Download Hub' on the book's product page: https://www.routledge.com/Human-Microan-atomy-Cell-Tissue-and-Organ-Histology-with-Celebrity-Medical/Stricker/p/book/9780367364571.

CELL AND TISSUE HISTOLOGY—OVERVIEW OF HISTOLOGY, CELL STRUCTURE, AND TISSUES

PART ONE

Introduction to Human Microanatomy and its Associated Techniques

---

PREVIEW

1.1 **Human microanatomy** and the general organization of this textbook

1.2 **Cells, size scales**, and **limits of resolution** in **light microscopy (LM)** vs. **electron microscopy (EM)**

1.3 Imaging **live whole mounts** by **LM** and **preparing fixed sections** for **LM** and **TEM**

1.4 **Staining sections, histological artifacts,** and **interpreting microanatomical images**

1.5 Identifying **specific components** in histological samples: **polarization microscopy, autoradiography,** and **indirect immunofluorescence**

1.6 Methods for **reconstructing three-dimensional morphology** from the subcellular to organ level of organization

1.7 **Medical applications** of microanatomical analyses

1.8 Summary and self-study questions

1.1 HUMAN MICROANATOMY AND THE GENERAL ORGANIZATION OF THIS TEXTBOOK

During animal evolution, individual **cells** and their surrounding networks of molecules that constitute the **extracellular matrix** (**ECM**) have aggregated to form functionally integrated assemblages, called **tissues** (**Figure 1.1a**). Tissues, in turn, join together to create **organs** (**Figure 1.1a**), each of which comprises a discrete collection of interacting tissues with differing functional properties and developmental origins (Chapter 3). The structure and function of cells, tissues, and organs are the focus of **histology** (= Greek: "*histo*" [woven, web, tissue] + "*logos*" [study of]). Alternatively, because histological studies often rely on microscopes to analyze anatomical components, histology is also known as microscopic anatomy or **microanatomy**.

This book covers the fundamentals of **human microanatomy** from a biologically oriented point of view that incorporates **developmental** and **evolutionary** perspectives into its descriptions of functional morphology. Although medically relevant information is routinely presented, detailed depictions of pathological microanatomy are seldom included. Instead, this text aims to provide a concise account of the **normal structure and function** of cells, tissues, and organs. Accordingly, overviews of cell and tissue biology presented in Chapters 2 and 3 focus on essential background information needed for subsequent chapters covering specific cells and tissues in normally functioning organs. To help provide orientation for descriptions of organ-level microanatomy, many of the included micrographs have stitched together overlapping regions to generate wide-field panoramic views of whole organs or large portions of organs. Thus, owing to size limitations or the unavailability of suitable human material, some images depict the microanatomy of smaller animals such as mice, which nevertheless illustrate histological features resembling those of humans. Similarly, experimental analyses of tissue and organ

DOI: 10.1201/9780429353307-2

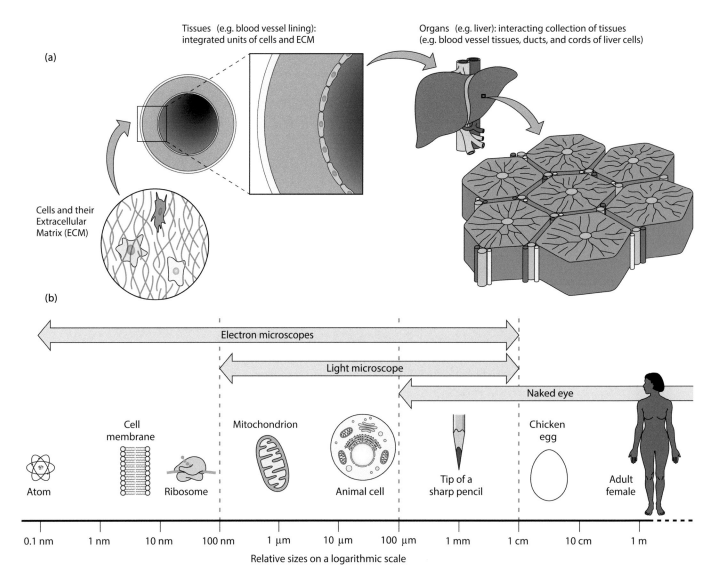

Figure 1.1 Introduction to microanatomy and its size scales. (a) Microanatomy (=histology) analyzes the functional morphology of (i) cells, (ii) tissues (=integrated assemblages of cells plus their extracellular matrix [ECM]), and (iii) organs (=discrete collections of interacting tissues). (b) Some biological structures with conserved size ranges include (i) cells (typically 10–50 μm in diameter), (ii) mitochondria (~0.5–1 μm wide), (iii) ribosomes (~25 nm in diameter), and (iv) cell membranes (=plasma membranes) (7 nm thick).

functioning that are highlighted throughout the text often involve animal models that can provide key insights into how the human body works.

In addition to conventional presentations of microanatomy, short **medical histories** are also included here as a way of lending "human faces" to histology. For such histories, 80 deceased individuals ranging from world-renowned icons to far less-famous figures have been selected in order to assemble a diverse collection of celebrities. Each of these accounts contains non-medically related information that helps flesh out its celebrity beyond simply being the bearer of a specific pathology, and by linking medical conditions to people from varied walks of life, this ancillary material is meant to provide broad context for accompanying descriptions of normal histology.

The rest of this chapter focuses on size scales, limits of resolution, and basic techniques used in microanatomy. In particular, two main methods traditionally employed in microanatomical studies—**light microscopy** (**LM**) and **electron microscopy** (**EM**)—are described while also summarizing some of the more modern ways of analyzing cell, tissue, and organ histology. Such introductory material is not intended to present in-depth explanations of the highlighted techniques but rather to offer some practical insights into the capabilities and limitations of these methods that can help with interpretations of microanatomical images included throughout the text.

1.2 CELLS, SIZE SCALES, AND LIMITS OF RESOLUTION IN LIGHT MICROSCOPY (LM) VS. ELECTRON MICROSCOPY (EM)

Recent estimates indicate that the adult human body is composed of **~30 trillion cells** plus a similar number of symbiotic bacteria and other microbes that constitute the body's **microbiome**. Although there is no universally accepted definition of what comprises distinct cell types, a figure of ~200 kinds of human cells is often quoted, with red blood cells accounting for the vast majority of human cells, followed distantly by such examples as bone marrow cells, endothelial cells of blood vessels, and neural cells (Table 1.1).

In spite of such diversity, human cells are fairly uniform in size, with most of their diameters measuring **~10–50 μm**. Some cells such as neurons in the sciatic nerve can extend narrow axonal projections up to a meter in length. However, surface-to-volume ratios required for efficient diffusion-based processes constrain the thickness of cells. This, in turn, limits typical cellular widths to ~100–120 μm, which is the size of fully grown human eggs and large versions of dorsal root ganglion cells near the spinal cord. Conversely, diameters of ~5 μm in a few human cells, such as sperm or granular cells of the brain represent the minimum size for housing a nucleus plus some cytoplasmic organelles. For context, the **50-μm** value for the upper end of typical cell sizes approximates the average width of **scalp hairs** in adult humans. Moreover, subcellular components with highly conserved sizes across numerous cell types include such examples as: (i) **mitochondria**, which are usually **~0.5–1 μm wide**, (ii) **ribosomes** with their **~25-nm diameters**, and (iii) **cell membranes** (=plasma membranes) that are routinely **7 nm thick** (**Figure 1.1b**).

A single human hair placed on a contrasting background can be easily **detected** by an unaided human eye. However, if two closely positioned hairs are separated from each other by only 50 μm, human vision cannot discriminate between two hairs vs. only a single thick hair being present, because the eye cannot **resolve** such narrow spacing. Put in another way, 50 μm is beyond the eye's **limit of resolution**, which is the ability to discern two closely apposed objects as being separate. Although the limit of resolution for human eyes depends on factors, such as illumination intensity and object-to-retina distance, **0.1–0.2 mm** (100–200 μm) represents a general approximation of this limit, which, in turn, prevents unaided eyes from resolving structures in the size range of not only cellular components but also most cells.

Thus, the exploration of cellular and subcellular morphology had to wait until the 17th century, when newly invented **light microscopes** (**LMs**) allowed pioneering microscopists to discover unicellular microbes and various features of plant and animal cells (Box 1.1 A Brief History of Biological Microscopy) (**Case 1.1 Pioneering Histologist, Marcello Malpighi**). Subsequent advances in optical physics provided the theoretical framework that helped drive further improvements in microscope design, fabrication, and capabilities. For example, during the late 1800s, German physicist Ernst Abbe defined the **minimal resolvable distance**, **d**, between two closely placed

TABLE 1.1 HUMAN CELL TOTALS

Cell type	Estimated % of ~30 trillion cells in adult human
Erythrocytes	84
Platelets	4.9
Bone marrow cells	2.5
Vascular endothelial cells	2.1
Lymphocytes	1.5
Hepatocytes	0.8
Neurons and glia	0.6
Epidermal cells	0.5
All other cells	~3

Source: Data from Sender, R et al. (2016) Revised estimates for the number of human and bacteria cells in the body. *PLoS ONE Biology* 14: doi:10.1371/journal.pbio.1002533, reproduced under a creative commons license.

BOX 1.1 A BRIEF HISTORY OF BIOLOGICAL MICROSCOPY

Although lens-containing eyeglasses were in use by the year 1300, it was not until 1595 that Dutch spectacle makers Hans and Zacharias Jansen invented a **compound microscope** with its dual lenses—an **objective lens** for imaging the specimen and an eyepiece **ocular lens** for magnifying the image. The actual capabilities of this first microscope remain unknown, as no descriptions of its images have been discovered. Thus, today, Italian astronomer/physicist, Galileo (1564–1642), and English scientist, Robert Hooke (1635–1703), are commonly viewed as the key developers of compound microscopes for the versions that they built, used, and documented during the 17th century (**Figure a and b**). In addition, while Hooke was examining

various biological specimens, Dutch cloth merchant Antonie van Leeuwenhoek (1632–1732) fabricated hundreds of **simple microscopes**. Such microscopes had only a single imaging lens and yet outperformed early compound microscopes until eventually the degrading effects of lens aberrations inherent in compound microscopes began to be corrected in the early 19th century. Thus, with resolving capacities of ~1 µm, van Leeuwenhoek's microscopes allowed him to describe, for example, **bacteria** and **sperm** cells. Following these early years of microscopy, various advances over the last three centuries have helped shape the current capabilities of biological imaging. Some of these **milestones are summarized in Figure c.**

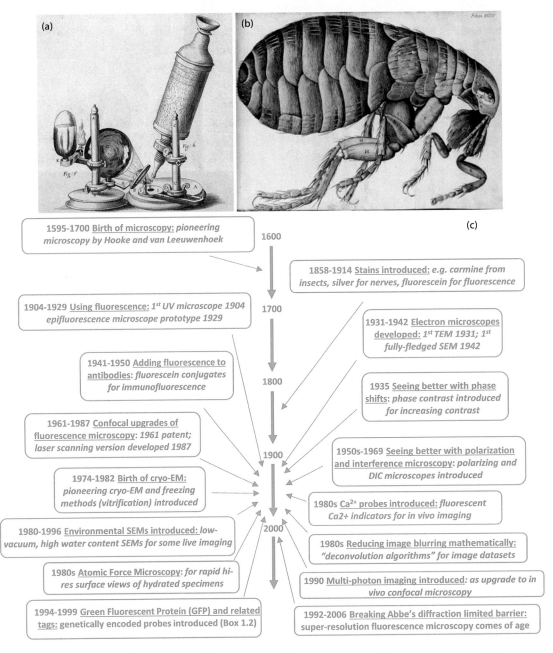

(a)

(b)

(c)

1595-1700 <u>Birth of microscopy:</u> *pioneering microscopy by Hooke and van Leeuwenhoek*

1600

1858-1914 <u>Stains introduced:</u> *e.g. carmine from insects, silver for nerves, fluorescein for fluorescence*

1904-1929 <u>Using fluorescence:</u> *1st UV microscope 1904 epifluorescence microscope prototype 1929*

1700

1931-1942 <u>Electron microscopes developed:</u> *1st TEM 1931; 1st fully-fledged SEM 1942*

1941-1950 <u>Adding fluorescence to antibodies:</u> *fluorescein conjugates for immunofluorescence*

1800

1935 <u>Seeing better with phase shifts:</u> *phase contrast introduced for increasing contrast*

1961-1987 <u>Confocal upgrades of fluorescence microscopy:</u> *1961 patent; laser scanning version developed 1987*

1900

1950s-1969 <u>Seeing better with polarization and interference microscopy:</u> *polarizing and DIC microscopes introduced*

1974-1982 <u>Birth of cryo-EM:</u> *pioneering cryo-EM and freezing methods (vitrification) introduced*

1980s <u>Ca²⁺ probes introduced:</u> *fluorescent Ca2+ indicators for in vivo imaging*

1980-1996 <u>Environmental SEMs introduced:</u> *low-vacuum, high water content SEMs for some live imaging*

2000

1980s <u>Reducing image blurring mathematically:</u> *"deconvolution algorithms" for image datasets*

1980s <u>Atomic Force Microscopy:</u> *for rapid hi-res surface views of hydrated specimens*

1990 <u>Multi-photon imaging introduced:</u> *as upgrade to in vivo confocal microscopy*

1994-1999 <u>Green Fluorescent Protein (GFP) and related tags:</u> *genetically encoded probes introduced (Box 1.2)*

1992-2006 <u>Breaking Abbe's diffraction limited barrier:</u> *super-resolution fluorescence microscopy comes of age*

Box 1.1 (a, b) *From R. Hooke's Micrographia 1665: a drawing of his microscope and a flea that he had examined.* **(c)** *A summary of major breakthroughs in biological imaging over the last several centuries.*

CASE 1.1 PIONEERING HISTOLOGIST, MARCELLO MALPIGHI

(1628–1694)

"... to no one ... observing the structure of the human body does Nature seem to have revealed her secrets as fully as to her beloved Malpighi"

(From a 1668 letter written by a member of the Royal Society of London extolling M.M.'s masterful research)

Often referred to as the "Father of Histology", Marcello Malpighi was a 17th-century physician and scientist who used early versions of microscopes to conduct pioneering studies of microanatomy. Malpighi was born in present-day Italy and after obtaining degrees in medicine and anatomy was offered a professorship at the University of Pisa in 1656. While there and at other positions during his nearly 40-year career, Malpighi made key discoveries regarding the microscopic structure

and function of organs not only in adult animals and plants but also in developing embryos, perhaps becoming best known for describing the circulation of red blood cells through capillaries. In recognition of his groundbreaking work, Malpighi had several structures named in his honor, including Malpighian corpuscles in the kidney and spleen. Late in life, Malpighi had to deal with years of poor health as well as a fire at his home that destroyed his manuscripts and equipment, before eventually dying of a stroke in 1694.

objects as **d = 0.61 λ/NA**, where 0.61 is a constant; λ is the wavelength of the illuminating source (i.e., visible light for LM); and NA (<u>n</u>umerical <u>a</u>perture) is a measure of the light-gathering capacity of lenses. In order to enhance resolution so that smaller d spacings can be resolved, Abbe's equation reveals that either λ in the numerator **must be decreased** or **NA** in the denominator **must be increased**. For visible light, the effective lower limit of λ that can be readily transmitted through glass-lens-containing LMs is ~500 nm (~0.5 μm). Similarly, the highest NA of typical LM lenses is on the order of 1.4. Thus, the **limit of resolution for a conventional LM** is calculated as d ~ 0.61 × 0.5 μm/1.4, which corresponds to ~**0.2 μm** (=200 nm)—a value that is also referred to as the diffraction-limited resolving capacity of standard LMs. Even though in recent years, various modes of super-resolution LMs (pg. 19) have been able to resolve structures substantially smaller than the value defined by Abbe's equation, analyses of subcellular features were for several centuries constrained by the ~0.2-μm limit that light diffraction imposes. Thus, LM was able to resolve 0.5-wide mitochondria by the end of the 19th century, but smaller structures like ribosomes or cell membranes remained beyond the resolving capacity of LMs, even in <u>light micrograph</u> images (also called LMs) produced by modern-day versions of conventional light microscopes.

As an alternative to light-based methods, **electron microscopes** (**EMs**) (Box 1.1) began in the 1930s to image specimens via accelerated beams of **electrons**. To minimize the scattering effects of gas molecules in air, such beams are focused by electromagnetic lenses within evacuated chambers, and due to their short wavelengths, electrons reduce the λ value in Abbe's equation to yield an ~1000-fold increase in resolving capacity over that of a typical LM. Thus, depending on the particular EM mode that is used, **resolutions** for biological samples slightly above or below **0.2 nm** (=0.0002 μm) can be achieved. With this increased resolving capacity, EMs are able to analyze subcellular (=ultrastructural) features, including, for example, the precise morphologies of ribosomes and cell membranes. However, because electrons, rather than light rays, are used to produce EM images, the numerous hues of color that can be captured by LM are lacking. Instead, EM images characteristically appear as shades of gray (i.e., black-and-white images), unless a false pseudocoloring is computationally added to the original monochromatic version.

Two basic kinds of EMs—**transmission electron microscopes** (**TEMs**) and **scanning electron microscopes** (**SEMs**)—are used to generate ultrastructural images, which are likewise termed TEMs and SEMs for transmission electron micrographs and scanning electron micrographs, respectively (**Figure 1.2a–e**). To produce TEMs, the **electron beam is transmitted through the sample**, thereby requiring thin and dehydrated specimens for effective beam propagation. Accordingly, TEM samples are typically cut on a highly precise sectioning device (=**microtome** or **ultramicrotome**) as ~70-nm **thin sections** that are gathered on a perforated TEM **grid** (**Figure 1.3a–d**). Prior to being inserted in the evacuated TEM chamber, grid-supported thin sections are stained with heavy metals like lead and uranium to increase specimen contrast, thereby yielding light vs. dark areas in 2-dimensional (2D) TEM images (**Figure 1.2b**).

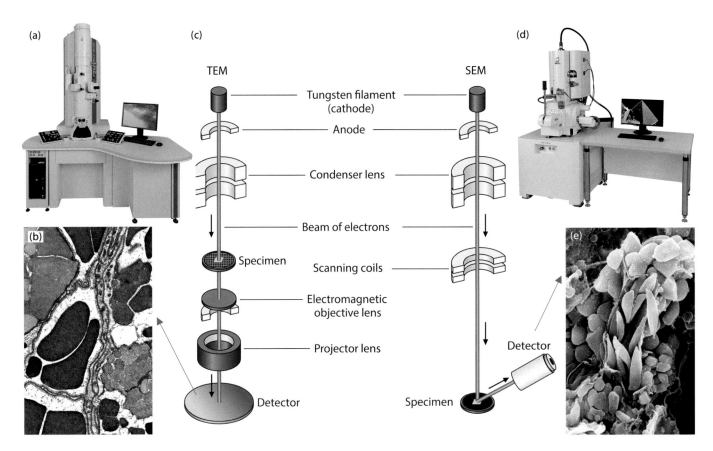

Figure 1.2 Transmission electron microscopy (TEM) vs. scanning electron microscopy (SEM). (a–e) As illustrated in the EM schematics and representative images of glandular cells, TEMs such as the JEOL JEM-2100Plus in figure (a) utilize an evacuated chamber with electromagnetic focusing lenses to transmit an electron beam through thin specimens, thereby generating 2D images (b). Alternatively, SEMs like the JEOL JSM-IT800 in figure (d) scan the beam over specimen surfaces to produce 3D images with enhanced depth of field (e). EM photos courtesy of JEOL USA; (b) 8,000X; (e) 3,200X.

Alternatively, **scanning electron microscopes scan electrons over the surface** of specimens to generate 3-dimensional (3D) images which not only enhance depths of focus but also provide larger fields of view than those obtained by TEMs (**Figure 1.2e**). For SEM imaging, biological samples are routinely dehydrated via a method, called **critical point drying (CPD)** in order to avoid distortions caused by other modes of removing water. The dried specimens are then coated with heavy metals before being placed in the SEM's evacuated chamber and scanned in an orderly fashion by the electron beam. Such scanning dislodges relatively low-energy electrons, called secondary electrons (SEs), which are collected by a secondary electron detector and converted into surface images. In addition, with proper ancillary equipment, SEMs can collect higher-energy backscattered electrons (BSEs) that are primarily emitted by heavier atoms within the sample, thereby allowing identifications of subregions exhibiting differing atomic weights. Similarly, with energy-dispersive X-ray (EDX) spectroscopy, SEM can quantify elemental compositions in different regions of probed samples (**Figure 1.4**).

Each of these EM modes has its own **advantages** and **disadvantages** (Table 1.2). For example, TEM offers a resolving capacity down to the atomic level and can thus provide the most precise imaging of subcellular and molecular architecture. However, except in such cases as when films of molecules are placed on special grids, conventional TEM typically requires specimens to be thinly sectioned, and such sectioning can be technically difficult to achieve. Moreover, TEM sections are restricted to relatively narrow and thin subregions of the specimen, thereby complicating elucidations of 3D relationships across broad expanses of cells and tissues. Conversely, SEM generates easy-to-interpret 3D-like images of comparatively wide fields of view. In addition, although most samples examined by either TEM or SEM are dead and dehydrated, some SEMs can allow hardy hydrated specimens to be imaged live over short timeframes by: (i) viewing specimens coated with a protective "nanosuit" of particles in conventional SEMs or (ii) imaging specimens in specialized **environmental**

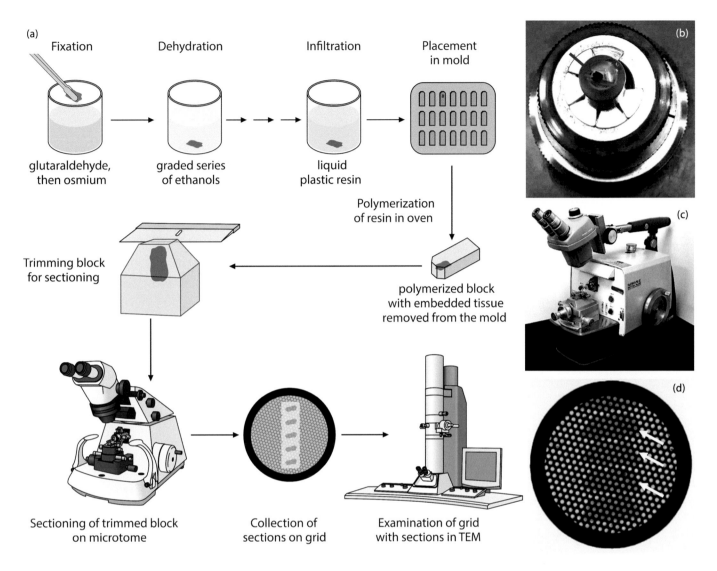

Figure 1.3 Specimen preparation for transmission electron microscopy (TEM). *(a–c) Following fixation, dehydration, and infiltration of small samples, hardened plastic blocks containing a specimen (b, red arrow) are trimmed for sectioning with a microtome (c). (d) The resulting ~70-nm-thick sections (white arrows) are then collected on grids for imaging in a TEM.*

or **atmospheric SEMs** that possess sample chambers with higher pressure and water vapor content than in standard SEMs.

However, most SEMs can neither match the sub-0.2-nm resolution that is obtained by TEM nor easily examine internal compartments within cells and tissues. Thus, TEM and SEM are often used in conjunction with each other to obtain complementary views of cell, tissue, and organ ultrastructure. In addition, hybrid techniques such as scanning transmission electron microscopy (STEM), freeze-etching methods, or SEM combined with serial thin sectioning (pg. 21) can supplement data obtained from traditional EM modes.

As an alternative to EM and its general requirements for examining dehydrated samples under a vacuum, **atomic force microscopy (AFM)** allows simplified imaging of surface structures not only in air but also in aqueous solutions, thereby enabling cellular details to be assessed in their hydrated states. During AFM, a flexible cantilevered probe with a very sharp tip is scanned near untreated sample surfaces while reflected laser light from the top of the probe is sent toward a position-sensitive photodetector (**Figure 1.5a**). As the probe encounters surface irregularities, corresponding deflections in the laser beam are converted into sub-nanometer-resolution topographical images (**Figure 1.5b, c**).

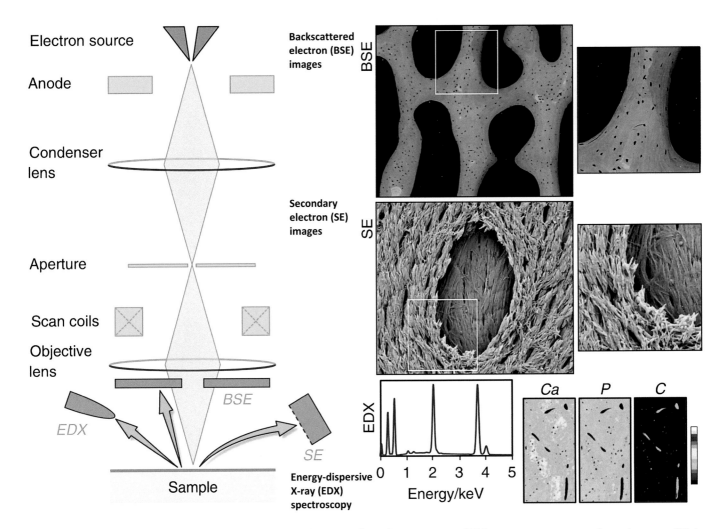

Figure 1.4 Scanning EM (SEM) capabilities. *As shown in these bone images, SEMs can capture secondary electrons (SEs) for traditional surface views or detect backscattered electrons (BSEs) as signals emitted by heavier atoms. In addition, SEMs with energy-dispersive X-ray (EDX) spectroscopy are able to analyze such components as calcium (Ca), phosphorus (P), and even carbon (C). (From Shah, FA et al. (2019) 50 years of scanning electron microscopy of bone—a comprehensive overview of the important discoveries made and insights gained into bone material properties in health, disease, and taphonomy. Bone Res 7:15. https://doi.org/10.1038/s41413-019-0053 and reproduced under a CC BY 4.0 creative commons license.)*

TABLE 1.2 COMPARISON OF TEM VS. SEM

Feature	TEM	SEM
Resolution	High (<0.2 nm)	Typically lower than most TEMs
Field of view	Considerably smaller (a maximum of few millimeters but typically much less)	Large (up to a 1 cm or even more with ancillary attachments)
Type of individual image that is produced	2D	3D-like (i.e., with greater depth of focus)
Ability to view internal cell components	Yes via sectioning	Yes but requires fracturing or ancillary devices inside SEM
Ability to view live, hydrated specimens	Not generally possible with currently available TEMs	Yes with certain kinds of SEMs (e.g., environmental or atmospheric SEMs)
Capacity for backscattered electron detection	Not common	Commonly available
Ease of specimen preparation	Relatively difficult to obtain thin sections	Easier to image whole mounted samples
Energy dispersive X-ray spectroscopy of elemental composition	Yes but not common	Commonly used

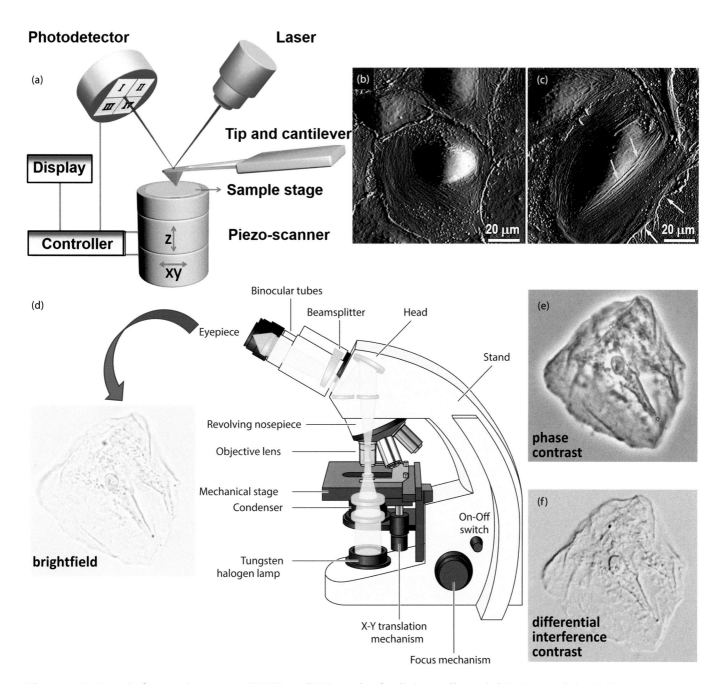

Figure 1.5 Atomic force microscopy (AFM) and LM modes for living cells and thin tissue slabs. (a) *By scanning a cantilevered tip over samples while using a position-sensitive photodetector to collect reflected laser light, AFM can convert tip deflections into images of living, hydrated samples at atmospheric pressure without the fixation, dehydration, and evacuation steps typically required in EM analyses. (**b, c**) AFM images of before (b) and after (c) stimulation of lung cells show formation of stress fibers (c, yellow arrows) near the nucleus. (**d–f**) Figure (d) diagrams the general optical path of a conventional light microscope, whereas images of a live human cheek cell are shown using conventional brightfield optics (d) vs. phase contrast (e) or differential interference contrast (f) methods for enhancing contrast (500X). ([a]: From Zhou, L et al. (2017) Progress in the correlative atomic force microscopy and optical microscopy. Sensors 17, 938. doi:10.3390/s17040938, reproduced under a CC BY 4.0 creative commons license; [b, c]: From Wang, X et al. (2018) The significant role of c-Abl kinase in barrier altering agonists-mediated cytoskeletal biomechanics. Sci Rep 8. doi:10.1038/ s41598-018-19423-w, reproduced under a CC BY 4.0 creative commons license.)*

1.3   IMAGING LIVE WHOLE MOUNTS BY LM AND PREPARING FIXED SECTIONS FOR LM AND TEM

Some live samples can be examined as unsectioned whole mounts using the standard **brightfield** (**BF**) mode of LM imaging that simply illuminates the specimen with unprocessed light (**Figure 1.5d**). However, because morphological details are difficult to discern in most living material, BF microscopy is often supplemented with **phase-contrast** or **differential interference contrast** (**DIC**) optics that employ specialized objective lenses and condensers to alter phase and interference properties of light, which, in turn, generates higher-contrast images (**Figure 1.5e, f**).

Alternatively, living cells and tissues can be labeled with non-toxic (=vital) types of **fluorescent dyes** that absorb specific wavelengths of light and rapidly emit longer wavelengths for detection by **fluorescence LM**. In its simplest form, which is referred to as **wide-field fluorescence microscopy**, both in- and out-of-focus emissions from a wide range of focal planes combine to produce relatively low-resolution images. However, in recent years, wide-field optics have been greatly improved by the introduction of **confocal microscopy** (**CM**) (Box 1.1) and related non-linear imaging modes, such as **multiphoton microscopy** (**MPM**) that allow specimens to be precisely imaged without requiring physical sectioning of the sample (**Figure 1.6a–i**). Confocal microscopy uses a pinhole aperture within the optical path to help eliminate out-of-focus rays emanating from the sample, whereas in MPM, two or more longer-wavelength light rays from a rapidly pulsed laser converge to excite fluorophores only in thin regions of the sample, thereby generating highly localized fluorescent emissions at the plane of focus. These non-invasive methods not only produce high-resolution 2D optical sections (Box 1.2 Imaging Physiological Processes and Ultrastructural Features in Live Cells via Fluorescent Probes) but can also render 3D images from stacks of 2D sections (**Figure 1.6c, d**), which in the case of MPM can penetrate ~1 mm into some living samples before image quality becomes substantially degraded by overlying tissues (**Figure 1.6f, g**).

For bulky or dense materials that are beyond the capabilities of contrast-enhanced or CM/MPM modes of LM, samples are routinely subjected to a series of processing steps in order to generate **fixed** (i.e., dead) **sections** that can be **stained** for subsequent examination by brightfield and/or fluorescence microscopy. To provide expedited processing as is often required in pathology laboratories, pieces of biological samples are simply fixed by rapid freezing and then cut in a refrigerated microtome (=**cryostat**) (**Figure 1.7a–d**). Such sectioning typically yields suboptimal morphological preservation. However, by employing quick freezing without chemical fixatives, cryo-sectioning helps retain the proper reactivities of proteins to antigen-recognizing **antibodies** so that specific sample components can be identified via **immunocytochemistry** protocols (pg. 18).

As opposed to such methods, traditional modes of preparing sections for LM tend to focus on preserving morphological features without necessarily stabilizing antigenicity for immunocytochemistry. Following the same general kind of protocol that is illustrated for TEM processing in **Figure 1.3**, morphology-centered LM methods typically begin with the incubation of up to ~2 cm-wide pieces of tissues and organs in an aldehyde-containing **fixative**, such as a buffered solution of formaldehyde. This step kills the specimen and ideally fixes its various cellular and ECM components not only in configurations that resemble their states prior to fixation but also in forms that help maintain sample integrity during subsequent processing. The fixed sample is then gradually **dehydrated** via a graded series of water-absorbing solutions like ethanol in order to minimize distortions while freeing up space for a liquefied **embedding medium**. After the dehydrated sample is treated with a transitional solvent such as xylene, liquid embedding material is infiltrated into the spaces that were once occupied by water molecules in order to provide support for sectioning. Traditionally, LM samples are infiltrated with molten **paraffin wax**, which, after being cooled, solidifies within and around the specimen to yield firm **paraffin blocks**. Because the sticky nature of paraffin helps link together successive sections into long **ribbons of serial sections**, paraffin sectioning is well suited for 3D reconstructions of sectioned tissues. Such serial sections are typically cut at a thickness of ~6–8 μm on a paraffin microtome that is equipped with a long steel knife for sampling large portions of the specimen (**Figure 1.7e**). Most paraffin sections fail to provide precise views of cellular details. However, methods have been devised for higher resolution imaging either by viewing paraffin sections via SEM or by first imaging such sections with LM before conducting SEM analyses, thereby allowing correlative LM and EM views of sectioned tissues (**Figure 1.7e–h**).

Alternatively, after using different kinds of fixatives and transitional fluids on smaller samples up to ~2 mm-wide, specimens can be embedded in hard **plastic resins** that allow thinner sectioning than is achieved

BOX 1.2 IMAGING PHYSIOLOGICAL PROCESSES AND ULTRASTRUCTURAL FEATURES IN LIVE CELLS VIA FLUORESCENT PROBES

As a way of supplementing microscopical analyses of general cellular morphology, various non-toxic fluorescent dyes can be used in conjunction with fluorescence imaging to examine **specific physiological processes** and **ultrastructural features in living cells**. Such dyes are loaded into cells either by microinjection or by simple incubation in the case of membrane-permeable forms, thereby allowing parameters such as membrane voltages and ion concentrations to be monitored via changes in the probe's fluorescence. For example, fluorescent Ca^{2+} indicators can track intracellular **calcium dynamics**, particularly when used with **confocal microscopy**, where a confocal aperture placed in the optical path excludes out-of-focus light to yield more precise optical sections in bulky cells such as eggs (**Figure a**). Alternatively, **genetically encoded calcium indicators** (**GECIs**) can be expressed in live cells and used in Ca^{2+} imaging studies without having to pre-load low-MW fluorophores. The development of such GECIs can be traced back to pioneering work done on **Green Fluorescent Protein** (**GFP**) that is encoded by the gfp gene discovered in jellyfish. After cloning the gene and genetically modifying the spectral properties of GFP, such enhanced gfps were linked to other genes encoding a wide variety of non-fluorescent proteins. In this way, gfp-containing recombinant genes have been able to provide readily imaged fluorescent tags of gene expression, thereby revolutionizing cell biology and earning for their key developers the 2008 Nobel Prize in Chemistry. Following the discovery of jellyfish GFP, additional fluorescent proteins with wide-ranging spectra (**Figure b**) have been found in other marine invertebrates. The efficacy of these new probes has been further optimized by genetic engineering, thereby greatly expanding the capabilities of genetically encoded fluorescent tags. For example, by linking these genes to genes encoding proteins with localization signals for specific components of the cell, multi-spectral fluorescence microscopy allows **several organelles** to be simultaneously monitored in living cells (**Figure c**). Similarly, various types of genetically encoded fluorescent probes can be used to track not only calcium concentrations but also numerous other kinds of functional and structural properties in living cells from the cellular level up to the entire organism.

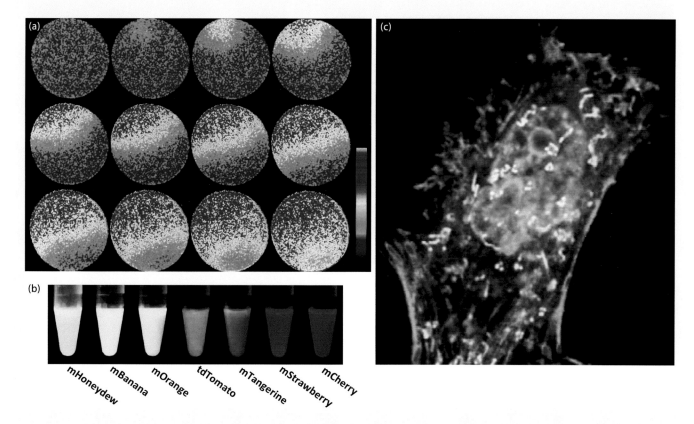

Box 1.2 (a) Time-lapse confocal images of a calcium wave spreading across a starfish egg during fertilization (150X). *(b)* Some of the yellow to red fluorophores used in cell biology. *(c)* A multi-labeled living human HeLa cell, showing nuclear histones (blue), actin (green), mitochondria (yellow), Golgi (orange), and focal adhesions (red). *([b]: From Shaner, NC et al. (2004) Improved monomeric red, orange and yellow fluorescent proteins derived from Discosoma sp. red fluorescent protein. Nat Biotech 22: 1567–1572, reproduced with publisher permission; [c]: From Chudakov, DM et al. (2010) Fluorescent proteins and their applications in imaging living cells and tissues. Physiol Rev 90: 1103–1163, reproduced with publisher permission.)*

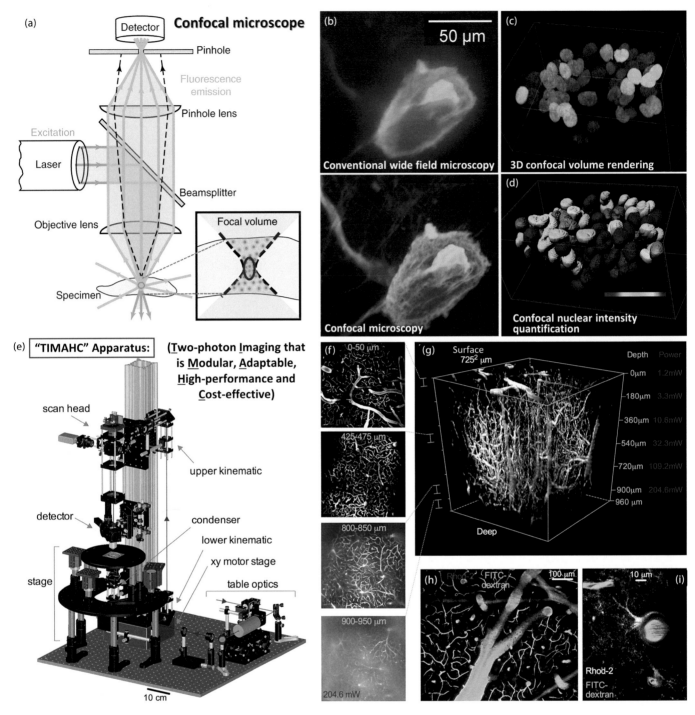

Figure 1.6 Conventional fluorescence microscopy and enhanced fluorescence imaging modes—confocal and two-photon microscopy. (**a**) Confocal microscopy is able to use a pinhole aperture in front of the detector to help eliminate out-focus-emissions (dotted lines) arising from the focal volume above or below optical planes of interest (red oval). (**b**) For example, confocal imaging of a neuron provides enhanced views of morphology compared to conventional wide-field fluorescence microscopy. (**c, d**) As shown for a cultured "mini-organ" organoid with fluorescently labeled nuclei, confocal datasets can also be converted into 3D volumetric renderings and quantifications of fluorescence intensities. Figure (**e**) depicts a TIMAHC (Two-photon Imaging that is Modular, Adaptable, High-performance and Cost-effective) type of two-photon microscope. (**f, g**) As examples of two-photon microscopy capabilities, TIMAHC can optically section deep within the mouse brain to generate 3D reconstructions of neural blood vessels (FITC labeled) plus associated astrocyte glial cells (Rhod-2 labeled). (**h, i**) Two-photon microscopy also allows more precise calcium imaging (i) than obtained by wide-field fluorescence microscopy (h). ([a, c, d]: From Jonkman J et al. (2020) Tutorial: guidance for quantitative confocal microscopy. Nature Protocols 15: 1585–1611, reproduced with publisher permission; [b]: From Martial, FP and Hartell, NA (2012) Programmable illumination and high-speed, multi-wavelength, confocal microscopy using a digital micromirror. PLoS ONE 7(8): e43942. doi:10.1371/journal.pone.0043942,

Figure 1.6 (Continued)   *reproduced under a CC BY creative commons license; [f–i]: From Rosenegger, DG et al. (2014) A high performance, cost-effective, open-source microscope for scanning two-photon microscopy that is modular and readily adaptable. PLoS ONE 9(10): e110475. doi:10.1371/journal.pone.0110475 reproduced under a CC BY 4.0 creative commons license.)*

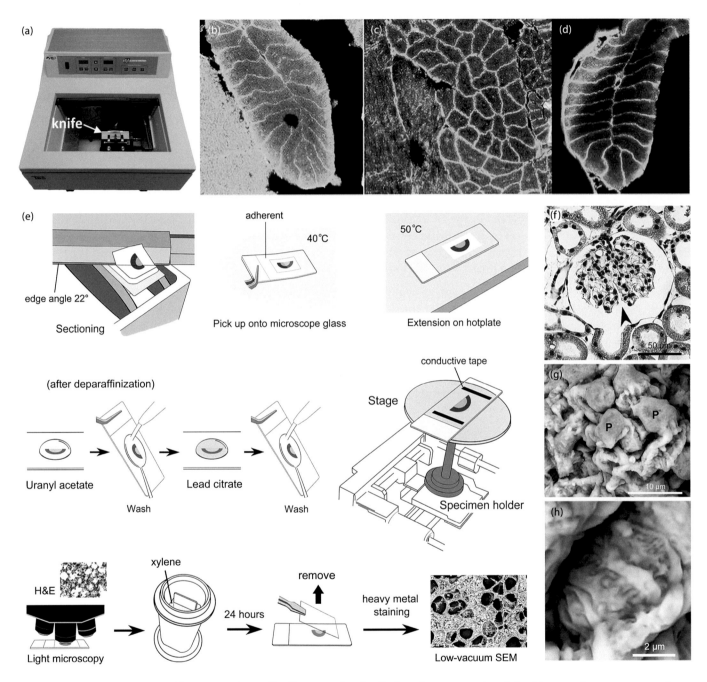

Figure 1.7 Cryo- and paraffin sectioning of bulky samples for light microscopy. (a–d) *A refrigerated cryostat type of microtome (a) can rapidly produce frozen sections, such as shown here for fruit fly muscle in a wild type specimen (b) vs. either in a mutant with disordered muscle patterning (c) or in a mutation rescue that restores normal patterning (d) (60X).* **(e–h)** *As a mainstay of histological analyses, samples embedded in paraffin wax provide views of general tissue morphology but often fail to preserve the cellular details visible in plastic-embedded specimens. However, paraffin sections can be stained with heavy metals and processed for SEM, either following initial LM evaluations (e, lowest row of figures) or without previous LM imaging (e, upper two rows of figures). Such techniques allow high-resolution views, such as shown in a conventionally prepared paraffin section of the kidney (f) vs. SEM images of comparable paraffin sections viewed by SEM at progressively higher magnification (g, h). ([b–d]: Images courtesy of Drs. R Cripps and T Lovato; [e–h]: From Sawaguchi, A et al. (2018) Informative three-dimensional survey of cell/tissue architectures in thick paraffin sections by simple low-vacuum scanning electron microscopy. Sci Rep 8:7479. doi:10.1038/s41598-018-25840-8, reproduced under a CC BY 4.0 creative commons license.)*

with paraffin blocks. Such plastic-embedded samples are cut using ultra-sharp glass or diamond knives for producing not only 0.5–1 μm thick plastic sections for LM but also 70-nm thin sections that are examined by **TEM**. Compared to paraffin sections, properly prepared plastic sections tend to yield much clearer images of subcellular architecture. However, even if collected in ribbons of serial sections, reconstructing broad expanses of morphology is often challenging, and thus paraffin sectioning remains a mainstay of histological analyses at the organ level of organization.

1.4   STAINING SECTIONS, HISTOLOGICAL ARTIFACTS, AND INTERPRETING MICROANATOMICAL IMAGES

For most morphological analyses, sections are **stained** with one to several kinds of colored dyes before being examined by BF light microscopy. With paraffin sections, the workhorse staining solution is a mixture of **hematoxylin** and **eosin**, commonly referred to as **H&E**. After being properly processed, hematoxylin functions as a **basic dye** with a net **positive charge**, whereas eosin is an **acidic dye** with an overall **negative charge** (**Figure 1.8a**). In general, negatively charged materials in sections bind to positively charged basic dyes and hence are termed **basophilic** ("basic-loving") **substances**. Conversely, positively charged **acidophilic** ("acid-loving") **substances** bind to the negative charge of acidic dyes. Accordingly, the negatively charged phosphate groups of nucleic acids contribute to the basophilia of nuclei, which are stained magenta to dark-purple by different **hematoxylins**, whereas positively charged amino acids of cytoplasmic and ECM proteins account for much of the **acidophilia** that is rendered pink by **eosin** (**Figure 1.8a, b**)

Because H&E staining can only categorize regions based on their overall charge, other stains were developed over the last century and a half (Box 1.1) with the aim of more precisely characterizing cell and tissue compositions. Such staining recipes constituted the field of **histochemistry**, which has been largely supplanted in recent years by the more specific and reliable antibody-based identifications employed in **immunocytochemical analyses** (pg. 18). However, several histochemical stains are still used today to augment standard H&E staining of paraffin sections. For example, silver-containing solutions help visualize reticular fibers in the ECM (Chapter 3), whereas other staining methods can be used to detect macrophage cells that engulf materials via the process of phagocytosis (Chapter 2) (**Figure 1.8c, d**).

Stained sections can often be categorized by their orientation relative to the following **three main axes**, with some of the synonyms used in human anatomy for these axes being noted in parentheses: (1) top (=superior) vs. bottom (=inferior); (2) left vs. right; and (3) front (=ventral, anterior) vs. back (=dorsal, posterior) (**Figure 1.8e**). Thus, in adults, the **transverse** plane cuts across the body's long axis to generate a **top** and **bottom** component. Alternatively, the **longitudinal** plane cuts along the long axis parallel to the dorsal-ventral axis to form **left** vs. **right** sides, whereas the **frontal** plane also sections the long axis but is perpendicular to the dorsal-ventral axis to generate **front** vs. **back** components. It should be noted that in other animals, anterior vs. posterior typically refer to head vs. tail regions, and in human development, cephalic vs. caudal are often used to describe the embryonic head vs. tail ends.

For relatively straight tubular organs, transverse sections tend to have a rounder profile than the elongated outlines of longitudinal and frontal sections occurring toward the middle of the organ. However, longitudinal vs. frontal planes are hard to tell apart without clear indicators of front from back or left from right. Moreover, tubular organs, such as the small intestine can be highly folded into differing configurations, thereby yielding difficult-to-interpret profiles. This is particularly problematic in **grazing** and **oblique** sections, which either barely sample the specimen or are well out of alignment with a major sectioning axis, respectively, thereby generating oddly shaped profiles in section (**Figure 1.8f**).

Along with the confounding effects of section plane orientations, morphologies can also be misinterpreted owing to abnormal features that are generated by the processing protocol. Such preparatory **artifacts** range from global changes, such as overall **shrinkage** of tissues to much more localized alterations within particular regions of the section. For example, without proper fixation, lipids normally contained within fat cells can be leached out during paraffin processing, thereby generating artifactual intracellular spaces. Similarly, improperly mounted specimens can contain bubbles, debris, or folds that might be confused as constituting parts of the tissue, although careful through-focusing will often reveal that such contaminants lie above or below, rather than within, the specimen.

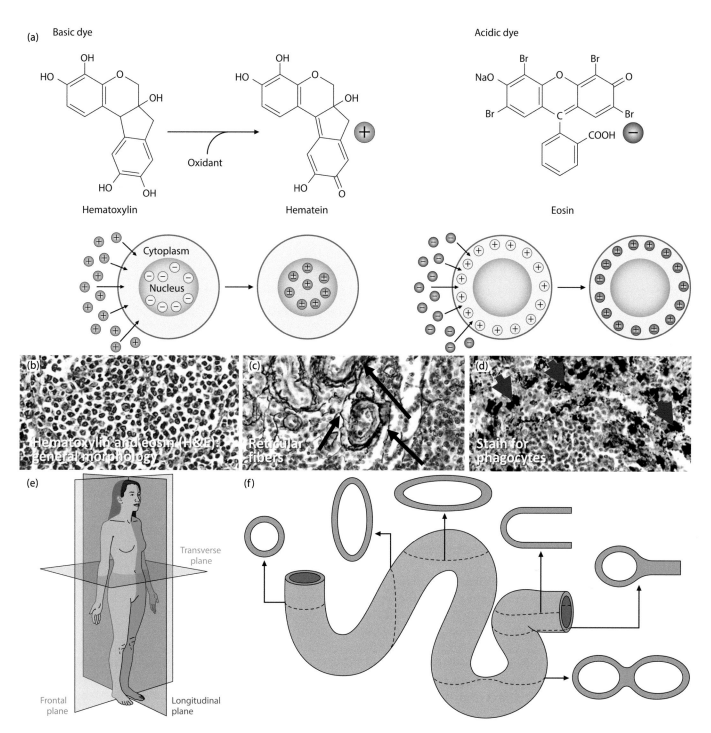

Figure 1.8 Staining LM sections with hematoxylin and eosin (H&E) or with more specialized histochemical reactions plus section plane orientations. (a) *Hematoxylin and eosin (H&E) is the most commonly used combination stain for viewing general tissue morphology. Oxidized hematoxylin acts as a positively charged basic dye that binds negatively charged basophilic substances, such as nucleic acids in nuclei, whereas eosin is a negatively charged acidic dye that binds positively charged molecules like cytoplasmic proteins. (**b–d**) Paraffin sections of the spleen are treated here with H&E (b) vs. silver staining for reticular fibers (c, black arrows) vs. a protocol for labeling phagocytic cells (d, red arrows) (30X). (**e**) The three main section planes—transverse, longitudinal, and frontal—are diagrammed for an adult human. (**f**) Without changing sample positioning within the microtome, orientations of section planes can greatly vary over relatively short distances, particularly in irregularly configured specimens, such as the intestines, thereby generating variable profiles that are difficult to relate to whole specimen morphology.*

In spite of these potential complications, histological and ultrastructural features can be properly identified in microanatomical images with some practice, particularly when considering **microscopy-** and **size-related benchmarks**. For example, determining whether the image is an LM or EM may help to ascertain if an organ, tissue, cell, or set of subcellular structures is depicted, because low-magnification LMs are capable of exhibiting entire organs of small animals, whereas EMs tend to encompass only subcellular to tissue levels of organization (**Figure 1.1**). Accordingly, given that visible light used in LM can produce colors, whereas TEM and SEM yield black-and-white images, full-color images with comparatively low resolution generally correspond to LMs. Conversely, higher-resolution monochromatic images tend to be EMs, although LMs can certainly be rendered in monochrome, and EMs may be digitally altered with pseudocolors. After taking such steps in the identification process, features such as nuclei inside cells vs. fiber bundles in the ECM can help discriminate intracellular vs. extracellular compartments in LMs, whereas distinctive intracellular organelles like mitochondria vs. banded collagen fibrils present in the ECM can play similar roles when assessing EMs.

1.5 IDENTIFYING SPECIFIC COMPONENTS IN HISTOLOGICAL SAMPLES: POLARIZATION MICROSCOPY, AUTORADIOGRAPHY, AND INDIRECT IMMUNOFLUORESCENCE

Along with histochemical stains that continue to be used today, several other methods can provide helpful information regarding sample composition. For example, **polarization microscopy** (**PM**) typically uses a pair of orthogonally arranged polarizing filters in LMs in order to assess heterogeneous kinds of polarized light refraction as the sample is rotated around the microscope's optical axis. Thus, in regions with either crystals or highly aligned materials, PM is able to detect **birefringence** arising from dual refractive indices in these ordered sites, which, in turn, can appear bright white to dark depending on the particular orientation of the birefringent material relative to polarized light. As one of the first applications of PM in living cells, birefringent **microtubule** arrays in the meiotic **spindle** (Chapters 2 and 18) of marine invertebrate eggs were detected during the last half of the 20th century, and today PM facilitates the selection of optimally differentiated human eggs for use in fertility clinics (**Figure 1.9a, b**).

To elucidate dynamic processes at the microanatomical level, **autoradiography** was developed in the 1950s as a method that incorporated radioactively labeled molecules into living specimens for tracking radiation emissions over time. Such analyses employ sections of **serially fixed** replicates of specimens that are preserved at multiple timepoints after radioisotope introduction. Isotope-containing sections are then coated with a radiation-sensitive photo emulsion that traditionally requires exposures lasting days to weeks in order to detect radiation-induced silver grains in the emulsion. Alternatively, newer phosphor imaging modes can provide enhanced sensitivity with greatly reduced exposure times. In either case, autoradiographic series can elucidate time-dependent molecular dynamics, which are often difficult to infer from conventional fixations. For example, by using a radiolabeled precursor to ECM molecules, it was determined that extracellular components are secreted both by **eggs** and by their surrounding **follicle cells** (Chapter 18), based on the presence of radioactive signals on the interior and exterior of the **zona pellucida** coat that envelops mammalian eggs (**Figure 1.9b, c**).

However, safety concerns and prolonged exposure times have led to a decline in autoradiographic usage over the last few decades, while at the same time **immunocytochemical** methods have become widely adopted for antibody-based monitoring of antigenic components such as proteins. As the most commonly applied mode of immunocytochemistry, **indirect immunofluorescence** is a dual-antibody method that typically utilizes sections or permeabilized cells that are readily penetrated by large antibody molecules. Initially, such specimens are incubated in a solution containing unlabeled **primary antibody,** which had been raised against the particular antigenic protein of interest (**Figure 1.9d, e**). After the primary antibody binds to antigen-containing sites, a fluorescently labeled **secondary antibody** raised against the primary antibody is added. By attaching numerous secondary antibodies to each primary antibody, secondary antibody fluorescence allows visualization of antigen/primary-antibody complexes while also increasing signal-to-noise ratios for enhanced sensitivity (**Figure 1.9d, e**).

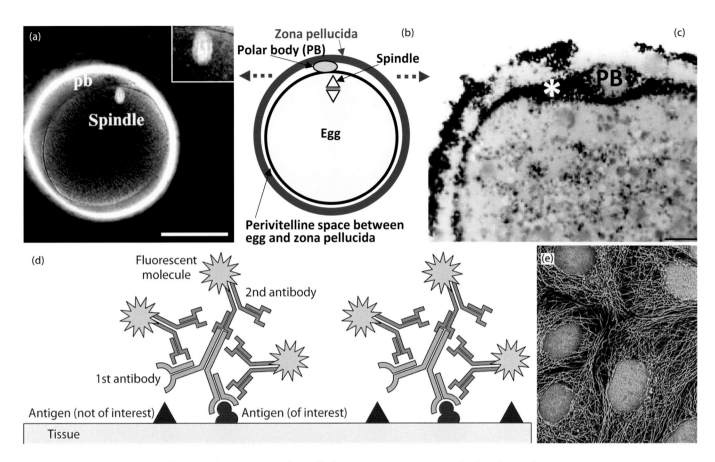

Figure 1.9 LM methods for analyzing specific cellular components—polarization microscopy, autoradiography, and immunocytochemistry. *(**a, b**) Polarization microscopy tracks the birefringence of highly ordered materials, such as microtubule arrays in the spindle of human eggs, as shown in (a) (scale bar = 50 μm). (**b, c**) Autoradiography of radioactively labeled substances allows dynamic processes to be tracked in serially fixed sample replicates. For example, as shown in (c), black deposits (asterisk) representing precursors to ECM products localize to both the external and internal sides of the zona pellucida coat of a pig egg, indicating that these substances were secreted from within and outside the egg (1,000X). (**d, e**) Indirect immunofluorescence employs a fluorescently tagged secondary antibody to visualize antigen/primary-antibody complexes, thereby identifying specific cellular components, such as shown in figure (e) for microtubules (green) and Golgi bodies (orange) in human HeLa cells with counterstained blue nuclei (1,000X). ([a]: From Tomari, H et al. (2018) Meiotic spindle size is a strong indicator of human oocyte quality. Reprod Med Biol 17: 268–274, reproduced with publisher permission; [c]: From Flechon, JE et al. (2003) The extracellular matrix of porcine mature oocytes: origin, composition and presumptive roles. Reprod Biol and Endocrinol 1, http://www.rbej.com/ content/1/1/124, reproduced with publisher permission; [e]: A public domain image generated by the National Institutes of Health of the federal government and available at https://en.wikipedia.org/wiki/HeLa#/media/File:HeLa-I.jpg.)*

1.6   METHODS FOR RECONSTRUCTING THREE-DIMENSIONAL MORPHOLOGY FROM THE SUBCELLULAR TO ORGAN LEVEL OF ORGANIZATION

As one of the most difficult tasks of microanatomical analyses, elucidating 3D morphology from 2D images can benefit from several imaging modes that span the subcellular to organ levels of organization. Toward the atomic end of this spectrum, **cryo-electron microscopy** (**cryo-EM**) uses rapid methods of specimen freezing to avoid the damaging effects of ice crystal formation and places these frozen samples on specialized grids in liquid-nitrogen-cooled TEMs (**Figure 1.10a, b**). By tilting the grids at differing angles and taking multiple images (**Figure 1.10c**), stacks of cryo-EM images can be assembled into sub-nanometer-resolution 3D renderings of biological materials unaffected by chemical fixations (**Figure 1.10d**). Similarly, several forms of **super-resolution LM**—also known

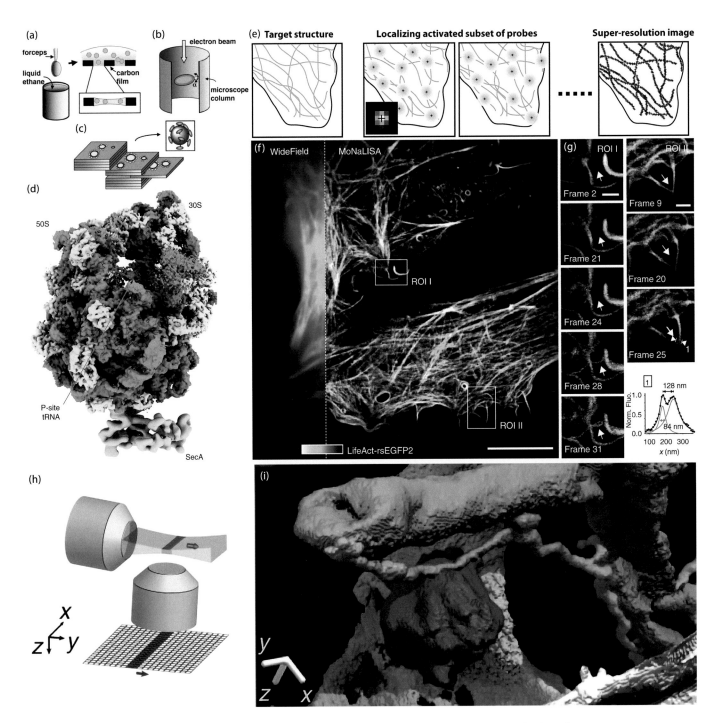

Figure 1.10 Reconstructing subcellular 3D microanatomy via cryo-EM, super-resolution fluorescence microscopy, and light-sheet microscopy. (a–d) In cryo-EM, rapidly frozen samples collected on specialized grids are imaged after being tilted at differing angles in a refrigerated TEM to generate high-resolution 3D reconstructions, such as the bacterial ribosome and associated molecules shown in figure d. **(e)** In one version of super-resolution microscopy, emissions of specialized fluorophores can be rapidly switched from on to off to allow precise imaging of cells with much greater resolution than is achieved in conventional wide field microscopy. **(f, g)** For example, such techniques can provide enhanced images of living epithelial cells (f, scale bar = 10 μm) and higher magnification regions of interest (ROIs) of their individual pseudopodial processes, as generated by a "MoNaLISA" version of super-resolution microscopy (g, scale bars = 1 μm). **(h)** In axially swept light-sheet microscopy (ASLM), specimens are probed with a sheet of excitation light, as an orthogonally oriented objective synchronously scans fluorescent emissions over serial bands of a photodetector for rapid 2D/3D imaging. **(i)** A 3D reconstruction of a chemically cleared portion of bone marrow subjected to ASLM

Figure 1.10 (Continued) shows a hematopoietic stem cell (magenta) protruding into a sinusoidal capillary (gray) with an arteriole (orange) and nerve fiber (green) nearby (scale bar = 2.5 μm). ([a–c]: From Fogarty, KH et al. (2011) New insights into HTLV-1 particle structure, assembly, and Gag-Gag interactions in living cells. Viruses 3: 770–793, reproduced under a CC BY 3.0 creative commons license; [d]: From Wang, S et al. (2019) The molecular mechanism of cotranslational membrane protein recognition and targeting by SecA. Nat Struct Mol Biol 26: 919–929, reproduced with publisher permission; [e]: From Huang, B et al. (2009) Super-resolution fluorescence microscopy. Annu Rev Biochem 78: 993–1016, reproduced with publisher permission; [f, g]: From Masullo, LA et al. (2018) Enhanced photon collection enables four dimensional fluorescence nanoscopy of living systems. Nat Commun 9. doi: 10.1038/s41467-018-05799-w, reproduced under a CC BY 4.0 creative commons license; [h, i]: From Chakraborty, T et al. (2019) Light-sheet microscopy of cleared tissues with isotropic, subcellular resolution. Nature Methods 16: 1109–1113, reproduced with publisher permission.)

as fluorescence nanoscopy—can provide 3D reconstructions of living cells at effective resolutions of a few dozen nanometers or even lower in the case of molecular solutions. A key factor in achieving this greatly enhanced resolving capacity is the use of small fluorescent molecules whose emission can be reversibly turned on or off, thereby allowing individual molecules in the on state to be discriminated from nearby neighbors in the off state for markedly improved resolution (**Figure 1.10e–g**). Alternatively, light-sheet fluorescence microscopy (LSFM) optically sections fluorescent specimens by scanning samples with a sheet of excitation light as the fluorescence emission is separated 90° away from the exciting rays for detection. As an upgrade to conventional LSFM, axially swept light-sheet microscopy (ASLM) scans the narrow waist of the excitation sheet in synchrony with a serially progressing band within the photodetector for rapid generation of high-resolution datasets that can be rendered in 3D (**Figure 1.10h, i**).

In addition to light microscopic methods, **high- and ultrahigh-voltage EMs** (**HVEMs**, **UHVEMs**) use up to 1000 kV accelerating voltages that substantially exceed the 60–100 kV range employed by most conventional TEMs. Such energetic EM beams are able to penetrate thick sections or even whole cells and can be combined with grid tilting to provide stereoscopic micrographs, thereby simplifying 3D reconstructions compared to conventional TEM methods, which have to piece together numerous thin sections.

Because thin sectioning for TEM is not only laborious but also prone to the loss or mis-registration of sections, several improvements have been introduced in recent years. One of these, termed **serial block-face electron microscopy**, places a microtome inside an SEM for cutting serial thin sections of plastic-embedded specimens that had been pre-stained with heavy metals before embedding (**Figure 1.11a–c**). After each thin section is produced and discarded, the SEM scans the newly cut block surface and generates a digital image based on backscattered electrons emitted by the metal-impregnated specimen. Eventually, a series of such images is computationally reconstructed into 3D renderings without having either to collect sections or to image each one by TEM. Alternatively, instead of a microtome, **focused ion beam SEM** (**FIB-SEM**) uses an ion beam as a way of finely milling new surfaces on the plastic-embedded block (**Figure 1.11d**).

For larger-scale views spanning from organs to whole-organism levels, microanatomical imaging often relies on two main types of non-invasive scanning methods: (i) magnetic-wave-based **magnetic resonance imaging** (**MRI**) (**Figure 1.12a–c**); and (ii) X-ray scans as part of a general process, called computerized axial tomography (CAT), or simply **computed tomography** (**CT**) (**Figure 1.12d–i**). In typical medical applications, these two methods provide 2D sections of whole living organs or even entire humans at resolutions on the order of ~200 μm and ~5–150 μm, respectively, with both being able to assemble digitized sections into informative 3D reconstructions. For example, subsets of through-focus MRIs can be projected on each other to convert 2D images into 3D-appearing renderings (**Figure 1.12a, b**). Alternatively, the entire dataset of 2D **pixels** can be transformed into 3D **voxels**, thereby allowing full 3D renderings to be cleaved and viewed along various imaging planes (**Figure 1.12c**). Similarly, CT scans can be partitioned into differing views, such as the exterior surface vs. the internal skeletal system (**Figure 1.12d, e**). In addition, industrial-grade versions of CT scanners, termed **micro-** and **nano-CTs**, are capable of achieving **sub-micron resolution** in fixed samples up to ~20 cm wide, thereby allowing CT scanning to provide volumetric renderings from the organ to subcellular level (**Figure 1.12f–i**).

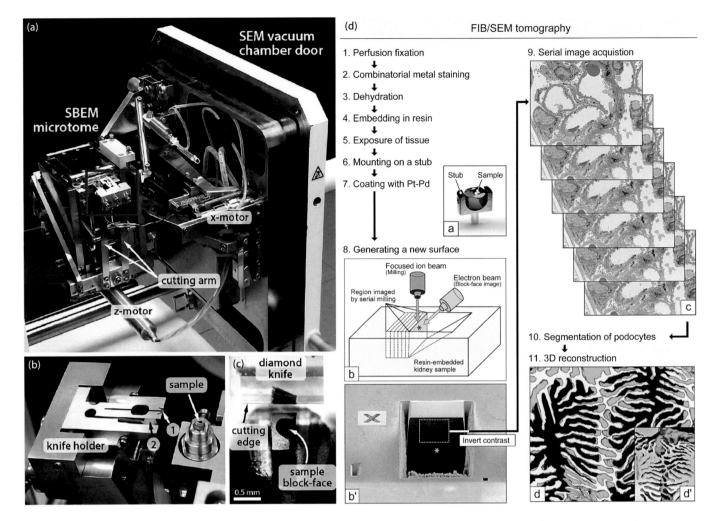

Figure 1.11 Reconstructing 3D ultrastructure via serial block face SEM and focused ion beam SEM. (a–c) *As an upgrade over serial thin-sectioning analyzed by TEM, a microtome placed within an SEM chamber allows thin sections to be removed from a metal-impregnated, plastic-embedded block. Backscattered electrons (BSEs) emitted from each newly exposed block face can then be used to generate a series of SEM images for reconstituting 3D ultrastructure. (**d**) Instead of cutting sections, a focused ion beam can mill the block face for BSE imaging used in 3D ultrastructural reconstructions, as illustrated here with kidney cells. ([a–c]: From Titze, B., and Genoud, C. (2016) Volume scanning electron microscopy for imaging biological ultrastructure. Biol Cell 108: 307–323, reproduced with publisher permission; [b]: From Miyaki, T et al. (2020) Three-dimensional imaging of podocyte ultrastructure using FE-SEM and FIB-SEM tomography. Cell Tissue Res 379: 245–254, reproduced under a CC BY 4.0 creative commons license.)*

1.7   MEDICAL APPLICATIONS OF MICROANATOMICAL ANALYSES

Although genomic sequencing has revolutionized the diagnosis of diseases and the individual tailoring of treatment options, **microanatomical methods** continue to be used in various fields of **medicine**. For example, automated **LM** is widely employed for differential counts of cell types in blood smears. Moreover, for *in vivo* studies, thin and flexible **endoscopes** can be enhanced with ancillary capabilities, such as confocal optics for microanatomical analyses of the digestive tract or other body regions exposed by keyhole-sized incisions. In addition, LM modes, such as **multiphoton microscopy** and **light-sheet microscopy** either individually or in tandem allow relatively thick slabs of fluorescently labeled tissues to be rapidly imaged. Similarly, high-resolution **MRIs** and **CT scans** are routinely employed *in vivo* to identify various pathologies, such as tumorous growths, diseased blood vessels, and damaged skeletomuscular components, thereby greatly aiding subsequent treatment options.

For fixed materials, conventional **TEM** allows confirmation of such cellular pathologies as Kartagener's syndrome (Chapter 3), whereas **indirect immunofluorescence** of sections can supplement serum-based diagnoses

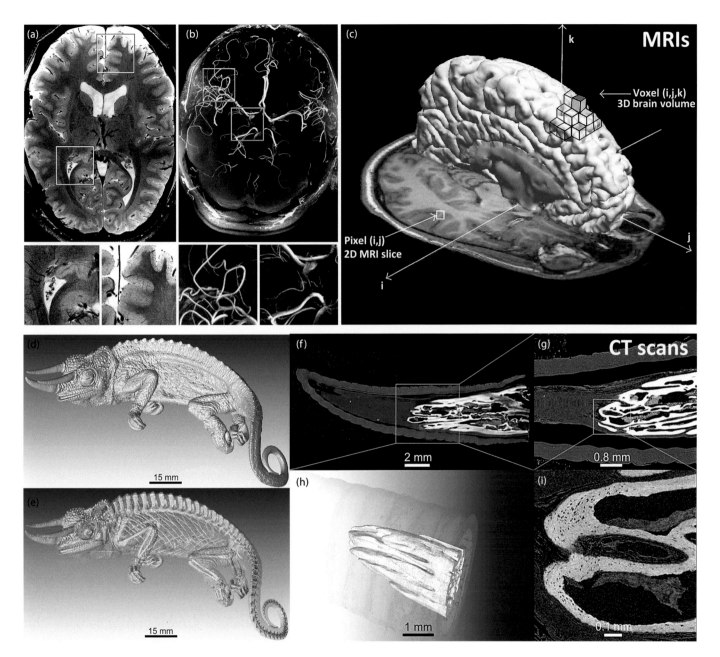

Figure 1.12 Reconstructing 3D microanatomy from the tissue to whole body level with MRI (magnetic resonance imaging) and CT (computed tomography) scanning. (a, b) *Two-dimensional MRI scans of the human brain can be assembled into 3D-like projected series, as shown here for neural vasculature. (c) As illustrated here for an MRI-scanned brain, 2D pixels can be rendered into 3D voxels, thereby allowing computational sectioning through various planes of the brain. (**d–i**) 2D (f, g, i) and 3D reconstructions (d, e, h) of high-resolution CT scans of an adult chameleon lizard illustrate microanatomy from the subcellular to whole-body level organization. ([a, b]: From Stucht, D et al. (2015). Highest resolution in vivo human brain MRI using prospective motion correction. PLoS ONE 10: e0133921. doi:10.1371/ journal.pone.0133921, reproduced under a CC BY 4.0 creative commons license; [c]: From Despotovic, I et al. (2015) MRI segmentation of the human brain: challenges, methods, and applications. Comput Math Meth Med: http://dx.doi. org/10.1155/2015/450341, reproduced under a CC BY 4.0 creative commons license; [d–i]: From du Plessis, A et al. (2017) Laboratory x-ray micro-computed tomography: a user guideline for biological samples. GigaScience 6: 1–11, reproduced under a CC BY 4.0 creative commons license.)*

and provide important spatial information for differentiating disease subtypes. Moreover, in order to analyze the systemic effects of candidate drugs, the pharmaceutical industry sometimes employs **whole-body autoradiography** on sections of entire animals that were obtained using very large microtomes.

Using these and other techniques, microanatomical studies take a multipronged approach in analyzing the structural bases of subcellular to organ-level functioning. In doing so, modern microanatomy represents a key nexus for integrating information at several levels of biological organization. Accordingly, the rest of this text supplements medically relevant material with various biology-based correlates of microanatomy that collectively aim to present a novel and cogent overview of human microanatomy.

1.8   SUMMARY–MICROANATOMY AND ITS ASSOCIATED TECHNIQUES

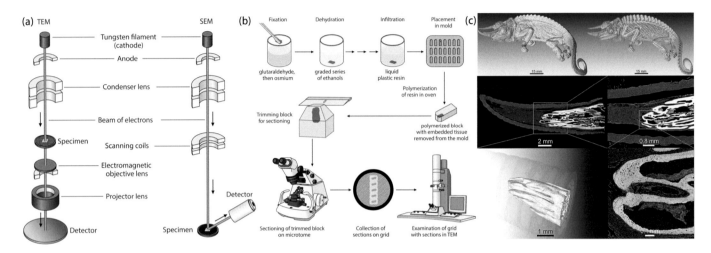

Pictorial Summary Figure Caption (a) *See Figure 1.2c;* ***(b)*** *See Figure 1.3a;* ***(c)*** *See Figure 1.12d–i*

Microanatomy (=histology) analyzes the functional morphology of (i) **cells**, (ii) **tissues**, which are integrated assemblages of **cells** and their **extracellular matrix (ECM)**, and (iii) **organs**, which comprise discrete collections of interacting tissues. Human cells typically measure **~10–50 μm** and contain uniformly sized components like **cell membranes (7 nm)**, **ribosomes (~25 nm)**, and **mitochondria (0.5–1 μm** wide). Unaided human eyes can only **resolve** closely apposed objects as being separate down to spacings of **~100–200 μm**. Thus, it was not until the first functional **light microscopes (LMs)** were developed in the 17th century that subcellular microanatomy could be analyzed. Conventional LMs increase resolving capacity to **~0.2 μm** as defined by **Abbe's equation: d = 0.61 λ/NA** (d = minimum resolvable distance; λ = wavelength of light [~0.5 μm]; NA = the numerical aperture, or light-gathering capacity, of imaging lenses [maximum ~1.4]—i.e., 0.61 × 0.5 μm/1.4 = ~0.2 μm). For **~0.2-nm resolutions**, **electron microscopes (EMs)** use **electron beams** with shorter wavelengths than light and thus can resolve subcellular components like ribosomes and membranes.

Two main EMs—**transmission (TEM)** and **scanning (SEM)**—provide complementary views of cell ultrastructure (**Figure a**). **TEMs** transmit electrons through thin samples for precise 2D images of restricted regions, whereas **SEMs** scan electrons over samples for 3D surface images of larger fields of view. Supplementing standard TEM/SEM analyses are: (i) **high voltage EM** (for transmitting beams through thicker samples), (ii) **cryo-EM** (flash freezing without chemical fixations for high-resolution analyses of native-state molecules), and (iii) various hybrid methods, such as **serial block-face EM** and **focused-ion beam SEM**, which places a sectioning apparatus (=**microtome**) or ion beam milling device in an SEM to expose new surfaces on plastic-embedded samples so that the SEM can image 2D and 3D regions of specimens without having to use a TEM.

Some cell and tissue samples can be examined as live **whole mounts** via LM using: (i) standard (**brightfield**) optics; (ii) contrast-enhancing **phase** or **differential interference microscopy**; or (iii) **fluorescence microscopy** of non-toxic fluorescent dyes, particularly with **confocal** or **multiphoton** methods for enhanced optical sectioning and 3D imaging.

More commonly, microanatomical samples are **fixed**, **dehydrated**, and **embedded** in **paraffin wax** or **plastic resin** for sectioning on a **microtome**. Paraffin allows serial-sectioning of relatively large (cm sized) samples at

~6–8 µm thickness. Conversely, plastic sections can be cut much thinner (~70 nm for TEM) but are limited to smaller samples (~mm wide) (**Figure b**). Sections can range from **transverse** (cut across the long axis into top/bottom sides) to **longitudinal** (along the long axis into left/right sides) to **frontal** (along the long axis into front/back sides) and are routinely stained with colored dyes for LM.

Basic dyes have a net + charge and bind **basophilic** substances like negatively charged phosphates of nucleic acids; **acidic dyes** have a − charge and attach to **acidophilic** materials like positively charged proteins in the cytoplasm/ECM. For more specific staining than routine **hematoxylin** & **eosin** (**H&E**) dyes, samples can be subjected to **histochemical** stains or to **immunocytochemistry** (e.g., **indirect immunofluorescence** with unlabeled **primary antibodies** and fluorescently labeled **secondary antibodies** typically used on cells or on frozen sections cut with a refrigerated microtome [**cryostat**]).

Additional microanatomical methods include: (i) **polarization microscopy**, which detects **birefringence** from crystals or other highly ordered arrays (e.g., microtubules); and (ii) **autoradiography**, which tracks radioactively labeled molecules across serially fixed specimens at cellular to whole-body levels.

Methods for **reconstructing 3D** microanatomy include: (i) **cryo-EM**, **super-resolution LM**, or **atomic force microscopy** (**AFM**) for reconstructing molecules to cells; (ii) **multiphoton** and **light-sheet fluorescence microscopy** for optical sectioning and reconstructions of cells and tissues; (iii) **micro-** and **nano-computerized tomography** (**CT**) for tissue to whole-body reconstructions (**Figure c**); and (iv) **magnetic resonance imaging** (**MRI**) for organ/tissue to whole body reconstructions.

SELF-STUDY QUESTIONS

1. Which major kind of electron microscope provides 3D images of surfaces?
2. T/F Decreasing the wavelength of illumination increases the resolving capacity of a microscope.
3. What kind of section cuts perpendicular to the long axis of an organ?
4. Which of the following detects sample birefringence?
   A. Cryo-EM
   B. Indirect immunofluorescence
   C. H&E stained paraffin sections
   D. Polarization microscopy
   E. Multiphoton microscopy
5. About how thick are typical paraffin sections?
   A. 70 nm
   B. 0.5–1 µm
   C. 6–8 µm
   D. 70 µm
   E. 100–200 µm
6. About how big are most human cells?
   A. 7 nm
   B. 25–30 nm
   C. 100 nm
   D. 10–50 µm
   E. 200–500 µm
7. T/F The limit of resolution for a standard LM is ~0.2 µm.
8. T/F Cell membranes are about 200 nm thick.
9. T/F Material processed for paraffin embedding is routinely dehydrated before sectioning.
10. What does the "H" of "H&E" represent?
11. Which of the following would bind an acidic dye such as eosin?
    A. Positively charged substances
    B. Basophilic substances
    C. Negatively charged phosphate groups of nucleic acids
    D. ALL of the above
    E. NONE of the above

12. Which of the following can non-invasively section living samples for 3D reconstructions?
    A. Confocal fluorescence microscopy
    B. MRI scans
    C. Multiphoton fluorescence microscopy
    D. Light-sheet fluorescence microscopy
    E. ALL of the above
13. T/F The maximum NA of a LM lens is ~100.

"EXTRA CREDIT" For each term on the left, provide the BEST MATCH on the right (answers A-F can be used more than once)

14. Standard LM_____   A. Photo emulsion
15. Autoradiography_____   B. Primary Ab
16. Contrast-enhancing LM_____ C. Secondary Ab
17. Microtome within SEM_____   D. Brightfield
18. Fluorescently labeled molecule used in indirect   E. DIC
    immuno-fluorescence_____   F. Serial block-face EM

19. Compare and contrast TEM vs. SEM.
20. Discuss Abbe's equation, factors affecting resolution, and imaging capabilities in LMs vs. EMs.

ANSWERS

1) SEM; 2) T; 3) transverse; 4) D; 5) C; 6) D; 7) T; 8) F; 9) T; 10) hematoxylin; 11) A; 12) E; 13) F; 14) D; 15) A; 16) E; 17) F; 18) C; 19) The answer should include among other pertinent topics: size/thickness of samples; beam paths; 2D vs. 3D images; surface vs. internal views of samples; possibility of imaging live samples; hybrid techniques like serial block-face EM; 20) The answer should include and clearly define such terms as: wavelength of illumination; numerical aperture (NA); colored vs. monochromatic images; fixed vs. living specimens; dehydration/vacuum; super-resolution LM; various LM methods for tracking specific components in samples; modes of reconstructing 3D morphology

Primer of Cellular Ultrastructure

2.1 INTRODUCTION TO CELLS AND THEIR GENERAL FEATURES

Although various cell types have diversified by evolving specialized adaptations for particular functions, all cells retain a core set of unifying features. For example, cells are invariably delimited by a bilayered **cell membrane** (=**plasmalemma**, **plasma membrane**). Moreover, cells of **eukaryotes**, which range from protists to humans, characteristically possess a double-membrane-bound **nucleus** containing **DNA** with **genes** that not only control cellular processes prior to cell division but also provide functional continuity once daughter cells are generated. Between the plasmalemma and nucleus of eukaryotic cells, the **cytoplasm** usually possesses a wide variety of membrane-bound **organelles**, such as Golgi bodies, endoplasmic reticulum (ER) arrays, and mitochondria as well as non-membrane-bound **inclusions** like ribosomes, lipid droplets, and cytoskeletal elements. Thus, eukaryotes differ fundamentally from archean and bacterial **prokaryotes**, which lack nuclei and most other organelles. For pedagogical purposes, components of a generalized human cell are summarized in **Figure 2.1a** and Table 2.1, whereas features of specialized cell types are listed in Table 2.2.

DOI: 10.1201/9780429353307-3

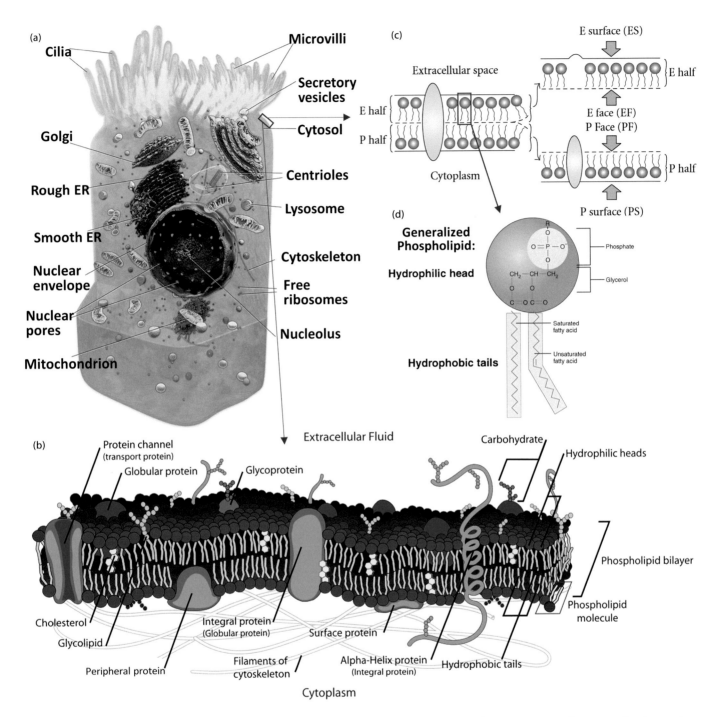

Figure 2.1 Introduction to cells and cell membranes. (a) *This cut-away view shows common features of eukaryotic cells. (**b**) The fluid-mosaic model maintains that cell membranes are dynamic structures (i.e., fluid) with numerous components (i.e., mosaic), such as phospholipids, proteins, and carbohydrate moieties. (**c**) Membranes comprise an external E half (= E leaflet) plus an internal P half (=P leaflet). (**d**) Membrane leaflets contain a backbone of phospholipids, with each phospholipid comprising (i) a hydrophilic head at the leaflet periphery and (ii) a pair of hydrophobic tails extending from the head toward the membrane center. ([a]: From Blausen.com staff (2014) "Medical gallery of Blausen Medical 2014" WikiJournal of Medicine 1 (2): 10. doi:10.15347/wjm/2014.010 reproduced under CC0 and CC SA creative commons licenses; [b]: From public domain drawing by Mariana Ruiz available at https://commons.wikimedia.org/wiki/File:Cell_membrane_detailed_diagram_en.svg; [c]: From Robenek, H et al. (2011) Topography of lipid droplet-associated proteins: insights from freeze-fracture replica immunogold labeling. J Lipids Volume 2011: doi:10.1155/2011/409371 reproduced under a CC BY 3.0 creative commons; [d]: From OpenStax College (2013) Anatomy and Physiology. OpenStax. http://cnx.org/content/col11496/latest, reproduced under a CC BY 4.0 creative commons license.)*

TABLE 2.1 FUNCTIONAL MICROANATOMY OF GENERALIZED, NON-EPITHELIAL CELLS

Organelle/inclusion	Major function(s)
Cell membrane	Selective barrier separating intra- and extracellular milieus
Nucleus	Contains genetic material for cell functioning and post-division continuity
Nuclear envelope	Selective barrier separating cyto- and nucleoplasm
Nucleolus	Ribosomal subunit production
Speckles/Cajal bodies	mRNA splicing
Golgi bodies	Protein packaging; nexus between endo- and exocytotic pathways
Endoplasmic reticulum	Protein synthesis; calcium ion storage; peroxisome biogenesis
Ribosomes	Protein synthesis
Lysosomes	Enzymatic degradation of residual products
Peroxisomes	Hydrogen peroxide catabolism and synthesis
Mitochondria	ATP production; driver of intrinsic apoptotic pathway
Cytoskeleton	Structural support; intracellular trafficking; signal transduction

TABLE 2.2 SOME SPECIALIZED CELL TYPES

Cell type	Major function(s)
Epithelial	Lining cavities and covering body surfaces; gland formation
Muscle	Locomotion/motility
Neuron	Action potential generation/propagation to control various target cells
Glial cells	Support and maintenance of neurons
Sensory cells	Perception of external or internal cues
Fibroblasts	Connective tissue fiber formation
Adipocytes	Energy storage
Chondrocytes	Cartilage formation
Osteoblasts	Bone formation
Osteoclasts	Bone resorption
Red blood cells	Gas exchanges
White blood cells	Anti-pathogenic reactions/immune defenses
Macrophages	Phagocytosis
Stem cells	Replenishment/proliferation of body components
Eggs and sperm	Fertilization

To provide essential background information for subsequent chapters on tissue and organ microanatomy, the following sections describe common features of cellular ultrastructure from the plasmalemma to nucleus. More specialized adaptations, such as brush borders, motile cilia, intercellular junctions, and basement membranes are typically restricted to epithelial cells and are thus covered in the next chapter.

2.2   THE CELL MEMBRANE: MULTIFACETED GATEKEEPER THAT REGULATES TRANSMEMBRANE MOLECULAR FLOW VIA SEVERAL MODES (PASSIVE AND FACILITATED DIFFUSION VS. ACTIVE TRANSPORT)

According to the **fluid mosaic model** of plasmalemma organization, cell membranes are dynamic (fluid) and multipartite (mosaic) structures composed of two halves—an **inner** and an **outer leaflet**—which are situated toward the intra- vs. extracellular sides of the cell, respectively (**Figure 2.1b, c**). The outer leaflet is also referred to as the E-half for "external", whereas the inner leaflet is termed the P-half, because the cytoplasm was historically designated as "protoplasm".

Forming the backbone of membrane leaflets are numerous **phospholipids** (**Figure 2.1d**), each of which contains: (i) a **hydrophilic** ("water-loving") head that is situated at leaflet peripheries near the aqueous milieus of the cytoplasm or extracellular matrix (ECM) (**Figure 2.1b, c**) and (ii) two **hydrophobic** ("water-repelling") **fatty acid tails** that partition away from the cytoplasm and ECM to form opposing arrays facing the membrane center (**Figure 2.1b, c**). Although there are numerous membrane phospholipids that vary in their head and/or tail compositions, a few common examples, such as **phosphatidylcholine** (**PC**), **sphingomyelins** (**SMs**), **phosphatidylserine** (**PS**), and **phosphatidylethanolamine** (**PE**) predominate across various cell types. PC and SM phospholipids are more concentrated in the outer leaflet, whereas PS and PE forms tend to be enriched in the inner leaflet (**Figure 2.2a, b**). These asymmetric distributions optimize membrane functioning at the extra- vs. intracellular surfaces and are established by ATP-requiring **flippase** enzymes. To stabilize these plasmalemmal components for EM imaging, cells are routinely fixed with the heavy metal **osmium**, which preferentially binds leaflets' phospholipid heads more avidly than their hydrophobic core. Thus, sectioned membranes exhibit a **trilaminar** configuration of two dense laminae sandwiched around a lighter central core (**Figure 2.2c**).

Along with phospholipid tails, the hydrophobic core contains other lipids, such as the sterol **cholesterol**. Each cholesterol molecule contains four rigid rings that reduce the mobility of adjacent phospholipids. Thus, in cells at normal or elevated temperatures, cholesterol helps maintain membrane rigidity (**Figure 2.2d**). Conversely, at low temperatures, intervening cholesterol molecules prevent phospholipids from crystallizing, thereby keeping cold membranes from becoming overly rigid. As opposed to such beneficial effects, excessive membrane cholesterol can promote various pathologies, such as the incorporation of toxic metabolic byproducts that drive the neurodegeneration of Alzheimer's disease (**Figure 2.2e**) (Chapter 9).

Scattered among membrane lipids are numerous **proteins**, which are broadly classified as **integral** vs. **peripheral**. Integral proteins are firmly bound to the internal hydrophobic core of the membrane and in most cases span the entire membrane (**Figure 2.3a–c**). Such **transmembrane** integral proteins serve various functions, including as selective **channels** that facilitate molecular transport (**Figure 2.3b, c**). In addition, transmembrane **receptors** like **receptor tyrosine kinases** (**RTKs**) (**Figure 2.3a**) can trigger important intracellular processes by binding activating molecules, called **ligands** (**Figure 2.3c**). Conversely, compared to integral proteins, **peripheral proteins** attach more loosely to membrane components. For example, proteins of the **heterotrimeric G protein** superfamily transiently bind to transmembrane receptors, called **G protein coupled receptors** (**GPCRs**) (**Figure 2.3d**), and following stimulation of GPCRs, G proteins dissociate from the membrane to initiate signaling cascades within cells (**Figure 2.3d**).

Cell membranes also contain **carbohydrates** that tend to be attached to outer leaflet phospholipid heads or to the extracellular domains of transmembrane proteins. At such sites, carbohydrates can help link cells to the ECM and provide key signals regarding cellular identities as is the case with blood cell glycosylations (Chapter 6).

Along with their heterogeneous composition, cell membranes are fluid structures as evidenced by classical experiments where membranes labeled with fluorescent probes were attached to unlabeled membranes and

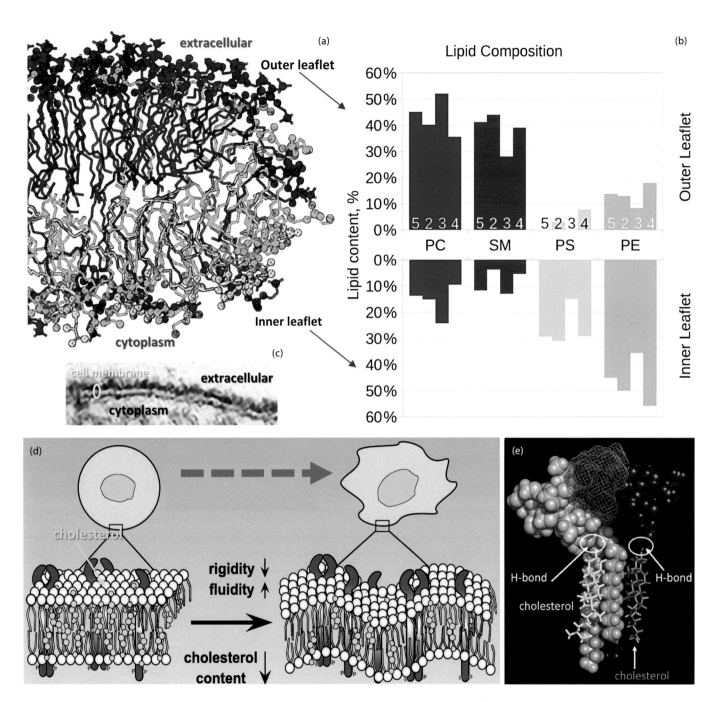

Figure 2.2 Membrane phospholipids and cholesterol. (*a*) *Of four common phospholipid types—phosphatidylcholine (PC, blue), sphingomyelins (SMs, red), phosphatidylserine (PS, yellow), and phosphatidylethanolamine (PE, green)—PC and SMs are enriched in the outer leaflet, whereas PS and PE are more concentrated in the inner leaflet, with this asymmetry helping to optimize outer vs. inner membrane functioning. (**b**) Graphs of different models (labeled 5, 2, 3, 4) estimating PC, SM, PS, and PE content in normal outer vs. inner leaflets. (**c**) The fixative osmium tetroxide preferentially binds phospholipid heads rather than tails. Thus, membranes appear trilaminar in EMs and measure ~7 nm from edge to edge. (**d**) Decreased cholesterol content leads to greater membrane fluidity and reduced rigidity. (**e**) Excessive cholesterol (yellow and green molecules) enhances the binding of neurotoxic β-amyloid peptides (=brown fishnet molecules) to nerve cell membranes during Alzheimer's disease progression. ([a, b]: From Rivel, T et al. (2019) The asymmetry of plasma membranes and their cholesterol content influence the uptake of cisplatin. Sci Rep 9:5627. https://doi.org/10.1038/s41598-019-41903-w reproduced under a CC BY 4.0 creative commons license; [c]: From Cardoso, MS et al. (2013) Identification and functional analysis of Trypanosoma cruzi genes that encode proteins of the glycosylphosphatidylinositol biosynthetic pathway. PLoS Negl Trop Dis 7(8): e2369. doi:10.1371/journal. pntd.0002369 reproduced under a CC BY creative commons license; [d]: From Zhang, J et al. (2019) Cholesterol content in cell*

Figure 2.2 (Continued)   *membrane maintains surface levels of ErbB2 and confers a therapeutic vulnerability in ErbB2-positive breast cancer. Cell Commun Signaling 17. https://doi.org/10.1186/s12964-019-0328-4 with modified labeling reproduced under permission of a CC BY 4.0 creative commons license; [e]: From Fantini, J et al. (2013) Cholesterol accelerates the binding of Alzheimer's β-amyloid peptide to ganglioside GM1 through a universal hydrogen-bond-dependent sterol tuning of glycolipid conformation. Front Physiol 4. doi: 10.3389/fphys.2013.00120 reproduced under a CC BY 3.0 creative commons license.)*

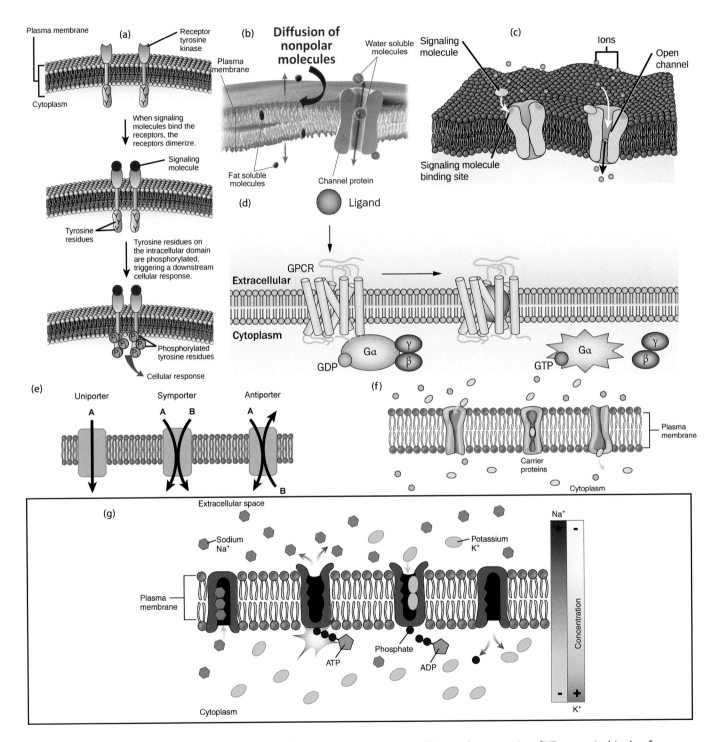

Figure 2.3 Membrane proteins and molecular transport across cell membranes. (a–d) *Two main kinds of membrane proteins are integral (a–c) vs. peripheral (d) proteins, with integral proteins being more firmly attached within the membrane than are peripheral proteins. Integral proteins include: (i) voltage-gated ion channels for water-soluble molecules (b) and (ii) receptors such as receptor tyrosine kinases (a) and ligand-gated channels (c). (d) G-proteins are*

Figure 2.3 (Continued) *examples of peripheral proteins that reversibly bind to receptors and transduce key intracellular signals. (**e**) Polar molecules that do not readily cross membranes typically utilize protein transporters which can be subclassified as: (i) uniporters (i.e., moving a single molecule type), (ii) symporters (i.e., transporting multiple molecule types in the same direction), and (iii) antiporters (i.e., moving different molecules in opposite directions). (**f**) For facilitated diffusion, transporters do not require energy input. (**g**) During active transport, transporters need ATP-based energy in order to function. ([a, c, f]: From OpenStax College (2013) Anatomy and Physiology. OpenStax http://cnx.org/content/col11496/latest, reproduced under a CC BY 4.0 creative commons license; [b]: From Blausen.com staff (2014) "Medical gallery of Blausen Medical 2014" WikiJournal of Medicine 1 (2): 10. doi:10.15347/wjm/2014.010 reproduced under CC0 and CC SA creative commons licenses; [d]: From Neumann, E et al. (2014) G protein-coupled receptors in rheumatology, Nat Rev Rheumatol 10: 429–436 with publisher permission; [e]: From OpenStax College (2013) Biology for AP courses. OpenStax. http://cnx.org/content/col11496/latest reproduced under a CC BY 4.0 creative commons license; [g]: From public domain drawing by Mariana Ruiz Villarreal available at https://commons.wikimedia.org/wiki/File:Scheme_sodium-potassium_pump-en.svg.)*

found over time to distribute their fluorescent tags throughout the fused membranes. These and other results indicate that membrane lipids and proteins can readily undergo **rotational motion** around an axis perpendicular to the membrane and **translational diffusion** along the plane of each leaflet, with the rates of movement being influenced by such factors as temperature, phospholipid types, and cholesterol content.

Collectively, plasmalemmal components form a **hydrophobic** and **semi-permeable** barrier that partitions the cell from its ECM while also partially restricting transmembrane molecular flow. In the case of **passive diffusion**, small molecules cross membranes in response to concentration and charge gradients without added energy input or other forms of facilitation. The rates of diffusion largely depend on the **size** and degree of **polarity** of the traversing molecule. **Nonpolar molecules** such as O_2, CO_2, and many organic (i.e., carbon-based) compounds share electrons fairly equally throughout the molecule, thereby reducing their solubility in water and hence conferring hydrophobicity. Conversely, **polar molecules** like water, ethanol, and charged ions distribute their electrons asymmetrically and thus are hydrophilic. Accordingly, when encountering hydrophobic plasmalemmata, small non-polar molecules are generally capable of crossing membranes by passive diffusion (**Figure 2.3b**), whereas charged molecules and high-molecular-weight polar, or even nonpolar, molecules do not readily diffuse across membranes. Instead, their translocation requires integral protein **transporters**, which can be subdivided into: (i) **uniporters** that transport a single type of molecule across the membrane; (ii) **symporters** that allow different types of molecules to cross the membrane in the same direction; and (iii) **antiporters** that move different molecules across the membrane in opposing directions (**Figure 2.3e**).

Transporter-mediated translocations can utilize either **facilitated diffusion** or **active transport** modes, which are distinguished based on their energy requirements. **Facilitated diffusion**—for example, when **water** moves through its transporter channels without added energy input—simply relies on the conducive internal milieu of the transporter and needs no supplemental energy to translocate impermeable molecules (**Figure 2.3f**). Conversely, **active transport**, such as when ions are pumped against concentration gradients to establish resting membrane potentials (Chapter 9), requires energy input for ATP-mediated pumps (**Figure 2.3g**). Alternatively, for macromolecules and large materials that are not easily translocated by either of these two modes, intracellular incorporation is accomplished via several endocytic pathways, as described further below.

2.3 ENDOCYTOSIS: PHAGOCYTOSIS VS. CLATHRIN-MEDIATED AND CLATHRIN-INDEPENDENT PINOCYTOTIC MODES PLUS THE ROLES OF LYSOSOMES

Endocytosis has historically been classified into **phagocytosis**, which involves the uptake of large particles or even whole cells vs. **pinocytosis**, where extracellular molecules dissolved in fluids are incorporated (**Figure 2.4a**). More recently, several subtypes of pinocytosis have been recognized, including: (1) **macropinocytosis** involving large cellular extensions, called pseudopodia; (2) **clathrin-mediated endocytosis** that utilizes intracellular **clathrin** proteins plus extracellular receptors for a more selective uptake of materials; (3) **caveolin-dependent endocytosis** whose incipient vesicles (=**caveolae**) are enriched in the protein **caveolin**; and (4) **clathrin-/caveolin-independent pathways** (**Figure 2.4b**).

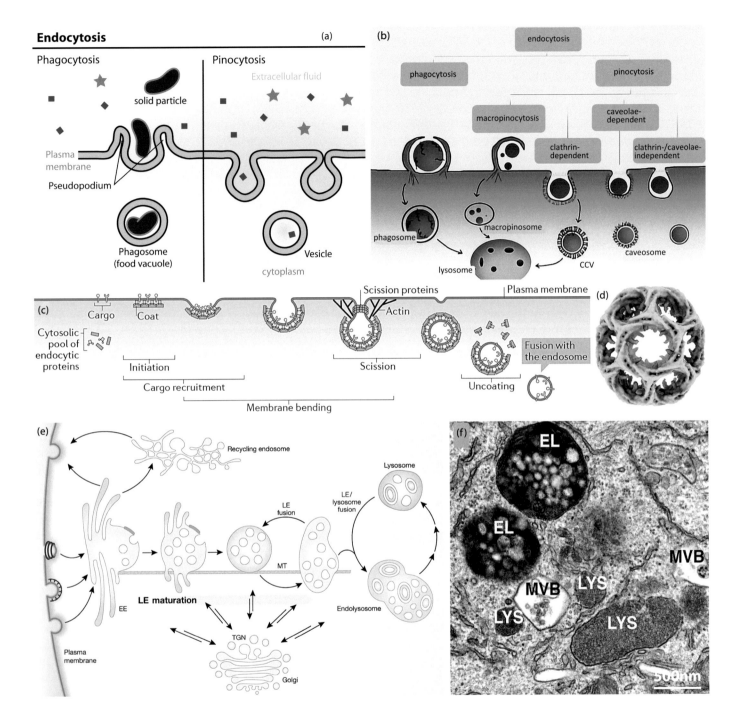

Figure 2.4 Endocytosis. *(a, b) Endocytosis can be classified as phagocytosis (incorporation of large particles or even cells) vs. pinocytosis (uptake of fluid and small particles), with pinocytosis being further distinguished based on clathrin- or caveolae-mediated pathways (b). (c, d) Clathrin can form a geodesic cage around endocytic vesicles during specific particle uptake, termed receptor-mediated endocytosis (outer diameter of cage in d = 80 nm). (e, f) Early endosomes (EEs) mature while fusing to Golgi-derived vesicles to become multivesicular bodies (MVBs). These, in turn, fuse with lysosomes (LYS) and thereby become endolysosomes (ELs) where catabolism and recycling can occur. ([a]: From public domain drawing by Mariana Ruiz Villarreal available at https://commons.wikimedia.org/wiki/File:Endocytosis_types.svg; [b]: From Kuo, L et al. (2013) The endocytosis and intracellular fate of nanomedicines: implication for rational design. Asian J Pharm Sci 8: 1–10 reproduced with publisher permission; [c]: From Kaksonen, M and Roux, A (2018) Mechanisms of clathrin-mediated endocytosis. Nat Rev Mol Cell Biol 10: doi:10.1038/nrm.2017.132 reproduced with publisher permission; [d]: From Paraan, M et al. (2020) The structures of natively assembled clathrin-coated vesicles. Sci Adv 6: eaba8397 reproduced under a CC BY 4.0 creative commons license; [e]: From Huotari, J and Helenius, A (2011) Endosome maturation. EMBO J 30: 3481–3500 reproduced with publisher permission; [f]: From Truschel, ST et al. (2018) Age-related endolysosome dysfunction in the rat urothelium. PLoS ONE 13: e0198817 https://doi.org/10.1371/journal.pone.0198817 reproduced under a CC BY 4.0 creative commons license.)*

CASE 2.1 HENRI DE TOULOUSE-LAUTREC'S LYSOSOMES

Artist
  (1903–1941)

  "Of course one should not drink much, but often."

  (T-L's view of drinking)

Henri de Toulouse-Lautrec was a French artist whose life was dominated by congenital skeletal disease and alcoholism. While growing up to a height of only 1.5 meters, Toulouse-Lautrec took solace in art and alcohol, feeling at home in the nightlife of cabarets and brothels. As he documented that world, Toulouse-Lautrec drank heavily, which eventually contributed to his early demise at age 36. Various lines of evidence suggest Toulouse-Lautrec suffered from pycnodysostosis, a rare autosomal recessive type of lysosomal storage disease

involving the cathepsin K enzyme in osteoclastic bone cells. In healthy bones, osteoclasts use lysosomal cathepsin K to digest organic matrix components during bone resorption, whereas in pycnodysostosis, the altered enzyme generates hypermineralized and poorly growing bones. Thus, inbreeding by his parents who were first cousins and both heterozygous for mutated cathepsin K yielded the lysosomal dysfunction underlying Toulouse-Lautrec's skeletal defects.

At the onset of clathrin-mediated endocytosis, plasmalemmal invaginations at incipient pinocytotic sites become coated on their intracellular side with clathrin molecules that interlink with each other to form a cage around the enlarging **clathrin-coated pit** (**Figure 2.4c, d**). During this process, extracellularly positioned receptors in the pit bind specific ligands, thereby enhancing the selectivity of molecular uptake in what is referred to as **receptor-mediated endocytosis**. After further invagination of the CCP, circumferentially arranged **dynamin** proteins constrict the neck of the CCP to dislodge it from the plasmalemma. This abscission process yields an internalized **clathrin-coated vesicle** (**CCV**) with its enclosed material being referred to as **cargo**. Cargo-carrying CCVs represent one kind of endocytically derived vesicles, which are referred to as **endosomes** in general, or **early endosomes** directly after their incorporation into the cell. Each CCV endosome eventually sheds its clathrin coat while also uncoupling its enclosed ligands from their receptors so that unbound receptors can be recycled for additional use. The remaining endosome fuses with Golgi-derived vesicles and eventually matures into a **multivesicular body** (**MVB**) (**Figure 2.4e, f**). MVBs can then combine with Golgi-derived organelles, called **lysosomes**, which contain acid hydrolase enzymes in a luminal pH of ~4.5, and following fusion, the hybrid **endolysosome** digests its cargo either for recycling purposes or for exocytosis of waste products. Lysosomes can also fuse with **autophagosomes** containing materials that arise from the cell itself, thereby helping to clear debris generated by turnover processes. Accordingly, human pathologies, termed **lysosomal storage diseases**, result from inadequate lysosomal functioning, which allows a buildup of toxic substances, thereby compromising cellular functions (**Case 2.1 Henri de Toulouse-Lautrec's Lysosomes**).

2.4 GOLGI APPARATUS: MULTIFUNCTIONAL NEXUS BETWEEN ENDO- AND EXOCYTIC PATHWAYS

Since its discovery in the 19th century by Italian anatomist Camillo Golgi, the **Golgi apparatus** (=Golgi complex) has been shown to play various roles, including: (i) relaying signals in intracellular cascades; (ii) providing a supplemental store of calcium ions; and (iii) helping to nucleate microtubules (pg. 45). However, the Golgi's major function is to modify proteins and lipids made by the **ER**. Such modifications include **glycosylations** that add carbohydrate moieties to ER-derived cargo. The Golgi-processed molecules can then be either delivered to **endocytic pathways** as transport vesicles and lysosomes or sent as vesicles to the plasmalemma for release into the ECM during a process termed **exocytosis**.

To perform these functions, the Golgi apparatus comprises curved stacks of membrane-bound sacs (=**cisternae**) that are closely associated with the ER (**Figure 2.5a**). Each stack, or **Golgi body**, usually contains ~5–10 cisternae that are polarized into three main compartments: (i) the **cis** (=same side) compartment closest to the ER, (ii) the **trans** (=opposite side) region farthest from the ER, and (iii) an intervening **medial** component between the cis and trans compartments (**Figure 2.5b**).

Neighboring Golgi cisternae are linked together via **GRASPs** (Golgi reassembly stacking proteins), which not only attach to neighboring cisternal membranes but also interlock with each other to keep the Golgi intact

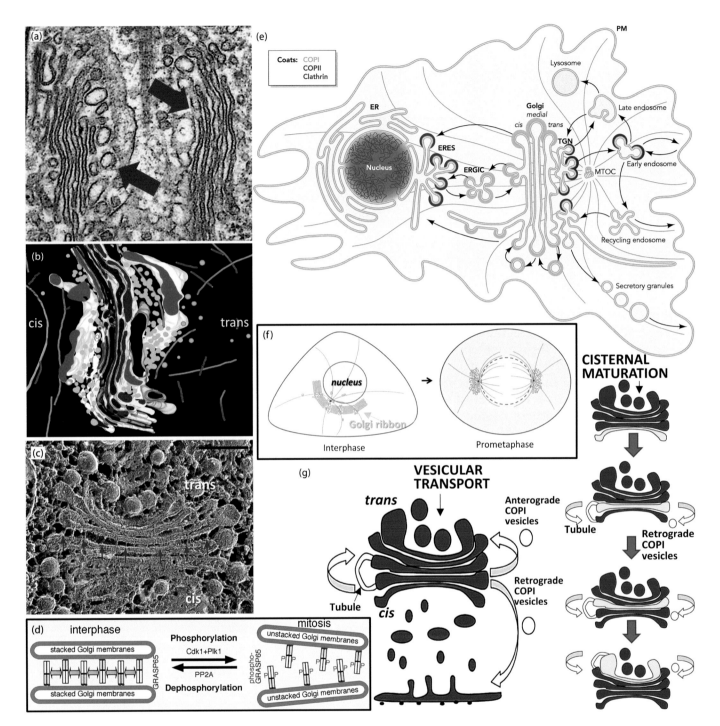

Figure 2.5 Golgi apparatus. (a, b) *Figure (a) shows a TEM of two Golgi bodies (red arrows) (12,000X), whereas a 3D reconstruction in (b) depicts the Golgi's cis- (closest to ER) vs. trans sides (farthest from ER). Microtubules are displayed in fuchsia, and clathrin-coated vesicles are green.* **(c)** *A freeze etched view of Golgi cisternae shows GRASP connections (blue arrows) in a non-dividing cell (scale bar = 100 nm).* **(d, e)** *Prior to division, unphosphorylated GRASPs keep Golgi bodies intact. However, during division, phosphorylation by kinases such as Cdk1 and Plk1 allow disruption of the Golgi and a spreading of Golgi components throughout the cell (d, e). Conversely, dephosphorylation by the phosphatase PP2A after division allows Golgi reassembly (d). Figure e summarizes the Golgi's endo- and exocytic roles: trafficking of Golgi vesicles is mediated by two main coat protein complexes (COPI and COPII), with COPII proteins initially assembling around vesicles that pinch off the ER at discrete ER exit sites (ERESs). COPII vesicles move in an anterograde direction along microtubules (blue lines) to reach an ER-to-Golgi intermediate complex (ERGIC). The ERGIC has a heterogeneous collection of COPI vesicles, some of which recycle material in a retrograde direction back to the ER. After materials move from the cis to*

Figure 2.5 (Continued) *medial to trans Golgi, clathrin coating in the trans Golgi network (TGN) can target vesicles to the endolysosome system rather than to the cell membrane for exocytosis. (**f**) Golgi stacks of non-dividing mammalian cells are typically linked together to form elongated ribbons with incompletely understood functions. (**g**) Movement of materials through the Golgi can involve anterograde vesicular transport, the maturation of cisternae as they move from a cis to trans position, and the direct trafficking of large molecules via tubules interconnecting nearby cisternae. ([b]: From Martinez-Martinez, N et al. (2017). A new insight into the three-dimensional architecture of the Golgi complex: Characterization of unusual structures in epididymal principal cells. PLoS ONE 12: e0185557 reproduced under a CC BY 4.0 creative commons license; [c, d]: From Zhang, X and Wang, Y (2016) GRASPs in Golgi structure and function. Front Cell Dev Biol 3: doi:10.3389/fc reproduced under a CC BY 4.0 creative commons license; [e]: From Szul, T and Sztul, E (2011) COPII and COPI traffic at the ER-Golgi interface. Physiology 26: 348–364 reproduced with publisher permission; [f]: From Saraste, J and Prydz, K (2019) A new look at the functional organization of the Golgi ribbon. Front Cell Dev Biol 7: doi: 10.3389/fcell.2019.00171 reproduced under a CC BY 4.0 creative commons license; [g]: From Martinez-Menarguez, JA (2013) Intra-Golgi transport: roles for vesicles, tubules, and cisternae. ISRN Cell Biol: http://dx.doi.org/10.1155/2013/126731 reproduced under a CC BY 3.0 creative commons license.)*

prior to cell division (**Figure 2.5c–e**). However, as cells prepare to divide, GRASP chains are depolymerized by phosphorylating **kinase** enzymes. Such phosphorylations disassemble the Golgi before it re-assembles again via **phosphatase**-induced de-phosphorylations of GRASPs as daughter cells reform their Golgi bodies (**Figure 2.5d-f**). GRASP-mediated stacking of Golgi cisternae minimizes transport distances and ensures that processing steps are carried out in the proper sequence. Conversely, human pathologies such as **Alzheimer's disease** (**AD**) (Chapter 9) are characterized by Golgi fragmentation due to GRASP over-phosphorylation, leading to accelerated and inaccurate trafficking through disordered Golgi assemblages (**Figure 2.5d**) (**Case 2.2 Ronald Reagan's Golgi Bodies**). In addition to such roles, GRASPs and other proteins such as **golgins** can also assemble multiple Golgi stacks into **ribbon-like** arrays, whose functions have yet to be fully elucidated (**Figure 2.5f**).

Along with its endocytic roles involving production of lysosomes and other vesicles that fuse with endosomes, the Golgi also interacts with the ER during **exocytosis**. Thus, ER-derived vesicles that move in an **anterograde** direction toward the cell periphery are able to combine with the Golgi before being directed to the plasmalemma for exocytosis (**Figure 2.5e, g**). As summarized in **Figure 2.5e**, such trafficking involves two main kinds of proteinaceous coats that cover ER vesicles, called **coat protein complex II** (**COPII**) and **coat protein complex I** (**COPI**). COPII proteins assemble around vesicles containing newly synthesized protein, lipid, or complex carbohydrate cargos that are pinched off the ER at discrete ER exit sites (**ERESs**) (**Figure 2.5e**). The COPII-coated vesicles can then move to an **ER-to-Golgi intermediate complex** (**ERGIC**), which contains material

CASE 2.2 RONALD REAGAN'S GOLGI BODIES

40th President of the United States
(1911–2004)

"Honey, I forgot to duck."

(R.R. joking to his wife after being shot in 1981, years before his Alzheimer-induced forgetfulness set in)

As the 40th U.S. president, Ronald Reagan had prepared for a career in politics by working as a radio sportscaster and actor. After serving as Governor of California, Reagan used his mastery of the media to win two presidential terms. Already 69 years old when he first took office, Reagan's vigor was severely tested only a few months into his presidency after he was shot by a would-be assassin. The bullet lodged near his left lung and caused substantial internal bleeding. Although Reagan initially quipped with his wife and physicians, he was, in fact, gravely wounded, and only timely emergency care prevented his death. Following the shooting, Reagan remained relatively healthy until 1985, when he underwent surgery for colon cancer.

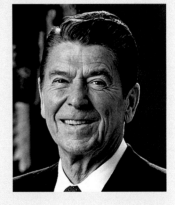

Reagan recovered from the operation and survived nearly 20 more years before succumbing to Alzheimer's disease (AD). Some recent analyses of AD focus on Golgi bodies, which among other functions serve to modify proteins and lipids. In neurons of AD patients, Golgi bodies often undergo fragmentation thereby contributing to increased levels of abnormally modified proteins. Thus, along with other processes typically associated with the disease (Chapter 9), Reagan's terminal decline most likely involved a breakdown in his neuronal Golgi bodies.

newly incorporated from COPII-coated vesicles as well as a heterogeneous collection of **COPI-coated** vesicles. Some of the COPI vesicles mediate the **retrograde** return to the ER of improperly folded proteins or Golgi-derived components, thereby allowing the clearing of defective products and the recycling of materials used in trafficking. Alternatively, other COPI vesicles are transported via **microtubules** (**Figure 2.5e**) in an **anterograde** direction to the cis-Golgi. The particular directionality of COPI vesicle movements depends on specific adaptor proteins, transmembrane receptors, and microtubule-associated motors.

Following fusion with the cis-Golgi, COPI-delivered materials can make their way to the trans face both by additional vesicular fusions at the Golgi periphery and by **cisternal maturation**, in which cis cisternae mature into trans configurations while moving further away from the ER (**Figure 2.5g**). In addition, tubules directly connecting neighboring cisternae can facilitate the intra-Golgi transport of large molecules like collagen precursors, which are not easily accommodated by canonical Golgi trafficking (**Figure 2.5g**). Following arrival in the **trans-Golgi network** (**TGN**), **CCVs** tend to target materials to the endolysosomal system used in endocytosis, whereas alternative vesicle types can be moved to the plasmalemma for exocytosis (**Figure 2.5e**).

2.5   ENDOPLASMIC RETICULUM (ER): SHEET (ROUGH, WITH RIBOSOMES) VS. TUBULAR (SMOOTH, WITHOUT RIBOSOMES) ER REGIONS PLUS STEPS IN PROTEIN SYNTHESIS

The **ER** is an expansive network of membranes comprising two main parts: (i) the double-membrane **nuclear envelope** (**NE**) that surrounds the nucleus (**Figure 2.6a**) and (ii) a **peripheral ER** that is contiguous with the NE and distributed throughout the cytoplasm as flattened **sheets** and rounded **tubules** (**Figure 2.6a, inset**). The nuclear envelope regulates the bidirectional transport of RNAs and proteins between the nuclear and cytoplasmic compartments, whereas the peripheral ER plays multiple key roles, including protein and lipid synthesis, calcium ion storage, and mediating certain types of stress reactions (Box 2.1 Cystic Fibrosis and Cellular Stress Responses Involving the Endoplasmic Reticulum) (**Case 2.3 Alice Martineau's Endoplasmic Reticulum (ER) Networks**). Each flattened **sheet** of the peripheral ER has an internal lumen that typically measures ~50 nm wide and has attached on its cytoplasmic surface individual **ribosomes** (**Figure 2.6b**) or clustered ribosomes, called **polysomes**. Conversely, ribosomes tend to be lacking on curved ER **tubules** (**Figure 2.6a, inset; c, d**). Thus, sheet vs. tubular ER regions were historically termed rough vs. smooth ER based on their ribosomal content.

ER sheets provide stable and expansive lumens for use in **protein synthesis**, whereas curved tubules are optimized for synthesizing certain **lipids** and for use in **signaling pathways**. The overall proportion of sheet-to-tubular regions tends to be maintained in fully differentiated cell types, with secretory cells such as pancreatic acinar cells containing predominantly ER sheets for abundant protein exocytosis, whereas muscle and epithelial cells that secrete relatively few proteins possess mainly tubular ER. Some ER proteins such as the general translocator **Sec61** are expressed in both sheet and tubule membranes (**Figure 2.6d, e**), whereas proteins like **CLIMP63** are localized to sheets. Conversely, **atlastins** and certain **reticulons** are characteristic of ER tubules (**Figure 2.6d**).

Protein translation involving ribosome-studded ER sheets is routinely initiated in the cytoplasm by individual **free ribosomes** and **polysomes** that are unattached to the ER (**Figure 2.6f**). Each ~25-nm-wide ribosome comprises a **large** and a **small ribosomal subunit** between which an mRNA strand can be translated into protein. Cytoplasmic proteins complete their translation on ribosomes unattached to the ER, whereas ribosomes forming membrane-associated or secreted proteins are **recruited** to the ER so that nascent proteins can be translocated into the ER lumen. Such recruitment and translocation occur via several pathways, with one of the most intensively studied modes being mediated by large ribonucleic protein complexes, called **signal recognition particles** (**SRPs**). After an SRP attaches to the appropriate **signal sequence** on a nascent polypeptide chain, the SRP binds to an SRP receptor on the ER membrane and becomes anchored to a nearby ER **translocon** complex (**Figure 2.6f**). Such docking can alter the translocon's conformation to facilitate import of the elongating protein into the ER lumen. Subsequently, the ribosome, mRNA, and signal sequence are released for re-use in the cytoplasm, and eventually protein synthesis is completed within the ER lumen while also incorporating post-translational modifications like glycosylations and proper protein folding (**Figure 2.6f**).

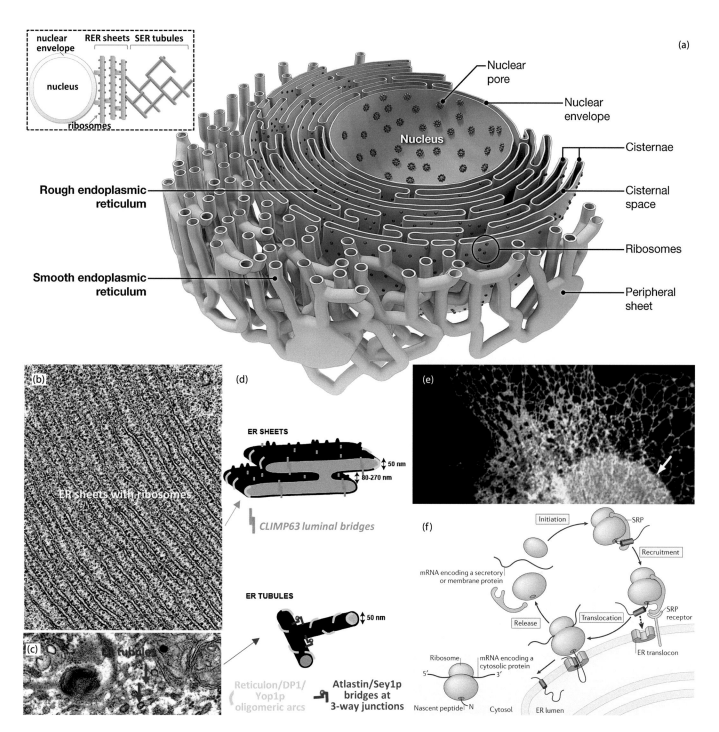

Figure 2.6 Endoplasmic reticulum (ER). *(a–c) Contiguous with the nuclear envelope, the peripheral ER comprises sheets with ribosomes (b) (15,000X) and tubules lacking ribosomes (c) (18,000X). (d) ER sheets are characterized by proteins such as CLIMP63, whereas tubules are enriched in atlastins and reticulons. (e) A doubly labeled ER in a cultured cell shows sheet-specific fluorescence (green) near the nucleus, more peripherally located tubule-associated fluorescence (red), and an overlap region of both signals (yellow). (f) Translation of cytosolic proteins occurs on free ribosomes (lower left), whereas ribosomes translating secretory or membrane proteins can contain a signal recognition particle (SRP) that allows ribosome docking on the ER and completion of translation within the ribosomal lumen. ([a]: From Goyal, U and Blackstone, C (2013) Untangling the web: mechanisms underlying ER network formation. Biochimica et Biophysica Acta 1833: 2492–2498, reproduced with publisher permission; [a, inset]: From Yamanaka, T and Nukina, N (2018) ER dynamics and derangement in neurological diseases. Front Neurosci 10. doi: 10.3389/fnins.2018.00091 reproduced under a CC BY 4.0 creative commons license; [d]: From Romero-Brey, I and Bartenschlager, R (2016) Endoplasmic reticulum: the favorite intracellular niche for viral replication and assembly. Viruses 8. doi:10.3390/v8060160 reproduced under a CC BY 4.0*

Figure 2.6 (Continued) *creative commons license; [e]: From Shibata, Y et al. (2006) Rough sheets and smooth tubules. Cell 126: 435–439, reproduced with publisher permission; [f]: From Reed, DW and Nicchitta, CV (2015) Diversity and selectivity in mRNA translation on the endoplasmic reticulum. Nat Rev Mol Cell Biol 16: doi:10.1038/nrm3958 reproduced with publisher permission.)*

BOX 2.1 CYSTIC FIBROSIS AND CELLULAR STRESS RESPONSES INVOLVING THE ENDOPLASMIC RETICULUM

Although mutations in the **cystic fibrosis transmembrane conductance receptor** (**CFTR**) trigger abnormal ionic fluxes that are the main drivers of **cystic fibrosis** (**CF**), cellular stress reactions related to defective CFTR expression also contribute to CF pathogenesis. In particular, improperly folded forms of CTFR proteins can accumulate in the ER of certain cells, such as **macrophages**, and thereby elicit an **unfolded protein response** (**UPR**) that has initially evolved as a corrective mechanism for eliminating proteins that are occasionally assembled improperly. However, with a continual accumulation of incorrectly folded CFTR protein, CF macrophages are re-programmed toward a deleterious ("M1") phenotype that chronically activates the ER membrane protein IRE1α (Inositol-Requiring Enzyme-1α) and, in turn, results in

the nuclear incorporation of a transcription factor, called XBP1 (X-box Binding Protein 1). In response to XBP1, genes encoding compounds that promote a variety of destructive processes collectively termed inflammation are transcribed, thereby allowing high levels of pro-inflammatory compounds to be exocytosed (**Figure a**). In addition, in normal macrophages, CFTR typically protects cells against bacterial toxins like lipopolysaccharide that bind to plasmalemmal proteins, called Toll-like receptors (TLRs) and via such binding upregulate inflammatory pathways. Conversely, in CF patients, who are prone to excessive infections owing to the thickened mucus that they produce in the absence of CFTR, TLR-mediated signaling further stimulates inflammation, thereby exacerbating ER-enhanced CF pathogenesis.

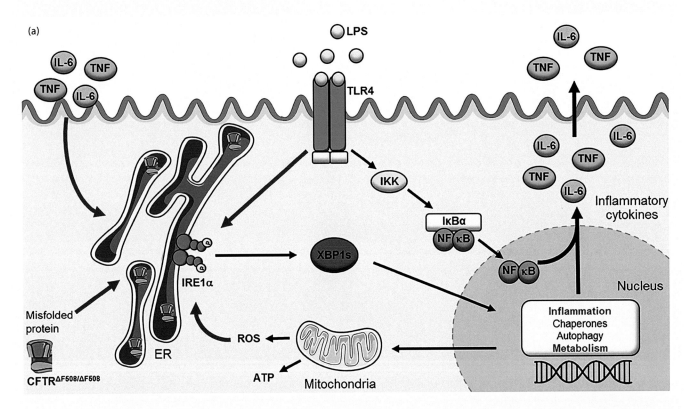

Box 2.1 (a) *Mutations of the cystic fibrosis transmembrane conductance regulator (CFTR) cause (i) misfolded protein accumulation, (ii) endoplasmic reticulum (ER) stress, and (iii) activation of inositol-requiring enzyme 1-α (IRE1-α) pathways that regulate unfolded protein response (UPR) genes, thereby exacerbating cystic fibrosis pathogenesis. LPS= lipopolysaccharide; ROS= reactive oxygen species; TLR= Toll-like receptor. (From Lara-Reyna, S et al. (2019) Metabolic reprogramming of cystic fibrosis macrophages via the IRE1α arm of the unfolded protein response results in exacerbated inflammation. Front Immunol 10: doi: 10.2289/fimmu.2019.01789 reproduced under a CC BY 4.0 creative commons license.)*

CASE 2.3 ALICE MARTINEAU'S ENDOPLASMIC RETICULUM (ER) NETWORKS

Model/Singer/Songwriter
  (1972–2003)

"The last thing I want is for people to see me as a victim."

(A.M. commenting in an interview about her illness)

Alice Martineau was born in London, England with cystic fibrosis (CF), which afflicts about 1 in 3,000 Caucasian babies. In spite of her disease, Martineau graduated from college, became a model, and released an album of her music before dying at age 30. For patients like Martineau, research is beginning to elucidate the far-reaching effects of CF beyond dysfunction of cystic fibrosis transmembrane conductance regulator (CFTR)

channels in cell membranes. For example, improperly folded CFTR proteins cause ER-associated stress reactions, called unfolded protein responses (UPRs), which exacerbate CF (Box 2.1). Accordingly, after being placed on a triple-transplant list to replace her lungs, heart, and liver, Martineau succumbed to widespread destruction triggered by CF, including damage presumably involving ER networks of her cells.

2.6 MITOCHONDRIA: ATP-GENERATING POWERHOUSES AND THEIR ROLES IN APOPTOTIC FORMS OF PROGRAMMED CELL DEATH

After evolving from ancient bacteria that were incorporated into unicellular progenitors of eukaryotes, dual-membrane-bound **mitochondria** have come to contain four main compartments: (1) an **outer mitochondrial membrane** (**OMM**); (2) an **inner mitochondrial membrane** (**IMM**) that is elaborated into inwardly directed folds, called **cristae**; (3) an intervening **intermembrane space**; and (4) an innermost gel-like **matrix** with mitochondrial genes, ribosomes, and soluble enzymes that are collectively surrounded by the IMM (**Figure 2.7a–d**). Via these components, mitochondria can modulate Ca^{2+} **signaling** pathways, **reactive oxygen species** (**ROS**) levels, and **redox** (reduction-oxidation) types of chemical reactions, while also carrying out their major role of **generating** a sufficient supply of **ATPs** for energy-consuming cellular processes.

As the most efficient mechanism by which eukaryotic cells synthesize ATP, **oxidative phosphorylation** (**OXPHOS**) begins in the mitochondrial matrix with **citric acid cycle** enzymes converting breakdown products of carbohydrates, lipids, and proteins into **NADH** (nicotinamide adenine dinucleotide bound to hydrogen) and other types of **electron carriers** (**Figure 2.7e**). After being shuttled to the IMM, these carriers pass their electrons to protein complexes constituting the **electron transport chain** (**ETC**), whose abundance is optimized via the increased surface area of IMM cristae (**Figure 2.7f**). Additional electron transfers and energy-releasing redox reactions by ETC components end with molecular oxygen as the terminal electron acceptor, and along the way, the ETC pumps **protons** (H^+ **ions**) into the intermembrane space for storing potential energy in a chemiosmotic gradient across the IMM. This gradient is tightly linked to ATP production, because intermembrane protons normally return to the matrix by flowing through IMM channels, called **ATP synthases**, which efficiently couple such ion movements to ATP synthesis (**Figure 2.7f**).

In humans, **mitochondrial DNA** (**mtDNA**) retained during evolution from ancient bacterial symbionts encodes 13 ETC genes that are essential for ATP synthesis. Normally, proper mtDNA expression is maintained by several kinds of nuclear genes. However, mutations causing **mitochondrial DNA depletion syndromes** (**MDDSs**) compromise mtDNA expression, thereby leading to widespread pathogenesis and lethality, owing to insufficient ATP production (**Case 2.4 Charlie Gard's Mitochondria**).

Although most mitochondria are cylindrical and measure ~0.5–1 µm wide by a few micrometers long, some can form 20-µm-long tubules, branched networks, or even annular organelles (**Figure 2.7d**). Variability in mitochondrial sizes and shapes depends on differing levels of **fission** vs. **fusion** events that are mediated by such proteins as **dynamin-related protein 1** (**Drp1**) vs. the **mitofusins Mfn1/2**, respectively (**Figure 2.7g**). Fission subdivides large mitochondria into smaller units that can facilitate not only the distribution of mitochondria into daughter progeny following cell divisions but also the movement of mitochondria to cellular regions with increased ATP demands. Conversely, fusions are required for generating larger mitochondria and for tethering mitochondria to components such as the ER. Accordingly, abnormally fused mitochondria due to mitofusin

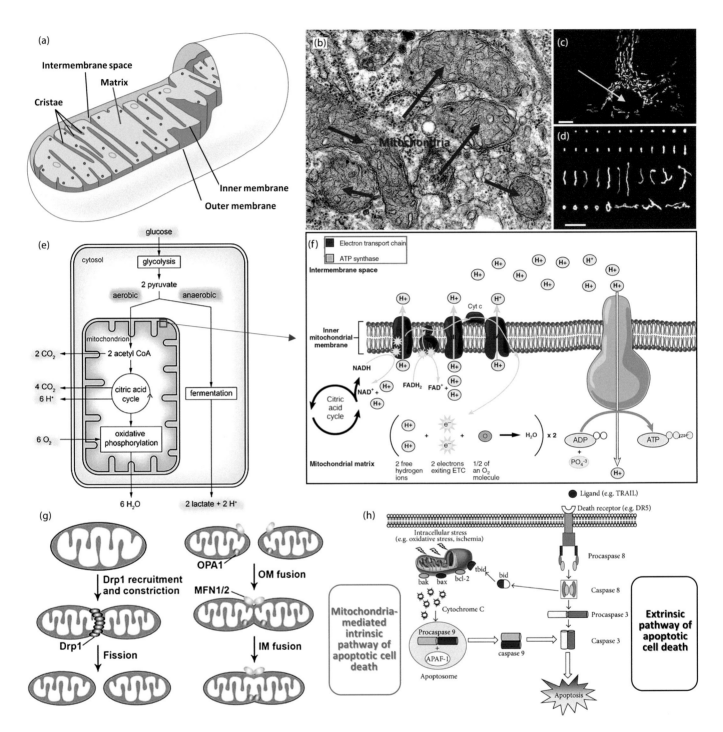

Figure 2.7 Mitochondria. (*a, b*) *Mitochondria are double-membrane-bound and typically rod-shaped, as shown in a cut-away diagram (a) and TEM (b, red arrows) (16,000X). (**c**) Most mitochondria are spread throughout the cytoplasm, as shown in (c) where mitochondrial fluorescence is lacking only in the central nucleus (arrow) of a smooth muscle cell; scale bar = 10 μm. (**d**) Mitochondria are typically ~0.5–1 μm wide and up to a few micrometers long, although considerable morphological variability can also occur, as illustrated by mitochondria from cultured smooth muscle cells; scale bar = 10 μm. (**e**) Mitochondria use oxidative phosphorylation (OXPHOS) to produce ATPs more efficiently than is achieved by anaerobic fermentation. (**f**) During OXPHOS, the citric acid cycle generates electron carriers that donate electrons to the inner mitochondrial membrane's electron transport chain (ETC). The ETC pumps protons (H+ ions) into the intermembrane space so that protons returning to the matrix through ATP synthases can be coupled to ATP production. (**g**) Mitochondria can change shape and size by DRP1-mediated fission processes and by fusions of their outer and inner membranes (OM, IM) involving mitofusins (MFNs). (**h**) Under cellular stress, mitochondria can release cytochrome c to form parts*

Figure 2.7 (Continued) *of apoptosomes that mediate cell death via an intrinsic mode of apoptosis. ([a, f]: From OpenStax College (2013) Anatomy and Physiology from OpenStax College (2013) Anatomy and Physiology. OpenStax http://cnx.org/content/col11496/latest, reproduced under a CC BY 4.0 creative commons license; [c, d]: From McCarron, JG et al. (2013) From structure to function: mitochondrial morphology, motion and shaping in vascular smooth muscle. J Vasc Res 50: 357–371 reproduced under a CC BY 3.0 creative commons license; [e]: From Kienenger, J et al. (2018) Microsensor systems for cell metabolism – from 2D culture to organ-on-chip. Lab Chip 18: DOI: 10.1039/c7lc00942a reproduced under a CC BY 3.0 creative commons license; [g]: From Cai, Q a Cai, Q and Tammenini, P (2019) Alterations in mitochondrial quality control in Alzheimer's disease. Front Cell Neurosci 10. doi:10.3389/fncel.2016.00024 reproduced under a CC BY 4.0 creative commons license; [h]: From Loreto, C et al. (2014) The role of intrinsic pathway in apoptosis activation and progression in Peyronie's disease. J Biomed Biotechnol 2014: doi:10.1155/2014/616149 reproduced under a CC BY 3.0 creative commons license.)*

mutations are linked to various pathologies, including a neurodegenerative disease, called Charcot-Marie-Tooth type 2A neuropathy.

As opposed to their roles in maintaining homeostasis, mitochondria also react to various cytotoxic stresses to trigger a form of programmed cell death, called **apoptosis** (**Figure 2.7h**). As a key driver of the intrinsic apoptotic pathway, the protein **cytochrome c** normally accumulates within mitochondria to form part of the ETC. In healthy cells, cytochrome c is kept from escaping through OMM channels into the cytoplasm. However, in response to apoptotic stress, such channels become reconfigured, thereby allowing cytochrome c to enter the cytoplasm. Such cytoplasmic cytochrome c associates with **Apaf-1** (apoptotic protease promoting factor 1) to help organize multimeric Apaf-1/cytochrome c arrays, called **apoptosomes** (**Figure 2.7h**). These, in turn, activate **caspase** enzymes that trigger the characteristic shrinkage and blebbing of cells as they die by apoptosis. Via such mechanisms, apoptosis ensures that overly stressed cells can be eliminated in a programmed fashion that helps prevent deleterious outcomes. For example, apoptosis is used to eliminate aged oocytes, which if allowed to fertilize could generate defective embryos (Chapter 18).

2.7 FUNCTIONAL MORPHOLOGY OF PEROXISOMES

Peroxisomes (=microbodies) are so-named, based on their ability both to break down and to synthesize **hydrogen peroxide** (H_2O_2) (**Figure 2.8a**). Similar to mitochondria, peroxisomes can process reactive oxygen species such as H_2O_2 and modulate redox reactions in various cell types (**Figure 2.8b**). However, unlike mitochondria, peroxisomes are: (i) invariably bounded by a single membrane, (ii) often endowed with a **crystalline core** that

---

CASE 2.4 CHARLIE GARD'S MITOCHONDRIA

Dying Infant Who Was Subject of a Legal Dispute (2016–2017)

"...all those who provided second opinions and the consultant instructed by the parents in these proceedings share a common view that further treatment would be futile."

(From a judge's ruling stating that additional therapies would not be able to save C.G.'s life)

Charlie Gard was an infant suffering from a fatal mitochondrial disorder who gained international fame during court cases adjudicating his right to die. Born in London, England, Gard was brought back to the hospital when it became clear he was failing to thrive. Tests revealed that Gard suffered from mitochondrial DNA depletion syndrome (MDDS) due to a mutated gene required for proper mitochondrial

DNA expression. Soon after his diagnosis, Gard's condition substantially deteriorated, and in the absence of approved treatments, hospital officials proposed taking him off life support. Gard's parents opposed the move and launched an internet campaign to raise money for an experimental therapy offered in the United States. The standoff was brought to court and during the initial hearing as well as throughout subsequent appeals, Gard's case received worldwide attention. Eventually, the High Court ruled that the extremely limited chance a therapy might help Gard was not worth the pain and suffering it would cause, and after being removed from life support, Gard died due to damage triggered by his dysfunctional mitochondria.

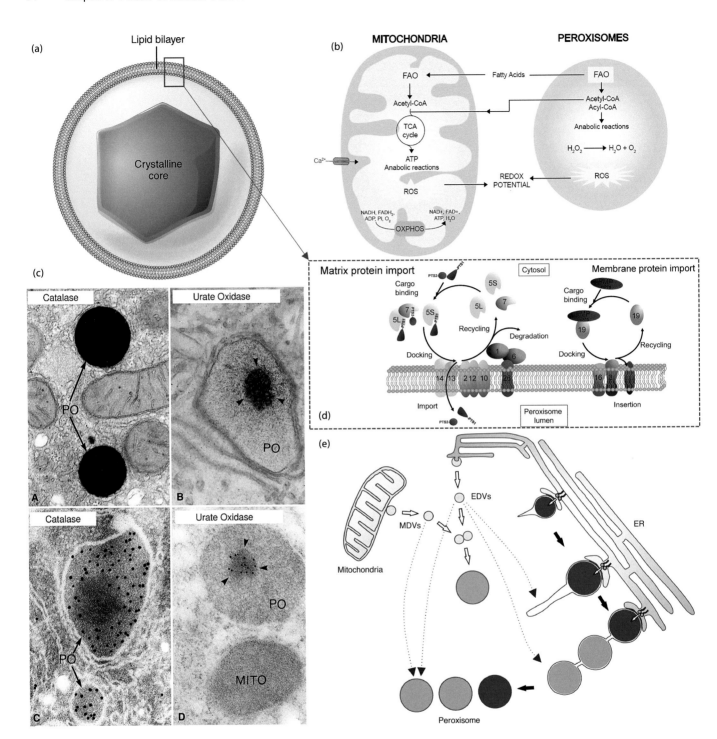

Figure 2.8 Peroxisomes. (**a**) *Spherical and often with a crystalline core, peroxisomes are so named for their ability to generate and break down hydrogen peroxide (H$_2$O$_2$).* (**b**) *Peroxisomes resemble mitochondria in several regards but lack double membranes or their own DNA and protein-synthesizing machinery.* (**c**) *When assayed by histochemistry (upper row) or immunocytochemistry (lower row), some peroxisomal enzymes such as H$_2$O$_2$-degrading catalase occur throughout the peroxisome, whereas others like urate oxidase are localized to the peroxisome (PO) crystalline core (c-A 17,000X; c-B 50,400; c-C 42,800; c-D: 36,600X).* (**d**) *Numerous kinds of peroxisome-specific peroxin (PEX) proteins (numbered in drawing) help incorporate proteins into the peroxisomal lumen or membrane.* (**e**) *Peroxisome biogenesis can involve either an ER-mediated pathway, where additional peroxisomes bud from a pre-existing one that arose in contact with the ER or a de novo mode generated from mitochondria- and ER-derived vesicles (MDVs, EDVs). ([a]: From OpenStax College (2013) Anatomy and Physiology from OpenStax College (2013) Anatomy and Physiology. OpenStax http://cnx.org/content/col11496/latest, reproduced under a CC BY 4.0 creative commons license; [b]: From Dosil, SG et al.*

Figure 2.8 (Continued) *(2020) The swing of lipids at peroxisomes and endolysosomes in T cell activation. Int J Mol Sci 21: doi:10.3390/ijms21082859 reproduced under a CC BY 4.0 creative commons license; [c]: From Schrader, M and Fahimi, HD (2008) The peroxisome: still a mysterious organelle. Histochem Cell Biol 129: 421–440 reproduced with publisher permission; [d]: From Waterham, HR et al. (2016) Human disorders of peroxisome metabolism and biogenesis. Biochimica Biophysica Acta 1863: 922–933 reproduced with publisher permission; [e]: From Costello, JL and Schrader, M (2018) Unloosing the Gordian knot of peroxisome formation. Curr Opin Cell Biol 5: 50–56 reproduced under a CC BY 4.0 creative commons license.)*

contains enzymes like urate oxidase (**Figure 2.8c**), and (iii) characterized by organelle-specific proteins, called **peroxins** (**PEXs**), which can help incorporate proteins into the peroxisomal membrane and lumen (**Figure 2.8d**). Via such attributes, peroxisomes generally serve to protect cells from cytotoxic stresses, thereby helping to prevent various diseases and aging-related pathogenesis.

Since peroxisomes lack nucleic acids or protein-synthesizing machinery, all their membrane proteins and luminal contents are formed elsewhere and incorporated into differentiating peroxisomes. During such differentiation, two distinct biogenic modes can be utilized (**Figure 2.8e**). In the case of **ER-based biogenesis**, an initial peroxisome forms in contact with the ER, and following influx of some ER-derived lipid and protein components, the peroxisome **elongates** and **buds off** additional peroxisomes (**Figure 2.8e**). Alternatively, during **de novo synthesis**, new peroxisomes are not derived from pre-existing ones but instead develop individually from **ER-derived vesicles** (**EDVs**) or **mitochondria-derived vesicles** (**MDVs**) (**Figure 2.8e**). Of these two pathways, formation on the ER and subsequent budding appears to be the chief biogenetic mode used by most cells.

2.8 THE CYTOSKELETON (MICROTUBULES, MICROFILAMENTS, AND INTERMEDIATE FILAMENTS): CELLULAR SCAFFOLDING, RAILROAD TRACKS, AND MOTILITY MEDIATORS PLUS SEPTIN CYTOSKELETAL ELEMENTS

The **cytoskeleton** is an intracellular proteinaceous network that is composed mainly of **microtubules, microfilaments**, and **intermediate filaments** (**Figure 2.9a**). At only 6–7 nm wide, microfilaments (=MFs, actin filaments) are the thinnest cytoskeletal elements and comprise solid rope-like helices of polymerized **actin** molecules (**Figure 2.9b–d, f, h**). Conversely, microtubules (MTs) are the largest cytoskeletal components measuring ~20–25 nm wide, with each MT forming a hollow array of α- and β-tubulin polymers (**Figure 2.9b, e–g**). Like MFs, intermediate filaments (IFs) are helically wound filaments lacking a central lumen and are so-named owing to their ~10–12-nm size falling in between those of MTs and MFs (**Figure 2.9a**). However, unlike actin-based MFs, IFs are composed of various non-actin proteins, including cytoplasmic **keratins** and nuclear **lamins** (Table 2.3). Collectively, networks generated by these three cytoskeletal elements provide structural support while also mediating cellular movements, intracellular trafficking, and signaling cascades.

As the building blocks of **microfilaments (MFs)**, actin molecules occur as globular monomers, called **G-actin**, which are typically bound to **adenosine diphosphate** (**ADP**) to reduce the likelihood of spontaneous polymerization. However, in the presence of **actin-binding-proteins** (**ABPs**) that facilitate the exchange of ATP for ADP, polymerization of ATP-bound G-actin occurs at the MF + **end** to generate filamentous **F-actin** (**Figure 2.9b**). Conversely, loss of G actin at the MF – **end** reduces MF length (**Figure 2.9b**). Additions and subtractions of actin underlie the dynamic process of actin **treadmilling** in which continually refreshed MFs can drive cellular shape changes, migration, as well as endo- and exocytic processes. MF dynamics also help mediate various physiological processes including myofibril contractions (Chapter 8), daughter cell separations during cell divisions, and even some nuclear reactions in the case of intranuclear actin filaments (Box 2.2 Intranuclear Actin Dynamics during the Cell Cycle).

During **microtubule** formation, **heterodimers of α- and β-tubulin** are linked together as 13 protofilament polymers arranged around the central microtubular lumen (**Figure 2.9b**). Such polymerization occurs in discrete sites within the cytoplasm, termed **microtubule-organizing centers** (**MTOCs**), which typically comprise a pair of cylindrical **centrioles** plus additional **pericentriolar proteins**, collectively constituting a **centrosome**. Both centrosomal centrioles contain a pinwheel array of nine MT triplets. In addition, the older **mother** centriole of

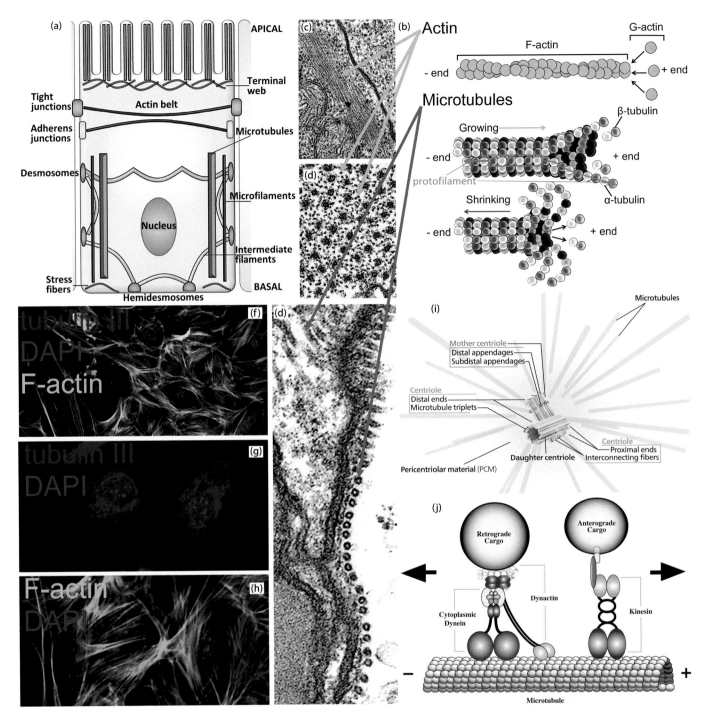

Figure 2.9 Cytoskeleton I: microfilaments (MFs) and microtubules (MTs). (**a**) Cells contain three main kinds of cytoskeletal components: 6–7 nm-wide MFs, 20–25 nm-wide MTs, and 10–12 nm-wide intermediate filaments. (**b–d**) MFs are composed of a double helix of actin polymers, whereas each microtubule comprises 13 protofilament chains of αβ-tubulin heterodimers arranged around a central lumen. (c) 18,000X; (d) 90,000X. (**e**) As shown in a glandular cell, MTs appear hollow both in longitudinal (upper) vs. transverse (lower) sections (91,000X). (**f–h**) Doubly labeled testicular cells show MTs (red) near the nucleus and MFs (green) in more peripheral locations. (**i**) Microtubule-organizing centers (MTOCs) typically comprise two centrioles plus pericentriolar material that together constitute a centrosome from which MTs arise. (**j**) Microtubule-associated proteins like dyneins help move vesicular cargo along MTs in a retrograde direction toward the MT's – end, whereas most kinesins transport materials in an anterograde direction toward the MT's + end. ([a]: From Chifflet, S and Hernandez, JA (2012) The plasma membrane potential and the organization of the actin cytoskeleton of epithelial cells. Int J Cell Biol 2012: doi:10.1155/2012/121424 reproduced under a CC BY 4.0

Figure 2.9 (Continued) *creative commons license; [b]: From Mostowy S (2014) Multiple roles of the cytoskeleton in bacterial autophagy. PLoS Pathog 10: e1004409. doi:10.1371/journal.ppat.1004409 reproduced under a CC BY 4.0 creative commons license; [f–h]: From Raju, D et al. (2015) Accumulation of glucosylceramide in the absence of the beta-glucosidase GBA2 alters cytoskeletal dynamics. PLoS Genet 11(3): e1005063 doi:10.1371/journal.pgen.1005063 reproduced under a CC BY 4.0 creative commons license; [i]: By Kelvinsong from https://commons.wikimedia.org/ wiki/File:Centrosome_(borderless_version)-en.svg reproduced under a CC BY 3.0 creative commons license; [j]: From Duncan, JE and Goldstein, LSB (2006) The genetics of axonal transport and axonal transport disorders. PLoS Genet 2: e124. DOI: 10.1371/journal.pgen 0020124 reproduced under a CC BY creative commons license.)*

each centrosome projects short appendages, whereas the more-recently generated **daughter** centriole lacks such structures (**Figure 2.9i**).

Within the pericentriolar material of the centrosome, **ϒ-tubulin** and associated proteins help initiate α-/β-tubulin dimer polymerization as new α-/β-dimers bound to **GTP (guanosine triphosphate)** are added at the distal + end of the nascent MT, thereby extending it outward from the centrosome. Conversely, the – end of the MT that remains anchored in the centrosome represents the site where **GDP (guanosine diphosphate)-bound** dimers are actively removed from the MT. With balanced additions and removals at the + and – ends, MTs can remain at relatively constant lengths. However, via a process of dynamic instability, MTs can also rapidly cycle between periods of intense polymerization vs. depolymerization, thereby dramatically altering MT configurations, particularly as cells transition to and from cycles of division.

Although all MTs consist of α-/β-tubulin dimers, considerable heterogeneity exists, owing to multiple tubulin isotypes and their post-translational modifications. Such alterations not only affect growth rates and overall stabilities of MTs but also modulate the kinds of ATP-dependent molecular motors that associate with MTs, thereby influencing the directionality of cargo movements. For example, most **kinesin** motors move vesicles and organelles in an anterograde direction toward the + ends of MTs, whereas **dyneins** plus cofactors such as dynactin mediate retrograde movement toward MT – ends (**Figure 2.9j**).

In addition to dissimilarities in size and protein composition, **intermediate filaments** (**Figure 2.10a, b**) differ from MTs and MFs in several key ways. For example, owing to their strong intramolecular bonds and lack of treadmilling, IFs can withstand harsh extraction protocols or intense mechanical loads that readily degrade MTs and MFs. Moreover, with over 70 IF genes constituting six main families (Table 2.3), IFs undergo a complex multi-step assembly process that begins with the translation of individual IF monomers, each of which contains a central α-helical rod flanked by globular head and tail regions (**Figure 2.10c**). Monomers then form dimers and tetramers before becoming loosely packed into a short filamentous array, called a **unit length filament** (**ULF**). Each ULF, in turn, links end-to-end with other ULFs to form a ~10–12 nm-wide mature cytoplasmic IF (**Figure 2.10c–e**), whereas **nuclear IFs** coat the inside of nuclear envelopes (pg. 50).

TABLE 2.3 SIX MAIN TYPES OF HUMAN INTERMEDIATE FILAMENTS

Type	Examples	~MWs (kD)	Typical expression sites
I	Acidic keratins (28 cytokeratins and hair keratins)	40–65	Epithelial cells
II	Basic keratins (26 cytokeratins and hair keratins)	45–70	Epithelial cells
III	Vimentin, desmin, glial fibrillary acidic protein (GFAP), peripherin		Mesenchymal cells, muscle cells, glial cells, peripheral neurons
IV	Neurofilament proteins, nestin, synemins	65–200	Neurons and a few other cells
V	Lamins	60–75	Nuclear lamina
VI	Phakinin, filensin	40–50 after processing	Eye lens

BOX 2.2 INTRANUCLEAR ACTIN DYNAMICS DURING THE CELL CYCLE

Unlike either MTs that are exclusively cytoplasmic or intranuclear IFs that are restricted to the nuclear lamina, recent evidence suggests that a system of **actin MFs** is distributed throughout the nucleoplasm of certain cell types. Actin filaments are typically difficult to detect in somatic cell nuclei, owing to their reduced concentrations and more rapid turnover rates compared to counterparts in the cytoplasm. However, in addition to actin itself, a full complement of actin-binding proteins like **cofilin** have been identified in nuclei. Moreover, when various probes for visualizing actin are concentrated in

the nucleus via nuclear localization sequences, dynamically changing individual filaments and bundles of filaments can be observed (**Figure a**). Although the precise roles of nuclear actin remain to be fully elucidated, results from experimental manipulations reveal that these nuclear filaments can aid in (i) transducing plasmalemmal signals in response to such ECM stimuli as serum, LPA (lysophosphatidic acid), or attachments of fibronectin; (ii) mediating chromatin reorganizations involving cyclin-dependent kinase 1; and (iii) repairing double-stranded DNA breaks (DSBs) (**Figure a**).

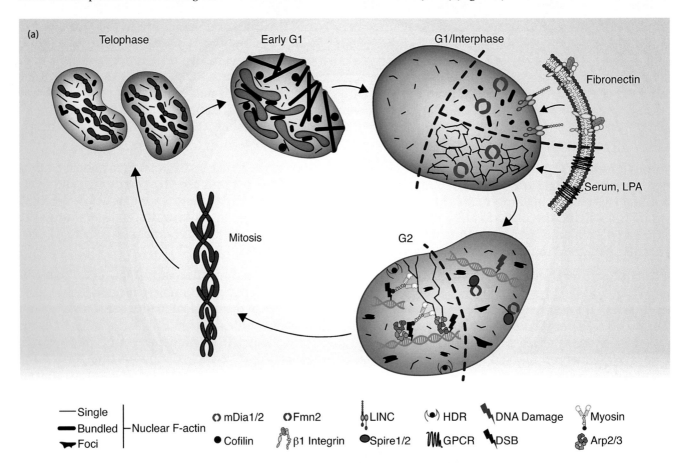

Box 2.2 (a) Cell cycle-specific nuclear F-actin assembly. *From telophase to early G1, newly formed daughter cell nuclei contain both single and bundled filamentous actin that help decondense chromatin and expand nuclear volume before such intranuclear actin filaments undergo a disassembly process mediated by cofilin. Subsequently, re-assembly of F-actin can be triggered by different stimuli, such as lysophosphatidic acid (LPA), fibronectin (involving mDia1/2), or DNA damage (involving Fmn2 with Spire1/2). Before dividing cells enter mitosis, nuclear actin filaments are cleared via pathways that include Arp2/3 and nuclear myosins. (From Plessner, M. and Grosse, R. (2019) Dynamizing nuclear actin filaments. Curr Opin Cell Biol 56: 1–6 reproduced under a CC BY 4.0 creative commons license.)*

Along with the three primary components of the cytoskeleton, humans and most other eukaryotes produce a fourth kind of cytoskeletal element composed of GTP-binding proteins, called **septins**. Such kidney-bean-shaped proteins interact with each other via their GTP-binding domains (**G interfaces**) and N- and C-terminal regions (**NC interfaces**) to form heteromeric oligomers, which can further assemble into filamentous and ring-like arrays (**Figure 2.10f, g**). These cytoskeletal proteins generally serve as septate barriers within various cell types, thereby facilitating such processes as cell blebbing, endocytosis, locomotion, and meiotic divisions during polar body formation (Chapter 18) (**Figure 2.10g**).

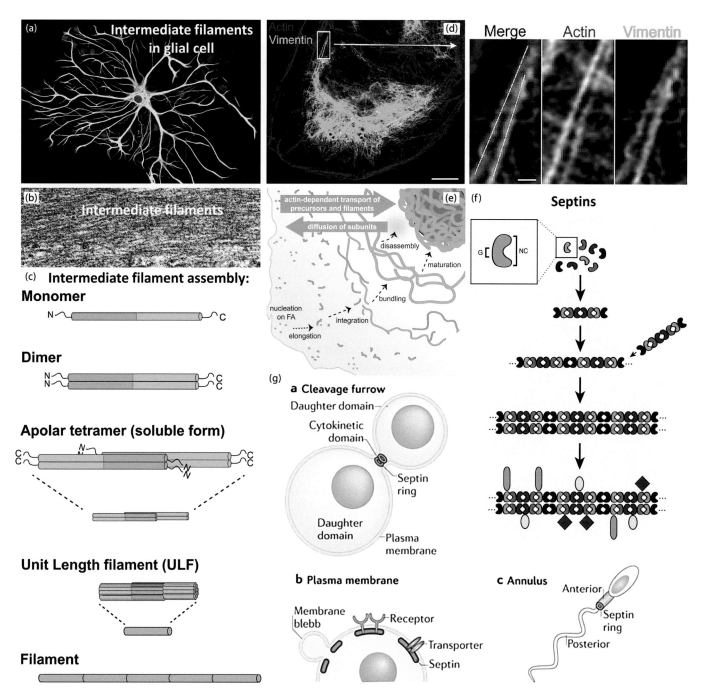

Figure 2.10 Cytoskeleton II: intermediate filaments (IFs) and septins. (a, b) *Intermediate filaments (IFs) are visible in an astrocyte imaged by immunofluorescence microscopy (a: 200X) and in an epithelial cell sectioned for TEM (b: 25,000X). (**c**) IFs develop from filamentous monomers that are eventually assembled into mature 10–12 nm-wide filaments. (**d**) Maturing IFs made of proteins such as vimentin often interact with MFs during differentiation (scale bars, left/right: 10 μm/1 μm). (**e**) After such interactions, IFs can become concentrated near the nucleus. (**f**) As a fourth cytoskeletal element, septins form heteromeric polymers from kidney-bean shaped monomers, each of which has a GTP-binding (G) interface and a N-/C-terminal (NC) interface that interact with similar interfaces of neighboring monomers during polymerization. (**g**) Filamentous and ring-shaped arrays of septins can function in various processes, including in unorthodox cell divisions, cell membrane changes, and locomotion. ([a]: From original work of G Shaw downloaded from https://upload.wikimedia.org/wikipedia/commons/6/63/Astrocyte5.jpg and reproduced by permission of a CC BY SA creative commons license; [c] and [e]: From Roberts, A et al. (2016) Intermediate filament dynamics: what we can see now and why it matters. Bioessays 38: 232–243 reproduced with publisher permission; [d]: From Jiu, Y et al (2015). Bidirectional*

Figure 2.10 (Continued) *interplay between vimentin intermediate filaments and contractile actin stress fibers Cell Rep 11: 1511–1518 reproduced with publisher permission; [f]: From Estey, MP et al. (2011) Septins. Cur Biol 21: doi: https://doi. org/10.1016/j.cub.2011.03.067 reproduced with publisher permission; [g]: From Mostowsky, S and Cosartt, P (2012) Septins: the fourth component of the cytoskeleton. Nat Rev Cell Mol Biol 13: doi:10.1038/nrm3284 reproduced with publisher permission.)*

2.9   THE NUCLEUS: ITS DELIMITING NUCLEAR ENVELOPE WITH PORE COMPLEXES AND LAMINA SURROUNDING NUCLEOPLASM WITH CHROMATIN ORGANIZED INTO CHROMOSOMES FOR GENE EXPRESSION

Each non-dividing human cell typically possesses a single **nucleus** with gene-containing **chromosomes**. However, multinucleated skeletal muscle cells and binucleated heart and liver cells also occur in the body. Conversely, fully differentiated platelets and red blood cells lack a nucleus.

Nuclei range from round to decidedly elongated forms (**Figure 2.11a, insets**), but most are oblong and possess a smooth periphery. Alternatively, in some white blood cells, the nucleus is distinctly indented at one to multiple sites to form nuclear **lobes**, which can increase flexibility for squeezing through narrow passageways. In spite of such morphological diversity, nuclei are fairly uniform in length, typically measuring ~5–20 µm long, and all are delimited by a double-membrane **nuclear envelope** that is ~20–30 nm wide.

The two membranes of the NE comprise an **outer nuclear membrane** (**ONM**) and **inner nuclear membrane** (**INM**), which are spanned by numerous **nuclear pore complexes** (**NPCs**) and underlain by a meshwork of IF proteins, called the **nuclear lamina** (**Figure 2.11a–e**). The ONM contains its own unique proteins like **nesprins** and is separated by an intermembrane space from the INM, which possesses such INM-specific components as **SUN-domain-containing** proteins. Collectively, nesprin and SUN proteins constitute l̲inker of n̲ucleoskeleton and c̲ytoskeleton (**LINC**) complexes that connect the nuclear lamina with the cytoskeleton (**Figure 2.11c**), thereby allowing mechanical forces and their accompanying signals to be transduced across the cell.

Each of the several thousand **NPCs** in a typical NE measures ~120 nm from edge to edge and contains at its center an ~5 nm-wide trans-envelope channel (**Figure 2.11b**). Although molecules smaller than 40 kDa can readily traverse NPCs, larger macromolecules must be actively translocated. Such translocation involves importin and exportin carriers, which interact both with NPC components and with either nuclear localization signals (NLSs) or export-related sequences on cargo molecules. In this way, cytoplasmic proteins are imported into the nucleus for structural support and transcriptional processes, whereas mRNAs, tRNAs, and ribosomal subunits are exported from the nucleus into the cytoplasm for translating proteins.

As a reticular network coating the inner surface of the NE, the **nuclear lamina** contains filaments formed from two main kinds of **lamins**, termed **A** vs. **B types** (**Figure 2.12a, b**). B-type lamins occur in nearly all cells and tend to be the sole lamins found in either stem cells or those undergoing differentiation, whereas A-type lamins characteristically co-occur with B lamins in differentiated cells. In addition to its mechanical functions, the nuclear lamina can help modulate gene expression by: (1) anchoring components to the nuclear periphery; (2) sequestering proteins needed for proper transcription; and (3) participating in signaling pathways both via LINC-mediated connections to the cytoskeleton and via the modulation of molecular influx through NPCs.

Medial to the nuclear lamina, the **nucleoplasm** that fills each nucleus contains various discrete units, including one to several relatively large **nucleoli** plus other smaller intranuclear components, such as **nuclear speckles** and **Cajal bodies** (**Figure 2.12c–i**). Nucleoli typically comprise (i) fibrillar, (ii) dense fibrillar, and (iii) granular component subregions (**Figure 2.12c, d, g,** and **i**), which collectively allow nucleoli to produce **ribosomal subunits** for assembly into cytoplasmic ribosomes. In addition, nucleoli disassemble during cell divisions, and their contents end up coating condensed chromosomes to form a substantial amount of the chromosomal volume (**Figure 2.12h**). Unlike nucleoli, speckles and Cajal bodies typically interact with **spliceosomes** in the processing of pre-mRNAs into mRNA (**Case 2.5 Lou Gehrig's Nuclei**). Alternatively, some Cajal bodies associate with nucleoli (**Figure 2.12g**) and may coordinate certain functions shared by these two nuclear subunits.

Aside from its scattered components, the nucleoplasm is filled with a mixture of gene-bearing DNA plus DNA-associated proteins, which is collectively termed **chromatin**. Chromatin is initially organized into ~10-nm-wide **nucleosomes**, each of which is composed of double-stranded DNA wrapped around an octameric complex of **histone**

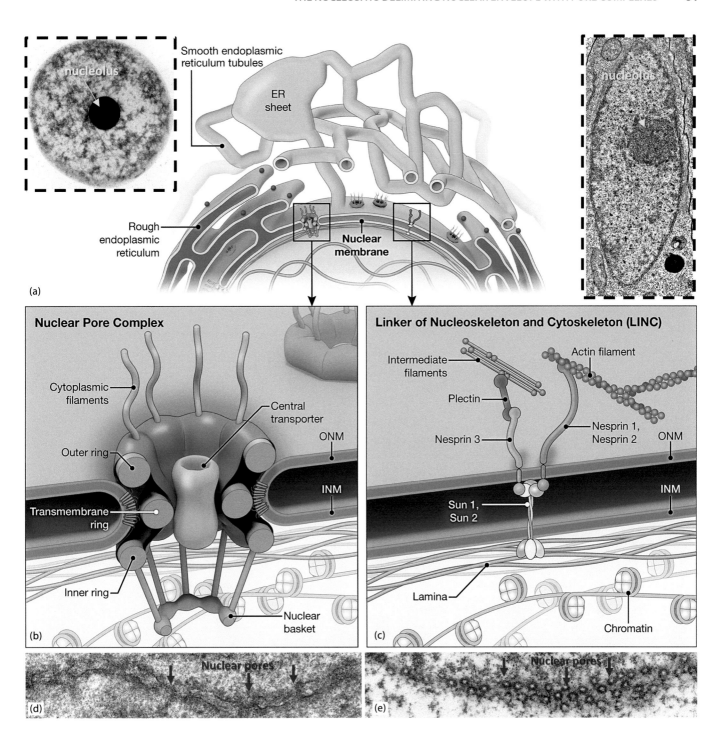

Figure 2.11 Nucleus I: nuclear envelope. (a) *Nuclei range from round (left inset of whole-mounted oocyte nucleus viewed without sectioning by high voltage EM [HVEM] [800X]) to elongated (right inset of conventional TEM of epithelial cell [2,800X]). However, all are bounded by a double-membrane nuclear envelope that contains nuclear pore complexes and is contiguous with the ER.* **(b–e)** *Each nuclear pore complex (as indicated by red arrows in longitudinal [d, 66,000X] and tangential [e, 33,000X] HVEM sections) comprises numerous nucleoporin proteins and a central ~5-nm-wide lumen that can regulate molecular transport into and out of the nucleus. Sun and nesprin proteins (c) of the inner nuclear membrane (INM) and outer nuclear membrane (ONM), respectively, connect the nuclear lamina underlying the nuclear envelope with the cytoplasmic cytoskeleton as part of a Linker of Nucleoskeleton and Cytoskeleton (LINC) network. ([a–c]: From Goyal, U and Blackstone, C (2013) Untangling the web: mechanisms underlying ER network formation. Biochimica et Biophysica Acta 1833: 2492–2498 reproduced with publisher permission.)*

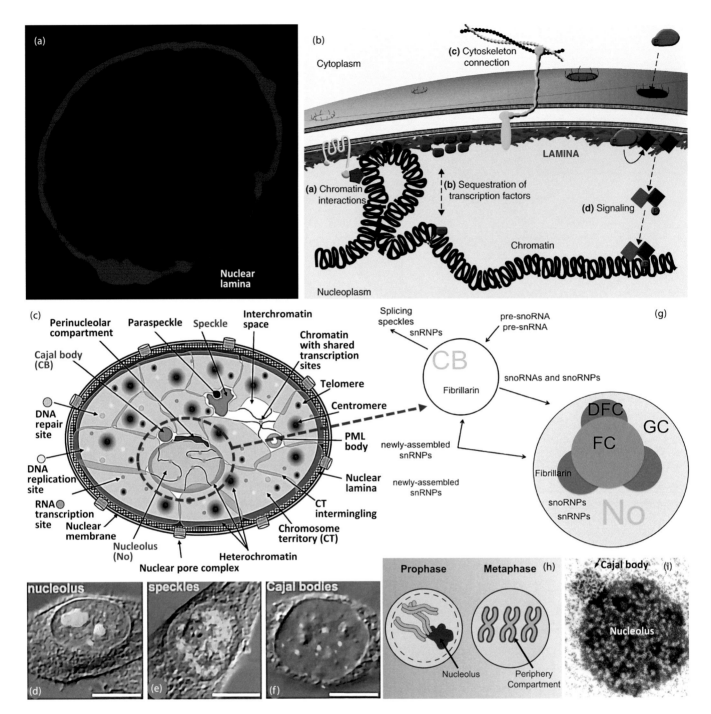

Figure 2.12 Nucleus II: nuclear lamina and nucleoplasmic components. (*a*) *The nuclear lamina lines the inner surface of nuclear envelopes, as illustrated by the red fluorescence of a human osteosarcoma cell nucleus probed for lamins A/C.* (***b***) *Along with providing structural support, the nuclear lamina can help modulate gene expression by (i) binding chromatin and sequestering components needed for transcription; (ii) altering transport through nuclear pore complexes; and (iii) regulating signal transduction via the LINC network.* (***c–i***) *Major intranuclear components include: (1)* nucleoli *(figures c, d) with their fibrillar (FC), dense fibrillar (DFC), and granular (GC) components (figure g), which collectively function in producing ribosomal subunits and coating metaphase chromosomes (figure h); (2)* speckles *(figures c, e), which help splice genes into alternative mRNAs; and (3)* Cajal bodies *(figures c, f, g, i), which similarly participate in gene splicing but can also coordinate with nucleolar functioning by occurring either near (figure g) or attached to nucleoli (figure i, arrow shows Cajal body bound to a nucleolus). Scale bars d–f = 10 µm. ([a, b]: From Dittmer, T, and Misteli, T (2011) The lamin protein family. Genome Biol 12: http://genomebiology.com/2011/12/5/222 reproduced with publisher permission; [c–f]: From Hemmerich, P et al. (2011) Dynamic as well as stable protein interactions contribute*

Figure 2.12 (Continued) *to genome function and maintenance. Chrom Res 19: 131–151 reproduced with publisher permission; [g]: From Trinkle-Mulcahy, L and Sleeman, JE (2017) The Cajal body and the nucleolus: "In a relationship" or "It's complicated"? RNA Biol 14: 739–751 reproduced with publisher permission; [h]: From Booth, DG et al. (2016) 3D-CLEM reveals that a major portion of mitotic chromosomes is not chromatin. Mol Cell 64: 790–802 reproduced under a CC BY 4.0 creative commons license; [i]: From Pena, E (2001) Neuronal body size correlates with the number of nucleoli and Cajal bodies, and with the organization of the splicing machinery in rat trigeminal ganglion neurons. J Comp Neurol 430: 250–63 reproduced with publisher permission.)*

proteins (**Figure 2.13a**). When extracted from nuclei and prepared for EM, nucleosomes can exhibit a **beads-on-a-string** configuration, with linear stretches of DNA connecting neighboring nucleosomes (**Figure 2.13a**). However, nucleosome-containing chromatin does not typically remain as beaded strings **in vivo** but instead becomes highly folded upon itself in secondary and tertiary levels of organization to form **chromosomes** (**Figure 2.13b**). Correlative sequencing and imaging analyses have revealed that instead of being randomly distributed, chromosomes are organized into distinct chromosomal territories, whose 3D-architecture and interactions with neighboring territories help modulate patterns of gene expression (**Figure 2.13b**). In addition, chromosomal chromatin can range from relatively loose configurations, termed **euchromatin**, to more tightly condensed arrangements, called **heterochromatin** (**Figure 2.13c**). Heterochromatic chromatin tends to be transcriptionally inactive, since its tight packing blocks DNA from accessing transcription factors and polymerases needed for gene expression, whereas euchromatic chromatin can be transcribed in the presence of proper regulators.

In addition to the canonical organization of chromatin occurring throughout each chromosome, the termini of chromosomes form unique **telomeres** (**Figure 2.13d**), which comprise looped DNA plus protective complexes of proteins, called **shelterins** (**Figure 2.13d**). A key component of these complexes is a telomerase reverse transcriptase (TERT) that utilizes telomerase RNA (TER) to add DNA to the ends of the telomere (**Figure 2.13e**). In so doing, telomeres not only allow full DNA replication at chromosomal tips but also prevent termini from being sensed as double-strand breaks (DSBs) that can elicit deleterious cellular responses.

2.10   MITOSIS VS. MEIOSIS, THE CELL CYCLE, AND STRUCTURAL REORGANIZATIONS DURING MITOTIC STAGES (PRO-, META-, ANA-, TELOPHASE)

Cellular division can occur via **mitosis** or **meiosis**, with mitosis being used by all cells in the body except for meiotically dividing cells that produce egg and sperm **gametes** (**Figure 2.14a–f**). Non-gamete-forming cells are termed **somatic cells** and typically possess a **diploid (2n)** amount of DNA, where n represents the total

CASE 2.5 LOU GEHRIG'S NUCLEI

Baseball Player and Celebrity ALS sufferer
(1903–1941)

"Fans, for the past two weeks you have been reading about the bad break I got. Yet today I consider myself the luckiest man on the face of this earth."

(From Lougherig.com: text of L.G.'s farewell speech given a few weeks after being diagnosed with ALS)

One of baseball's greatest players, Lou Gehrig was the first baseman for the 1923-1939 New York Yankees before being diagnosed with amyotrophic lateral sclerosis (ALS), a motor neuron disorder that today is commonly referred to as Lou Gehrig's disease. During his Hall-of-Fame career as a prolific hitter, "Iron Horse" Gehrig played in 2,130 consecutive games until increased muscle weakness triggered by ALS forced his retirement and not long thereafter led to his death in 1941. Unfortunately, as in Gehrig's time, most ALS patients today die within a few years after their diagnosis, typically from

respiratory failure. However, as new ALS genetic mutations are identified, optimism for finding effective treatments is increasing. For example, mutation of C9orf72, which is the most common genetic defect linked to ALS, occurs specifically in the gene's first intron. Normally, RNA transcribed from the intron is spliced from the initial transcript before undergoing degradation. However, with mutated C9orf72, the modified intronic RNA accumulates intranuclearly or is exported and translated into protein, thereby promoting neurodegeneration. Thus, although too late to help Gehrig extend his playing streak and overall lifespan, new therapies targeting such nuclear abnormalities may well help future ALS sufferers.

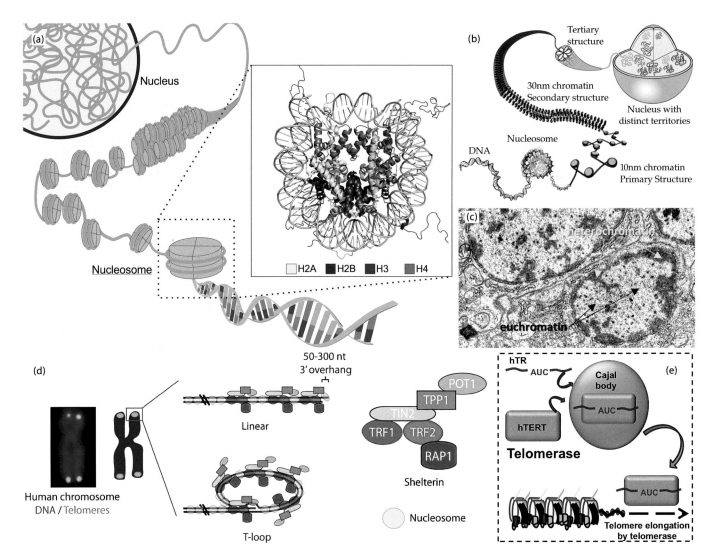

Figure 2.13 Nucleus III: chromatin. (a) *Chromatin is a complex of double helical DNA bound to proteins, which are most commonly histones. In its primary structure, chromatin forms ~10 nm-wide bead-like units, called nucleosomes, each of which comprises eight histones (H2A, H2B, H3, and H4 histone pairs) around which the DNA double helix is wrapped. (**b**) Such "beads-on-a-string" arrangements of nucleosomes are further folded on each other at secondary and tertiary levels of packing to form chromosomes that occur within discrete chromosomal territories in intact nuclei. (**c**) Morphologically, chromatin ranges from transcriptionally inactive, tightly packed heterochromatin to more loosely arranged euchromatin that is capable of being transcribed (2,200X). (**d**) The tips of chromosomes contain unique units, called telomeres, which comprise looped DNA configurations associated with shelterin protein complexes. Telomeres allow terminal DNA to be fully replicated while also protecting against deleterious reactions potentially launched toward double-strand breaks. (**e**) Telomeres are generated and maintained within Cajal bodies by telomerase enzyme complexes. In humans, such processes involve human telomerase reverse transcriptase (hTERT) that utilizes single-stranded human telomerase RNA (hTR) to elongate DNA in telomeres. ([a]: From Okinawa Institute of Science and Technology, reproduced under a CC BY 2.0 creative commons license; [b]: From Iyer, BVS et al. (2011) Hierarchies in eukaryotic genome organization: Insights from polymer theory and simulations. BMC Biophysics 4: http://www.biomedcentral.com/2046-1682/4/8 reproduced under a CC BY 2.0 creative commons license; [d]: From Cesare, AJ and Karlsleder, J (2012) A three-state model of telomere control over human proliferative boundaries. Curr Opin Cell Biol. 24: 731–738 reproduced with publisher permission; [e]: From Novo, CL and Londono-Vajello, JA (2013) Telomeres and the nucleus. Sem Cancer Biol 23: 116–124 reproduced with publisher permission.)*

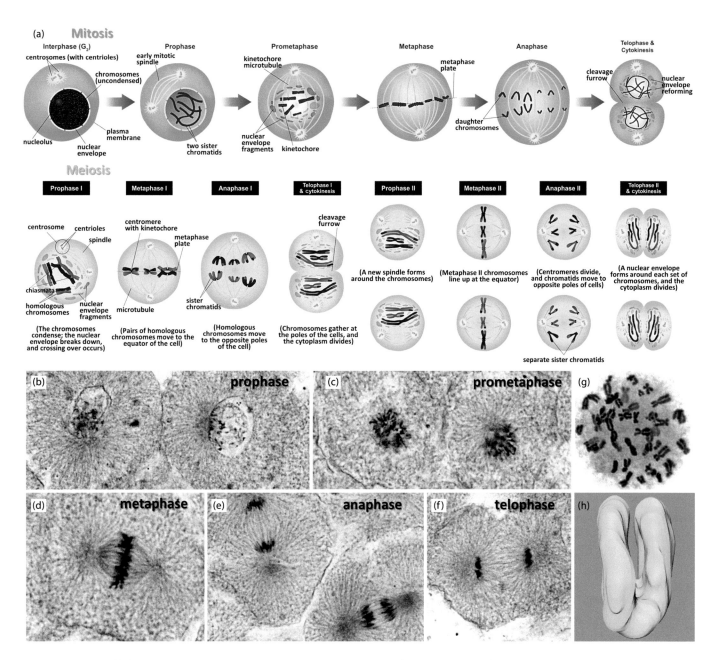

Figure 2.14 Cell division. (*a*) *Somatic cells—which are all cells in the body except for those that form gametes (sperm and eggs)—divide by mitosis, whereas developing sperm and eggs divide via meiosis. As opposed to the single division of mitosis that yields diploid (2n) daughter cells, meiosis has two divisions, thereby reducing eggs and sperm to a haploid (1n) state so that diploidy can be restored after fertilization. (**b-f**) The major stages of cell division (pro-, prometa-, meta-, ana-, and telophase) are illustrated in sections of a fish embryo, which has rapid and synchronous divisions, thereby often allowing more than one example to be observed in a high-magnification field of view (1,500X). (**g**) A stained spread of a mitotic human cell shows 46 duplicated chromosomes (4,500X). (**h**) A 3D reconstruction of serial block-face SEM images through a metaphase human chromosome shows two sister chromatids attached to each other. ([a]: From Cell division diagrams from A Zifan downloaded from https://commons.wikimedia.org/wiki/File:MitosisStages.svg and https://commons.wikimedia.org/wiki/File:MeiosisStages.svg, both reproduced under a CC BY 4.0 creative commons license; [h]: From Booth, DG et al. (2016) 3D-CLEM reveals that a major portion of mitotic chromosomes is not chromatin. Mol Cell 64: 790–802 reproduced under a CC BY 4.0 creative commons license.)*

number of chromosomes in an egg or sperm, which in humans equals 23. Thus, somatic cells normally possess **23 chromosome pairs** (i.e., 46 chromosomes), with each of the 23 pairs comprising a **maternal** vs. **paternal** chromosome derived post-fertilization from the egg vs. sperm, respectively (**Figure 2.14g**).

Before initiating mitosis, somatic cells duplicate their DNA so that after mitosis divides the cell into two, each daughter cell will end up with the full 2n genetic complement that was present before cell division. Accordingly, pre-mitotic DNA duplication yields a **4n** amount of DNA distributed across 46 **duplicated chromosomes**, with each chromosome consisting of two identical conjoined halves, called **sister chromatids** (**Figure 2.14a, g, h**). In preparation for cell division, each 4n cell disassembles its nucleus and progresses through four main mitotic stages: (i) **prophase**; (ii) **metaphase**; (iii) **anaphase**; and (iv) **telophase** (**Figure 2.14a–f**). Following telophase, the dividing cell separates into two mono-nucleated daughter cells, with one chromatid from each chromosome being distributed into each diploid progeny.

Similarly, DNA is also duplicated prior to meiosis during gamete formation. However, because meiosis involves two cell divisions (meiosis I/meiosis II) without another intervening round of DNA replication, fully formed gametes are **haploid** (**1n**) and thus contain only half the DNA content of diploid somatic cells (**Figure 2.14a**). This, in turn, allows diploidy to be restored when eggs are fertilized by sperm (Chapters 17 and 18). Moreover, meiosis also increases genetic diversity compared to clonal products generated by mitosis, owing to crossing-over and independent segregation processes that occur during meiosis I (**Figure 2.14a**).

Dividing cells undergo cyclic reorganizations during a **cell cycle**, which is broadly divided into **M-phase** and **interphase** (**Figure 2.15a**). For somatic cells, mitotic M-phase begins with **karyokinesis** (=mitosis proper), which comprises prophase through telophase. M-phase then ends with **cytokinesis**, which separates the mitotic cell into two daughter cells. Following cytokinesis, each cell transitions into **interphase** prior to the next M-phase (**Figure 2.15a**). Some cells can also exit the cell cycle and enter a quiescent G_0 state before returning from G_0 **arrest** to resume the cell cycle (**Figure 2.15a**).

Similar to the substages of M-phase, interphase can be further subdivided into three distinct phases: (i) G_1 (Gap 1) **phase**, which represents a period of growth between M-phase and DNA replication; (ii) **S-phase**, in which DNA synthesis duplicates chromosomes in preparation for division; and (iii) G_2 (Gap 2) **phase**, which is a second growth period between S- and M-phases (**Figure 2.15a**). To cull out defective cells, quality-control **checkpoints** assess the fidelity of such features as DNA replication and mitotic spindle structure. Following successful passage through these checkpoints, subsequent cell cycle progression is driven in large part by **cyclin-dependent kinases** (**CDKs**) that bind **cyclin** proteins (**Figure 2.15a**). In particular, activation of the **CDK1/cyclin B** heterodimer initiates M-phase and was historically, called maturation-promoting factor (MPF) for its roles in oocyte maturation (Chapter 18). However, now this mediator of G_2-**M transitions** is often referred to as M-phase promoting factor due to its roles played in both meiotic and mitotic M-phases.

After entry into M-phase, **prophase** is characterized by the formation of a cytoplasmic **mitotic apparatus** that mediates segregation of chromosomes into daughter cells (**Figure 2.15b**). Prophase cells also undergo **nuclear envelope breakdown** (**NEB**), **nuclear lamina depolymerization**, and **nucleolar dissolution**, thereby exposing chromosomal chromatin as it becomes further condensed and coated with nucleolar-derived materials (**Figure 2.12h**). Each condensed chromosome has a **centromere** that connects the sister chromatids of the duplicated chromosome. Such centromeric regions are characterized by unique centromeric proteins (CENPs) (**Figure 2.15c**) and are coated on both sides by a bilayered complex of proteins, called the **kinetochore** (**Figure 2.15d, e**). The **inner kinetochore** interacts with the centromere, whereas the **outer kinetochore** binds MTs (**Figure 2.15e**).

In preparation for mitosis, the original **centrosome** of the cell duplicates and matures during interphase before the two resulting centrosomes start nucleating MTs of the nascent mitotic apparatus in prophase (**Figure 2.15f**). Some of these MTs can capture chromosomes by binding to their kinetochores. Via the coordinated actions of MTs, associated motor proteins, and additional cellular components, captured chromosomes are gathered into the inter-centrosomal cytoplasm during the culmination of prophase, which is sometimes viewed as its own distinct phase, termed **prometaphase**.

Following prophase, **metaphase** chromosomes become aligned at a **metaphase plate** that typically occurs near the cell center so that equal-sized daughter cells can be generated (**Figure 2.15b, d**). At this point, the mitotic apparatus consists of: (1) a **centrosome** at each cellular pole; (2) **astral MTs** that radiate from centrosomes toward

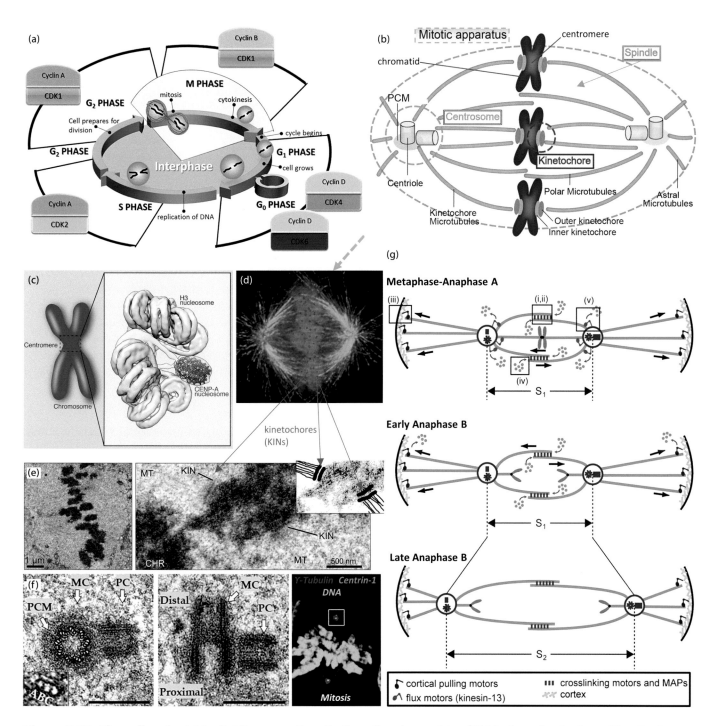

Figure 2.15 The cell cycle. (**a**) *In dividing somatic cells, the cell cycle consists of (i) M-phase, beginning with mitosis proper (=karyokinesis, i.e., prophase through telophase) followed by cytokinesis in which daughter cells separate, and (ii) interphase comprising G1 (first growth period), S (DNA synthesis to duplicate chromosomes), and G2 (second growth period) phases. Non-dividing cells enter a quiescent G_0 phase, whereas progression through the various cell cycle phases requires successful passage of checkpoints as well as stimulation by different heterodimers composed of cyclin-dependent kinases (CDKs) and their cyclin protein binding partners. (**b**) The mitotic apparatus allows dividing somatic cells to provide daughter cells with a diploid set of chromosomes, each of which had duplicated prior to mitosis to contain two sister chromatids attached to each other by a centromere that is coated on each side by an inner or outer kinetochore. The metaphase mitotic apparatus comprises: (i) two centrosomes, each with a pair of centrioles and pericentriolar material (PCM) from which microtubules (MTs) arise; (ii) astral MTs that extend toward the cell periphery; and (iii) a mitotic spindle occurring between the centrosomes and consisting mainly of kinetochoral MTs that attach to kinetochores plus non-kinetochoral MTs such as polar MTs that do not attach to kinetochores. (**c**) Cryo-EM reveals*

*Figure 2.15 (Continued) unique centromeric chromatin such tri-nucleosome units containing CENP-A protein. (**d**) An immunolabeled mitotic apparatus at metaphase shows chromosomes in blue, MTs in green, and kinetochores in pink. (**e**) Low- and high-magnification TEMs show chromosomes and their associated kinetochores, to which microtubules (MTs) attach. (**f**) The centrosome consists of pericentriolar material with proteins, such as Centrin-1 that surround a pair of centrioles. Each centriole contains nine triplets (a–c) of MTs, with the younger centriole initially forming perpendicular to the older mother centriole (MC) as a procentriole (PC) that matures into the daughter centriole; scale bars = 400 nm. (**g**) During poleward movement of chromosomes, coordinated MT polymerization/depolymerization at several sites [i–v] translocate chromosomes toward the centrosomes during Anaphase A while the inter-centrosomal spacing (S1) remains the same. During Anaphase B, inter-centrosomal spacing increases (S2) as centrosomes move toward the cell periphery via pulling from astral MT depolymerization and pushing due to non-kinetochoral MT polymerization in the spindle. ([a]: From Garcia-Reyes, B et al. (2018) The emerging role of cyclin-dependent kinases (CDKs) in pancreatic ductal adenocarcinoma. Int J Mol Sci 19: doi:10.3390/ijms19103219 reproduced under a CC BY 4.0 creative commons license; [b]: From Ito, KK et al. (2020) The emerging role of ncRNAs and RNA-binding proteins in mitotic apparatus formation. Non-coding RNA 6: doi:10.3390/ncrna6010013 reproduced under a CC BY 4.0 creative commons license; [c]: From Takizawa, Y et al. (2019) Cryo-EM structures of centromeric tri-nucleosomes containing a central CENP-A nucleosome. Structure 28: 44–53.e3 doi: 10.1016/j.str.2019.10.016 reproduced with publisher permission; [d]: From public domain image downloaded from https://commons.wikimedia. org/wiki/File:Kinetochore.jpg; [e]: From Samoshkin, A et al. (2009) Human condensin function is essential for centromeric chromatin assembly and proper sister kinetochore orientation. PLoS ONE 4: e6831. doi:10.1371/journal.pone.0006831 reproduced under a CC public domain creative commons license; [f]: From Sullenberger, C et al. (2020) With age comes maturity: biochemical and structural transformation of a human centriole in the making. Cells 9: doi:10.3390/cells9061429 reproduced under a CC BY 4.0 creative commons license; [g]: From Scholey, JM et al. (2016) Anaphase B. Biology 2016, 5, 51; doi:10.3390/biology5040051 reproduced under a CC BY 4.0 creative commons license.)*

the plasmalemma; and (3) a central **spindle** between the centrosomes comprising **kinetochoral MTs** that attach to kinetochores plus **non-kinetochoral MTs** not bound to kinetochores (**Figure 2.15b**).

During **anaphase**, the mitotic spindle begins to break apart each duplicated chromosome at its centromeric region before pulling detached chromatids toward opposite cellular poles (**Figure 2.15g**). Such movements involve coordinated polymerizations and depolymerizations of MTs resulting in two stages of chromosomal translocations: (1) **anaphase A** that moves chromatids closer to their nearest centrosome while the centrosomes remain at a constant distance apart from each other; and (2) **anaphase B**, in which chromatids continue to move toward the poles as the two centrosomes are spread apart (**Figure 2.15g**). Collectively, such processes segregate diploid sets of chromosomes toward opposite cellular sides prior to the onset of cytokinesis.

In preparation for daughter cell formation, an actomyosin-based **contractile ring** of circumferentially arranged **MFs** forms in the cortical cytoplasm within a plane perpendicular to the long axis of the spindle (**Figure 2.16a**). The contractile ring then constricts telophase cytoplasm to produce a tubular intercellular **midbody** that eventually splits and thereby generates two new cells at the end of M-phase (**Figure 2.16a**).

After M-phase is completed, some cells can be triggered to undergo additional mitoses. A key enabler of repetitive cell cycling is the presence of sufficient **telomere** caps on chromosomes. In most cell types, telomere lengths are gradually reduced during cell divisions (**Figure 2.16b**) until insufficient amounts of telomeres and increased genomic instability prevent mitosis and promote senescence. Similarly, inadequate telomere maintenance and/or excessive rates of telomere loss are linked to various pathologies, including heart disease, cancers, and premature aging (see next section). Accordingly, methods such as TERT replacement/activation are being tested as disease therapies and as ways of extending overall longevity (**Figure 2.16c**) (**Case 2.6 Frances Oldham Kelsey's Telomeres**).

2.11 SOME CELLULAR DEFECTS IN HUMAN DISEASES: LYSOSOMAL STORAGE DISEASES AND LAMINOPATHIES

Lysosomal storage diseases (**LSDs**) arise from a wide range of genetic mutations that result in malfunctioning lysosomes and hence the buildup of wastes normally degraded by lysosomal processes. For example, mucopolysaccharidosis (MPS) kinds of LSDs like Hunter syndrome (MPS type II) involve defective lysosomal enzymes needed to break down glycosaminoglycan (GAG) (=mucopolysaccharide) types of carbohydrates and

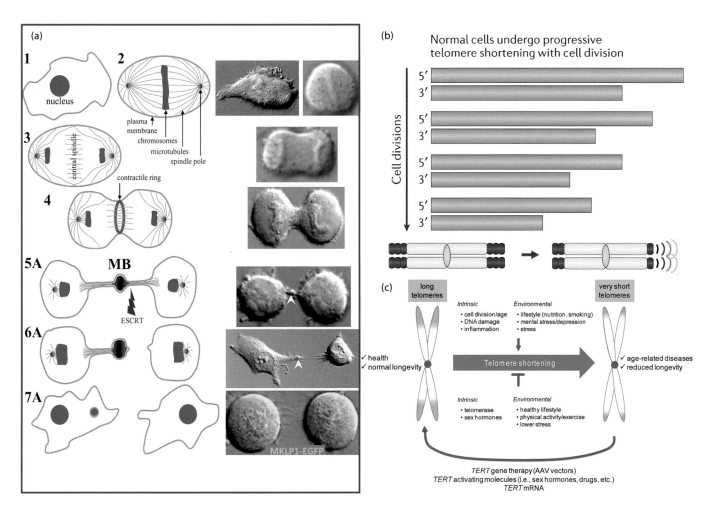

Figure 2.16 Cytokinesis and telomere loss. *(**a**) As illustrated in dividing human HeLa cells, constriction of an actomyosin contractile ring circumferentially arranged around the spindle of a telophase-stage cell forms a transient midbody (MB) connection. Such midbodies contain components, such as MKLP1 before abscission triggers like ESCRT enable separation of the two daughter cells. (**b**) With each cell division, telomere length normally decreases until overly shortened telomeres in senescent cells prevent further mitoses. (**c**) Various therapies focusing on activating or replacing telomerase reverse transcriptase (TERT) are being tested as ways of both treating aging-related diseases and increasing overall longevity. ([a]: From Antanaviciute, I et al. (2018) Midbody: from the regulator of cytokinesis to postmitotic signaling organelle. Medicina 54: doi:10.3390/medicina54040053 reproduced under a CC BY 4.0 creative commons license; [b]: From Shay, JW and Wright, WE (2019) Telomeres and telomerase: three decades of progress. Nat Rev Gen 20: 299–319 reproduced with publisher permission; [c]: From Bar, C and Blasco, MA (2016) Telomeres and telomerase as therapeutic targets to prevent and treat age-related diseases.F1000Research 5: 89 https://doi.org/10.12688/f1000research.7020.1 reproduced under a CC BY 4.0 creative commons license.)*

thus affect GAG-rich tissues, such as cartilage and skin. Although individually rare, LSDs have a collective incidence of ~1:5,000 births, and symptoms typically present by early childhood. Current treatments often utilize either replacement therapies to bolster the affected enzymes or substrate reductions to reduce the buildup of products caused by defective lysosomes.

Laminopathies are diseases involving dysfunctional nuclear laminas mainly caused by abnormal forms of the alternatively-spliced lamins A and C produced from the lamin A gene. Such defects are linked to a wide range of disorders, including a type of muscular dystrophy (Chapter 8), several lipodystrophies (Chapter 5), and a rare form of accelerated aging, termed **Hutchinson-Gilford progeria syndrome** (**HGPS**). Most HGPS cases are due to an abnormally spliced lamin A, termed **progerin**, which results in the early onset of such age-related pathologies as cataracts (Chapter 19), atherosclerosis (Chapter 7), and accumulated telomere loss. Because effective treatments are currently lacking, HGPS patients usually succumb to cardiovascular disease in early adulthood.

CASE 2.6 FRANCES OLDHAM KELSEY'S TELOMERES

Pharmacologist
    (1914–2015)

"Her exceptional judgment in evaluating a new drug for safety for human use has prevented a major tragedy of birth deformities in the United States."

(President Kennedy during a speech honoring F.O.K. for blocking the sale of a birth-defect-causing drug)

Born in Canada, Frances Oldham earned a Ph.D. in Pharmacology and joined the faculty of the University of Chicago, before marrying a colleague, Prof. F. Kelsey. In 1960, Oldham Kelsey was hired by the U.S. Food and Drug Administration (FDA) and assigned an application for the sale of thalidomide, a sedative that was marketed in many countries as a treatment for morning sickness during pregnancy. Oldham Kelsey detected red flags in the application and repeatedly asked for data to verify the drug's safety. During the year-long delay caused by these requests, thalidomide was linked to major birth defects, like the deformation

of arms and legs into flipper-like appendages. Because of such harm, the drug was eventually banned. For helping to prevent additional births of "thalidomide babies", Oldham Kelsey received a presidential award in 1962 and was named to the Order of Canada in 2015, just before her death at 101. Related to her long lifespan are findings that the telomeric ends of chromosomes shorten during aging and that delaying this process may prolong cellular lifespans. Thus, the long-lived pharmacologist should be remembered not only for putting the brakes on thalidomide but also for the apparently decelerated rate at which her telomeres shortened.

2.12   SUMMARY—CELLULAR ULTRASTRUCTURE

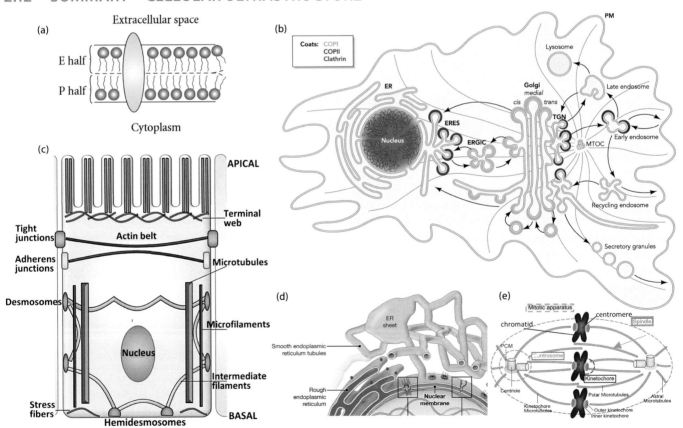

***Pictorial Summary Figure Caption: (a)**: see figure 2.1c; **(b)**: see figure 2.5e; **(c)**: see figure 2.9a; **(d)**: see figure 2.11a; **(e)**: see figure 2.15b*

The 7 nm-thick **cell membrane** is a multifaceted and dynamic bilayer of **phospholipids** supplemented with additional lipids (e.g., cholesterol), **proteins** (integral vs. peripheral), and **carbohydrates**. Each membrane comprises an **E**- (outer) and **P**- (inner) half, which collectively provide a semi-permeable barrier between the ECM and cytoplasm (**Figure a**). Small non-polar molecules (e.g., O_2) can **passively diffuse** across membranes, whereas large and/or polar molecules require either: (i) **facilitated diffusion** via membrane **transporter** proteins or (ii) ATP-mediated energy input to drive **active transport**. For other forms of intake, cells use two main kinds of **endocytic** processes—**phagocytosis** (particle/whole cell incorporation) vs. **pinocytosis** (fluid uptake) to generate incorporated **endosomes**. Pinocytotic endosomes form via pathways that are dependent or independent of **clathrin**, and such structures often fuse with **lysosomes**, which are Golgi-derived sacs of acid-optimized enzymes that digest/recycle wastes (**Figure b**).

The **Golgi** comprises stacks of ~5–10 flat sacs with their **cis** side closest to the ER and **trans** side farthest from the ER (**Figure b**). Golgi sacs help process proteins and lipids and are normally held together by **GRASP** proteins. The **ER** comprises the nuclear envelope plus peripheral components consisting of: (i) flat **sheets** with attached **ribosomes** for protein synthesis (rough ER) and (ii) curved **tubules** lacking ribosomes (smooth ER), used in cell signaling and lipid synthesis. **Anterograde** trafficking moves vesicles and organelles from the ER toward the cell periphery, often using **COPII coating proteins**. **Retrograde** Golgi-to-ER trafficking uses certain kinds of **COPI** coats. Golgi vesicles can participate in **endocytosis** (e.g., via lysosomes) or **exocytosis** (= secretion) (**Figure b**).

Peroxisomes are membrane-bound spheres with enzymes involved in maintaining cellular health, particularly via **catalase**-mediated breakdown of toxic H_2O_2. **Mitochondria** have outer and inner membranes plus protons concentrated in the intermembrane space for **ATP production**. The inner membrane possesses large folds (**cristae**) and surrounds a matrix that contains its own DNA. When stressed, mitochondrial membranes release **cytochrome c** into the cytoplasm to trigger **apoptotic** death.

The **cytoskeleton** (**Figure c**) comprises: (i) **microtubules** (**MTs**) (~20-nm **tubulin** polymers); (ii) **MFs** (~6-nm **actin** helices); (iii) **intermediate filaments** (**IFs**) (~10-nm helices of various proteins, such as keratins and lamins); and (iv) **septin** rings/filaments; MTs/MFs undergo treadmilling (i.e., adding new monomers to their + end and removing old monomers from their – end). The MT – end occurs in a microtubule organizing center (centrosome). The **centrosome** has two centrioles plus pericentriolar material. **Kinesin** vs. **dynein motors** typically move cargo toward the + vs. the – MT end.

The **nucleus** is bounded by a double-membrane ~25 nm-wide **nuclear envelope** with **nuclear pores**. Subjacent to the nuclear envelope is an IF-containing **nuclear lamina** that connects to the cytoskeleton via **LINC complexes** (**Figure d**). Nucleoplasm is filled with **chromatin** (a mixture of DNA + protein that forms **chromosomes**) but also has substructures (e.g., ribosome-forming **nucleoli** and **splicing machinery**). Heterochromatin is compacted and transcriptionally inactive. Euchromatin is loosely packed and potentially transcribed. Some nuclei contain **actin filaments** that can modulate gene expression.

Somatic cells divide via **mitosis**, whereas developing eggs and sperm divide via **meiosis**. **M-phase** of the mitotic **cell cycle** is separated from the next M phase by **interphase** (=G1, S, G2). Cell cycle progression is driven by cyclin-dependent kinases (e.g., CDK1 at G2/M transitions). Before mitosis, chromosomes and the centrosome replicate, and in prophase, dual centrosomes begin to set up the **mitotic apparatus** of: (i) two centrosomes, (ii) MTs, and (iii) duplicated chromosomes with sister chromatids and kinetochores coating centromeres (**Figure e**). Chromatids are moved to opposite poles during anaphase by polymerization/depolymerization of MTs, and after telophase, cells are separated during **cytokinesis** by constriction of a MF-containing contractile ring.

Some cell defects in human diseases: (i) **lysosomal storage diseases** (e.g., Hunter syndrome)—buildup of toxic wastes in various cells due to lysosome dysfunction; (ii) **laminopathies:** e.g., **Hutchinson-Gilford progeria syndrome**—improperly spliced lamin (**progerin**) accelerates aging; (iii) **unfolded protein responses** to misfolded proteins in ER can exacerbate diseases like CF; (iv) **mitochondrial DNA depletion syndromes**: mtDNA is not properly expressed thereby causing mitochondrial dysfunction; (v) **hyperphosphorylation of GRASPs** can cause Golgi disassembly, which is linked to disorders like Alzheimer's disease; (vi) rapid attrition of **telomeres** (endcap chromatin of chromosomes) correlates with early cell aging/death.

SELF-STUDY QUESTIONS

1. Which cellular component makes ribosomal subunits?
2. T/F Mitochondria release cytochrome c during apoptosis.
3. The MTOC contains which end of a microtubule?
4. When does the nuclear envelope break down during cell division?
   A. Prophase
   B. Metaphase
   C. Anaphase
   D. Telophase
   E. Cytokinesis
5. About how wide is the nuclear envelope?
   A. 7 nm
   B. 25 nm
   C. 100 nm
   D. 7 μm
   E. 30 μm
6. Which kind of cellular intake can involve clathrin?
   A. Passive diffusion
   B. Facilitated diffusion
   C. Active transport
   D. Pinocytosis
   E. Phagocytosis
7. T/F Heterochromatin tends to be transcriptionally active.
8. T/F A centrosome contains two centrioles.
9. T/F O_2 can passively diffuse across cell membranes.
10. In which specific phase of the cell cycle is DNA duplicated?
11. Which of the following contains large amounts of catalase?
    A. Mitochondria
    B. Golgi
    C. ER
    D. Nucleus
    E. Peroxisome
12. Which of the following is NOT a feature of intermediate filaments (IFs)?
    A. ~10–12-nm width
    B. Heterogeneous protein composition
    C. Nuclear lamins
    D. Cytoplasmic keratins
    E. Actin
13. T/F. The contractile ring is composed of microtubules.

"EXTRA CREDIT" For each term on the left, provide BEST MATCH on the right (answers A-F can be used more than once)

14. COP II___	A. Retrograde movement
15. Dynein____	B. Abnormal lamin
16. GRASP___	C. ER
17. Progerin___	D. Golgi cisternae linker
18. CDK1_____	E. G2/M transition
	F. Anterograde movement

19. Describe nuclear architecture, including major functions for each listed component.
20. Discuss the cell cycle for mitotically active cells.

ANSWERS

1) nucleolus; 2) T; 3) minus; 4) A; 5) B; 6) D; 7) F; 8) T; 9) T; 10) S; 11) E; 12) E; 13) F; 14) F; 15) A; 16) D; 17) B; 18) E; 19) The answer should include functional morphology of nuclear envelopes, pores, laminas, LINCs, chromatin, chromosomes, nucleoli, speckles; 20) The answer should include and define clearly such terms as: M phase, interphase, G1, S, G2, G0, checkpoints, cyclin dependent kinases, centromeres, mitotic spindle, metaphase plate, kinetochore, non-kinetochore MTs, astral MTs, anaphase A vs. B, telophase, cytokinesis

Overview of Embryology and the Four Basic Tissue Types with Emphasis on Epithelium and Connective Tissue Proper

PREVIEW

3.1 Highlights of **human development: fertilization, implantation, placenta formation**, and **embryonic/fetal development**

3.2 General properties of the four basic tissue types—**epithelium, muscle, nervous tissue**, and **connective tissue**

3.3 Unifying features of **epithelial cells: tight packing, apical-basal polarity, intercellular junctions**, and **basement membranes**

3.4 **Functional microanatomy** of common **apical elaborations** of epithelia: **glycocalyx, brush border, stereovilli**, and **cilia**

3.5 Epithelial **junctions: zonula occludens, zonula adherens, desmosome, gap junction,** and **hemidesmosome**

3.6 Classifying **lining epithelia: simple** vs. **stratified** plus unusual epithelial types—**pseudostratified** and **transitional**

3.7 **Glandular epithelia: uni-** vs. **multi-cellular** glands plus **endocrine glands** (without ducts) vs. **exocrine glands** (with ducts)

3.8 **Embryonic connective tissues: mesenchymal CT** and **mucous CT (Wharton's jelly)**

3.9 **Adult CT proper: cells** (e.g., fibroblasts, macrophages, mast cells), **fibers (collagen, reticular, elastic)** and **ground substance** plus classifications based on fiber densities (**loose** vs. **dense**) and arrangements (**irregular** vs. **regular**)

3.10 The **triple helical organization** of **collagens** and the formation of **collagen fibrils** in health and disease

3.11 **Elastic fibers** and **Marfan syndrome**

3.12 Some disorders of epithelia and CT proper: **Kartagener's syndrome** and **Alport syndrome**

3.13 Summary and self-study questions

3.1 HIGHLIGHTS OF HUMAN DEVELOPMENT: FERTILIZATION, IMPLANTATION, PLACENTA FORMATION, AND EMBRYONIC/FETAL DEVELOPMENT

Interpreting adult histology often requires an understanding of how tissues and organs develop. Thus, this chapter begins with a brief summary of human **embryology**, and each subsequent chapter also considers key developmental highlights relevant to the particular tissues and organs that are discussed.

DOI: 10.1201/9780429353307-4

Human **embryogenesis** is initiated by **sperm fertilizing** an **egg** within the woman's **oviduct** to yield a diploid unicellular **zygote** (**Figure 3.1a**). Following mitotic divisions of the zygote, the resulting **embryo** is transported to the **uterus**, where it develops into a **blastocyst** with a thin outer layer of **trophoblast cells** surrounding a more medially positioned **inner cell mass** (**ICM**) (**Figure 3.1a**). The blastocyst then hatches from its protective coat and **implants** in the uterine lining at about a week post-fertilization (pf). The implanted blastocyst's trophoblast cells combine with maternal tissues to begin forming a **placenta** that serves as a nourishing organ where gas and nutrient exchanges occur. Conversely, the **ICM** generates both the **embryo proper** and its surrounding extraembryonic membranes, such as the fluid-filled **amnion** that protects the embryo. Following 8 weeks of successful pf development, each embryo (**Figure 3.1b**) is referred to as a **fetus**. The fetus continues to develop until birth, which in a **full-term pregnancy** of three ~3-month-long **trimesters** occurs at around 9 months pf. Alternatively, because the precise timing of fertilization is difficult to pinpoint, pregnancies are often described in terms of **gestational ages**, in which each developmental event is calculated relative to the onset of the last menstrual period before fertilization. Such calculations add roughly 2 weeks to the actual pf timings of embryogenesis, and thus, for example, implantation at ~1 week pf would occur at a gestational age of ~3 weeks (**Case 3.1 Pioneering Experimental Embryologist, Hilde Mangold**).

Throughout these developmental timelines, future adult tissues and organs are assembled from three **primary germ layers**, called **ectoderm**, **mesoderm**, and **endoderm**, which the embryo establishes as part of a gastrulation process during the 2nd and 3rd weeks pf (**Figure 3.1c, d**). Ectoderm in the outer portion of the embryo forms the **epidermis** of the skin as well as the **nervous system** while also producing a **neural crest** component of migratory cells that dissociate from the future brain and spinal cord regions. Such neural crest ectoderm generates not only components of the peripheral nervous system but also non-neural structures scattered throughout the body, such as pigment-forming cells of skin (Table 3.1). **Mesoderm** sets up internal to the epidermis and differentiates into: (i) muscles, (ii) linings of coelomic compartments (i.e., secondary body cavities outside the gut), (iii) tissues in organs like blood vessels, kidneys, and gonads, (iv) specialized connective tissues (skeletal elements, fat, and blood), and (v) connective tissue proper (e.g., tendons and dermis of skin). **Endoderm** is the innermost germ

---

CASE 3.1 PIONEERING EXPERIMENTAL EMBRYOLOGIST, HILDE MANGOLD

(1898–1924)

"Hers is one of the very few doctoral theses in biology that have directly resulted in the awarding of a Nobel Prize."

(From Scott Gilbert's Developmental Biology, 6th Ed. text describing H.M's research)

Not long after Marie Curie became the first woman to win a Nobel Prize in 1903, Hilde Mangold conducted pioneering embryological experiments that helped Hans Spemann win his 1935 Nobel Prize for Physiology or Medicine. For such studies, Mangold removed the dorsal blastopore lip from the embryo of one newt species and put the excised region in an embryo of a different species of newt (see thumbnail sketch above). Without effective antibiotics to aid post-surgical survival, Mangold had to conduct 250 of these xenografts to obtain five usable chimeras that allowed her to conclude that the transplants induced cells in the recipient embryo to form a second neural tube. After writing up her groundbreaking work, which was published with Spemann as first author, Mangold died that same year in a house fire. Since Nobel prizes are not awarded posthumously, Mangold could not share Spemann's prize, but today the dorsal blastopore lip is sometimes designated the Spemann-Mangold organizer in honor of her key contributions.

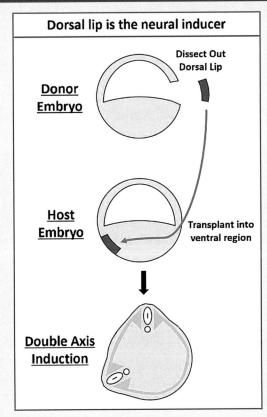

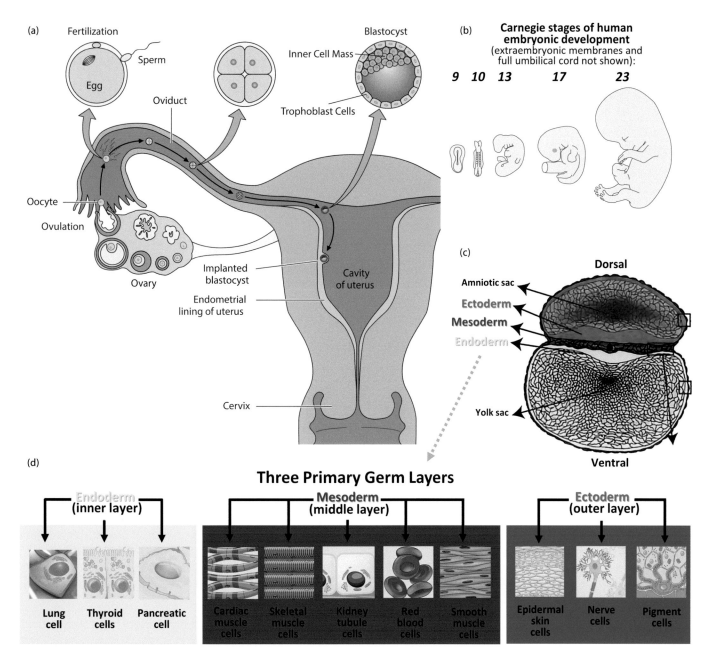

Figure 3.1 Introduction to human development. (*a*) *Following fertilization of an egg in the oviduct, the dividing embryo is moved into the uterus to become a blastocyst comprising: (i) outer trophoblast cells that generate part of the placenta, and (ii) inner cell mass (ICM) cells that form the embryo proper. The blastocyst implants in the endometrial lining of the uterus at ~7 days post-fertilization. (**b**) Carnegie stages of embryonic development end with stage 23, which is completed after 8 weeks post-fertilization as the embryo becomes a fetus. (**c**) Early post-implantation development involves three primary germ layers—ectoderm, mesoderm, and endoderm—forming within the embryo plus the production of surrounding extraembryonic membranes like the amnion. (**d**) Ectoderm gives rise to the nervous system, epidermis of skin, and some other non-neural cells like pigment cells that are derived from neural crest ectoderm; endoderm generates much of the digestive and respiratory systems plus some outgrowths of the embryonic foregut such as the thyroid. Mesoderm forms the rest of the body, including muscles, urogenital tissues, blood, and other connective tissues. ([b]: From photographs in Bleyl, SB and Schoenwolf, GC (2017) What is the timeline of important events during pregnancy that may be disrupted by a teratogenic exposure? Teratology Primer, 3rd Edition, https://www.teratology.org/primer/Teratogenic-Exposure.asp; [c]: From: Isaza-Restrepo, A et al. (2018) The peritoneum: beyond the tissue—a review. Front Physiol 9: doi: 10.2338/fphys.2018.00738 reproduced under a CC BY 4.0 creative commons license; [d]: From OpenStax College (2013) Anatomy and Physiology. OpenStax http://cnx.org/content/col11496/latest, reproduced under a CC BY 4.0 creative commons license.)*

TABLE 3.1 NEURAL CREST SUBTYPES AND THEIR DERIVATIVES

Neural crest subtype	Tissues	Cell types	
Cranial	Craniofacial skeleton Skin Cornea Connective tissue	Fibroblasts Melanocytes Adipocytes Osteocytes Odontoblasts Chondroblasts	Neurons Glia Mesenchymal cells Myocytes Pericytes
Cardiac	Branchial arches Cardiac septum Parasympathetic cardiac Ganglia	Myocytes Pericytes Neurons Glia	
Vagal	Enteric nervous system	Neurons Glia	
Trunk	Peripheral nervous system Skin Adrenal medulla	Neurons Glia Melanocytes Chromaffin cells	

Source: From Delaney SP et al. (2014) The neural crest lineage as a driver of disease heterogeneity in Tuberous Sclerosis Complex and Lymphangioleiomyomatosis. Front Cell Dev Biol 2: doi: 10.3389/fcell.2014.00069 reproduced under a CC BY creative commons license.

layer and along with producing much of the digestive system also forms associated organs like the thyroid, parathyroids, thymus, and lungs (**Figure 3.1d**).

As **fundamental building blocks** of the embryo, primary germ layers generate parts of each adult organ. Thus, within any single organ, diverse derivatives, such as ectodermal nerves, mesodermal muscles, and endodermal lining layers arise from more than a single germ layer. Accordingly, innervated blood vessels with both ectodermal and mesodermal derivatives can be viewed as full-fledged organs, whereas small nerves or tendons, formed exclusively from ectoderm or mesoderm, respectively, would be classified as tissues.

To help compare rapidly vs. slowly developing species, embryos of humans and other vertebrates have been assigned to a series of **23 Carnegie stages** that classifies development according to keystone morphological features rather than absolute timelines (**Figure 3.1b**). Some notable post-gastrulation stages, as well as their general timings and features in humans, include: **stage 9** (end of 3rd week pf: neural tube and somite formation begins [Chapter 9]); **stage 11** (4th week pf: heart begins beating; neural crest is forming from the neural tube); **stage 13** (end of 4th week pf: neural tube closes; liver formation begins; embryo has head, trunk, and tail); **stage 17** (end of 6th week pf: pigmented eyes are present; olfactory and auditory organs begin to form); **stage 23** (end of 8th week pf and the termination of embryonic development; essentially all external features of the future baby are discernable). Organs continue to develop during the fetal phase that extends through the 2nd and 3rd trimesters until birth, which occurs at ~37–39 weeks pf for a full-term pregnancy.

3.2   GENERAL PROPERTIES OF THE FOUR BASIC TISSUE TYPES— EPITHELIUM, MUSCLE, NERVOUS TISSUE, AND CONNECTIVE TISSUE

The **cells** and **extracellular matrix** (**ECM**) of tissues have evolved varied compositions and configurations, thereby generating numerous kinds of tissues. Nevertheless, these highly divergent tissue types can be classified into four basic categories: **epithelium**, **connective tissue**, **muscle**, and **nervous tissue** (**Figure 3.2a–e**).

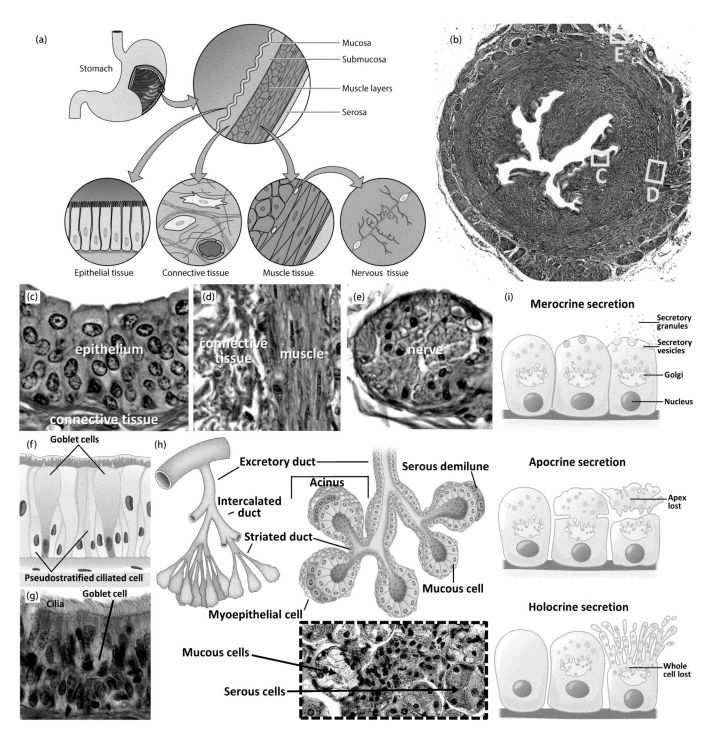

Figure 3.2 The four basic tissue types and an introduction to epithelium. (*a*) Distributed in different locations of large organs, such as the stomach are all four basic tissue types—epithelium, connective tissue, muscle, and nerve. (*b–e*) The four basic tissues in a transversely sectioned urethra are shown at low (b: 12X) and higher magnifications (c: 420X; d: 330X; e: 300X). (*f–h*) Epithelia line cavities such as the trachea (f, g; g: 270X), cover the body surface, and form glands, which in the case of exocrine glands like salivary glands secrete their products via ducts (h: 130X). (*i*) Glandular cells typically undergo merocrine secretion where little of the cell is lost during exocytosis. Conversely, the apex or whole cell is lost during apocrine vs. holocrine secretion, respectively. ([f, i]: From OpenStax College (2013) Anatomy and Physiology. OpenStax http://cnx.org/content/col11496/latest, reproduced under a CC BY 4.0 creative commons license; [h]: From Andreasen, S (2018) Molecular features of adenoid cystic carcinoma with an emphasis on microRNA expression. J Pathol Microbiol Immunol 126 (Suppl. 140): 7–57 reproduced with publisher permission.)

CASE 3.2 ALTHEA GIBSON'S LIGAMENTS

Tennis Player and Golfer
(1927–2003)

"Shaking hands with the Queen of England was a long way from being forced to sit in the colored section of the bus going into downtown Wilmington, North Carolina."

(A.G. after winning the Wimbledon tennis championship)

Before her death at age 76 due to pneumonia, Althea Gibson was a pioneering African-American tennis player and golfer, who had to battle both her opponents and segregation. Growing up, Gibson excelled at several sports but decided to focus her efforts on tennis and quickly became a world-class player. Gibson dominated the American Tennis Association in the 1940s but was initially excluded from elite tournaments that allowed only white players. Eventually, however, she won the French Open and followed that victory with wins at the Wimbledon and U.S. National championships, all of

which were firsts for a person of color. Even though Gibson was the U.S. Female Athlete of the Year in 1957 and 1958, she never received the money-making opportunities given to white players. Thus, in search of more lucrative paychecks, she left tennis and became the first African-American woman to play professional golf. As a golfer, Gibson could not match her tennis achievements. Nevertheless, it is a testament to her remarkable athleticism and competitive drive that she was able to train a complex array of muscles, tendons, and ligaments in order to master the two very different swinging motions required for playing both golf and tennis at a professional level.

Epithelium typically comprises a **sheet** of integrated cells that either line internal cavities (**Figure 3.2f, g**) or cover the body's surface in the case of the epidermis. In addition to such sheet-like configurations, epithelia also form secretory **glands**—both unicellular and multicellular types—the latter of which may possess distinct **ducts** for draining glandular secretions (**Figure 3.2h**). At the cellular level, glandular epithelia typically exocytose their products via a **merocrine** form of secretion, where secretory vesicles fuse with the plasmalemma thereby minimizing cellular loss. Alternatively, some glandular cells either lose much of their upper (=apical) cytoplasm (e.g., mammary gland cells) or all of the cell (e.g., sebaceous gland cells) during **apocrine** vs. **holocrine** modes of exocytosis, respectively (**Figure 3.2i**). Both lining and glandular epithelia share in common several unifying features, such as a tightly packed organization with relatively little intervening ECM, which, in turn, provides effective boundary layers.

Connective tissue (**CT**) is the most heterogeneous of the four basic tissue types and generally serves to interconnect body components via its well-developed ECM. The variable nature of CT ECMs makes it difficult to generate a single universally accepted classification. However, CTs are often split into **embryonic** forms that function *in utero* vs. **adult** CTs that reach their mature configurations after birth and are further divided into **CT proper** vs. **specialized CTs** (**Figure 3.3a**). Adult CT proper more closely resembles generalized CTs in basal animal lineages and comprises loosely arranged versions like organ-suspending mesenteries or densely configured examples such as **ligaments** (**Figure 3.3b**) (Case 3.2 Althea Gibson's Ligaments). Alternatively, specialized CTs are **cartilage**, **bone**, **fat**, and **blood** (**Figure 3.3c–j**), which have evolved highly modified forms for additional functions beyond those of CT proper.

Muscles are composed of contractile cells, which can be either **striated** or **smooth** based on whether or not their intracellular **myofilaments** used in contractions form serially repeated striations (**Figure 3.4a–c**). Via their myofilament-mediated contractions, muscles function mainly in motility-related processes, such as bodily locomotion, blood pumping, and material propulsion along the digestive tract.

Nervous tissue is characterized by hardly any ECM interspersed among its two cell types: (i) **neurons** that propagate electrical impulses (=**action potentials**, **APs**) along their cell membranes, and (ii) supportive **glial cells** that help maintain neuronal viability but are typically unable to transmit APs themselves (**Figure 3.4d, e**). Collectively, nervous tissue within the brain, spinal cord, and peripheral nervous system work together to modulate various physiological processes, thereby coordinating with other regulators, such as hormones in controlling appropriate responses to stimuli that impinge upon, and arise within, the body.

Since specialized CTs, muscles, and nervous tissue are also covered in separate chapters, only cursory introductions are provided here. Conversely, without additional coverage provided elsewhere, this chapter includes further details regarding **epithelia** and **CT proper**.

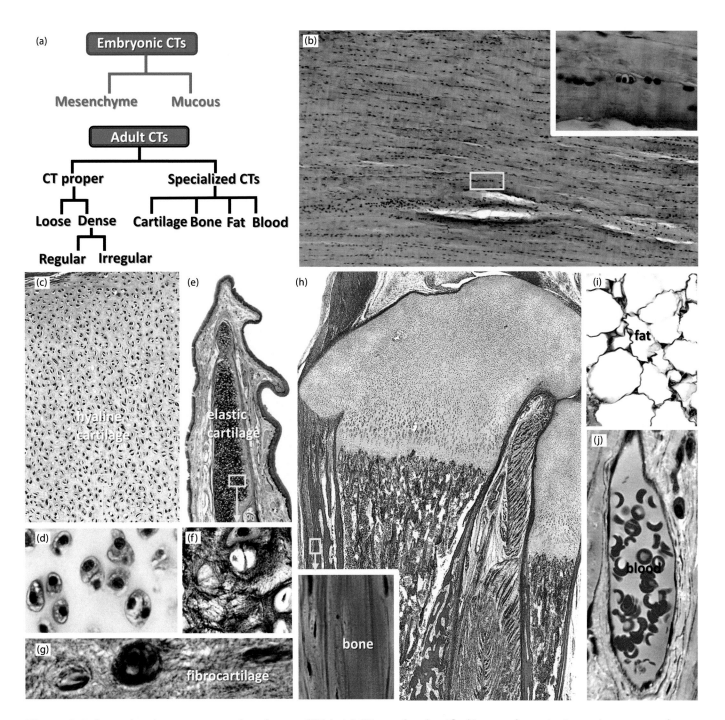

Figure 3.3 Introduction to connective tissues (CTs). (*a*) *CTs can be classified into embryonic tissues (mucous and mesenchyme) that function exclusively before birth vs. adult tissues that mature after birth and are further subdivided into generalized CT proper (with loosely vs. densely arranged fibers) vs. more highly modified specialized CTs (cartilage, bone, fat, and blood).* (*b*) *A section through a ligament shows an example of dense regular CT proper (170X; inset: 40X).* (*c–j*) *Examples of specialized CTs are illustrated in sections of cartilage (c: 45X; d: 230X; e: 25X; f: 230X; g: 250X), bone (h: 20X; inset: 200X), fat (i: 160X), and blood (j: 560X).*

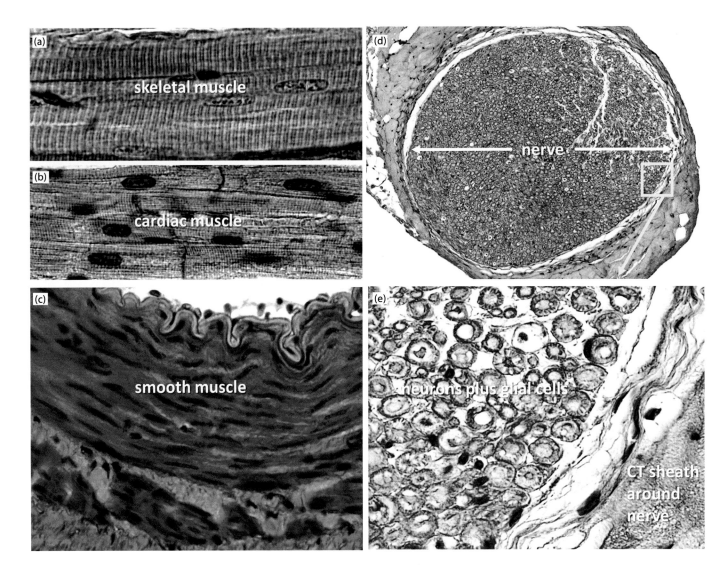

Figure 3.4 Introduction to muscle and nervous tissue. *(a–e) Along with epithelium and CT, muscle (skeletal [a: 700X], cardiac [b: 500X], and smooth [c: 250X]) plus nerve (d: 60X; e: 400X) constitute the four basic tissue types of the body.*

3.3   UNIFYING FEATURES OF EPITHELIAL CELLS: TIGHT PACKING, APICAL-BASAL POLARITY, INTERCELLULAR JUNCTIONS, AND BASEMENT MEMBRANES

Epithelial cells are characterized by their: (1) **tight packing** and thus minimal intervening ECM; (2) **apical-basal polarity** such that their **apex** that is directed toward either an internal lumen or the outer environment differs markedly from their **base** that overlies subjacent tissues; (3) **intercellular junctions** that hold neighboring cells together and help maintain epithelial functioning; and (4) a **basement membrane** (~basal lamina), which is a thin layer of collagen and other components lying directly beneath epithelia (**Figure 3.5a–h**). Although such features serve as hallmarks of epithelia, a few exceptions can occur. For example, some lymph and blood vessel linings lack a basement membrane (Chapter 7). Conversely, several kinds of non-epithelial cells are also known to exhibit cellular polarization, intercellular junctions, and/or basement membrane-like sheaths.

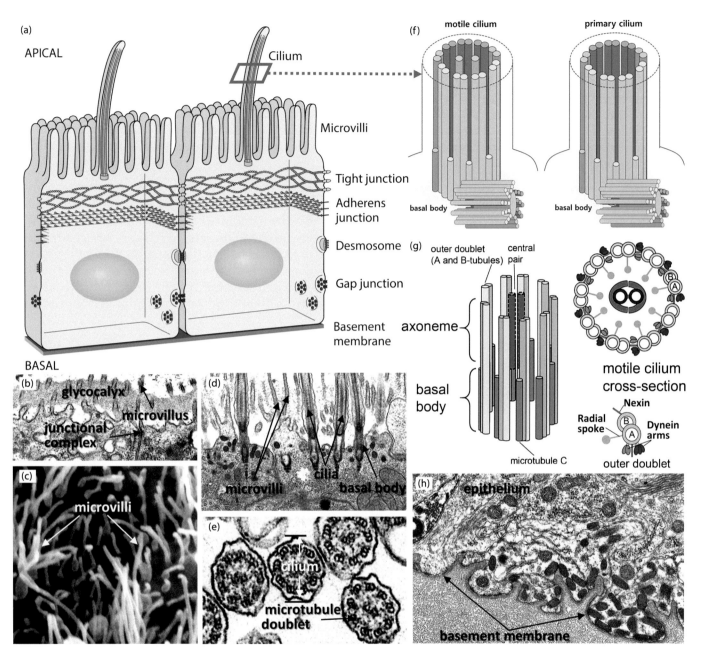

Figure 3.5 Epithelium: apical-basal polarity. (a) *Epithelial cells are: (1) tightly packed (i.e., with little intervening ECM); (2) interconnected by several types of junctions; (3) polarized so that their apices differ from their bases; and (4) underlain by a basement membrane. (**b–g**) Apical elaborations of typical epithelial cells include: (i) microfilament-containing microvilli (b: 16,000X; c: 15,000X) that lie subjacent to a glycocalyx; (ii) junctional complexes; and (iii) cilia that contain at least nine pairs of microtubules (d: 15,000X; e: 80,000X). (**f, g**) Motile cilia have a central pair of microtubules plus dynein arms and radial spokes attached to their peripheral microtubules, which collectively enable ciliary motility. Alternatively, immotile primary cilia lack a central pair and complete set of spokes and dynein arms, thereby adapting such organelles to sensory, rather than locomotory, functions. (**h**) The basal surface of each epithelium rests on a thin basement membrane that, in turn, lies above the rest of the subepithelial extracellular matrix (8,500X). ([f]: From https://www.ncbi. nlm.nih.gov/books/NBK373370/: Park, JH et al. Polycystic Kidney Disease, Chapter 15 Autosomal dominant polycystic kidney disease induced by ciliary defects reproduced under a CC BY 4.0 creative commons license; [g]: From Fabczak, H and Osinka, A (2019) Role of the novel Hsp90 co-chaperones in dynein arms' preassembly. Int J Mol Sci. 20: doi:10.3390/ ijms20246174 reproduced under a CC BY 4.0 creative commons license.)*

3.4   FUNCTIONAL MICROANATOMY OF COMMON APICAL ELABORATIONS OF EPITHELIA: GLYCOCALYX, BRUSH BORDER, STEREOVILLI, AND CILIA

The **glycocalyx** is an extracellular coat of mainly glycosylated proteins that covers the apices of many epithelia (**Figure 3.5b**). Particularly well-developed glycocalyces that can reach widths of several micrometers occur in the gut lining, pulmonary air sacs, and next to blood vessel lumens. Functionally, the glycocalyx can modulate access to plasmalemmal receptors while also generating forces that help shape specialized domains within underlying cell membranes.

Beneath the glycocalyx, the epithelial apex often forms finger-like projections, called **microvilli**, which typically measure ~0.1 µm wide by ~0.5–1 µm tall. Such elaborations tend to be sparsely arranged (**Figure 3.5c**), but in the small intestine and proximal tubules of the kidney, abundant microvilli form a distinct unit, called the **brush border**. Within each microvillus are vertically oriented actin **microfilaments**, which undergo treadmilling via actin additions and removals, thereby either maintaining or altering microvillar lengths, depending on the relative rates of these processes. Conversely, the basal ends of microvillar microfilaments are anchored in a several-micrometer-wide band of cytoskeletal elements and associated proteins that constitute a **terminal web**. Together, microvilli and terminal webs form dynamic organelles that greatly increase surface area for optimizing absorptive and secretory processes.

As opposed to normal-sized counterparts, elongated microvilli are formed by several types of epithelial cells, including some lining cells in the male reproductive tract (Chapter 17). Before their ultrastructure had been elucidated, such elaborations had been termed stereocilia in the mistaken belief that they were non-motile cilia. However, given their microvillar organization and lack of cilia-like features, these elaborations are more aptly designated **stereovilli**. In addition to scattered types of stereovilli occurring along epithelial apices, hair cells of the inner ear contain discrete clusters of these organelles (Chapter 20). In general, the high surface-to-volume ratio of stereovilli in lining cells aids absorption and secretion, whereas hair cell stereovilli serve sensory roles.

Along with microvilli, epithelial apices can produce at least one **cilium**, which is a cylindrical organelle containing an orderly array of **microtubules (MTs)** that are attached to an underlying **basal body** centriole (**Figure 3.5d, e**). Two main types of cilia—**primary** vs. **motile**—are recognized. It is believed that most epithelial cells form a single non-motile primary cilium (**Figure 3.5f**). Typically measuring ~0.2–0.3 wide by ~2–10 µm long, the elongated shaft (=**axoneme**) of primary cilia lacks the complete intracellular machinery required for active movements. Instead, primary cilia help sense and transduce signals during development or following cellular differentiation. Consequently, dysfunctional primary cilia are linked to developmental defects and/or post-natal diseases termed **sensory ciliopathies** (Box 3.1 Primary Cilia in Health and Disease).

Alternatively, ciliated cells in the epithelial lining of respiratory passageways, female reproductive tract, and cerebral-spinal-fluid-containing cavities of the brain and spinal cord have multiple **motile cilia** extending apically from each cell. Similar in width and length to those of a primary cilium, such cilia are capable of rapidly bending back and forth, owing to **axonemal** components that are lacking in primary cilia. Thus, in comparing transverse sections of motile vs. primary cilia, the axoneme in both cases contains nine peripheral pairs of MTs, with each pair comprising an "A" and a "B" MT before eventually transitioning into nine triplets of MTs within the subjacent centriole (**Figure 3.5g**). However, in typical motile cilia, the axoneme contains an additional central pair of MTs to provide a 9+2 MT arrangement, whereas primary cilia generally lack central MTs and thus have a 9+0 configuration (**Figure 3.5f**). Moreover, unlike motile forms, primary cilia lack components associated with peripheral MTs, such as nexin linkages, a full complement of radial spokes, and key structures, termed the inner and outer **dynein arms** (**Figure 3.5g**). ATPase activity in dynein arms provides the driving force for MTs translocating past each other during a **sliding microtubule mechanism** of ciliary beating used in producing power- and recovery strokes. Such movements, in turn, allow ciliated epithelia to: (i) clear debris from respiratory passageways, (ii) sweep oocytes into the oviduct and translocate fertilized embryos from the oviduct to the uterus, or (iii) circulate cerebral spinal fluid. Similarly, the solitary **flagellar** tail of each sperm (Chapter 17) has 9+2 MTs, dynein arms, and a sliding MT mechanism for sperm movements, albeit with different waveform dynamics than are observed in beating cilia.

BOX 3.1 PRIMARY CILIA IN HEALTH AND DISEASE

Except for a few non-ciliated cells like T lymphocytes, various cell types in mammals form a single **primary cilium** that projects from an anchoring basal body. The elongated shaft (=**axoneme**) of each cilium contains nine peripheral pairs of microtubules (MTs) but lacks a central pair plus other key components of motile cilia. Thus, primary cilia have a 9+0 MT organization and are non-motile. Within their stationary axonemes, primary cilia carry out abundant anterograde **intraflagellar transport** (IFT) that moves molecules toward the ciliary tip as well as retrograde IFT that sends molecules back toward the basal body.

IFT and membrane receptors enable primary cilia to serve as sensory organelles and transducers of key signals during development and post-differentiation maintenance of cellular functioning. When properly stimulated, signaling cascades in primary cilia help mediate such developmental processes as eye morphogenesis, mid-axial partitioning, and digit formation.

Conversely, based on congenital diseases termed **sensory ciliopathies**, defects in primary cilia can result in wide-ranging pathogenesis, including retinal dystrophy, skeletal malformations, and kidney defects. For example, primary cilia genes, called Pkd1 and Pkd2, are mutated in **polycystic kidney disease** (PKD). Such genes encode **PKD1** (=polycystin 1, PC1) and **PKD2** (=polycystin 2, PC2) proteins in the ciliary membrane that allow Ca^{2+} influx into renal tubule cells (**Figure a, b**). According to one model, the normal influx of Ca^{2+} occurs in concert with the activation of **epidermal growth factor** (**EGFR**) in cells with normal primary cilia, thereby blocking "cystogenic signals" that cause cysts to form in renal tissues. Conversely, without functional PDK1/PDK2 proteins, insufficient Ca^{2+} influx allows formation of pathological cysts, which degrade renal function and underscore the roles of dysfunctional primary cilia in disease progression.

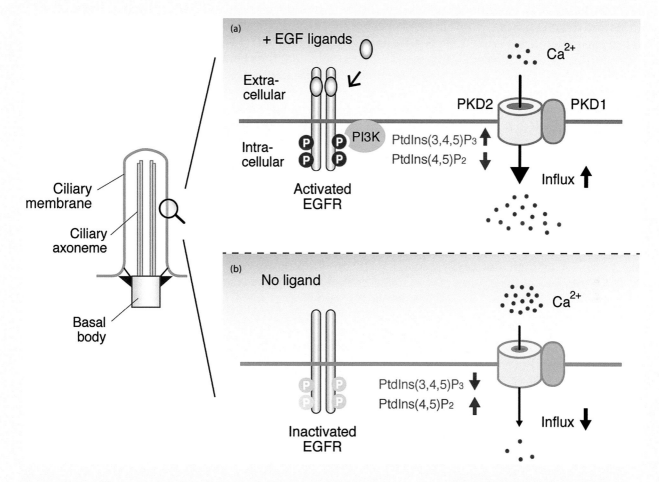

Box 3.1 (a, b) *By epidermal growth factor (EGF)-associated activation of Ca^{2+} influx through polycystic kidney disease (PKD) channels, primary cilia may block the transmission of cystogenic signals that lead to cyst formation in renal tissues during PKD. (Diagram from: Nishimura, Y et al. (2019) Primary cilia as signaling hubs in health and disease. Adv Sci 6: doi: 10.1002/advs.201801138 reproduced under a CC BY 4.0 creative commons license.)*

3.5 EPITHELIAL JUNCTIONS: ZONULA OCCLUDENS, ZONULA ADHERENS, DESMOSOME, GAP JUNCTION, AND HEMIDESMOSOME

Several kinds of intercellular junctions not only help keep epithelia intact but also serve other functions, such as in signaling pathways and in providing a physical barrier between the apical and basal compartments flanking the epithelium (**Figure 3.6a**). The apical-most junction of epithelial cell membranes is called a **zonula occludens (ZO)** (**Figure 3.6b**). Each ZO extends continuously around the entire cell in a belt-like fashion (**Figure 3.6c**) and is associated on its cytoplasmic side with **microfilaments**. The outer leaflets of cell membranes in ZOs essentially fuse at discrete sites enriched in proteins like occludins to restrict transepithelial flow of molecules (**Figure 3.6a–f**). Thus, ZOs are also referred to as **tight junctions** and are common in locations, such as the **blood-brain barrier** (Chapter 9) that serve to minimize the flow of potentially deleterious molecules between sub-epithelial tissues and lumens overlying epithelia. Directly beneath the ZO, a **zonula adherens** (ZA or adherens junction) also forms a continuous belt-like junction with attached microfilaments that help hold epithelial cells together (**Figure 3.6a–c, h**). However, compared to ZOs, ZAs have a larger intercellular space in which glycoproteins, such as **cadherins** provide greater adhesion. Moreover, instead of acting as physical barriers, ZAs often help transduce mechanical forces to signaling pathways within epithelia.

Along with zonula-type junctions, many epithelial cells possess two kinds of button-shaped connections—**gap junctions** and **desmosomes**—which instead of extending around the entire cell are restricted to discrete sites in the lateral plasmalemma (**Figure 3.6b, c, g, h**). Each **gap junction (GJ)** (=a nexus or communicating junction) is a quasi-crystalline cluster of channels that enable direct molecular flow between neighboring cells. The individual channels within a gap junction are assembled from six **connexin** proteins that create a heximeric hemichannel, called a **connexon**, which, in turn, lines up with a corresponding connexon in the neighboring membrane to yield a complete channel (**Figure 3.6g**). Such conduits allow small (<1 kDa) molecules, such as ions and second messengers like cAMP to flow between cells, thereby providing chemo-electric coupling in epithelia as well as in other GJ-generating cells (Box 8.2).

Alternatively, each **desmosome** (=macula adherens) contains numerous membrane proteins and attached **intermediate filaments** (**Figure 3.6a–c, h**). Collectively, these thickened membrane sites provide strong **spot welds** for epithelial integrity and are particularly abundant in epidermal cells (Chapter 10). Similarly, **hemidesmosomes (HDs)** are linked to intermediate filaments at discrete membrane sites. However, instead of being shared between two cells, HDs attach to integrin proteins along the basal surface of individual cells, where they help anchor the epithelium, especially in tissues subjected to high shear forces, such as the epidermis (**Figure 3.6h**). Accordingly, defects in epidermal HD functioning and/or in associated collagens of the ECM can lead to epidermolysis bullosa types of skin diseases that are characterized by epithelial fragility and blistering detachments of the epidermis. Some of the key characteristics of epithelial junctions are summarized in Table 3.2.

3.6 CLASSIFYING LINING EPITHELIA: SIMPLE VS. STRATIFIED PLUS UNUSUAL EPITHELIAL TYPES—PSEUDOSTRATIFIED AND TRANSITIONAL

Different morphological types of lining epithelia are distinguished based on the number of their cellular layers and on the shape of their apical-most cells. Thus, a **simple epithelium** comprises a single layer of cells that all touch the underlying basement membrane (**Figure 3.7a**). Such unilaminar epithelia are further classified as **squamous**, **cuboidal**, or **columnar** for cells that are: (i) flattened along the apical-basal axis, (ii) essentially as tall as they are wide, or (iii) elongated from base to apex, respectively. As examples of these three types, the endothelial lining of blood vessels is **simple squamous**, whereas renal tubules tend to be lined by a **simple cuboidal epithelium** as opposed to the **simple columnar** lining of the intestines. Alternatively, in a **stratified epithelium**, multiple layers are present, as not all cells touch the basement membrane (**Figure 3.7b**). By far the most common type of stratified epithelium is the stratified squamous epithelium of the epidermis, oral cavity, esophagus, and anal canal, whereas stratified cuboidal and stratified columnar epithelia are encountered in such restricted locations as certain glandular ducts and ocular tissues, respectively.

In addition, two epithelial subtypes—**pseudostratified** and **transitional**—possess unusual characteristics. A pseudostratified epithelium lines regions like respiratory passageways and is so named because it falsely

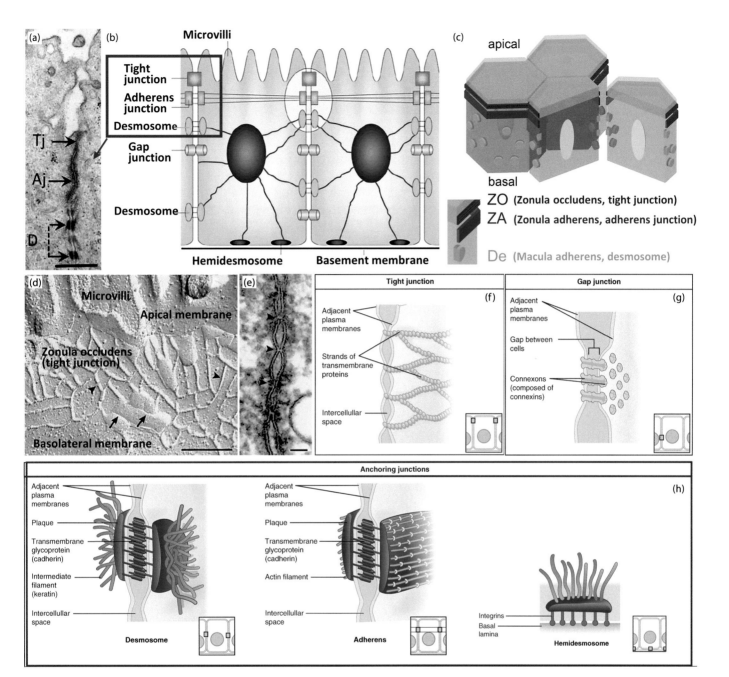

Figure 3.6 Epithelium: intercellular junctions. *(**a, b**) Of the five major intercellular junctions in epithelia, tight junctions (TJs, zonulae occludentes), adherens junctions (zonulae adherentes), and desmosomes tend to be more apically located (a, scale bar = 500 nm, b), whereas gap junctions and hemidesmosomes occur laterally and basally, respectively, in epithelial cells. (**c**) Zonular junctions form continuous belt-like structures around epithelial cells, whereas macular junctions like desmosomes, hemidesmosomes, and gap junctions comprise discrete patches within epithelial cell membranes. (**d–f**) As shown in a freeze-etch EM image (d, scale bar = 200 nm), TEM (e, scale bar = 50 nm), and diagram (f), neighboring cells connected by a tight junction have discrete sites (arrows, e) where essentially fused outer leaflets restrict transepithelial molecular fluxes. (**g**) Each gap junction (GJ) consists of a highly ordered array of transcellular channels (connexons), which are composed of connexin proteins. Aligned connexons form conduits through which small molecules like ions can pass, thereby coupling neighboring cells electrically and functionally. (**h**) Three anchoring junctions—desmosomes, adherens junctions, and hemidesmosomes—serve to maintain epithelial integrity and are associated on their intracellular side with intermediate filaments (desmosomes, hemidesmosomes) or microfilaments (adherens junctions). ([a]: From Prozorowska, E et al. (2019) Ultrastructural study of uterine epithelium in the domestic cat during prenatal development. Theriogenology 130: 49–61 reproduced with publisher permission; [b, d, e]: From Tsukita,*

Figure 3.6 (Continued) *S et al. (2001) Multifunctional strands in tight junctions. Nat Rev 2: 285–293 reproduced with publisher permission; [c]: From Mueller H-AJ (2018) More diversity in epithelial cell polarity: A fruit flies' gut feeling. PLoS Biol 16(12): e3000082. https://doi.org/10.1371/journal.pbio.3000082 reproduced under a CC BY 4.0 creative commons license; [f–h]: From OpenStax College (2013) Anatomy and Physiology. OpenStax http://cnx.org/content/col11496/latest, reproduced under a CC BY 4.0 creative commons license.)*

appears to be multilayered owing to the staggered positioning of its nuclei. However, with relatively short cells that are interspersed among taller ones, pseudostratified is, in fact, a simple epithelium whose cells all touch the basement membrane (**Figure 3.7A4**). Conversely, the transitional epithelial lining of the urinary tract is a stratified epithelium with multiple distinct layers, but is so-named, because its apical-most cells transition from cuboidal to squamous as the lumen fills with urine and thus stretches the epithelium into a more flattened state (**Figure 3.7B4**).

3.7   GLANDULAR EPITHELIA: UNI- VS. MULTI-CELLULAR GLANDS PLUS ENDOCRINE GLANDS (WITHOUT DUCTS) VS. EXOCRINE GLANDS (WITH DUCTS)

In addition to sheet-like configurations, epithelia also form glands that can comprise scattered **unicellular** secretory units (e.g., mucous goblet cells and neuroendocrine cells). Alternatively, two types of multicellular glands—**endocrine** vs. **exocrine**—are also epithelial derivatives. During development, each endocrine gland typically invaginates from a surface epithelium and eventually loses its duct before associating with blood vessels and secreting its **hormones** into the bloodstream (Chapter 15). Conversely, exocrine glands remain connected to internal cavities or the exterior environment via a **duct**. Such glandular ducts branch or do not branch in **compound** vs. **simple** ducts, respectively (**Figure 3.7c**), and via these conduits, exocrine glands deliver various secretory products, including enzymes, lubricants, and anti-microbial agents. Exocrine glands are further classified based on the morphology of their secretory regions, which range from a spherical acinus (=alveolus) to a more cylindrical tube, with such tubes in some cases being wound into coils (**Figure 3.7c**). For example, sweat glands of the skin are **simple coiled tubular** glands, where the duct is unbranched and the tubular secretory region is coiled.

3.8   EMBRYONIC CONNECTIVE TISSUES: MESENCHYMAL CT AND MUCOUS CT (WHARTON'S JELLY)

The most common CT in embryos and fetuses comprises relatively dense arrays of irregularly shaped mesodermal cells, called **mesenchymal cells**. Such cells can differentiate into various cell types and secrete key molecules to regulate the functioning of nearby cells (**Figure 3.8a**). Mesenchymal CTs are spread throughout the developing

TABLE 3.2 INTERCELLULAR JUNCTIONS

Junction	Major protein types	Associated filaments
Zonula occludens	Claudin, occludin, zonulin	Actin
Zonula adherens	Cadherin, catenin	Actin
Macula adherens	Desmoglein, desmocollin, desmoplakin	Intermediate
Gap junctions	Connexin	
Hemidesmosome	Integrin, plectin, tetraspanin	Intermediate

Source: Based on data from Lee JY et al. (2018) Molecular pathophysiology of epithelial barrier dysfunction in inflammatory bowel disease. Proteomes 6, 17: doi:10.3390/proteomes6020017 reproduced under a CC BY creative commons license.

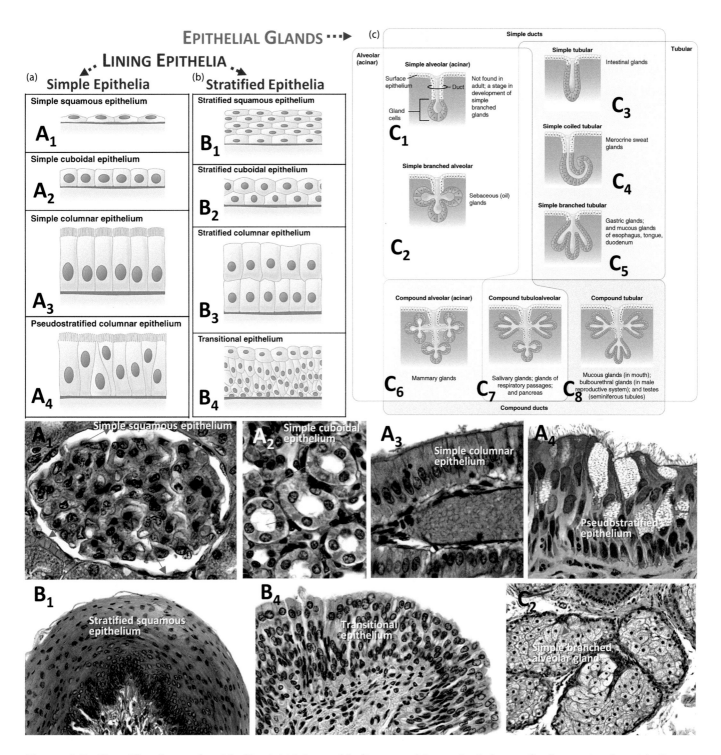

Figure 3.7 Classifications of epithelia. (a) *Lining epithelia comprising a single layer of cells are simple epithelia, whose cells range from thin (=squamous) to medium height (=cuboidal) to tall (=columnar).* **(b)** *Conversely, multilayered epithelia are stratified. The pseudostratified epithelium of the respiratory tract is a simple epithelium, and the transitional epithelium of the urinary tract changes from stratified squamous to stratified cuboidal depending on urine content. Examples **A1–B4**: A1—a simple squamous epithelium (arrows) surrounds renal corpuscles of kidney (300X); A2—a simple cuboidal epithelium lines kidney tubules (arrows) (330X); A3—a simple columnar epithelium (arrow) lines much of the digestive tract (300X); A4—a pseudostratified epithelium lines the trachea (300X); B1—a stratified squamous epithelium lines the esophagus (150X); B4—transitional stratified epithelium lines the ureter (200X).* **(c)** *As opposed to endocrine glands that lack ducts, exocrine glands derived from epithelia can have unbranched ducts (simple glands) or branched ducts (compound glands) with secretory portions that range from*

Figure 3.7 (Continued) *tubular to alveolar (=acinar) and can be either unbranched or branched; C2—sebaceous glands (arrows) are examples of a simple branched alveolar gland (85X). ([a–c]: From OpenStax College (2013) Anatomy and Physiology. OpenStax http://cnx.org/content/col11496/latest, reproduced under a CC BY 4.0 creative commons license.)*

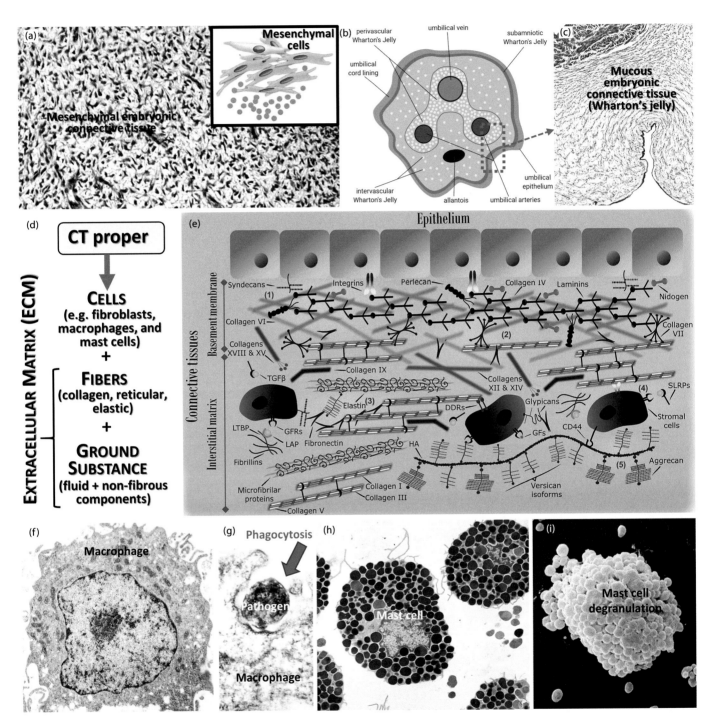

Figure 3.8 Embryonic connective tissues (CTs) and adult CT proper. *(a) Non-specialized CTs in embryos and fetuses consist mostly of mesenchyme with fairly dense arrays of stellate cells that can secrete growth factors (inset, yellow spheres) to influence differentiation of nearby cells (100X). (b, c) As opposed to mesenchyme, Wharton's jelly in the umbilical cord comprises loosely arranged cells with a mucopolysaccharide-rich ECM (60X). (d) Adult CT proper consists of: (1) various cells including fibroblasts, macrophages, and mast cells; (2) fibers (collagen, reticular, and elastic), and (3)*

Figure 3.8 (Continued) *ground substance (=fluid plus non-fibrous components such as numerous glycoproteins and proteoglycans). Together, fibers and ground substance constitute the ECM of CT proper. (e) The composition of CT proper can vary dramatically not just throughout the body, but even at different sites within the same organ, for example when comparing basement membranes under epithelia vs. the interstitial ECM of organ stroma interspersed among epithelia. Figures f (6,800X) and g (15,000X) show TEMs of macrophages in a quiescent state vs. in the process of phagocytosing a pathogen. (h, i) Mast cells are characterized by numerous metachromatic granules that they secrete via a degranulation process (h: 2,300X; I: 2,500X). ([a, inset]: From de la Torre, P et al. (2019) Human placenta-derived mesenchymal stromal cells: a review from basic research to clinical applications. DOI: 10.5772/intechopen.76718 reproduced under a CC BY 3.0 creative commons license; [b]: From Stefańska, K et al. (2019) Human Wharton's jelly—cellular specificity, stemness potency, animal models, and current application in human clinical trials. J Clin Med 9(4), 1102: https://doi.org/10.3390/jcm9041102 reproduced under a CC BY 4.0 creative commons license; [e]: From Theocharis, AD et al. (2019) The extracellular matrix as a multitasking player in disease. FEBS Lett 286: doi:10.1111/febs.14818 reproduced with publisher permission; [f, g]: From Dedonder, SE et al. (2012) Transmission electron microscopy reveals distinct macrophage- and tick cell-specific morphological stages of Ehrlichia chaffeensis. PLoS ONE 7: e36749 doi:10.1371/jounal.pone0036749 reproduced under a CC BY license; [h, i]: Images courtesy of Dr. V. Borelli; from Borelli, V et al. (2018) The secretory response of rat peritoneal mast cells on exposure to mineral fibers. Int J Environ Res Public Health 15: doi:10.3390/ijerph15010104 reproduced under a CC BY 4.0 creative commons license.)*

fetus, except within the umbilical cord that joins the fetus with the placenta (**Figure 3.8b, c**) (Chapter 18). Instead, the umbilicus possesses a loose arrangement of cells plus scattered wisps of mucilaginous material, collectively called mucous CT, or **Wharton's jelly**. By providing a pliable and supportive CT, Wharton's jelly helps prevent umbilical vessels from kinking during pregnancy and can contain pluripotent cells that offer researchers an accessible supply of embryonic stem cells for supplementing adult stem cell sources.

3.9   ADULT CT PROPER: CELLS (E.G., FIBROBLASTS, MACROPHAGES, MAST CELLS), FIBERS (COLLAGEN, RETICULAR, ELASTIC) AND GROUND SUBSTANCE PLUS CLASSIFICATIONS BASED ON FIBER DENSITIES (LOOSE VS. DENSE) AND ARRANGEMENTS (IRREGULAR VS. REGULAR)

As in other tissue types, the CT proper of adults is composed of three elements: (1) **cells**, (2) **fibers**, and (3) a **ground substance** comprising fluid plus non-fibrous molecules. Ground substance and fibers, in turn, constitute the **ECM** (**Figure 3.8d**). CT proper cells are further classified as **transient cells** like white blood cells that enter and exit the CT proper in response to pathogens or other cues vs. **fixed cells**, such as **fibroblasts** that spend essentially their entire lifetimes within CT proper. The proportions and compositions of the three CT proper components can vary widely not only across the body but even within micro-niches of the same tissue (**Figure 3.8e**). Given this complexity, the following introduction focuses only on generalizations that are applicable to most examples of adult CT proper.

Fibroblasts tend to be the most common fixed cells of CT proper and as discussed further below are the key producers of CT fibers. In addition, two other frequently encountered fixed cells are **macrophages** and **mast cells**. Derived from monocytes, macrophages function in host defenses like phagocytosis, thereby acquiring such distinguishing features as well-developed arrays of lysosomes, abundant endocytic vesicles, and irregular peripheries as they engulf large materials or even whole cells (**Figure 3.8f, g**). Depending on their specific origin sites, differentiation patterns, and localization with the body, macrophages are assigned different names as part of a wide-ranging **mononuclear phagocyte system** (**MPS**) (Table 3.3).

Unlike macrophages, mast cells are derived from non-monocyte precursors and are filled with **metachromatic granules** that can cause dyes like toluidine to shift from their normal hues to different colors (**Figure 3.8h, i**). Such color-shifting **metachromasia** is due to granular components like heparin and chondroitin sulfate generating dense accumulations of dye molecules that, in turn, alter the chromatic properties of the stain. Along with these constituents, mast cells also release other bioactive compounds like histamine that can either help modulate normal homeostasis or contribute to wide-ranging pathogenesis, including acute systemic responses that lead to a life-threatening condition, termed **anaphylaxis** (Box 3.2 Stinging Insects, Mast Cells, and Anaphylaxis).

TABLE 3.3 MONONUCLEAR PHAGOCYTE SYSTEM

Prenatal/adult source	Phagocyte type	Location/cell name
Yolk sac	Microglia	Central nervous system
	Tissue macrophages	Several tissues
	Langerhans cells	Skin
Fetal liver	Alveolar macrophages	Lungs
	Langerhans cells	Skin
	M-CSF*-driven tissue resident macrophages	Spleen
		Liver/Kupffer cells
		Bone/osteoclasts
		Kidney/mesangial cells
	IL4**-driven tissue resident macrophages	Several tissues
Adult bone marrow	Inflammatory macrophages	Various sites
	Inflammatory resolution-promoting macrophages	Various sites
	Atherosclerotic plaque	Arteries
	Intestinal macrophages	Intestines
	Microglia	Central nervous system
	Langerhans cells	Skin

Source: Based on data from Fejer, G et al. (2015) Self-renewing macrophages–a new line of enquiries in mononuclear phagocytes. Immunobiology 22: 169–174 reproduced with publisher permission.
Abbreviations: *M-CSF = macrophage colony-stimulating factor; **IL-4 = interleukin 4.

BOX 3.2 STINGING INSECTS, MAST CELLS, AND ANAPHYLAXIS

Insects such as bees and wasps can sting humans and inject venoms that cause allergic reactions. In most cases, such envenomation generates a relatively mild response that remains localized to the skin surrounding the sting site. However, in ~0.5–8% of adults who are stung, a more systemic reaction occurs, with up to 40% of these cases rapidly progressing to a life-threatening condition, called **anaphylaxis**. As is the case with anaphylaxis triggered by such stimuli as snake venoms or food and drug allergies, the acute, system-wide effects of insect stings pose significant risks to susceptible individuals.

Anaphylaxis is predominantly mediated by granule-containing CT cells, called **mast cells**. Such cells secrete compounds that normally generate localized responses to allergens but in imbalanced cases are able to trigger the systemic effects of anaphylaxis. Upstream steps in mast cell stimulation typically involve an initial allergen encounter causing helper T cells to release cytokines (Chapter 7) that, in turn, activate **IgE antibody** production by B cells (**Figure a**). Secreted IgE antibodies can then bind to mast cells via high-affinity **FcεRI** receptors that are attached to the mast cell plasmalemma. Subsequent allergen crosslinking of attached IgEs triggers a Ca^{2+} rise that promotes mast cell degranulation and the release of bioactive compounds. For example, **histamine** released at high levels can trigger anaphylactic responses, such as widespread skin reactions, airway constriction, and hypotension. To counteract these effects, patients prone to anaphylaxis are advised to carry a ready source of epinephrine (e.g., "EpiPens") for rapidly stimulating adrenergic receptors throughout the body (Chapter 9).

Life-threatening anaphylactic responses to localized insect envenomation are difficult to understand in an evolutionary context. However, there is evidence that non-lethal forms of anaphylaxis can enhance subsequent survival rates during repeated challenges with broadly distributed venoms, such as those delivered via the bloodstream by snake or scorpion bites. Thus, although maladaptive when directed against the restricted venom deliveries of insect stings, the capacity of anaphylaxis to generate wide-ranging and rapid responses may help counteract more systemic insults that the body occasionally encounters.

(a)

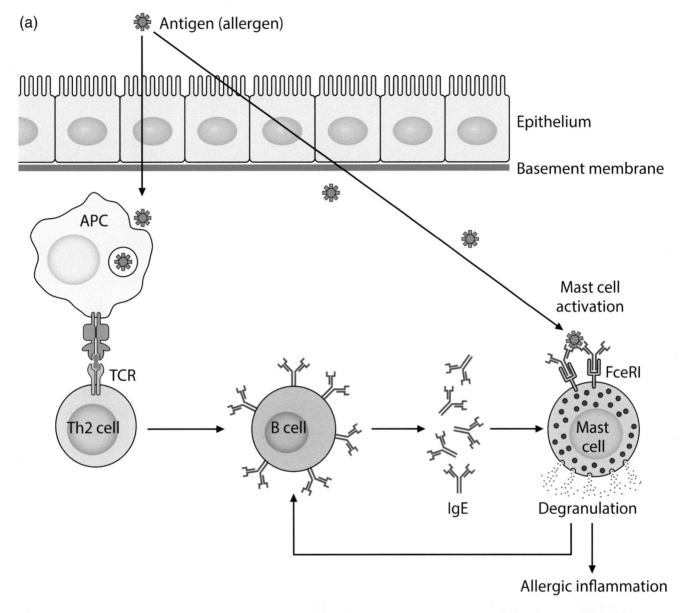

Box 3.2 (a) *In response to antigens such as found in bee venom, antigen-presenting cells (APCs) trigger T helper (Th)-mediated B cell secretion of IgE antibodies. When challenged again by the antigen, antibody crosslinking of Fc receptors on mast cells stimulates degranulation as a key driver of allergic inflammation and potentially the onset of anaphylaxis. (Diagram based on Amin, K (2012) The role of mast cells in allergic inflammation. Resp Med 106: 9–14 reproduced with publisher permission.)*

Surrounding cells of CT proper, the ECM typically has one to three kinds of ropelike strands, called **collagen**, **reticular**, and **elastic fibers** (**Figure 3.9a–d**). Collagen and reticular fibers are composed of the protein **collagen** (**Case 3.3 Cesar Chavez's Collagen Fibers**), which comprises numerous subtypes within various CTs (**Figure 3.9e**). Alternatively, elastic fibers consist mainly of **elastin** protein.

The three CT proper fibers are generally 0.2–20 μm wide and may form relatively sparse collections in **loose** (=areolar) **CT proper** (**Figure 3.9b**) vs. more tightly packed arrays in **dense CT proper**, which is further subdivided

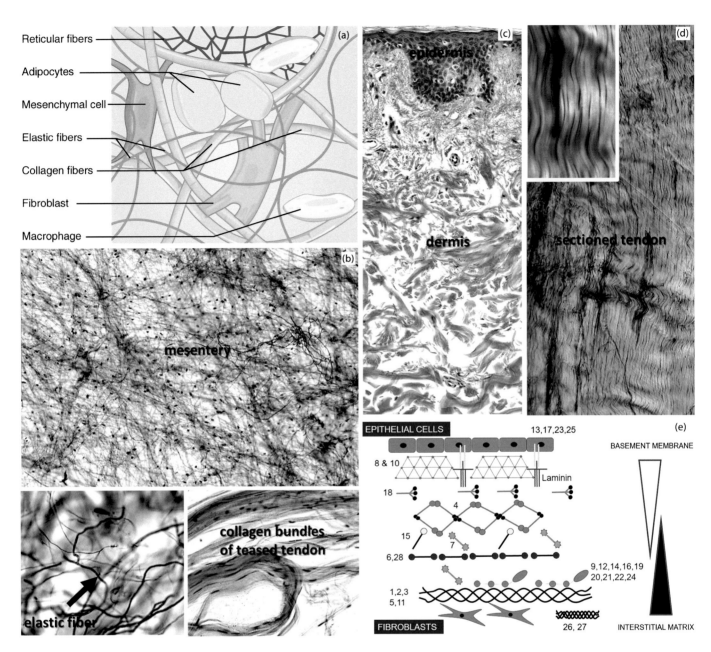

Figure 3.9 Loose vs. dense CT proper. (a–d) *The fibers of loose (=areolar) CT are not as densely packed (a, b: 85X; b, left inset: 480X; b, right inset: 225X) as they are in dense CT proper (c: 150X; d: 240X, inset 400X). The arrow in the left inset of (b) points to a branched elastic fiber within a whole mount of a mesentery, whereas the right inset shows bundles of collagen fibers teased from a tendon and viewed in a whole mount. Dermis of skin is an example of dense irregular CT, whereas the tendon shown sectioned at low and higher magnifications in (d) is a dense regular CT. (e) The various kinds of collagen (types 1–28) can show gradients of distribution relative to the basement membranes underlying epithelia. ([a]: From OpenStax College (2013) Anatomy and Physiology. OpenStax http://cnx.org/content/col11496/latest, reproduced under a CC BY 4.0 creative commons license; [e]: From Karlsdal, MA et al. (2017) The good and the bad collagens of fibrosis–their roles in signaling and organ function. Adv Drug Del Rev 121: 43-56 reproduced with publisher permission.)*

into **dense irregular** vs. **dense regular** types based on the uniformity of fiber orientations (**Figure 3.9c, d**). Examples of loose CT proper include subepithelial CTs in digestive tract linings. Conversely, dense irregular CT proper occurs in various locations like the dermis of skin, whereas dense regular CT proper is found in such sites as tendons. Bathing CT proper fibers is a **ground substance** derived from blood plasma, which CT cells

---

CASE 3.3 CESAR CHAVEZ'S COLLAGEN FIBERS

Activist for Migrant Farm Worker Rights
(1927–1993)

"This is a man who refuses to eat so that all of us can continue to eat."

(From a documentary film on C.C.'s
life called "Cesar's Last Fast")

Cesar Chavez rose to fame during the 1960s through 1980s by leading the United Farm Workers (UFW) union in efforts to secure better pay and working conditions for various workers in the United States. After his family lost its homestead during the Great Depression, Chavez moved from farm to farm finding jobs as a temporary laborer. Motivated by injustices he witnessed, Chavez began organizing fellow workers, and following the chartering of the UFW as an official union in 1966, he led its members and supporters during numerous protests, boycotts, and strikes aimed at raising the standard of

living for migrant farm workers. As a part of such efforts, Chavez often fasted for more than several weeks, not only for the publicity it brought but also for what he called the personal spiritual transformation it triggered. When Chavez died at age 66 of undisclosed causes, some medical historians speculated his habit of long-term fasting had hastened his demise, based on the deleterious effects that prolonged starvation can have on various organs, including the liver and kidneys. At the cellular level, collagen fibers often break down more rapidly than do other proteinaceous components in animal models subjected to sustained starvation, suggesting that Chavez probably lost disproportionately large amounts of his collagen fibers while conducting his prolonged fasts.

---

modify by secreting additional soluble substances (e.g., glycoproteins, proteoglycans, and bioactive molecules like growth factors).

3.10 THE TRIPLE HELICAL ORGANIZATION OF COLLAGENS AND THE FORMATION OF COLLAGEN FIBRILS IN HEALTH AND DISEASE

As keystone glycoproteins that have played crucial roles in animal evolution, **collagens** (from the Greek terms for "glue" and "to give rise to") are the most abundant proteins in the human body. Each collagen molecule comprises three polypeptide **alpha-chains** that are wound together as a **triple helix** (**Figure 3.10a**). Within these alpha-collagen chains are numerous Gly-Xaa-Yaa motifs, where a **glycine** (**Gly**) is often flanked by a **proline** (**Pro**) and a **hydroxyproline** (**Hyp**) at the X and Y positions (**Figure 3.10a**). The unique structural properties of these triple helices not only endow collagen with **great tensile strength** for withstanding mechanical loads but also provide **increased resistance to proteolysis and wide-ranging pHs**, collectively making collagens durable ECM components that can survive in some cadavers for centuries.

Vertebrates produce 28 distinct types of collagens, which differ in their alpha-chain amino acids, post-translational modifications, and patterns of super-molecular assembly (Table 3.4). **Types I**, **II**, **III**, and **IV** collagens account for ~90% of the body's collagen and because of their abundance were the first to be identified biochemically. Type I is by far the most widely distributed collagen in adults and represents the predominant collagen in CT proper, bones, teeth, and fibrocartilages. Type II and III collagens are present mainly in other kinds of cartilage vs. lymphoid and vascular tissues, respectively, whereas type IV collagen is a major component of basement membranes where it forms a mesh-like network. In addition to being concentrated in different sites across the entire body, collagens can also be variably distributed within micro-niches of the same organ (**Figure 3.9e**).

As opposed to 21 kinds of non-fibrillar collagens that comprise either network-forming examples like type IV collagen or three additional classes of collagens (Table 3.4), type I, II, III, V, XI, XXIV, and XXVII collagens are assembled into thread-like structures, called **fibrils**. Such fibrils usually measure ~20–100 nm in diameter and can aggregate together to yield ~1–20 μm-wide **collagen fibers**. **Fibrillar collagens** are often produced by **fibroblasts** (**Figure 3.10b–e**), although other cell types (e.g., osteoblasts, odontoblasts, chondrocytes, and smooth muscle cells) are also capable of forming collagen fibrils. Because fibril formation (=**fibrillogenesis**) has been most intensively analyzed in type I fibrils of tendon fibroblasts, the following account summarizes tendon

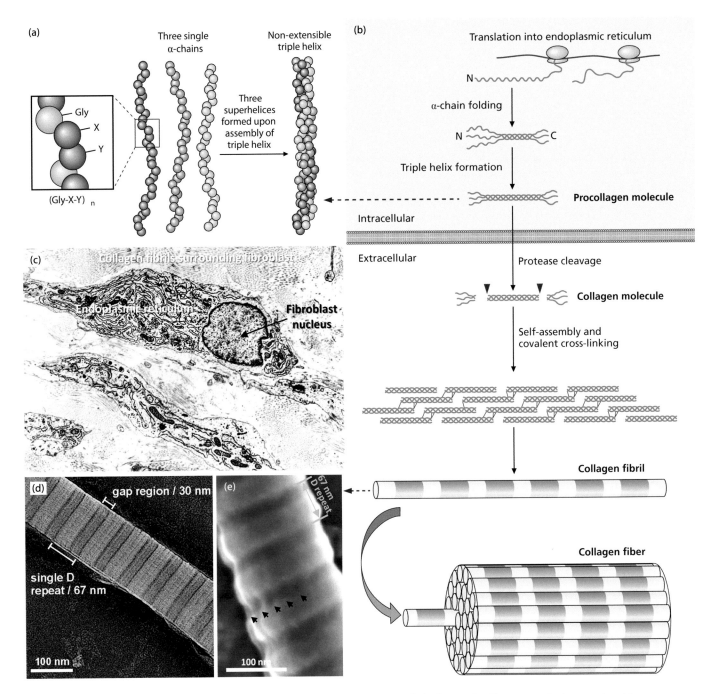

Figure 3.10 Collagen fibrillogenesis. (a) *Collagen forms stable triple helices of collagen α-chains, each with numerous Gly-X-Y motifs.* **(b, c)** *Formation of banded collagen fibrils such as those composed of type I collagen begins within the endoplasmic reticulum (ER) of fibroblasts (c: 1,750X) as translation of collagen α-chains and their post-translational modifications allow the formation of triple-helical procollagen molecules with globular N- and C-termini. A key post-translational modification of each α-chain is the hydroxylation of lysines and prolines. Such hydroxylations require vitamin C as a cofactor and thus without sufficient vitamin C intake can result in a CT disease, called scurvy. There is evidence that some cleavage of the globular telopeptide ends of procollagen can begin within the fibroblast (see* **Figure 3.11a***) before eventually being completed by proteases in the ECM following procollagen exocytosis. The cleaved collagen molecules can then polymerize into orderly arrays of collagen that are stabilized by lysyl-oxidase-mediated cross linkages (red bonds, middle image) between lysine and hydroxylysine residues of neighboring fibrils, which, in turn, stack together to form a collagen fiber.* **(d, e)** *Each fibril has a repeated banding pattern of 67 nm, as shown in high-magnification EMs. ([a]: From Fidler, AL et al. (2018) The triple helix of collagens – an ancient protein structure that enabled animal multicellularity and tissue evolution. J Cell Sci 131: jcs203950 doi: 10.1242/jcs.203950; [b]:*

Figure 3.10 (Continued) *From Yamauchi, M and Sricholpech, M (2012) Lysine post-translational modifications of collagen. Essay Biochem 52: 113–133; [c]: Image courtesy of Dr. M. Morroni; [d, e]: From Greiner, JFW et al. (2019) Natural and synthetic nanopores directing osteogenic differentiation of human stem cells. Nanomed Nanotech Biol Med 17: 319–328 reproduced with publisher permission.)*

TABLE 3.4 COLLAGEN CLASSES AND TYPES

Class	Type	Distribution (examples)
Fibril-forming (Fibrillar)	I	Bone, skin, tendon, ligaments, cornea
	II	Cartilage, vitreous humor in the eyes
	III	Skin, blood vessels
	V	Bone, dermis, co-distribution with type I
	XI	Cartilage, intervertebral discs, co-distribution with type II
	XXIV	Bone, cornea
	XXVII	Cartilage
Fibril-associated collagens with interrupted triple helices (FACIT)	VII	Bladder, dermis
	IX	Cartilage, cornea
	XII	Tendon, dermis
	XIV	Bone, dermis, cartilage
	XVI	Kidney, dermis
	XIX	Basement membrane
	XX	Cornea of chick
	XXI	Kidney, stomach
	XXII	Tissue junctions
	XXVI	Ovary, testis
Network-forming	IV	Basement membrane
	VI	Muscle, dermis, cornea, cartilage
	VIII	Brain, skin, kidney, heart
	X	Cartilage
	XXVIII	Dermis, sciatic nerve
Membrane-associated collagens with interrupted triple helices (MACIT)	XIII	Dermis, eyes, endothelial cells
	XVII	Hemidesmosomes in epithelia
	XXIII	Heart, retina
	XXV	Heart, testis, brain
Multiple triple-helix domains and interruptions (MULTIPLEXINs)	XV	Capillaries, testis, kidney, heart
	XVIII	Liver, basement membrane

Source: From Samad, NABA and Sikarwar, AS (2016) Collagen: new dimension in cosmetic and healthcare. Int J Biochem Res & Rev 14: 1–8 reproduced under a CC BY 4.0 creative commons license.

fibrillogenesis and contrasts this process with **reticular fibril** formation before ending with comments about diseases involving abnormal collagen fibrils.

Following the transcription and translation of type I collagen genes in tendon fibroblasts, newly translated alpha chains must undergo several post-translational modifications in the **ER** for proper chain folding and triple helix formation. For example, some **lysines** and **prolines** undergo hydroxylation to become hydroxylysines and hydroxyprolines, with such hydroxylating enzymes requiring **vitamin C** as a cofactor.

Appropriately modified alpha chains are then assembled into triple helices that possess at both their C- and N-terminal ends a globular domain (=propeptide) that helps prevent complete polymerization within the fibroblast. At this stage, the triple helix plus its terminal peptides is termed **procollagen**. Because of their large size, procollagens are secreted via an unusual exocytic pathway that can involve mega-vesicle production and/ or direct fusions of ER and Golgi compartments.

During post-Golgi trafficking, procollagen is sent via microtubules to the plasmalemma, and after secretion into the ECM, further stages in fibrillogenesis involve removal of procollagen's N- and C-terminal propeptides by extracellular N- and C-proteinases. Such cleavage forms an ~300-nm-long triple-helical molecule, termed **collagen** (**Figure 3.10b**). Individual collagens are then assembled into numerous highly ordered chains with regularly arranged gaps and overlaps among neighboring molecules, which, in turn, are stabilized via inter-chain crosslinking of collagen lysine and hydroxylysine residues by **lysyl oxidase** (**LOX**) enzymes (**Figure 3.10b**). Although individual collagen fibrils can vary in diameter from 10 to 500 nm, each fibril typically exhibits a banding pattern of ~67-nm periodicity, owing to the regularly arranged gaps and overlap regions that are repeated along the fibril (**Figure 3.10b, d, e**).

Previously, it was thought type I procollagen is cleaved only after being exocytosed and that once cleavage is completed, subsequent assembly into fibrils takes place without substantial influence from the synthesizing fibroblast. However, based on serial block-face SEM and correlative biochemical analyses, some cleavage of procollagen termini can occur in Golgi-derived vesicles, thereby generating intracellular vesicles that contain ~28 nm-wide nascent collagen fibrils prior to fibroblast exocytosis (**Figure 3.11a**). Moreover, fibroblasts often contain grooves and ridges that facilitate fibril formation. In particular, discrete actin-containing projections of the fibroblast periphery, called **fibripositors**, appear to orchestrate not only the elongation of a nascent fibril but also its alignment with nearby fibrils (**Figure 3.11a, b**).

Reticular fibers are the first CT fibers to form during development and in adults are distinguished by their more **restricted distribution**, occurring in such locations as blood vessels and the reticular framework of lymphoid organs like the spleen (Chapter 7). These fibers are also characterized by their: (i) **diverse collagen composition**, given that each fibril not only contains **type III collagen** but can also include other collagens like **types I, IV,** and **V (Figure 3.11c–e)**; (ii) **slender dimensions**, with reticular fibrils generally measuring only ~20–40 nm in diameter and generating fibers <2 μm wide; and (iii) **ability to bind silver**, owing to non-type-I collagens and associated carbohydrates having argyrophilic ("silver-loving") properties. Accordingly, because reticular fibers are often difficult to detect in conventionally stained sections, treatment with silver solutions can aid in their identification.

As might be expected given the complex pathways underlying collagen fibrillogenesis, various defects in collagen fibrils have been identified as drivers of human CT diseases. For example, insufficient dietary intake of vitamin C can impair production of hydroxyprolines and hydroxylysines, thereby leading to **scurvy**, which is characterized by bruising, bleeding gums, and overall weakness. Scurvy was once a common cause of death for those lacking citrus fruits or other vitamin C sources in their diets, until, by trial and error, foods that effectively treat and/or prevent scurvy began to be discovered. In fact, British sailors, who used daily rations of lime juice to ward off scurvy, were often referred to as "limeys" well before the role of vitamin C in procollagen hydroxylation had been elucidated.

Alternatively, as examples of heritable diseases affecting fibrillar collagens or associated subtypes (Table 3.5), a heterogeneous group of disorders, called **Ehlers-Danlos syndromes**, are characterized by joint hypermobility, excess skin extensibility, and CT fragility. Such conditions are attributable to various disruptions in fibrillogenesis, ranging from point mutations in glycines blocking triple helix formation to inadequate propeptide cleavage reducing collagen production.

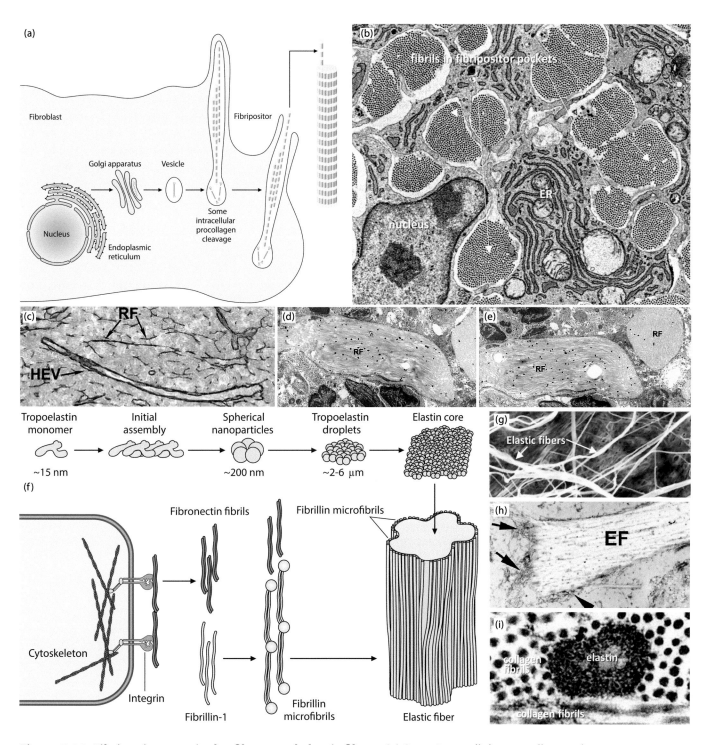

Figure 3.11 Fibripositors, reticular fibers, and elastic fibers. *(a) Some intracellular procollagen cleavage can occur in fibroblasts prior to the extracellular completion of fibril assembly, which, in turn, may be aided by regions of fibroblasts, called fibripositors. (b) In this TEM of a mouse embryonic fibroblast, developing fibrils (arrows) occur within fibripositor invaginations (10,000X). (c–e) Based on LM and EM immunostaining of monkey lymph nodes, reticular fibers (RFs) are heterogeneous ECM components, containing for example type I collagen (c, d) in addition to type III collagen (e); (c, d: 5,000X). (f) Elastic fibers are composed of an elastin core surrounded by a sheath of fibrillin microfibrils that are assembled together into fibers by means of heterogeneous networks of extracellular components, including fibronectin fibrils. (g–i) An SEM (g) and two TEMs (h: longitudinal, i: transverse) show elastic fibers with their peripheral sheath of fibrillin microfibrils (h, arrows) (35,000X). ([a]: From Kalson, NS et al. (2013) Nonmuscle myosin II powered transport of newly formed collagen fibrils at the plasma membrane. PNAS 110 (49) E4743–E4752; https://doi.org/10.1073/pnas.1314348110;*

Figure 3.11 (Continued) *[b]: TEM courtesy of Dr. K. Kadler; [c–e]: From Sobocinski, GP et al. (2010) Ultrastructural localization of extracellular matrix proteins of the lymph node cortex: evidence supporting the reticular network as a pathway for lymphocyte migration. BMC Immunol 11: http://www.biomedcentral.com/1471-2172/11/42 reproduced under a CC BY 2.0 creative commons license; [f]: From Kanta, J (2016) Elastin in the liver. Front Phys 7: doi:10.3389/phys.2016.00491 reproduced under a CC BY 4.0 creative commons license; [g]: From Tavakoli, J and Costi, JJ (2017) Development of a rapid matrix digestion technique for ultrastructural analysis of elastic fibers in the intervertebral disc. J Mech Behav Biomed Mat 71: 175–183 reproduced with publisher permission; [h]: From Ozturk, N et al. (2013) Pressure applied during surgery alters the biomechanical properties of human saphenous vein graft. Heart Vessels 28: 237–245 reproduced with publisher permission; [i]: From Lewis, PN et al. (2016) Three-dimensional arrangement of elastic fibers in the human corneal stroma. Exp Eye Res 146: 43–53 reproduced under a CC BY 4.0 creative commons license; image courtesy of Drs. P. Lewis, T. White, and K. Meek.)*

3.11 ELASTIC FIBERS AND MARFAN SYNDROME

As a means of enhancing the elasticity of such structures as large arteries, lungs, elastic cartilage, and dermal layers of the skin, CTs contain ~0.2-2 µm-wide **elastic fibers** (**Figure 3.11f–i**). When fully formed, elastic fibers are able to branch and assemble into meshed networks (e.g., in skin) or into fused sheets (e.g., in arteries), with each fiber possessing an **elastin core** surrounded by **fibrillin-containing microfibrils** (**Figure 3.11f**).

Over 90% of each elastic fiber is composed of the protein **elastin** that is produced from midgestation through adolescence by such cells as fibroblasts, chondrocytes, and smooth muscle cells. Rich in **lysine** plus unusual lysine derivatives such as **desmosine**, mature elastin begins its biosynthesis as soluble **tropoelastin** monomers. Such tropoelastins become cross-linked at the extracellular surface of the secretory cell via steps requiring **LOX** enzymes. The cross-linked tropoelastin aggregates are then translocated to a network of fibrillin microfibrils that is generated extracellularly by numerous interacting components, including LOXs and other key elements like **fibronectin fibrils** (**Figure 3.11f**). When fully assembled, elastin polymers form the core of each elastic fiber, and a sheath of ~10 nm-wide **fibrillin microfibrils** coats the core (**Figure 3.11f**). Owing to their high hydrophobicity and abundant crosslinking, the half-life of elastic fibers can exceed 70 years. However, rapid elastic fiber breakdown can occur in response to either excessive activities of elastases like **MMP-9** (**matrix metalloproteinase 9**) and/or the mechanical rupturing of fragile elastic fibers that occurs in CT diseases, such as **Marfan syndrome** (**MFS**).

TABLE 3.5 HUMAN DISEASES INVOLVING MUTATIONS OF FIBRILLAR OR FIBRILLAR-ASSOCIATED COLLAGENS

Collagen type affected	Disease
I	Osteogenesis imperfecta, Ehlers-Danlos syndrome type VII
II	Several chondrodysplasias, osteoarthritis
III	Ehlers-Danlos syndrome type IV, aortic aneurysms
V	Ehlers-Danlos syndrome types I and II
VII	Epidermolysis bullosa dystrophica
IX	Multiple epiphyseal dysplasia (MED)

Source: From Karlsdal, MA et al. (2017) The good and the bad collagens of fibrosis–their roles in signaling and organ function. Adv Drug Del Rev 121: 43–56 reproduced with publisher permission.

CASE 3.4 NICCOLO PAGININI'S ELASTIC FIBERS

Violinist
   (1782–1840)

"...one of the audience ...had distinctly seen, while I was playing my variations, the devil at my elbow, directing my arm and guiding my bow."

   (N.P. recounting to a friend how an audience member had seen the devil guiding N.P. as he played his violin)

Niccolo Paginini played violin with such virtuosity that some concertgoers believed he had made a pact with the devil. As a child prodigy, Paginini quickly outperformed his teachers before eventually becoming a tall and gaunt man who could use his extremely flexible hands in ways that revolutionized violin playing. By 1835, Paginini stopped giving concerts due to ill health, and in 1840, he died after refusing his last rites, further fueling tales of his ties to the devil. In recent years, several medical histories have attributed Paginini's

body morph and flexible fingers to Marfan syndrome, a genetic disorder that alters elastic fibers in connective tissues. Along with such marfanoid attributes, Paginini also had vocal cord dysfunction and reportedly died of internal hemorrhaging, as might be expected of a Marfan patient. Without genetic tests to rule out other proposed causes, such as Ehlers-Danlos disease, the etiology of Paginini's unique physical attributes remains unclear, although abnormal elastic fiber formation certainly offers a plausible alternative to explanations citing satanic-derived powers.

   MFS is a heritable disorder, which may have affected such celebrities as the violinist Niccolo Paganini (**Case 3.4 Niccolo Paginini's Elastic Fibers**) and the 16th U.S. President Abraham Lincoln (although see **Case 9.2** for an alternative diagnosis). Mutations in the fibrillin 1 (FBN1) gene required for proper microfibril formation drive MFS pathogenesis, which in typical cases involves such symptoms as skeletomuscular abnormalities (e.g., an extremely tall and thin body with spider-like digits and hyperextensible joints) plus cardiovascular defects that may ultimately prove to be fatal, for example via tearing of the aorta (=aortic dissection). Given that FBN1 normally has binding sites for **transforming growth factor-beta** (**TGF-β**), one model proposes that properly functioning microfibrils sequester enough TGF-β to prevent overstimulation of nearby CT cells and thereby keep excessive amounts of MMP-9 from being secreted. Conversely, with defective microfibrils that are formed during MFS, constitutive stimulation by non-sequestered TGF-β may trigger MMP-9 hypersecretion and the concomitant breakdown of elastic fibers.

3.12 SOME DISORDERS OF EPITHELIA AND CT PROPER: KARTAGENER'S SYNDROME AND ALPORT SYNDROME

Kartagener's syndrome (KS) involves mutations in genes that encode proteins of the **outer dynein arms** and/or other axonemal structures of motile cilia and flagella. Thus, ciliated epithelia that line conducting passageways of the respiratory system are often incapable of effectively clearing mucus and debris, thereby triggering chronic respiratory tract diseases. In addition, KS can cause infertility due to impaired motility of sperm and oviductal cilia. Furthermore, ~50% of KS patients suffer from **situs inversus** (i.e., left-right switching of organ positions). This is because motile cilia in the nodal region of developing mammalian embryos normally move fluids with left-side-specific signaling molecules toward the future left side to ensure proper left-right polarity. Thus, without such movements, successful outcomes of left-right determination become randomized.

   Alport syndrome (**AS**) is a heritable **basement membrane disease** that involves mutated **glycines** in alpha chains of **type IV collagens**. Such mutations compromise triple helix formation, thereby leading to defective basement membranes and concomitant pathogenesis in organs, such as the kidneys, eyes, and cochleas. To treat AS, new therapies such as molecular chaperones for facilitating triple helix formation are being evaluated in order to augment blood pressure medicines (Chapter 16) that are used in mitigating renal failure and cardiac arrest.

3.13 SUMMARY—EMBRYOLOGY AND THE FOUR BASIC TISSUE TYPES WITH EMPHASIS ON EPITHELIUM AND CONNECTIVE TISSUE PROPER

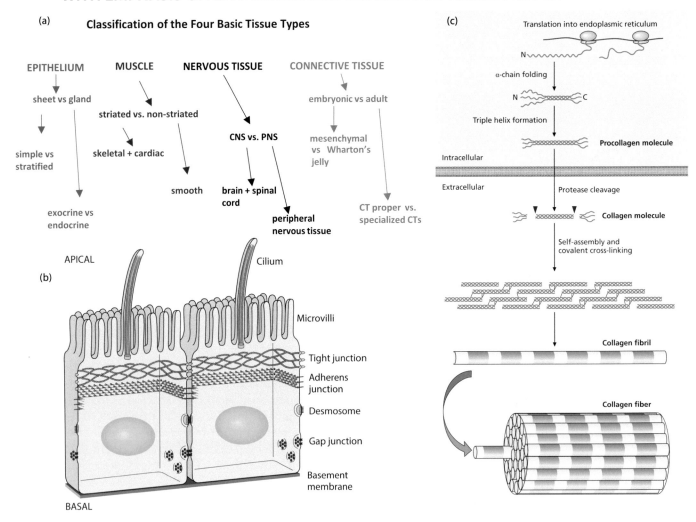

***Pictorial Summary Figure Caption: (a)**: Classification of major tissue types; **(b)**: see figure 3.5a; **(c)**: see figure 3.10b.*

In humans, **eggs** fertilized by **sperm** produce **zygotes** that divide to become bipartite **blastocysts**, each containing an outer **trophoblast** layer and an **inner cell mass (ICM)**. Blastocysts **implant** in the uterine lining and use trophoblast cells to help form **placental** tissues while ICM cells produce the **embryo**. During the 2nd and 3rd weeks, embryonic **gastrulation** yields three primary germ layers—(i) **ectoderm**, (ii) **mesoderm**, and (iii) **endoderm**—which produce, respectively: (i) nervous system, epidermis, and neural crest cells, (ii) non-gut-related internal organs (e.g., muscles and connective tissues), and (iii) much of the digestive system plus related organs (e.g., respiratory tract and thyroid). After gastrulation, the first functional organ is the heart, which beats by the end of the 4th week. By week 8, most organs have begun forming, and at the end of the 8th week, the embryo is termed a **fetus**. Fetal organogenesis continues through the **1st trimester** (first 3 months) as well the 2nd and 3rd trimesters until birth occurs at ~9 months. Adult tissues can vary greatly based on differences in their **cells** and **extracellular matrix (ECM)** but are generally classified into **four basic types**: epithelium, **muscle**, **nervous tissue**, and **connective tissue (CT)** (**Figure a**).

Epithelia are **sheet-like** or **glandular**. Sheets can be single- (**simple**) or multi-layered (**stratified**) with **squamous** (flat), **cuboidal** (medium height), or **columnar** (tall) cells. Glands can have ducts (**exocrine**) or lack them (**endocrine**). Muscles are either **striated** with myofilaments that form striations (skeletal and cardiac muscle) or **non-striated** (smooth muscle). Nervous tissue comprises **neurons** and supporting **glia cells** within the brain, spinal cord, and peripheral nervous system. CTs are subdivided into **embryonic** (mesenchymal [most common] vs. **Wharton's jelly** [restricted to the umbilical cord]) vs. **adult** (=**CT proper** [e.g., dermis and tendons] or **specialized CTs** [cartilage, bone, fat, and blood]).

Hallmark features of epithelial cells are: (1) **tight packing**, (2) **apical-basal polarity**, (3) **intercellular junctions**, and (4) a **basement membrane (Figure b)**. Epithelial apical elaborations include: (i) an outer **glycocalyx** coat; (ii) **microvilli** (**MV**) (=microfilament-containing finger-like projections), which sometimes form a dense array, called the **brush border**, in organs like the intestines, or generate elongated versions (=**stereovilli**), for example in auditory hair cells; and (iii) **microtubule** (**MT**)-**containing cilia**. **Motile cilia**: (i) contain a 9+2 arrangement of MTs with **dynein arm** ATPases for ciliary beating; (ii) exhibit a restricted distribution (e.g., in the respiratory tract); (iii) occur in multiple numbers per epithelial cell; and (iv) often serve to remove debris and pathogens. **Primary cilia**: (i) have 9+0 MTs without dynein arms; (ii) are commonly formed by epithelial cells; (iii) occur one cilium per cell; and (iv) comprise non-motile sensors. Typical **intercellular junctions** from apex to base are—(1) **zonula occludens** (belt-like tight junctions [TJs] with MFs for sealing off the apex from the base); (2) **zonula adherens** (belt-like with MFs for signaling); (3) **macula adherens** (=**desmosomes**; with intermediate filaments; non-belt-like; form strong spot welds); (4) **gap junctions** (with **connexon** channels for cell-cell communication); and (5) **hemidesmosomes** that keep the base attached to the **basement membrane**. Two unusual epithelia are—(1) **pseudostratified** (e.g., in the respiratory tract)—appears stratified, but is simple; and (2) **transitional** (e.g., lining the bladder)—changes from stratified **cuboidal** to stratified **squamous** as the lumen fills with urine.

CT proper has (i) **cells**: **transient** (e.g., wandering blood cells) vs. **fixed** (resident in CT; e.g., fiber-producing **fibroblasts**, phagocytic **macrophages**, and **mast cells** whose degranulation can trigger **anaphylaxis** in unchecked allergic reactions), (ii) **ground substance** (=ECM fluid + non-fibrous solutes), and (iii) **ECM fibers**— **collagen**, **reticular**, and **elastic**—the first two of which are composed of triple-helical **collagen**; the third fiber type contains extensible **elastin** with **desmosine** amino acids plus surrounding **microfibrils** containing **fibrillin**. As the most common collagen in the body, **type I collagen** can be secreted by fibroblasts to form **fibrils** (typically 20–100 nm wide) in the ECM that are bundled into ~2–20 μm-wide fibers. Stages in fibrillogenesis include (**Figure c**): (1) **translation** of three **alpha-collagen** chains; (2) **post-translational modifications** of alpha chains (e.g., Lys, Pro hydroxylation); (3) three alpha chains forming triple-helix **procollagen** with terminal **propeptide** globular domains; (4) **procollagen secretion** and **propeptide cleavage** to yield **collagen**; and (5) **alignment** and **crosslinking** of collagens to form fibrils.

Some epithelial and CT proper **disorders**: (1) **Kartagener's syndrome**: dynein arm defects that compromise ciliary movement; (2) **Alport syndrome**: defective type IV collagen in basement membranes; (3) **scurvy**: improper collagen fibrillogenesis due to lack of vitamin C needed for Pro/Lys hydroxylations; (4) **Ehlers-Danlos syndromes**: various malfunctions of collagen fibrillogenesis (e.g., improper propeptide cleavage); and (5) **Marfan syndrome**: defective elastic fibers due to mutations in fibrillin 1 gene needed for microfibril production.

SELF-STUDY QUESTIONS

1. What is the most common type of collagen in humans?
2. T/F Gap junctions occur only in epithelial cells.
3. Which portion of the blastocyst forms part of the placenta?
4. Which involves defective elastic fibers?
   A. Scurvy
   B. Marfan syndrome
   C. Alport syndrome
   D. Kartagener's syndrome
5. Which of these epithelia consists of a single layer of cells?
   A. Stratified squamous
   B. Transitional
   C. Stratified cuboidal
   D. Pseudostratified
   E. NONE of these

6. What are stereovilli?
   A. Immobile cilia
   B. Primary cilia
   C. Motile cilia
   D. Microvilli
   E. Microtubule-containing apical elaborations
7. T/F The heart begins to beat by the end of the 4th week of development.
8. T/F Exocrine glands lack ducts.
9. T/F Fat is a type of CT proper.
10. What are the four basic tissue types?
11. When is the embryonic to fetal transition in humans?
    A. End of 1st week post-fertilization (pf)
    B. During 2nd/3rd weeks pf
    C. During 4th week pf
    D. End of 8th week pf
    E. End of 2nd trimester
12. What is the approximate width of typical type I collagen fibrils?
13. T/F. Ground substance includes ECM fibers.

    "EXTRA CREDIT" For each term on the left, provide the BEST MATCH on the right (answers A-G can be used more than once)

14. Vitamin C deficiency___   A. Connexon
15. Wharton's jelly____   B. Desmosine
16. Motile cilia___   C. Dynein arms
17. Basement membrane___   D. Type IV collagen
18. Covers epithelial   E. Umbilical cord
    cell microvilli___   F. Scurvy
                                               G. Glycocalyx

19. Describe the highlights and general timelines of human development covered in this chapter.
20. Describe in proper order the steps in type I collagen fibril formation noting precisely where such steps occur

ANSWERS

1) type I; 2) F; 3) trophoblast cells; 4) B; 5) D; 6) D; 7) T; 8) F; 9) F; 10) epithelium, muscle, nervous tissue, connective tissue; 11) D; 12) 50-150 nm 13) F; 14) F; 15) E ; 16) C; 17) D; 18) G; 19) The answer should include: fertilization, cleavage, blastocyst formation, implantation, placenta formation, gastrulation, embryo to fetal transition, organogenesis; trimesters, full term pregnancy; 20) The answer should include and clearly define such terms as: alpha collagen genes transcription and translation, post-translational modifications; procollagen secretion, procollagen propeptide cleavage; collagen alignment/ crosslinking; fibril to fiber production

Cartilage and Bone

4.1 INTRODUCTION TO GENERAL FUNCTIONS AND DEVELOPMENTAL TIMELINES OF SKELETAL TISSUES—CARTILAGE AND BONE

In addition to forming blood cells (Chapter 6) and storing key elements like calcium and phosphorus, the adult **skeleton** protects and supports body components while also providing a firm foundation for muscle attachments. Such protective and supportive roles are carried out by two interacting skeletal tissues—**cartilage** and **bone**. Cartilage is a pliable and well-hydrated type of specialized connective tissue that can cushion bone joints. In addition, cartilage also protects respiratory passages and provides a transient template during *in utero* development as certain kinds of bone undergo **endochondral** ("internal cartilage") **bone formation**. Conversely, to increase skeletal rigidity, bone is a hardened connective tissue that contains a **mineral phase** of calcium phosphate crystals referred to as carbonated hydroxyapatite, or simply **apatite**. In addition to apatite, bone possesses **cells** plus a predominantly collagenous **organic matrix**, both of which remain after decalcification protocols remove the mineral phase.

As the first embryonic cartilage to form, neural-crest-derived **Meckel's cartilage** begins differentiating during the 4th–5th week of development and eventually gives rise to the incus and malleus bones of the ear (Chapter 20)

DOI: 10.1201/9780429353307-5

while also helping the lower jaw to undergo proper bone formation (=**ossification**). Similarly, calcified bone tissue first appears in jaw- and collar bones by the end of the 6th week, whereas parts of the vertebral column and skull can take up to 15 weeks before ossifying and thus represent some of the last embryonic structures to begin organogenesis.

4.2   BASIC HISTOLOGY OF THE THREE CARTILAGE TYPES—HYALINE, ELASTIC, AND FIBROCARTILAGE

The three main types of cartilage—**hyaline**, **elastic**, and **fibrocartilage** (**Figure 4.1a**)—share various structural and functional properties. For example, all cartilages lack blood vessels. Consequently, molecular exchanges tend to occur via long-range diffusion, thereby requiring a well-hydrated ECM. In addition, although rigid bones can only enlarge **appositionally** by adding to its edges, cartilages are able to expand not only by appositional growth at their peripheries but also by **interstitial** growth, in which new cartilage is added within the interior of previously formed cartilage. Moreover, all three cartilage types possess relatively spherical cells, called **chondrocytes**, which in conventionally prepared LM sections tend to reside in discrete spaces, termed **lacunae** (**Figure 4.1b**). However, such lacunae are actually artifacts of suboptimal fixation and processing methods, since samples that are properly prepared for EM lack lacunae and instead contain chondrocytes that are directly surrounded by a delicate ECM (**Case 4.1 Christiaan Barnard's Cartilage**).

As opposed to such unifying features, the three cartilage types can be distinguished based on their coloration, mechanical properties, ECM composition, and the presence vs. absence of a distinct boundary layer, called the **perichondrium** (**Figure 4.1c**). For example, **hyaline cartilage**, which is named for its translucent appearance, is a moderately flexible and grayish tissue with a homogeneous matrix, in which thin fibers composed predominantly of **type II collagen** blend imperceptibly with other ECM components (**Figure 4.1d–g**). Hyaline cartilage is usually encased by a perichondrium whose cartilage-forming chondrogenic layer abuts existing cartilage. However, a subset of hyaline cartilage, called **articular cartilage**, coats the ends of bones at skeletal joints (=articulations) and lacks a discrete perichondrium. **Elastic cartilage** resembles non-articular hyaline cartilage in being surrounded by a perichondrium and containing type II collagen that appears relatively non-fibrous when viewed with conventional methods. However, compared to hyaline cartilage, living elastic cartilage is more flexible and has a yellower hue owing to abundant ECM **elastic fibers**, which are readily apparent when treated with elastin-specific stains (**Figure 4.2a–d**). Conversely, as a comparatively inflexible and whitish cartilage, **fibrocartilage** lacks a perichondrium and comprises a tough, fibrous ECM that contains mainly **type I collagen** fibers (**Figure 4.2e**).

CASE 4.1 CHRISTIAAN BARNARD'S CARTILAGE

Surgeon
(1922–2001)

"On Saturday, I was a surgeon in South Africa, very little known. On Monday, I was world renowned."

(C.B. on his whirlwind celebrity after performing the first human-to-human heart transplant in 1967)

Born in South Africa, Christiaan Barnard obtained his medical degree and carried out numerous organ transplants before gaining worldwide fame when he performed the first human-to-human heart transplant. In 1955, Barnard was recruited to the United States, where he developed cutting-edge techniques in the field of organ transplantation. After returning to South Africa, Barnard continued to hone his skills by performing heart transplants in dogs. In 1964, the first xenograft of a chimpanzee heart into a human patient was attempted in the United States but quickly ended when the heart stopped beating after 90 min. With this backdrop of limited success, Barnard forged ahead with transplanting a human heart

into a patient in 1967, and although the recipient died 18 days later, the surgery was hailed as a major breakthrough. Subsequent improvements that Barnard and others made to the original surgical techniques have dramatically increased survival times so that some patients can now live for decades following a heart transplant. As for Barnard's last years, the groundbreaking surgeon had to retire from conducting operations in 1983 due to the breakdown of his articular cartilage caused by arthritis, and in 2001, cartilage also played a role in his death, when he succumbed to a severe asthma attack, unable to open his cartilage-encased airways.

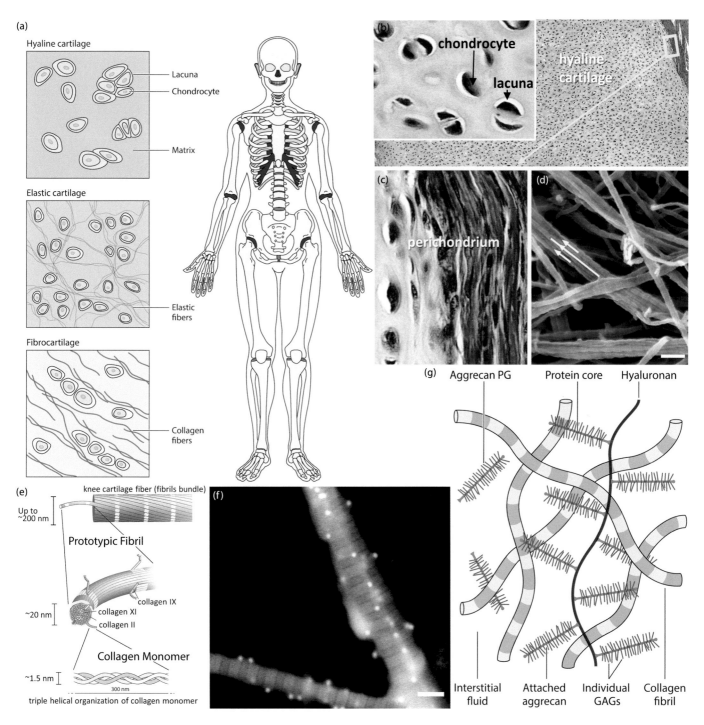

Figure 4.1 Introduction to cartilage types and hyaline cartilage. (*a*) *The human body produces three main types of cartilage: hyaline (e.g., bone articulations), elastic (e.g., ears), and fibrocartilage (e.g., intervertebral discs).* (*b*) *As the most widely distributed cartilage, hyaline cartilage contains rounded chondrocytes, which in conventional paraffin sections occur within artifactual spaces (=lacunae) caused by the processing protocol (200X; inset 20X).* (*c*) *A perichondrium surrounds hyaline cartilages, except those found at bone articulations (225X).* (*d, e*) *High-resolution SEMs of type II collagen fibrils in articular hyaline cartilage reveal ~20 nm-wide prototypic fibrils (arrows), which are highly aligned within a banded collagen fiber measuring up to ~200 nm wide.* (*f*) *Immuno-EM shows gold-conjugated antibodies (white dots) labeling type II collagen in articular cartilage fibrils.* (*g*) *Numerous proteoglycans are scattered among collagen fibrils in hyaline cartilage. Scale bars = 100 nm. ([d–f]: From Gottardi, R et al. (2016) Supramolecular organization of collagen fibrils in healthy and osteoarthritic human knee and hip joint cartilage. PLoS ONE 11: doi:10.1371/journal. pone.0163552 reproduced under a CC BY 4.0 creative commons license.)*

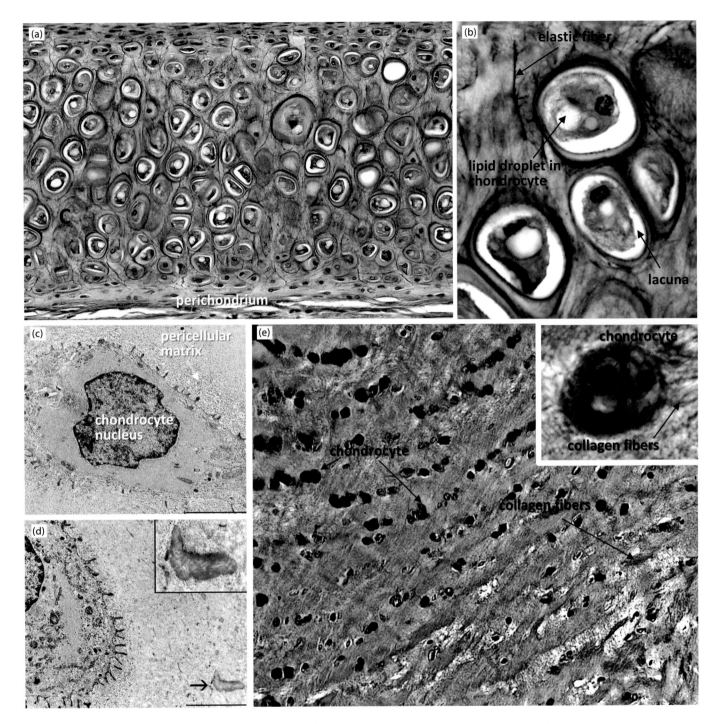

Figure 4.2 Elastic cartilage and fibrocartilage. (**a, b**) Elastic cartilage such as shown here from a cat resembles hyaline cartilage in possessing a perichondrium but differs in having elastic fibers (a: 200X; b: 425X). (**c**) TEM of rabbit ear elastic cartilage shows the lack of an artifactual lacuna around a chondrocyte and instead the presence of pericellular matrix with fine, non-fibrous components (scale bar = 5 μm). (**d**) A higher-magnification TEM shows an elastic fiber (arrow, d, and further magnified in inset) in the territorial matrix surrounding the circum-chondrocytic pericellular matrix (scale bar = 2 μm). (**e**) Fibrocartilage lacks a perichondrium and contains type I collagen fibers in its ECM (220X; inset: 500X). ([c, d]: From Rosa, RG et al. (2014) Growth factor stimulation improves the structure and properties of scaffold-free engineered auricular cartilage constructs. PLoS ONE 9: e105170: doi:10.1371/journal.pone.0105170 reproduced under a CC BY 4.0 creative commons license.)

The three cartilage types are also differentially distributed throughout the body. As the **most widely dispersed type**, hyaline cartilage forms the cartilage models of endochondrally developing bones and persists as articular cartilage in freely mobile joints of the adult skeleton. Moreover, hyaline cartilage contributes to less mobile cartilaginous joints, such as the **symphysis pubis** of the pelvic girdle (**Figure 4.3a–d**) and provides support for the nose and larger respiratory passageways. Alternatively, elastic cartilage occurs in the epiglottis, Eustachian tubes, and external ear, where its elastic fibers allow considerable malleability and support. Fibrocartilage, on the other hand, is found in scattered tendons subjected to excessive mechanical loading and in parts of the **symphysis pubis** as well as in **intervertebral discs** that cushion spinal vertebrae (**Figure 4.3a–i**) (**Case 4.2 Elizabeth Taylor's Intervertebral Discs**). Such discs contain a central lighter-staining **nucleus pulposus** that is surrounded by a dense sheath of **fibrocartilage**, called the **annulus fibrosus** (**Figure 4.3e–i**). Collectively, the histological features and typical locations of the three main cartilage types are summarized in Table 4.1.

4.3   FUNCTIONAL MORPHOLOGY OF CHONDROCYTES, COLLAGEN FIBRILS, AND PROTEOGLYCANS

As the cells that secrete and recycle the ECM macromolecules of cartilage, **chondrocytes** possess a well-developed synthetic machinery of Golgi bodies and ER as well as appropriate catabolic organelles such as lysosomes. In addition, chondrocytes often store energy in lipid droplets (**Figure 4.2b**) to help sustain their high metabolic activities.

Each chondrocyte is encased by a thin sheath of non-fibrous ECM, called the pericellular matrix (**Figure 4.2c**). This sheath is surrounded by a territorial matrix that contains ECM fibers (**Figure 4.2d, inset**), which, in turn, is separated from the territorial matrices of other chondrocytes by interterritorial matrix. The pericellular and territorial matrices associated with each chondrocyte contain specific growth factors and other regulatory molecules that can help modulate chondrocytic differentiation as well as the rates at which differentiated chondrocytes endocytose and secrete macromolecules.

A key secretory product of chondrocytes is **collagen**, which is assembled within the ECM into fibrils or fibers that collectively account for ~60% of cartilage's dry weight. Except for fibrocartilage where **type I collagen** predominates, >90% of cartilaginous collagen comprises **type II collagen** (**Figure 4.1d–f**), which is supplemented by minor amounts of other types (e.g., IX and XI). In adult articular cartilage (**Figure 4.1d–f**), the diameters of collagen fibrils and fibers average ~20 nm and ~200 nm, respectively, and such collagens occur within a well-hydrated ECM. In addition to collagens, several non-fibrous macromolecules are secreted by chondrocytes into the ECM, thereby generating a gel-like **ground substance**.

CASE 4.2 ELIZABETH TAYLOR'S INTERVERTEBRAL DISCS

Actress
   (1932–2011)

"I would rather be a symbol of a woman who makes mistakes, perhaps, but a woman who loves."

(E.T. reflecting on her passionate and stormy life)

Elizabeth Taylor was a British-American movie star, whose tumultuous personal life helped fuel her worldwide celebrity. At age 12, Taylor starred in *National Velvet* as a jockey who rides in a prestigious horse race. After that breakout role, Taylor went on to win two Oscars and reached peak notoriety in 1963 due to a highly publicized affair she conducted with her co-star, Richard Burton. Taylor and Burton subsequently divorced their spouses and married each other, which ended up being the fifth in her series of eight marriages. As her acting career waned, Taylor helped raise funds for AIDS research at

a time when the disease was widely stigmatized, and she continued such philanthropy as well as several business ventures while suffering from ill health throughout much of her life. One of her most debilitating maladies was caused during the filming of *National Velvet*, when she fractured her back during a horse riding accident, thereby necessitating surgery on her intervertebral discs. After the operation, Taylor still experienced chronic back pain that contributed to her alcoholism and prescription drug addiction, and although she ultimately died of congestive heart failure, intervertebral disc issues long plagued her iconic, but stormy, life.

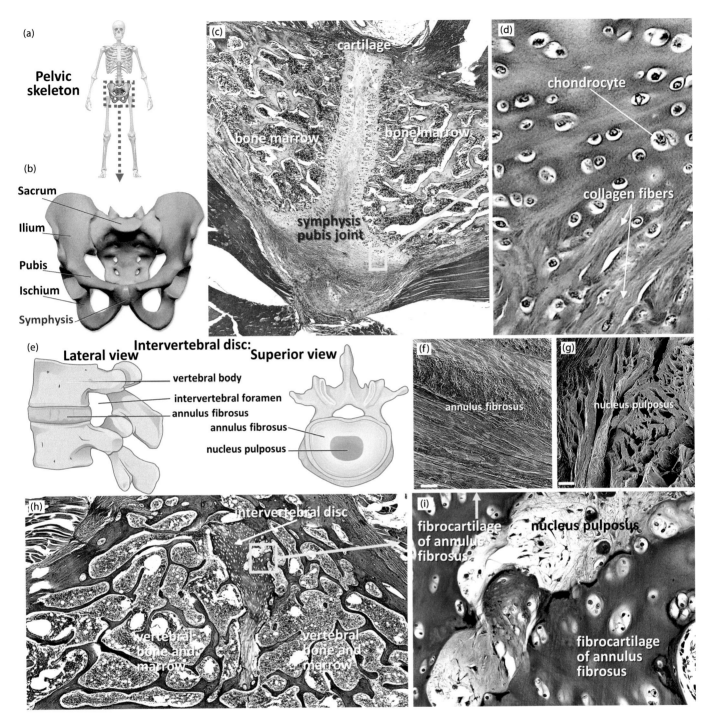

Figure 4.3 Fibrocartilage in the symphysis pubis joint and intervertebral discs. *(a–d) The symphysis pubis joint that connects bones of the pelvis contains fibrocartilage (c: 20X; d: 200X). (e) Each intervertebral disc comprises a peripheral annulus fibrosus and a central nucleus pulposus, which collectively serve to cushion vertebral bones. (f–i) The annulus fibrosus contains fibrocartilage plus highly oriented bundles of collagen fibers (f, g scale bars = 100 μm; h: 15X; i: 150X). ([a, b]: From Blausen.com staff (2014) "Medical gallery of Blausen Medical 2014" WikiJournal of Medicine 1 (2): 10. doi:10.15347/wjm/2014.010 reproduced under CC0 and CC SA creative commons licenses; [e]: From OpenStax College (2013) Anatomy and Physiology. OpenStax. http://cnx.org/content/col11496/latest reproduced under a CC BY 4.0 creative commons license; [f, g]: From Fontes, RVB et al. (2019) Normal aging in human lumbar discs: An ultrastructural comparison. PLoS ONE 14(6): e0218121: https://doi.org/10.1371/journal.pone.0218121 reproduced under a CC BY 4.0 creative commons license.)*

TABLE 4.1 COMPARISON OF CARTILAGES

Cartilage	Predominant collagen	Perichondrium	Major sites in body
Hyaline	Type II	Yes (except in articular)	Bone articulations, around trachea and bronchi
Elastic	Type II	Yes	Ears, epiglottis
Fibrocartilage	Type I	No	Symphysis pubis, intervertebral discs

The major constituents of cartilaginous ground substance are chondrocyte-derived glycosylated proteins, called **proteoglycans** (**PGs**). Each PG comprises a protein core bound to one or more chains of repeated disaccharides, with each chain being termed a **glycosaminoglycan** (**GAG**) (**Figure 4.1g**). Thus, one to more than 100 GAGs are connected via linker regions to the core protein, and a single attached GAG can comprise dozens of polymerized disaccharides.

GAGs can also contain sulfate groups at variable positions along each disaccharide polymer. Based on their specific disaccharide composition and sulfation pattern, sulfated cartilaginous GAGs include **chondroitin sulfate** (**CS**), **dermatan sulfate** (**DS**), **heparan sulfate** (**HS**), and **keratan sulfate** (**KS**). For example, of the three most abundant sulfated PGs in cartilage—**aggrecan**, **biglycan**, and **decorin**—aggrecan has more than 100 attached CS and KS GAGs, whereas biglycan and decorin are each typically bound to two or one DS GAG(s), respectively. Sulfate groups of cartilaginous GAGs are water-retaining **hygroscopic sites** that keep cartilage hydrated so that it can function as an effective cushion. In addition to such roles, aggrecan can form large aggregates of more than 1000 individual molecules, which are linked together by the non-sulfated GAG **hyaluronic acid** (=**hyaluronan**) (**Figure 4.1g**). Aggrecan aggregates allow cartilage to withstand high compressive loads, while also helping to block calcification by inhibiting apatite growth.

Although PGs can have long half-lives of up to 25 years, chondrocyte-mediated lysis and endocytosis of PGs regularly occur as part of continual turnover processes both during normal upkeep and during aging- or pathogenic-driven re-organizations. For such turnover, chondrocytes secrete PG-targeting enzymes, such as **matrix metalloproteinases** (**MMPs**) that are normally kept in check by specific **tissue inhibitors of metalloproteinases** (**TIMPs**) so that the rates of PG synthesis and degradation are properly balanced.

4.4 SYNOVIAL JOINTS AND ARTICULAR CARTILAGE IN HEALTH AND DISEASE

Skeletal elements are joined by three main types of articulations—**fibrous**, **cartilaginous**, and **synovial**. Fibrous joints, such as occur between skull bones, lack a specialized cavity and comprise collagen fibers that allow little bone movement. Similarly, relatively immobile cartilaginous joints also lack a cavity but consist either entirely of hyaline cartilage or a mixture of hyaline- and fibro-cartilage. Alternatively, synovial joints such as in knees and hips contain a fluid-filled cavity encapsulated by a **synovium** (=synovial membrane), which collectively allow a wide range of joint movements (**Figure 4.4a–e**). The synovium typically comprises a surface **intima** that contains fibroblast-like **synoviocytes** plus an underlying **subintima** of relatively loose connective tissue. Surface synoviocytes secrete macromolecules, such as hyaluronan and lubricin that lubricate bone termini to reduce friction during joint movements.

Contiguous with the synovium is **articular cartilage** that comprises three major strata—the **superficial**, **middle**, and **deep zones** (**Figure 4.4d, e**). Flattened chondrocytes and nearby collagen fibers tend to be tangentially oriented in the superficial zone before transitioning through the middle zone and becoming rounded cells associated with radially arranged fibers in the deep zone (**Figure 4.4e**). Functionally, the superficial zone protects and maintains underlying portions of articular cartilage while also resisting sheer forces generated during bone movements. The thickened middle zone forms the first line of resistance to compressive loads, and such resistance is augmented by the numerous vertically arranged collagen fibers in the deep zone.

Along with these major regions, the articular superficial zone in post-adolescents is overlain by a thin surface layer of tangentially aligned collagen fibers, called the **lamina splendens**. Conversely, the lower edge of adult

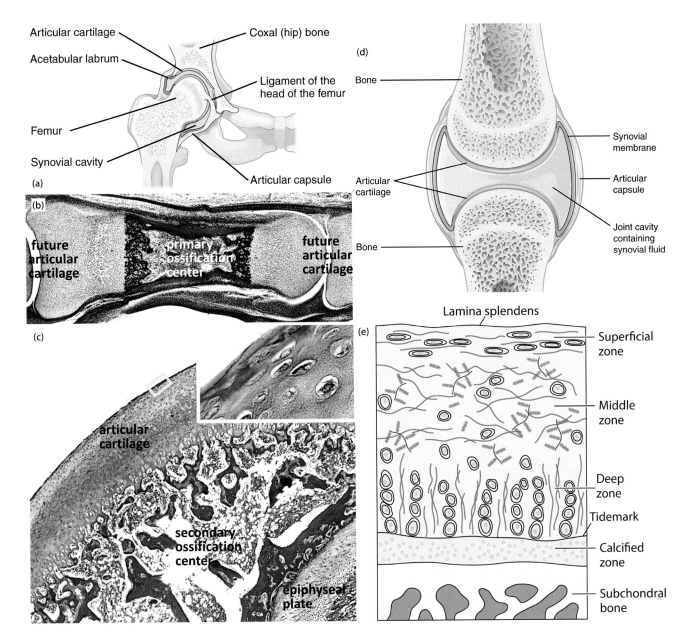

Figure 4.4 Articular cartilage of synovial joints. (a) *As shown in this diagram of the hip, synovial joints possess a fluid-filled synovial cavity in which bone termini are coated with articular hyaline cartilage, thereby enabling a wide range of movements. (**b, c**) Articular cartilage begins to develop in the fetus, such as shown here at low magnification of finger bone articulations (b: 10X) and lacks an overlying perichondrium (c: 25X; inset: 300X). (**d**) The synovium of synovial joints is contiguous with articular cartilage. (**e**) Orientations of chondrocytes and ECM fibers change from tangential to radial from the superficial to deep zones of articular cartilage. ([a, d]: From OpenStax College (2013) Anatomy and physiology. OpenStax: http://cnx.org/content/col11496/latest reproduced under a CC BY 4.0 creative commons license.)*

articular cartilage forms an undulating **tidemark** line at the border between un-calcified cartilage and underlying calcifying cartilage that abuts ossified bone. The precise functions of the lamina splendens and tidemark have not been fully elucidated, but both components are the subject of investigations that aim to optimize therapies for joint-targeting diseases.

As the most common form of inflammatory joint disease (=arthritis), **osteoarthritis (OA)** typically involves wear-and-tear degradation without a major autoimmune component. Since chondrocytes have little or no regenerative capacity to counteract OA-induced PG deficits, loss of cartilage and joint functioning are common outcomes of advanced OA. Alternatively, **rheumatoid arthritis (RA)** is an autoimmune disease in which the synovium undergoes marked swelling and re-organizations, with synoviocytes typically secreting proteases and inflammatory cytokines that exacerbate the effects of RA, thereby accelerating joint damage.

4.5 CALCIFIED CARTILAGE VS. BONE: TWO DIFFERENT TYPES OF MINERALIZED SKELETAL TISSUES

Adult cartilage is normally non-calcified. Conversely, developing bones that undergo **endochondral ossification** initially calcify their cartilage before ossification proceeds (pg. 113). Accordingly, malnourished children lacking enough vitamin D for calcium absorption from their diet can develop a bone disease, called **rickets**, in which inadequate cartilage calcification and impaired bone ossification lead to short stature and other skeletal deformities.

In developing skeletons, calcified cartilages differ substantially from bones with respect to their organic matrix compositions, vascularization patterns, and mechanical properties. Similarly, although many invertebrate animals produce calcified structures, such products are also structurally and functionally distinct from vertebrate bones. Thus, bones represent key evolutionary innovations providing both protective skeletons and effective scaffolds for muscles, thereby facilitating wide-ranging adaptive radiations among vertebrate lineages (Box. 4.1 Invertebrate Calcified Structures, Cartilage, and the Evolution of Vertebrate Bones).

4.6 HISTOLOGICAL PREPARATIONS OF BONE: DECALCIFIED VS. GROUND SECTIONS

Unlike routine processing carried out with non-calcified cartilage or other soft tissues, the hardened nature of bones (**Figure 4.5a**) requires specialized procedures to generate finely cut pieces. Thus, along with non-destructive imaging methods like MRIs and CT scans, physical sectioning of bone is typically achieved using either: (i) **ground sections** where calcified bone is mechanically abraded to a thin slab that retains the mineral phase (**Figure 4.5b, c**), or (ii) **decalcification** protocols that allow de-mineralized bone to be embedded and sectioned (**Figure 4.5d, e**). Such methods yield complementary views of bone, since

---

BOX 4.1 INVERTEBRATE CALCIFIED STRUCTURES, CARTILAGE, AND THE EVOLUTION OF VERTEBRATE BONES

About 1.5 billion years ago, the Earth underwent major tectonic upheavals that triggered increased depositions of calcium (Ca^{2+}) into seawater. Given that high intracellular Ca^{2+} levels are generally toxic, it has been speculated that some primitive eukaryotes adapted to such changes by precipitating Ca^{2+} in materials containing mainly **calcium carbonate ($CaCO_3$)**, with various modern-day invertebrate groups continuing to form a wide range of calcified products (**Figures a–c**). A few extant invertebrates also have **chondrocytes** for secreting **cartilage,** which contains predominantly **type I** and **V collagens** and probably initially arose to help support feeding organs. Conversely, in vertebrates, chondrocyte-derived cartilage is characterized by **type II collagen** and is the sole, or predominant, tissue in the non-bony skeletons of agnathans (e.g. lampreys) and cartilaginous fish (e.g. sharks).

In addition to cartilage, all other extant vertebrates also produce **osteoblast-derived bone**, which is a vascularized connective tissue enriched in **type I collagen** and $CaPO_4$ crystals of **apatite.** Vertebrate bones appear to have arisen >400 million years ago, based on heterostracan ostracoderm fossils with a **dermal exoskeleton** of acellular bones that were sometimes capped with dentin or enamel (**Figure d**). More complex **endochondrally formed** bone evolved in bony fish and eventually formed a calcified **endoskeleton** in the trunk and in parts of the head in humans and other tetrapod ("four-legged") species. Conversely, tetrapod exoskeletons became reduced mostly to the head and developed via **intramembranous ossification**, which appears to be the most recently acquired mode of bone development. Among vertebrate lineages, **apatite** has evolved as the major $CaPO_4$ of bone, apparently because compared to $CaCO_3$, apatite is more stable at reduced pH. Accordingly, it has been speculated that such stability allowed vertebrates to produce well-vascularized skeletons for vigorous muscle contractions, whose associated lactic acid buildup and extracellular pH acidification, in turn, might have degraded a $CaCO_3$-based skeleton.

A key innovation after the acquisition of chondrocytes was the evolution of osteoblasts. Based on analyses of extant vertebrates, chondrocyte differentiation requires the transcription factor **Sox9** (SRY box 9), whereas osteoblast differentiation is driven by **Runx2** (Runt-related transcription factor 2; = Core-binding factor subunit alpha 1, Cbfa1) (**Figure e**). Thus, it is hypothesized that Runx2 and other important osteogenic proteins evolved >400 mya to enable dermal bone formation before eventually apatite-containing bone arose to provide additional benefits over invertebrate calcified products and cartilaginous skeletons, thereby allowing widespread vertebrate radiations.

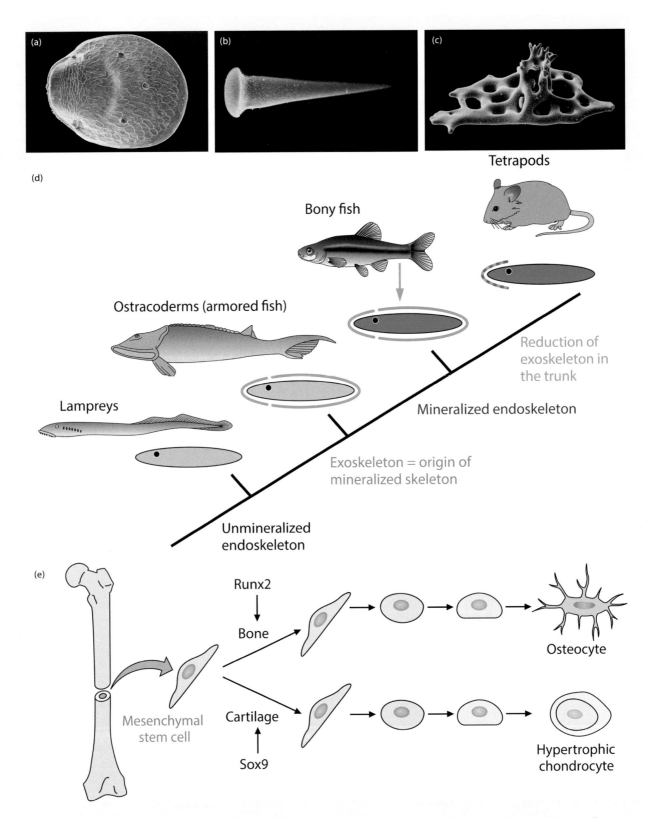

Box 4.1 (a) *SEM of a calcium carbonate shell from a juvenile brachiopod invertebrate (100X); (**b**) SEM of a calcium phosphate stylet made by a nemertean worm invertebrate (300X); (**b**) SEM of a calcium carbonate ossicle from the body of a sea cucumber invertebrate (350X); (**d**) Possible stages in vertebrate bone evolution. (Adapted from Shimada, A et al. (2013) Trunk exoskeleton in teleosts is mesodermal in origin. Nat Comm: doi: 10.1038/ncomms2643); (**e**) regulation of cartilage vs. bone formation. (Adapted from Wagner, DO and Aspenberg, P (2011) Where did bone come from? An overview of its evolution. Acta Orthoped 82: 393–398.)*

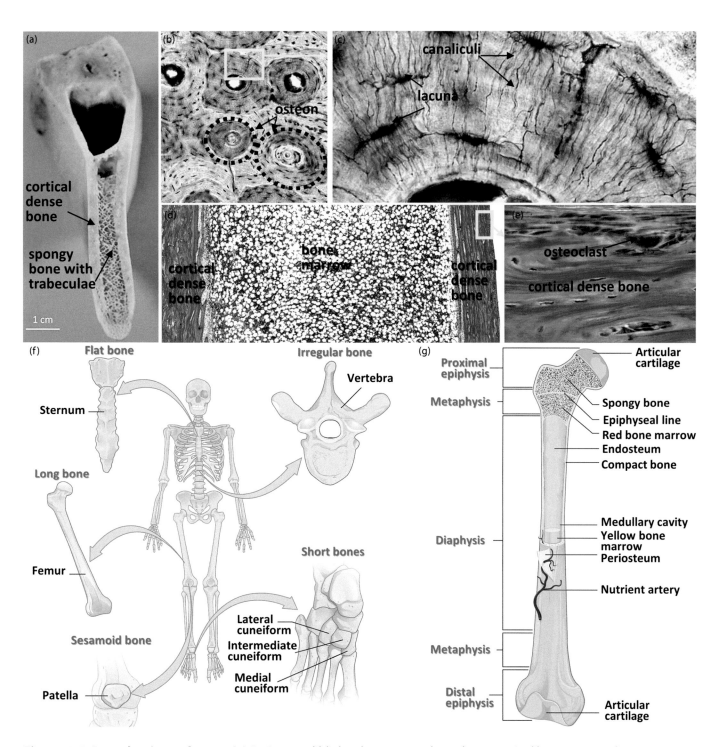

Figure 4.5 Introduction to bones. (*a*) A pig mandible has been cut to show dense cortical bone surrounding spongy bone with its trabeculae separated by large spaces. (*b, c*) For traditional modes of preparing bone sections, bones can be ground into fine slabs to view mineralized features, such as osteons with their lacunae and canaliculi (b: 100X; c: 320X). (*d, e*) Alternatively, bones can be decalcified before sectioning, allowing cells and the organic matrix in bone and marrow to be analyzed in the absence of the mineral phase (d: 15X; e: 250X). (*f, g*) Anatomically, most bones are classified as flat or long, with long bones having three main regions: the epiphyses (heads), diaphysis (shaft), and metaphyses (transition zones between the heads and shaft). ([a]: From Ben-Zvi, Y et al. (2017) 3D architecture of trabecular bone in the pig mandible and femur: inter-trabecular angle distributions. Front Mats 4: doi: 10.3389/fmats.2017.00029 reproduced under a CC BY 4.0 creative commons license; [f, g]: From OpenStax College (2013) Anatomy and physiology. OpenStax: http://cnx. org/content/col11496/latest reproduced under a CC BY 4.0 creative commons license.)

decalcified sections show cells and organic matrices in the absence of the mineral phase, whereas ground sections preserve the mineral phase, but owing to heat and friction generated during the grinding process typically destroy cells and organic matrix components. In addition to such traditional modes, various modern methods, such as either tissue clearing protocols in conjunction with light sheet microscopy or cryo-fixations followed by FIB-SEM serial sectioning are currently being used to analyze the 3D organization of bone components at high resolution.

4.7 DEFINING OSTEOLOGICAL TERMS: BONES AS ORGANS VS. BONE AS A TISSUE

The various terms used in the study of bones (=**osteology**) are particularly complex because they encompass anatomical features of **bones as organs** as well as histological characteristics of **bone tissue** occurring within each bone organ. As examples of organ-related terms, the slightly more than 200 bones in the adult human skeleton can be divided into two main anatomical types—**flat** (e.g., skull bones) vs. **long** (e.g., hand digits and leg bones), with a few others being more appropriately classified as short (e.g., carpals of wrists), irregular (e.g., vertebrate), or sesamoid (e.g., patella) (**Figure 4.5f**). In a typical long bone, the shaft (=**diaphysis**) accounts for most of the bone's length and connects at each of its ends to a transitional zone (=**metaphysis**), which terminates in a rounded head (=**epiphysis**) (**Figure 4.5g**). In addition, long bones tend to have a central **marrow** cavity that is surrounded by a cortical **collar** of bone.

The macroscopic structure of bone tissue visible by the naked eye comprises two main morphotypes: **cancellous** vs. **compact** (**Figure 4.6a–c**). Cancellous (=trabecular, spongy) bone is found in such sites as metaphyses and is characterized by relatively large spaces intermixed among its calcified **trabeculae** (=spicules). Conversely, the collars of bone shafts are examples of compact (=dense, cortical) bone that lack large cavities. As measured by weight, long bones comprise predominantly compact bone, whereas bones such as vertebrae are composed mostly of cancellous bone.

At a microanatomical level, bone tissue can be arranged in two ways—developing **woven** bone vs. adult **lamellar** bone (**Figure 4.6a, b**). In developing bones as well as in adult bones that are initiating fracture repair, bone tissue lacks a well-defined sub-structuring and is thus referred to as woven (=non-lamellar) bone. Conversely, after early childhood, bone normally comprises numerous ~5–15 μm-wide lamellae, and hence is termed lamellar bone. The basic building blocks of lamellae are mineralized collagen fibrils, and within each lamella, alternating alignments of collagen fibrils and apatite crystals can generate a re-enforced organization that provides added strength. Multiple lamellae often become grouped together to generate cylindrical assemblages, called **osteons** (=**Haversian systems**), which are separated from each other by a few intervening layers of bone, called **interstitial lamellae** (**Figures 4.5b, c and 4.6c**). Interstitial lamellae are, in turn, often traversed by **Volkmann canals** (=perforating canals) that form perpendicularly or obliquely oriented connections between adjacent osteons (**Figure 4.6c**). In addition to its internal osteons, living bone is covered on its outer surface by a protective epithelium-like layer, called the **periosteum**, whereas a similar **endosteum** lines inner bone surfaces adjacent to the marrow cavity (**Figure 4.6c, d**).

Each Haverisan system has running down its length a medially situated conduit, called the **central** (=Haversian) **canal**, which in living bones contains blood vessels, lymphatics, and nerves (**Figure 4.6c**). Surrounding the central canal are several to more than a dozen concentrically arranged lamellae of calcified bone, and scattered among the lamellae are ovoid spaces, termed **lacunae**. Each lacuna houses an **osteocyte** type of bone cell that extends cellular processes toward its neighbors within thin channels in bone, called **canaliculi** (**Figure 4.5c**). Lamellae, canaliculi, and lacunae are visible in ground sections but are lost following decalcification.

4.8 THE MINERALIZED PHASE AND ORGANIC MATRIX COMPONENTS OF BONE

The **mineral phase** of bones typically constitutes ~**60–70%** of each bone's wet weight vs. ~**30–40** wt% for the bone's **organic matrix plus water**. Nearly all of the mineral consists of a carbonated form of calcium phosphate crystal that is often designated **hydroxyapatite [Ca$_{10}$(PO$_4$)$_6$(OH)$_2$]**. However, bone crystals are probably better referred to as biological **apatite**, given their unique ionic composition and crystallographic properties that facilitate rapid deposition and resorption processes.

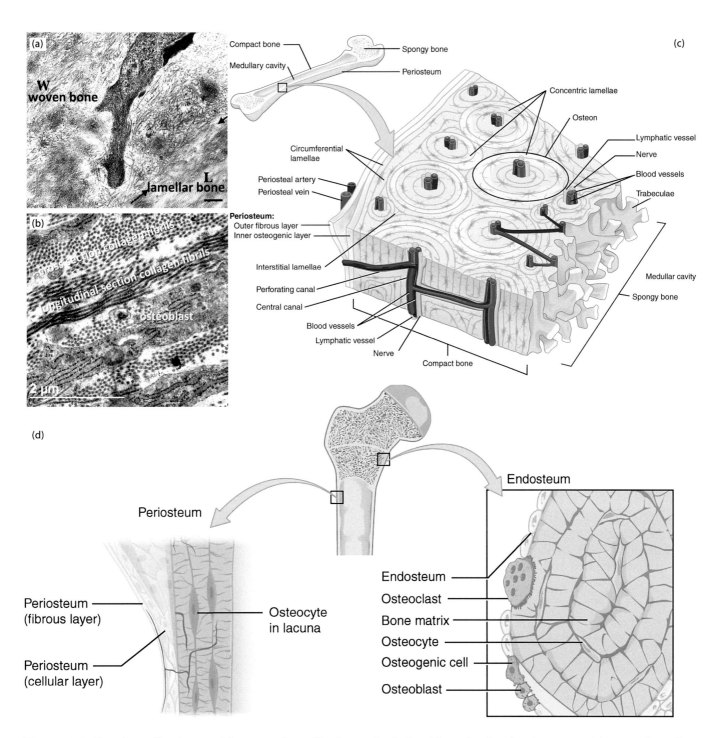

Figure 4.6 **Non-lamellar (woven) bone vs. lamellar bone**. (*a, b*) *Fetal bone begins development with a non-lamellar (=woven) arrangement of mineralized collagen fibrils before eventually being converted into lamellar bone, which is the only type of bone in adults except for sites initiating fracture repair. (**c**) Lamellae of compact bone are mostly organized into distinct cylindrical units, called osteons (=Haversian systems), each of which comprises multiple concentric layers arranged around a central (=Haversian) canal containing blood vessels and nerves. Neighboring osteons are separated from each other by interstitial lamellae that are traversed by orthogonally oriented Volkmann's (perforating) canals. (**d**) The external periphery and inner surfaces of bones are covered by periosteum and endosteum layers, respectively. ([a, b]: From Shapiro, F and Wu, JY (2019) Woven bone overview: structural classification based on its integral role in developmental, repair and pathological bone formation throughout vertebrate groups. Eur Cells Mat 38: 137–167 reproduced under a CC BY 4.0 creative commons license; images courtesy of Dr. F. Shapiro and reproduced by permission of a creative commons license; [c, d]: From OpenStax College (2013) Anatomy and Physiology. OpenStax. http://cnx.org/ content/col11496/latest reproduced under a CC BY 4.0 creative commons license.)*

Similar to a mineral phase dominated by apatite, ~90% of bone's organic matrix consists of **type I collagen** that forms apatite-nucleating fibrils. In addition, the matrix contains additional collagens (e.g., III, V, X) and various **non-collagenous proteins** (**NCPs**) that can promote or inhibit apatite mineralization. Notable NCPs include **osteocalcin, bone sialoprotein** (**BSP**), **osteonectin**, as well as osteonectin-like proteins of the **secretory calcium binding phosphoprotein** (**SCPP**) family that is a hallmark of bone-forming vertebrates (Box 4.1).

4.9   BONE CELLS AND OSTEOBLASTIC BONE DEPOSITION VS. OSTEOCLASTIC BONE RESORPTION WITH AN OVERVIEW OF OSTEOPOROSIS

Based on their cellular lineages and functional properties, four types of resident bone cells are recognized: (1) **osteoclasts**, (2) **osteoblasts**, (3) **bone lining cells**, and (4) **osteocytes** (**Figure 4.7a**). Osteoclasts are ~100–200 μm-wide multinucleated cells derived from pluripotent **hematopoietic stem cells**. After being delivered to bone via blood vessels, individual osteoclast precursors fuse and differentiate into functional osteoclasts under the influence of regulatory molecules that are secreted mainly by osteocytes but also to a lesser extent by osteoblasts and bone lining cells (Box 4.2 Roles of Osteoblasts and Osteocytes during Osteoclastic Differentiation and Activation). When fully activated, osteoclasts serve to break down bone, with such **resorption** mediating normal bone replenishment and the homeostatic regulation of blood calcium levels (see below).

By contrast, bone-secreting **osteoblasts** are derived from pluripotent **mesenchymal stem cells** (**MSCs**). Each mononucleated osteoblast contains abundant Golgi and ER both for **depositing bone** and for **modulating osteoclasts** via such secretions, such as **macrophage colony stimulating factor** (**M-CSF**), **osteoprotegerin** (**OPG**), and **receptor activator of nuclear factor kappa B ligand** (**RANKL**) (Box 4.2). After laying down bone, osteoblasts can (i) die via **apoptosis**, (ii) form **bone lining cells**, or (iii) develop into **osteocytes**. Osteoblasts that do not apoptose while remaining at the bone surface become **bone lining cells** in the **periosteum** and **endosteum** (**Figure 4.7a, b**). Lining cells possess few synthetic organelles, and prior to the remodeling of bone must first pull away from the bone surface so that bone resorption and deposition can be carried out by osteoclasts and osteoblasts, respectively.

As opposed to differentiating into surface lining cells, most non-apoptosing osteoblasts deposit bone and are thus translocated beneath the bone surface as they become encased in ossifying bone tissue. Such interiorly positioned osteoblasts transition into long-lived **osteocytes**, each of which resides in a distinct lacunar space within bone (**Figure 4.7c, d**). Osteocytes are characterized by dozens of elongated cytoplasmic processes that

---

BOX 4.2 ROLES OF OSTEOBLASTS AND OSTEOCYTES DURING OSTEOCLASTIC DIFFERENTIATION AND ACTIVATION

Hematopoietic stem cells entering the marrow cavity of bones are triggered to differentiate as **osteoclast precursors** by osteoblast-secreted stimuli like **macrophage colony-stimulating factor** (**MCSF**) (**Figure a**). During differentiation, such pre-osteoclasts fuse to form a multinucleated osteoclast, which, in turn, is further activated mainly by osteocytes in two major ways: (i) by increased secretion of **receptor activator of nuclear factor kappa B ligand** (**RANKL**) and (ii) by decreased exocytosis of **osteoprotegerin** (**OPG**). The binding of osteocyte-derived RANKL to RANK receptors on osteoclasts activates osteoclast precursors, whereas OPG acts as a soluble decoy that sequesters RANKL and prevents osteoclastic activation via RANK/RANKL (**Figure a, b**). Once activated, key markers of osteoclastic functionality are lysosomal **tartrate-resistant acid phosphatases** (**TARPs**). Such TARPs are secreted at the ruffled border of activated osteoclasts to modulate bone matrix in several ways, including by dephosphorylating **osteopontin**

and **bone sialoprotein**, thereby breaking these proteins' bonds with integrins at the sealing zone. This cleavage allows the osteoclast to reposition itself within Howship's lacuna and establish a new seal during bone remodeling.

As a key driver of osteoclast activation, parathyroid hormone (PTH) is secreted by parathyroid glands in response to low blood calcium levels (Chapter 15). However, osteoclasts lack functional **PTH receptors** (**PTHRs**), and instead PTH binds to PTHRs on osteocytes and osteoblasts to increase and decrease RANKL and OPG, respectively. This, in turn, upregulates osteoclastic resorption so that blood calcium balance is restored. Counterintuitively, low doses of recombinant PTH are used to treat osteoporosis, which rather than triggering more osteoclastic activity actually serve to upregulate bone deposition, perhaps by blocking osteocytic secretions of **sclerostin**, a potent inhibitor of osteoblastic bone formation (**Figure b**).

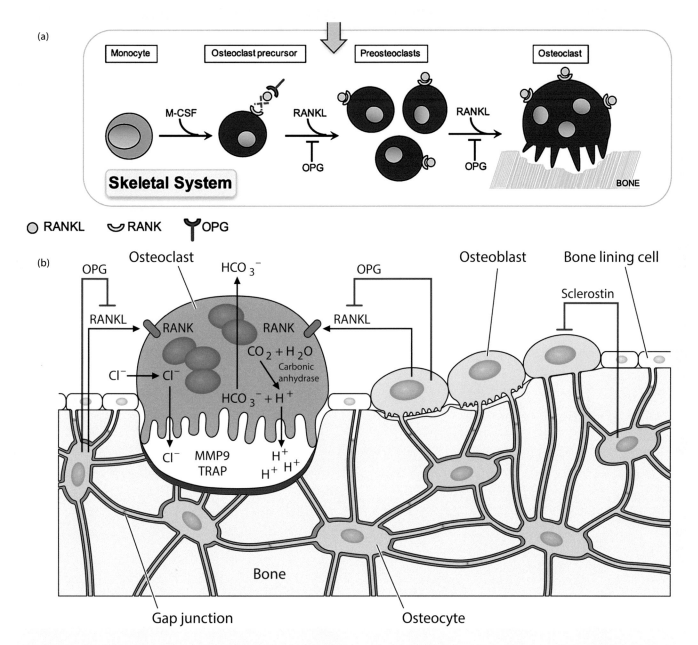

Box 4.2 *(**a**) Osteoclastic activation via the RANKL/RANK system and its inhibition by osteoprotegerin (OPG) (Adapted from Titanji, K (2017) Beyond antibodies: B cells and the OPG/RANK-RANKL pathway in health, non-HIV disease and HIV-induced bone loss. Front Immunol 8: doi:10.3389/fimmuu.201710851 reproduced under a CC BY 4.0 creative commons license); (**b**) Osteoclast-mediated resorption of bone and its regulation by osteoblasts and osteocytes. (Adapted from Florencio-Silva, R et al. (2015) Biology of bone tissue: structure, function, and factors that influence bone cells. Biomed Res Int: http://dx.doi.org/10.1155/2015/421 reproduced under a CC BY 3.0 creative commons license.)*

extend through thin **canalicular** channels within the bone (**Figure 4.7e–h**). At their termini, such processes connect via **gap junctions** to neighboring processes as well as to surface osteoblasts and lining cells, thereby forming a syncytium that facilitates molecular exchanges throughout avascular regions of bone. In addition, the osteocytic syncytium helps coordinate bone's two main activities— **osteoblastic bone deposition** and **osteoclastic bone resorption** (**Figure 4.8a**).

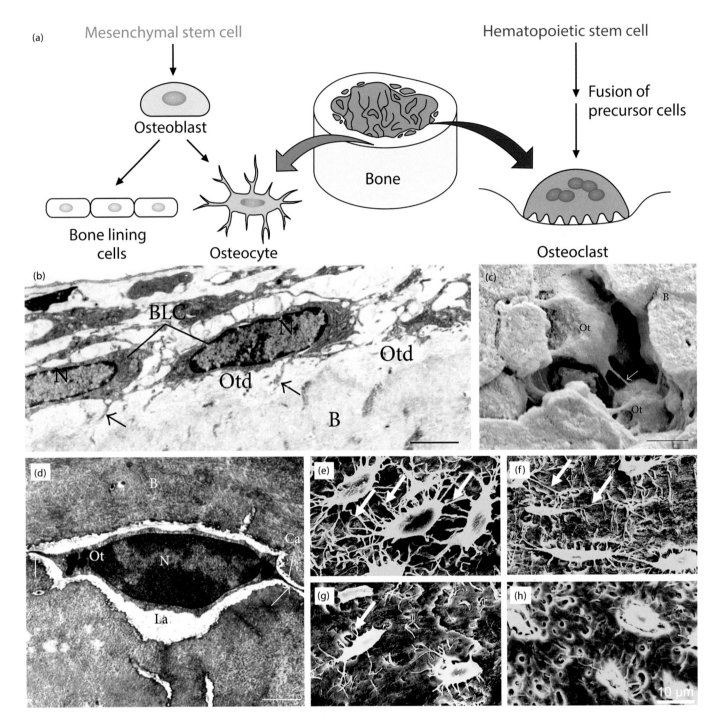

Figure 4.7 Bone cells. *(a) Mesenchymal stem cells give rise to osteoblasts, which, in turn, can differentiate into either bone lining cells of the periosteum and endosteum or long-lived osteocytes within bone lacunae. Hematopoietic stem cells form osteoclast precursors, which fuse to form multinucleated osteoclasts. (**b**) A TEM shows bone lining cells (BLCs) covering unmineralized matrix (=osteoid, Otd) and subjacent mineralized bone (B). (**c**) In this pseudocolored SEM of bone, two osteocytes (Ots) are linked by cellular processes (arrow) that are joined by gap junctions. (**d**) A TEM shows an intra-lacunar (La) osteocyte (Ot) with its processes (arrows) that extend into canalicular (Ca) spaces among the bone (B) matrix (scale bars = 2 μm). (**e–h**) SEMs show aging-related loss of osteocytes and their processes (arrows) as human middle ear bones become hypermineralized, thereby filling lacunae and canaliculi with bone. ([a]: From OpenStax College (2013) Anatomy and Physiology. OpenStax. http://cnx.org/content/col11496/latest reproduced under a CC BY 4.0 creative commons license; [b–d]: From Florencio-Silva, R et al. (2015) Biology of bone tissue: structure, function, and factors that influence bone cells. Biomed Res Int: http://dx.doi.org/10.1155/2015/421746 reproduced under a CC BY 3.0 creative*

Figure 4.7 (Continued) *commons license; [e–h]: From Rolvien, T et al. (2018) Early bone tissue aging in human auditory ossicles is accompanied by excessive hypermineralization, osteocyte death and micropetrosis. Sci Rep 8: doi:10.1038/ s41598-018-19803-2 reproduced under a CC BY 4.0 creative commons license.)*

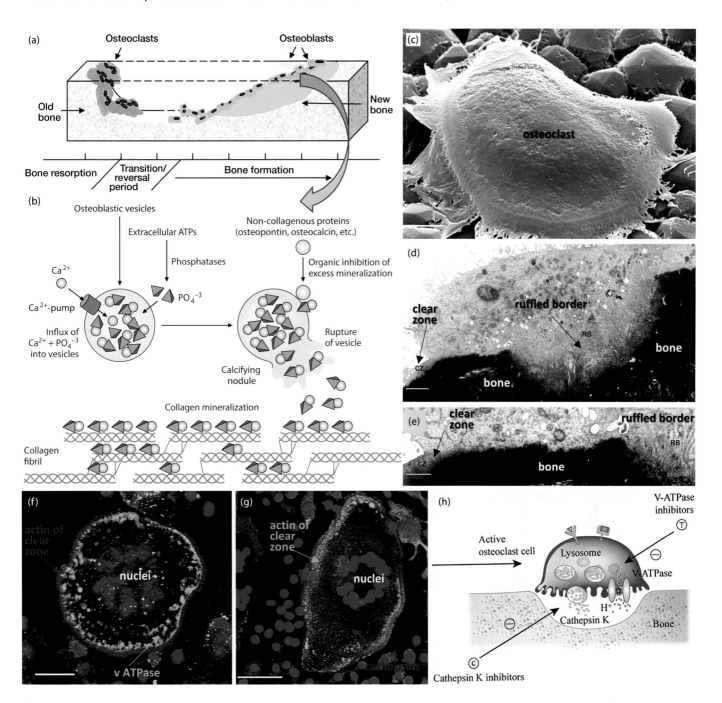

Figure 4.8 Bone deposition by osteoblasts vs. bone resorption by osteoclasts. *(a) New bone is continually deposited by osteoblasts, while old bone is resorbed by osteoclasts, thereby balancing blood calcium levels (see also* **Box. 4.2***). (b) According to one model of bone deposition, osteoblastic vesicles accumulate calcium and phosphate ions through pumps and membrane channels to reach supersaturating levels before apatite crystals rupture the vesicle and are seeded on collagen fibrils of the bone matrix. Such processes are regulated by various non-collagenous proteins in the ECM. (c–e) Osteoclasts can be grown on surfaces containing the glycoprotein vitronectin, which binds to osteoclastic integrins, thereby stimulating: (i) osteoclast adhesion, (ii) clear zone formation, and (iii) development of a ruffled border in preparation for bone resorption. (scale bars: c = 5 μm; d = 1 μm; e = 2 μm). (f, g) During osteoclast activation, a ring of actin in the clear zone forms around the cell periphery to seal the osteoclast and allow the formation of a ruffled border,*

Figure 4.8 (Continued) *in which protons pumped by ATPases and cathepsins released from lysosomes demineralize apatite and break down matrix components, respectively (scale bars: f = 5 μm; g = 10 μm).* **(h)** *Therapies targeting osteoclastic ATPases and cathepsins can be used to reduce bone loss. ([a]: From Roberts, S et al. (2016) Ageing in the musculoskeletal system. Cellular function and dysfunction throughout life. Acta Orthopedica 87: 15–25 reproduced with publisher permission; [c–g]: From Szewczyk, KA et al. (2013) Distinctive subdomains in the resorbing surface of osteoclasts. PLoS ONE 8(3): e60285 https://doi.org/10.1371/journal.pone.0060285 reproduced under a CC BY creative commons license; [h]: From Bi, H et al. (2017) Key triggers of osteoclast-related diseases and available strategies for targeted therapies: A review. Front Med 4:234. doi: 10.3389/fmed.2017.00234 reproduced under a CC BY 4.0 creative commons license.)*

Osteoblasts can upregulate their **deposition of bone** in response to such stimuli as elevated blood calcium levels (Chapter 15) by secreting into the ECM various organic matrix components like type I collagen and non-collagenous proteins. Following collagen fibrillogenesis, the organic matrix comprises a non-calcified **osteoid**, which can undergo a two-step mineralization process involving a **vesicular** and a **fibrillar** phase. According to one model of bone formation, vesicular-phase osteoblasts release into osteoid **matrix vesicles** (**MVs**) containing phosphatases that generate phosphates from ECM sources. Such phosphates plus extracellular Ca^{2+} are then accumulated in the vesicular lumen to reach supersaturated levels of Ca^{2+} and PO_4^{-3}. This, in turn, triggers apatite crystal formation, thereby rupturing the vesicle and allowing further apatite crystallization on collagen fibrils of bone tissue (**Figure 4.8b**). An alternative theory maintains that vesicles within osteoblasts initially generate non-crystalline **amorphous calcium phosphate** (**ACP**). ACP is then released from vesicles to initiate the extracellular fibrillar phase, in which ACP crystallizes into apatite. In either case, apatite crystals form orderly arrays within spaces along type I collagen fibrils, and the various steps in mineralization can either be enhanced by matrix components, such as osteonectin and SCPPs or inhibited by constituents like osteopontin.

To counterbalance bone deposition, activated **osteoclasts** (Box 4.2) begin **bone resorption** by anchoring to the bone surface via an actin- and integrin-rich ring, called the **sealing-** or **clear-zone** (**Figure 4.8c–h**). Medial to this zone, the osteoclastic apical membrane forms long microvilli that constitute a **ruffled border** (Box 4.2) (**Figure 4.8d, e**). Using protons generated via osteoclastic **carbonic anhydrase** enzymes plus energy supplied by mitochondrial ATPs, vacuolar **H^+-ATPases** on the ruffled border pump out H^+ ions to produce an acidic milieu that demineralizes bone (**Figure 4.8f, h**). In addition, **cathepsin** enzymes in osteoclastic lysosomes are exocytosed to digest the organic matrix (**Figure 4.8g, h**), collectively carving out surface depressions, called **Howship's lacunae**. Given the key roles that these osteoclastic components play in bone resorption, H^+-ATPases and cathepsins are the targets of several therapies that aim to prevent excessive demineralization caused by overly active osteoclasts (**Figure 4.8h**). Conversely, defective osteoclastic resorption can lead to reduced Howship lacuna production and an excessively mineralized condition, referred to as **osteopetrosis**.

Bones are continually replenished at an average rate of ~10% of the adult skeleton each year. Such replenishment, or **bone remodeling**, involves numerous integrated collections of osteoclasts, osteoblasts, bone lining cells, and osteocytes, with each collection constituting a **basic multicellular unit** (**BMU**). In bones of children and young adults, BMUs normally allow osteoblastic deposition to exceed osteoclastic resorption until overall bone mass eventually peaks in 30–40 year-olds. Thereafter, resorption begins to dominate, and bone mass declines. Compared to men, women tend to undergo more rapid bone loss, particularly after menopause, when **estrogen** production falls (Chapter 18). Such post-menopausal acceleration in bone loss is due to the fact that estrogen normally serves to retard bone demineralization via several mechanisms, including by inhibiting osteoblastic/osteocytic apoptosis and by downregulating osteoclastic activation via modulation of OPG and RANKL levels (Box 4.2).

In cases where osteoclastic osteolysis greatly outpaces osteoblastic deposition, excessive holes form within the mineral phase and cause bones to become brittle as part of a common aging-related pathology, called **osteoporosis**. In between its holes, the remaining osteoporotic bone may have a normal mineral content and thus differs from bones weakened by **osteomalacia,** where excessive pores are typically lacking and the osteoid is inadequately mineralized. Along with prescribing exercise plus calcium and vitamin D supplements, osteoporosis treatments often aim to reduce bone resorption by **estrogen replacement therapies** or **bisphosphonate** drugs that reduce osteoclast numbers.

Approximately 7 million people in the United States suffer a **bone fracture** each year. During repair of these fractures, the periosteum produces a mass of chondrocytes (=**callus**) that links the broken ends of the bone before becoming replaced by bone. Bone formed by osteoblasts in hard calluses initially resembles embryonic woven bones and is subsequently remodeled into lamellar bone.

4.10 BONE DEVELOPMENT: INTRAMEMBRANOUS OSSIFICATION DURING EXOSKELETON FORMATION VS. ENDOCHONDRAL OSSIFICATION WITH CARTILAGE MODELS AND EPIPHYSEAL PLATES DURING ENDOSKELETON FORMATION

Vertebrate bones evolved as components of two skeletal networks—an **endoskeleton** vs. **exoskeleton**—which differ in their positioning relative to the body surface and in how they are formed during fetal development (Box 4.1). Endoskeletal bones of humans tend to occur fairly deeply embedded in the body within the limbs, trunk, spine, and parts of the head. Moreover, such bones typically develop via an **endochondral** mode of ossification that involves the formation of **cartilage models** and **epiphyseal plates**. Conversely, exoskeletal bones, such as in the skull, usually occur closer to the body surface and tend to undergo **intramembranous ossification** without cartilage models or epiphyseal plates.

Intramembranously formed bones that form parts of the skull, face, and collarbones begin as condensations of pluripotent mesenchymal cells that differentiate into **osteoblasts**. Such cells exocytose matrix components that subsequently become mineralized by apatite crystals to form irregular bone trabeculae (**Figure 4.9a–c**) (**Case 4.3 Harriet Tubman's Skull Bones**).

As opposed to intramembranous ossification, endochondral bone formation begins with the formation of a miniature cartilaginous version of the bone, called the **cartilage model**. Such models initially comprise a solid mass of hyaline cartilage that subsequently cavitates and undergoes ossification (**Figure 4.9d**). In typical long bones, these re-organizations are initiated by chondrocytic hypertrophication in the center of the model, thereby resulting in swollen cells that secrete matrix metalloproteinases to break down the surrounding cartilaginous matrix. Such matrix dissolution combined with the apoptosis of hypertrophied chondrocytes generates a central cavity in the model. Concurrently, the perichondrium around the future diaphysis is converted into a periosteum with osteoblasts that secrete a thin **bone collar** at the diaphyseal periphery (**Figure 4.9d**).

Subsequent steps in ossification require a rich **vascular supply**. Blood vessels penetrating the bone collar deliver bone cell precursors into the diaphyseal marrow cavity while helping to organize this region into a

---

CASE 4.3 HARRIET TUBMAN'S SKULL BONES

Civil Rights Activist
   (1822?–1913)

"I was the conductor of the Underground Railroad for eight years, and I...never ran my train off the track and I never lost a passenger."

(H.T. on her successes while guiding slaves to freedom using the Underground Railroad network of pro-abolitionists)

Born in Maryland as a slave named Araminta Ross, Harriet Tubman escaped slavery in 1849 and guided slaves to freedom via an Underground Railroad network of people who opposed slavery. After the Civil War, Tubman campaigned for women's rights while also having to deal with the aftereffects of a skull injury sustained as a young woman, when she was struck with a heavy object thrown by an irate slave-owner. The blow fractured intramembranously formed bone in her skull,

causing chronic headaches, seizures, and periods of drowsiness. In the late 1890s, Tubman underwent surgery to relieve pressure on her brain, enduring the operation without anesthesia while biting down on a bullet, before recovering from the surgery and eventually dying of

pneumonia in 1913. In honor of her acts of valor, Tubman was selected in 2016 to be the first woman to have her likeness put on U.S. paper currency, a picture that no doubt will fail to show the troublesome skull fracture that plagued her throughout her adult life.

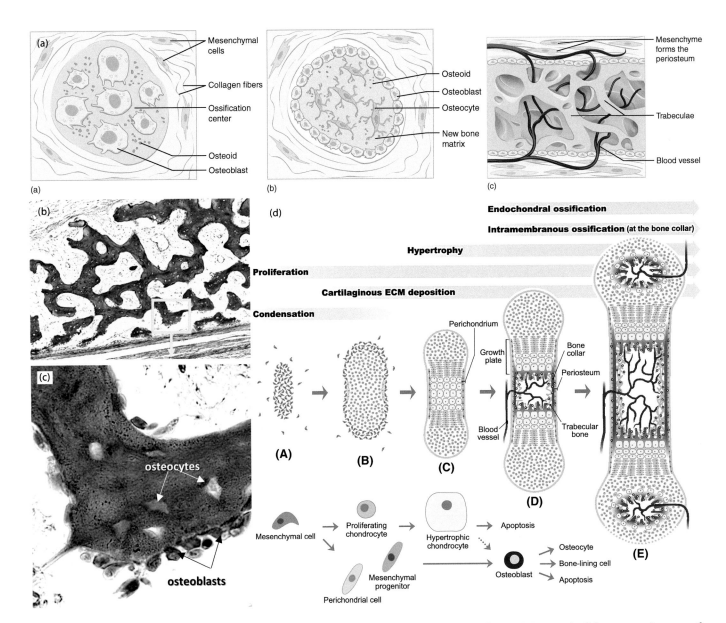

Figure 4.9 Modes of bone formation: intramembranous vs. endochondral. (a–c) *Some skull bones and parts of collarbones form via intramembranous ossification, in which osteoblasts deposit bone without first forming a cartilage model. (b) 20X; (c) 200X. (**d**) Other bones of the body, such as appendicular long bones develop via endochondral ossification. During such ossification, mesenchyme cells lay down a miniature cartilage model of the bone before apoptosis generates a primary ossification center in the bone shaft. Blood vessels then invade the primary ossification center to allow bone formation in the fetus. After birth, a vascularized secondary ossification center forms in each bone head so that an epiphyseal growth plate becomes situated between the primary and secondary ossification centers. ([a]: From OpenStax College (2013) Anatomy and Physiology. OpenStax. http://cnx.org/content/col11496/latest reproduced under a CC BY 4.0 creative commons license; [d]: From Egawa, S et al. (2014) Growth and differentiation of a long bone in limb development, repair and regeneration. Develop Growth Differ 56: 410–424 reproduced with publisher permission.)*

primary ossification center prior to birth (**Figure 4.9d**). After birth and continuing through puberty, blood vessels also infiltrate and organize a similar region within each epiphysis. These **secondary ossification centers** lie beneath articular cartilage (**Figure 4.9d**) and are separated from the primary ossification center by an intervening mass of cartilage, called the **epiphyseal plate**. When actively elongating in a typical long bone, the epiphyseal plate adds new bone material toward the shaft of the bone, thereby allowing the articular cartilage

CASE 4-4 KING TUT'S FOOT BONES

Egyptian Pharaoh
(~1342–1325 B.C.)

"We were astonished by the beauty and refinement of the art displayed by the objects surpassing all we could have imagined."

(Archeologist Howard Carter on finding the magnificent treasures buried in King Tut's tomb)

Once a relatively obscure pharaoh who ruled in Egypt from ~1332 to 1323 B.C., Tutankhamun became famous after his tomb was discovered and its treasures sparked worldwide interest in the now iconic King Tut. Tutankhamun began his reign at around age 9 and died about 10 years later. Based on analyses of his mummified body, the short-lived king had a frail body and a foot with necrotic bone degradation resembling Kohler's disease, type II. Such foot pathologies would have necessitated the use of a walking aid, and consistent with that view, numerous canes were found in Tutankhamun's tomb. Recent genetic testing of his remains and those of immediate family members indicate Tutankhamun was the son of King Akhenaten and one of Akhenaten's sisters. Such inbreeding presumably contributed to the young pharaoh's early demise, which, according to recent theories, was due to multiple ailments. For example, it has been speculated that damage to Tutankhamun endochondrally formed bones in combination with a case of malaria ultimately proved to be lethal.

in the bone head to remain uncalcified and thus capable of sliding past its neighboring bone during articulating movements (**Case 4.4 King Tut's Foot Bones**).

Traditionally, the epiphyseal plate has been viewed as comprising multiple horizontal zones from the top to bottom of the plate. However, functionally, two main regions predominate within the plate: (1) **immature cartilage** comprising an upper **resting zone** of quiescent chondrocytes plus a lower **proliferative zone** of dividing chondrocytes, and (2) a subjacent region of **mature cartilage** with a **hypertrophic zone**, where each chondrocyte enlarges before **calcifying** and undergoing **apoptosis** near the metaphyseal border (**Figure 4.10a–d**).

In response to stimulatory cues like growth hormone that increase during puberty, quiescent immature chondrocytes at the top of the plate are triggered to divide in the proliferative zone, thereby driving more advanced stages toward the bottom of the plate. Such mitoses often form a vertical stack of chondrocytes, which in turn grow much larger in the hypertrophic zone before eventually apoptosing as the cartilage becomes calcified (**Figure 4.10e, f**). At the end of adolescence, closure of the epiphyseal plate converts the former cartilaginous plate into bone, thereby leaving behind an **epiphyseal** (fusion) **line** of more compact bone that links spongy bone in the epiphysis with that in the metaphysis. In this way, closed (inactive) vs. open (active) epiphyseal plates can be used forensically to help date recovered bones relative to puberty.

4.11   SOME CARTILAGE AND BONE DISORDERS: RELAPSING POLYCHONDRITIS AND PAGET'S DISEASE

Relapsing polychondritis (**RP**) is a rare autoimmune disease of children and adults that is characterized by sporadic inflammation affecting either cartilage (particularly in the ears and nose) or other proteoglycan-rich tissues in organs such as eyes. Disease etiology is incompletely understood, and in many cases, such relapses self-resolve with only mildly debilitating symptoms. However, inflammation of respiratory tract cartilage can have

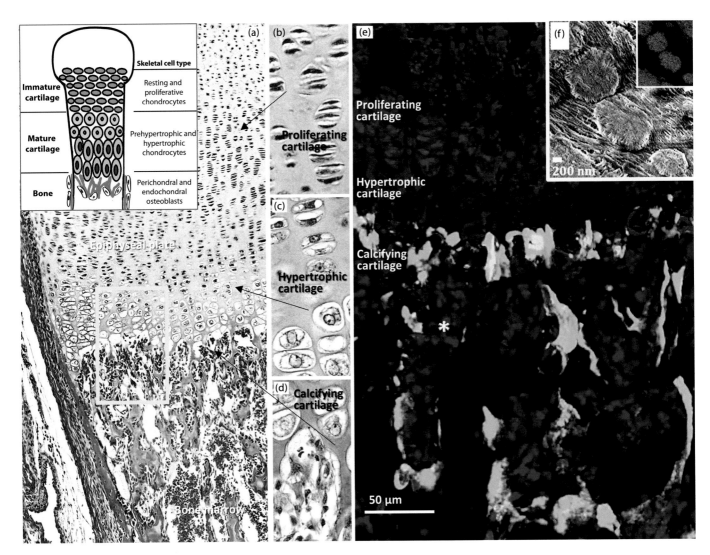

Figure 4.10 Functional microanatomy of the epiphyseal plate. *(a–d) The epiphyseal plate of growing endochondral bones comprises two main regions: (1) an upper area of immature cartilage with chondrocytes that can be stimulated from a resting state to undergo mitoses within the proliferative zone (figure b); and (2) a subjacent area of mature cartilage, where chondrocytes undergo hypertrophy (figure c) followed by apoptosis as the matrix becomes calcified (figure d). (a) 25X; (b, c, d) 250X. (e) Calcein-green fluorescence of a tibial epiphyseal plate in the area comparable to that outlined by the yellow box in (a) shows new mineralization directly beneath the border between proliferating and hypertrophic chondrocytes. An asterisk marks blood vessels interspersed among and eventually replaced by ossified bone within developing spicules at the top of the primary ossification center. (f) High magnification cryo-SEM shows several mineralized particles embedded in collagen fibrils near a hypertrophic chondrocyte (not visible at this magnification). The backscattered electron (BSE) image shown in the inset confirms that these particles are mineralized. ([a, inset]: From Gomez-Pico, P and Eames, BF (2015) On the evolutionary relationship between chondrocytes and osteoblasts. Front Gen: doi: 10.23389/fgene.2015.00297 reproduced under a CC BY 4.0 creative commons license; [e, f]: From Haimov, H et al. (2020) Mineralization pathways in the active murine epiphyseal growth plate. Bone 130: https://doi.org/10.1016/bone.2019.115086 reproduced with publisher permission.)*

fatal consequences, and to prevent such outcomes, treatments rely on immunosuppressants or even surgical intervention.

Paget's disease (PD) causes fragile and misshapen bones to form due to excessive osteolysis. PD is distinguished from osteoporosis by the markedly increased numbers and sizes of osteoclasts that characterize PD. Causes of PD remain unknown, although a virus-induced etiology is suggested for at some cases of PD. As with osteoporosis, treatments for PD often include bisphosphonates to reduce osteoclast numbers.

4.12 SUMMARY—CARTILAGE AND BONE

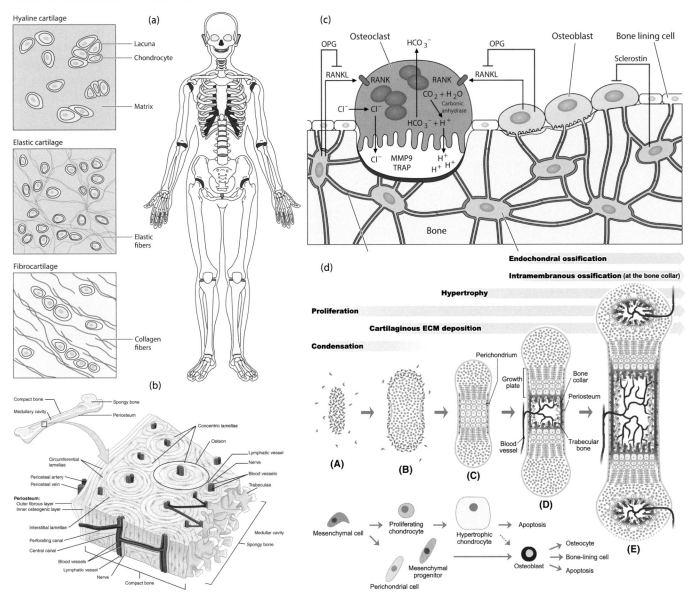

Pictorial Summary Figure Caption (a): *see figure 4.1a;* ***(b):*** *see figure 4.6c;* ***(c):*** *see box 4.2b;* ***(d):*** *see figure 4.9d.*

Pliable **cartilage** and rigid **bone** are **skeletal connective tissues**. Of the three main **cartilage** types—**hyaline** (e.g., bone heads), **elastic** (e.g., epiglottis), and **fibrocartilage** (e.g., intervertebral discs)—hyaline is the most widely distributed (**Figure a**). Each cartilage type in adults lacks blood vessels and contains **chondrocytes** in artifactual spaces, called lacunae. The ECM of cartilage is enriched in **collagen** (mainly **type II** in hyaline and elastic vs. mostly **type I** in fibrocartilage) plus glycosylated proteins, termed **proteoglycans** (**PGs**). PGs have a protein core that is typically attached to sulfated disaccharide chains, called **glycosaminoglycans** (**GAGs**), which attract water and allow cartilages to act as cushions. Chondrocytes not only secrete PGs, collagens, and the elastic fibers of elastic cartilage but also break down these ECM components with matrix metalloproteinases (MMPs), collagenases, and elastases. A cellular **perichondrium** surrounds elastic- and most hyaline cartilages but is lacking in fibrocartilage, which is also distinguished from the other two cartilage types by its prominent **collagen fibers**. Hyaline cartilage coating the heads of bones in **synovial joints** is called **articular cartilage**, which is lubricated by synovial fluid but can become degraded and/or calcified during **osteoarthritis**. Cartilage normally calcifies during **endochondral bone formation**, but such calcification and subsequent bone formation are impaired in children with **rickets**, typically due to insufficient vitamin D intake for calcium absorption.

Bone is ~**60–70% mineralized** mostly in the form of a $CaPO_4$ crystal called **apatite** vs. ~**30–40% organic matrix** containing mainly **type I collagen** plus water. To obtain conventional sections, bone is either (1) **decalcified** before sectioning, which removes apatite while retaining the organic matrix and cells of bone, or (2) prepared as **ground sections**, which retain apatite but destroy organic matrix components and cells. Key bone terms include: (1) **long** (e.g., leg bones with **epi-, meta-, diaphysis** regions [**Figure b**]) vs. **flat** (e.g., skull); (2) **compact** (e.g., peripheral bone collar) vs. **cancellous** (e.g., near marrow cavity with hardened **trabeculae** separated by large spaces); and (3) **woven** (non-lamellar bone formed in fetuses or during fracture repair in adults) vs. **lamellar** (with bone **lamellae** arranged as **osteon** units, the most complex of which comprise **Haversian systems**). Each Haversian system has (i) a **central canal** with blood vessels; (ii) concentric **bone lamellae** with interspersed **lacunae**, each of which contains an **osteocyte**; and (iii) thin channels (=**canaliculi**) interconnecting lacunae and containing osteocyte cell processes that are linked by **gap junctions** to form a syncytial network (**Figures b, c**). The four bone cells are: (1) multinucleated **osteoclasts** that resorb bone and are derived from hematopoietic stem cells; (2) mesenchymal-stem-cell-derived **osteoblasts** that deposit matrix (=**osteoid**), and after osteoid is mineralized can become either **osteocytes** or **bone lining cells**; (3) **osteocytes** comprising long-lived cells within bone lacunae that help maintain bone; and (4) **bone lining cells** that form the **periosteum/endosteum** on outer/ inner bone surfaces.

Although invertebrates can make cartilage and calcified products, bones are vertebrate innovations, with proteins like **Runx2** and **SCPP family** members having evolved for **osteoblast functioning**. Once formed, **bone** is continually **remodeled** to keep blood Ca^{2+} levels balanced. When circulating Ca^{2+} is too low, **parathyroid hormone** is secreted, causing osteocyte-derived **RANKL** to bind to **RANK** receptors on osteoclasts, thereby triggering acid and enzyme secretion at the osteoclastic **ruffled border**. This results in surface depressions (**Howship's lacunae**) as bone resorption raises blood Ca^{2+} levels (**Figure c**). Conversely, excessive osteoclastic activity can lead to the fragile bones of **osteoporosis**. During development, bone can form either by **endochondral ossification** (with **cartilage models** and epiphyseal growth plates such as occur in long bones [**Figure d**]) or by **intramembranous ossification** (with no cartilage models/epiphyseal plates; e.g., in some skull bones).

Some cartilage/bone diseases: **relapsing polychondritis**—sporadic cartilage inflammation, particularly in ears; **Paget's disease**—similar to osteoporosis but characterized by greater numbers of large osteoclasts; **osteomalacia**—weak bones due to inadequate **osteoid mineralization; osteopetrosis**—over-mineralized bone due to insufficient osteoclast activity.

SELF-STUDY QUESTIONS

1. What is the most common type of cartilage in humans?
2. T/F Some invertebrates form bone.
3. Approximately what percentage of bone is mineralized?
4. Which disease involves excessive bone mineralization?
   A. Osteoporosis
   B. Rickets
   C. Osteomalacia
   D. Osteopetrosis
5. Which of the following occurs in high amounts in the epiglottis?
   A. Hyaline cartilage
   B. Fibrocartilage
   C. Elastic cartilage
   D. ALL of the above
   E. NONE of the above
6. Which occurs in high amounts in most cartilages?
   A. Proteoglycans
   B. Type II collagen
   C. Glycosaminoglycans (GAGs)
   D. ALL of the above
   E. NONE of the above

7. T/F Cancellous bone contains numerous trabeculae.
8. T/F Osteocytes are interconnected by gap junctions.
9. T/F Long bones develop via intramembranous ossification.
10. Which two kinds of cells can osteoblasts become?
11. What is the name of the spaces that contain cell bodies of osteocytes?
    A. Canaliculi
    B. Central canals
    C. Lacunae
12. Which of the following is made by osteocytes?
    A. RANK
    B. RANKL
    C. Ruffled border
    D. ALL of the above
    E. NONE of the above
13. T/F Ground sections of bone typically contain cells and organic matrix.

"EXTRA CREDIT" For each term on the left, provide the BEST MATCH on the right (answers A-G can be used more than once)

14. Breaks down bone___   A. Osteoclast
15. A type of hyaline cartilage___   B. Apatite
16. Major bone mineral___   C. $CaCO_3$
17. Osteoblast marker___   D. Articular
18. Contains bone lining cells____   E. Runx2
                                                 F. Periosteum
                                                 G. Fibrocartilage

19. Describe bone functional morphology from anatomical to microanatomical (histological/cytological) levels.
20. Describe steps in endochondral ossification and the types of bones that develop by such methods.

ANSWERS

1) hyaline 2) F; 3) 60-70%; 4) D; 5) C; 6) D; 7) T; 8) T; 9) F; 10) osteocytes and bone lining cells; 11) C; 12) B; 13) F; 14) A; 15) D; 16) B; 17) E; 18) F; 19) The answer should include among other pertinent topics: long vs. flat; epi-, meta-, diaphysis of long bones; cancellous vs. compact; woven vs. lamellar; Haversian system components, periosteum/endosteum; osteoid, mineral phase; 20) The answer should include such terms as: cartilage model; vascularization; primary/secondary ossification centers; epiphyseal plates and its zones; conversion of calcified cartilage into ossified bone.

Adipose Tissue

5.1 INTRODUCTION TO ADIPOSE TISSUES: DYNAMIC CONNECTIVE TISSUES WITH WHITE, BROWN, AND BEIGE ADIPOCYTES

From yeast to humans, eukaryotes store surplus energy within cytoplasmic **lipid droplets (LDs) (Figure 5.1a)**, which in vertebrates can be produced by specialized cells, called **adipocytes** (=lipocytes). Adipocytes combine with blood vessels, nerves, and interspersed **stromal** cells to form a connective tissue, called **adipose tissue (fat)**. Two major kinds of adipose tissue—**white adipose tissue (WAT)** and **brown adipose tissue (BAT)**—are distinguished, based on their morphologies, developmental modes, and functional properties (**Figure 5.1b, c**). As the most common fat in humans and other vertebrates, WAT is characterized by unilocular **white adipocytes**, each of which possesses a single large LD containing **triglyceride** types of lipids. Scattered patches of **BAT** have also evolved in humans and most other mammals but not in non-mammalian vertebrates. Such fat contains multilocular **brown adipocytes** with many small triglyceride-containing LDs, which unlike energy-storing LDs in WAT, primarily serve to **generate heat**. In addition, a third mammalian fat cell, called **beige adipocyte**, shares features of both white and brown adipocytes.

Traditionally, fat was often viewed as a relatively uniform and simple store of energy that can mechanically cushion and thermally insulate organs. However, more recent analyses have revealed that adipose tissues differ markedly across body regions while also playing far more complex roles than were previously appreciated. For example, healthy WAT depots are now known to release various beneficial compounds that broadly help

DOI: 10.1201/9780429353307-6

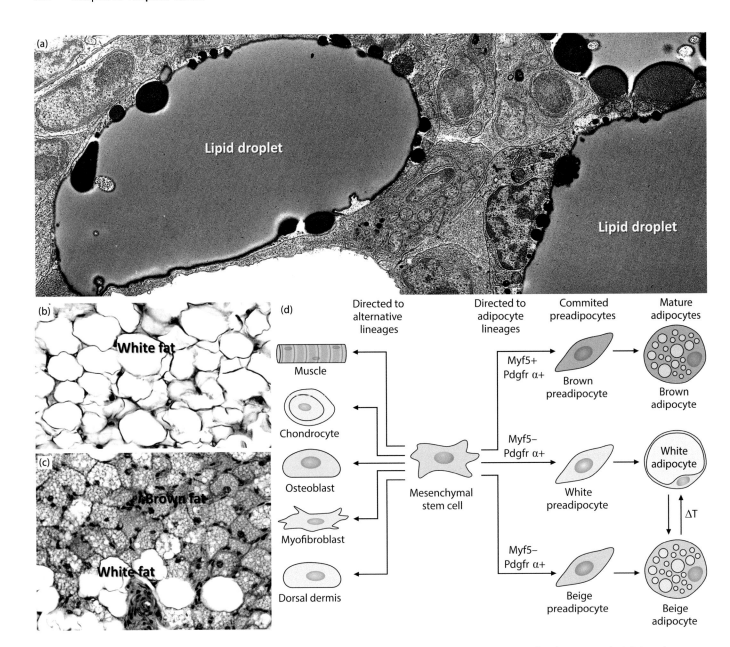

Figure 5.1 Introduction to adipose tissue (fat). (a) *As highly conserved energy stores of eukaryotes, lipid droplets are formed by numerous cells as exemplified in this TEM of a developing invertebrate embryo (3,300X). (**b, c**) Vertebrates have evolved two major kinds of specialized adipocytes for producing lipids—unilocular adipocytes of white fat in which a single large lipid droplet fills most of the adipocyte (b, 300X) and multilocular adipocytes of brown fat, each of which contain multiple smaller lipid droplets (c, 250X). Spaces in adipocytes represent artifacts arising from the dissolving of lipids by conventional preparatory techniques. (**d**) Mesenchymal stem cells can differentiate into various non-adipocyte cell types, or they can express platelet-derived growth factor-alpha receptor (Pdgfra) to become white, brown, or beige types of adipocytes. The particular adipocyte type depends on whether or not other key determinants, such as myogenic factor 5 (myf5) are co-expressed with Pdgfra. In addition, a population of beige adipocytes can arise via temperature-induced conversion of previously differentiated white adipocytes. ([d]: From Shapira, S and Seale, P (2018) Transcriptional control of brown and beige fat development and function. Obesity 27: doi:10.10002/oby.22334.)*

BOX 5.1 ADIPOSE TISSUE AND TRYPANOSOME PARASITES OF HUMANS

Humans are parasitized by two **trypanosome protozoans**: (1) *Trypanosoma cruzi*, the causative agent of **Chagas disease** which affects ~6–7 million people mainly in Latin and South America, and (2) *T. brucei*, whose infections occur in sub-Saharan Africa, resulting in an estimated 10,000 new cases of **human African trypanosomiasis** (**HAT**) each year. Although trypanosomes also invade other body regions, such as the central nervous system to trigger the lethal symptoms of "African sleeping sickness", human-infecting species tend to form particularly high parasite loads in **adipose tissues**. During the life cycle of *T. cruzi*, intracellular parasitic stages are typically found within adipocytes and preadipocytes of subcutaneous WAT, whereas *T. brucei* remains extracellular and most often resides within blood vessels and the interstitial stroma of visceral WAT (**Box 5.1 figure**).

Trypanosomes are transmitted into the bloodstream of humans by the biting of parasite-carrying insects. Both trypanosome species exit blood vessels and colonize their proper intra- or extracellular niches in fat, which in turn helps to sustain systemic infections that can range from mildly incapacitating to fatal. The establishment of trypanosomes in fat may be facilitated by at least **four non-mutually exclusive mechanisms**: (i) parasite delivery throughout the body, but **enhanced survival in adipose tissues**, due to beneficial secretions and/or protective immune responses in fat; (ii) **specific attraction of parasites to adipose tissues**; (iii) **increased likelihood of parasites escaping from blood vessels in fat** due to frequent vascular remodeling; and/or (iv) **immune responses that promote parasite entry into fat**. Accordingly, given that adipose tissues can form safe havens for trypanosomes, further analyses of these potential explanations may yield alternative therapies to help combat trypanosome infections of humans.

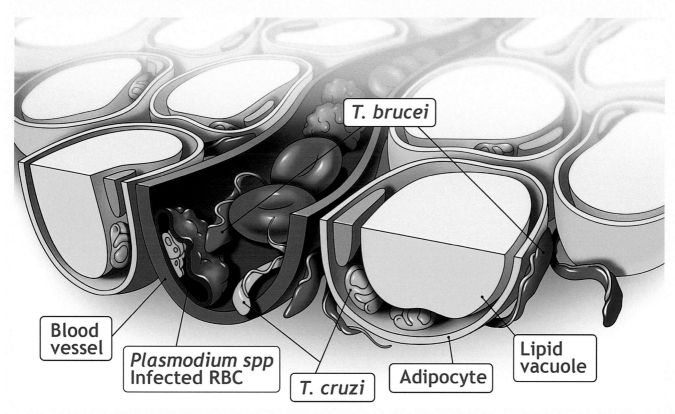

Box 5.1 *White adipose tissue can serve as a reservoir for Trypanosoma parasite infections in humans. (From Tanowitz, HB et al. (2017) Adipose tissue—a safe haven for parasites? Trends Parasit 33: 276–284 reproduced with publisher permission.)*

in regulating systemic homeostasis. Conversely, along with facilitating certain kinds of parasitic infections (Box 5.1 Adipose Tissue and Trypanosome Parasites of Humans), WAT can produce a wide array of deleterious compounds leading to **chronic inflammation** and associated maladies, such as hypertension (Chapter 7), cardiovascular disease (Chapter 7), and diabetes mellitus (Chapter 15), which collectively constitute **metabolic syndrome** disorder. In addition to such WAT-mediated effects, brown and beige fat depots not only generate adaptive temperature rises but also secrete broadly beneficial molecules that help combat obesity-related pathogenesis. Thus, contrary to previous views, adipose tissues are multifaceted and dynamic connective tissues that can serve both positive and negative functions throughout the body.

5.2   VARIED PATHWAYS OF ADIPOCYTE DEVELOPMENT

Substantial numbers of adipocytes begin to develop from mesodermal and neural crest sources by the 14th week of gestation, with BAT and WAT adipocytes being produced by different precursor cells (**Figure 5.1d**). Classical (=constitutive) **brown adipocytes** arise from bipotent stem cells that express a marker, called **myogenic factor 5 (Myf5)**. Upon further stimulation, these Myf5$^+$ stem cells become BAT **pre-adipocytes** that express **platelet-derived growth factor receptor alpha (PDGFRα)** before differentiating into functional brown adipocytes (**Figure 5.1d**). Alternatively, **white adipocytes** are generally derived from Myf5$^-$/PDGFRα$^+$ precursors, and **beige adipocytes** usually arise either from Myf5$^-$/PDGFRα$^+$ precursors in fetuses or from recruitable types of brown cells in adults. In addition, some adult white adipocytes can transdifferentiate into beige adipocytes (**Figure 5.1d**).

As fat develops within the fetus, major BAT aggregations localize to the neck, clavicular, interscapular, perirenal, and axillary regions, whereas WAT becomes much more widely distributed throughout the body and head. Directly following birth, WAT deposits continue to expand via the lipid-mediated hypertrophy of already-existing adipocytes, and following puberty, additional cellular proliferation also increases the quantities of both pre-adipocytes and functional adipocytes in WAT. These increased cell numbers tend to set by early adulthood the total quantity of adipocytes that will occur throughout pre-senescent life.

Unlike the general post-natal expansion of WAT, BAT was thought to be essentially lost after birth in humans and was generally viewed as an adaptation of small-bodied and hibernating mammals. However, recent studies have identified pockets of active BAT or BAT-like tissues that are maintained in humans throughout life (pg. 131).

5.3   WHITE ADIPOSE TISSUE (WAT): MAJOR FAT DEPOTS CONTAINING WHITE ADIPOCYTES, NEURONS, AND STROMAL VASCULAR CELLS (SVCS)

In various mammals, **white adipose tissue** (WAT) can be arranged into specialized deposits, such as the humps of camels that provide energy stores for prolonged desert journeys while also segregating insulation from the body core to avoid over-heating. Although similar specializations are lacking in humans, WAT nevertheless forms functionally distinct sub-compartments across different regions of the human body. Collectively, these WAT depots account for ~8–18% of the total weight in men vs. ~14–28% in women, with at least some of the extra fat in females having been selected to accumulate in the breasts, buttocks, and hips both as mating attractants and as energy reserves for child rearing.

Two major subdivisions of WAT are traditionally recognized: (i) **subcutaneous WAT** (=**sWAT**, or subcutaneous adipose tissue, SAT) occurring below the skin proper (**Figure 5.2a, b**), and (ii) **visceral WAT** (=**vWAT**, or visceral adipose tissue, VAT) surrounding internal organs (**Figure 5.2c, d**) such as the intestines, which are covered by a fat-rich layer, called the **omentum.** The sWAT vs. vWAT dichotomy is often predicative of disease outcome, since excessive vWAT particularly distributed as omental fat is generally associated with increased pathogenesis. Conversely, extra subcutaneous adiposity tends to be less deleterious than added vWAT fat.

In both WAT subtypes, **white adipocytes** measure ~25–250 μm wide, with each adipocyte possessing a single large LD that restricts the nucleus and residual cytoplasm to a narrow peripheral rim (**Figure 5.3a**). Because lipids are normally extracted during conventional histological processing, WAT typically appears as a reticulum of spaces where the LDs were once located. Alternatively, with lipid-stabilizing treatments such as **osmium tetroxide** fixation, LDs can be retained in whole mounts and sections (**Figure 5.3b, c**).

Scattered among WAT adipocytes are **stromal vascular cells** (SVCs) that predominantly comprise blood vessel lining cells plus surrounding fibroblasts, pre-adipocytes, white blood cells, and macrophages (**Figure 5.3c, d**). Along with **autonomic neurons** (Chapter 9) that innervate WAT, SVCs are also surrounded by a meshwork of collagen fibers that permeate and envelop WAT.

Visceral WAT depots in obese individuals undergo deleterious structural and functional reorganizations. For example, unhealthy increases in WAT reduce adipose vasculature while increasing hypoxia and fibrosis. In

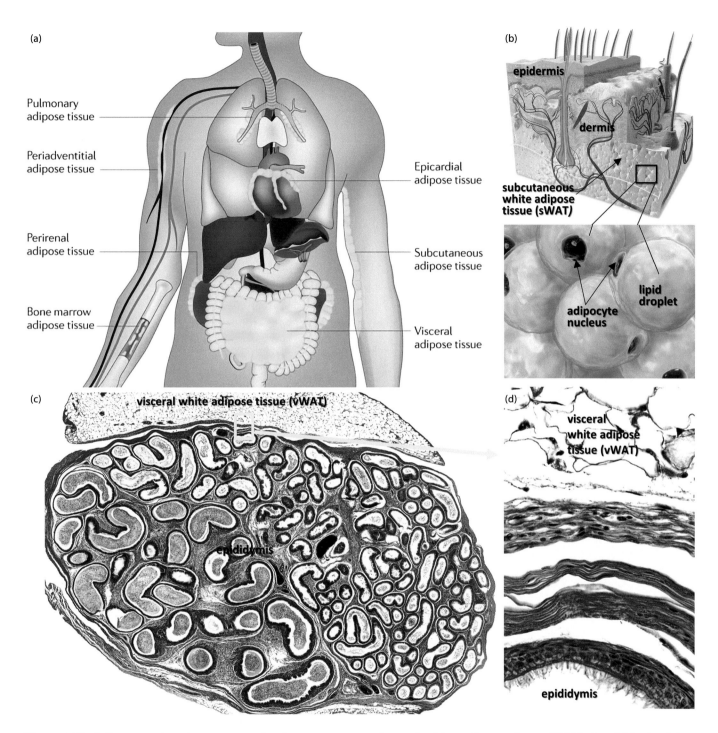

Figure 5.2 Adipocyte development and subcutaneous vs visceral white adipose tissue (WAT). (a–d) *The body's two main depots of WAT are: subcutaneous WAT (sWAT) occurring directly below the dermis of skin (a, b) vs. visceral WAT (vWAT) that is associated with various internal organs, such as shown here for the epididymis in the male reproductive system (c: 12X; d: 200X). ([a]: From Ouchi, N et al. (2011) Adipokines in inflammation and metabolic disease. Nat Rev Immunol 11: 85–97 reproduced with publisher permission; [b]: From Blausen.com staff (2014) "Medical gallery of Blausen Medical 2014" WikiJournal of Medicine 1 (2): 10. doi:10.15347/wjm/2014.010 reproduced under CC0 and CC SA creative commons licenses.)*

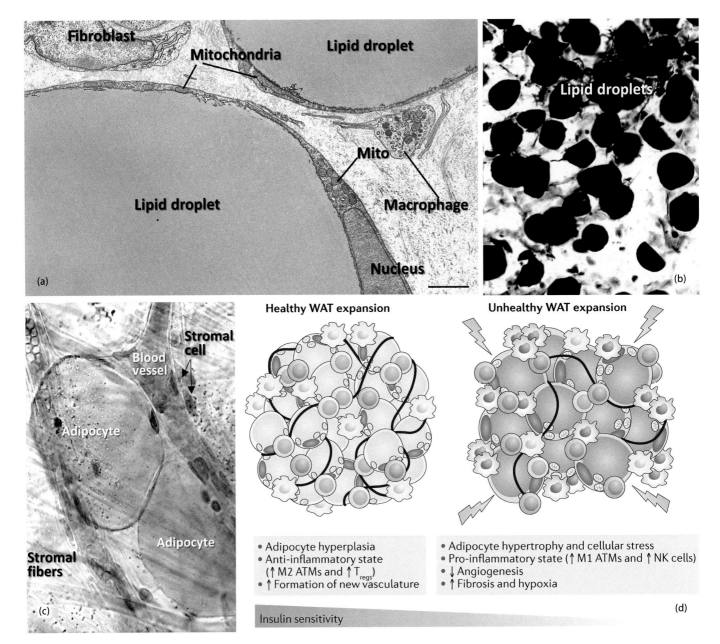

Figure 5.3 *Microanatomy of white adipocytes and their surrounding stroma.* (*a*) *As shown in this TEM of mouse adipose tissue, the large lipid droplet in each fully developed white adipocyte is surrounded by a thin peripheral rim of cytoplasm containing the nucleus and a few other organelles like mitochondria. The stromal ECM around adipocytes possesses blood vessels and various cells, such as fibroblasts and macrophages; scale bar = 2 μm. (**b**) Lipids in adipocytes are normally lost during conventional processing for light microscopy (**Figure 5.1b, c**) but can be maintained via treatment with osmium tetroxide, which is reduced to a black state in stabilized lipids (50X). (**c**) A lightly osmicated whole mount of adipose tissue shows two adipocytes and their surrounding stroma (260X). (**d**) During pathological reorganizations of white adipose tissue (WAT) in obese individuals: (1) fewer but larger (hypertrophied) adipocytes occur per unit volume; (2) macrophages convert to a more pro-inflammatory M1 phenotype (pink cells); (3) vascularization decreases, thereby promoting fibrosis and hypoxia; and (4) sensitivity to insulin is reduced. ([a]: From Cinti, S (2005) The adipose organ. Prostaglandins Leuk Essent Fatty Acids 73: 9–15 reproduced with publisher permission; [d]: From Kusminski, CM et al. (2016) Targeting adipose tissue in the treatment of obesity-associated diabetes. Nat Rev Drug Disc 15: 639–660 reproduced with publisher permission.)*

addition, obesity tends to change the number and types of **macrophages** in WAT (**Figure 5.3d**). Thus, in lean individuals, WAT macrophages have a predominantly **M2** phenotype that secretes anti-inflammatory molecules, whereas unhealthy WAT expansion can cause macrophages to transform into **M1** cells that release pro-inflammatory compounds (Chapter 7). Moreover, obesity is often associated with decreased **insulin sensitivity** in which circulating insulin fails to trigger its normal effects in target cells (Chapter 15), further exacerbating metabolic syndrome pathologies.

5.4   TRIGLYCERIDE METABOLISM IN WAT PLUS DISEASES SUCH AS ECTOPIC FAT DEPOSITION AND METABOLIC SYNDROME

Although normally functioning tissues, such as muscles, cartilage, and liver can also accumulate small amounts of lipids, WAT represents the major site where excess energy is stored in LDs. Such LDs are generated by the process of **lipogenesis** and later mobilized via **lipolysis** when energy is required. In **white adipocytes** undergoing lipogenesis, LDs accumulate **triglycerides** (**TGs**) (=triacylglycerols, TAGs), each of which is composed of three **fatty acids** (**FAs**) that are attached via ester bonds to a **glycerol** backbone (**Figure 5.4a–c**). For TG production during lipogenesis, blood-derived **glucose** can be converted by adipocytes into glycerol and **free fatty acid** (**FFA**) molecules via *de novo* synthetic pathways (**Figure 5.4c**). Alternatively, TGs can also circulate in the bloodstream in the form of large transport molecules, called **chylomicra** and **very low density lipoproteins** (**VLDLs**), which are made mainly by the intestines and liver, respectively. To yield smaller molecules that can cross adipocyte membranes, such transporters are first enzymatically digested by **lipoprotein lipase** (**LPL**), which is produced by adipocytes and attached to the lining of nearby blood vessels (**Figure 5.4a**). In this way, LPL-mediated hydrolysis can yield FFAs and glycerol that diffuse into adipocytes and supplement de-novo-generated pools formed from glucose. FFAs and glycerol are then esterified to form the TGs in adipocyte LDs (**Figure 5.4b, c**).

Conversely, in response to negative energy balances that arise between meals and after marked exertion, LDs in white adipocytes undergo lipolysis (**Figure 5.4d–f**). Lipolysis can be triggered by the hormone **glucagon** (Chapter 15) or by **norepinephrine** released from nearby neurons binding to **β3-adrenergic receptors** (**β3-ARs**) on adipocyte plasmalemmata (**Figure 5.4d, e**). Following β3-AR stimulation, a cAMP rise in adipocytes activates **protein kinase A** (**PKA**), which in turn phosphorylates **perlipin 1** (**Plin1**) proteins at the LD surface to activate **adipose triglyceride lipase** (**ATGL**), the enzyme that initiates lipolysis (**Figure 5.4e**). PKA-mediated phosphorylation also activates **hormone-sensitive lipase** (**HSL**), which helps fully hydrolyze TGs into glycerol and FFAs (**Figure 5.4e, f**). Glycerol can then be extruded for processing within the liver, whereas FFAs are used for ATP production by mitochondrial β-oxidative pathways either within the adipocyte itself or in other tissues following FFA export. In addition, lipolysis can also modulate glucose uptake and protein synthesis in adipocytes (**Figure 5.4f**).

The balance between lipogenesis and lipolysis in WAT is mainly controlled by pancreatic secretions of **insulin** (Chapter 15). Thus, high insulin levels circulating in the bloodstream after a meal can bind insulin receptors on adipocyte plasmalemmata to increase glucose uptake and upregulate lipogenesis while also downregulating lipolysis (**Figure 5.4c**). Conversely, during fasting, insulin levels drop, as glucagon (Chapter 15) and adrenergic stimulation rise, thereby promoting cAMP-induced lipolysis. Collectively, insulin-mediated processes help keep circulating FFAs below pathologically elevated levels (=hypertriglyceridemia), whereas obesity with its expanded adipose depots and increased tendency for lipolysis tends to elevate blood-borne lipid levels. This in turn can promote **ectopic fat deposition** in non-adipose tissues, particularly in the case of fatty liver disease (Chapter 13).

In addition to hypertriglyceridemia and ectopic fat deposition, obesity is typically linked to metabolic syndrome. In particular, obese people with vWAT containing large (**hypertrophic**) adipocytes are prone to developing metabolic syndrome pathologies (**Case 5.1 William Taft's Adipose Tissue**), whereas increased numbers of smaller (**hyperplastic**) adipocytes are characteristic of expanded vWAT depots in **metabolically healthy obese** (**MHO**) patients. Such adipose configurations help MHO individuals to maintain relatively normal blood lipid levels and a lower incidence of metabolic syndrome.

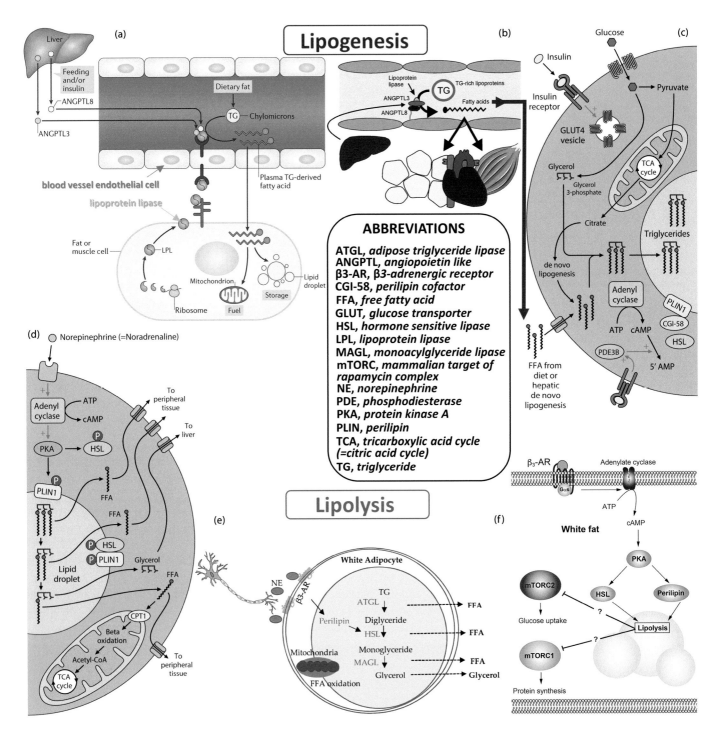

Figure 5.4 Pathways of lipogenesis and lipolysis in white adipocytes. (**a**) Lipoprotein lipase (LPL) made in adipocytes is transferred to vascular lining cells where its activity is regulated by angiopoietin like proteins (ANGPTLs) made in the liver. (**b**) Following meals, LPL serves to break down chylomicra and VLDL transporter complexes that carry triglycerides (TGs) in the bloodstream, thereby releasing glycerol and free fatty acids (FFAs) from TG catabolism to be taken up by adipocytes. (**c**) Glycerol and FFA are re-combined into TGs and are supplemented by de novo TGs synthesis from incorporated glucose to form large lipid droplets in adipocytes. (**d–f**) Between meals as energy needs rise, norepinephrine (NE), epinephrine, and glucagon can stimulate cAMP elevations in white adipocytes. This activates adipocyte protein kinase A (PKA), which stimulates hormone sensitive lipase (HSL) and perilipin (Plin) to promote both lipolysis of lipid droplets and mitochondrial ATP production from FFAs via β-oxidation pathways. ([a]: From Kersten, S (2013) Angiopoietin-like 3 in lipoprotein metabolism. Nat Rev Endocrinol 13: 731–739 reproduced with publisher permission; [b]: From Kersten, S (2014) Physiological regulation of lipoprotein lipase. Biochim Biophys Acta 1841 919–33

Figure 5.4 (Continued) reproduced with publisher permission; [c, d]: From Rutkowski, JM et al. (2015) The cell biology of fat expansion. J Cell Biol 208: 501–512; [e]: From Zhu, Q et al. (2019) Neuroendocrine regulation of energy metabolism involving different types of adipose tissues. Int J Mol Sci 20: 2707. doi:10.3390/ijms20112707 reproduced under a CC BY 4.0 creative commons license; [f]: From Chia, LY et al. (2019) Adrenoceptor regulation of the mechanistic target of rapamycin in muscle and adipose tissue. Br J Pharmacol 176:2433–2448 reproduced with publisher permission.)

CASE 5.1 WILLIAM TAFT'S ADIPOSE TISSUE

27th President of the United States
(1857–1930)

"How's the horse?"

(A friend's reply to a telegram that the obese W.T. sent proclaiming he was feeling in great shape after riding a horse 25 miles to the top of a 3500-ft-tall mountain)

Elected U.S. President in 1909, William Taft was an athletic but obese man who tended to eat too much when stressed, thereby causing his weight at one point to reach 340 pounds. Taft's corpulence was often the subject of jokes, particularly when news broke that a custom-made bathtub was installed in the White House, supposedly because Taft had gotten stuck in a conventional tub. Following his four years in the White House, Taft became the only ex-president to be appointed to

the Supreme Court, where he served as Chief Justice. As might be expected, Taft suffered from various weight-related maladies, including cardiovascular disease, gout, and a tendency to fall asleep during the day, presumably owing to obesity-induced sleep apnea. Eventually, such disorders led to the corpulent ex-president's death due to arteriosclerosis, a disease which no doubt was exacerbated by his abundant supplies of adipose tissues.

5.5   WAT SUB-COMPARTMENTS AND THEIR VARIOUS SECRETIONS

In the 1990s, it was demonstrated that WAT secretes into the bloodstream a protein, called **leptin**, which among other roles targets hypothalamic neurons that reduce food intake. Since then, adipose tissues have been shown to produce additional bioactive compounds that collectively constitute a subset of cell-signaling cytokines, termed **adipokines**. Of the hundreds of human adipokines that have been identified, most can be subdivided into proteins that promote normal homeostasis vs. those that are associated with pathogenesis. Alternatively, **anti-** vs. **pro-inflammatory adipokines** are sometimes recognized, based on their relationships to chronic inflammation.

As summarized in Table 5.1, the cellular sources of WAT adipokines vary from **adipocytes** themselves to **stromal cells**, such as **macrophages**. In addition to blood-borne adipokines, adipocytes and stromal cells of WAT also secrete bioactive compounds like **growth factors** to affect nearby tissues in healthy and diseased states. Selected examples of localized WAT secretory products and blood-borne adipokines are summarized below.

Dermis

Dermal WAT (**dWAT**) surrounding **hair follicles** (**HFs**) of the skin undergoes cyclic changes that are coupled to phases in hair growth (Chapter 10). Dermal fat cycling is largely controlled by HF-derived lipogenic activators, such as bone morphogenetic protein (BMP). In addition, the dWAT adipokine **adiponectin** promotes **wound healing** in skin, and in response to infections, dWAT depots secrete **antimicrobial peptides** to help ward off invading bacteria.

Bone marrow

Occurring in the cavities of bones, two types of **bone marrow** (Chapter 6) contain bone-marrow-specific adipose tissue (**bmAT**). Fetal bmAT secretes a variety of compounds for either up- or downregulating the rates of blood cell formation in neighboring cells (Chapter 6).

TABLE 5.1 SOURCES AND FUNCTIONS OF KEY ADIPOKINES

Adipokines	Primary source(s)	Function
Leptin	Adipocytes	Control of appetite through the central nervous system
Resistin	Peripheral blood mononuclear cells (human) adipocytes (rodent)	Promotes insulin resistance and inflammation
RBP4	Liver, adipocytes, macrophages	Implicated in systemic insulin resistance
Lipocalin 2	Adipocytes, macrophages	Promotes insulin resistance
ANGPTL2	Adipocytes, other cells	Local and vascular inflammation
TNF	Stromal vascular fraction cells, adipocytes	Inflammation, antagonism of insulin signaling
IL-6	Adipocytes, stromal vascular fraction cells, liver, muscle	Varies with source and target tissue
IL-18	Stromal vascular fraction cells	Broad-spectrum inflammation
CCL2	Adipocytes, stromal fraction cells	Monocyte recruitment
CXCL5	Stromal vascular fraction cells (macrophages)	Antagonism of insulin signaling
NAMPT (=visfatin)	Adipocytes, macrophages, other cells	Monocyte chemotactic activity
Adiponectin	Adipocytes	Insulin sensitizer, anti-inflammatory
SFRP5	Adipocytes	Suppression of pro-inflammatory signaling

Source: From Ouchi, N et al. (2011) Adipokines in inflammation and metabolic disease. Nat Rev Immunol 11:85–97 reproduced with publisher permission.
Abbreviations: ANGPTL2, angiopoietin-like protein 2; CCL2, CC-chemokine ligand 2; CXCL5, CXC-chemokine ligand 5; IL, interleukin; NAMPT, nicotinamide phosphoribosyltransferase; RBP4, retinol-binding protein 4; SFRP5, secreted frizzled-related protein 5.

Mammary glands

Mammary gland WAT (mgWAT) forms around epithelial tissues that secrete milk during lactation (Chapter 18). Such mgWAT reaches peak lipid content during pregnancy before markedly diminishing during post-natal lactation as lipids are transferred to mammary gland secretory cells. mgWAT normally secretes various adipokines and other molecules to maintain normal glandular functioning. Conversely, a subset of mgWAT adipocytes, termed cancer-associated adipocytes (CAA), can nourish tumor cells, thereby facilitating breast cancer metastases (Chapter 18).

Gonads

Fat deposits affect **gonadal** processes, such as the onset of the first menstrual cycle (=menarche) (Chapter 18), as corpulent girls tend to have earlier menarches. Conversely, lean individuals, such as athletes and those with severely reduced caloric intake due to malnutrition or illnesses, such as anorexia nervosa have later-onset or fully blocked menarche. As a part of this regulatory process, increased secretion of **leptin** by adipocytes in corpulent girls facilitates the production of gonadotropin-releasing hormones by the hypothalamus (Chapters 15 and 18), which in turn accelerates menarche onset.

5.6 BROWN ADIPOSE TISSUE (BAT) AND ITS UCP1-MEDIATED MODE OF HEAT PRODUCTION

In response to decreased body temperatures, mammals can raise their basal metabolic rate (BMR) and undertake counteractive measures such as shivering, both of which are energetically expensive processes requiring recurrent food intake to continue long term. Alternatively, heat production by **BAT** is also used by most mammals to regain thermal neutrality, particularly in hibernating species that do not feed. As the chief source of non-shivering thermogenesis, BAT is also present in newborn babies, which are unable to shiver.

Classical brown adipocytes can be distinguished from white adipocytes by their: (i) smaller size (typically, only 15–60 μm wide); (ii) more rounded, centrally located nucleus; and (iii) multiple relatively small LDs surrounding the nucleus (**Figure 5.5a–c**). In addition, brown cells have abundant cytochrome-rich mitochondria, which along with enhanced vascularization of the surrounding stroma confer an overall brownish hue to BAT. In many cases, the outer membranes of these BAT mitochondria fuse with ER membranes, thereby facilitating Ca^{2+} delivery, which in turn can specifically activate BAT, but not WAT, adipocytes (**Figure 5.5d**).

In addition to such morphological features, BAT is characterized by its unique developmental sequence (**Figure 5.1d**) as well as by the types of proteins it produces after differentiating. In particular, the inner membrane of BAT mitochondria contains high amounts of **uncoupling proteins** (**UCPs**). UCP1 is so named because it is the major protein for uncoupling ATP production from the proton pool that the mitochondrial electron transport chain (ETC) normally pumps into intermembrane spaces (**Figure 5.5e**). Prior to such uncoupling, UCP1 normally remains inactive owing to its inhibition by purine nucleotides in mitochondria. However, in response to lipolysis-mediated signals like elevated FFAs, UCP1 becomes activated and provides an alternative channel for the return of intermembrane protons into the mitochondrial matrix (**Figure 5.5e**). This proton leak pathway reduces the number of protons that can drive ATP-synthase-mediated synthesis of ATPs, thereby allowing energy present in the proton gradient formed by ETC pumping to be dissipated as heat instead of being stored in ATP molecules (**Figure 5.5e, f**).

5.7 BEIGE FAT, WAT BROWNING, AND USING BROWN AND BEIGE ADIPOSE TISSUES TO TREAT OBESITY AND RELATED DISEASES

In addition to BAT clusters that occur in newborns, scattered brown-like adipocytes also accumulate among white adipocytes in some adult WAT depots (**Figure 5.5g–i**). Histological evidence for such **beige cells**, which are also called **brite** (<u>br</u>own <u>i</u>n whi<u>te</u>) **adipocytes**, dates back several decades. However, the presence of beige fat in adult humans has only gained widespread acceptance over the last decade when whole-body imaging studies helped identify active beige cells generated by various stimuli, including cold, vigorous exercise, and excessive adrenergic input. Subsequent analyses showed that like classical BAT, beige cells can express UCP1 and generate heat, while also helping to reduce obesity and associated pathogenesis.

Even though beige cells broadly resemble classical brown adipocytes, the two cell types differ in key ways. For example, the size of beige adipocytes is intermediate between those of brown and white adipocytes (Box 5.2 Activating Thermogenic Fat to Treat Obesity and Its Associated Pathologies). In addition, brown adipocytes are derived from Myf5$^+$ precursors, whereas beige cells arise from Myf5$^-$ precursors or from the transdifferentiation of white adipocytes. Along with other modes of increasing beige fat within WAT, such conversion from white adipocytes contributes to an overall process, referred to as the **browning of WAT**. Moreover, unlike brown cells, a subset of beige cells lacks the ability to produce UCP1 and yet can still generate heat by incompletely defined mechanisms.

Fat burning by brown and beige fat not only decreases adiposity but also provides benefits beyond simple weight loss. Some of the beneficial effects of adipokines produced by BAT and browned WAT include reducing insulin resistance (Chapter 15), increasing tissue vascularization, and even upregulating restorative sleep patterns. Accordingly, activations of BAT and browned WAT are currently being investigated as potential treatments for obesity and its related diseases (Box 5.2).

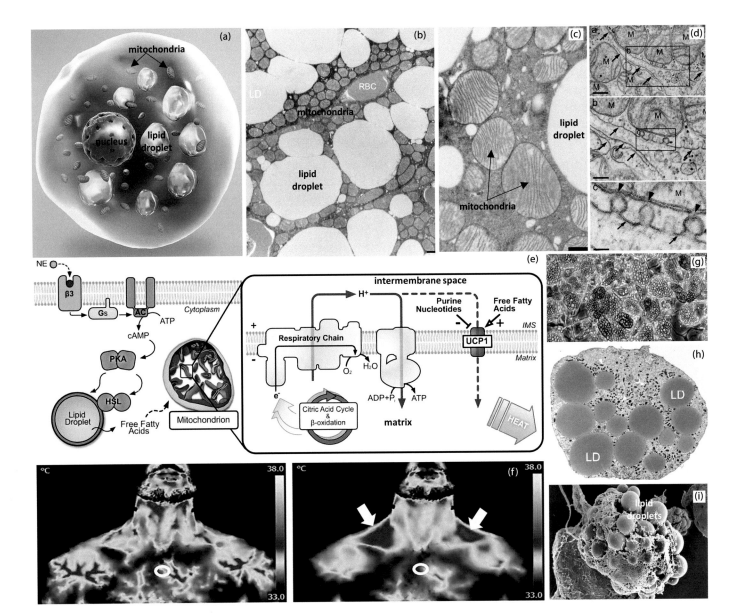

Figure 5.5 Thermogenic brown and beige adipocytes. (*a–c*) *Adipocytes in brown adipose tissue (BAT) are characterized by numerous small lipid droplets surrounded by specialized thermogenic mitochondria that serve to generate heat (b: 4,000X; c: 17,000X). (**d**) Fusion of BAT mitochondria with nearby ER membranes facilitates influx of calcium that helps to activate thermogenic properties of brown adipocytes. (**e**) Following stimulation by norepinephrine (NE), uncoupling proteins (UPCs) in BAT inner mitochondrial membranes provide leak channels for protons that had been concentrated in the intermembrane space by the electron transport chain. By passing through UCPs rather than through ATP synthase channels, protons fail to drive ATP synthesis and instead radiate heat. (**f**) The typical trigger for BAT thermogenesis is cold, but in this live thermal imaging of the upper thorax, BAT heat production increases (white arrows pointing to red sites of higher temperatures) following the ingestion of caffeine in coffee. (**g–i**) Beige adipocytes, shown here by phase contrast (g: 270X), TEM (h: 1,100X), and SEM (i: 640X), have structural and functional properties that are intermediate between classic white and brown adipocytes. ([a]: From https://www.scientificanimations.com/ reproduced by permission of CC-BY-SA 4.0 creative commons license; [b, c]: From Gonzalez-Hurtado, E et al. (2018) Fatty acid oxidation is required for active and quiescent brown adipose tissue maintenance and thermogenic programing. Mol Metab 7: 45–56 reproduced with publisher permission; [d]: From De Meis, L et al. (2010) Fusion of the endoplasmic reticulum and mitochondrial outer membrane in rat brown adipose tissue: activation of thermogenesis by Ca²⁺. PLoS ONE 5(3): e9439 reproduced under a CC BY 4.0 creative commons license; [e]: From Crichton, PG et al. (2017) The molecular features of uncoupling protein 1 support a conventional mitochondrial carrier-like mechanism. Biochemie 134: 35–50 reproduced with publisher permission; [f]: From Velickovic, K et al. (2019) Caffeine exposure induces browning features in adipose*

Figure 5.5 (Continued) *tissue in vitro and in vivo. Sci Rep 9:9104 | https://doi.org/10.1038/s41598-019-45540-1 reproduced under a CC BY 4.0 creative commons license; [g–i]: From Singh, AM et al. (2020) Human beige adipocytes for drug discovery and cell therapy in metabolic diseases. Nat Comm 11: https://doi.org/10.1038/s41467-020-16340-3 reproduced under a CC BY 4.0 creative commons license.)*

5.8   REMODELING OF ADIPOSE TISSUES DURING NORMAL AGING

During aging, organs like the thymus, skeletal muscles, and bones tend to lose substantial mass. Conversely, fat often increases in volume while also becoming markedly redistributed, with subcutaneous WAT often being re-allocated to visceral WAT locations, particularly in the abdomen. In most adults, fat depots peak by about **age 70**, and until that time, increased adiposity maintains or increases overall weight as other tissues undergo reductions in mass. Conversely, toward the end of life, both body weight and adiposity steadily decline, with the decrease in adipose tissues becoming especially noticeable in such regions as the eye sockets, cheeks, arms, and calves.

5.9   SOME ADIPOSE TISSUE DISORDERS: DYSLIPIDEMIA AND LIPODYSTROPHIES

Dyslipidemia is a common disorder characterized by abnormal concentrations of lipids circulating in the blood. Approximately one-third of adults in industrialized countries have some form of dyslipidemia, with the most common suite of abnormalities comprising elevated triglycerides (**TGs**), high amounts of low-density lipoproteins (**LDLs**), and/or low amounts of high-density lipoproteins (**HDLs**). Elevated TGs can contribute to increased amounts of circulating LDLs termed **bad cholesterols**, owing to their ability to trigger vascular plaques that promote heart attacks and strokes (Chapter 7). Conversely, HDLs tend to enhance the excretion of excess cholesterol via liver-produced bile (Chapter 13) and hence are referred to as **good cholesterols**.

Unlike normal aging-related fat loss, diseases that pathologically reduce adipose tissues either throughout the body or in discrete regions are referred to as generalized- or partial-**lipodystrophies**, respectively. Some generalized forms of these rare diseases can be acquired through auto-immune reactions. Alternatively, mutations in such genes as **Plin1** can interfere with proper adipocyte functioning and contribute to several partial forms of lipidodystrophy, some of which have benefited from treatments with recombinant **leptin**.

BOX 5.2 ACTIVATING THERMOGENIC FAT TO TREAT OBESITY AND ITS ASSOCIATED PATHOLOGIES

Mitochondria in brown and beige adipocytes not only dissipate heat via the activation of **uncoupling protein 1** (**UCP1**) but also burn fat to promote leanness. Conversely, reduced UCP1 activation in obese individuals contributes to excess WAT deposition and weight gain (**Box 5.2 figure**). Thus, to combat obesity, attempts are being made to upregulate UCP1 activity by enhancing already-existing UCP1 pools and/or by generating new UCP1 in added stores of brown and beige fat.

Most obesity treatments involving brown or beige fat have taken two major approaches for upregulating UCP1-mediated thermogenesis—**sympathetic neural stimulation** (**SNS**) and **cold** (**Box 5.2 figure**). Generalized SNS can cause serious side-effects, such as elevated blood pressures and heart rates, which precludes its use as a safe activator of brown and beige fat. However, the more specific **β3-AR agonist** mirabegron has been shown to raise BAT thermogenesis without marked effects on cardiovascular functioning in short-term trials. Similarly, cold is a well-established trigger of BAT activity in mice, and "cold vests" are being sold for fat burning in adult humans. However, such treatments have failed to yield marked WAT browning in humans, suggesting that enhanced cold challenges may be needed. For example, dietary supplements, such as the **capasicins** of peppers and **catechins** of teas have shown some promise in helping to activate additional thermogenic fat depots in lean individuals. Moreover, even normal doses of caffeine contained in coffee can produce heat within thermogenic fat (see **Figure 5.5f**). Alternatively, to induce more thermogenic fat in obese individuals, it may be possible to **downregulate** WAT-specific genes while also **upregulating genes** that promote either the determination of brown and beige pre-adipocytes or the differentiation of already determined pre-adipocytes along brown and beige pathways. In addition, beige adipocytes lacking UCP1 expression can generate heat via unidentified mechanisms that in turn could provide another way of reducing obesity.

Along with its anti-obesity effects, thermogenic fat also secretes beneficial molecules, such as **adiponectin**, **interleukin-6 (IL-6)**, and **fibroblast growth factor 21 (FGF21)** to improve heart and liver functioning while also enhancing glucose metabolism. In particular, secretions derived from brown and beige fat help ensure proper insulin production by the pancreas while further guarding against hyperglycemia by enhancing glucose uptake in peripheral tissues.

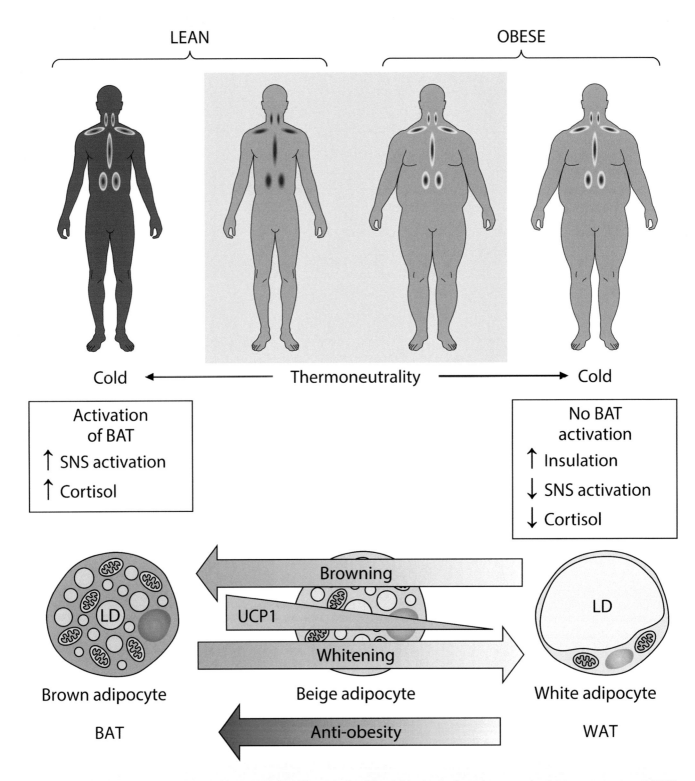

Box 5.2 *Brown adipose tissue (BAT) is more readily activated by cold, cortisol, and sympathetic nervous system (SNS) stimulation in lean vs. obese individuals, thereby helping to reduce adiposity. (Figure adapted based on illustrations in: (i) Symonds, ME et al. (2018) Recent advances in our understanding of brown and beige adipose tissue: the good fat that keeps you healthy. F1000 Res: doi: 10.12688/f1000research.14585.1. www.nature.com/naturecommunications and (ii) Ro, S-H et al. (2019) Autophagy in adipocyte browning: emerging drug target for intervention in obesity. Front Phys 10: doi:10.3389/fphys.00022, with both articles being reproduced under CC BY 4.0 creative commons licenses.)*

5.10   SUMMARY—ADIPOSE TISSUE

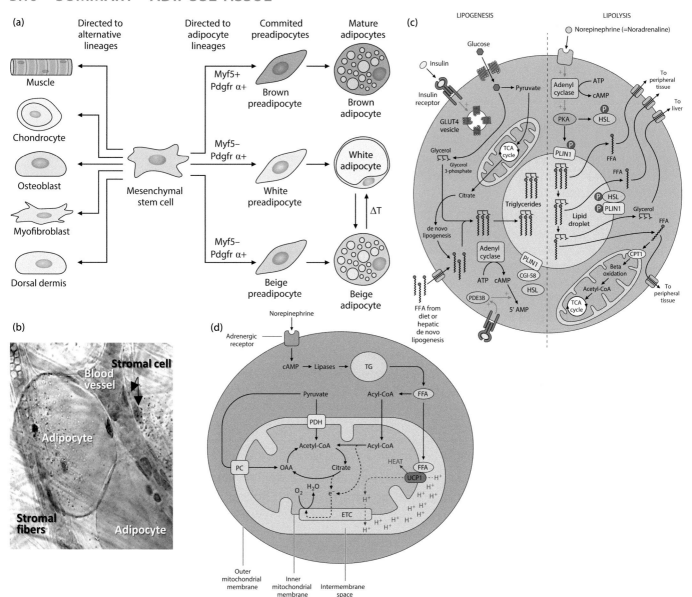

Pictorial Summary Figure Caption: (***a***): *see* figure 5.1d; (***b***): *see* figure 5.3c; (*c*): *see* figure 5.4c, d; (d): *Summary of UCP-1-mediated heat production in mitochondria of brown/beige adipocytes.*

Two main **adipose tissues** (**fats**) are (i) **WAT**, the major lipid **energy store** of the body, and (ii) scattered **BAT**, which **generates heat**. Along with regulating energy and heat levels, fat can play active roles in maintaining health or triggering pathogenesis. Three **adipocyte** types—**white**, **brown**, and **beige**—differ in morphology and body positioning. **White fat cells** occur in two main WAT depots—**subcutaneous** (sWAT) vs. **visceral** (vWAT)—with added sWAT generally posing less health risks than excessive vWAT. Reaching diameters of up to 250 μm wide, white adipocytes have a single large **lipid droplet (LD)** encased by a thin peripheral rim of cytoplasm. **Brown adipocytes**, which occur in scattered BAT depots mainly in the neck and upper thorax, are the smallest fat cells and possess a central nucleus surrounded by numerous mitochondria and LDs. **Beige adipocytes**, which are intermediate to white and brown cells both in overall size and in their numbers of LDs and mitochondria, can replace white adipocytes during a process, called the **browning of WAT**. In the fetus, the three adipocyte types arise via **alternative pathways** that yield different **pre-adipocyte** precursors (**Figure a**). Along with adipocytes, pre-adipocytes, stem cells, and **neurons**, adult fat contains various **stromal vascular cells** (**SVCs**) (**Figure b**). Such SVCs include blood vessel cells, fibroblasts, leukocytes, and macrophages some of which can produce cell signaling proteins, called **adipokines**.

Adipocytes and SVCs in normal WAT release into blood beneficial adipokines that can aid mammary gland function, wound healing, and adaptations to severe energy deficits. Conversely, deleterious adipokines promote diseases like **ectopic fat deposition** (lipid accumulation in non-adipose tissues) and **metabolic syndrome** (hypertension, cardiovascular disease, and diabetes). Harmful adipokines from hypertrophic and hypoxic vWAT in obese individuals have **pro-inflammatory** effects that can promote metabolic syndrome by increasing **insulin resistance** (i.e., reduced target cell reactions to insulin). Conversely, **brown** and **beige cells** often produce beneficial adipokines and thus are potential targets for obesity-related diseases.

During post-meal **lipogenesis**, adipocyte LDs store surplus energy in **triglycerides (TGs)**, each comprising **three fatty acids (FAs)** bound to **glycerol (Figure c)**. In addition to blood-borne **glucose** that is converted into glycerol and **free fatty acids (FFAs)** via *de novo* pathways in adipocytes, lipoprotein carriers of TGs in blood are hydrolyzed into FFAs and glycerol by **lipoprotein lipase (LPL)**, which is secreted by adipocytes and attached to blood vessel linings.

Between meals, **lipolysis** hydrolyzes TGs in **adipocyte LDs** to release stored energy (**Figure c**). Such lipolysis often begins by neurons secreting **noradrenaline** to cause cAMP-mediated **protein kinase A (PKA)** activation in adipocytes. PKA phosphorylates: (i) **perlipin 1 (Plin1)** at the LD surface to activate **adipocyte triglyceride lipase (ATGL)**, the initial catalyst of lipolysis, and (ii) **hormone-sensitive lipase (HSL)**, which completes lipolysis to yield FFAs and glycerol. FFAs from hydrolyzed TGs are sent to adipocyte mitochondria or exported to other tissues to generate ATPs via mitochondrial β-oxidation. The balance between lipogenesis and lipolysis is largely controlled by circulating **insulin** levels. High insulin promotes lipogenesis, while also blocking lipolysis. Insulin resistance of metabolic syndrome reduces insulin's effectiveness and allows blood TGs to exceed normal levels (=**hypertriglyceridemia**).

Unlike in WAT adipocyte mitochondria that produce large amounts of ATPs, **brown** and **beige** mitochondria use **uncoupling protein 1 (UCP1)** to disconnect ATP production from proton gradients generated by the mitochondrial **electron transport chain (ETC)**, thereby allowing energy dissipation as **heat (Figure d)**. **Cold** is the major activator of brown and beige adipocytes. Accordingly, thermogenic BAT helps newborns survive before they gain the ability to shiver. Adult humans retain some BAT while also **browning** WAT with beige cells. Cold causes noradrenaline release and lipolysis in brown and beige cells to yield FFAs that activate UCP1 on adipocyte inner mitochondrial membranes.

Some adipose tissue disorders: **dyslipidemia**: abnormal blood lipid levels: e.g., high TGs, low HDLs, high LDLs; **lipodystrophies**: general or partial fat loss over and above normal aging-related reductions in fat depots.

SELF-STUDY QUESTIONS

1. What is the most common type of fat in humans?
2. T/F White fat cells produce large amounts of UCP1.
3. Name a lipolytic hormone that is directly activated by PKA.
4. Which is NOT a feature of beige adipocytes?
    A. Intermediate in size between white and brown fat cells
    B. More mitochondria than in white fat cells
    C. UCP1 expression
    D. ALL of the above
5. In triglycerides (TGs), what are bound to glycerol?
    A. Glucose
    B. Fatty acids
    C. HDLs
    D. LDLs
    E. Uncoupling protein 1
6. Where is Perilipin 1 (Plin1) found in high amounts?
    A. On endothelia of blood vessels in fat
    B. In adipose tissue stroma
    C. At the surface of adipocyte lipid droplets
    D. Within adipocyte mitochondria
    E. Circulating in the blood

7. T/F Hypoxic vWAT with large adipocytes tends to reduce chances of developing diseases like metabolic syndrome.
8. T/F Excess fat in subcutaneous WAT is generally a greater health risk than added adiposity in visceral WAT.
9. T/F Non-mammalian vertebrates form functional BAT.
10. Which fixative is used to stabilize lipids?
11. Which cells synthesize and secrete large amounts of lipoprotein lipase in adipose tissues?
    A. Endothelial cells
    B. Pericytes
    C. Adipocytes
    D. ALL of the above
    E. NONE of the above
12. Which of the following would trigger lipolysis?
    A. Noradrenaline (norepinephrine)
    B. FFAs
    C. PKA activation in adipocytes
    D. cAMP rise in adipocytes
    E. ALL of the above
13. T/F. Adiponectin is a generally beneficial type of adipokine.

    "EXTRA CREDIT" For each term on the left, provide the BEST MATCH on the right (answers A-F can be used more than once)

14. Breaks down VLDLs___	A. UCP1
15. Proton leak channel____	B. PKA
16. Directly activated by cAMP___	C. ETC
17. Directly activates UCP1___	D. FFAs
18. Pumps protons into inter-membrane space___	E. ATGL
	F. LPL

19. Compare and contrast WAT vs. BAT.
20. Describe lipogenesis and lipolysis in white adipocytes.

ANSWERS

1) white adipose tissue (WAT); 2) F; 3) hormone sensitive lipase (HSL); 4) D; 5) B; 6) C; 7) F; 8) F; 9) F; 10) osmium tetroxide; 11) C; 12) E; 13) T; 14) F; 15) A; 16) B; 17) D; 18) C; 19) The answer should include among other pertinent topics: which kinds of vertebrates that generate such fat types; sizes/nuclear positioning/LD numbers in adipocytes; typical locations in adult body; UCP1 expression; major functions in health and disease, ; 20) The answer should include such terms as: insulin, glucose, de novo lipogenesis; transport lipoproteins; lipoprotein lipase; exogenous FFAs/glycerol, TG; LD; noradrenaline, cAMP, PKA, ATGL, HSL, FFAs, mitochondrial B-oxidation

TISSUE AND ORGAN HISTOLOGY—BLOOD, CIRCULATORY AND LYMPHATIC SYSTEMS, MUSCLES, NERVOUS SYSTEM, AND SKIN

PART TWO

Blood

6.1 INTRODUCTION TO BLOOD

Blood is a specialized connective tissue comprising individual cells suspended in a **liquid ECM** that serves to maintain systemic homeostasis in various ways. For example, blood supplies cells with **oxygen** and **nutrients** while also conveying **carbon dioxide** and **soluble wastes** to the lungs, kidneys, and liver for removal from the body. In addition, blood distributes hormones, facilitates thermoregulation, helps repair tissues, and contributes to defenses against pathogens via its widespread delivery of immune cells and molecules (Chapter 7).

Following centrifugation, blood is partitioned into three layers: (1) an uppermost liquid fraction, which is called **plasma** before a clotting reaction occurs vs. **serum** after clotting reagents are removed from plasma; (2) a large mass of red blood cells (**erythrocytes**) in the bottom half of the tube; and (3) a thin intervening **buffy coat** comprising thrombocytes (**platelets**) and white blood cells (**leukocytes**) (**Figure 6.1a–c**). Five types of leukocytes are present in blood: **neutrophils, eosinophils, basophils, lymphocytes**, and **monocytes**. Of these, only neutrophils, lymphocytes, and monocytes occur at high enough concentrations to be commonly observed scattered among numerous erythrocytes and platelets in low-magnification views of normal blood smears (**Figure 6.1b**) (Table 6.1).

DOI: 10.1201/9780429353307-8

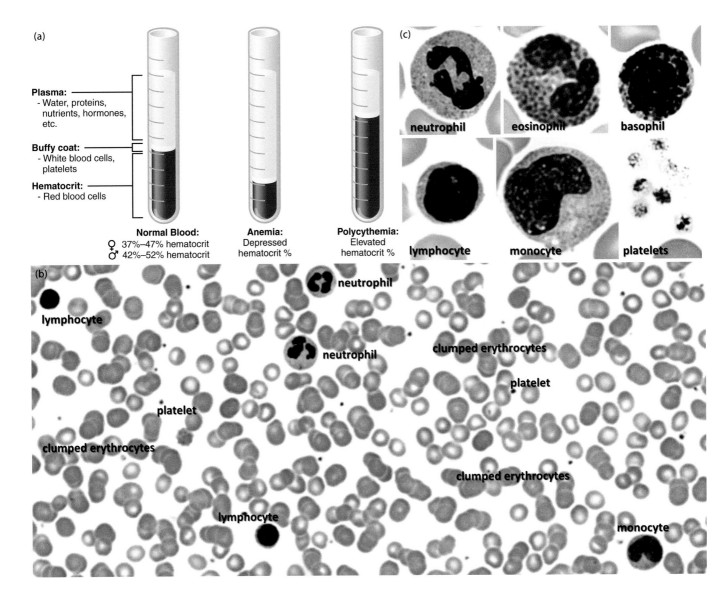

Figure 6.1 Introduction to blood. *(a) Centrifuged blood contains a fluid plasma fraction toward the top of the tube, packed red blood cells at the bottom, and a thin buffy coat comprising white blood cells and platelets in between those two fractions. The percent volume of red blood cells in centrifuged blood (=hematocrit) is typically about 40–50% but is lower than normal in anemias and elevated in polycythemias. (b, c) A low magnification (800X) view of a blood smear shows a few commonly encountered white blood cells (=leukocytes) and platelets scattered among numerous red blood cells (erythrocytes), whereas figure c depicts higher magnification (2,000X) images of the five leukocytes plus some platelets. ([a]: From OpenStax College (2013) Anatomy and Physiology. OpenStax. http://cnx.org/content/col11496/latest reproduced under a CC BY 4.0 creative commons license.)*

Erythrocytes, leukocytes, and platelets of adults are produced mainly within specialized tissue of bone cavities, called **marrow**. Once formed, blood cells combine with plasma to yield a fairly uniform range of cellular concentrations that are nevertheless affected by such factors as altitude and overall hydration levels. Accordingly, **hematocrit**, which is the volume percent of centrifuged blood occupied by packed erythrocytes normally varies from ~37% to 47% for women vs. ~42% to 52% for men, whereas pathologies involving abnormally low or high hematocrits are classified as **anemias** or **polycythemias**, respectively (**Figure 6.1a**). Given such values and the fact that the buffy coat constitutes ~1% of packed volume, plasma accounts for ~50–60% of adult blood volumes.

Integrally associated with blood is **lymph**, a colorless fluid that lacks erythrocytes and circulates throughout the body within **lymph vessels** (Chapter 7). Lymph is produced from ECM fluids, which in turn are derived from plasma that continually leaks from small blood vessels. Since large lymph vessels eventually connect with veins

TABLE 6.1 BLOOD CELLS

Cell	Granulocyte/ agranulocyte	Diameter (µm)	Nucleus	Number/µl of blood: avg (typical range)	~Average of total nucleated leukocytes (%)
Erythrocyte	—	7–8	None	5.2 million (4.4–6 million)	—
Neutrophil	Granulocyte	10–13	Multilobed	4,150 (1,800–7,300)	59
Lymphocyte	Agranulocyte	8–12	Spherical	2,185 (1,500–4,000)	31
Monocyte	Agranulocyte	15–30	Indented	455 (200–950)	7
Eosinophil	Granulocyte	10–15	Bilobed	165 (0–700)	2
Basophil	Granulocyte	6–12	Irregular	44 (0–150)	0.6
Platelet	—	1–3	None	350,000 (150,000–500,000)	—

Source: Concentration data from OpenStax College (2013) Anatomy and Physiology. OpenStax. http://cnx.org/content/col11496/latest reproduced under a CC BY 4.0 creative commons license.

supplying the heart, lymph not only serves to transport immune molecules and cells (Chapter 7) but also helps replenish blood by returning lost fluid to the bloodstream.

6.2 THE FLUID COMPONENT OF BLOOD

As the straw-colored fluid of blood, ~90–95% of **plasma** comprises water that is normally maintained at a slightly alkaline pH of 7.4. Dissolved within plasma are numerous small molecules that account for ~1% of total plasma volume. Some of these molecules (e.g., oxygen, amino acids, vitamins, lipids, glucose, and electrolytes such as sodium, potassium, and calcium) serve to nourish cells, whereas others consist of low-molecular-weight wastes and metabolic byproducts such as carbon dioxide, uric acid, and lactic acid.

The remaining ~7% of non-aqueous plasma comprises over 280 identified **plasma proteins**. A few of these are clotting factors, enzymes, and other bioactive molecules, whereas the rest belong to three main categories of proteins, which in decreasing order of abundance are: (1) **albumins**; (2) **globulins** (α-, β-, and γ-globulins); and (3) **fibrinogen**. Except for immunoglobulin types of γ-globulins that are made by activated B lymphocytes (Chapter 7), these three major plasma protein types are predominantly synthesized by hepatocytes in the liver, and because albumins and globulins are retained in serum after fibrinogen and other clotting-related components are removed, such liver-derived constituents are referred to as **serum proteins**. Owing to their high concentrations, serum proteins contribute to the osmotic pressure of blood (Chapter 16), thereby helping to regulate blood volume, whereas fibrinogen plays a key role in the formation of blood clots (pg. 149).

6.3 STRUCTURE AND FUNCTION OF ERYTHROCYTES

Erythrocytes are the most abundant cells in humans, totaling ~25 trillion of the ~30 trillion cells in a typical adult. Unlike nucleated versions in other vertebrates, each fully formed erythrocyte in adult mammals lacks a nucleus and comprises a biconcave disc that is filled with gas-carrying **hemoglobin** (**Hb**) molecules (**Figure 6.2a–c**). In humans, erythrocytes typically measure ~7–8 µm wide and ~2–3 µm thick at their periphery but only ~0.7–1 µm thick in their indented central region. Such morphology optimizes surface area for molecular exchanges while also keeping erythrocytes small enough to navigate through narrow capillaries.

Of the various components of erythrocyte cell membranes, the most abundant glycoprotein is called **band 3** based on its position in polyacrylamide gels. Band 3 is a 100-kD anion channel that takes up Cl^- in exchange for bicarbonate (HCO_3^-) following: (i) diffusion of CO_2 into erythrocytes, (ii) the coupling of CO_2 and H_2O to form carbonic acid (H_2CO_3), which is catalyzed by **carbonic anhydrase**, and (iii) a rapid dissociation of carbonic acid into bicarbonate and H^+ ions (**Figure 6.2d–f**). By such means, band 3 delivers HCO_3^- to the plasma and prevents the swelling and lysis of erythrocytes while also helping to modulate blood pH.

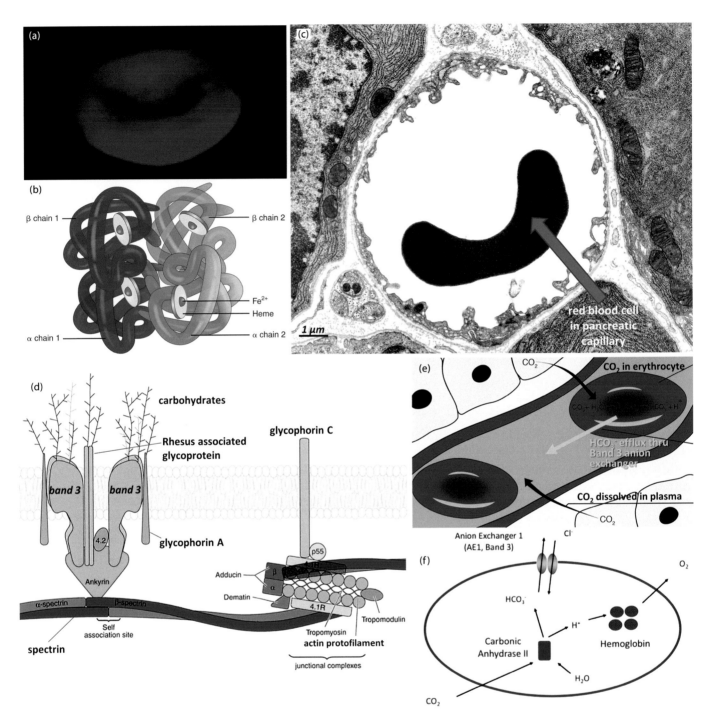

Figure 6.2 Erythrocytes. *(a–c) A confocal 3D reconstruction shows the concave nature of anucleate human erythrocytes (a: 6,700X), which are packed with the gas carrier hemoglobin (b) and thus appear electron-dense in TEMs (c: 7,300X). (d) Erythrocyte plasma membranes contain band 3 anion exchange channels plus glycophorins and are underlain by a cytoskeleton that mainly comprises actin and spectrin filaments. (e, f) Band 3 allows bicarbonate ion efflux from erythrocytes following carbonic-anhydrase-mediated condensation of water and CO_2 to form carbonic acid. ([a]: From Tutwiler, V et al. (2018) Shape changes of erythrocytes during blood clot contraction and the structure of polyhedrocytes. Sci Rep 8: 17907: doi:10.1038/s41598-018-35849-8 reproduced under a CC BY 4.0 creative commons license; [b, e]: From OpenStax College (2013) Anatomy and physiology. OpenStax: http://cnx.org/content/col11496/latest reproduced under a CC BY 4.0 creative commons license; [d]: From Louisa Howard, released into the public domain and downloaded from https://commons.wikimedia.org/wiki/File:A_red_blood_cell_in_a_capillary,_pancreatic_tissue_-_TEM.jpg; [d]: From An, X and Mohandas, N (2008) Disorders of red cell membranes. Br J Haematol 141: 367–375 reproduced with publisher permission; [f]: From Reithmeier, RAF et al. (2017) Band 3, the human red cell chloride/bicarbonate anion exchanger (AE1,SLC4A1), in a structural context. Biochim et Biophys Acta 1858: 1507–1532 reproduced with publisher permission.)*

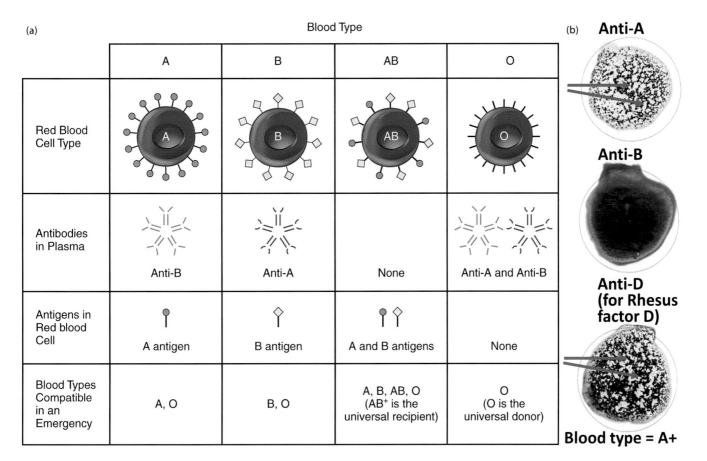

	Blood Type				
(a)	A	B	AB	O	(b) **Anti-A**
Red Blood Cell Type	A	B	AB	O	
Antibodies in Plasma	Anti-B	Anti-A	None	Anti-A and Anti-B	**Anti-B**
Antigens in Red blood Cell	A antigen	B antigen	A and B antigens	None	**Anti-D (for Rhesus factor D)**
Blood Types Compatible in an Emergency	A, O	B, O	A, B, AB, O (AB⁺ is the universal recipient)	O (O is the universal donor)	**Blood type = A+**

Figure 6.3 Blood types. *(a) Human blood types have differing antigens (A, B, A and B, or neither A nor B [=O]) plus varying Rhesus (Rh) factors on their erythrocyte surfaces and hence contain differing antibody types in their plasma. Accordingly, transfusions require compatible blood with the proper antibodies in order to avoid deleterious agglutination reactions. (b) After a drop of sample blood is placed in three wells coated with anti-A, anti-B or anti-Rhesus factor D antibodies, blood agglutination (arrows) occurs only in the anti-A and anti-D wells indicating the tested blood is A-positive (type A with Rhesus factor). ([a, b]: From OpenStax College (2013) Anatomy and Physiology. OpenStax. http://cnx.org/content/col11496/latest reproduced under a CC BY 4.0 creative commons license.)*

Other common plasmalemmal glycoproteins include transmembrane **glycophorins** that help prevent circulating erythrocytes from sticking to vessel walls or other suspended cells. In addition, antigenic carbohydrates and proteins on erythrocyte membranes constitute the **ABO** and **Rh** (=Rhesus) **blood group systems**. Rh components help preserve membrane integrity, whereas A and B antigens apparently evolved as mimics of viral epitopes, thereby enabling the pre-formation of circulating antibodies (Chapter 7) (**Figure 6.3a**). However, such antibodies complicate blood transfusions as improperly matched blood types can generate erythrocyte clumping (=agglutination) and rupturing (=hemolysis). For example, a patient with type A blood has anti-B plasma antibodies that agglutinate transfused type B or type AB blood due to B antigens that the incompatible blood introduces (**Figure 6.3b**).

Beneath the erythrocytic plasmalemma is a **cytoskeleton** comprising mainly **actin** and **spectrin** molecules (**Figure 6.2d**) that help resist shear forces while also maintaining cellular flexibility during passage through capillaries whose lumen may be smaller than the erythrocyte's diameter. However, as erythrocytes age or are subjected to pathologies, such as **sickle cell anemia** (Box 6.1 Sickle Cell Anemia and Malaria), flexibility and optimal morphology are compromised (**Case 6.1 Miles Davis's Erythrocytes**), thereby stimulating the clearing of defective cells by phagocytosis in organs, such as the liver and spleen (Chapter 7).

During the later stages of red blood cell formation (=**erythropoiesis**) (pg. 160), erythrocytes lose their membrane-bound cytoplasmic organelles and nuclei via autophagy and exocytosis in order to make more room for **Hb**, which accounts for ~97% of the enucleated erythrocyte's dry weight. An individual Hb molecule in adults consists of two **α**- and typically two **β-globin proteins** plus four **heme** coordination complexes (**Figure 6.2b**),

BOX 6.1 SICKLE CELL ANEMIA AND MALARIA

With ~7–8 million cases worldwide and particularly high levels in sub-Saharan Africa (**Figure a**), **sickle cell anemia** (**SCA**) is a type of sickle cell disease (SCD) that causes **hemoglobin** (**Hb**) dysfunction when inherited in the homozygous state (**Figure b, c**). Thus, mutations in both of Hb's β-globin alleles cause normally monomeric Hb molecules to **polymerize at low oxygen levels**

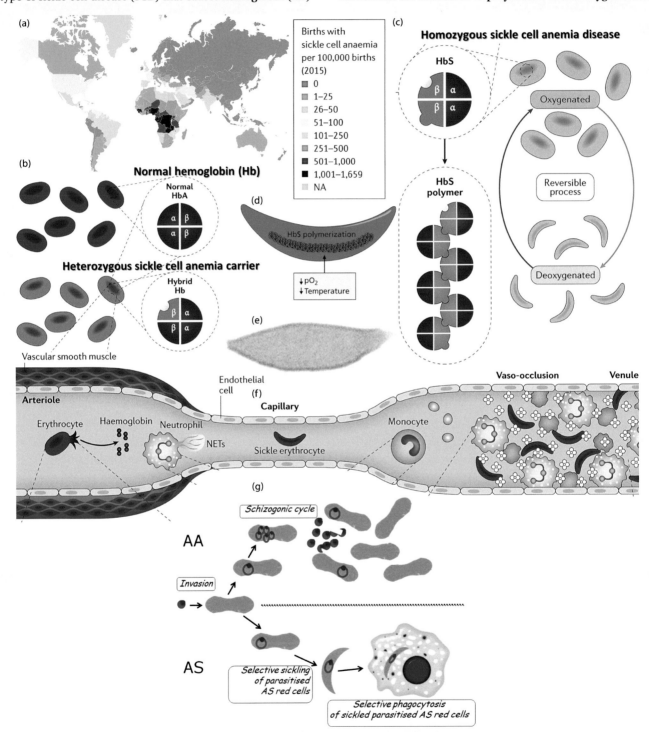

Box 6.1 *Sickle cell disease: prevalence, cellular pathogenesis, and heterozygote advantage for resisting malaria. ([**a–c, f**]: From Kato, GJ et al. (2018). Sickle cell disease. Nat Rev Dis Primers 4: doi:10.1038/nrdp.2018.10 reproduced with publisher permission; [**g**]: From Luzzatto, L (2012) Sickle cell anaemia and malaria. Mediterr J Hematol Infect Dis 4: doi 10.4048/MJHID. 2012.065 reproduced under a CC BY 2.0 creative commons license.)*

into filaments, thereby often converting discoidal erythrocytes into shapes resembling a **sickle** (**Figures c–e**). Such sickled cells are prone to hemolysis and are targeted by inflammatory reactions that can collectively clog blood vessels (**Figure f**). SCA-induced ischemia often compromises organ functioning and thereby causes ~50–90% of African children with SCA to die before age 5. With such high infant mortality rates, it was initially difficult to explain how SCA continued to be passed on. However, SCA co-occurs with an especially lethal form of **malaria** caused by the unicellular parasite **Plasmodium falciparum**, which is delivered to human hosts by biting mosquitos. After infecting host erythrocytes, *Plasmodium* parasites develop into extraerythrocytic stages that can eventually be transmitted to another mosquito as part of the malarial life cycle (**Figure g**).

The enhanced lethality of *P. falciparum* led to the hypothesis that heterozygotic carriers of the sickle cell trait are conferred **protection against infections**. Indeed, carriers with a single mutated allele have 50–80% less parasites loads than cohorts who lack the mutation, apparently due to: (i) lower parasite uptake into erythrocytes; (ii) reduced parasite reproduction; and/or (iii) increased clearance of infected erythrocytes by phagocytosis (**Figure g**).

CASE 6.1 MILES DAVIS'S ERYTHROCYTES

Trumpeter
(1926–1991)

"Miles Davis played a crucial and inevitably controversial role in every major development in jazz since the mid-'40s, and no other jazz musician has had so profound an effect on rock."

(From a biography of M.D. in the *Rolling Stone Encyclopedia of Rock and Roll*, Simon Schuster, 2001)

Called the "Picasso of Jazz" and "Prince of Darkness" for his innovative style and self-destructive habits, Miles Davis was a ground-breaking jazz trumpeter who helped shape 20th century music. Born and raised in Illinois, Davis started his career in the 1940s with the Charlie Parker quintet. After giving hit-or-miss performances due to heroin abuse, Davis overcame his addiction to launch a successful solo career in the 1950s. For a five-year period at the end of the 1970s, Davis retired from music with failing health. Although this silent

period was partly due to depression exacerbated by alcohol and drugs, Davis also suffered from numerous ailments, such as vocal cord polyps, osteoarthritis, and sickle cell anemia, a disease that can trigger sharp pain when abnormally shaped erythrocytes become lodged in narrow vessels (Box 6.1). Davis revived his musical career in the 1980s. However, his health continued to deteriorate, and eventually he died at age 65. Davis has been widely hailed as a musical genius, whose downward spiral was fueled by substance abuse and chronic pain, some of which was due to his sickled erythrocytes.

with each complex comprising a **porphyrin** molecule linked to a central iron atom (Fe^{2+}) that binds gas molecules (O_2, CO_2, and NO). Hb bound to O_2 is the chief mode by which oxygen is supplied to body tissues. Conversely, relatively little of total CO_2 is bound to Hb compared to the large amounts of CO_2 plus bicarbonate ions that are dissolved in plasma. As a potent vasodilator, NO can also attach to heme groups as part of a long-range delivery system for tissues with reduced oxygen levels (=hypoxia). Normally, such gases reversibly bind heme groups. Conversely, **carbon monoxide** (**CO**) remains tightly bound to Hb and thus can eventually block oxygen delivery, thereby contributing to fatal carbon monoxide poisoning.

6.4 PLATELETS: MULTIFUNCTIONAL CELLULAR FRAGMENTS OF MEGAKARYOCYTES INVOLVED IN HEMOSTASIS AND OTHER BIOLOGICAL PROCESSES

Platelets arise in bone marrow as anucleate cellular fragments that are derived from precursor cells, called **megakaryocytes** (**Figure 6.4a–c**). As the name "large nucelus cell" suggests, each megakaryocyte possesses an oversized, multilobed nucleus that undergoes multiple endomitoses without concomitant nuclear divisions to yield 2–32X the normal 2n amount of DNA in diploid cells. Such polyploidy allows megakaryocytes to amplify genes needed for expressing the numerous kinds of proteins found in each mature platelet.

From a typical megakaryocyte, a few thousand platelets are generated via the process of **thrombopoiesis** that begins with each megakaryocyte forming an internal network of membranes, called the demarcation membrane system (DMS). Such internal membranes facilitate the extension of elongated pseudopodia into large bone capillaries, termed **sinusoids** (**Figure 6.4a**). Intra-sinusoidal pseudopodia can branch and reach

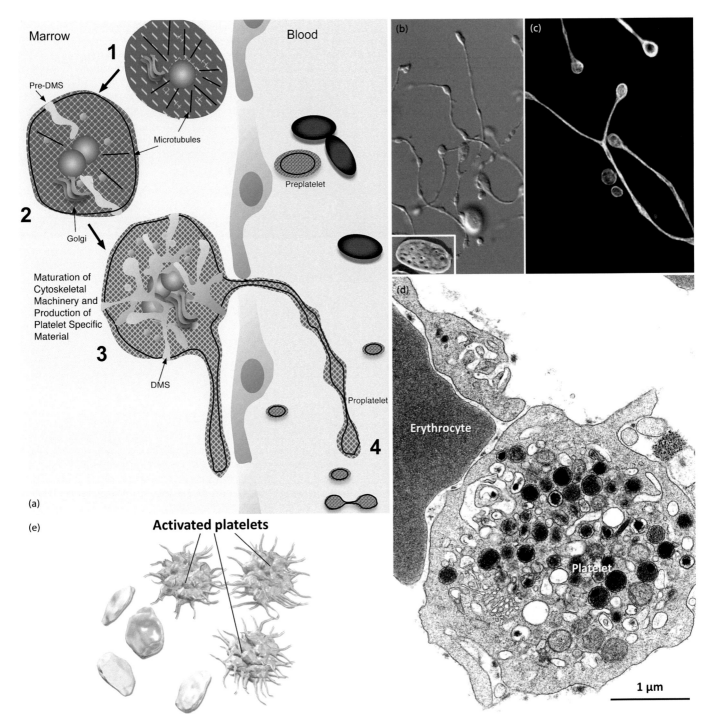

Figure 6.4 Platelets. *(a) Platelets are formed in bone marrow from large megakaryocytes, which generate an internal demarcation membrane system (DMS) and microtubular network that collectively allow slender proplatelet pseudopodia to extend into sinusoidal blood vessels. Preplatelets pinch off from proplatelets and mature into platelets (red discs). (b) DIC optics show proplatelets with swellings that will form individual preplatelets. These in turn mature to become functional platelets (inset). (c) Immunofluorescence microscopy reveals brightly staining microtubules in proplatelets. (d) When fully formed, platelets contain numerous granules with various bioactive molecules. (e) A drawing of mature platelets shows the marked morphological changes that occur when unstimulated platelets (left) undergo activation (right). ([a–c]: From Italiano, JJ and Hartwig, J (2015) Production and destruction of platelets. Intech Open access: DOI: 10.5772/60678 reproduced under a CC BY 3.0 creative commons license. [d]: TEM image courtesy of Drs. H. Jastrow and H. Wartenberg reproduced with permission from Jastrow's Electron Microscopic Atlas (htttp://www.drjastrow.de); [e]: From Blausen.com staff (2014) "Medical gallery of Blausen Medical 2014". WikiJournal of Medicine 1 (2): 10. doi:10.15347/ wjm/2014.010 reproduced under CC0 and CC SA creative commons licenses.)*

collective lengths of several hundred micrometers, thereby becoming slender **proplatelets** with numerous discrete knob-like swellings where individual platelets will be generated (**Figure 6.4a–c**). Eventually, nearly all of megakaryocyte except for its multilobed nucleus is released into the sinusoid as a reticulum of proplatelets. These reticula utilize microtubule-mediated reorganizations to shed pre-platelets, which mature into disc-like platelets (**Figure 6.4d**) that circulate throughout the body or are stored in sites such as the spleen.

Each discoidal platelet measures ~1–3 μm wide and lives ~1–2 weeks. Unlike enucleated erythrocytes which lack synthetic organelles, platelets possess Golgi bodies, an endoplasmic reticulum network, and mitochondria, enabling these cellular fragments to be metabolically active. In addition to its organelles, each platelet contains three types of granules: (1) **α-granules** that possess numerous large molecules, such as **von Willebrand factor** (**vWF**); (2) a few dense **δ- granules** that contain mainly small molecules like as **ADP**, serotonin, and calcium ions; and (3) scattered **lysosomes** with **hydrolase enzymes** (**Figure 6.4d**). In inactive platelets, the granule-containing cytoplasm is surrounded by circumferentially arranged **microtubules** that promote a smooth disc-like morphology for reducing platelet aggregation. However, during activation, platelets dramatically change shape, as described further below (**Figure 6.4e**).

Most of the contents of platelet granules function in **blood clotting** (=blood coagulation), thereby helping to stop **hemorrhages** from damaged vessels during a multifaceted dampening process, called **hemostasis**. At the onset of bleeding, hemostasis is initiated by upstream vasoconstriction to reduce blood flow. Platelets near the wound site are then **activated** by various stimulatory molecules, such as **collagen**, **integrins**, and **ADP**, which become exposed by breaches in the vessel lining (**Figure 6.5a**). Such stimulants can bind G protein coupled receptors (GPCRs) on platelet membranes, thereby initiating positive feedback loops in which **arachidonic acid** derived from the cleavage of plasmalemmal phospholipids becomes converted into **thromboxane A_2** (**TXA_2**), a potent agonist of further platelet activation.

Following activation, platelets attach to the vessel wall via plasmalemmal proteins that bind both **vWF** and **collagen** molecules in the exposed ECM (**Figure 6.5a**). Secreted α-granules from activated platelets provide additional vWF for adhesion, whereas dense granule exocytosis delivers **ADP**, serotonin, and Ca^{2+} ions for activating and aggregating nearby platelets. During such activation, platelets undergo dramatic **actin filament** and **microtubule remodeling** that help generate multiple pseudopodia (**Figure 6.5a, b**). Such extensions can trap and pack platelets over the wound site, thereby forming a **platelet plug** that diminishes bleeding (**Figure 6.5a**).

The platelet plug is then enlarged and strengthened by a multi-step cascade involving **clotting factor** proenzymes and associated bioactive molecules that collectively form a fibrous network over the plug (**Figure 6.5a–c**). Such pro-coagulants normally circulate in plasma and are supplemented by secretions from either activated platelets (e.g., Factors V, XI, and XIII plus prothrombin) or cells in the damaged blood vessel (e.g., **tissue factor** [=Factor III]). Two pathways—**intrinsic** and **extrinsic**—can be activated prior to clotting (**Figure 6.5d**). The intrinsic pathway is stimulated by exposed ECM components at the wound site, whereas the extrinsic pathway is launched by tissue factor secreted from blood vessel endothelial cells in response to extravascular trauma. Some of the initial reactions in the intrinsic and extrinsic pathways require **vitamin K** as a cofactor and thus are targets of "anti-vitamin K" forms of anticoagulant drugs. Ultimately, both cascades converge on a **final common pathway** that converts inactive **prothrombin** into active **thrombin**, which in turn produces insoluble **fibrin** from soluble **fibrinogen** molecules dissolved in plasma. The generated fibrin strands entangle and crosslink erythrocytes and platelets in the platelet plug (**Figure 6.5c**), thereby forming a **blood clot** (=**thrombus**), as immune cells are recruited to the site for wound healing.

Normally, clotting is beneficial in preventing blood loss, but under pathological conditions, platelets can contribute to cases of **thrombosis**, where a blood clot does not simply plug the damaged endothelium but also protrudes substantially into the vessel lumen (**Case 6.2 Yasser Arafat's Platelets**). Large thrombi can in turn occlude blood flow, and the resultant ischemia may cause serious, or even fatal, effects. In particular, **deep vein thromboses** (**DVTs**) can dislodge as free-floating masses, called (thrombo-) **emboli**, which may block smaller-bore vessels, thereby potentially causing fatal embolic suffocation, heart attacks, or strokes. For these reasons, various anticoagulant medicines ("blood thinners") have been developed to combat pathological thrombi. For example, originally sold at high concentrations as a rat poison that functions by triggering excessive bleeding, **warfarin** (=Coumadin) was later approved for anticoagulative purposes, owing to its ability to inactivate vitamin-K-requiring clotting reactions. Alternatively, drugs like **aspirin** inhibit **cyclooxygenase** (**Cox**) enzymes needed to produce **TXA_2** from **arachidonic acid**, thereby reducing platelet activation and aggregation.

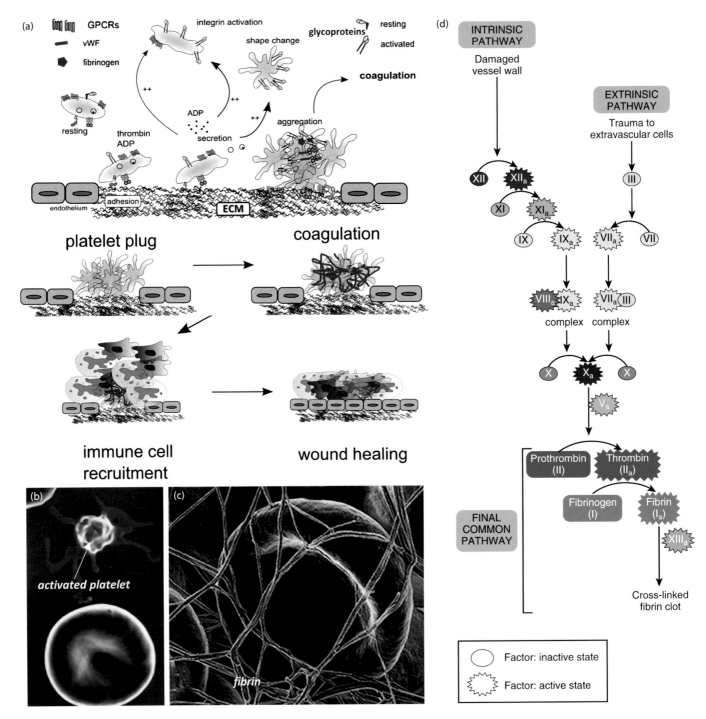

Figure 6.5 Blood clotting. (*a*) *To dampen bleeding from wounds, platelets react to exposed ECM components, such as collagen and von Willebrand's factor (vWF), thereby undergoing activation processes that involve shape changes, granule secretion, and positive feedback loops for further activation of neighboring platelets. After binding to the wound site to form a platelet plug, platelets produce a fibrin clot and attract immune cells to promote wound healing. (**b, c**) SEMs show an activated platelet and fibrin mesh near an erythrocyte (b: 4,400X; c: 8,800X). (**d**) The intrinsic pathway of blood clotting is activated via ECM molecules exposed by wounds, whereas the extrinsic pathway is activated by the release of tissue factor by endothelial cells in response to external damage. Both pathways converge on a common pathway that ultimately forms soluble fibrin from fibrinogen. ([a]: From Golebiewska, EM and Poole, AW (2015) Platelet secretion: from haemostasis to wound healing and beyond. Blood Rev 29: 153–162 reproduced under a CC BY 3.0 creative commons license; [b]: From Kell, DB and Pretorius, E (2018) No effects without causes: the iron dysregulation and dormant microbes hypothesis for chronic, inflammatory diseases. Biol Rev: doi: 10.1111/brv.12407 reproduced under a CC BY 4.0 creative commons license; [c]: From Bester, J et al. (2013) High ferritin levels have major effects on the morphology of erythrocytes*

Figure 6.5 (Continued) *in Alzheimer's disease. Front Aging Neurosci 5: doi: 10.3389/fnagi.2013.00088 reproduced under a CC BY 3.0 creative commons license; [d]: From OpenStax College (2013) Anatomy and Physiology. OpenStax. http://cnx. org/content/col11496/latest reproduced under a CC BY 4.0 creative commons license.)*

CASE 6.2 YASSER ARAFAT'S PLATELETS

Palestinian Leader
(1929–2004)

"Palestine is the cement that holds the Arab world together, or it is the explosive that blows it apart."

(Y.A. in a 1974 Time Magazine article discussing the volatility of Palestinian-Israeli relations)

The man widely known as Yasser Arafat was born to Palestinian parents in Egypt, where he grew up and studied engineering at Cairo University. As a student, Arafat began supporting Arab nationalism and opposing the state of Israel, a stance he continued to take as the co-founder and leader of a Palestinian group, called Fatah. In 1988, Arafat acknowledged Israel's right to existence as part of negotiations for Arab-Israeli co-existence, and for such efforts, he was co-awarded the 1994 Nobel Peace Prize. During a staff meeting in 2004, Arafat vomited due to what was initially thought to be the flu, but after his condition worsened, he was transported to a French hospital, where he soon lapsed into a coma and died at age 77. Arafat's cause of death has remained controversial,

as some speculate that after documented assassination attempts, Israeli agents succeeded in poisoning him with radioactive polonium. Initial assays of Arafat's tissues were consistent with polonium poisoning, but other tests tended to contradict those findings. Moreover, physicians at his French hospital diagnosed Arafat as having an idiopathic case of disseminated intravascular coagulation, where platelets formed numerous clots throughout his body. Accordingly, French authorities recently ruled against poisoning as triggering Arafat's death and instead cited natural causes, exacerbated by such pathologies as widespread blood clots triggered by over-active platelets.

6.5 LEUKOCYTES: GRANULOCYTES (NEUTROPHILS, EOSINOPHILS, AND BASOPHILS) VS. AGRANULOCYTES (LYMPHOCYTES AND MONOCYTES)

At collective concentrations ~750X lower than that of erythrocytes (Table 6.1), the far less common **leukocytes** of human blood are grouped into two classes—**granulocytes** that contain cell-specific **secondary granules** in their cytoplasm vs. **agranulocytes** that lack such granules (**Case 6.3 Sadako Sasaki's Leukocytes**) (**Figure 6.6a-l**). Granulocytes are ephemeral cells that typically live less than a few days. In decreasing order of abundance, such cells comprise **neutrophils**, **eosinophils**, and **basophils**. Following activation in peripheral tissues, granulocytes broadly function in responses that help clear the body of pathogens and physical damage. Conversely, agranulocytes consist of **lymphocytes** and **monocytes**, both of which can live for weeks to months,

CASE 6.3 SADAKO SASAKI'S LEUKOCYTES

Casualty of the Hiroshima atomic bombing
(1943–1955)

"This is our cry. This is our prayer. Peace in the world."

(Translation of an inscription on a statue of S.S. in Hiroshima Peace Memorial Park)

As an enduring symbol of nuclear warfare's destruction, Sadako Sasaki became widely known for folding 1,000 origami paper cranes while battling terminal cancer caused by the bombing of Hiroshima, Japan during World War II. In 1945, two-year-old Sasaki was playing a mile from ground zero when an American plane dropped an atomic bomb on the city. Although shock waves from the blast tossed her about, Sasaki amazingly showed no external signs of major damage and initially exhibited normal health. However, in 1955, Sasaki developed a pronounced swelling of her neck as well as red and purple patches on her leg. Tests showed that Sasaki's leukocyte

count was greatly elevated, indicating that she, like many other irradiated children, was suffering from leukemia, a cancer characterized by leukocyte overproduction. After being admitted to the hospital for blood transfusions and palliative care, Sasaki was told by her roommate of a legend that whoever folds 1,000 origami paper cranes would be granted a wish. Although continuing to deteriorate during her months-long hospitalization, Sasaki nevertheless was able to find enough paper scraps and strength to fold 1,000 cranes, before eventually dying at age 12 due to her leukocyte-related cancer.

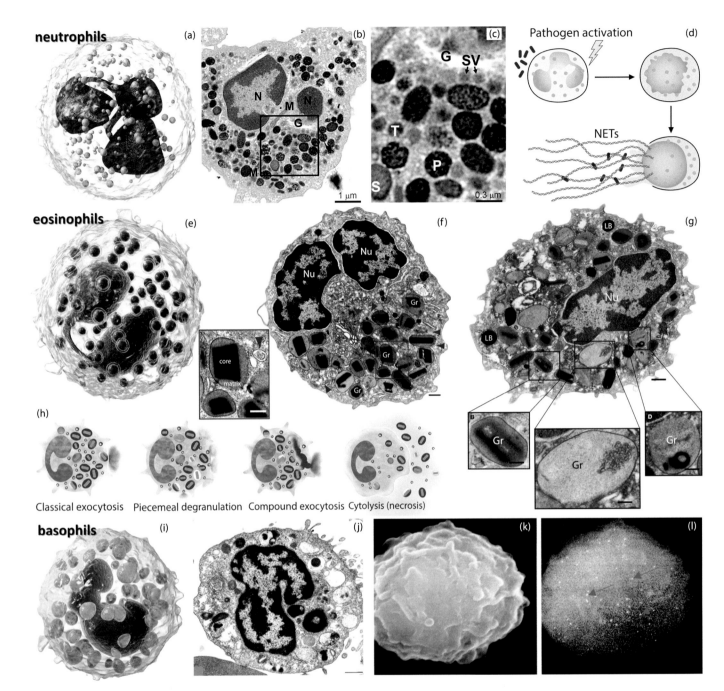

Figure 6.6 Granulocytic leukocytes. (a, b) *Neutrophils, which are also called polymorphonuclear neutrophils (PMNs, polys), possess a multilobed nucleus of varying morphology that often appears as comprising separate compartments in TEMs. (**c**) The three granule types of neutrophils—primary (P), secondary (S), and tertiary (T)—occur in this TEM near Golgi bodies (G) and secretory vesicles (SVs). (**d**) Neutrophils often form the first line of cellular defense against pathogens, typically phagocytosing bacteria (red rods) or encasing them in neutrophil extracellular traps (NETs) composed of exocytosed chromatin with bactericidal products such as myeloperoxidases. (**e, f**) Eosinophils contain a bilobed nucleus (Nu), lipid bodies (LBs), and large secondary granules, each of which in unstimulated eosinophils (F) comprises a crystalline core surrounded by a more electron lucent matrix (scale bar = 500 nm; inset: 300 nm). (**g, h**) Via various modes of secretion (figure h), eosinophilic granules (Gr) can ward off pathogens, as illustrated in figure g for a piecemeal mode of degranulation by a stimulated eosinophil (scale bar = 500 nm; insets: 170 nm). (**i–l**) As the smallest and most rare granulocytes, basophils have large basophilic granules that they can release in response to IgE binding to cell surface receptors, as illustrated by a conventional SEM (k) and correlative backscattered electron image (l) of anti-IgE-FcεRI complexes that were labeled with gold-conjugated protein A (arrows); scale bar in (j) = 1 μm; (k, l) = 3,500X. ([a, e, i]: From Blausen.com staff (2014) "Medical gallery of Blausen Medical 2014". WikiJournal of Medicine 1 (2): 10. doi:10.15347/wjm/2014.010 reproduced under CC0 and CC SA creative commons licenses; [b, c]: From Sheshachalam, A et al. (2014)*

Figure 6.6 (Continued) Granule protein processing and regulated secretion in neutrophils. Front Immunol 5: doi: 10.3389/fimmu.2014.00448 reproduced under a CC BY 4.0 creative commons license; [e–h]: From Spencer, LA et al. (2014) Eosinophil secretion of granule-derived cytokines. Front Immunol 5: doi: 10.3389/fimmu.2014.00496 reproduced under a CC BY 4.0 creative commons license; [j–l]: From Kepley, CL et al. (1998) The identification and characterization of umbilical cord blood-derived human basophils. J Leukoc Biol 64: 474–483 reproduced with publisher permission.)

with monocytes differentiating into long-lived **macrophages**. Lymphocytes perform various immune functions, whereas monocyte-derived macrophages are mainly involved in phagocytosing pathogens and debris (Table 6.1).

Neutrophils

Polymorphonuclear neutrophils (=PMNs, polys, or **neutrophils**) measure ~10-13 µm in diameter and are the most abundant leukocyte, representing ~55–70% of white cell totals (Table 6.1). Aside from their common occurrence in blood, neutrophils can also be distinguished by their multi-lobed nucleus of variable shapes (**Figure 6.6a**). Surrounding the nucleus, the cytoplasm contains three types of granules—**primary**, **secondary**, and **tertiary** (**Figure 6.6b, c**). **Primary** (=azurophilic) **granules** stain with azure dyes and contain **antimicrobial agents**, such as **defensins** and ROS-generating **myeloperoxidase** (**MPO**) to help kill microbial invaders. Although primary granules are particularly abundant in neutrophils, such azurophilic granules are also found in the cytoplasm of all other leukocytes, including both granulocytes and agranulocytes. Conversely, neutrophil-specific **secondary granules** appear light pink when stained with neutral dyes and contain various **bactericidal proteins**. Such proteins include **lysozyme**, which hydrolyzes the cell wall of gram-positive bacteria, and **lactoferrin**, which sequesters iron to limit bacterial growth. **Tertiary granules** comprise a subpopulation of secondary granules, which contain ECM-digesting enzymes like **gelatinase B** that help neutrophils migrate through tissues.

Neutrophils released from the bloodstream into peripheral tissues can detect and **phagocytose** invasive microbes before internally releasing granular contents that then fuse with intracellular endosomes containing the incorporated microorganisms. Alternatively, neutrophils are also able to destroy pathogens in the ECM either via an external **degranulation** process or by the production of **neutrophil extracellular traps** (**NETs**) comprising DNA strands with attached histones plus granule-derived proteins like lactoferrin and MPO (**Figure 6.6d**). Collectively, the intra- and extracellular modes of pathogen elimination used by neutrophils often provide the first line of defense against invading microorganisms and can contribute to much of the yellowish exudate (=**pus**) that forms at the site of infected wounds.

Eosinophils

At ~10–15 µm wide, eosinophils are slightly larger than neutrophils but far less common, typically accounting for only 1–5% of total leukocytes (Table 6.1). Moreover, eosinophils have a bilobed nucleus, rather than the multilobar type of neutrophils (**Figure 6.6e**). In addition to **azurophilic** granules with unique components like T-lymphocyte modulators, eosinophils contain abundant secondary granules. Compared to their counterparts in neutrophils, the redder secondary granules of eosinophils are also larger, and in TEMs, each granule is distinctly bipartite, with a crystalline core surrounded by an outer matrix (**Figure 6.6f**). Within the granular core, the most abundant component is **major basic protein** (**MBP**), which increases membrane permeability in pathogens to trigger their death. Conversely, components of the granular matrix include **eosinophil peroxidase** (**EPO**), which like neutrophilic MPOs destroys various pathogens via ROS generation. With this toolkit, eosinophils can engulf and kill endocytosed pathogens by an intracellular release of MBP and other proteins into endocytic vesicles. Alternatively, via several degranulation modes, exocytosis of granules into the ECM helps destroy various pathogens including even **worm parasites** (**Figure 6.6g, h**).

Basophils

As the smallest granulocytes and least common of all leukocytes, **basophils** measure only ~6–12 µm wide and usually constitute less than 1% of total white cell counts (Table 6.1) (**Figure 6.6i, j**). Unlike other granulocytes, each basophil contains large secondary granules that stain blue to purple with various basic dyes. Such **metachromatic** shifting of colors resembles that exhibited by **mast cell** granules and coincides with other lines of evidence that basophils and mast cells diverged from a common cell type while maintaining similar functional properties (Box 6.2 Basophils vs. Mast Cells). As with mast cells, metachromasia of basophil granules

BOX 6.2 BASOPHILS VS. MAST CELLS

Cells with similar secretory products to those made by vertebrate **basophils** and **mast cells** occur in tunicate invertebrates (Phylum Chordata, Subphylum Tunicata) **(Figure a)**.

Such findings indicate basophils and mast cells arose from an ancestral cell type that was present before the divergence of the subphylum Vertebrata from the rest of the phylum Chordata.

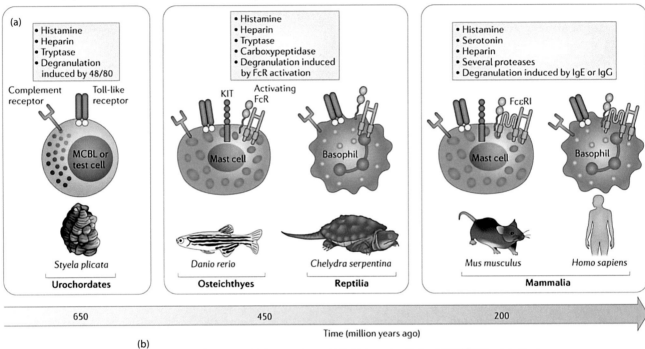

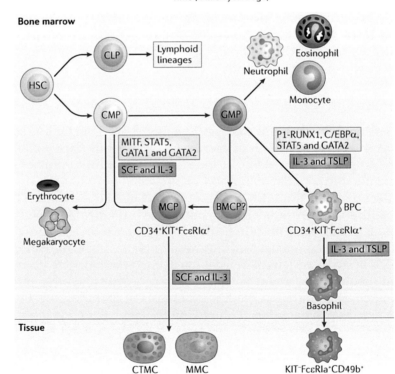

Box 6.2 (a, b) *Evolutionary trends and developmental pathways in basophils vs. mast cells. (Figures from: Voehringer, D (2013) Protective and pathological roles of mast cells and basophils. Nat Rev Immunol: doi:10.1038/ nri/3427 reproduced with publisher permission.)*

Although extant basophils and mast cells share in common several key structural and functional features like **metachromatic granules** that mediate allergic reactions, both cell types arise via alternative pathways involving discrete suites of differentiating cues (**Figure b**). In addition, basophils mature in bone marrow before being transported via blood to peripheral tissues, where they typically live less than 3 days. Conversely, mast cells exit the marrow as immature cells that subsequently differentiate in peripheral tissues during a more prolonged life span that lasts weeks to months. Moreover, differentiated mast cells diverge into two functionally different subclasses— (1) **connective tissue mast cells** (**CTMCs**) which occur mainly in the skin, peritoneal cavity, intestinal submucosae, and around blood vessels vs. (2) **mucosal mast cells** (**MMCs**) that tend to occur near epithelial cells of mucosal tissues in the lungs and intestines. Collectively, such contrasting patterns indicate that basophils and mast cells are distinct cell types that operate in similar ways within their own specialized niches in the body.

is mainly due to sulfated molecules like **chondroitin sulfate** and **heparin**. Such products are released along with **histamine** and other cellular secretions in response to IgE antibodies binding to basophil surface receptors (**Figure 6.6k, l**). Collectively, this exocytosis contributes to allergic reactions, including those that can lead to life-threatening anaphylaxis (Chapter 3).

Lymphocytes

The smallest and most common of the two agranulocytic leukocytes are **lymphocytes**, which typically constitute ~20–40% of white cell totals and measure about 8–12 μm in diameter (Table 6.1). An additional distinctive feature of lymphocytes is a large spherical nucleus that occupies much of the cell's interior (**Figure 6.7a**). Thus, in conventionally prepared blood smears, only a thin rim of peripheral cytoplasm remains visible, although fully hydrated lymphocytes viewed by atmospheric SEM can exhibit broad pseudopodia (**Figure 6.7b**).

Among numerous lymphocyte subtypes with distinctive molecular markers and specific functions, two main classes can be recognized (see also Chapter 7): (1) a relatively small number of **innate lymphocytes** (e.g., **natural killer cells** [NKs] and **innate lymphoid cells** [ILCs]), which detect general motifs in broad groups of encountered pathogens and can mount a rapid, but often non-specific and short-lived, anti-pathogenic response vs. (2) the more common **adaptive lymphocytes** (**B** and **T lymphocytes**), which react in a highly specific fashion to pathogens by means of **B cell antibodies** and **T cell receptors** (**TCRs**) (Chapter 7). After being produced in bones, B lymphocytes are delivered via blood and lymph to various lymphoid organs and other peripheral tissues, whereas T lymphocytes are typically transported to the **thymus** for further differentiation. Lymphocyte subtypes are not readily distinguished morphologically in conventional preparations and instead are discriminated based on their molecular markers or distributions in tissues where certain lymphocytes are known to predominate (Chapter 7). Thus, further discussions of lymphocyte nomenclature, features, and functions are deferred to next chapter's coverage of the lymphatic system.

Monocytes

As the largest of all leukocytes, **monocytic** agranulocytes measure ~15–30 μm in diameter and constitute ~5–10% of the white blood cells in normal blood smears (Table 6.1). Monocytes are also distinguished by their distinctly indented nucleus and abundant cytoplasm, which often contains peroxisomes (**Figure 6.7c**). Following export into blood from their sites of origin in bones, monocytes can progress through several distinct stages (classical, intermediate, and non-classical) (**Figure 6.7d**), which are characterized by their differing molecular markers and functional roles. Circulating monocytes are eventually recruited to peripheral tissues, where they differentiate into various members of a **mononuclear phagocyte system** (**MPS**) comprising such cells as **macrophages** and bone **osteoclasts**. Although monocytes and lymphocytes both form in bones and circulate in the blood, monocytes are products of common myeloid progenitor cells, whereas lymphocytes are derived from common lymphoid progenitors (**Figure 6.7e**).

6.6 INTRODUCTION TO BLOOD CELL FORMATION (HEMATOPOIESIS): ADULT BONE MARROW WITH ITS HEMATOPOIETIC STEM CELLS (HSCS) PLUS DIFFERING HEMATOPOIETIC SITES DURING DEVELOPMENT

In the fetus and following birth, the medullary cavities of bones become filled with **marrow** that contains blood-cell-forming progenitors, called **hematopoietic stem cells** (**HSCs**), which are able to form blood cells during the process of **hematopoiesis** (**Figure 6.8a–h**). HSCs in marrow tend to aggregate around large discontinuous

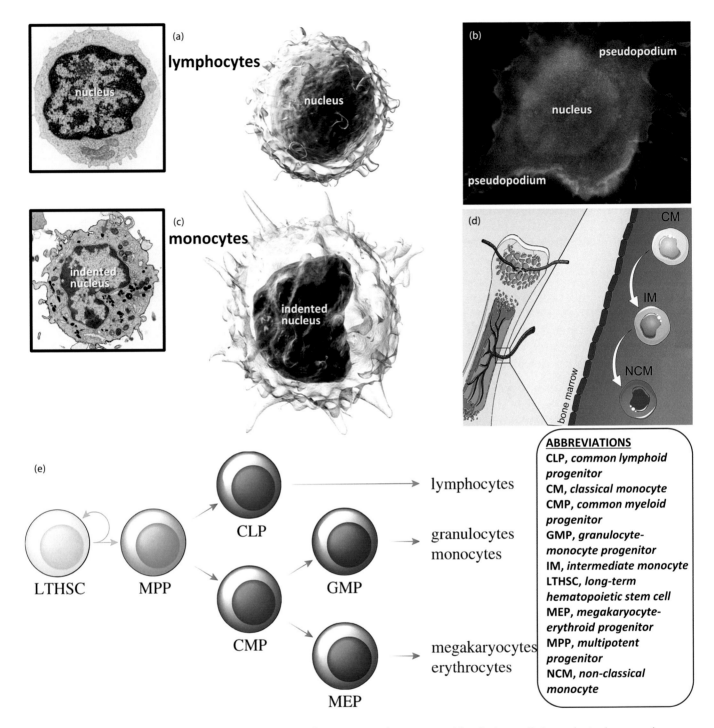

Figure 6.7 Agranulocytic leukocytes. (a, b) *Lymphocytes are characterized by their small size, relatively smooth profiles, and high nuclear:cytoplasmic ratios (a, TEM inset: 3,600X), although when viewed hydrated and attached to a substrate in an atmospheric SEM, broad pseudopodia can be seen (b: 5,000X).* **(c)** *Monocytes are the largest leukocytes and typically contain an indented nucleus as well as peroxisomes (c: TEM inset, arrows) (1,750X).* **(d)** *Based on molecular markers, monocytes can differentiate from classical monocytes (CMs) into non-classical monocytes (NCM) via an intermediate (IM) monocyte stage.* **(e)** *Stages in blood cell differentiation. ([a, c drawings]: From Blausen.com staff (2014) "Medical gallery of Blausen Medical 2014". WikiJournal of Medicine 1 (2): 10. doi:10.15347/wjm/2014.010 reproduced under CC0 and CC SA creative commons licenses Blausen.com staff (2014) reproduced by permission of creative common licenses; [a, TEM inset]: From Kaido, M et al. (2017) Investigation of morphological changes for the discrimination of nucleated red blood cells and other leukocytes in Sysmex XN hematology analyzer scattergrams using transmission electron microscopy. Practical Lab Med 8: 70–76 reproduced under a CC BY 4.0 creative commons license; [b]: From*

Figure 6.7 (Continued) *Murai, T et al. (2013) Ultrastructural analysis of nanogold-labeled cell surface microvilli in liquid by atmospheric scanning electron microscopy and their relevance in cell adhesion Int. J. Mol. Sci. 14, 20809–20819; doi:10.3390/ijms141020809 reproduced under a CC BY 3.0 creative common license; [c, TEM inset]: From Kang, Y-H et al. (1990) Ultrastructural and immunocytochemical study of the uptake and distribution of bacterial lipopolysaccharide in human monocytes. J Leukoc Biol 48: 316–332 reproduced with publisher permission; [d]: From Kapellos, TS et al. (2019) Human monocyte subsets and phenotypes in major chronic inflammatory diseases. Front Immunol 10: doi: 10.3389/fimmu.2019.02035 reproduced under a CC BY 4.0 creative common license; [e]: From Wiesner, K et al. (2018) Haematopoietic stem cells: entropic landscapes of differentiation. Interface Focus 8: 20180040 reproduced under a CC BY 4.0 creative commons license.)*

sinusoidal capillaries, with each stem cell occupying discrete **niches** that optimize blood cell proliferation (**Figure 6.8e**). In addition to nerves, blood vessels, and HSCs, marrow also contains **mesenchymal stem cells** (**MSCs**) that generate non-hematopoietic cells, such as chondrocytes, osteoblasts, and adipocytes.

Regions of active hematopoiesis in adults are referred to as **red marrow**, owing to their high numbers of differentiating erythrocytes (**Figure 6.8a, b**), whereas less active sites contain **yellow marrow** with abundant unilocular fat (**Figure 6.8c, d**). To supply enough blood cells for rapid post-natal growth, nearly all marrow in infants is hematopoietically active red marrow. Over time, much of this marrow becomes replaced with yellow marrow, so that adult red marrow is restricted to the axial skeleton and a few pockets in long bones like the femur and humerus.

Prior to blood cells forming in bone marrow, a **primary wave** of hematopoiesis starts about 2–3 weeks post-fertilization outside the embryo proper within the **yolk sac** to generate nucleated erythrocytes, monocytes, and megakaryocytes (**Figure 6.9a**) (see also Chapter 18 regarding nucleated fetal blood cells). Subsequently, such hematopoiesis is supplanted by a **definitive hematopoietic wave** involving **HSCs**, which can differentiate into all adult blood cell types, including enucleated erythrocytes (**Figure 6.9a**). By around 9–11 weeks of gestation, HSCs colonize mainly the **liver** before eventually migrating to developing **marrow** regions and the **thymus**, with post-natal hematopoiesis continuing predominantly within bone marrow throughout life.

Although adult blood cell formation occurs mainly in bone marrow, lymphocytes also proliferate in the **thymus** and secondary lymphoid organs (e.g., lymph nodes and spleen) (Chapter 7). Moreover, megakaryocytes of humans and other mammals normally migrate to blood vessels in adult lungs, where additional **pulmonary thrombopoiesis** occurs. Alternatively, as a compensatory mechanism for diseases that reduce blood cell production (**Case 6.4 Eleanor Roosevelt's Bone Marrow**), various marrow-derived progenitor cells are transported to organs, such as the liver, lymph nodes, spleen, and thymus for **extramedullary hematopoiesis (EMH).**

CASE 6.4 ELEANOR ROOSEVELT'S BONE MARROW

First Lady, Activist, Diplomat
(1884–1962)

"At all times, day by day, we have to continue fighting for freedom of religion, freedom of speech, and freedom from want…"

(E.R. in one of her newspaper columns, called My Day, which she wrote for several decades)

As First Lady of the United States from 1933 to 1945, Eleanor Roosevelt was a civil rights advocate, who continued her activism during her post-White-House years by taking on various diplomatic roles. After being orphaned at age 9, Roosevelt was raised by her grandmother before marrying her distant cousin Franklin Delano Roosevelt (FDR). While serving as First Lady, Roosevelt played a key role in publicly promoting FDR's policies. She also remained FDR's wife until he died in office, even though the two were estranged from each other, with both having engaged in extramarital affairs. Following

FDR's death, Roosevelt was often asked to run for public office and thereby capitalize on her perennial status as Most Admired Woman in various polls. However, she declined such requests and instead served as a delegate to the United Nations, while also championing for human rights around the world. In 1960, Roosevelt was hit by a car, and after receiving steroid treatments, she developed aplastic anemia that was exacerbated by an activated case of latent tuberculosis that had been residing within her bone marrow. Ultimately, the inability of her marrow to form enough blood cells contributed to her death by cardiac failure in 1962.

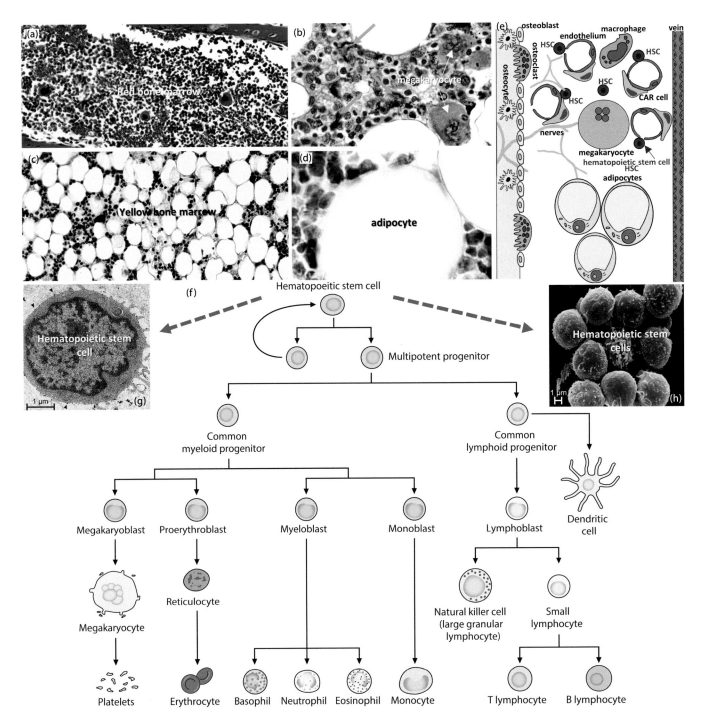

Figure 6.8 Introduction to hematopoiesis. (a, b) *Adult blood cell formation (=hematopoiesis) occurs mainly in red marrow of axial and long bones (a: 50X). Such marrow contains numerous developing blood cells along with megakaryocytes and phagocytic macrophages (arrows, b) plus relatively few adipocytes (200X). (**c, d**) Aging red marrow can be converted into yellow marrow (c: 40X) with fewer hematopoietic cells and more abundant adipocytes (d: 320X). (**e**) During hematopoiesis, hematopoietic stem cells (HSCs) occupy discrete marrow niches that often contain macrophages, megakaryocytes, CXCL12-Abundant Reticular (CAR) cells, and sinusoidal endothelia. Aged marrow with added adiposity is depicted toward the bottom of the figure. (**f–h**) Figure f summarizes general stages of hematopoiesis, whereas (g) and (h) show EMs of isolated HSCs that had been labeled for HSC-specific surface markers (g, arrowheads) and then subjected to a pull-down purification step. ([e]: From Al-Drees, MA et al. (2015) Making blood: The haematopoietic niche throughout ontogeny. Stem Cell Int 2015: Article ID 571893 http://dx.doi.org/10.1155/2015/571893 reproduced under a CC BY 3.0 creative commons license; [g, h]: From Neumueller, J et al. (2015) Development of myeloid dendritic cells under the influence of sexual hormones visualized using scanning and transmission electron microscopy. Intech Open: http://dx.doi.org/10.5772/62310 reproduced under a CC BY 3.0 creative commons license.)*

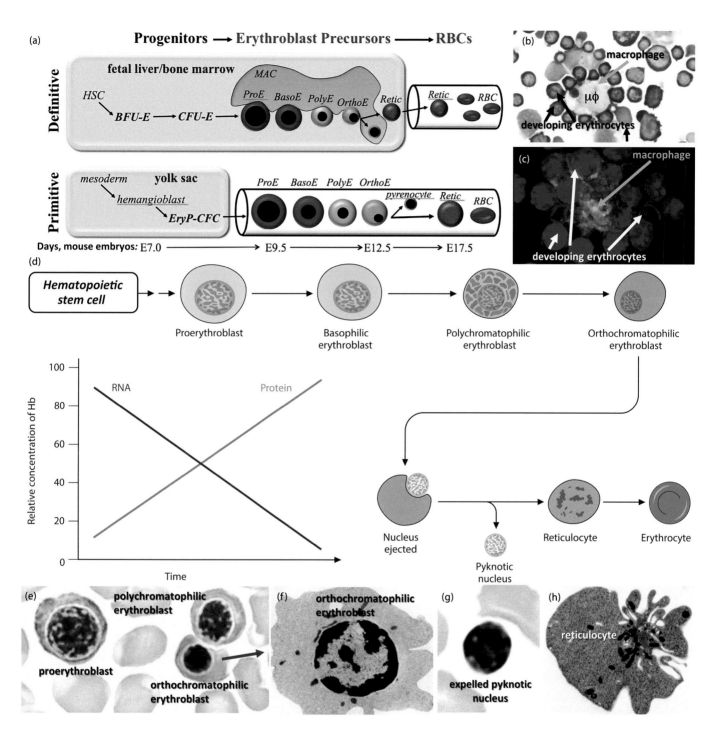

Figure 6.9 Erythropoiesis: the formation of red blood cells. (a) Before birth, a primary wave of erythropoiesis within the yolk sac initially generates erythrocytes. Subsequently, a definitive erythropoietic wave initiated in the liver and later transitioning into bone marrow utilizes blood islands with a central macrophage (MAC) that assists the differentiation of erythrocytes. **(b, c)** Conventional LM (b) and fluorescence (c) images depict the central macrophage with its surrounding developing erythrocytes in a blood island during definitive erythropoiesis. **(d–h)** Each hematopoietic stem cell becomes a nucleated proerythroblast (ProE) with a basophilic cytoplasm arising from abundant basophilic globin mRNAs for hemoglobin (Hb) and relatively low levels of acidophilic globin protein. Within blood islands, such cells transition from basophilic (BasoE) to polychromatophilic (PolyE) to orthochromatophilic (OrthoE) erythroblasts (d–f) as acidophila from translated globin proteins overwhelms the basophilia of untranslated globin mRNAs. Orthochromatophilic (=acidophilic, eosinophilic) erythroblasts expel their nucleus (d, g) and most organelles to make more room for hemoglobin, thereby becoming an irregularly shaped reticulocyte (d, h), which over several days matures into a biconcave mature erythrocyte (b: 1,000X; c: 1,400X; e: 2,000X; f: 6,000X; g: 3,500X; h: 6,000X). ([a]: From Palis, J (2014) Primitive and definitive

Figure 6.9 (Continued) *erythropoiesis in mammals. Front Physiol 5: doi: 10.3389/fphys.2014.00003 reproduced under a CC BY 4.0 creative commons license; [b]: From Palis, J (2017) Interaction of the macrophage and primitive erythroid lineages in the mammalian embryo. Front Immunol 7: doi: 10.3389/fimmu.2016.00669 reproduced under a CC BY 4.0 creative commons license; [c]: From Giger, KM and Kalfa, TA (2015) Phylogenetic and ontogenetic view of erythroblastic islands. BioMed Res Int 2015: http://dx.doi.org/10.1155/2015/873628 reproduced under a CC BY 3.0 creative commons license; [f, h]: From Kaido, M et al. (2017) Investigation of morphological changes for the discrimination of nucleated red blood cells and other leukocytes in Sysmex XN hematology analyzer scattergrams using transmission electron microscopy. Practical Lab Med 8: 70–76 reproduced under a CC BY 4.0 creative commons license.)*

6.7   RED BLOOD CELL FORMATION (ERYTHROPOIESIS): PRO-ERYTHROBLAST → BASOPHILIC ERYTHROBLAST → POLYCHROMATOPHILIC ERYTHROBLAST → ACIDOPHILIC ERYTHROBLAST → RETICULOCYTE → ERYTHROCYTE

Red blood cells normally circulate in adults for **100–120 days** before being removed by macrophages located mainly in the liver and spleen. To balance this loss, new erythrocytes are formed by an **erythropoietic process** that also spans ~110 days. Such **erythropoiesis** begins with **hematopoietic stem cells** (**HSCs**) forming multipotential progenitors (MPPs) that can become either red or white blood cells (**Figure 6.8f**). Erythroid-inducing stimuli subsequently cause MPPs to differentiate into: (i) **megakaryocytes**, which form platelets, and (ii) **pro-erythroblasts**, which produce erythrocytes (**Figure 6.8f**).

During the definitive wave of hematopoiesis, developing erythrocytes organize into radially arranged units, called **erythroblastic** (=blood) **islands** (**Figure 6.9a–c**). Within each erythroblastic island, a central **macrophage** nurse cell interacts with nearby pro-erythroblasts and more advanced erythropoietic stages via various secreted proteins (**Figure 6.9a–c**). In response to macrophage-mediated cues, proeythroblasts form **basophilic erythroblasts**, so-named because their cytoplasm stains with basic dyes, owing to their numerous basophilic mRNA transcripts for globin proteins of Hb (**Figure 6.9d, e**). Basophilic erythroblasts become **polychromatophilic** (=multicolored) **erythroblasts** that are filled with both globin mRNAs and acidophilic globin proteins, thereby conferring mixed baso- and acidophilia. In **acidophilic** (=orthochromatophilic) **erythroblasts** that develop from polychromatophilic erythroblasts, nearly all of the globin mRNA has been translated into protein, thereby generating the same cytoplasmic acidophilia as exhibited by mature erythrocytes (**Figure 6.9d**).

Up to this point in erythropoiesis, developing erythrocytes are nucleated (**Figure 6.9d–f**). However, in addition to degrading their cytoplasmic organelles, acidophilic erythroblasts condense their nuclei and eject such **pyknotic** structures plus degenerated cytoplasmic organelles from the cell (**Figure 6.9g**). The resulting enucleated cells possess a reticulum of residual ER profiles, leading to their designation as **reticulocytes** (**Figure 6.9h**). After being released into the bloodstream, reticulocytes mature over several days into concave erythrocytes.

Although erythropoiesis also depends on localized modulators in erythroid-specific niches of adult bone marrow, the overall stimulator of erythrocyte production is a glycoprotein hormone, called **erythropoietin** (**Epo**). Epo is made mainly by peritubular fibroblasts in the **kidney** (Chapter 16) in response to decreases in blood oxygenation. After its transport via blood to bone marrow, Epo increases erythrocytic differentiation by binding to receptors on pro-erythroblast progenitors. Accordingly, recombinant human erythropoietin (rhEpo) and related erythropoiesis-stimulating agents (ESAs) are used to treat insufficient erythrocyte production. Such agonists can similarly provide a means of **blood doping** for athletes who want to supplement their erythrocyte numbers and oxygen-carrying capacity.

6.8   AN OVERVIEW OF WHITE BLOOD CELL FORMATION: THE PRODUCTION OF MYELOBLASTS AND LYMPHOBLASTS DURING LINEAGE-SPECIFIC DIFFERENTIATION OF LEUKOCYTES

As with erythropoiesis, white blood cell formation (=leukopoiesis) begins in adult marrow with the conversion of long-term populations of self-replicating HSCs into multipotent progenitors (MPPs). However, instead of differentiating along an erythroid pathway, some MPPs are stimulated by alternative cues and eventually form two leukocyte-committed precursor stages, called the **granulocyte-monocyte progenitor** (**GMP**) and **common lymphocyte progenitor** (**CLP**) (**Figure 6.7e**). From GMPs, unipotent intermediary stages, called **myeloblasts**, develop, and under appropriate stimulation, each differentiates into a granulocyte, monocyte, or mast cell (Box 6.2)

(**Figure 6.8f**). Alternatively, CLPs can generate either non-monocyte-derived dendritic cells or **lymphoblasts** (**Figure 6.8f**), which form several lymphocyte subtypes (Chapter 7).

Unlike kidney-derived Epo that stimulates erythropoiesis mainly via its binding to plasmalemmal receptors on precursors of pro-erythroblasts, distinct types of glycoprotein **colony stimulating factors** (**CSFs**) are produced by various tissues throughout the body to enhance multiple steps in **granulocyte** and **macrophage** formation. For example, CSFs are capable of: (i) stimulating precursor cell divisions, (ii) promoting myeloid maturation, (iii) enhancing mature cell functioning, and (iv) blocking apoptosis. Alternatively, lymphocyte production (=lymphopoiesis) is controlled by an array of relatively large **growth factors** and smaller **cytokines**, some of which are covered further in Chapter 7.

6.9 SOME BLOOD DISORDERS: HEMOPHILIAS AND ANEMIAS

Hemophilias are diseases in which excessive bleeding occurs due to impaired clotting capabilities. Hemophilia can be acquired, for example from insufficient intake of **vitamin K** needed for clotting reactions. More commonly, however, hemophilia is inherited. For example, **hemophilia A** involves mutations in the clotting factor VIII gene. Depending on the particular mutation, symptoms can range from mild—with increased injury-induced bruising and slower clotting reactions— to severe, involving spontaneous internal bleeding. Disease progression can often be controlled via transfusions of clotting factors, with the current hope of finding an outright cure through targeted gene therapy.

Anemias are characterized by the reduced oxygen-carrying capacity of blood owing to insufficient levels of normally functioning erythrocytes. Anemia can result from excessive bleeding, inadequate production of functional erythrocytes, or an increased breakdown of erythrocytes (=**hemolytic anemias**). Along with sickle cell anemia (Box 6.1), **thalassemia** is a common form of hemolytic anemia that affects nearly 300 million people worldwide, with α- and β-thalassemias arising from mutations in Hb's α- and β-globin chains, respectively. Depending on the particular mutations, symptoms can range from general fatigue to life-threatening, and treatments usually involve frequent blood transfusions.

6.10 SUMMARY—BLOOD

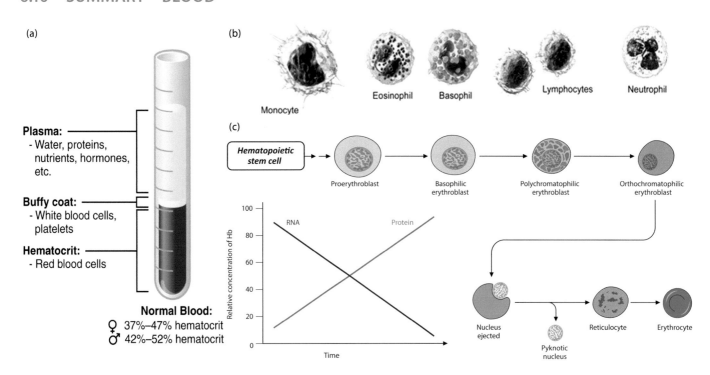

Pictorial Summary Figure Caption: (***a***) *see figure 6.1a;* (***b***) *Blood cells from Blausen.com staff (2014) "Medical gallery of Blausen Medical 2014". WikiJournal of Medicine 1 (2): 10. doi:10.15347/wjm/2014.010 reproduced under CC0 and CC SA creative commons licenses;* (***c***) *see figure 6.9d.*

Blood transports gases, nutrients, soluble wastes, bioactive molecules, and blood cells. For such functions, blood comprises: **plasma** fluid (=**serum** after removing clotting elements), red blood cells (**erythrocytes**), white blood cells (**leukocytes**), and **platelets**. After centrifugation, plasma is in the top layer (~50–60% of blood volume), leukocytes and platelets are in middle **buffy coat** (~1%), and erythrocytes occur at the bottom, thereby constituting the **hematocrit** of blood (~40–50%) (**Figure a**).

Plasma is ~90–95% water plus **small solutes, wastes**, and **proteins** (e.g., **albumins, globulins**, and **fibrinogen**), and some of this plasma leaks from the bloodstream into ECM compartments to be converted into lymph and eventually returned to the heart.

Erythrocytes constitute 99+% of all blood cells, measure **~7–8 μm** wide, have a concave discoidal shape, and lack a nucleus or synthetic organelles. Instead, erythrocytes are filled with **hemoglobin (Hb)** for carrying O_2 (and to a lesser extent, other gases). Each Hb molecule typically consists of: two **α-globin** proteins, two **β-globin** proteins, and **four heme groups** for reversibly binding gas molecules. The erythrocyte plasmalemma and underlying **cytoskeleton** normally maintain integrity and flexibility. However, during **sickle cell anemia**, mutated **β-globin** allows Hb to **polymerize**, thereby deforming erythrocytes into a sickle shape and promoting hemolytic breakdown with accompanying vessel clogging.

Platelets are anucleate cellular fragments that arise from **proplatelets** generated by **megakaryocytes**. Activated platelets can attach to damaged vessel walls via several molecules (e.g., **von Willebrand factor**) and can stop bleeding by producing **blood clots (thrombi)** comprising aggregated platelets, erythrocytes, and a **fibrin** meshwork that collectively plug breaks in the endothelial lining. Platelet activation involves **thromboxane A2**, and some steps in clotting reactions require **vitamin K**. Such processes can be downregulated by **aspirin** and **warfarin** types of anticoagulants, respectively, to prevent pathological clotting (**thrombosis**) that occludes vessels.

Leukocytes comprise: (i) **granulocytes (neutro-, eosino-**, and **basophils**) with cell-specific **secondary granules** plus **azurophilic granules** vs. (ii) **agranulocytes (lymphocytes** and **monocytes**) that have only azurophilic granules (**Figure b**). Typical fractions of white blood cell counts consist of neutrophils (50–70%), lymphocytes (20–40%), monocytes (5–10%), eosinophils (1–5%), and basophils (<1%). **Neutrophils** have multi-lobed nuclei plus pink granules with bactericidal agents (e.g., **myeloperoxidase**) and can phagocytose pathogens or trap them in the ECM via **NETs** (neutrophil extracellular traps). Neutrophils are abundant in **pus** of infected wounds. **Eosinophils** have bipartite secondary granules with components like **major basic protein** to kill pathogens, including worm parasites. **Basophils** possess **metachromatic** secondary granules and use molecules like **histamine** in allergic responses. **Lymphocytes** are the smallest leukocytes. Two main kinds of lymphocytes (**T** and **B**) are used in immune responses. **Monocytes** are the largest leukocytes and can differentiate into **macrophages**.

The sites of blood cell formation (**hematopoiesis**) change in the embryo and fetus before ending up in **bone marrow** and the **thymus** before birth. Hematopoiesis continues mainly in marrow through adulthood. Blood cells arise from **hematopoietic stem cells (HSCs)** in **red marrow** that is often replaced by less-active, fat-containing **yellow marrow**. Depending on stimulatory cues, HSCs differentiate along: i) an erythroid line to form **erythrocytes** plus platelet-producing **megakaryocytes** or (ii) **myeloblast** vs. **lymphoblast** lineages that generate granulocytes, monocytes, and mast cells vs. lymphocytes.

Red blood cell formation (**erythropoiesis**) occurs mainly in response to **erythropoietin (Epo)** hormone from the kidney that causes **pro-erythroblasts** to develop through stages that switch from basophilic to acidophilic cytoplasm as Hb's globin mRNAs are translated into globin proteins (**Figure c**). **Acidophilic** (=orthochromatophilic) **erythroblasts** lose their nucleus and other organelles for maximum Hb storage. The enucleated immature **reticulocyte** enters the bloodstream to mature into an erythrocyte that lives for **~110 days** before being cleared by phagocytes in the spleen and liver.

Some blood disorders: hemophilias cause excessive bleeding due to impaired blood clotting; **anemias** result in reduced oxygen-carrying capacity due to excessive bleeding, insufficient erythropoiesis, and/or erythrocytic lysis (e.g., hemolytic anemia like **sickle cell anemia** and **thalassemias**).

SELF-STUDY QUESTIONS

1. Which hormone acts as an overall stimulator of erythropoiesis?
2. T/F The buffy coat contains leukocytes and platelets.
3. Which gene is mutated in sickle cell anemia?
4. Which is filled almost entirely with untranslated globin mRNA?
    A. Acidophilic erythroblast
    B. Basophilic erythroblast
    C. Polychromatophilic erythroblast
    D. Erythrocyte
    E. Reticulocyte
5. Which is a major component of pus in infected wounds?
    A. Monocytes
    B. Myeloblasts
    C. Lymphoblasts
    D. Neutrophils
    E. Basophils
6. Which cells differentiate into macrophages?
    A. Neutrophils
    B. Monocytes
    C. Lymphocytes
    D. Eosinophils
    E. Megakaryocytes
7. Which cells give rise to platelets?
8. T/F Lymphocytes are the most common agranulocytes.
9. Which is the least common leukocyte?
    A. Lymphocyte
    B. Neutrophil
    C. Eosinophil
    D. Basophil
    E. Monocyte
10. Which circulating blood cell has a multi-lobed nucleus?
11. Which cells produce large amounts of myeloperoxidase?
    A. Erythrocytes
    B. Basophils
    C. Monocytes
    D. Eosinophils
    E. Neutrophils
12. Which of the following is NOT a common plasma protein?
    A. Albumin
    B. Glycophorin A
    C. Globulin
    D. Fibrinogen
    E. ALL of the above
13. T/F. Most erythrocytes live for a total of 14 days.

"EXTRA CREDIT" For each term on the left, provide the BEST MATCH on the right (answers A-F can be used more than once)

14. Metachromasia _____   A. Neutrophil
15. Major basic protein_____   B. Eosinophil
16. NET _____   C. Basophil
17. Reticulocyte_____   D. Platelet
18. von Willebrand factor _____ E. Monocyte
                                   F. Erythrocyte

19. Describe platelet formation and blood clotting.
20. Describe basic steps in red blood cell formation from *in utero* through adulthood.

ANSWERS

1) erythropoietin; 2) T; 3) β-globin; 4) B; 5) D; 6) B; 7) megakaryocytes; 8) T; 9) D; 10) PMN; 11) E; 12) B; 13) F; 14) C; 15) B; 16) A; 17) F; 18) D; 19) The answer should include among other pertinent topics: bone marrow, megakaryocyte, sinusoid, proplatelet, platelet granules/cytoskeleton, clotting factors, platelet activation, platelet attachment, von Willebrand factor, platelet plug, erythrocytes, fibrin, thrombus. 20) The answer should include and clearly define such terms as yolk sac, liver, bone marrow (red vs. yellow), HSCs, MPPs, erythroid lineage, Epo, pro-, basophilic-, polychromatophilic-, acidophilic erythroblasts, reticulocytes, maturation, erythrocyte

Circulatory and Lymphatic Systems

7.1 INTRODUCTION TO CIRCULATION

Blood and **lymph** are circulated throughout the body by two interrelated networks—the **circulatory** (=blood vascular) and **lymphatic systems**. Collectively, such circulation helps maintain homeostasis by: (i) facilitating gas, nutrient, and waste exchanges; (ii) regulating ECM fluid levels; and (iii) distributing various cells and bioactive molecules for normal physiological responses and for reactions to toxins and pathogens.

In addition to blood, the **circulatory system** comprises **arteries, capillaries, veins**, and the **heart**. The heart pumps blood through arteries to supply beds of interconnected capillaries, whereas capillaries return blood via veins to the heart. During this transport, plasma that had leaked into the ECM can be taken up by lymph vessels and converted into lymph. Along with lymphatic vessels and fluid, the **lymphatic system** comprises: (i) **non-encapsulated tissues**, such as bone marrow, tonsils, and more diffuse arrays in the mucosal lining of the

DOI: 10.1201/9780429353307-9

digestive and respiratory tracts; and (ii) three types of **encapsulated lymphoid organs—lymph nodes** plus the **spleen** and **thymus**.

7.2   BLOOD CIRCULATION IN THE FETUS VS. ADULT PLUS MORPHOLOGY OF THE HEART WITH ITS TWO ATRIA AND TWO VENTRICLES

The heart develops from both mesodermal and neural crest sources of differentiating cells. Such cardiogenic precursors initially form two separate primordia that subsequently fuse to produce a rhythmically contracting tube by 4 weeks post-fertilization (**Figure 7.1a**). By the 6th–8th week of embryogenesis, the tubular heart **loops** back on itself and undergoes a **septation** process to yield a four-chambered organ comprising two upper **atria** that receive blood plus two lower **ventricles** that pump blood from the heart (**Figure 7.1a**).

Before birth, the lungs are non-functional, and gas exchanges as well as nutrient loading occur in the placenta (Chapter 18). Thus, a fetal shunt, called the **ductus arteriosus**, connects the pulmonary arterial trunk with the aorta, thereby diverting lung-targeted blood to the placenta (**Figure 7.1b, c**). In addition, because little blood returns to the heart from the lungs, the median cardiac septum forms a hole, called the **foramen ovale**, which allows direct blood flow between the atria, thereby counteracting blood volume imbalances caused by the ductus arteriosus (**Figure 7.1a–c**). Soon after birth, both fetal structures normally close on their own (**Figure 7.1b**), although surgical intervention may be needed to correct defects, such as a patent foramen ovale ("hole in the heart"). Following post-natal remodeling, the right side of the heart functions in **pulmonary circulation**, sending de-oxygenated blood from the right ventricle via pulmonary arteries to the lungs for oxygenation (**Figure 7.1d**). Oxygen-rich blood then returns via pulmonary veins to the left atrium (**Figure 7.1d**). Conversely, the left-side **systemic circulation** pumps oxygenated blood from the left ventricle into the aorta to supply all of the body except for gas-exchange sites in the lungs before returning deoxygenated blood to the right atrium via the inferior and superior **venae cavae** (**Figure 7.1d**).

From the inside outward, the wall of the adult heart comprises three main layers— **endocardium**, **myocardium**, and **epicardium** (**Figure 7.2a–e**) (**Case 7.1 Frederick Douglass's Heart**). The endocardium lines heart chambers with a simple squamous epithelium that overlies a thin sub-epithelial layer of connective tissue. Beneath the endocardium, the muscle-containing myocardium constitutes the bulk of the heart wall and is particularly thick in the left ventricle where stronger contractions are needed to propel blood throughout the body. For such roles, myocardial myocytes are densely packed and interconnected by **intercalated discs** (Chapter 8) that help coordinate heartbeats and maintain myocyte integrity during repeated contraction cycles (**Figure 7.2f, g**). Overlying the myocardium, the epicardium is also referred to as the **visceral pericardium**, which in turn is overlain by a thin pericardial cavity and an outermost parietal pericardium (**Figure 7.2a**).

CASE 7.1 FREDERICK DOUGLASS'S HEART

Abolitionist and Orator
    (1818–1895)

    "The life of a nation is secure only while the nation is honest, truthful, and virtuous."

    (From an 1885 speech by F.D.)

Before becoming the most famous and influential African-American activist in 19th-century America, Frederick Douglass learned to read and write while enslaved in Maryland. At age 20, Douglass was able to escape and settle in Massachusetts. Drawing upon his oratorical skills, Douglass rallied against the injustices of slavery and the lack of voting rights for women. To disseminate his views, Douglass gave numerous speeches and published several books as well as a series of progressive newspapers that emphasized a call for equal rights. Following the adoption of the 13th and 15th amendments that outlawed slavery and gave African-American men voting rights, Douglass continued his support for women's suffrage, and in 1895, after receiving a standing ovation at the National Council of Women in Washington DC,

Douglass returned home, where he died of a heart attack at age 77. Thus, the eloquent activist, who spoke out for justice throughout his life, was eventually silenced by his ailing heart.

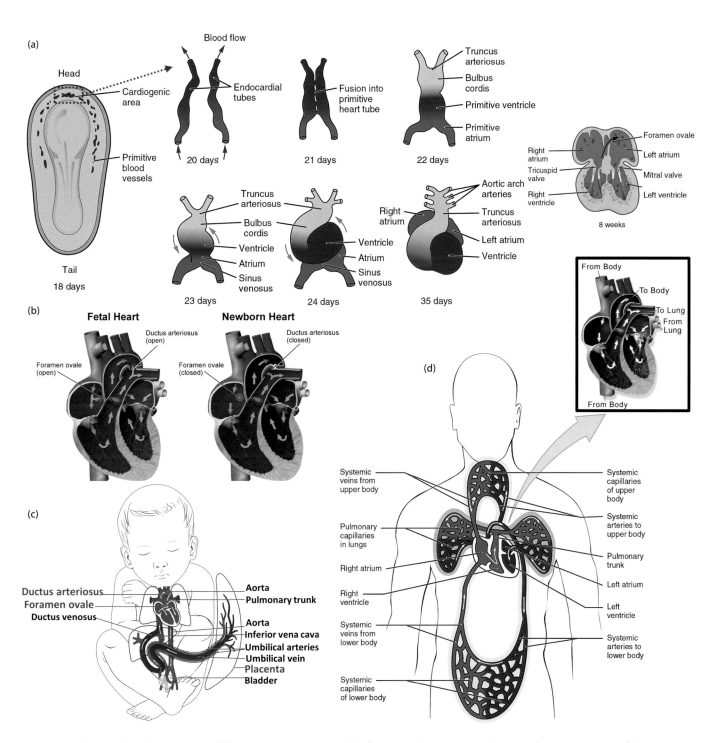

Figure 7.1 Heart development. (a) *The heart develops mainly from cardiogenic mesoderm at the anterior end of the embryo and transitions via the fusion of two primordia into a single tube. After beating begins by the end of the 4th week, the tubular heart loops back on itself to form a four-chambered organ with two upper atria and two lower ventricles by 8 weeks.* **(b, c)** *The fetal heart has a ductus arteriosus that shunts blood from the pulmonary arteries to the aorta so that most of the blood bypasses the non-functional lungs and is sent to the placenta for nutrient, waste, and gas exchanges with the maternal circulation. To compensate for the pressure imbalance caused by the ductus arteriosus, a foramen ovale allows blood to move directly from the right to left atrium (see also figure a).* **(d)** *Both the ductus arteriosus and foramen ovale normally close soon after birth so that blood can be directed to alveoli of the lungs via the pulmonary circulation and circulated throughout the rest of the body via the aorta and systemic circulation. Atria receive blood from veins, whereas ventricles expel blood into arteries. ([a, c, d]: From OpenStax College (2013) Anatomy and physiology. OpenStax. http://cnx.org/content/col11496/latest reproduced under a CC BY 4.0 creative commons license; [b, d, inset]: From Blausen.com staff (2014) Medical gallery of Blausen Medical 2014. Wiki J Med 1 (2): 10: doi:10.15347/wjm/2014.010 reproduced under CC0 and CC SA creative commons licenses.)*

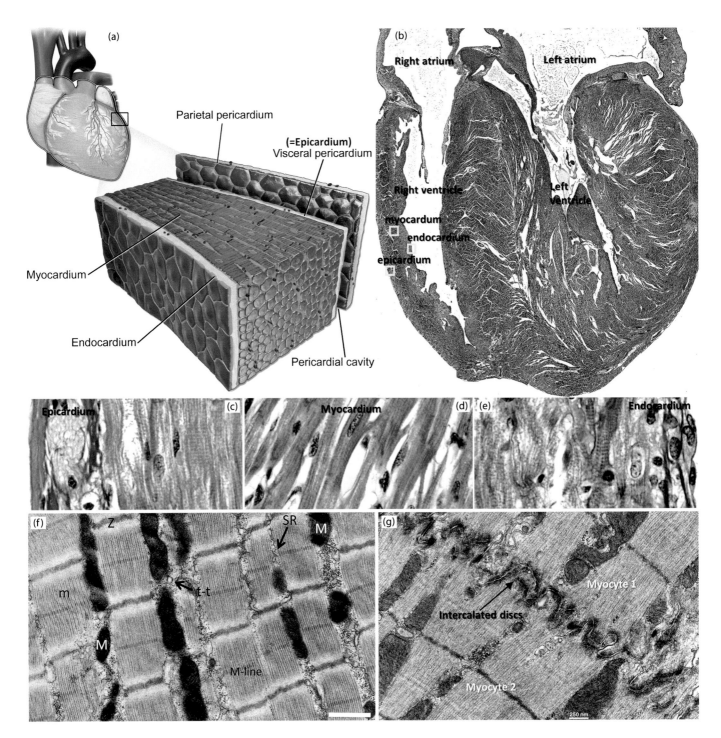

Figure 7.2 The heart wall. (*a–e*) *The heart is covered by a pericardial sac and has a wall composed of three layers—epicardium (=visceral layer of the pericardium), myocardium, and endocardium. (a) 6X; (c–e) 600X. (**f, g**) The myocardium constitutes most of the cardiac wall and contains cardiac myocytes that are held together at their ends by intercalated disc junctions. (f) m = myofilaments; M = mitochondria; SR = sarcoplasmic reticulum; t-t = transverse tubule; scale bar = 1 μm. ([a]: From Blausen.com staff (2014) "Medical gallery of Blausen Medical 2014". Wiki J Med 1 (2): 10. doi:10.15347/wjm/2014.010 reproduced under CC0 and CC SA creative commons licenses; [f]: From Pinali, C and Kimotto, A (2014) Serial block face scanning electron microscopy for the study of cardiac muscle ultrastructure at nanoscale resolutions. J Mol Cell Cardiol 76: 1–11 reproduced with publisher permission; [g]: From TEM images courtesy of Drs. H. Jastrow and H. Wartenberg reproduced with permission from Jastrow's Electron Microscopic Atlas (htttp://www.drjastrow.de).)*

7.3 ACCESSORY CARDIAC STRUCTURES: VALVES AND A CONDUCTING SYSTEM COMPRISING THE SA NODE, AV NODE, BUNDLE OF HIS, AND PURKINJE FIBERS FOR PROPAGATING ELECTRICAL IMPULSES

The heart contains four **valves**. Two of these guard the atrioventricular (AV) openings, whereas the other two occur at ventricular outlets leading into the aorta or the pulmonary arterial trunk on the left and right sides of the heart, respectively (**Figure 7.3a–d**). Heart valves comprise flap-like leaflets, each of which contains a connective tissue core covered by endocardium (**Figure 7.3c, d**). Retrograde flow pushing against the flaps shuts the valve to prevent backflow, thereby ensuring unidirectional transport from atria into ventricles and from ventricles into the aorta and pulmonary arteries. The left-side AV valve contains only two flaps and is referred to as the **bicuspid (=mitral) valve**, whereas the other three valves have three flaps to form **tricuspid valves** (**Figure 7.3b**).

The two AV valves are anchored in a fibrous collagenous skeleton that also insulates the upper heart from the lower half so that atria and ventricles can contract at different times. In addition, each valve is tethered to the ventricular wall via tough collagenous strands, called **chordae tendineae**, which normally help keep valves from folding into the atria (**prolapsing**), thereby ensuring efficient pumping. However, mitral valve prolapse affects ~2% of the population and may eventually cause chordae rupturing with accompanying heart failure.

In addition to supporting structures, the heart has a **conducting system** that comprises: (1) two **nodes** of cells that can self-generate action potentials (APs) (Chapter 9), and (2) a network of conducting cells terminating in **Purkinje fibers** that propagate node-produced APs to the myocardium for triggering contractions (**Figure 7.3e–g**). Situated in the right atrial wall near its connection to the superior vena cava, the **sinoatrial (SA) node** serves as the heart's primary **pacemaker** by generating ~70 APs/min. This baseline firing rate can be slightly reduced by parasympathetic innervation (Chapter 9). Conversely, sympathetic nerves (Chapter 9) not only cause more rapid heartbeats but also increase the force of contractions, mainly via myocyte **β-adrenergic receptors**. Thus, **beta-blocker** drugs targeting these receptors are used to treat various cardiovascular diseases.

A small portion of SA node output is sent via **Bachmann's bundle** to the left atrium, whereas the majority of APs are relayed to the **atrioventricular (AV) node** in the right AV wall (**Figure 7.3e**). The AV node functions as a relay station as well as a backup pacemaker that can replace a non-functional SA node. Due to the collagenous skeleton separating the atria and ventricles, AV node output can only reach the ventricles via the **bundle of His**, which is a tract of transmitting cells in the median septum. This bundle bifurcates into the right and left ventricles to terminate as ramifying **Purkinje fibers** situated mainly between the endo- and myocardium (**Figure 7.3f**). Purkinje fibers are modified myocytes with reduced perinuclear myofibrils (**Figure 7.3g**) as well as unique ion channels that allow more rapid AP transmissions.

7.4 MICRO- VS. MACROVASCULATURE AND REDUCED BLOOD FLOW VELOCITIES IN CAPILLARIES

Based on overall size, blood vessels can be classified into two major groups: (i) **microvasculature** that requires microscopy in order to be observed vs. (ii) **macrovasculature** that can be seen with a magnifying glass or unaided eye. The smallest microvasculature vessels consist of ramifying networks of **capillaries** plus **arterioles** and **venules** that lead into, and away from, capillary beds. Conversely, the largest macrovasculature vessels are the **venae cavae** and **aorta**.

Given the diameter of individual vessels, it might be expected that blood flow velocities would increase in microvasculature compared to macrovasculature in the same way that placing a thumb over a garden hose reduces the hose's effective diameter and thereby elevates flow rates. However, blood flow velocity (**V**) is inversely proportional to the **total cross-sectional area (CA)** of vessels in which blood travels (**V = 1/total CA**). Thus, even though each microvessel is small compared to large vessels, interconnections between microvessels in and around capillary beds generate a large total cross-sectional area (**Figure 7.4a**). This property, along with other factors, such as added friction of blood cells against microvessel walls, causes blood to flow slowly through microvasculature (**Figure 7.4a**), thereby facilitating gas, nutrient, and waste exchanges in capillaries.

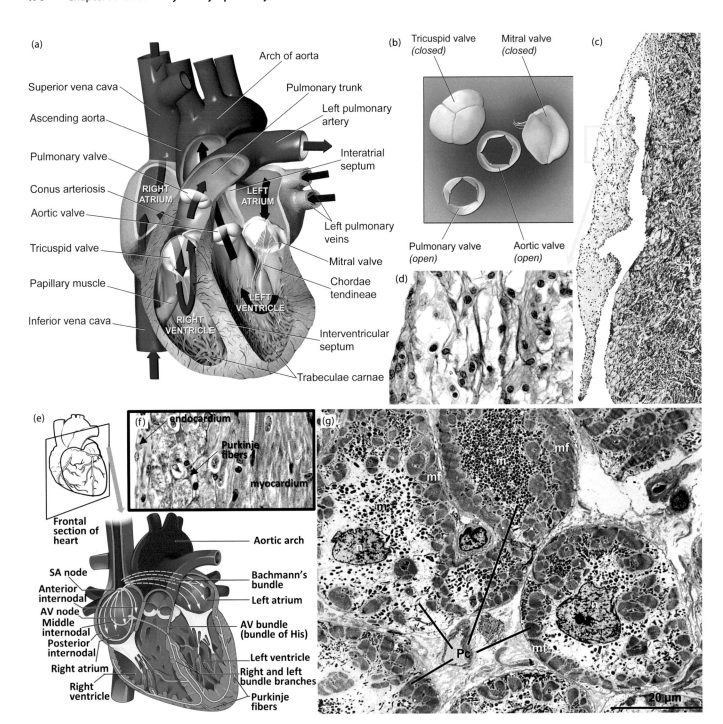

Figure 7.3 Cardiac valves and conducting system. *(a, b) The atrioventricular (AV) passageways and the openings from the ventricles leading into the pulmonary trunk and the aorta are guarded by valves that ensure one-way blood flow. Except for the dual-leaflet mitral valve between the left ventricle and the aorta, cardiac valves have three leaflets and hence are tricuspid. (c, d) Each valve has a connective tissue core and is tethered by tough strands, called chordae tendinae (see figure (a)); c: 45X; d: 250X. (e, f) To coordinate cardiac contractions and relaxations, a conducting system composed of modified myocytes, called Purkinje fibers (cells), occurs mainly at the endo-/myocardium interface. Such fibers receive rhythmic stimulation from a pacemaker region in the right heart, called the sinoatrial (SA) node, which has a backup center in the atrioventricular (AV) node. (g) Purkinje cells (Pc) are characterized by (i) their larger size, (ii) reduced number of myofibrils (mf), which occur mainly around the cell periphery, and (iii) abundant mitochondria (m) surrounding the nucleus (n) in the cell center. ([a, b]: From Blausen.com staff (2014) Medical gallery of Blausen Medical 2014. Wiki J Med 1 (2): 10. doi:10.15347/wjm/2014.010 reproduced under CC0 and CC SA creative commons licenses; [e]: From OpenStax College (2013) Anatomy and Physiology. OpenStax: http://cnx.org/content/col11496/latest reproduced under a CC BY 4.0 creative commons license; [f]: From Mitrofanova, LB et al. (2014) Evidence of specialized tissue in human*

Figure 7.3 (Continued) *interatrial septum: histological, immunohistochemical and ultrastructural findings. PLoS ONE 9: doi:10.1371/journal.pone.0113343 reproduced under a CC BY 4.0 creative commons license.)*

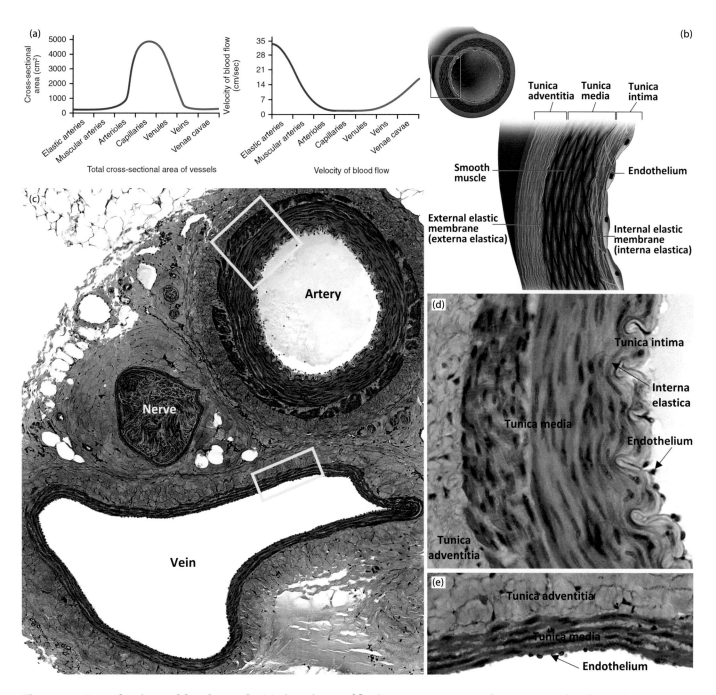

Figure 7.4 Introduction to blood vessels. (*a*) *The velocity of fluid movement is inversely proportional to the total cross-sectional area of the tube in which fluid is flowing. Because of their large total cross-sectional area, interconnected capillaries, arterioles, and venules of capillary beds have relatively low blood flow velocities thereby optimizing gas and nutrient exchanges in capillaries. (**b**) The walls of arteries and veins comprise three layers: (i) tunica intima (simple squamous endothelium and subendothelial connective tissue plus a peripheral interna elastica in arteries); (ii) tunica media (smooth muscle layers plus an externa elastica in arteries); and (iii) tunica adventitia (connective tissue, nerves, and a vasa vasorum of supporting blood vessels in the case of larger vessels). (**c–d**) Compared to corresponding veins, arteries tend to have: (i) thicker walls, (ii) a more scalloped intima due to their interna elasticas, and (iii) a wider lumen, which does not collapse as readily during fixation. (a) 60X, (d, e) 250X. ([a]: From OpenStax College (2013) Anatomy and Physiology. OpenStax. http://cnx.org/content/col11496/latest reproduced under a CC BY 4.0 creative commons license; [b]: From Blausen.com staff (2014). "Medical gallery of Blausen Medical 2014". WikiJournal of Medicine 1 (2): 10. doi:10.15347/wjm/2014.010 reproduced under CC0 and CC SA creative commons licenses.)*

7.5   THREE PARTS OF A TYPICAL BLOOD VESSEL WALL—TUNICA INTERNA (=INTIMA), TUNICA MEDIA (=MEDIA), TUNICA EXTERNA (=ADVENTITIA)—PLUS CAPILLARY TYPES AND ARTERIES VS. VEINS

The lumen of all blood vessels is lined by a simple squamous epithelium, called the **endothelium**. In vessels larger than capillaries, the endothelium forms the innermost component of three concentrically arranged layers (or **tunics**)—the **tunica interna (=intima)**, **tunica media (=media)**, and **tunica externa (=adventitia)** (**Figure 7.4b–e**). The **intima** comprises an endothelium plus a thin sub-endothelial connective tissue sheath. In small to large arteries, the intima is bounded peripherally by an elastic-fiber-containing **interna elastica**, which is generally lacking in arterioles and is invariably absent in venous vessels. Peripheral to the intima, the media contains **smooth muscle** arrays ranging from one layer in arterioles and venules to numerous layers in larger vessels. Moreover, in medium to large arteries, the outermost muscle layer is bordered by an **externa elastica** of elastic fibers. Surrounding the media is a connective tissue **adventitia** with parasympathetic nerve fibers (Chapter 9) that promote muscular relaxation. Conversely, sympathetic nerves (Chapter 9) in the adventitia trigger vasoconstriction and hence increase peripheral resistance, thereby causing elevated blood pressure (=**hypertension**). In larger vessels, the adventitia also contains its own set of smaller blood vessels (=**vasa vasorum**), which facilitate molecular exchanges in a thickened wall whose functions could not be maintained simply by diffusion from the vessel's central lumen.

The **aorta** and other **large arteries** have abundant **elastic fibers** in their tunica media and are thus referred to as **elastic arteries** (**Figure 7.5a, b**). Such fibers not only re-enforce an arterial wall that is subject to relatively high pressures but also aid in vessel recoiling during pulsatile propulsions of blood from the heart (**Case 7.2 Margaret Thatcher's Arteries**). Intravascular elastic fibers and smooth muscle layers become progressively less abundant as large arteries transition into arterioles (**Figure 7.5c–e**).

Arteries can be distinguished from similarly sized veins by a scalloped endothelial profile due to their **interna elasticas** and multiple smooth muscle layers. Alternatively, veins lack interna elasticas and possess fewer smooth muscle layers than do cognate arteries. Thus, the thinner walls of veins tend to collapse in conventional preparations, although the lumens of veins can be preserved in an expanded state by injecting blood vessels with resins and then digesting tissues surrounding the polymerized vascular casts (**Figure 7.5f, g**) (**Case 7.3 Richard Nixon's Veins**). In addition, veins are distinguished from arteries not only in having **valves** that ensure unidirectional blood flow but also in failing to undergo pathological thickening of their intimas as occurs in **atherosclerotic** arteries (Box 7.1 Cell Biology of Atherosclerosis).

Capillaries differ from other blood vessels in containing just an endothelium plus associated **pericytes**, which control vascular permeability and help repair damaged capillaries (**Figure 7.6a, b**). Based on overall size and

---

CASE 7.2 MARGARET THATCHER'S ARTERIES

British Prime Minister
(1925–2013)

"While the conventional, political dangers...appear to be receding, we have all recently become aware of another insidious danger. It is the prospect of irretrievable damage to the atmosphere, to the oceans, to earth itself."

(M.T. in a 1989 speech to the United Nations warning about destruction of the environment)

Nicknamed the "Iron Lady" for her dogged determination, Margaret Thatcher was the first female Prime Minister of the United Kingdom and the longest-serving head of British government during the 20th century. Born in Lancashire, England, Thatcher received a Bachelor's degree in Chemistry at Oxford University while conducting her thesis research under the supervision of future Nobel Laureate Dorothy Hodgkin. In her first chemistry job interview, Thatcher was

rejected for being *"headstrong, obstinate and dangerously self-opinionated"* but was later hired as a chemist at another company to work on emulsifiers for ice cream. After earning a law degree, Thatcher went on to win local and national elections, including for Prime Minister. Drawing upon her scientific training, Thatcher sounded early wake-up calls about impending environmental destruction, although critics argue

she later reversed her views to side with climate-change deniers. As she aged, Thatcher began suffering minor strokes and ischemia-induced dementia. In the end, the iron-willed ruler of the United Kingdom died at age 87 due to a stroke, as her arteries became incapable of maintaining cerebral functions.

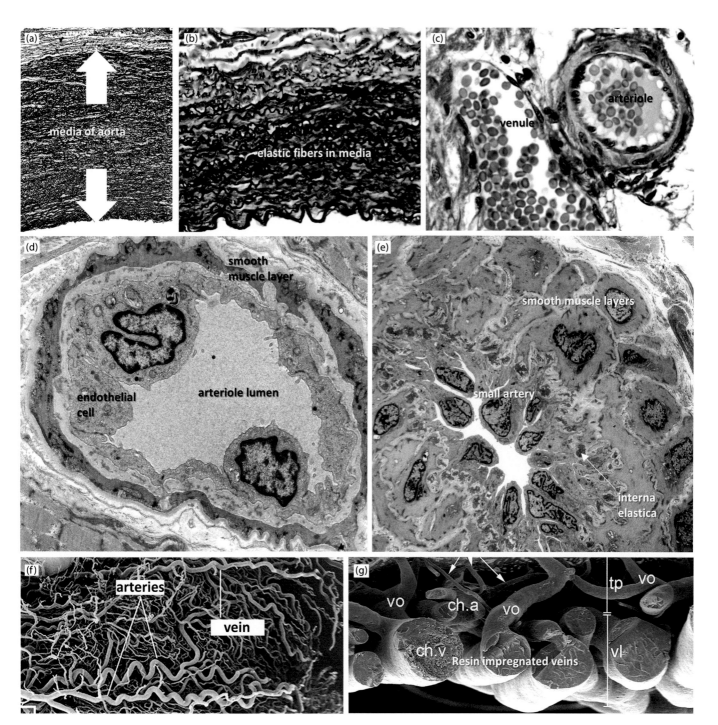

Figure 7.5 Arteries, veins, arterioles, and venules. (a, b) *The largest arteries, such as the aorta and its offshoots are characterized by abundant elastic fibers in their media, providing elasticity for high pressure blood flow from the heart. (a) 20X; (b) 150X. (**c–e**) The smallest blood vessels—arterioles and venules (c, 350X)—typically have but a single layer of smooth muscle, and in the case of arterioles, the interna elastica is either absent or poorly developed (d, 2,700X) compared to small arteries with their multiple smooth muscle layers (e, 800X). (**f, g**) Veins often collapse during conventional preparation techniques (see* **Figure 7.4b**) *but can be kept in an expanded state by injecting plastic resin and subsequently digesting surrounding tissues after resin polymerization, as shown in these vascular corrosion casts of a horse's eye ((f) 60X; (g) 140X). ([de]: From Buckley, AF and Bossen, EH (2013) Skeletal muscle microvasculature in the diagnosis of neuromuscular disease. J Neuropathol Exp Neurol 72: 906–918, reproduced with publisher permission; images courtesy of Dr. A. Buckley. [f, g]: From Ninomiya, H and Inomata, T (2014) Functional microvascular anatomy of the horse eye: a scanning electron microscopic study of corrosion casts. Open J Vet Med 4: 91–101, reproduced under a CC BY 4.0 creative common license.)*

CASE 7.3 RICHARD NIXON'S VEINS

37th U.S. President
   (1913–1994)

   "Well, I am not a crook."

   *(R.N. regarding his alleged criminal acts during the
                                        Watergate scandal)*

As Richard Nixon was beginning his second term as U.S. President in 1972, the arrest of five burglars at the Democratic National Committee office in a Washington D.C. complex, called Watergate, hardly seemed noteworthy. However, mounting evidence began to connect the Watergate break-in and associated crimes to the White House, and after several of his aides were sent to prison, Nixon eventually resigned rather than face an impeachment vote. During the unfolding crisis, Nixon abused alcohol and according to several accounts became deranged and suicidal while wandering the White House at night. In addition, Nixon was also

dealing with recurring phlebitis, an inflammation of veins that is capable of generating fatal blood clots particularly when blood pools in the legs during excessive standing. In spite of his phlebitis, Nixon refused hospitalization and two months before his resignation went instead on an arduous tour of the Middle East where he chose to be on his feet many hours a day. Based in part on Nixon's own accounts of his deep despair, it has been speculated that while trying to redeem his legacy during the Middle East trip, Nixon had also hoped his diseased veins might serve as his "pistol in the drawer" and thereby provide him with a way out of his problems.

endothelial morphology, capillaries are classified into three subtypes—**continuous**, **fenestrated**, and **sinusoidal** (=discontinuous). As the most common morphotype in the body, **continuous** capillaries measure only ~5–10 μm wide and thus necessitate that erythrocytes typically move in a single file in order to squeeze through the constricted lumen. Such narrow passageways can in turn cause major problems for sickle cell patients whose abnormally shaped erythrocytes may clog capillary flow (Chapter 6). The endothelium of continuous capillaries lacks pores but often possesses numerous caveolae for aiding trans-endothelial transport. Moreover, in organs requiring separation between blood and surrounding tissues (e.g., blood-brain-, blood-testis-, or blood-thymus barriers), continuous capillary endothelia are sealed by abundant tight junctions (**Figure 7.6c, d**). **Fenestrated** capillaries resemble continuous capillaries in size and are common in sites, such as renal glomeruli (Chapter 16) where rapid trans-endothelial fluid flow is required. To facilitate such flow, each endothelial cell is perforated by ~50–100 nm-wide pores (=**fenestrations**) (**Figure 7.6a**), which in some cases can be closed by a modified membrane, called a **diaphragm**. **Sinusoidal** capillaries, such as found in the liver, spleen, and bone marrow, reach diameters two to five times greater than those of continuous and fenestrated capillaries. Moreover, along with fenestrations, the endothelial cells of sinusoids rest on a patchy basement membrane and are separated from each other by large gaps (**Figure 7.6a**). Such discontinuities facilitate movements of small cells to and from the lumen, thereby enabling rapid exchanges with the ECM.

7.6   UNUSUAL BLOOD FLOW PATTERNS: CAPILLARY BYPASSES AND DOUBLE-CAPILLARY PORTAL SYSTEMS

Unlike conventional arteriole-to-capillary flow patterns, **metarterioles** can reversibly divert blood to venules via thoroughfare **shunts**, which along with permanently fused arteriole-to-venule **anastamoses** serve to bypass capillaries (**Figure 7.6e**). Reversible shunt bypasses are abundant in the dermis of the hands and feet where they can use muscular sphincters for temporarily bypassing capillaries, which in turn reduces heat loss by avoiding blood being spread throughout the large surface areas of capillary beds.

   As opposed to capillaries bypasses, several body regions (e.g., liver, kidney, and hypothalamic-hypophyseal tract) transport blood from one capillary bed into a second set of capillaries before reaching the heart (**Figure 7.6f**). Such double-capillary circulations constitute **portal systems** and provide separations for distinct functional specializations, such as divergent hormone deliveries or different kinds of blood processing within the two capillary beds.

7.7   LYMPHATIC SYSTEM DEFENSES AGAINST PATHOGENS: HUMORAL VS. CELL-MEDIATED MODES PLUS INNATE VS. ADAPTIVE IMMUNITY

Barriers provided by skin and mucosal linings (Chapters 10 and 12) help protect internal body compartments from pathogens, such as disease-causing viruses, bacteria, and parasites (**Figure 7.7a**). However, for deleterious invaders that have breached these physical blockades, cells of the lymphatic system can launch

BOX 7.1 CELL BIOLOGY OF ATHEROSCLEROSIS

As the most common subtype of arteriosclerosis (hardening of the arteries), **atherosclerosis** involves the formation of plaques (**atheromas**) within the tunica intima and a concomitant reduction in blood flow as the enlarging intima occludes the arterial lumen (**Figure a–c**). Atheromas may also rupture and become plugged by fibrin-containing blood clots, with such complicated plaques providing dual threats of not only atheroma-induced occlusion but also the dislodging of clots

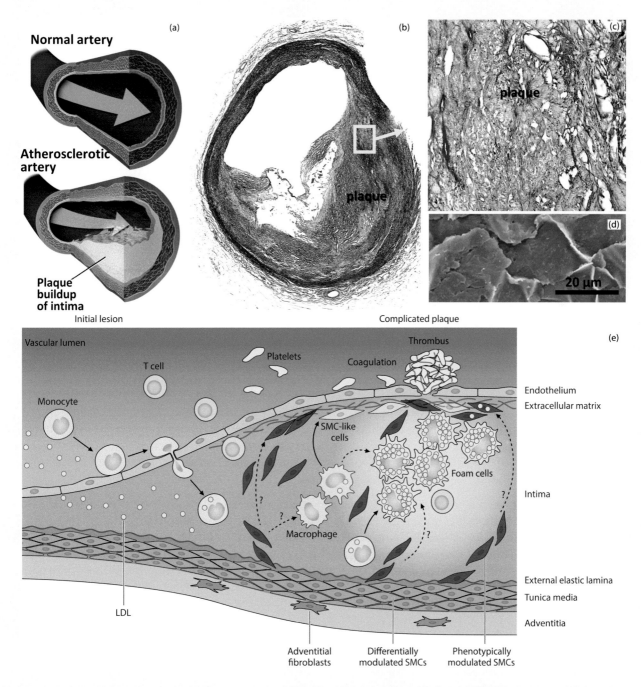

Box 7.1 *Atherosclerosis ((**a**) from Blausen.com staff (2014) "Medical gallery of Blausen Medical 2014". WikiJournal of Medicine 1 (2): 10. doi:10.15347/wjm/2014.010 reproduced under CC0 and CC SA creative commons licenses); (**b, c**) Paraffin section of atherosclerotic artery; (**d**) SEM of human calcified plaque (from: Curtze, SC et al. (2016) Step down vascular calcification analysis using state-of-the-art nanoanalysis techniques. Sci Rep 6: DOI: 10.1038/srep23285 reproduced under a CC BY 4.0 creative commons license); (**E**) (Adapted from Badimon et al. (2009) Lipoproteins, platelets, and atherothrombosis. Rev Exp Cardiol 62: 1161–1178.)*

and their subsequent blockage of smaller-bore vessels via thrombo-embolisms. Moreover, calcified atheromas (**Figure d**) negatively affect the biomechanical properties of elastic arteries, with the collective pathogenesis of atherosclerosis enhancing the chances of developing ischemia-mediated heart attacks and strokes.

Atheroma initiation is associated with excessive amounts of circulating **low-density lipoproteins** (**LDLs, bad cholesterol**) (**Figure e**). Such LDLs are able to breach an artery's endothelial lining particularly at susceptible sites, such as bends or branch points that are associated with unorthodox blood flow patterns. Intima-concentrated LDLs can then help trigger inflammation by reducing nitric oxide levels that normally inhibit atherogenesis. Elevated LDL levels also attract to the intima various cell types. For example, monocyte-derived macrophages are recruited to phagocytose LDLs, and the resulting lipid-containing **foam cells** secrete pro-inflammatory molecules that further thicken the intima. Conversely, **smooth muscle cells** (**SMCs**) and SMC-like myofibroblasts from the media can help form a cap over the atheroma (**Figure e**) for limiting atheroma growth. In some cases, an intact cap keeps the artery from progressing into an acutely diseased state. Alternatively, if a lesion forms in the cap, complicated plaque development can lead to blood clots and high-risk pathogenesis.

anti-pathogenic **immune responses** to the pathogen's **antigens**, which are defined as molecules that elicit host **defense reactions**, such as **antibody** production. Based on the sites where anti-pathogenic immune cells mature, the lymphatic system is often divided into **primary lymphoid organs** (adult bone marrow and thymus) in which B and T lymphocytes (=**B** and **T cells**) form and mature vs. **secondary lymphoid organs** (lymph nodes, spleen, and non-encapsulated lymphatic tissues), where B and T cells from the marrow and thymus can interact with pathogens. Lymphoid organs and tissues are able to use widely dispersed secretory molecules like blood-borne antibodies to generate what are termed **humoral responses**. Alternatively, more localized defenses produced by certain lymphocytes, granulocytes, and macrophages constitute **cell-mediated responses**. Each integrated suite of anti-pathogenic defense reactions typically combines both humoral and cell-mediated responses, with the collective **immunity** supplied by such responses often being classified as **innate** vs. **adaptive** (**Figure 7.7a**).

Innate immunity is phylogenetically ancient and widely distributed across extant animal groups. As an inherited system that is already functional at birth, innate immunity provides initial anti-pathogenic defenses involving not only physical obstacles, such as skin and mucosal structures but also rapidly acting humoral and cell-mediated mechanisms. Cells contributing to innate immunity (e.g., granulocytes, macrophages, and natural killer [NK] cells) (**Figure 7.7a**) express a comparatively limited number of germline-derived receptors (e.g., **pattern recognition receptors** of neutrophils or **Toll-like receptors** of macrophages) that mediate relatively non-specific reactions to general groups of pathogens. In addition, innate immunity can often be launched within only a few hours after encountering pathogens but usually lacks a **memory** component from previous pathogen encounters for facilitating subsequent responses.

Conversely, **adaptive immunity** has evolved more recently and is traditionally viewed as an innovation of vertebrates. Full-fledged adaptive immunity is only acquired after post-natal exposures to pathogens, and compared to innate immunity, adaptive immunity provides greater specificity when targeting pathogens via its production of billions of versions of pathogen-reactive receptors. For T and B cells, respectively, these comprise: (i) **T-cell receptors** (**TCRs**) plus their co-receptor molecules vs. (ii) **B-cell receptors** (**BCRs**) consisting of **antibodies** plus associated membrane molecules (**Figure 7.7b, c**). Such diverse receptor arrays are produced in maturing T and B cells via random rearrangements of receptor molecules during a **somatic recombination** process. This allows T and B cells expressing the appropriate receptor for a particular pathogen to be amplified by mitotic divisions into expanded clones of suitably reactive cells. TCR and BCR production often requires antigen-presenting cells, such as dendritic cells that utilize **major histocompatibility** (**MHC**) proteins to help deliver antigens (**Figure 7.7b, d**). Due to their numerous mitoses and complex intercellular interactions, adaptive immune responses are typically slower in initiating than those triggered by innate immunity. However, adaptive responses can utilize memories of previous pathogen encounters to help adapt to subsequent exposures and thereby produce more accelerated and robust secondary reactions.

Although features such as these are traditionally used to distinguish innate vs. adaptive immunity (Table 7.1), several kinds of cells have attributes of both innate and adaptive immunity. For example, **innate lymphoid cells** (**ILCs**) lack diversified receptors but nevertheless possess other T-cell-like characteristics, thereby blurring the lines of a conventionally binary classification scheme.

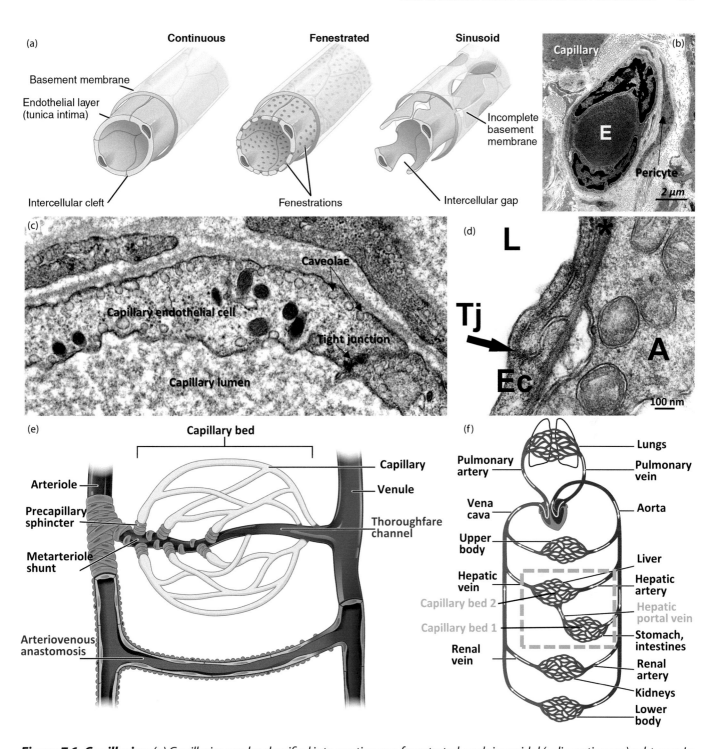

Figure 7.6 Capillaries. *(**a**) Capillaries can be classified into continuous, fenestrated, and sinusoidal (=discontinuous) subtypes. In all cases, capillaries lack the media and adventitia of larger vessels and instead comprise just an endothelium that is surrounded by regenerative cells, called pericytes. (**b**) As the most common morphotype, continuous capillaries measure only 5–10 μm wide. Thus, their lumen can be entirely filled by a single erythrocyte (labeled as "E"). (**c**) Continuous capillaries are characterized by an uninterrupted endothelium with numerous caveolae to assist trans-endothelial transport (27,500X). (**d**) In many organs, continuous endothelial cells (Ec) are interconnected by tight junctions (Tjs) to provide efficient seals between blood in the capillary lumen (L) and subendothelial compartments. In some cases, such as the blood-brain barrier, additional cells like astrocytes (A) directly beneath the endothelial basement membrane (asterisk) also contribute to an effective seal. (**e**) Although the typical circulatory pathway has arterioles delivering blood to a single capillary bed prior to blood reaching the heart by a venous return network, transitory shunts provided by metarterioles with muscular sphincters can bypass capillaries and send blood directly to venules via thoroughfare channels. Such reversible shunts supplement permanent arteriole-to-venule anastomoses and are often*

Figure 7.6 (Continued) *found in peripheral tissues like hands and feet, where bypassing capillary beds can minimize heat loss.* (***f***) *Alternatively, blood can pass through two capillary beds of a portal circulatory system, such as occurs with blood drained from digestive tract capillaries and subsequently delivered by the hepatic portal vein to sinusoidal capillaries in the liver. ([a, e, f]: From OpenStax College (2013) Anatomy and Physiology. OpenStax. http://cnx.org/content/col11496/latest reproduced under a CC BY 4.0 creative commons license; [b]: From Lam, MA et al. (2017) The ultrastructure of spinal cord perivascular spaces: Implications for the circulation of cerebrospinal fluid. Sci Rep 7: doi:10.1038/s41598-017-13455-4 reproduced under a CC BY 4.0 creative common license; [c, d]: From Buckley, AF and Bossen, EH (2013) Skeletal muscle microvasculature in the diagnosis of neuromuscular disease. J Neuropathol Exp Neurol 72: 906–918, reproduced with publisher permission; image courtesy of Dr. A. Buckley.)*

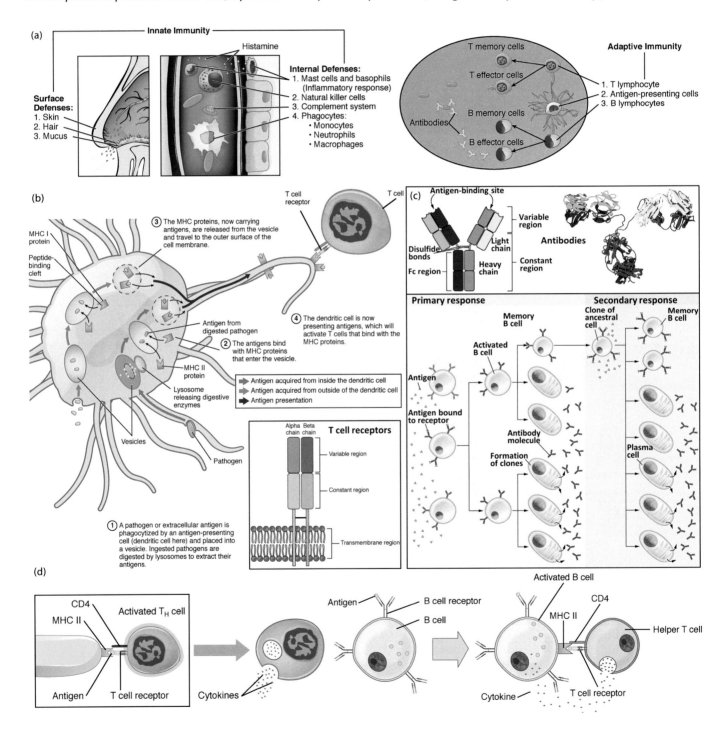

Figure 7.7 An introduction to immune functions of the lymphatic system. (***a***) *Immune responses to pathogens can utilize the more phylogenetically ancient and generalized modes of innate immunity that involves physical defenses like*

Figure 7.7 (Continued) *skin and mucosal linings as well as innate cells, such as mast cells, natural killer cells (NK cells), and phagocytes. In addition, vertebrates can launch the more recently evolved responses of adaptive immunity involving T and B lymphocytes (T cells and B cells), as diagrammed here for a germinal center region of lymph follicles in lymphatic organs like lymph nodes, tonsils, and the spleen. (**b**, **c**) T and B cells can be stimulated via pathogenic antigens delivered by antigen-presenting cells like dendritic cells to generate T-cell receptors (TCRs) and B-cell receptors (BCRs = antibodies). Along with co-receptor molecules, TCRs and antibodies provide highly specific responses to pathogens, which compared to innate defenses, often take longer to initiate owing to numerous mitoses of the clonal selection process. However, adaptive responses can include a memory component that allows more accelerated and robust secondary responses to previously encountered antigens. (**d**) As summarized here, interactions among antigen-presenting cells, B cells, and T cells can drive various adaptive immune responses. ([a–d]: From OpenStax College (2013) Anatomy and Physiology. OpenStax. http://cnx.org/content/col11496/latest reproduced under a CC BY 4.0 creative commons license.)*

TABLE 7.1 INNATE VS. ADAPTIVE IMMUNITY

Properties	Innate immunity	Adaptive immunity
Specificity	Non-specific	Antigen-specific
Action time	Quick	Slow, 2–6 days later than innate immune response
Persistence	Short	Long
Memory	No	Yes
Antigens	Conserved microbe specific molecules, such as LPS, glycans, microbial DNA	Diverse proteins, peptides, and carbohydrates
Receptors	Germ-line encoded	Encoded in gene segments, its diversity relies on rearrangements

Source: From Zhong, J et al. (2011) Innate immunity in the recognition of β-cell antigens in type 1 diabetes. In: Type 1 Diabetes—Pathogenesis, Genetics and Immunotherapy, David Wagner, IntechOpen, doi:10.5772/22264 reproduced under a CC BY 3.0 creative commons license.

7.8   LYMPH VESSELS FOR TRANSPORTING LYMPH AND REPLENISHING BLOOD FLUID

Lymph vessels begin in peripheral CT compartments as blind-ended **lymph capillaries** (=initial lymphatics) that become progressively larger as they extend from their origin site. The proximal wall of these originating vessels comprises overlapping, oak-leaf-shaped endothelial cells plus a patchy or absent basement membrane that collectively facilitate uptake of plasma-derived ECM fluid into the vessel lumen so it can be modified into **lymph** (**Figure 7.8a, b**). Lymph normally lacks erythrocytes and contains not only antigens derived from pathogens but also various immune cells (**Figure 7.8c, d**). In larger lymph vessels, the lining endothelium is surrounded by a few smooth muscle cells and protrudes periodically into the lumen to form **valves** (**Figure 7.8c, e**). Muscles and valves allow the efficient flow of lymph to and from **lymph nodes**, which in turn serve as in-line processing stations of the lymphatic system. Most of the lymphatic drainage from lymph nodes and other lymphatic organs eventually reach the largest lymph vessel—the left **lymphatic duct** (=**thoracic duct**)—which empties into the bloodstream near the superior vena cava. Thus, along with a minor amount delivered by the right lymphatic duct, lymph in the thoracic duct can restore fluid that had leaked from blood vessels into the ECM (**Figure 7.8a**).

7.9   ANTIGEN-PROCESSING LYMPH NODES: FUNCTIONAL MORPHOLOGY OF THEIR CORTEX, PARACORTEX, AND MEDULLA

The ~500–700 **lymph nodes** scattered throughout the adult body are encapsulated bean-shaped organs (**Figure 7.8f**), which occur in dense clusters within such regions as the groin, axillae, and neck. **Afferent** lymph vessels bring lymph into each node via its convex side, whereas **efferent** vessels transport lymph away from the

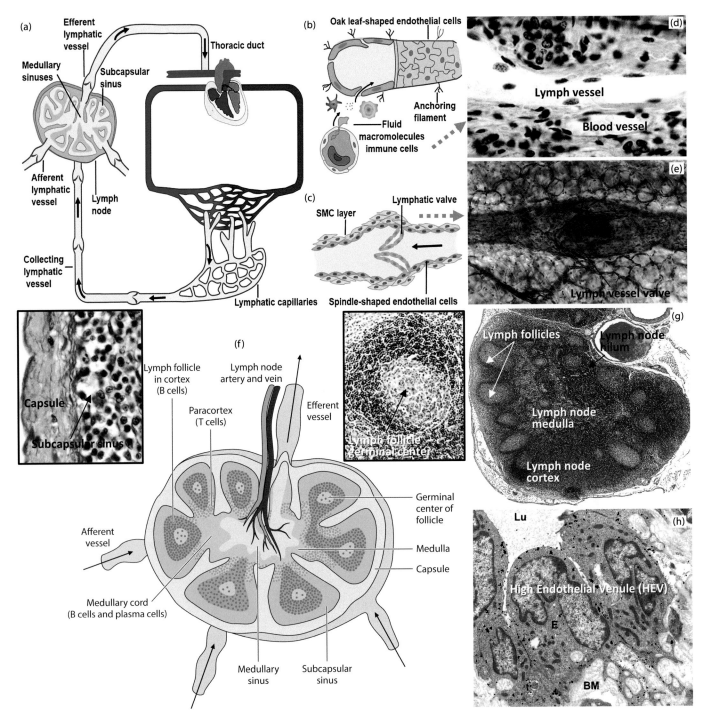

Figure 7.8 Lymph vessels and lymph nodes. *(a–e) Lymph vessels begin in peripheral tissues as blind-ended tubes that carry lymph, which is produced by the continual leakage of blood plasma into the ECM. Lymph contains immune cells and is transported unidirectionally owing to lymphatic valves before being transmitted through lymph nodes and eventually returned to the heart, mainly via the thoracic duct. (d) 230X; (e) 50X. (f, g) Each kidney-shaped lymph node receives lymph via afferent lymph vessels entering the nodal convex side and connects to an efferent lymph vessel at its indented hilum. From its periphery to its center, the node contains: (1) a connective tissue capsule; (2) subcapsular sinus; (3) a large cortex with numerous lymph follicles each of which may have a medial germinal center where B-cell mitoses occur; (4) a paracortical zone with abundant T cells; and (5) an inner medulla with blood vessels and activated T and B cells. (f, left inset) 180X; (f, right inset) 40X; (g) 10X. (h) High endothelial venules (HEVs) with cuboidal endothelial cells allow blood cells to enter and exit the lymph node tissues (1,700X). ([a–c]: From Shang, T et al. (2019) Pathophysiology of aged lymphatic vessels. Aging: doi: 10.18632/aging.10221 reproduced under a CC BY 4.0 creative commons license; [h]: From*

Figure 7.8 (Continued) *Sobocinski, GP et al. (2010) Ultrastructural localization of extracellular matrix proteins of the lymph node cortex: evidence supporting the reticular network as a pathway for lymphocyte migration. BMC Immunol 11: http://www.biomedcentral.com/1471-2172/11/42 reproduced under a CC BY 2.0 creative commons license.)*

indented concave side (=**hilum**) (**Figure 7.8g**). In addition, the hilum is where arterial blood enters, and venous blood exits, the node (**Case 7.4 Golda Meir's Lymph Nodes**).

Directly beneath its capsule, each lymph node has a **subcapsular sinus** that sends lymph from afferent vessels into inwardly directed channels traversing reticular tissue toward the nodal core (=**medulla**) (**Figure 7.8f**). With its lymph fluid and free-floating cells, the subcapsular sinus surrounds densely packed aggregations of mainly B cells that form individual **follicles** (=nodules) in the nodal **cortex** (**Figure 7.8g**). An ill-defined **paracortex** zone between the cortex and central medulla contains abundant T cells and dilated vessels with cuboidal lining cells, called **high endothelial venules** (**HEVs**) (**Figure 7.8h**). Medial to the paracortex, the **medulla** contains macrophages, activated T cells, and antibody-secreting B cells (=**plasma cells**) in medullary sinuses that connect with efferent lymph vessels draining the node (**Figure 7.8f**).

Such microanatomical adaptations allow lymph nodes to function like "antigen libraries". Thus, "books" comprising antigens and antigen-presenting dendritic cells derived from peripheral tissues can be browsed by a stream of visiting "patrons", which are unstimulated (=**naive**) **lymphocytes** capable of generating immune responses. To initiate their reactions, blood-borne forms of naive T and B cells undergo a process of **diapedesis** in which they exit the bloodstream by squeezing between endothelial cells of HEVs to interact with free antigens or antigen-loaded dendritic cells within nodal tissues. If newly arrived lymphocytes fail to recognize appropriate antigens within a few hours to days, the unstimulated T and B cells end up leaving the node via efferent vessels and are eventually returned to the bloodstream before recirculating to other lymphoid tissues and repeating the process. Alternatively, upon encountering a cognate antigen, a naive T or B cell becomes activated and can undergo clonal expansion to launch a full-fledged immune response.

Activation of immune cells is typically initiated in the paracortex, and in the case of T-cell activation, usually begins with TCR binding to an appropriate antigen that is presented by a dendritic cell. Subsequent T cell activation and proliferation can yield **effector T cells** (e.g., **helper T cells**) and **memory T cells**. One subpopulation of helper T cells, called **T follicular helper (TFH) cells**, is retained within the node to assist with B-cell activation. Similarly, **central memory T cells** (**TCMs**) continue to reside in the node to surveille re-exposures to antigen for rapid and robust secondary responses. Alternatively, other activated T cells exit nodes via efferent lymph vessels, and after circulating in blood, are able to utilize T-cell **homing molecules** for locating appropriate peripheral tissues to initiate immune responses at sites of infection.

Antigen-activated B cells in lymph nodes can secrete immunoglobulin antibodies during a two-step proliferative process that typically begins in cortical tissues surrounding follicles. Initial proliferation generates short-lived, dividing cells that produce antibodies which bind relatively weakly to antigenic **epitopes** (i.e., specific

CASE 7.4 GOLDA MEIR'S LYMPH NODES

Israeli Prime Minister
  (1898–1978)

"We've been waiting for 2,000 years. Is that hurrying?"

(G.M.'s 1948 reply when asked not to rush toward proclaiming Israel a sovereign state)

As the first and only female Prime Minister of Israel, Golda Meir had a grandmotherly countenance but was, in fact, an iron-willed leader who oversaw both covert operations and full military responses in defense of her country. Born in present-day Ukraine, Meir emigrated to the United States and briefly taught public school in Wisconsin before moving to Palestine in order to help establish the state of Israel. After being elected Israeli Prime Minister in 1969, Meir ordered secret

assassinations in retaliation for the 1972 massacre of Israeli Olympic athletes. In addition, during the prelude to the 1973 Yom Kippur War, Meir decided against a preemptive strike, and even though she secured a victory in less than 3 weeks, Meir resigned, in response to criticisms of her leadership. In the 1960s, Meir was diagnosed with lymphoma, a cancer involving enlarged lymph nodes due to an overproduction of lymphocytes. For years, Meir was able to manage the disease until her lymph-node-associated cancer eventually caused her death in 1978.

molecular sites that antibodies recognize). Some of the activated B cells formed during this initial phase can enter nodal follicles to launch a second, more persistent mode of antibody secretion. This second proliferative phase generates a discrete **germinal center** (**GC**) within the follicle (**Figure 7.8f**), with some of the newly formed B cells in GCs maturing into non-dividing, long-lived **plasma cells** that can sustain a prolonged secretion of strongly binding antibodies. As cell divisions expand follicular GCs, nodes can enlarge and yield the "swollen glands" that are palpable in areas, such as the neck and axillae. After their intrafollicular phase, some plasma cells typically accumulate in the medulla where they can secrete their antibodies into the bloodstream.

7.10   SPLEEN: RED PULP FOR BLOOD FILTERING PLUS SCATTERED WHITE PULP PATCHES FOR IMMUNE RESPONSES

The **spleen** begins to form during the 5th week of embryogenesis and eventually occupies the upper left quadrant of the abdominal cavity (**Figure 7.9a**). When fully developed, the reddish spleen is surrounded by a connective tissue capsule and consists mainly of erythrocyte-rich **red pulp** plus scattered patches of **white pulp** that are separated from red pulp by intervening **marginal zones** (**Figure 7.9b–e**). Red pulp carries out vascular-related functions, whereas white pulp and marginal zones serve immune-related roles analogous to those performed by lymph nodes and non-encapsulated lymphatic tissues.

Splenic functioning relies on a complex vascular supply in which branches of the **splenic artery** (**Case 7.5 James Garfield's Spleen**) eventually transition into **central arterioles** penetrating white pulp patches and nearby marginal zones (**Figure 7.9f**). Some of the marginal-zone blood flows directly into nearby sinusoids and venules of red pulp via a rapid **closed** pathway involving intact blood vessels. Other marginal blood percolates more slowly through an **open** route of unlined spaces interspersed among cellular **cords** in adjacent red pulp. Such splenic cords comprise mainly: (i) fibroblasts that produce **reticular fibers** for structural support (**Figure 7.9c**), and (ii) phagocytic **macrophages** plus other stromal cells such as lymphocytes (**Figure 7.9d**). After traversing the unlined channels of the splenic cords, blood can enter red pulp sinusoids, whose discontinuous endothelia are reinforced by circumferentially arranged ECM hoops (**Figure 7.9f, g**). Open circulation involving macrophage-rich splenic cords facilitates the major function of red pulp as a key sifter of blood. Thus, free-floating debris and pathogens in blood are constantly phagocytosed by macrophages associated with splenic cords. Similarly, aged erythrocytes lacking the flexibility to squeeze between sinusoidal cells interspersed among splenic cords are engulfed by macrophages (**Figure 7.9h**), and much of their hemoglobin is recycled. Conversely, plasma and blood cells that are able to enter sinusoids are drained by the splenic vein that can return blood to the heart.

As opposed to red pulp, each **white pulp** region of the spleen comprises: (i) a T-cell-rich **periarteriolar lymphatic sheath** (**PALS**) surrounding a central arteriole, and (ii) one or a few **B-cell follicles** in the cortical region between the PALS and marginal zone (**Figure 7.9f**). These two white pulp components resemble lymph node paracortical and cortical regions, respectively, in their immune-related roles. However, the HEVs that deliver blood cells to nodal tissues are lacking in white pulp, as blood cells can percolate through white pulp from open-circulation sources. In addition, some marginal zone components have no counterpart in lymph nodes. For example, **marginal zone B cells** (**MZB cells**) are unique B cells that can react in a T-cell-independent manner to produce most of the body's IgM antibodies (**Figure 7.9f**).

7.11   THYMUS: PERIVASCULAR SPACE (PVS) IN SEPTA THAT PARTITION THE THYMIC EPITHELIAL SPACE (TES) INTO LOBULES, EACH OF WHICH COMPRISES A CORTEX AND MEDULLA WHERE PROCESSES OF T-CELL MATURATION OCCUR

The **thymus** develops mainly from pharyngeal pouch derivatives that fuse during the 8th week of embryogenesis to form a single bilobed organ in the mediastinum region above the heart (**Figure 7.10a**). During development, pouch derivatives generate the major scaffolding of the **thymic epithelial space** (**TES**), in which **T cells** (=**thymocytes**) derived from bone marrow undergo maturation (=**thymopoiesis**). The thymus also develops a non-thymopoietic compartment, called the **perivascular space** (**PVS**). Together with the TES, the PVS yields an average thymic weight of ~20 g at birth before eventually decreasing to <15 g in 80-year-olds during an aging-related loss in functionality, called **thymic involution** (Box 7.2 Evolution of Thymic Involution).

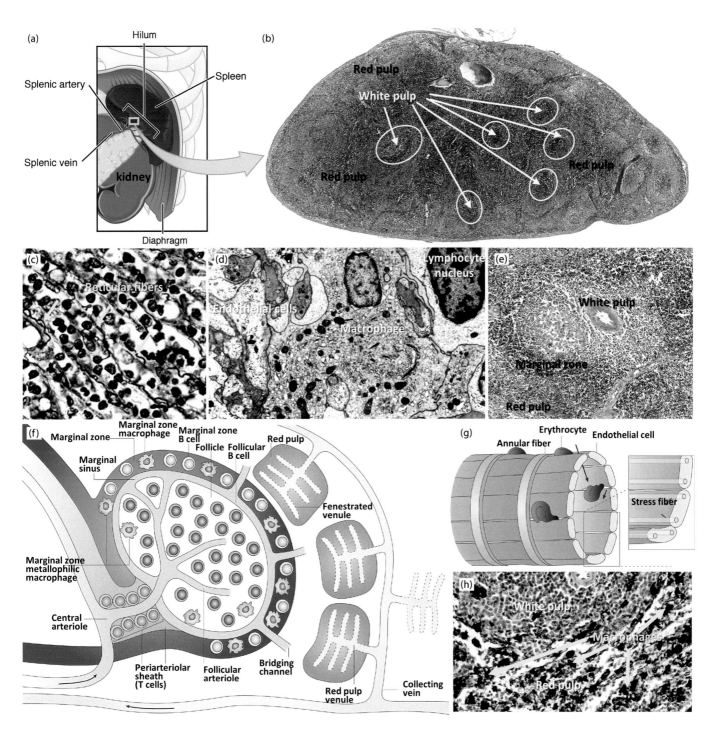

Figure 7.9 The spleen. (**a**, **b**) The spleen occurs in the upper left abdomen and comprises mostly red pulp with interspersed patches of white pulp (b: 8X). (**c**) Red pulp contains abundant reticular fibers (arrows) that bind sliver (c: 250X). (**d**) Surrounding red pulp endothelia are macrophages and other stromal cells like lymphocytes (3,200X). (**e, f**) A transitional marginal zone occurring between white and red pulp contains unique macrophages and B cells, whereas white pulp is characterized by a periarteriolar sheath of T lymphocytes plus B-cell follicles (e: 70X). (**g**, **h**) Blood from white pulp can reach sinusoids directly via lined vessels or indirectly by percolating through the stroma in closed vs. open pathways, respectively. Senescent erythrocytes that cannot squeeze through sinusoidal endothelial are phagocytosed by red pulp macrophages (h, arrows; 125X). ([a]: From OpenStax College (2013) Anatomy and Physiology. OpenStax. http:// cnx.org/content/col11496/latest reproduced under a CC BY 4.0 creative commons license; [d]: From Image courtesy of Dr. M. Morroni; [f]: From Pillai, S and Cariappa, A (2009) The follicular versus marginal zone B lymphocyte cell fate decision. Nat Rev Immunol 9: 767–777 doi: 10.1038/nri2656 reproduced with publisher permission; [g]: From Mebius, RE and Kraal, G (2005) Structure and function of the spleen. Nat Rev Immunol doi:10.1038/nri1669 reproduced with publisher permission.)

CASE 7.5 JAMES GARFIELD'S SPLEEN

20th U.S. President
   (1831–1881)

"Tell her I am seriously hurt; how seriously I cannot yet say."

(J.G.'s message to his wife after he had been shot by an assassin)

After winning the U.S. presidency in 1880, James Garfield had to deal with an out-of-control spoils system that required the White House open its doors several hours a day to receive long lines of people seeking governmental positions. One especially demanding applicant, named Charles Guiteau, was convinced Garfield owed him a prestigious appointment, and after being banned from the White House, Guiteau shot Garfield in the back at a local train station. The wounded president was rushed to the White House, where doctors repeatedly probed for the bullet with their unwashed fingers thereby undoubtedly helping to spread a systemic infection. The bullet was never located, even after the inventor of the telephone,

Alexander Graham Bell, volunteered to use a metal detector prototype he had built specifically for Garfield. Just before dying 11 weeks after being shot, Garfield called out in pain, and post-mortem analyses indicated that an aneurysm of his splenic artery was the immediate cause of the president's death. Ironically, before being hanged, Guiteau may have gotten it right when he stated that the doctors had killed the president, since, in retrospect, the actual gunshot damage itself was probably non-lethal. Instead, Garfield's burst splenic artery was most likely the result of widespread sepsis exacerbated by the doctors' unsanitary treatments.

In each young thymus, the surrounding thymic capsule invaginates a few thin **septa** with blood vessels, thereby generating a relatively restricted PVS compartment with little adipose tissue (**Figure 7.10b–e**). These PVS septa in turn partition the well-developed TES of both thymic lobes into multiple **lobules**, each of which comprises: (i) a more densely staining outer **cortex** with its high concentration of T cells, and (ii) a lighter inner **medulla** with sparser T cells plus scattered **Hassall's corpuscles** of keratinizing cells that aid the differentiation of **regulatory T cells** (**Tregs**) (**Figure 7.10f, g**). The continuous capillary endothelia in the cortex contain tight junctions contributing to a **blood-thymus barrier** that prevents pathogens from entering sites where T-cell maturation is initiated. Conversely, medullary capillaries are connected to high endothelial venules that allow the influx of T cells derived from bone marrow.

Most of the **stromal** scaffolding in functioning lobules comprises relatively lightly staining and large **thymic epithelial cells** (**TECs**) that develop from endoderm, thereby differing from smaller and denser T cells derived from mesoderm (**Figure 7.10d, inset**). TECs produce **reticular fibers** for structural support and secrete the hormone **thymosin** to help regulate T-cell functions. Via thymosin and other secreted factors, TECs coordinate with stromal cells like macrophages and dendritic cells to modulate two key thymic processes: (i) **positive selection** during the production of functional T cells for counteracting pathogens, and (ii) **negative selection** involving the elimination of T cells that are potentially directed against the body's own components.

Both of these thymic processes begin with T-cell progenitors arising in the bone marrow and circulating via the bloodstream to lobular **cortices**. Under the influence of TECs and additional cells, maturing T cells not only undergo somatic recombination to produce a diverse array of TCRs but also begin expressing the TCR co-receptors **CD4** and **CD8** (**Figure 7.10h**). These **double-positive T cells** are then tested by TECs plus other stromal cells that utilize plasmalemmal **MHC** proteins to present small peptides of potential pathogens to the TCRs on maturing T cells (**Figure 7.10h**). Failure to recognize and bind the presented peptide can trigger apoptosis of the T cell. Conversely, antigen binding causes clonal amplification and further differentiation to yield numerous T cells, each of which expresses a single type of TCR via the process of **positive selection**. During such selection, T cells downregulate co-receptors to become either: (i) a **CD4$^+$ cell** capable of forming several kinds of T helper cells, or (ii) a **CD8$^+$ cell** that can generate such effector cells as **cytotoxic T cells**. The single-positive CD4$^+$ and CD8$^+$ cells formed in the lobular cortex then migrate to the **medulla** where they are tested against **self-antigens** that are generated by one's own body. Such testing involves medullary TECs that express a key protein, called **AIRE** (**a**uto**i**mmune **re**gulator), which helps downregulate **autoimmune** reactions against autologous tissues (**Figure 7.10h**). Failure to bind with high avidity to self-antigens enables further T-cell maturation. After such differentiation, T cells can exit the thymus to populate T-cell niches of secondary lymphatic organs and peripheral tissues. Conversely, high-avidity binding to self-antigens normally triggers apoptosis of the T cell as

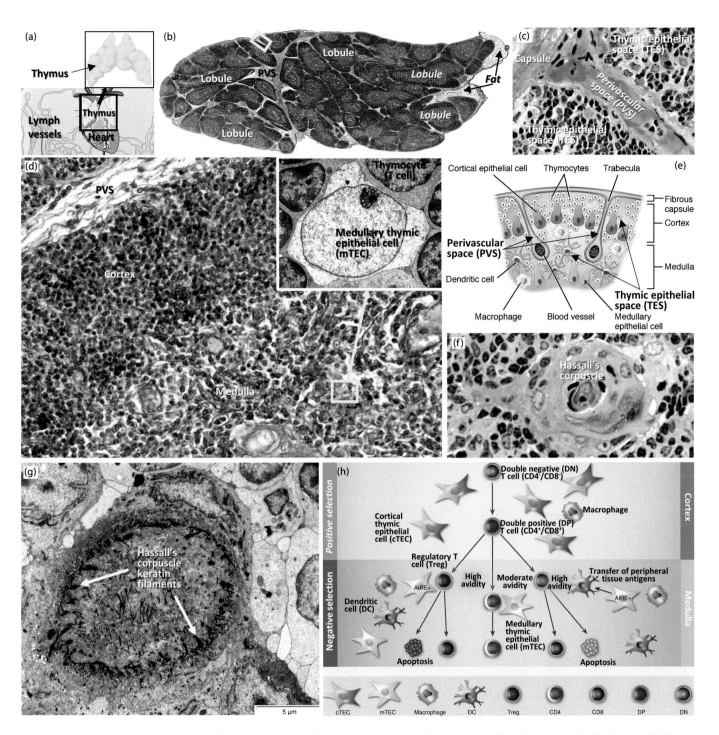

Figure 7.10 The thymus. (a) *The thymus is situated above the heart and comprises: (i) a thymic epithelial space (TES) in which T-cell maturation (thymopoiesis) occurs, and (ii) trabecular septa with blood vessels collectively constituting the perivascular space (PVS) of the thymus. (**b, c**) In young thymuses, relatively thin PVS septa divide the TES into thymopoietic lobules, each of which has an outer cortex and inner medulla. Conversely, in older thymuses, the expansive PVS contains abundant fat that replaces TES thymopoietic tissues during an aging-related process, called thymic involution (Box 7.2) (b: 10X, c: 250X). (**d, e**) Each thymic cortex contains a scaffold of cortical thymic epithelial cells, whereas medullas possess medullary thymic epithelial cells (mTECs) (d: 175X; d, inset: 2,300X). (**f, g**) In addition, the medulla has keratin-containing Hassall's corpuscles that promote regulatory T cell (Treg) differentiation (f: 300X; g: 2,300X). (**h**) During T-cell maturation, double-negative (DN) T-cell precursors that express neither CD4 nor CD8 become stimulated in the cortex during a positive selection process to become CD4+/CD8+ double-positive (DP) cells. DP cells are then transferred to the medulla, where they undergo a negative selection process modulated by medullary thymic epithelial cells (mTECs) that express AIRE (autoimmune regulator). Such AIRE+ mTECs ensure that T cells that might recognize self-antigens undergo apoptosis,*

Figure 7.10 (Continued) *thereby reducing the likelihood of generating autoimmune (AI) diseases. Prevention of AI pathogenesis is also aided by Tregs that are able to escape apoptosis. ([a, e]: From OpenStax College (2013) Anatomy and Physiology. OpenStax. http://cnx.org/content/col11496/latest reproduced under a CC BY 4.0 creative commons license; [d, inset and g]: From Ceylan, A and Alabay, B (2015) Ultrastructure of apoptotic T lymphocytes and thymic epithelial cells in early postnatal pig thymus. Turk J Vet Anim Sci 41: 613–620 images courtesy of Dr. A Ceylon; [h]: From Passos, GA et al. (2018) Update on Aire and thymic negative selection. Front Immunol 153: 10–20 reproduced with publisher permission.)*

BOX 7.2 EVOLUTION OF THYMIC INVOLUTION

Via its production of T cells, the **thymus** is a key defender against pathogens. Yet, in humans and many other vertebrates, the thymus normally loses thymopoietic tissue and overall functionality as part of an aging process, called **thymic involution**. During involution, active sites where T cells can mature in the **thymic epithelial space** become replaced by fat-rich **perivascular space** (**PVS**) tissue (**Figure a, b**). Based on weight loss, involution was once thought to start after adolescence. However, it is now clear that involution is initiated well before puberty, as reductions in thymic epithelial space

(TES) volume and maturing T-cell numbers begin in the first year of post-natal life and are accompanied by increases in non-thymopoietic PVS compartments (**Figure c**). Replacement of TES by PVS continues during aging and eventually results in decreased thymic functionality, thereby making the elderly more susceptible to diseases (**Figure d**).

Exactly why the apparently maladaptive process of thymic involution has evolved remains unknown. However, arguments along the lines of the **disposable soma theory** postulate that the human immune system was initially

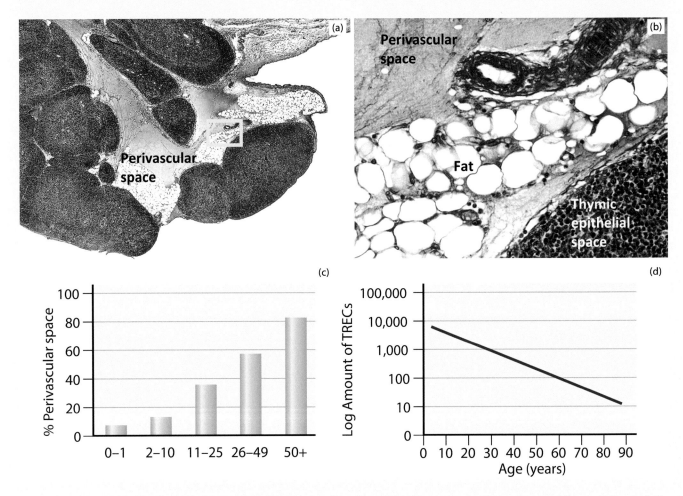

Box 7.2 *Thymic involution. (**a, b**) An involuting thymus showing fat in the perivascular space replacing thymopoietic tissue in the thymic epithelial space ((**c**) after Flores, KG et al. (1999) Analysis of the human thymic perivascular space during aging. J Clin Inves 104: 1031–1039, showing an increased perivascular space during involution); (**d**) after Lynch, HE et al. (2009) Thymic involution and immune reconstitution. Trends Immunol 30: 366–373, showing a graph of TREC (T-cell receptor excision circle) levels, indicating a loss of thymopoietic capacity during aging.*

adapted for lives lasting only a few decades. Accordingly, replacing energetically expensive tissues like the TES with non-thymopoietic PVS may have once provided a beneficial re-allocation of resources but now has become detrimental as longevity has increased. Alternatively, it has been argued that thymic involution has evolved as an **adaptive response**, because it mitigates two key problems related to aging: (1) the longer the thymus remains active, the more chances there are that invasive pathogens can infiltrate thymic tissues and interfere with T-cell defenses,

and (2) during aging, mutations progressively accumulate throughout the body, thereby increasing the likelihood that the aged thymus will be unable to recognize self-tissues and correctly protect them from autoimmune attacks. Accordingly, by generating an effective array of naive T cells and accumulating long-lived memory T cells, the young, pristine thymus may actually optimize life-long immune responses, compared to what would be generated by an older and more error-prone thymus that produces numerous defective T cells.

CASE 7.6 VIRGINIA ALEXANDER'S THYMUS

Physician and Healthcare Activist
(1899–1949)

"We will have to send physicians into sections which have no bright lights and … take public health across the railroad tracks, to serve those most in need of comfort and care."

(V.A. on the need for public health care in all corners of America)

Virginia Alexander was a pioneering African-American physician who sought to improve healthcare for the poor. After completing a residency in Obstetrics and Gynecology in Kansas City, Alexander moved to Philadelphia and in 1931 opened within her own home a clinic that offered free medical care to those who could not afford a doctor. Alexander went on to receive a Master's degree in Public Health and oversaw healthcare services for miners in Alabama during

World War II. While in Alabama, Alexander was diagnosed with the autoimmune disease systemic lupus erythematosus (SLE) and a few years later succumbed to SLE complications at age 49. Although still a puzzling syndrome, SLE is correlated with a smaller than normal thymus, and given that regulatory T lymphocytes (Tregs) from the thymus can help reduce autoimmune diseases in general, some recent types of treatment have proposed introducing into SLE patients more Tregs from harvested thymuses. Such a strategy begs the question: would supplementing a presumably defective thymus have helped stave off the lethal effects of Alexander's SLE and thereby prolonged her charitable career in public health?

part of a **negative selection** process that helps prevent **autoimmunity**. However, some CD4+ **Tregs** can avoid apoptosis and continue to differentiate in part due to secretions from medullary **Hassall's corpuscles**, and after gaining functionality, such cells further protect the body against autoimmune attacks. Accordingly, incomplete removal of self-reacting T cells and/or insufficient Treg production can promote autoimmune pathogenesis (**Case 7.6 Virginia Alexander's Thymus**).

7.12 NON-ENCAPSULATED SECONDARY AND TERTIARY LYMPHOID TISSUES

Unlike encapsulated lymphoid organs, **secondary lymphoid tissue** occurring in the skin and mucosal linings of the digestive, respiratory, and urogenital systems lack a connective tissue capsule and constitute a diffuse system that is collectively referred to as **mucosa-associated lymphoid tissue** (**MALT**). Along with the beneficial roles of MALT that are discussed in chapters covering organ systems with these tissues, various autoimmune diseases can also generate non-encapsulated lymphoid tissues at sites that normally lack a lymphoid component. Such **tertiary lymphoid tissues** may either supplement normal MALT by counteracting pathogenesis or contribute to disease progression via auto-reactive and pro-inflammatory processes.

7.13 SOME CIRCULATORY SYSTEM AND LYMPHOID ORGAN DISORDERS: HEART ATTACKS AND DIGEORGE SYNDROME

Heart attacks (=myocardial infarctions, MIs) typically occur due to atherosclerotic buildup in coronary arteries reducing blood flow to heart tissues. Such ischemia can compromise cardiac functions, and in non-fatal cases, damaged myocytes are eventually replaced by connective tissue scars. Although many MIs are accompanied by such symptoms as chest pain and dizziness, **silent** MIs can occur without any obvious warning. In fact, one analysis of middle-aged to elderly adults revealed that nearly 8% experienced a silent MI over the course of the several-year-long study.

As opposed to the abundant thymopoietic tissue that is formed in most newborns, several congenital diseases can hamper thymic development. For example, in **DiGeorge syndrome** (**DGS**), variable deletions of chromosome 22 cause thymic disorders ranging from nearly normal T-cell functioning to complete DGS (cDGS) where a thymus is absent (athymia). Current treatment for cDGS can utilize transplantation of cultured fetal thymic tissues to help restore at least some thymic function.

7.14 SUMMARY—CIRCULATORY AND LYMPHATIC SYSTEMS

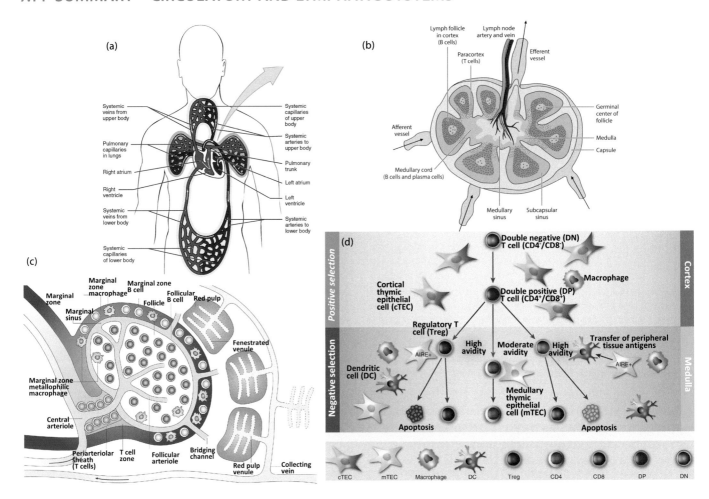

Pictorial Summary Figure Caption: *(**a**): see figure 7.1d; (**b**): see figure 7.8f; (**c**): see figure 7.9f; (**d**): see figure 7.10h.*

Circulation involves the **circulatory system** (heart, arteries, capillaries, and veins) and **lymphatic system** (vessels, tissues, and organs). The **heart** comprises two upper **atria** that receive blood from **veins** plus two lower **ventricles** that pump blood via **arteries** to **capillaries** for gas, nutrient, and waste exchanges. Plasma leaked from blood into the ECM can form **lymph** that both returns ECM fluid to the heart and supplies **lymph nodes** with **antigens** to elicit immune responses.

The **fetal heart** has a **ductus arteriosus** and **foramen ovale** that help re-route blood from the non-functional lungs to the placenta for molecular exchanges. After these structures close post-natally, the definitive circulation has a right-side **pulmonary** pathway (to and from the lungs) and left-side **systemic** pathway (throughout body, except for pulmonary alveoli) (**Figure a**). The **three layers of the heart wall** are the **endocardium** (with an endothelial lining), **myocardium** (cardiac myocytes), and **epicardium** (thin covering). One-way blood flow through the heart relies on **valves** tethered by **chordae tendinae** that help prevent abnormal valve folding (**prolapse**). The cardiac **conducting system** (**SA/AV node pacemakers**), **bundle of His**, **Purkinje fibers**) coordinates heart beats.

The velocity of blood flow through vessels is inversely related to the **total cross-sectional area** (**CA**) of vessels. Accordingly, flow through interconnected capillaries with **high total CA** is slow, thereby facilitating molecular exchanges.

The three layers (**tunics**) of non-capillary blood vessels are the: (1) **intima** (endothelium and subendothelial CT [plus a peripheral **interna elastica** in arteries]); (2) **media** (smooth muscle); and (3) **adventitia** (CT and nerves plus a **vasa vasorum** blood supply in larger vessels). **Capillaries** possess only an endothelium plus nearby **pericyte** cells and comprise three types: (1) **continuous** (most common type of capillary, typically measuring ~5–10 μm wide); (2) **fenestrated** (~5–10 μm wide and endowed with endothelial pores); and (3) **sinusoids** (much larger and more discontinuous than other capillaries so that cells can easily traverse the endothelium). **Arteries differ from veins** in possessing a thicker wall and an interna elastica, whereas veins have no interna elasticas but possess valves. Unusual circulatory pathways include: (1) **capillary shunts** that provide direct arteriole-to-venule connections, thereby bypassing capillaries; and (2) **portal systems** that transport blood through two capillary beds before it returns to the heart.

The **lymphatic system** comprises: (i) **primary lymphoid organs** (bone marrow and thymus), which make **B** and **T lymphocytes** (=B and T cells), and (ii) **secondary lymphoid organs** (lymph nodes, spleen, and non-encapsulated lymphatic tissues), where B and T cells can interact with antigens. Immune responses can use: (1) secreted molecules like **antibodies** during **humoral defenses** for wide-ranging effects and/or (2) more localized cell-pathogen interactions during **cell-mediated defenses**. **Innate immunity** is widespread in animals; functional at birth; rapid and generalized; and based on limited receptors without a memory component. **Adaptive immunity** is vertebrate-specific; fully functional post-natally; slower in onset; and able to provide specific anti-pathogenic responses utilizing diverse B/T-cell receptors with a memory capacity.

Lymph nodes are bean-shaped. **Afferent** lymph enters the node's convex side, whereas **efferent** lymph exits the concave **hilum** (**Figure b**). Unstimulated (**naive**) T and B cells reach nodes via **diapedesis** in **high endothelial venules** (**HEVs**). The **cortex** at the periphery of the node has **follicles** where B cells can proliferate in **germinal centers**. The **paracortex** underlying the cortex has T cells that can be activated by antigens. The **medulla** in the core of the lymph node contains macrophages, activated T cells, and long-lived antibody-secreting B cells (**plasma cells**).

The **spleen** comprises: (1) **red pulp** for removing old or damaged red blood cells; (2) scattered **white pulp** for immune responses; and (3) **marginal zones** between red and white pulp (**Figure c**). Blood entering white pulp via a **central arteriole** reaches the marginal zone and can percolate among red pulp **splenic cords** to enter **sinusoids**, whereas erythrocytes that cannot squeeze through sinusoidal gaps are endocytosed by splenic macrophages for hemoglobin recycling. White pulp consists of: (1) a T-cell-rich **periarteriolar lymphatic sheath** (**PALS**) around the central arteriole, and (2) **B-cell follicles**, which function like lymph node paracortices and cortices, respectively.

The **thymus** is composed of: (1) a **thymic epithelial space** (**TES**) comprising **thymopoietic** tissues where T-cell precursors from bone marrow mature, and (2) a non-thymopoietic **perivascular space** (**PVS**) that subdivides the TES into **lobules**, each with a **cortex** and **medulla**. The TES is well developed in newborns, but atrophies during aging (=**thymic involution**). T-cell maturation starts in the lobule cortex, as T cells interact with **thymic epithelial cells** (**TECs**) and express T-cell receptors (**TCRs**) plus **CD4** and **CD8** co-receptors. T cells with TCRs that recognize antigens can proliferate (**positive selection**) (**Figure d**). Positively selected T cells move to the medulla and are tested against self-antigens. Avidly binding T cells are normally removed by apoptosis (**negative selection**) to reduce the likelihood of launching autoimmune responses. However, some T cells utilize stimulation from **Hassall's corpuscles** to avoid apoptosis and become regulatory T cells (**Tregs**) that also help downregulate **autoimmune** diseases.

Some circulatory disorders: heart attacks: typically due to ischemia-induced myocardial damage; **atherosclerosis:** intima plaque buildup seeded by LDLs; **DiGeorge syndrome:** inherited thymus dysfunction.

SELF-STUDY QUESTIONS

1. Which chambers of the heart receive blood?
2. T/F CD4 is a TCR co-receptor in some T cells.
3. Where, exactly, are regions with PALS located?
4. Which is characteristic of innate immunity?
   A. Highly specific responses to each particular pathogen
   B. Found only in vertebrates
   C. With memory that adapts to previous encounters
   D. Already functional at birth
   E. Has highly diverse T and B cell receptors

5. Which is a common feature of veins?
    A. Comparatively thicker walls than nearby arteries
    B. Possession of an interna elastica
    C. A lumen that often collapses after fixation
    D. ALL of the above
    E. NONE of the above
6. Which makes up the bulk of the heart wall?
    A. Endothelium
    B. Endocardium
    C. Myocardium
    D. Epicardium
    E. Conductive system
7. T/F Blood flows more rapidly in capillaries than in the aorta.
8. T/F Afferent lymph vessels enter the hilum of a lymph node.
9. Where, exactly, does negative selection occur in the thymus?
    A. Cortex
    B. PVS
    C. Medulla
    D. Subcapsular sinus
    E. Lymphatic duct
10. What are the vasa vasorum?
11. How do most naive lymphocytes enter lymph nodes?
    A. Afferent lymph vessels
    B. Efferent lymph vessels
    C. Arteries
    D. High endothelial venules
    E. Directly through capsule
12. What is the name of a long-lived antibody-secreting B cell?
    A. Hassall's corpuscle
    B. Helper T cell
    C. Treg
    D. TEC
    E. Plasma cell
13. Chordae tendinae cells serve as the heart's primary pacemaker

"EXTRA CREDIT" For each term on the left, provide the BEST MATCH on the right (answers A-F can be used more than once)

14. T$_{reg}$ stimulator _____
15. Chordae tendineae _____
16. DiGeorge syndrome _____
17. Pericytes _____
18. Germinal center _____

A. Cardiac valve
B. Thymus dysfunction
C. Red pulp
D. Capillaries
E. B cell follicle
F. Hassall's corpuscle

19. Compare and contrast fetal vs. adult circulation.
20. Describe thymic structure and function.

ANSWERS

(1) left and right atria; (2) T; (3) in splenic white pulp directly around a central arteriole; (4) D; (5) C; (6) C; (7) F; (8) F; (9) C; (10) small vessels supplying blood to the adventitia of large blood vessels; (11) D; (12) E; (13) F; (14) F; (15) A; (16) B; (17) D; (18) E; (19) The answer should show blood flow among the following: right and left atria; right and left ventricles; pulmonary arteries; pulmonary veins; lungs, body (except for pulmonary alveoli), vena cavae, ductus arteriosus, foramen ovale; (20) The answer should include and define such terms as: capsule, TES, PVS, lobule, cortex, medulla, involution, adipose tissue, bone marrow, T cell precursors, TECs, TCRs, CD4+, CD8+, MHCs, stroma, positive selection, self-antigens, negative selection, apoptosis, Tregs, autoimmune disease

Muscles

8.1 INTRODUCTION TO MUSCLES: SKELETAL AND CARDIAC (STRIATED) VS. SMOOTH (NON-STRIATED)

Muscle comprises three main types— **skeletal**, **cardiac**, and **smooth**. **Skeletal muscle** is under voluntary control and typically attaches via tendons to the skeleton. Along with involuntarily regulated **cardiac muscles** of the heart, skeletal muscles are **striated**, owing to their highly aligned arrays of **myofilaments** in contractile units, called **sarcomeres**. Conversely, involuntary **smooth muscles** lack sarcomere-generated striations and are usually located in walls of blood vessels, digestive organs, urogenital components, and respiratory passageways.

Muscle cells play important ancillary roles, such as in glucose uptake (Chapter 15) and natriuretic peptide secretion (Chapter 16), but their primary function is to contract and thus mediate **motility**. In all muscles,

DOI: 10.1201/9780429353307-10

contractions require an **increase** in intracellular calcium ions (Ca^{2+}) that allow force-generating interactions to occur between **thin** and **thick** kinds of **myofilaments**, which contain **actin** and **myosin**, respectively. However, the three muscle types differ in their: (i) precise arrangement of myofilaments and associated membranous structures; (ii) modes of elevating calcium; and (iii) mechanisms of regulating actin-myosin interactions. Thus, this chapter focuses mainly on the functional microanatomy of skeletal muscle contractions and ends by summarizing distinctive features of such processes in cardiac and smooth muscle.

8.2   DEVELOPMENT OF A MULTINUCLEATED SKELETAL MUSCLE CELL (=MYOFIBER) FROM THE FUSION OF INDIVIDUAL MYOBLASTS

In 3-week-old embryos, mesodermal precursors to skeletal muscles begin to develop as mononucleated **myoblast** cells that fuse to form a multinucleated **syncytium** (=myotube) (**Figure 8.1a–c**). After differentiating functional sarcomeres, the myotube becomes a contractile muscle cell (=myocyte), which, given its syncytial nature, is often termed a **myofiber** in order to distinguish it from mononucleated cells (**Figure 8.1d**). Myofibers continue to grow during fetal development via further cellular fusions, whereas post-natal muscle enlargement involves mainly the hypertrophy of existing myofibers plus the production of a few new myofibers. Conversely, aging-related loss of skeletal muscles, called **sarcopenia**, usually starts in the fourth through fifth decade of life and eventually reduces muscle mass to ~50% of its peak levels, with fat replacing much of the lost muscle tissue.

8.3   ANATOMY OF SKELETAL MUSCLES, THEIR CONNECTIVE TISSUE SHEATHS, AND SATELLITE CELLS

Each adult has ~600 discrete skeletal muscles that can range widely from the ~1-mm-long stapedius muscle in the middle ear to ~60-cm-long sartorius muscles connecting hips to knees. In addition to sarcomere-containing myofibers, blood vessels, and nerves, each large muscle possesses three types of connective tissue sheaths, which from the muscle periphery to center are the **epi-**, **peri-**, and **endomysium** (**Figure 8.1d, e**). The **epimysium** surrounds the entire muscle and comprises dense regular connective tissue that may be overlain by a deep fascia capsule (**Figure 8.2a, b**). In cases where skeletal muscles connect to bones, the epimysium terminates in a **tendon** at the muscle's origin (proximal end) and insertion (distal end), with each myofiber of the muscle extending the entire inter-tendinal length and thus often measuring several centimeters long (**Figure 8.2c, d**). Medial to tendons, the epimysium invaginates as **perimysial sheaths** that surround myofiber clusters, called **fascicles** (**Figure 8.2a, b**). Each myofiber in a fascicle is covered by an **endomysium**, whose innermost layer abutting the fiber is a basement-membrane-like **external lamina**. At sites where motor neurons innervate myofibers, the external lamina plus its underlying plasmalemma form multiple invaginations. Such invaginations in turn associate with axonal termini and thereby collectively form an integrated neuro-muscular unit, called a **motor end plate** (see also **Figure 8.5c–f** and Chapter 9).

Approximately 3% of the nuclei scattered in and among adult myofibers belong to **satellite cells**, which occur directly next to myofiber plasmalemmata (**Figure 8.2e–g**). When activated by stimuli like hepatocyte growth factor (HGF), satellite cells can regenerate damaged myofibers. In addition, post-natal myofibers are unable to proliferate and only increase muscle size by growing larger. Conversely, satellite cells can divide and differentiate into a few new myofibers, thereby supplementing exercise-induced increases in myofiber volumes (**Figure 8.2g**).

8.4   MYOFIBER MICROANATOMY: T TUBULES, SARCOPLASMIC RETICULUM DOMAINS, AND MYOFILAMENT-CONTAINING MYOFIBRILS THAT COMPRISE A SERIES OF CONTRACTILE SARCOMERES

Each myofiber's plasmalemma, which is often termed the **sarcolemma** (**Figure 8.3a**), covers a thin rim of peripheral cytoplasm (=**sarcoplasm**) that contains myoblast-derived nuclei. In addition, at periodic intervals along the fiber, the surface plasmalemma invaginates toward the myofiber center as ~30 nm-wide **T tubules** (transverse tubules) that associate with a well-developed endoplasmic reticulum, called the **sarcoplasmic reticulum (SR)** (**Figure 8.3b**).

The SR functions primarily as a mobilizable store of Ca^{2+} and comprises both a **longitudinal** component and a **junctional** domain. The longitudinal SR extends along the length of the myofiber (**Figure 8.3a, b**) and is organized into elongated tubules with **sarco/endoplasmic reticulum Ca^{2+} ATPase (SERCA)** complexes that promote muscle relaxation by pumping Ca^{2+} from the sarcoplasm into the SR. Conversely, junctional SR forms

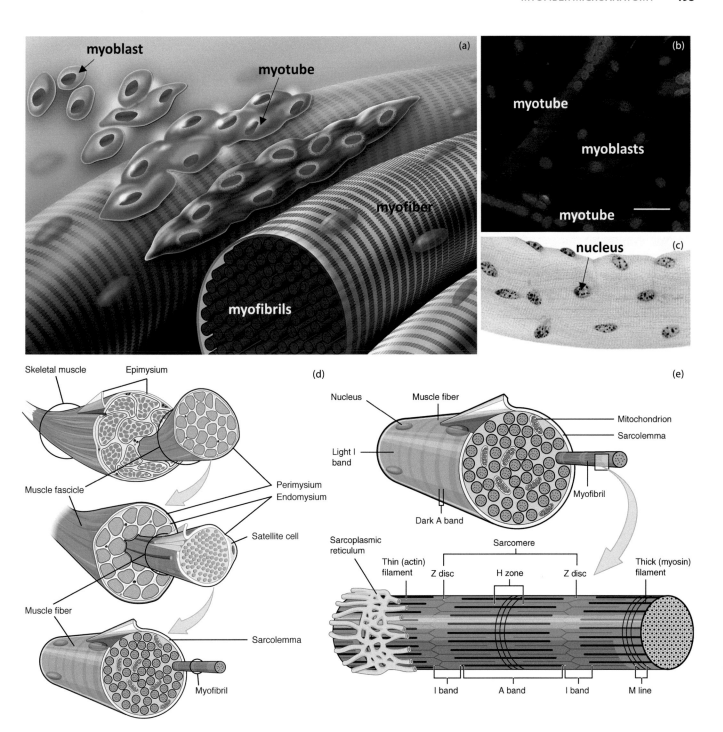

Figure 8.1 Skeletal muscle development and microanatomical organization. *(a, b) Skeletal muscles develop from the fusion of individual myoblast cells to form multinucleated myotubes that generate functional sarcomeres to become multinucleated muscle cells, called myofibers. Scale bar, b = 20 μm. (c) A DAPI-stained developing myofiber shows multiple nuclei derived from fused myoblasts (380X). (d) When fully developed, skeletal muscles are surrounded by an epimysium connective tissue sheath, which invaginates to form perimysia around bundles of myofibers, called fascicles. Each myofiber in a fascicle is surrounded by an endomysium and contains myofibril organelles in its cytoplasm. (e) Myofibrils comprise a chain of numerous functional units, called sarcomeres, each of which is bounded by neighboring Z discs that delimit lighter I bands flanking a central, darker-staining A band. ([a]: From public domain image from National Human Genome Research Institute, NIH https://www.flickr.com/photos/nihgov/38876545081/in/album-72157662951050375/; [b] From Jurdana, M et al. (2008) Neural agrin changes the electrical properties of developing human skeletal muscle cells. Cell Mol Neurobiol 29: 123–131 reproduced with publisher permission; [c] From Gnocchi, VF et al. (2011) Uncoordinated transcription and compromised muscle function in the Lmna-null mouse model of*

Figure 8.1 (Continued) *Emery-Dreifuss Muscular Dystrophy. PLoS ONE 6: doi:10.1371/journal.pone.0016651 reproduced under a CC BY 4.0 creative commons license; [d,e] From OpenStax College (2013) Anatomy and physiology. OpenStax. http://cnx.org/content/col11496/latest reproduced under a CC BY 4.0 creative commons license.)*

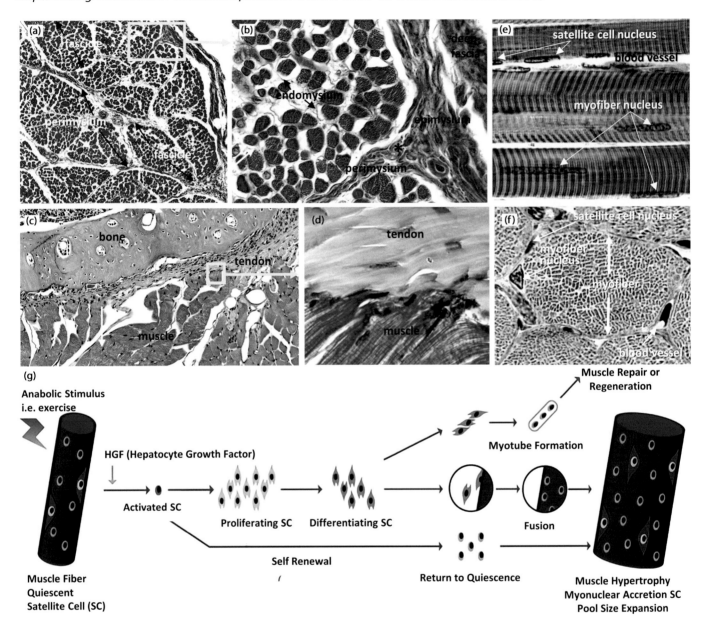

Figure 8.2 Skeletal muscle connective tissue sheaths, tendons, and satellite cells. *(a, b) Lying beneath deep fascia, the connective tissue sheaths of skeletal muscle from outside in are the epi-, peri-, and endomysium (asterisk in (b) shows an epimysium/perimysium connection). (a) 20X; (b) 100X. (c, d) Epimysia at the termini of myofibers can connect with tendons, thereby attaching muscles to bones. (c) 80X; (d) 350X. (e–g) Satellite cells occurring next to myofibers serve as stem cells that can replace myofibers and increase muscle bulk following exercise. (e) 650X; (f) 750X. ([g]: From Snijders, T et al. (2015) Satellite cells in human skeletal muscle plasticity. Front Physiol 6: doi: 10.3389/fphys.2015.00283 reproduced under a CC BY 4.0 creative commons license.)*

dilated sacs, called (terminal) **cisternae,** at the ends of longitudinal SR that associate with T tubules. Thus, at discrete sites along the length of the myofiber, a SR cisterna flanks each side of T tubules to form tripartite units, termed **triads**, which collectively serve to release SR Ca^{2+} for initiating muscle contractions (**Figure 8.3a, b**).

SR arrays are associated with cylindrical bundles of myofilaments, called **myofibrils**, which fill much of the cytoplasm and span the entire length of myofiber between its terminal connections with tendons (**Figure 8.3a–c**).

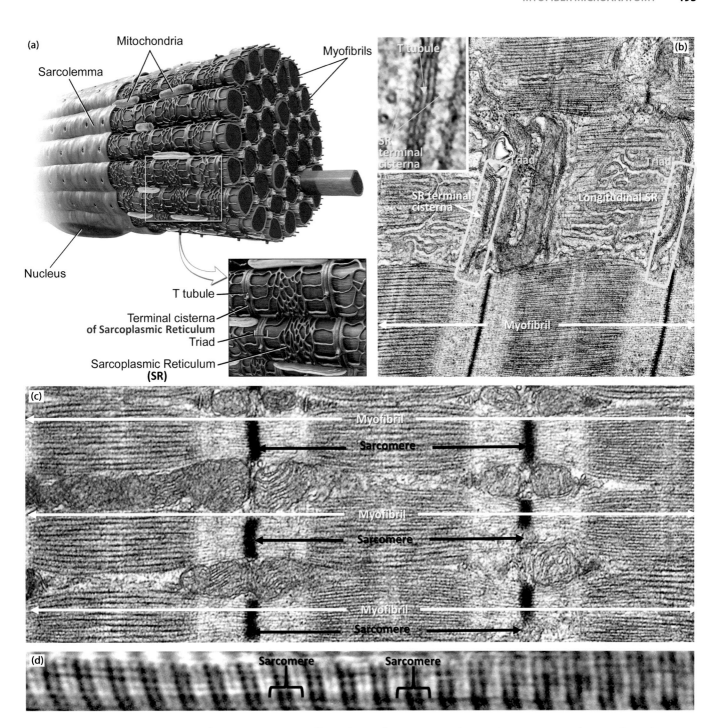

Figure 8.3 T tubules and myofibrils. (*a*) *The myofiber plasmalemma (sarcolemma) invaginates along its length to generate transversely oriented T tubules that extend toward centrally located myofibrils within the myofiber.* (*b*) *Each T tubule associates with two terminal cisternae of the smooth endoplasmic reticulum (=sarcoplasmic reticulum, SR) to form a triad that helps coordinate myofibril contractions throughout the myofiber (21,000X).* (*c*) *Myofibrils are composed of a chain of numerous sarcomeres, each of which measures ~2–3 μm long in relaxed muscles (31,000X).* (*d*) *Sarcomeres of neighboring myofibrils are normally highly aligned, thereby allowing their detection by light microscopy (4,000X).* ([a]: *From Blausen.com staff (2014) "Medical gallery of Blausen Medical 2014". WikiJournal of Medicine 1 (2): 10. doi:10.15347/ wjm/2014.010 reproduced under CC0 and CC BY SA creative commons licenses; [b,c]: From images courtesy of Dr. C. Franzini-Armstrong.)*

A typical myofibril measures ~1–2 µm in diameter and comprises a single chain of many thousands of **sarcomeres**, each of which consists of highly ordered myofilament arrays that generate alternating light and dark bands along the myofibril (**Figure 8.3c**). Similarly, each myofibril is precisely aligned with its neighbors so that their ordered sarcomeres yield the larger-scale, trans-myofiber striations that are visible by LM (**Figure 8.3d**).

8.5   SARCOMERE ORGANIZATION: THICK (MYOSIN) AND THIN (ACTIN) FILAMENTS ARRANGED INTO PERIPHERAL Z DISCS AND I BANDS SANDWICHED AROUND A CENTRAL A BAND

In relaxed muscles, each **sarcomere** measures ~2.5 µm long and comprises two **Z discs**, two **I bands**, and an **A band** with a central **H zone** (**Figure 8.4a**). The edges of the sarcomere are delimited by two adjacent **Z discs**, which abut lightly staining **I bands** (**Figure 8.4b**), and medial to the **I bands** is the darkly staining **A band**, whose central **H zone** contains in its middle an **M band** (=M line) (**Figure 8.4a, c**).

The staining and light-polarizing properties of sarcomere regions depend mainly on their **thick** and **thin filament** configurations. **Thick filaments** measure ~15 nm wide by 1.6 µm long and span from edge-to-edge of the A band. Each thick filament comprises staggered chains of few hundred **myosin heavy chain** (**MHC**) dimers, with each dimer resembling a pair of intertwined, miniature golf clubs (**Figure 8.4d**). The C-terminal portions of MHC dimers form coiled-coil, rodlike **tails** that constitute the filament shaft. Conversely, toward its N-terminal end, each MCH monomer has a short **neck** region with a pair of bound **myosin light chains** (**MLCs**) that help regulate filament functioning. Distal to the neck, the myosin monomer's N terminus forms a paddle-like **head** (=**crossbridge**), which, along with the neck, protrudes from the shaft (**Figure 8.4d**). In fully assembled thick filaments, myosin tails are polarized in opposite directions from the sarcomeric M band, and their protruding crossbridges occur helically wound around the filament except in a central bare zone where crossbridges are lacking (**Figure 8.4d**).

Unlike thick filaments, **thin filaments** are only ~6 nm wide by ~1 µm long and consist of non-myosin proteins. Most of the thin filament comprises a double helix of filamentous **actin** polymers, but situated along this actin backbone are **tropomyosin** and **troponin** proteins that respond to changes in Ca^{2+} levels, thereby regulating the ability of actin to bind myosin crossbridges (**Figure 8.4d**). Thin filaments attach in Z discs and extend across the entire I band as well as into the A band periphery before terminating at the H zone (**Figure 8.4d**). Thus, in relaxed sarcomeres, the I band contains **only thin filaments**, whereas the A band peripheries comprise **thick-thin overlap zones**, where each thick filament is surrounded by six thin filaments (**Figure 8.4a**). Conversely, the uncontracted H zone consists **entirely of thick filaments** and may exhibit a lighter **L zone** next to the M band, where myosin crossbridges are lacking (**Figure 8.4b, d**).

Aside from thick and thin filaments, each sarcomere has additional proteins that aid muscle functioning. For example, Z discs contain the actin-binding protein **α-actinin** that helps anchor thin filaments within the sarcomere. In addition, the protein **nebulin** attaches to Z discs and wraps around actin thereby stabilizing thin filaments. Similarly, the Ca^{2+}-sensitive protein **titin** connects Z discs to thick filaments and may also interact with thin filaments during active lengthening of the myofiber (=eccentric contractions) (Box 8.1 Thick Filaments: Titin-mediated Eccentric Contractions, Off vs. On States, and Muscle Evolution).

Z discs in peripheral myofibrils also extend various elements like **dystrophin, desmin**, and **actin** linkages to interact with the sarcolemma and surrounding ECM. Such interacting complexes collectively constitute **costameres**, which serve to transmit forces to neighboring connective tissues thereby helping to maintain proper myofiber functioning within muscles (**Figure 8.5a, b**). Accordingly, **dystrophin** mutations can result in muscular dystrophies (MDs) in which myofiber damage caused by aberrant costameres overwhelms satellite cell abilities to regenerate muscles (**Case 8.1 Darius Weems's Skeletal Muscles**).

8.6   EXCITATION-CONTRACTION COUPLING (ECC): NEUROMUSCULAR JUNCTIONS AND TRIAD NETWORKS FOR COORDINATING CONTRACTIONS

The term **excitation-contraction coupling** (**ECC**) refers to functional events occurring between neuronal excitation of the sarcolemma and subsequent contraction of the myofiber. Two key units—the **neuromuscular junction** (**NMJ**) and conducting networks of **triads**—are involved in initiating and propagating ECC processes. In addition to ECC descriptions provided here, further discussions of related topics are covered in the next chapter on the nervous system.

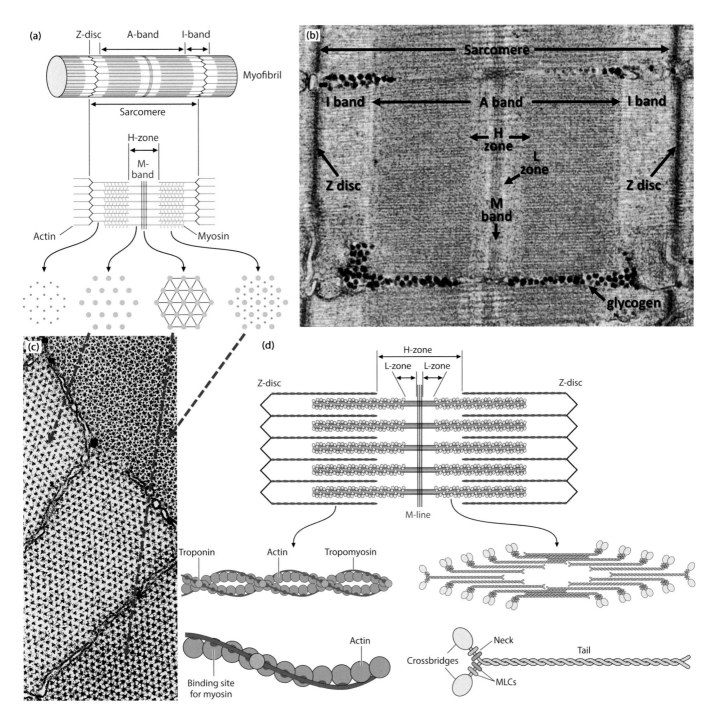

Figure 8.4 Sarcomere and myofilament microanatomy. (a–d) *Each sarcomere extends from one Z disc to its neighbor and has in its center an A band with myosin-containing thick filaments spanning the width of the band. Flanking the A band are I bands that contain actin thin filaments. The thin filaments extend from the Z disc into the periphery of the A band, where they form a thick-thin overlap area before terminating at the edge of the H zone. The H zone contains only thick filaments plus a central M band (M line) that interconnects them. Next to the M line is an L zone where thick filaments lack crossbridges. TEMs of frog muscle in figures b and c show sarcomeres in longitudinal (b, 42,000X) and transverse (c, 60,000X) section, with triads oriented over Z discs, rather than over A–I junctions, as occurs in adult mammalian myofibers (**Figure 8.3b**). Figure d shows that the double-helix of actin polymers within each thin filament associates with regulatory molecules consisting of tropomyosin filaments and globular troponin complexes spaced at intervals of seven actin monomers. Each thick filament comprises of a few hundred myosin heavy chain (MHC) dimers, whose C-termini form coiled-coil, rodlike tails within the filament shaft. Toward its N-terminal end, each MCH monomer has a short neck with paired myosin light chains (MLCs). Distal to the neck, the N terminus forms a paddle-like crossbridge head, which, along with the neck, protrudes from the shaft except in the central bare zone of the filament.*

Figure 8.4 (Continued) *([a,d]: Adapted from OpenStax College (2013) Anatomy and Physiology. OpenStax. http://cnx .org/content/col11496/latest and from Craig, R and Woodhead, JL (2006) Structure and function of myosin filaments. Curr Opin Struc Biol 16: 204–212; [b,c]: From images courtesy of Dr. C. Franzini-Armstrong.)*

BOX 8.1   THICK FILAMENTS: TITIN-MEDIATED ECCENTRIC CONTRACTIONS, OFF VS. ON STATES, AND MUSCLE EVOLUTION

Unlike passive shape changes caused by other muscles, muscle can actively generate force via: (1) a **static** (=isometric) process without joint/limb movement (e.g., pushing straight-armed against a wall); (2) a **dynamic** (=isotonic) **concentric contraction** where the muscle shortens as tension is generated (e.g., bringing a dumbbell toward the shoulder during a biceps curl); and

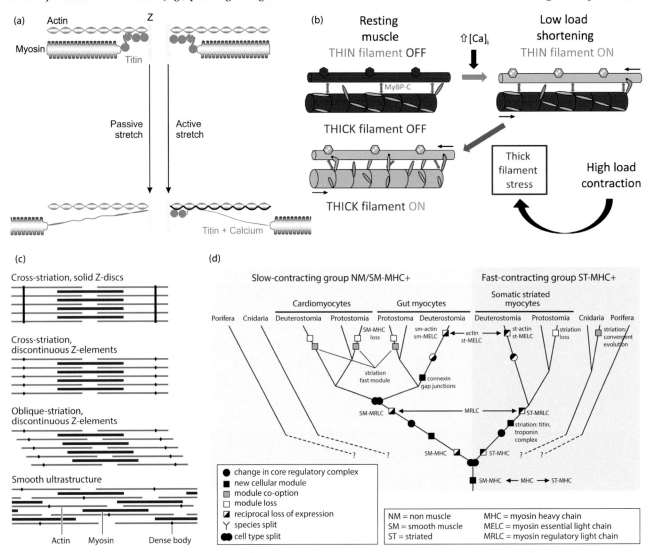

Box 8.1 *(a) Titin's potential role in active eccentric muscle stretching (from Herzog, W et al. (2015) A new paradigm for muscle contraction. Front Physiol 6: doi: 10.3389/fphys.2015.00174 reproduced under a CC BY 4.0 creative common license); (b) A diagram summarizing how thick filaments may help regulate contractions by placing their crossbridges in an active ON position for binding thin filaments (from Irving, M (2017) Regulation of contraction by thick filaments in skeletal muscle. Biophys J 113: 2579–2594 reproduced with publisher permission); (c, d) Aspects of muscle evolution in the animal kingdom depicting transitions from smooth muscle to more highly derived cross-striated muscle (note: these figures represent simplified versions of diagrams in Brunet, T et al. (2016) The evolutionary origin of bilaterian smooth and striated myocytes. eLife 5: doi: 10.7554/eLife.19607 reproduced under a CC BY 4.0 creative commons license.)*

(3) a **dynamic eccentric contraction** during which the muscle lengthens while generating force (e.g., moving a dumbbell away from the shoulder). Previously, eccentric contractions were difficult to explain via standard myosin-actin interactions. However, new evidence suggests thick-filament-associated **titin** molecules situated near Z discs can bind both calcium and thin filament actin to stiffen titin, thereby generating force during eccentric contractions (**Figure a**).

In addition to titin, striated **thick filaments** are attached to **myosin-binding protein C** (**MyBP-C**). Such proteins are believed to help thick filaments transition from an **Off state** in relaxed muscles, where most crossbridges are held close to the filament shaft, to an **On state** with more protruded crossbridges that facilitate actomyosin interactions (**Figure b**).

In **vertebrates**, sarcomeres with solid Z discs, unique thick filaments, and troponin-mediated contraction regulation are considered distinguishing characteristics of striated muscles. However, among **invertebrates**, such distinctions are not as clear-cut, as striated muscles can possess either **discontinuous** or **continuous Z discs** (**Figure c**). Moreover, in the flatworm *Schistosoma mansoni*, smooth muscle with **dense bodies** not only contains striated-like thick filaments but also expresses **tropomyosin** and the **TnT** subunit of **troponin**. Such features as well as the transitioning of thick filaments from OFF-to-ON states underscore the commonality of extant muscle types, which most likely diverged during evolution from smooth-muscle-like forms in basal animal groups (**Figure d**).

NMJs are synapses in motor end plates that develop between axon terminals and the sarcolemma, with the axon plasmalemma forming the **pre-synaptic membrane**, and the sarcolemma constituting the **post-synaptic membrane** (**Figure 8.5c, d**). Each motor axon that innervates myofibers typically divides into numerous branches. The dilated terminus of these branches (=end bulb or bouton) contains **synaptic vesicles** (**SVs**) filled with the neurotransmitter (**NT**) **acetylcholine** (**ACh**). Between the pre- and postsynaptic membranes is a **synaptic cleft** that in turn overlies an infolded sarcolemma. **ACh receptors** (**AChRs**) of the sarcolemma are concentrated near the top of these folds, whereas **voltage-gated sodium channels** (NaVs) occur deeper within the folds (**Figure 8.5e, f**).

As summarized in **Figure 8.6a**, NMJs transmit neuronal **action potentials** to the myofiber sarcolemma. Such transmission begins when an AP reaches the end bulb and opens **voltage-gated calcium channels** in the neuronal plasmalemma, thereby allowing external calcium ions to enter the axon and cause ACh exocytosis into the synaptic cleft. The binding of ACh to sarcolemmal AChRs opens nearby voltage-gated sodium channels, and the resulting influx of sodium ions generates a sarcolemmal AP that is transmitted along T tubules. After the AP spreads through T tubules, the **triad** network completes the final ECC steps by raising sarcoplasmic calcium levels via the release of Ca^{2+} ions that had been bound to such proteins like **calsequestrin** within the SR lumen (**Figure 8.6b**). In adult myofibers, triads tend to occur near the borders between sarcomeric A and I bands (**A–I junctions**) (**Figure 8.3b**), which in turn allows a coordinated release of Ca^{2+} throughout the relatively thick myofiber.

Calcium efflux from the SR involves two closely apposed channel types: (i) T tubule **L-type calcium channels** (**LTCCs**) and (ii) SR membrane **ryanodine receptor** (**RyR**) channels (**Figure 8.6b, c**). LTCCs and RyRs face each other across the narrow sarcoplasmic space that separates SR and T-tubule membranes, and extending into this space are the cytoplasmic domains of RyR tetramers, called RyR feet (**Figure 8.6b, c**). Via these channel arrays, APs propagated down T tubules cause conformational changes in LTCCs that open RyR channels, perhaps via LTCC-mediated displacements of nearby RyR feet. The opened RyRs then release their stored Ca^{2+} ions into the sarcoplasm, thereby initiating myofiber contraction, as described further below.

8.7 THE SLIDING FILAMENT MECHANISM OF SKELETAL MUSCLE CONTRACTION AND THE ROLES OF CA²⁺ AND ATP AS THIN FILAMENTS SLIDE PAST THICK FILAMENTS TO SHORTEN SARCOMERES

Based on careful measurements of myofiber sarcomeres, thin filaments do not shorten during muscle contraction, and thick filaments spanning A bands also remain the same size in relaxed vs. contracted muscles. Conversely, I-band and H-zone sizes decrease following contraction (**Figure 8.6d**). Such observations provided the basis for the **sliding-filament** mechanism of muscle contraction, which proposes that thin filaments are actively moved past thick filaments toward the central M band as myofibers contract. For sliding to proceed, **calcium-mediated** changes in the positioning of thin filament **regulatory proteins** must first expose crossbridge binding sites on actin (**Figure 8.6e**). Myosin crossbridges can then undergo **ATP-related cycles** of attachment, shape change, release, and repositioning to move thin filaments in a ratchet-like fashion, as described further below.

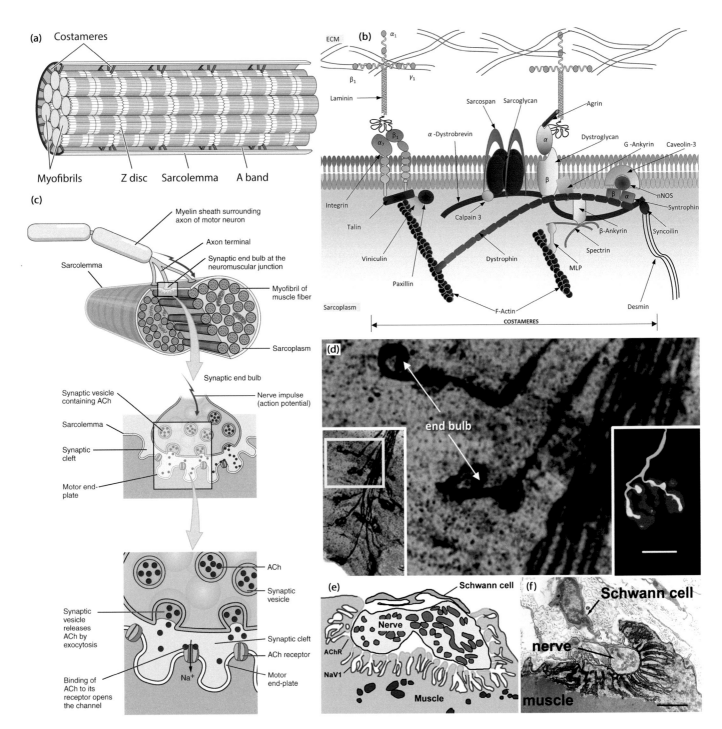

Figure 8.5 Costameres and neuromuscular junctions (NMJs). (*a, b*) *Costameres are complexes of filaments and associated proteins, such as dystrophins that link Z discs of peripheral myofibrils to the sarcolemma and surrounding ECM, thereby helping to transduce forces and signals between extra- and intracellular compartments of skeletal muscles. Such regions can break down following repeated cycles of contractions, particularly in cases of muscle degeneration due to mutations that cause muscular dystrophies. (*c–f*) As summarized in figure c, an excitation-contraction coupling (ECC) cascade links muscle membrane excitation to myofibril contractions. To activate such ECC pathways, motor neurons synapse with the sarcolemma at discrete neuromuscular junctions (NMJs) within a motor endplate (d). Each NMJ comprises the dilated terminus of an axon branch (=end bulb), which is separated from the invaginated sarcolemma by the synaptic cleft. After an action potential reaches the end bulb, voltage-gated calcium channels open, thereby triggering external calcium influx that drives exocytosis of acetylcholine-containing synaptic vesicles into the synaptic cleft. Binding of acetylcholine (ACh) to ACh receptors (AChRs) on the sarcolemma allows sodium influx through these receptors and through nearby voltage-gated sodium channel (NaVs) to initiate an action potential in the sarcolemma. The right inset in figure d shows a human end bulb*

Figure 8.5 (Continued) with the axon labeled green for synaptophysin and neurofilament protein. Postsynaptic AChRs are labeled red, and yellow represent overlaps of fluorescence (scale bar = 20 μm). A corresponding drawing and TEM are shown in e and f. Scale bar in f = 1 μm. ([a]: From Ervasti, JM (2003) Costameres: the Achilles' heel of herculean muscle. J Biol Chem 278: 13591–13594 reproduced under a CC-BY 4.0 creative commons license; [b]: From Mukund, K and Subramaniam, S (2019) Skeletal muscle: A review of molecular structure and function, in health and disease. WIREs Systems Biology and Medicine: DOI: 10.1002/wsbm.1462 reproduced under a CC BY 4.0 creative commons license; [c]: From OpenStax College (2013) Anatomy and Physiology. OpenStax. http://cnx.org/content/col11496/latest reproduced under a CC BY 4.0 creative commons license; [d]: From right inset; and [e,f]: From Slater, CR (2017) The structure of human neuromuscular junctions: some unanswered questions. Int J Mol Sci 18: doi:10.3390/ijms18102183 reproduced under a CC BY 4.0 creative commons license.)

CASE 8.1 DARIUS WEEMS'S SKELETAL MUSCLE

Disability Rights Activist and Rapper (1989–2016)

"I'm in a wheelchair but I'm just like a regular person."

(D.W. during a cross-country trip documenting accessibility issues and raising awareness of Duchenne's muscular dystrophy disease)

An American rap singer who suffered from Duchenne's muscular dystrophy (DMD), Darius Weems was the subject of an award-winning film that documented his cross-country journey to highlight accessibility rights. Weems was born in Georgia and became confined to a wheelchair by age 12 owing to DMD-induced skeletal muscle degeneration. Having not traveled beyond Georgia's borders, Weems and a group of friends embarked in 2005 on an 8000-mile road trip in a customized van with film equipment to shoot the

documentary *"Darius Goes West: The Roll of his Life"*. The crew filmed Weems as he navigated about in his wheelchair while visiting popular landmarks and lesser-known attractions. The oldest crew member was 24, and although they were novices, the young filmmakers' documentary won dozens of awards while also raising more than two million dollars for DMD research.  Subsequently, Weems went on to release two CDs of his rap songs and to begin an experimental treatment for DMD. However, in the end, the breakdown of skeletal muscles throughout his body and the various complications of DMD led to Weems's death in 2016 at age 27.

In relaxed myofibers, sarcoplasmic Ca^{2+} is kept at low levels mainly by SERCA pumps in the longitudinal SR, which in turn affects the precise positioning of **troponin** (**Tn**) and **tropomyosin** (**Tm**) regulatory proteins within thin filaments (**Figure 8.6e**). Troponin comprises scattered heterotrimeric complexes, each with three subunits (**TnT** [tropomyosin-binding], **TnI** [inhibitory], and **TnC** [calcium-binding]), whereas each tropomyosin is a ropelike double helix covering the outer side of seven actin molecules and linking end-to-end with other tropomyosins along the entire thin filament (**Figures 8.4d** and **8.6e**). At low sarcoplasmic Ca^{2+} levels, troponin and tropomyosin cover crossbridge binding sites on actin to prevent the binding of thick filaments to thin filaments (**Figure 8.6e**). However, in response to activated NMJs, Ca^{2+} ions released from triad SR cisternae bind to TnC, thereby causing a **conformational change** that shifts the Tn complex plus its attached Tm toward actin polymer centers and hence uncovers crossbridge binding sites on actin (**Figure 8.6e**).

Following exposure of binding sites for crossbridges, thick filaments can begin translocating thin filaments via an ATP-related **swinging crossbridge** mechanism (**Figure 8.7a, b**). In each thick filament myosin hexamer (=holoenzyme), such crossbridge movements involve: (i) an N-terminal pair of **crossbridges** with **ATPase activity** plus **actin** and **ATP binding sites**, collectively allowing crossbridges to act as actin-directed **molecular motors**; (ii) two pairs of **myosin light chains** that help optimize force generation during actin-myosin interactions; (iii) dual **necks** that function as flexible **lever arms** for amplifying crossbridge conformational changes; and (iv) a coiled-coil **tail**, which anchor necks and crossbridges to the thick filament (**Figure 8.7a**).

As summarized in **Figure 8.7b**, prior to thick and thin filament sliding, each **crossbridge** has bound an ATP and possesses a large **cleft** separating its upper and lower domains in a **pre-powerstroke** (**PPS**) configuration (see: "9 o'clock position" in **Figure 8.7b**). However, once calcium release exposes binding sites on actin, the crossbridge can **attach to actin**, and with actin serving as a cofactor for optimizing myosin ATPase activity, ADP and Pi are released from the crossbridge. Such release causes a conformational change that closes the crossbridge cleft and thereby re-positions the head to generate a crossbridge **powerstroke** that slides the attached thin filament medially. After the powerstroke is completed, the **crossbridge binds a new ATP molecule**, which **detaches** the crossbridge from actin as it undergoes a recovery stroke to restore its former PPS alignment.

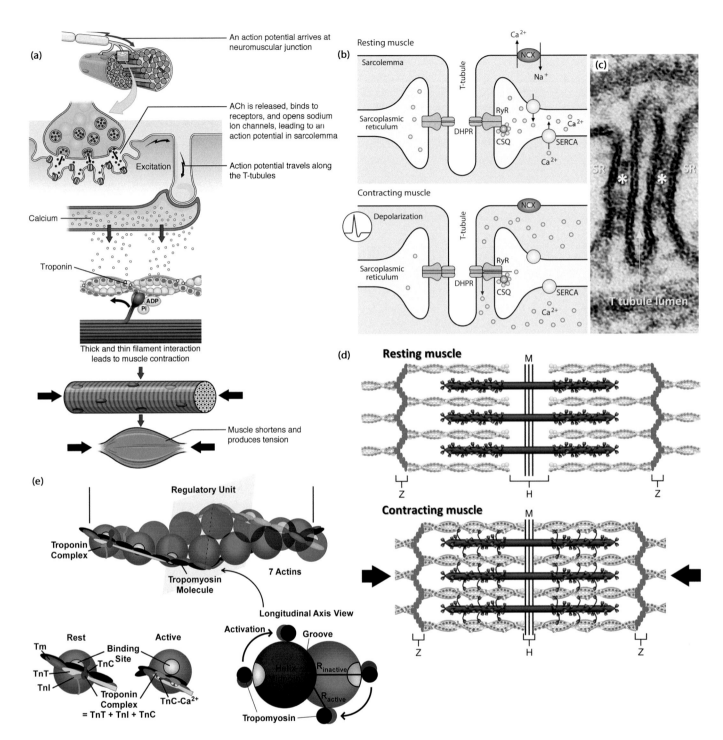

Figure 8.6 Excitation contraction coupling (ECC). (*a*) As a summary of the ECC process, acetylcholine (ACh) released at the NMJ triggers a sarcolemmal action potential that is propagated along T tubules, thereby releasing calcium ions from the sarcoplasmic reticulum (SR). This in turn allows interactions between thick and thin filaments for muscle contraction. (*b*) The depolarizing action potential in T tubules causes conformational changes in voltage-gated dihydropyridine receptor (DHPR) kinds of L-type calcium channels on the T tubule membrane. Such changes open SR ryanodine receptors (RyRs) whose end feet extend into the narrow space between T tubule and SR membranes (asterisks), thereby allowing the release of calcium ions that had been previously pumped into the SR lumen by SERCA pumps and bound to proteins, such as calsequestrin (CsQ). Following contraction, calcium is re-sequestered into the SR and extruded extracellularly by Na/Ca exchangers (NCXs), as DHPRs and RyRs return to pre-contraction states in the absence of high cytoplasmic calcium. (*c*) In this TEM of toadfish muscle, end feet of ryanodine receptors forms electron-dense material (asterisks) in the space between the T tubule and SR membranes (250,000X). (*d*) Based on measurements

Figure 8.6 (Continued) *of sarcomeres before and after contraction, neither thick nor thin filaments shorten during contraction. Instead, thin filaments slide medially past thick filaments toward the center of the sarcomere. Thus, as postulated by the sliding-filament mechanism of muscle contraction, the A band remains the same size, whereas contraction causes the I bands, H zone, and distance between Z discs to shorten. **(e)** As diagrammed in various orientations, calcium bound to the TnC subunit of the troponin complex causes a conformational change that moves troponin and its attached tropomyosin toward the central groove of the actin double-helix, thereby exposing binding sites on actin for thick filament crossbridges. ([a, d]: From OpenStax College (2013) Anatomy and Physiology. OpenStax. http://cnx.org/content/col11496/latest reproduced under a CC BY 4.0 creative commons license; [b]: From Lasa-Elgaressta, J et al. (2019) Calcium mechanisms in limb-girdle muscular dystrophy with CAPN3 mutations. Int. J. Mol. Sci. 20,4548: doi:10.3390/ijms20184548 reproduced under a CC BY 4.0 creative commons license; [c]: Image courtesy of Dr. C. Franzini-Armstrong; [e]: From Dupuis, LJ et al. (2016) Mechano-chemical interactions in cardiac sarcomere contraction: A computational modeling study. PLoS Comput Biol 12: doi:10.1371/journal.pcbi.1005126 reproduced under a CC BY 4.0 creative commons license.)*

In the absence of additional ATP production following death, thick and thin filaments remain attached, and the muscle generates post-mortem rigidity (**rigor mortis**). However, in living muscle with adequate ATP supplies, actin binding sites on the thin filament continue to be exposed, and the **swinging crossbridge cycle is repeated**, with each powerstroke moving the thin filament a few nanometers. Under normal conditions resulting in full contractions, repetitive ratcheting eventually contracts the sarcomere to ~70% of its resting length, which is essentially the width of the A band in each sarcomere.

At the end of a contraction, APs are no longer delivered to the NMJ, and T-tubule LTCCs revert to their former conformation so that SR RyRs close. Accordingly, as sarcoplasmic Ca^{2+} levels are returned to baseline levels by SERCA pumps and plasmalemmal channels, such as **Na/Ca exchangers** (**NCXs**) (**Figure 8.6b**), the Tn+Tm regulatory proteins shift back to mask actin binding sites for crossbridges so that thick and thin filaments can no longer interact.

8.8   SUBTYPES OF SKELETAL MUSCLE

Myofibers differ in two major functional properties—their rapidity of contraction and speed of fatigue. Contraction rates are determined mainly by the particular isoform of myosin heavy chains that the fiber contains, which has traditionally been used to separate fibers into **type 1** vs. **type 2** classes. Fatigability, on the other hand, is related to such factors as: (i) mitochondrial content; (ii) metabolic patterns (oxidative versus glycolytic); and (iii) overall muscle redness, which is due in large part to the levels of an oxygen-binding pigment, called **myoglobin**. Originally, myofibers were distinguished as **type 1 red fibers**, which are myoglobin-rich, comparatively slow to contract (slow-twitch), and able to sustain contractions for extended periods (fatigue-resistant) vs. **type 2 white fibers**, which have less myoglobin and can initiate contraction quickly (fast-twitch) before losing contractility rapidly (fatigable). These two fiber types explain the difference between the **dark breast meat** of ducks, which must fly long distances during migrations vs. the **white breast meat** of flightless domesticated turkeys, which no longer require abundant fatigue-resistant muscles to move their wings and instead have such dark meat confined to their legs for extended periods of standing. Accordingly, the enhanced myoglobin content of red fibers is related to their **oxidative** mode of metabolism, as opposed to **glycolytic** pathways used by classical white fibers with lower myoglobin levels.

After microdissection methods became more refined and additional parameters were assessed (e.g., ATPase activity, predominant metabolic mode, and precise speed of contraction), the classical view of a single type 2 white fiber needed to be revised to encompass three distinct subtypes of type 2 fibers that are produced by vertebrates (**Figure 8.8a**). Adding to this complexity, hybrids of the main fiber subtypes provide an even broader continuum than had been previously envisioned by the original red vs. white dichotomy.

8.9   MUSCLE SPINDLES: STRETCH-SENSITIVE SENSORY ORGANS IN SKELETAL MUSCLE THAT FUNCTION IN PROPRIOCEPTION

As components of a **proprioceptor** network that helps body parts orient relative to each other and to the surrounding environment, fusiform **muscle spindles** are embedded in virtually all skeletal muscles, where they provide sensory input regarding limb positioning and body movements (**Figure 8.8b**). Each spindle is ensheathed by a connective tissue capsule that is anchored to the surrounding epimysium. Inside the capsule, unmodified (=**extrafusal**) fibers can be distinguished from modified myofibers, termed **intrafusal fibers**, which possess

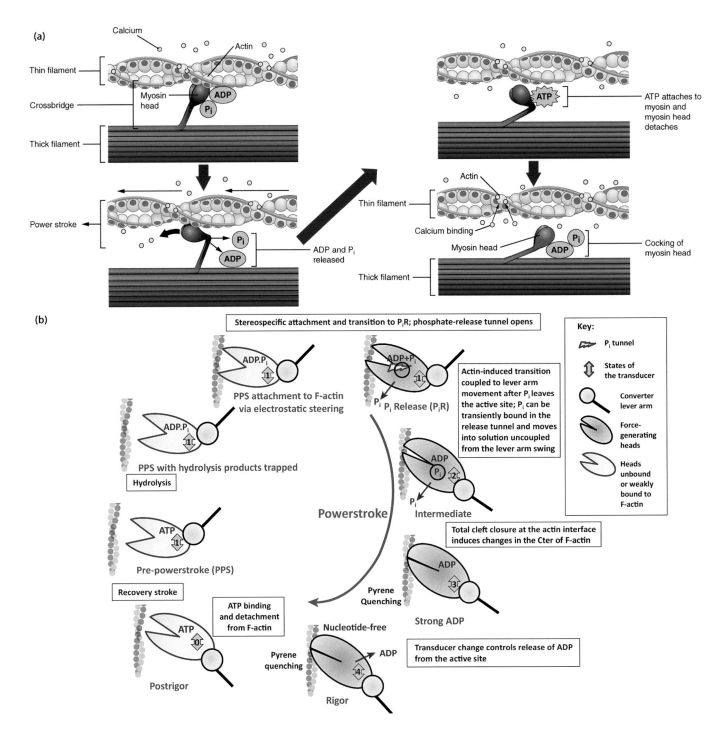

Figure 8.7 Molecular mechanisms of skeletal muscle contraction. (a, b) *After troponin binds calcium to expose crossbridge binding sites on thin filaments, thick filament attachment to actin increases ATPase activity in crossbridges so that ADP and Pi can be released. This causes a conformational change in the crossbridge allowing a powerstroke to move the thin filament toward the sarcomere center. The subsequent binding of a new ATP molecule to the crossbridge detaches the crossbridge and returns it to its pre-powerstroke (PPS) condition (9 o'clock position in diagram). Ratcheting of thin filaments past thick filaments can be repeated until the sarcomere is fully contracted, or until calcium is re-sequestered to obstruct crossbridge binding sites. ([a] From OpenStax College (2013) Anatomy and Physiology. OpenStax. http://cnx.org/content/col11496/latest reproduced under a CC BY 4.0 creative commons license; [b] From Houdusse, A and Sweeney, HL (2016) How myosin generates force on actin filaments. Trends Biochem Sci 41: 989–997 reproduced with publisher permission.)*

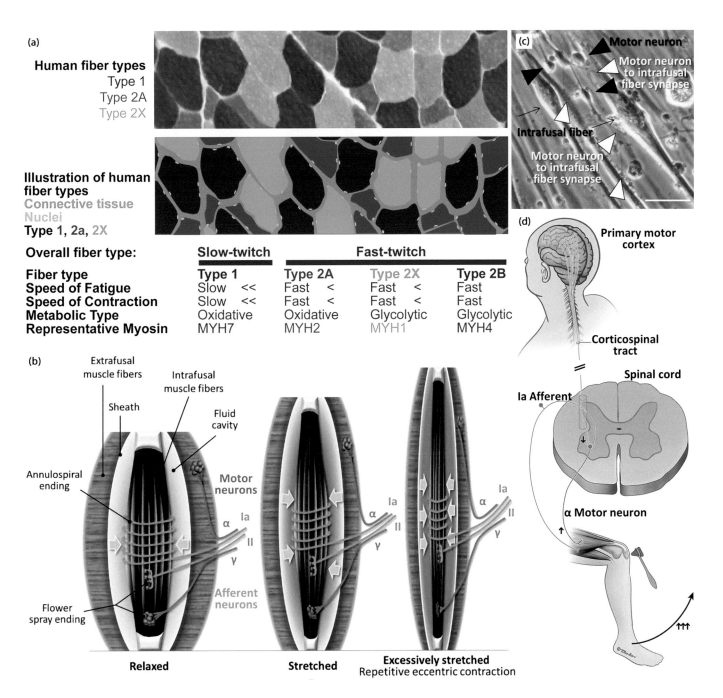

(a)

Human fiber types
Type 1
Type 2A
Type 2X

Illustration of human fiber types
Connective tissue
Nuclei
Type 1, 2a, 2X

(c)

Motor neuron
Motor neuron to intrafusal fiber synapse
Intrafusal fiber
Motor neuron to intrafusal fiber synapse

Overall fiber type:	Slow-twitch	Fast-twitch		
Fiber type	Type 1	Type 2A	Type 2X	Type 2B
Speed of Fatigue	Slow <<	Fast <	Fast <	Fast
Speed of Contraction	Slow <<	Fast <	Fast <	Fast
Metabolic Type	Oxidative	Oxidative	Glycolytic	Glycolytic
Representative Myosin	MYH7	MYH2	MYH1	MYH4

(d) Primary motor cortex

Corticospinal tract

Spinal cord

Ia Afferent

α Motor neuron

(b)

Extrafusal muscle fibers
Intrafusal muscle fibers
Sheath
Fluid cavity
Annulospiral ending
Motor neurons
α Ia
II
γ
Afferent neurons
Flower spray ending

Ia
α
II
γ

Ia
α
II
γ

Relaxed   **Stretched**   **Excessively stretched**
Repetitive eccentric contraction

Figure 8.8 Myofiber subtypes and muscle spindles. *(**a**) Each human skeletal muscle can contain three main subtypes of myofibers (1, 2a, 2X), which are distinguished based on such factors as speed of contraction, fatigability, myoglobin content, myosin isotypes, and metabolic pathways. (**b**) Within muscles are muscle spindle organs, which comprise a sheath of conventional extrafusal fibers that surround modified intrafusal fibers. Spindles connect to both motor and afferent neurons, thereby providing proprioceptive information regarding the contractile state of the muscle. (**c**) Synapses form between motor neurons and isolated intrafusal fibers in co-cultures of these two spindle components. Scale bar = 50 μm. (**d**) Muscle spindles also help prevent overstretching of muscles, as illustrated by the patellar tendon reflex, when tapping on the kneecap stretches upper leg muscles and causes a reflexive movement of the lower leg to compensate for such stretching. ([a]: From Talbot, J and Maves, L (2016) Skeletal muscle fiber type: using insights from muscle developmental biology to dissect targets for susceptibility and resistance to muscle disease. Wiley Interdiscip Rev Dev Biol 5: 518–534 reproduced with publisher permission; [b]: From Sonkodi, B et al. (2019) Have we looked in the wrong direction for more than 100 years? Delayed onset muscle soreness is, in fact, neural microdamage rather than muscle damage. Antioxidants 9: doi:10.3390/antiox9030212 reproduced under a CC BY 4.0 creative commons license; [c]: From Colon, A et al. (2017) Functional analysis of human intrafusal fiber innervation by human γ-motoneurons. Sci Rep 7: 17202,*

Figure 8.8 (Continued) *doi:10.1038/s41598-017-17382-2 reproduced under a CC BY 4.0 creative commons license; (d) From Florman, JE et al. (2014) Lower motor neuron findings after upper motor neuron injury: insights from postoperative supplementary motor area syndrome. Front Hum Neurosci 7: doi: 10.3389/fnhum.2013.00085 reproduced under a CC BY 3.0 creative commons license.)*

sarcomere-free central regions with aggregated nuclei (**Figure 8.8b**). Intrafusal fibers synapse with **afferent neurons** possessing mechanosensitive ion channels that can generate APs encoding proprioceptive information regarding surrounding muscle activities. Intrafusal fibers also synapse with motor neurons, which can fine-tune the spindle so that it functions properly over varying muscle lengths (**Figure 8.8c**). In addition, some spindle motor neurons synapse with extrafusal fibers as part of reflex arcs (Chapter 9) that help protect muscles from being overstretched. An example of this protective effect can be demonstrated by the patellar reflex (=**knee jerk**) test, where sharply tapping the patellar tendon of a flexed knee stretches quadriceps muscle spindles in the thigh, thereby reflexively causing quadriceps contraction and a kicking forward of the leg (**Figure 8.8d**).

8.10 CARDIAC MUSCLE—STRUCTURAL AND FUNCTIONAL DIFFERENCES BETWEEN CARDIAC AND SKELETAL TYPES OF STRIATED MUSCLE

Although cardiac muscle fundamentally resembles skeletal muscle in possessing sarcomeres that use a sliding-filament/swinging-crossbridge mode of contraction, several attributes of cardiac myocytes help differentiate these two types of striated muscles. Thus, as supplements to organ-level descriptions of the heart covered in Chapter 7, this section focuses on cellular, organellar, and molecular differences distinguishing cardiac myocytes from skeletal myofibers.

Cellular differences

During early development, cardiac myocytes possess a single nucleus and can actively divide to construct a functional heart. However, starting just before birth and continuing into early postnatal life, mononucleated cardiac cells terminally differentiate into functional myocytes that no longer divide. In many of these post-mitotic cells, the myocyte undergoes karyokinesis without cytokinesis to generate **two nuclei**, perhaps as an adaptation for high metabolic demands that require additional genetic material to be housed in an extra nucleus. Regardless of whether one or two nuclei are present, cardiac myocyte nuclei occur in the **center of the cell**, as opposed to the periphery, as in skeletal myofibers (**Figure 8.9a, b**). Moreover, unlike satellite cell contributions to adult myofiber regeneration, cardiac myocytes have relatively little capacity to regenerate.

When fully developed, each cardiac myocyte measures ~100–150 μm long and can form one to several **branches** along their lengths (**Figure 8.9a**). As adaptations for contractions that can exceed three billion cycles over the course of a lifetime, the terminal ends of myocytes have staircase-like intercellular junctions, called **intercalated discs (IDs)** (**Figure 8.9c–e**), which not only keep neighboring cells firmly attached to each other but also facilitate intercellular communication for coordinating contractions (**Figure 8.9d, e**). The transverse portions of IDs typically contain cardiac-specific adherens junctions that interconnect actin filaments in neighboring cells to help transmit contractile forces across the heart (**Figure 8.9e**). Conversely, longitudinal parts of IDs not only possess **desmosomes** that bind **desmin-containing intermediate filaments** to hold myocytes together but also contain **gap junctions** that allow the direct flow of small molecules for ionic coupling (**Figure 8.9e**) (Box 8.2 Cardiac Gap Junctions).

Organelle-related differences

Cardiac **T tubules** are substantially wider than their counterparts in skeletal muscle and form a network of both transversely oriented and axially arranged channels that collectively constitute a transverse-axial tubule system (TATS) (**Figure 8.9f–h**). In addition, rather than generating triads, cardiac myocytes produce **dyads** of a single SR cisterna next to its T tubule partner, with these dual units typically localizing over Z discs instead of A–I junctions, as in myofibers (**Figure 8.9h, i**).

Molecular differences

Contraction of cardiac cells functions mainly by a **calcium-induced calcium release (CICR)** mode of elevating sarcoplasmic calcium levels, in which **external Ca²⁺ ions** play a key role in excitation-contraction coupling. Thus, action potentials propagating down cardiac T tubules can open cardiac-specific LTCCs that allow abundant extracellular Ca²⁺ ions occurring within wide T tubules to flow into the cleft separating T tubule and SR membranes. Such external calcium binds to RyRs on the SR membrane and alters their conformation to

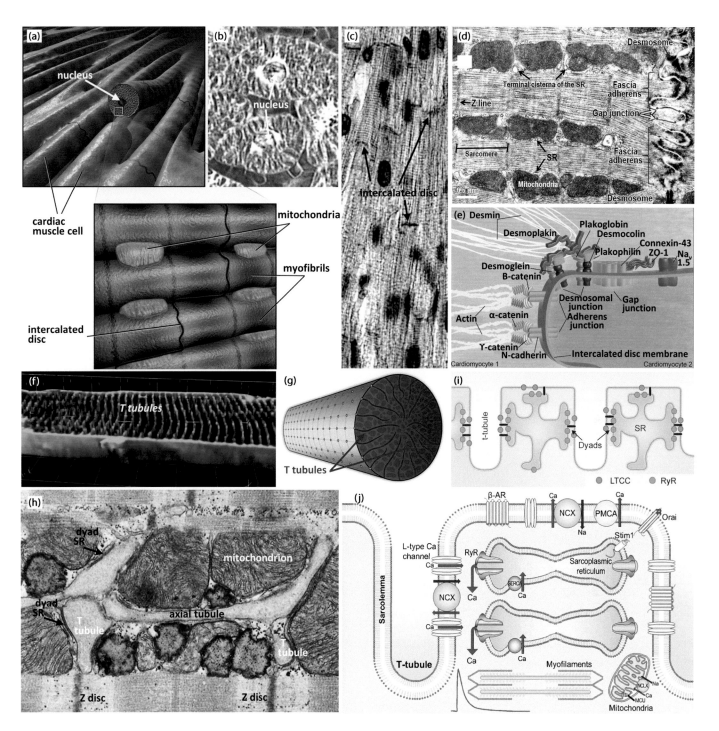

Figure 8.9 Cardiac myocytes. *(a–e) Cardiac myocytes have one, or sometimes two, central nuclei and are interconnected by staircase-like arrangements of junctions, called intercalated discs. (b) 675X; (d) 530X. (f–h) As shown in a 3D confocal reconstruction (f), diagram (g), and TEM (h, 13,000X), T tubules of cardiac cells are relatively wide and commonly joined together by axial tubules, thereby forming a well-developed transverse axial tubule system (TATS). (i, j) Each T tubule membrane has L-type calcium channels (LTCCs) and is associated with a single SR cisterna with ryanodine receptors (RyRs), thereby yielding a dyad over sarcomeric Z-discs. Contraction involves calcium release from SR stores as well as influx of calcium through plasmalemmal channels. At the end of contraction, calcium is re-sequestered into the SR by SERCA pumps and extruded into the ECM by Na/Ca Exchangers (NCXs) and plasma membrane calcium ATPases (PMCAs). ([a]: From Blausen.com staff (2014) "Medical gallery of Blausen Medical 2014". WikiJournal of Medicine 1 (2): 10. doi:10.15347/wjm/2014.010 reproduced under CC0 and CC BY SA creative commons licenses; [d]: From Yeh, C-H et al. (2019) Cisd2 is essential to delaying cardiac aging and to maintaining heart functions. PLoS Biol 17: e3000508. https:// doi. org/10.1371/journal.pbio.3000508 reproduced under a CC BY 4.0 creative commons license; [e]: From Manring, HR*

Figure 8.9 (Continued) *et al. (2018) At the heart of inter- and intracellular signaling: the intercalated disc. Biophys Rev 10: 961–971 reproduced with publisher permission; [f, g]: From Guo, A et al. (2013) Emerging mechanisms of T-tubule remodeling in heart failure. Cardiovas Res 98: 204–215 reproduced with publisher permission; [h]: From Fawcett, DW and McNutt, NS (1969) The ultrastructure of the cat myocardium. I. Ventricular papillary muscle. J Cell Biol. 42: 1–44 reproduced with publisher permission; [i]: From Jones, PP et al. (2018) Dyadic plasticity in cardiomyocytes. Front Physiol 9: doi: 10.3389/fphys.2018.01773 reproduced under a CC BY 4.0 creative commons license; [j]: From Eisner, DA et al. (2017) Calcium and excitation-contraction coupling in the heart. Circ Res 121: 181–195 reproduced with publisher permission.)*

BOX 8.2 CARDIAC GAP JUNCTIONS

A key feature of cardiac **intercalated discs (IDs)** are **gap junctions (GJs)**, which allow molecules <1000 MW to pass between neighboring myocytes via clusters of densely packed intercellular channels (**Figure a–d**). Each GJ channel is composed of a total of 12 **connexin** proteins, with 6 connexins forming within one cell membrane a half-channel unit, called a **connexon**, which

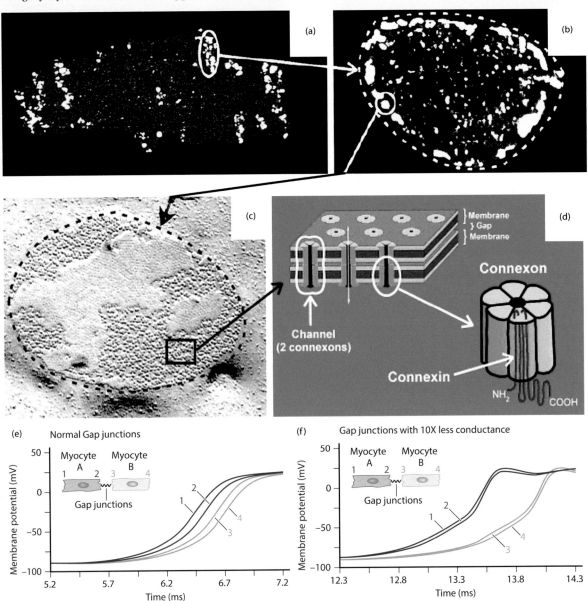

Box 8.2 (a–d) *Cardiac myocyte gap junctions (confocal microscopy [a, b]; freeze-fracture methods [c], and diagram [d]) (from Severs, NJ (2000) The cardiac muscle cell. BioEssays 22: 188–199 reproduced with publisher permission); (**e, f**) Computer modeling of accelerated impulse transmission in cardiac myocytes coupled with normal gap junctions vs. with abnormal coupling; based on data presented in Rohr (2004) (Rohr, S (2004) Role of gap junctions in the propagation of the cardiac action potential. Cardiovas Res 62: 309–322.)*

lines up with another 6-connexin-containing connexon in the neighboring cell membrane to form a full channel (**Figure d**). GJs of cardiac myocytes possess many hundreds of connexon-containing channels, with larger GJs tending to occur in a peripheral ring that surrounds smaller centrally located GJs (**Figure b**). Such arrangements help ensure faster AP propagations along the long vs. transverse myocyte axis.

The major role of cardiac GJs is to accelerate AP propagation within the myocardium. If GJs are obliterated, conduction rates are substantially reduced, mainly due to protracted intercellular transmissions (**Figure e, f**). Not surprisingly, cardiac GJ dysfunction is linked to various heart diseases. For example, arrhythmias in patients with Wolff-Parkinson White syndrome arise from added strands of GJ-rich tissue that bypass normal conducting pathways. Conversely, localized disordering of GJs and downregulation can contribute to heart attacks.

CASE 8.2 MOTHER TERESA'S CARDIAC MUSCLES

Catholic Nun and Saint
(1910–1997)

"I am grateful to receive it in the name of the hungry, the naked, the homeless, the crippled, the blind, the lepers, all those people who feel unwanted, unloved, uncared for throughout society..."

(M.T. on who would be cared for by the congregation she founded in what is now called Kolkata, India)

Mother Teresa was a nun who received the 1979 Nobel Peace Prize and posthumous canonization as Saint Teresa of Calcutta for her charitable work helping the downtrodden. Born in what is now North Macedonia, Teresa founded a congregation to help the sick and poor of Kolkata (=Calcutta), India, where she offered spiritual aid and shelter. Under Teresa's guidance, the original Calcutta unit expanded to over 500 affiliated

branches in numerous countries before she eventually stepped aside in 1997 due to heart disease. Such cardiac issues first surfaced in 1983, when Teresa suffered a heart attack while visiting the Pope. After another infarction four years later, Teresa was fitted with a pacemaker, and in 1996, while hospitalized for further cardiac surgery, Teresa became highly agitated, prompting an exorcism to be performed "to drive the devil from her heart". In spite of her surgery and the supposedly soothing effects of the exorcism, Teresa failed to recover from pervasive damage done to her cardiac muscles and died at age 87 due to congestive heart failure.

---

allow the release of stored SR calcium via CICR (**Figure 8.9i, j**). In addition, cardiac myocyte relaxation involves not only SERCA pumps but also considerable extrusion of Ca^{2+} ions into the external space by cardiac-specific plasma membrane calcium ATPases (PMCAs) and Na/Ca exchangers (NCXs) (**Figure 8.9j**). Accordingly, such channels are currently being targeted for potential therapies of heart disease, including certain forms of heart failure (**Case 8.2 Mother Teresa's Cardiac Muscles**).

8.11 SMOOTH MUSCLE CELLS (SMCS)—NO ALIGNED SARCOMERES BUT WITH THICK AND THIN FILAMENTS, DENSE BODIES (=Z DISC EQUIVALENTS), AND CONTRACTION REGULATION VIA MYOSIN LIGHT CHAIN KINASE

Collectively constituting ~1–2% of an adult's weight, **smooth muscle cells** (**SMCs**) are distributed throughout the body as components of circulatory, digestive, urogenital, respiratory, integumentary, and ocular organs (**Figure 8.10a, b**). Most SMCs are derived from mesenchymal populations of mesoderm, and following their initial differentiation *in utero*, some adult SMCs can be triggered to divide again, thereby contributing to the higher regenerative capacity of smooth vs. striated muscle.

In adults, spindle-shaped SMCs contain a single central nucleus that can be flanked by clustered mitochondria (**Figure 8.10a–d**). In addition, SMCs typically measure ~30 μm to a few hundred μm long and thus more closely resemble cardiac myocytes than skeletal myofibers.

However, unlike either type of striated muscle, fully differentiated SMCs lack regularly arranged sarcomeres that generate trans-cellular striations. Instead, each SMC has a more homogeneous distribution of **thick** and **thin myofilaments** throughout its myoplasm, which in turn is surrounded by a plasmalemma that lacks T tubule invaginations. In addition, scattered within the myoplasm are oblong-shaped **dense bodies**, which are interconnected by intermediate filaments that anchor this cytoplasmic network into **dense plaques** occurring directly beneath the plasmalemma (**Figure 8.10d, e**). Dense bodies contain actin-binding proteins, such as **α-actinin** that attach to the ends of thin filaments, thereby serving as the functional equivalents of striated

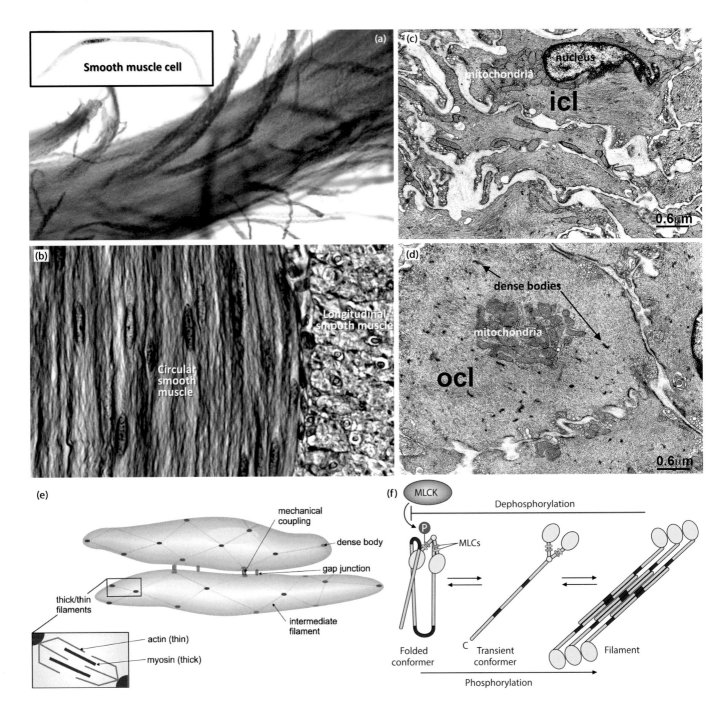

Figure 8.10 Smooth muscle cells. (**a**) Smooth muscle consists of spindle-shaped cells with a single central nucleus, as shown in stained whole mounts of a smooth muscle cell (SMC) (inset) and teased SMC bundle (300X). (**b**) Gut SMCs are shown in two different sectioning planes. 570X. (**c, d**) SMCs in the inner and outer portions of the circular layer (icl and ocl) of the gut wall have clustered mitochondria and Z-disc-equivalent dense bodies. (**e**) Cytoplasmic dense bodies connect via intermediate filaments with dense plaques in the SMC plasmalemma. During contractions, SMC thin filaments slide past thick filaments within mini-sarcomeres that are located between neighboring dense bodies. In conjunction with the intermediate filament network, such filament sliding serves to shorten the SMC in a purse-string-like fashion. (**f**) SMC contraction is not regulated by a troponin-based system. Instead, increased amounts of calcium ions in stimulated SMCs combine with calmodulin to activate myosin light chain kinase (MLCK), which phosphorylates myosin light chains in folded, inactive monomers of myosin. This phosphorylation causes monomer unfolding and assembly into thick myofilaments that can attach their crossbridges to nearby thin filaments. ([c, d]: From Traini, C et al. (2014) Inner and outer portions of colonic circular muscle: ultrastructural and immunohistochemical changes in rat chronically treated with otilonium bromide. PLoS ONE 9: doi:10.1371/journal.pone.0103237 reproduced under a CC BY 4.0 creative

Figure 8.10 (Continued) *commons license; [e] From MacIntyre, DA et al. (2007) Myometrial activation—coordination, connectivity, and contractility. Fetal Mat Med Rev: doi:10.1017/S0965539507002033 reproduced with publisher permission; [f] Adapted from Milton, DL et al. (2011) Direct evidence for functional smooth muscle myosin II in the 10S self-inhibited monomeric conformation in airway smooth muscle cells. PNAS 108: 1421–1426.)*

muscle Z discs. SMCs can be further distinguished from striated muscle cells by the absence of calcium-sensitive troponin subunits on thin filaments (Box 8.1), indicating ECC pathways and myosin-actin interactions in SMCs differ fundamentally from those used by striated muscle, as discussed further below.

Without the physical constraints of a sarcomere-based organization, SMCs are well adapted for functioning across a wide range of cellular lengths, particularly in highly distensible organs, such as the uterus and bladder. Moreover, SMC are able to undergo relatively short-lived (**phasic**) vs. prolonged (**tonic**) types of contractions in response to various stimuli (e.g., mechanical forces, neurotransmitters like ACh and noradrenaline, and/or diffusible factors, such as vasopressin, nitric oxide, and natriuretic peptides).

Prior to SMC contraction, intracellular Ca^{2+} rises are triggered by: (i) **external Ca^{2+} influx** through plasmalemmal channels; (ii) **internal calcium release** from the SR, which can involve inositol trisphosphate (IP_3)-mediated pathways; or (iii) **a combination of both mechanisms**. Such calcium elevations fail to expose crossbridge binding sites on thin filaments as in striated muscles. Instead, actin-myosin interactions are regulated by Ca^{2+} initially binding to **calmodulin (CaM)** molecules in the SMC myoplasm to form Ca/CaM complexes that in turn activate **myosin light chain kinase (MLCK)** (**Figure 8.10f**). Active MLCK then phosphorylates thick filament **myosin light chains (MLCs)**, thereby enabling actin-myosin binding in two major ways: (1) by enhancing the **ATPase activity** of thick filament crossbridges; and (2) by promoting the **unfolding of myosin molecules** from their inhibited state, thus allowing unfurled myosins to **polymerize** into new and/or longer thick filaments (**Figure 8.10f**).

Once activated by MLCK, thick filaments in SMCs lack a central bare region without crossbridges, which allows a broad range of filament sliding for thin filaments that are anchored in nearby **dense bodies.** Since SMC thin filaments possess continually exposed crossbridge binding sites unobstructed by tropomyosin/troponin complexes, thick filaments that are both activated by MLCK and primed by ATP binding are directly triggered to undergo a swinging-crossbridge-mediated sliding of thin filaments past thick filaments (**Figure 8.10e**). Collectively, such sliding occurs within "mini sarcomere-like" units in SMCs, thereby constricting the dense body network and folding the cell in a purse-string-like fashion to reduce its size. At the end of contraction, decreased sarcoplasmic calcium levels inactivate MLCK and drive myosin polymers toward less-active folded monomers (**Figure 8.10f**). The key roles played by MLCK are underscored by knock-outs of this kinase that block contraction. Conversely, overactive MLCK mediates chronic constrictions of SMCs in diseases, such as asthma (**Case 8.3 King Kamehameha IV's Smooth Muscles**). These and other distinguishing features of SMCs, cardiac myocytes, and skeletal myofibers are summarized in Table 8.1.

CASE 8.3 KING KAMEHAMEHA IV'S SMOOTH MUSCLES

Fourth Monarch of the Hawaiian Kingdom
 (1834–1863)

"I found he was the conductor and took me for somebody's servant just because I had a darker skin than he had."

(From K.K.IV's journal describing how he was nearly removed from a train by the conductor while visiting the U.S.)

Kamehameha IV was the fourth king of the United Hawaiian monarchy, who reigned from 1855 to 1863. As a well-educated and multilingual crown prince, Kamehameha IV traveled to Europe via the United States in 1849 as part of a diplomatic mission. While in America, the dark-skinned royal member was subjected to racist insults, which further contributed to his negative view of Americans at a time when their presence in Hawaii posed a growing threat to the kingdom's independence. After being crowned king in 1855, Kamehameha IV proposed a forward-thinking healthcare plan to deal with increases in foreigner-introduced diseases, such as smallpox that were laying waste to the native population, and in 1859 Queen's Hospital of Honolulu was opened. Unfortunately, such healthcare improvements failed to save the king's only son, who died at age 4 of an undisclosed illness. Similarly, before he could fully institute

his ambitious plan for Hawaii, Kamehameha IV succumbed at age 29 due to chronic asthma, a disease characterized by labored breathing due to overly constricted smooth muscle fibers in airway passages.

TABLE 8.1 COMPARISON OF SKELETAL, CARDIAC, AND SMOOTH MUSCLE

Property	Skeletal	Cardiac	Smooth
Cell name	Myofiber	Cardiac myocyte	Smooth muscle cell (SMC)
Locations in body	Typically attached to skeleton	Heart	Widespread (e.g., blood vessels; urogenital, digestive, and respiratory tracts)
Nucleus (number, position)	Numerous, peripheral	One or two, central	One, central
Length of cell	Up to ~60 cm	~100–150 μm	~30–300 μm
T-tubule invaginations of sarcolemma	Relatively narrow	Wide and supplemented with axial tubules	None
Triads/dyads	Triads over A–I junctions	Dyads over Z discs	None
Calcium rise required for contraction	Yes	Yes	Yes
Source of calcium rise	Mostly internal (release from sarcoplasmic reticulum)	Mix of external/internal	Mostly external with some internal
Thin (actin) and thick (myosin) filaments present	Yes	Yes	Yes
Thin filament attachment sites	Z discs	Z discs	Dense bodies
Aligned sarcomeres forming striations	Yes	Yes	No
Sliding of thick and thin filaments during contraction	Yes	Yes	Yes
Regulation of contraction	Troponin/tropomyosin	Troponin/tropomyosin	Myosin light chain kinase
Duration of contraction	From short to long	Short	From short to very long
Intercalated discs	No	Yes	No
Regenerative capacity	Limited	Essentially none	High

8.12   SOME MUSCLE DISORDERS: DUCHENNE MUSCULAR DYSTROPHY (DMD) AND MULTISYSTEMIC SMOOTH MUSCLE DYSFUNCTION SYNDROME (MSMDS)

As the most common of nine major muscular dystrophies, **Duchenne muscular dystrophy (DMD)** involves excessive striated muscle damage owing to **dystrophin** gene mutations that compromise dystrophin-mediated linkages between striated muscle and the ECM. Usually affecting males, DMD initially presents as generalized muscle weakness before steadily worsening until respiratory and cardiac insufficiencies eventually cause death, typically before age 30. Currently, there is no cure for DMD, although disease progression can be retarded by β-blocker drugs that treat heart disease (Chapter 7).

Multisystem smooth muscle dysfunction syndrome (MSMDS) is a rare genetic disease that involves mutations in the ACTA2 gene encoding the smooth-muscle-specific α2 isoform of actin. Young MSMDS patients often exhibit various smooth-muscle-related anomalies, such as aortic aneurysms and cerebrovascular disorders. Histologically, MSDS is characterized by fibrosis of smooth muscle cells within blood vessel walls, causing arteries to be less flexible and more prone to aneurysms. Currently, no cures for MSMDS are available, but surgical re-structuring of cerebral vessels may help prolong life.

8.13 SUMMARY—MUSCLES

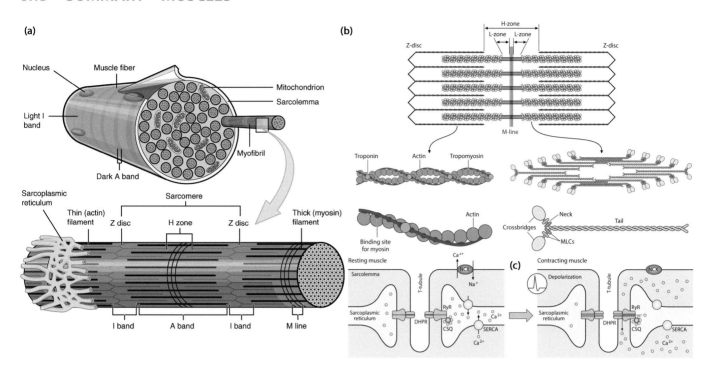

Pictorial Summary Figure Caption: (a)*: see figure 8.1d; (**b**): see figure 8.4d; (**c**): see figure 8.6b.*

The three main muscle types are **skeletal** and **cardiac** (=**striated muscle** with orderly arrays of **sarcomeres**) vs. **smooth** (=**non-striated** without aligned sarcomeres). All muscles **contract** using Ca^{2+} **elevations** that allow interactions between **actin thin filaments** and **myosin thick filaments**.

Skeletal muscle comprises multinucleated contractile **myofibers** that develop from the fusion of multiple myoblasts. Each muscle is surrounded by an **epimysium** CT sheath that forms **perimysia** around myofiber bundles (**fascicles**), with each myofiber being covered by an **endomysium**. The myofiber plasmalemma (**sarcolemma**) invaginates inward as transversely oriented **T tubules** that associate with **sarcoplasmic reticulum** (**SR**) cisternae to form **triads**, and most of the remaining cytoplasm is filled with rope-like **myofibrils** each comprising a chain of **sarcomeres** with ordered **thick** and **thin** filaments (**Figure a**). **Sarcomeres** measure 2–3 μm long in a relaxed state and comprise: (i) two **Z discs** (with **α-actinin** for binding actin); (ii) two **I bands** (with only thin filaments); and (iii) a medial **A band** (with thick filaments spanning edge-to-edge). The A band has a central **H zone** (only thick filaments), middle **M band** (interconnecting thick filaments), and peripheral **overlap zones** (where thick and thin filaments co-occur) (**Figure b**). Each **thin filament** consists of double-helical **actin** polymers with **troponin** and **tropomyosin** regulatory proteins along the actin backbone (**Figure b**). **Thick filaments** comprise mainly **myosin heavy chains** (**MHCs**) arranged as: (i) coiled-coil **tails** in the filament shaft; (ii) flexible **necks** with **myosin light chains** (**MLCs**); and (iii) **crossbridge heads** that bind actin (**Figure b**).

Neuromuscular junctions (**NMJs**) are axon-muscle **synapses** of **motor end plates** controlling myofiber contraction. **Acetylcholine** (**ACh**) released by the motor axon binds sarcolemmal **ACh receptors** to trigger an **action potential** (**AP**) that is transmitted down **T tubules** to cause Ca^{2+} **release** from SR cisternae flanking each T tubule (**Figure c**). Such Ca^{2+} release involves AP-mediated changes in T-tubule **L-type calcium channels** (LTCCs) opening **ryanodine receptors** (**RyRs**) in the SR.

As part of the **sliding filament/swinging crossbridge mode of contraction**, thick and thin filaments **do not shorten** during contraction. Instead, thin filaments **slide past** thick filaments. Thus, I bands and H zones shorten, as A bands stay the same size. Ca^{2+} from the SR binds **troponin** (Tn) to move **Tn** and attached **tropomyosin** (Tm), thereby exposing crossbridge binding sites on actin. Actin-bound crossbridges undergo **conformational changes** for **powerstrokes** that move thin filaments. Each crossbridge binds new ATP, detaches from actin, recovers its pre-powerstroke alignment, and re-binds actin. The cycle is then repeated until Ca^{2+} is returned to the SR by **SERCA** pumps, which in turn allows Tn and Tm to block crossbridge binding sites again.

Spindle organs in skeletal muscles aid proprioception and prevent overstretching. **Myofiber types** are classified by contraction speed, metabolic mode, MHCs, and ATPase activities.

Cardiac muscle uses the same sliding mechanism as in skeletal muscle but has mono- or bi-nucleated cardiac myocytes that: (1) are joined by **intercalated discs** with **gap junctions**; (2) have wider T tubules that form **dyads** (single SR per tubule); and (3) depend mainly on **calcium-induced calcium release (CICR)** from external calcium influx to open SR RyRs.

Smooth muscle cells (SMCs) in blood vessels and other organs lack ordered sarcomeres but still use Ca^{2+} **elevations, actin thin filaments,** and **myosin thick filaments** for contraction. Ca^{2+} released from the SR or brought in via plasmalemmal channels binds **calmodulin** to activate **myosin light chain kinase (MLCK)**. MLCK phosphorylates MLCs on myosin monomers to unfold them from an inactive state and form thick filaments that allow thin filament sliding between **dense bodies** (=Z-disc equivalents).

Some muscle disorders: **Duchenne muscular dystrophy (DMD)**: muscle cell damage due to the loss of **dystrophin** protein that helps link striated muscle plasmalemmata to ECM molecules; **multisystem smooth muscle dysfunction syndrome (MSMDS)**: rare disease that generally results in death owing to aneurysms and strokes arising from the weakening of vascular smooth muscle.

SELF-STUDY QUESTIONS

1. What are the Z-disc equivalents in smooth muscle cells?
2. T/F L-type calcium channels of striated muscle triads occur in T tubule membranes.
3. Which cells able to regenerate lost or damaged skeletal myofibers?
4. Which of the following is the largest?
    A. Myofilament
    B. Myofiber
    C. Myofibril
    D. Fascicle
    E. Triad
5. Which of the following is NOT a feature of cardiac myocytes?
    A. Calcium-induced calcium release
    B. Dyads
    C. Gap junctions
    D. Intercalated discs
    E. Multiple peripherally located nuclei
6. Which part of myosin hexamers occur within the thick filament shaft?
    A. Heads
    B. Necks
    C. Tails
    D. Myosin light chains
    E. Tropomyosin
7. Approximately how long are sarcomeres?
8. T/F Motor neurons release noradrenaline at neuromuscular junctions of skeletal myofibers.
9. T/F Each myofiber is directly surrounded by a sheath of connective tissue, called the epimysium.
10. What is the name of the encapsulated intramuscular sensory organs that help protect myofibers from overstretching?
11. Which is NOT a part of myofiber thin filaments?
    A. Actin
    B. Troponin
    C. Tropomyosin
    D. Crossbridges
    E. Crossbridge binding sites

12. Which region in relaxed sarcomeres contains only thin filaments?
   A. M band
   B. I band
   C. A band
   D. H zone
   E. NONE of the above
13. T/F. Ryanodine receptors occur in SR membranes.

"EXTRA CREDIT" For each term on the left, provide the BEST MATCH on the right (answers A-F can be used more than once)

14. Sequesters Ca^{2+} in SR _____
15. Part of dyad _____
16. Binds to tropomyosin _____
17. Found in dense bodies _____
18. Phosphorylates smooth muscle myosin monomer _____

   A. α-Actinin
   B. T tubule
   C. Troponin
   D. SERCA pump
   E. MLCK
   F. MHC

19. Describe the sliding-filament/swinging-crossbridge mechanism of skeletal muscle contraction from neural input to relaxation.
20. Describe contraction of smooth muscle cells.

ANSWERS

1) dense bodies; 2) T; 3) satellite cells; 4) D; 5) E; 6) C; 7) 2-3 µm; 8) F; 9) F; 10) muscle spindles; 11) D; 12) B; 13) T; 14) D; 15) B; 16) C; 17) A; 18) E; 19) The answer should include among other pertinent topics: NMJ, ACh, AChR, T tubule, LTCCs, RyRs, Ca^{2+} release, troponin, tropomyosin, crossbridge binding sites on actin, crossbridge ATPase, ATP, actin binding, powerstroke, recovery stroke, ratcheting, SERCA pumps; 20) The answer should include such terms as neural/hormonal stimulation, Ca^{2+} sources, calmodulin, MLCK, MLCs, self-inhibited states, myosin-actin interactions in "mini-sarcomeres", dense bodies

Nervous System

9.1 INTRODUCTION TO THE NERVOUS SYSTEM

The **nervous system** is organized into two main parts: (i) the **central nervous system (CNS)** consisting of the **brain** and **spinal cord** and (ii) the **peripheral nervous system (PNS)** comprising nervous tissue that occurs outside the CNS (**Figure 9.1a**). Such components are interconnected anteriorly and centrally by cranial and spinal nerves, respectively, whereas posteriorly, a horsetail-like collection of large nerves, called the **cauda equina**, links the CNS with the PNS in the pelvic and lower body regions (**Figure 9.1a, inset**).

Within the CNS and PNS, nervous tissue proper contains two cell types—**neurons** and **glial cells** (**Figure 9.1b**). Neurons are distinguished by their ability to transmit plasmalemmal electrical impulses, called **action potentials (APs)**. Conversely, glial cells, which collectively constitute the **neuroglia**, play key roles in supporting neuronal functions but are generally unable to produce APs (Box 9.1 Revising Neuroscience Dogma: Glial Cell Action Potentials and Neurogenesis in the Adult Brain). Previously, the human brain was thought to contain ~1 trillion glial cells and ~100 billion neurons, yielding a **glia-neuron ratio (GNR)** of ~10:1. However, recent analyses indicate roughly 60–90 billion neurons and only 40–90 billion glial cells occur in the human brain, resulting in a GNR of about 1:1.

DOI: 10.1201/9780429353307-11

Nervous System: CNS + PNS

Nervous Tissue: neurons + glial cells

Functional Neuron Types: Sensory-, Inter-, Motor Neurons

Development of the Nervous System

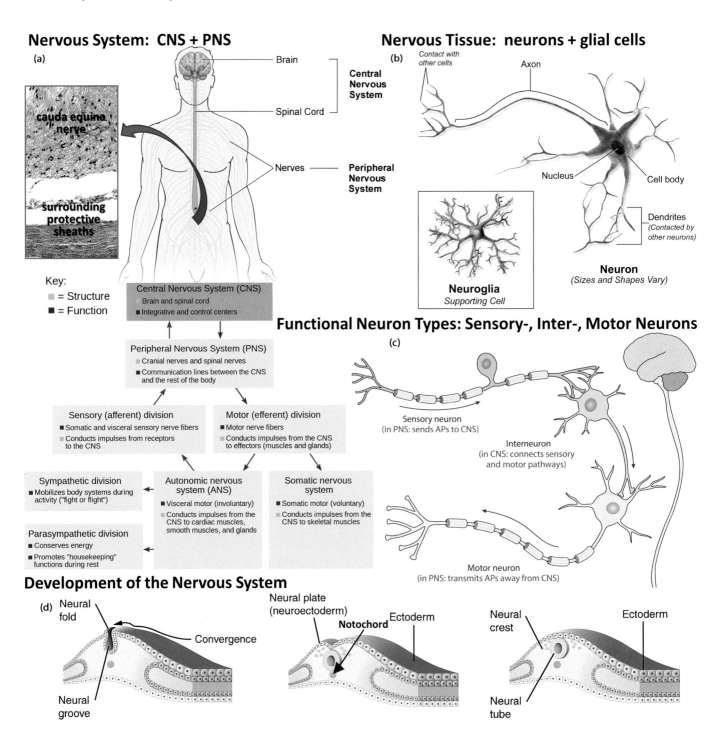

Figure 9.1 Introduction to the nervous system. (*a*) *The nervous system comprises: (i) a central nervous system (CNS) consisting of the brain and spinal cord, and (ii) a peripheral nervous system (PNS) of nervous tissue occurring external to the CNS. Functionally, the PNS is divided into: (i) the somatic system that voluntarily controls skeletal muscles vs. (ii) the autonomic system that involuntarily controls the heart, smooth muscles, and other tissues like glands. The autonomic system is further classified into sympathetic vs. parasympathetic subdivisions, which are adapted for fight-or-flight responses vs. resting housekeeping functions, respectively. Inset: cauda equina nerves are examples of links between the PNS and CNS (75X). (**b**) Nervous tissue proper consists of: (i) nerve cells (=neurons), which can generate neural impulses, called action potentials (APs), and (ii) supporting glial cells, which collectively constitute the neuroglia and are generally unable to generate APs. (**c**) PNS neurons are functionally classified as sensory (=afferent) if they send APs toward the CNS vs. motor (efferent) if they transmit APs away from the CNS. Interneurons in the CNS link sensory and motor pathways. (**d**) The nervous system develops from ectodermal folds that during a neurulation process converge, fuse, and separate from surface epidermis to form an internalized neural tube overlying the notochord. Neural crest cells migrate away from neural tube*

Figure 9.1 (Continued) *to form parts of the PNS and non-neural components in the body, such as pigment cells of skin and dentin-producing odontoblasts of teeth. ([a], drawing, and [d]: From OpenStax College (2013) Anatomy and Physiology. OpenStax. http://cnx.org/content/col11496/latest reproduced under a CC BY 4.0 creative commons license; [a]: bottom flowchart: From https://commons.wikimedia.org/wiki/File:NSdiagram.png which was assembled based on data available in Marieb's Human Anatomy & Physiology, 7th ed. New York: Pearson (2007) and reproduced by permission of a CC BY SA 3.0 creative commons license; [b]: From Blausen.com staff (2014) Medical gallery of Blausen Medical 2014. WikiJ Med 1 (2): 10. doi:10.15347/wjm/2014.010, reproduced under CC0 and CC BY SA creative commons licenses; [c] adapted from a CC BY SA creative commons image downloaded from https://maken.wikiwijs.nl/95908/flashmediaelement.swf#!page-3001026.)*

BOX 9.1 REVISING NEUROSCIENCE DOGMA: GLIAL CELL ACTION POTENTIALS AND NEUROGENESIS IN THE ADULT BRAIN

Traditionally, only neurons in nervous tissue were thought to generate action potentials (APs). However, more recent studies indicate that some cerebellar and hippocampal glial cells can also produce APs, which in turn may help modulate myelin formation or regulate AP transmission in nearby neurons.

Another long-held tenet was that CNS neurons are generally produced before birth or in some cases up to adolescence

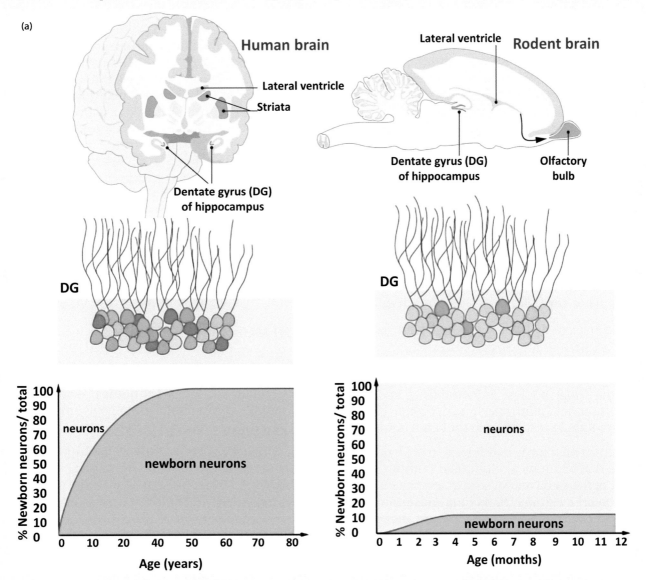

Box 9.1 (a) As an exception to neurons only being generated before adolescence, the hippocampal dentate gyrus (DG) continues to produce new neurons in adult mammals, including humans. (From: Ernst, A and Frisen, J (2015) Adult neurogenesis in humans—common and unique traits in mammals. PLoS Biol 10: e1002045 reproduced under a CC BY 4.0 creative commons license.)

before neurogenesis ceases. However, more recently, new neuron formation has been demonstrated in adult brains of various mammals, including humans. Such neurogenesis occurs in two major sites—(i) the **subventricular zone (SVZ)** near the lateral ventricle on each side of the cerebrum, and (ii) the **subgranular zone (SGZ)** directly beneath the granular layer of each hippocampal dentate gyrus (**Figure a**). The relative amounts of adult neuron production can differ in rodents vs. humans (**Figure a**). Nevertheless, pockets of new neuron formation in adult brains open the possibilities that such neurogenic capabilities might be used to help treat CNS disorders and injuries.

Within the CNS, each neuron connects with one or more other neurons, whereas PNS neurons not only connect with other neurons but may also join either with **sensory cells** that respond to external cues or with **target cells** that the neuron innervates. Junctions among neurons or between neurons and non-neural cells are called **synapses**, and such connections allow an upstream **pre-synaptic** neuron to communicate with its downstream **post-synaptic** partner. Via synapses and the neural circuits that they form, **sensory** (=afferent) **neurons** of the PNS direct APs toward the CNS, whereas PNS **motor** (=efferent) neurons transmit APs away from the CNS. In addition, as the most common functional neuron type, CNS **interneurons** relay APs between sensory and motor pathways (**Figure 9.1c**).

The nervous system begins to develop 3–4 weeks post-fertilization during a **neurulation** process that involves the movement of outer **ectoderm** into the interior of the embryo (**Figure 9.1d**). At the onset of neurulation, **neural plate** ectoderm on the dorsal embryonic surface generates **neural folds** that extend medially and fuse with each other while also sinking below the presumptive skin to yield a dorsal hollow **neural tube** that overlies a transient morphogenetic organizer, called the **notochord** (**Figure 9.1d**). **Neural crest cells** within the neural tube then migrate into the developing body to form PNS components and non-neural structures like pigment cells in skin and dentin-producing odontoblasts in teeth. Subsequently, the neural tube closes at each end, allowing the anterior **brain** and posterior **spinal cord** to develop. Failure to seal the anterior end causes **anencephaly**, a fatal lack of brain development, whereas incomplete posterior closures lead to **spina bifida** with varying degrees of health problems depending on the position and size of the residual opening (see **Case 10.5**).

When fully developed, the nervous system coordinates homeostatic activities and performs higher-order tasks, including cognition, learning, memory, and the integration of behavioral and emotional responses. Since these functions can be conducted on a conscious or unconscious level, the PNS is also viewed as comprising: (i) a **somatic nervous system** that can **voluntarily** control skeletal muscles and (ii) an **autonomic nervous system** that regulates **cardiac** and **smooth muscle**, **glands**, plus other tissues at an **involuntary** level (**Figure 9.1a**).

9.2   MULTIPOLAR NEURONS: THE MOST COMMON MORPHOLOGICAL TYPE OF NEURON

The cell body (=**soma** or perikaryon) of each neuron houses the nucleus and various cytoplasmic organelles, including Golgi bodies, mitochondria, and basophilic aggregates, called **Nissl bodies**, which comprise endoplasmic reticulum sheets and ribosomes (**Figure 9.2a, b**). In addition, intermediate filaments, microfilaments, and microtubules extend from the soma into cellular projections that are referred to as **neurites**.

Most neurons have multiple neurites and hence are classified morphologically as **multipolar neurons** (**Figure 9.2c**). In typical multipolar neurons, a single neurite process, called the **axon**, arises from the **axon hillock** of the soma and measures less than a few micrometers wide. Axons can extend for more than meter in length before eventually branching and terminating in bulbous tips, called **end bulbs** (=terminal boutons), which form the pre-synaptic ends of synapses. The soma also gives rise to **dendrite** forms of neurites that compared to axons tend to be shorter, more highly branched, and endowed with many small and dynamic protrusions, called **dendritic spines** (**Figure 9.2b**). Based on neurophysiological analyses, axons send APs **away from the soma**, whereas dendrites propagate their impulses **toward the soma**. In EMs, regions of neural tissue that contain high densities of neurites plus glial cells are referred to in general as **neuropil**.

As opposed to multipolar types, **bipolar neurons** in sensory organs have two neurites, whereas **pseudounipolar neurons** develop from bipolar neurons whose neurites fuse together to form a single T-shaped neurite that bifurcates near the soma (**Figure 9.2c**). Such neurons receive sensory input via one branch of the bifurcation

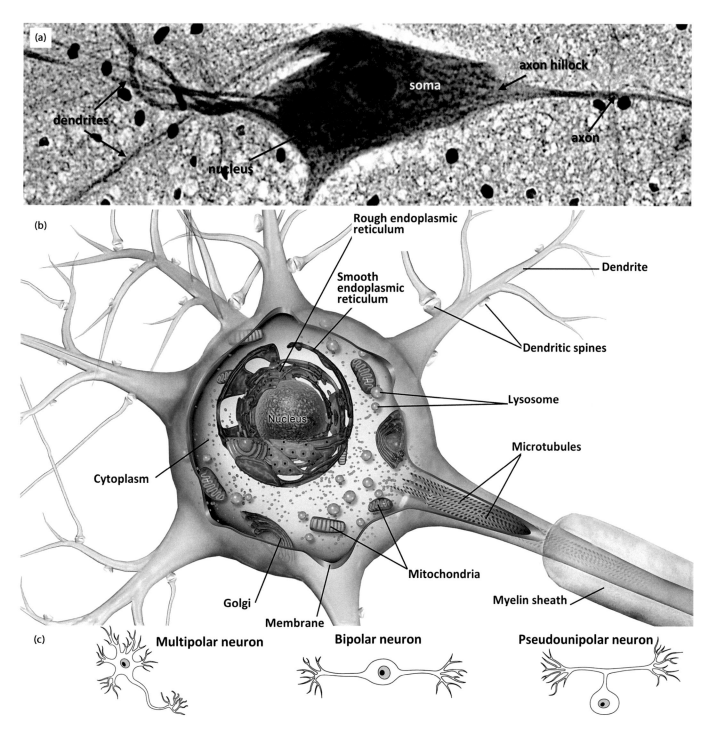

Figure 9.2 Neuron morphology. (*a*) *Neurons have a nucleus-containing cell body (=soma) from which neurite processes arise. Most of these neurites are dendrites that bring action potentials (APs) toward the soma vs. a single large axon type of neurite that transmits APs away from the soma (850X). (**b**) Each neuronal soma contains various cytoplasmic organelles, including collections of ribosomes and endoplasmic reticulum arrays, called Nissl bodies, and in many cases, the axon is surrounded by a glial-derived sheath of myelin. (**c**) Most neurons in the body have multiple neurites and are thus termed multipolar neurons. Bipolar neurons of sensory organs have two neurites, whereas a pseudounipolar neuron has a single short neurite that bifurcates into functional dendrite and axon branches. ([b]: From Blausen.com staff (2014) "Medical gallery of Blausen Medical 2014". WikiJournal of Medicine 1 (2): 10. doi:10.15347/wjm/2014.010 reproduced under CC0 and CC BY SA creative commons licenses; [c] from OpenStax College (2013) Anatomy and Physiology. OpenStax. http://cnx.org/content/col11496/latest reproduced under a CC BY 4.0 creative commons license.)*

and transmit impulses to the CNS via the other branch. Alternatively, **anaxonic neurons** (e.g., amacrine cells of the retina) lack a true axon and instead have dendrites that also function as axons in providing synaptic input for other cells.

9.3   HISTOLOGY OF GANGLIA AND NERVES IN THE PNS VS. THEIR CNS COUNTERPARTS

In the PNS, **ganglia** are encapsulated structures that contain neuronal cell bodies and ganglion-specific glial cells, called **satellite cells** (**Figure 9.3a–e**). Extending from ganglia are **nerves**, which are composed of multiple functional units, termed **nerve fibers**. Each nerve fiber comprises: (i) a **neuron** plus its **myelin sheath** if present (see next section), (ii) associated **glial cells**, and (iii) a thin connective tissue covering, called the **endoneurium** (**Figure 9.3e**). Three major nerve fiber types—**A**, **B**, and **C**—are classified based on their size, propagation rate, and degree of myelination (Table 9.1). Along with their nerve fibers, nerves also possess additional connective tissue sheaths similar in organization to those of skeletal muscles. Thus, **perineuria** surround bundles of nerve fibers, and an **epineurium** encases the entire nerve (**Figure 9.3a, b, d**).

Conversely, the CNS lacks abundant connective tissue. Thus, discrete encapsulated nerves are absent, and instead, closely apposed CNS axons are termed **neural tracts**. Similarly, collections of neuronal somata in the brain and spinal cord are referred to as **nuclei** rather than ganglia, although somata-rich regions at the base of the brain have historically been called **basal ganglia**.

9.4   MYELINATED VS. UNMYELINATED NEURONS OF THE PNS AND WHITE VS. GRAY MATTER OF THE CNS

Myelin is a complex mixture of lipids and proteins occurring within concentrically arranged membranes that can surround axons, thereby forming circum-axonal **myelin sheaths** (**Figure 9.4a–h**). In equal-sized myelinated vs. unmyelinated axons, myelin sheaths and their regularly arranged discontinuities serve to accelerate AP transmission rates compared to their unmyelinated counterparts, as discussed further below.

Myelin sheaths in the PNS are generated by glial cells, called **Schwann cells**. Each Schwann cell typically myelinates a single axon, with numerous Schwann cells occurring in a chain-like fashion down the length of axons. Similarly, **oligodendrocytes** (=oligodendroglial cells) are the CNS equivalents of Schwann cells, but unlike their PNS counterparts, each oligodendrocyte often branches and thus myelinates multiple axons. In unmyelinated nervous tissue, axons lack myelin sheaths and are instead loosely associated with non-myelinating glial cells. Accordingly, **gray matter** of the CNS appears gray in living brain and spinal cord samples, because it comprises mainly unmyelinated neurites. Conversely, CNS **white matter** is whiter, owing to abundant myelin around its predominantly myelinated axons.

During myelination, Schwann or oligodendroglial cells become obliquely wrapped around each axon multiple times (**Figure 9.4a**). Such wrappings begin at a starting point in the sheath, called the inner tongue (=inner mesaxon) and continue to the outer tongue farthest from the axon (**Figure 9.4d**). Concentrically arranged glial cell membranes and their intervening cytoplasm can generate up to a hundred myelin sheath layers, thereby increasing the overall width of the nerve fiber by ~50% beyond the diameter of its enclosed axon (**Figure 9.4b**). As myelination proceeds, most of the glial cytoplasm surrounding the axon is removed, allowing neighboring inner leaflets of glial membranes to fuse and thus form **major dense lines** of the sheath, which in turn are interspersed with **intraperiod spaces** containing outer leaflets of adjacent wraps (**Figure 9.4c**). In scattered sites along PNS myelin sheaths, some residual Schwann cell cytoplasm is retained, and the resulting separations of glial membranes form obliquely oriented incisures across the sheath, called **Schmidt-Lanterman clefts** (**Figure 9.4e**). These clefts facilitate transport of beneficial substances through the sheath during normal axonal maintenance and can be disrupted in neurodegenerative disorders, such as **Charcot-Marie-Tooth disease**.

As myelin sheaths form, distinct gaps in myelination, called **nodes of Ranvier**, develop every 0.1–1.5 mm down the length of the axon (**Figure 9.4f–h**), and as discussed further in the next section, such nodes help accelerate AP propagation. Most nodes are flanked by loosely arranged **paranodal sheaths** that transition into more tightly wound internodal sheaths between nodes (**Figure 9.4g, h**). These myelin configurations are normally maintained

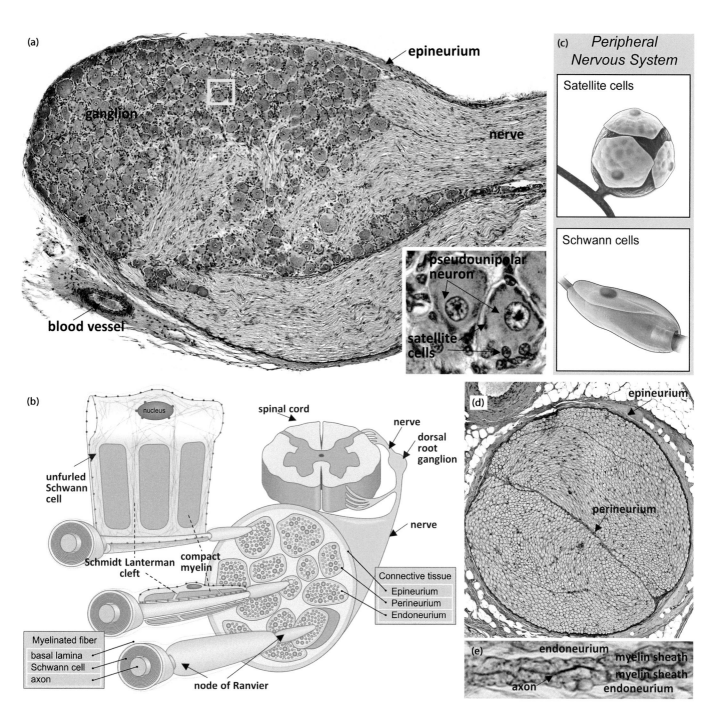

Figure 9.3 Ganglia, nerves, and their connective tissue sheaths. (*a, b*) *Collections of nerve cell bodies in the PNS are termed ganglia. In this section of a dorsal root ganglion (a), the ganglion is filled with unique satellite cell types of glial cells surrounding the large somata of pseudounipolar neurons. Nerves are ensheathed by epi-, peri-, and endo-neural layers of connective tissue (b). (a) 45X; (a, inset) 170X. (**c–e**) The axonal plasmalemma is typically encased by a sheath of myelin that is assembled by PNS glial cells, called Schwann cells (c, e). (d) 60X; (e) 480X. ([b]: From Belin, S et al. (2017) Influence of mechanical stimuli on Schwann cell biology. Front Cell Neurosci 11: doi: 10.3389/fncel.2017.00347 reproduced under a CC BY 4.0 creative commons license; [c]: From Blausen.com staff (2014) Medical gallery of Blausen Medical 2014. WikiJournal of Medicine 1 (2): 10. doi:10.15347/wjm/2014.010 reproduced under CC0 and CC BY SA creative commons licenses.)*

TABLE 9.1 CLASSIFICATION OF NERVE FIBERS

Fiber types	Function	Avg. fiber diameters (μm)	Avg. conductance velocity (m/s)
Aα	Primary muscle spindle afferents, motor to skeletal muscle	15	100 (70–120)
Aβ	Cutaneous touch and pressure afferents	8	50 (30–70)
Aγ	Motor to muscle spindle	5	20 (15–30)
Aδ	Cutaneous temperature and pain afferents	<3	15 (12–30)
B	Sympathetic preganglionic	3	7 (3–15)
C	Cutaneous pain afferents and sympathetic postganglionic	1	1 (02–2)

Source: From Innocenti, GM (2017) Network causality, axon computations, and Poffenberger. *Exp Brain Res* 235: 2349–2357 reproduced under a CC BY 4.0 creative commons license.

by Schwann or oligodendroglial cells, but in **demyelinating diseases**, myelin sheaths degenerate. For example, **multiple sclerosis** (**MS**) involves autoimmune-mediated attacks that leave behind numerous demyelinated lesions (=**sclerae**) within CNS white matter, thereby compromising neural function (**Case 9.1 Annette Funicello's Myelinated Nervous Tissue**).

9.5 STRUCTURAL BASES OF ACTION POTENTIAL (AP) GENERATION AND PROPAGATION

Like other **excitable cells**, such as gametes, myocytes, and sensory cells, stimulated neurons can rapidly change the distribution of ions across their plasmalemma to generate **action potentials** (**APs**). Before stimulation, the neuron's **resting membrane potential** is typically **−60–70 mV**, owing to more negative charge occurring inside the cell than outside it (**Figure 9.5a**). This difference depends on the active transport of sodium and potassium ions through plasmalemmal **Na⁺/K⁺ ATPases** that hydrolyze ATP and thereby supply the energy needed to pump out three Na⁺ ions in exchange for two K⁺ ions brought in (**Figure 9.5b**). Such ionic heterogeneity combines with other sources of enhanced intracellular negative charge to yield a negative membrane potential. Because this potential is not equal to 0, resting membranes are also termed **polarized**.

In sensory cells responding to cues, such as pressure, temperature, pain, and light, external stimuli can cause resting potentials to change from negative to positive values, thereby generating an AP (**Figure 9.5c–g**). APs are initiated by the opening of plasmalemmal Na⁺ channels that allow the influx of Na⁺ ions as they move down their concentration gradient from the ECM into the cytoplasm (**Figure 9.5c**). If enough Na⁺ channels open to surpass a **threshold potential** of −50–55 mV, further Na⁺ influx rapidly moves the potential toward 0 and thus **depolarizes** the membrane, as the positively going phase heads toward a non-polarized state of 0 potential. The positive charge added by Na⁺ influx not only causes a localized increase in membrane potential but also opens **voltage-gated Na⁺ channels** in adjacent downstream regions, thereby spreading the **depolarization** along the membrane. However, at a peak positive potential of about **+30 mV**, Na⁺ channels close, and K⁺ channels open. This stops Na⁺ influx while allowing K⁺ efflux into the ECM, collectively driving **repolarization** as the potential moves back toward a negative polarized value. Such processes often cause a slight overshoot beyond the original resting potential during a transient **hyperpolarization** phase before K⁺ channels close and a −60–70 mV potential is re-established (**Figure 9.5f–h**).

Along with depolarizations due to sensory cues, neurons can generate APs in response to **neurotransmitter** (**NT**) chemicals released at synapses. NT release may initially trigger only a small rise in potential at each synapse that connects to the post-synaptic neuron. However, after **summing together subthreshold responses** across multiple synapses, a **suprathreshold** stimulus can trigger full depolarization via an **all-or-none** mode of AP generation. Owing to heterogeneous ion channel distributions, this depolarization normally begins in the **initial segment** of the axon directly distal to the hillock (**Figure 9.2a**) before being transmitted along the axon.

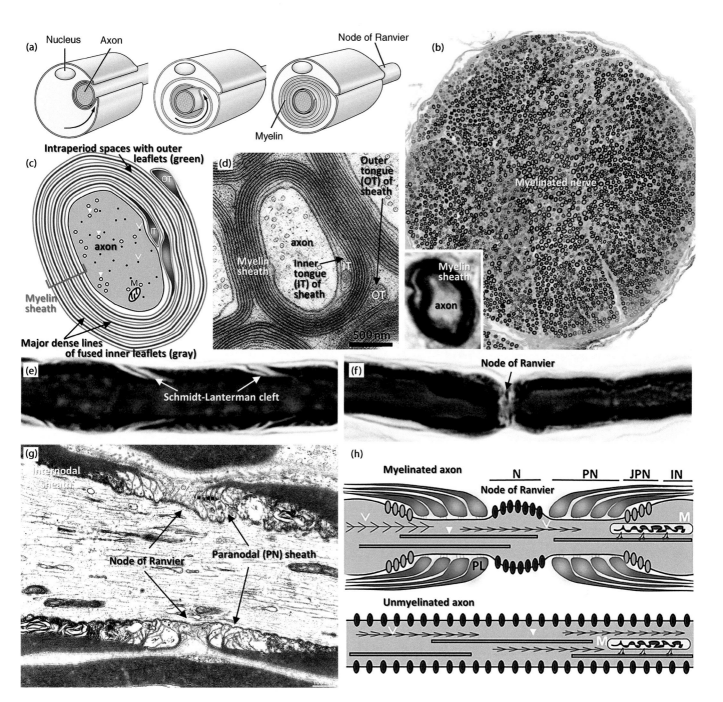

Figure 9.4 Myelin sheaths. (*a*) *Myelin sheaths are formed by either Schwann cells in the PNS or by oligodendrocytes in the CNS wrapping themselves multiple times around segments of axons. (**b–d**) Within each myelin sheath shown here in an LM (b) and TEM (d) of an osmicated nerve, the myelinating cell's membranes produce an orderly array of major dense lines interspersed with intraperiod spaces; (b) 120X; (b, inset) 2,200X. (**e**) As illustrated by an osmicated nerve fiber, Schmidt-Lanterman's clefts represent sites where the myelinating cell's cytoplasm is not fully resorbed, thereby leaving oblique incisures that facilitate axonal functioning. (**f–h**) Every 0.1–1.5 mm down the length of a peripheral nerve, there is a gap in the myelin sheath, called a node of Ranvier, which separates neighboring segments of internodal myelin (IN). Nodes have clustered sodium channels (figure h, red ovals) and are flanked by paranodal (PN) myelin, juxtaparanodal myelin (JPN), plus clustered potassium channels (figure h, blue ovals) that collectively accelerate action potential transmission along the length of the axon, compared to rates in unmyelinated fibers of similar size (e, f: 980X). ([a]: From OpenStax College (2013) Anatomy and Physiology. OpenStax. http:// cnx.org/content/col11496/latest reproduced under a CC BY 4.0 creative commons license; [c, d, h]: From Stassart, RM et al. (2018) The axon-myelin unit in development and degenerative disease. Front Neurosci 12: https://doi.org/10.3389/fnins.2018.00467 reproduced under a CC BY 4.0 creative commons license; [g]: TEM image courtesy of Drs. H. Jastrow and H. Wartenberg reproduced with permission from Jastrow's Electron Microscopic Atlas (htttp://www.drjastrow.de).)*

CASE 9.1 ANNETTE FUNICELLO'S MYELINATED NERVOUS TISSUE

Actress (1942–2013)

"Who's the leader of the Club that's made for you and me? M-I-C– K-E-Y– M-O-U-S-E!"

(A. F. as a Mouseketeer singing the Mickey Mouse song)

For kids in 1950s America, the crooning of the Mickey Mouse song by Annette Funicello and her fellow Mouseketeers was a Pied Piper's call to gather around the television. Capitalizing on her role in that popular show, Funicello appeared in other Disney-produced series, before recording pop songs and starring in films. In her late 40s, Funicello began suffering dizzy spells but for several years was able to conceal her symptoms from the public. However, to refute rumors that her repeated stumbles were due to alcoholism,

Funicello announced in 1992 that she had been diagnosed with multiple sclerosis (MS), an autoimmune disease that attacks myelin sheaths in the brain and spinal cord. After the announcement, Funicello used her celebrity status to advocate for MS research and education. Unfortunately, current drugs that can slow progression of certain MS forms were unavailable at the time, and following widespread damage to her myelinated nervous tissues, the former Musketeer lost her ability to walk and talk before eventually dying in 2013.

AP transmission rates down axons can be greatly increased by a surrounding myelin sheath that electrically insulates the axon. In particular, **nodes of Ranvier** that are interspersed within the myelin sheath interrupt this electrical insulation and represent sites where voltage-gated ion channels are aggregated (**Figure 9.4h**). Such nodal gaps and channel clusters allow **saltatory conduction**, whereby APs rapidly hop from node to node rather than propagating continuously along the axon, as in unmyelinated axons (**Figure 9.6a, b**).

In both myelinated and unmyelinated neurons, each AP typically persists for only a few milliseconds and may soon be followed by another depolarization during repetitive firing sequences. However, for about a millisecond after AP propagation, membrane sites that have just generated an AP are incapable of depolarizing during their **refractory period**. Such periods are due to Na$^+$ channels temporarily assuming an inactive conformation that prevents additional influx. Accordingly, **anesthetics** such as **lidocaine** can bind to voltage-gated Na$^+$ channels and thereby numb affected areas by causing a more prolonged blockage of the Na$^+$ influx needed for AP propagation.

9.6   THE FORM AND FUNCTIONING OF SYNAPSES: HOW NEURONS COMMUNICATE WITH EACH OTHER

Neural synapses are the sites where neurotransmitter-containing vesicles are exocytosed by the pre-synaptic neuron to affect the functioning of the post-synaptic neuron (**Figure 9.6c, d**). Morphologically, synapses can generally be classified as **axodendritic**, **axosomatic**, or **axoaxonic** for axons that synapse with a dendrite, soma, or axon, respectively. In addition, non-canonical neural synapses include axonal-glial connections or dendrites from retinal amacrine cells serving as axons. Regardless of their precise topologies, several thousand synapses can be established by each axon, and these can provide communication among multiple downstream neurons, thereby greatly increasing the complexity of wiring patterns in neural circuits.

Synapses are characterized by membrane-bound **NT vesicles** in the pre-synaptic neuron plus a **highly thickened** post-synaptic membrane, which in turn is separated from the pre-synaptic membrane by a narrow **synaptic cleft** (**Figure 9.6d**). Synaptic vesicles are formed in the soma, beginning with the production of either protein-based NTs in Nissl bodies or the processing of non-proteinaceous components by other organelles. After being packaged into membrane-bound vesicles, NTs are delivered to the synapse via axonal tracts of MTs associated with the motor protein **kinesin** that drives **anterograde** movement toward the thickened membrane at the axon terminus. This thickening contains specialized molecules needed for the docking, fusion, and release of NT vesicles. Such processes in turn depend on intracellular Ca^{2+} rises that are elicited when an AP reaches the axon terminus and triggers calcium influx through voltage-gated Ca^{2+} channels. Within microseconds of their influx, calcium ions attach to **synaptogamins**, allowing these calcium-dependent proteins to interact with a **SNARE** complex that promotes exocytosis of NTs for binding to their receptors in the post-synaptic membrane. Once synaptic transmission is completed, the pre-synaptic neuron can actively take up exocytosed NT components and then re-use materials within the soma, following **retrograde** transport via **dynein** motors (**Figure 9.6e**).

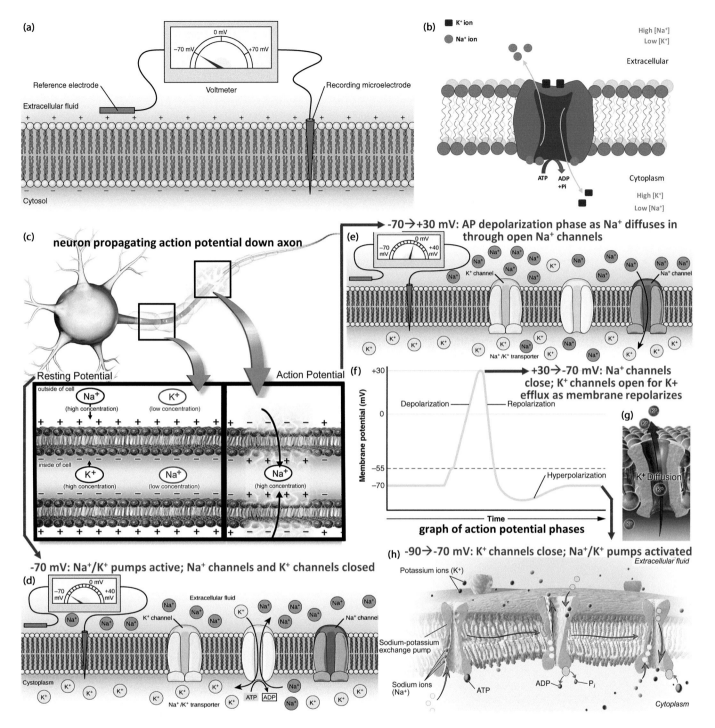

Figure 9.5 Generation of action potentials (APs). (a) *Mini electrodes that are placed directly inside and outside a cell can analyze the voltage difference across the cell's membrane prior to stimulation. Such resting membranes are polarized, since their membrane potentials do not equal zero but instead typically measure ~–70 mV. (**b**) The negative membrane potential is mainly due to Na⁺/K⁺ pumps that actively expel three sodium ions out for every two potassium ions brought in, thereby helping to make the inside of the cell more negative than the outside. (**c–h**) For neurons that are stimulated by such cues as light, pressure, pain, or neurotransmitters, the polarized membrane potential rapidly changes from –70 mV toward more positive values. During this positive-going phase, the membrane depolarizes as it first approaches, and then exceeds, 0 mV while generating an action potential. The increase in membrane potentials is due to stimulus-induced opening of sodium channels, thereby allowing positively charged sodium ions to rush into the cell and raise the membrane potential. At a peak positive value of ~30 mV, sodium channels close, and potassium channels open, thereby allowing the efflux of positively charged potassium ions as the membrane repolarizes toward the resting potential. A brief*

Figure 9.5 (Continued) *hyperpolarization at the end of the action potential can occur when the repolarized potential becomes more negative than the original –70 mV value before Na+/K+ pumps re-establish proper resting values. ([a]: From OpenStax College (2013) Anatomy and Physiology. OpenStax. http://cnx.org/content/col11496/latest reproduced under a CC BY 4.0 creative commons license; [b]: From Pivovarov, AS et al. (2019) Na+/K+-pump and neurotransmitter membrane receptors. Invert Neurosci 19: https://doi.org/10.1007/s10158-018-0221-7 reproduced under a CC BY 4.0 creative commons license; [c, g, h]: From Blausen.com staff (2014) "Medical gallery of Blausen Medical 2014". WikiJournal of Medicine 1 (2): 10. doi:10.15347/wjm/2014.010 reproduced under CC0 and CC BY SA creative commons licenses; [d–f]: From OpenStax Biology 2nd Edition, Biology 2e. OpenStax CNX. Sep 26, 2018 http://cnx.org/contents/8d50a0af-948b-4204-a71d-4826cba765b8@14.24 reproduced under a CC BY 4.0 creative commons license.)*

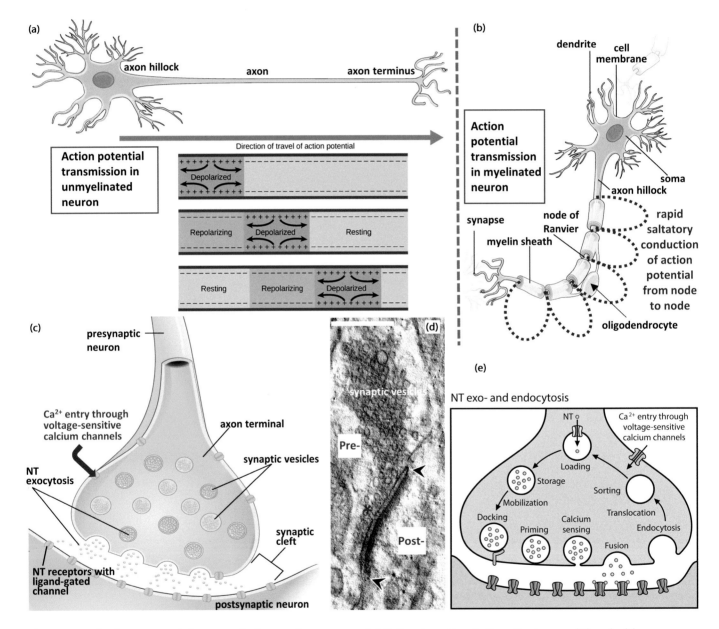

Figure 9.6 Action potential transmission and synapses. (**a**) Following stimulations that exceed thresholds needed for all-or-nothing transmission, action potentials are continuously propagated down an unmyelinated axon. (**b**) Myelinated axons can achieve faster transmission rates owing to saltatory conduction as depolarizations jump from node to node. (**c, d**) Action potentials reaching the axon terminus trigger calcium influx through voltage-gated calcium channels, which in turn drives the fusion and release of neurotransmitter-containing synaptic vesicles (asterisk) from the pre-synaptic neuron (Pre-) into the synaptic cleft, thereby allowing binding of neurotransmitter (NT) to receptors in

Figure 9.6 (Continued) *the post-synaptic neuron (Post-).* (*e*) *Each synapse has well-developed machinery for exocytosis and endocytosis; scale bar = 250 nm. ([a–c]: From OpenStax College (2013) Anatomy and Physiology. OpenStax. http:// cnx.org/content/col11496/latest reproduced under a CC BY 4.0 creative commons license; [d]: From Yakoubi, R et al. (2019) Ultrastructural heterogeneity of layer 4 excitatory synaptic boutons in the adult human temporal lobe neocortex. eLIFE8:e48373. doi: https://doi.org/10.7554/eLife.48373 reproduced under a CC BY 4.0 creative commons license; [e]: Adapted from Richmond, J. Synaptic function (December 7, 2007), WormBook, ed. The C. elegans Research Community, WormBook, oi/10.1895/wormbook.1.69.1,http://www.wormbook.org reproduced under a CC BY 2.5 creative commons license.)*

Accordingly, one class of anti-depressant drugs, called **selective serotonin reuptake inhibitors** (**SSRIs**), downregulates this recycling, thereby maintaining high levels of the NT **serotonin** in the synaptic cleft to help ameliorate depression.

As opposed to morphological classifications, synapses can be grouped into two functional types—**ionotropic** vs. **metabotropic**—based on the mechanism of ion channel gating in the post-synaptic neuron (**Figure 9.7a, b**). Thus, NT binding to **ionotropic receptors**, which are also called **ligand-gated channels**, directly opens the ion channel component of each receptor to depolarize or hyperpolarize the post-synaptic neuron. Conversely, the binding of NT to **metabotropic receptors** indirectly affects ion channel function by first stimulating a signaling cascade that subsequently alters ion channel gating. For example, after NT binding, **G protein coupled receptors** (**GPCRs**) initiate a **G-protein-mediated cascade** that modulates the gating properties of separate ion channels. Because of this two-step process, APs generated by metabotropic receptors often take longer to initiate and hence are sometimes also referred to as **slow synapses** compared to **fast synapses** using ionotropic receptors.

Alternatively, synapses can also be classified as **excitatory** or **inhibitory**, based on whether the transmitted NT makes the post-synaptic neuron more, or less, likely, respectively, to continue AP propagation (**Figure 9.7c, d**). Such opposing responses depend on the particular NTs and receptors that are utilized by each synapse and can differ among closely related NTs. For example, **aspartic** and **glutamic acid** types of amino acid NTs can bind receptors to excite post-synaptic neurons via the depolarization-associated opening of **Na$^+$ channels** (**Figure 9.7c, d**). Conversely, inhibitory NTs like **gamma-amino butyric acid** (**GABA**) and **glycine** can open **Cl$^-$ channels**, thereby resulting in post-synaptic neuron **hyperpolarization** (**Figure 9.7c, d**).

In addition to conventional signaling via exocytosed NTs, some neurons in the retina and several regions of the brain communicate via **electrotonic synapses** involving **gap junctions**. Such transmission is typically capable of bidirectional propagation, which allows a speedier spreading of impulses than can be achieved via unidirectional NT exocytosis.

9.7   THE CELL BIOLOGY OF NERVE REGENERATION

In mammals, only PNS neurons are normally capable of regenerating. However, after experimental grafting of PNS tissues into the CNS, neurons in the spinal cord and brain can undergo some regeneration, indicating that both intrinsic factors within the injured neuron itself and extrinsic factors in its surrounding environment affect nerve repair. Accordingly, various factors that promote or inhibit nervous tissue regeneration are being investigated in order to find ways of stimulating brain and spinal cord regeneration.

At the onset of myelinated nerve regeneration in the PNS, damaged neurons exhibit **retrograde modifications** upstream to the injury (**Figure 9.7e**). These include hypertrophy of the soma and loss of perinuclear Nissl bodies (=chromatolysis). Conversely, the distal end of the severed nerve undergoes **anterograde** (=**Wallerian**) **degeneration**, in which Schwann cells stop producing myelin and instead proliferate before transdifferentiating into repair cells that both phagocytose debris and recruit macrophages for additional phagocytosis. After axonal remnants and myelin sheath fragments are cleared by phagocytosis, reprogrammed Schwann cells can organize into cable-like arrays, called **bands of Bungner**. Such bands extend distally from the injury site to guide the regenerating axon as it attempts to re-form a functional connection with its target (**Figure 9.7e**). If these guiding bands are either improperly established or allowed to degrade over time, neuronal apoptosis increases, and regeneration fails. Alternatively, in some nerves that do not re-connect with their target cells, continued proliferation of surrounding nervous tissue can result in abnormal neural growths, termed **neuromas** (**Case 9.2 Abraham Lincoln's Putative Neuromas**).

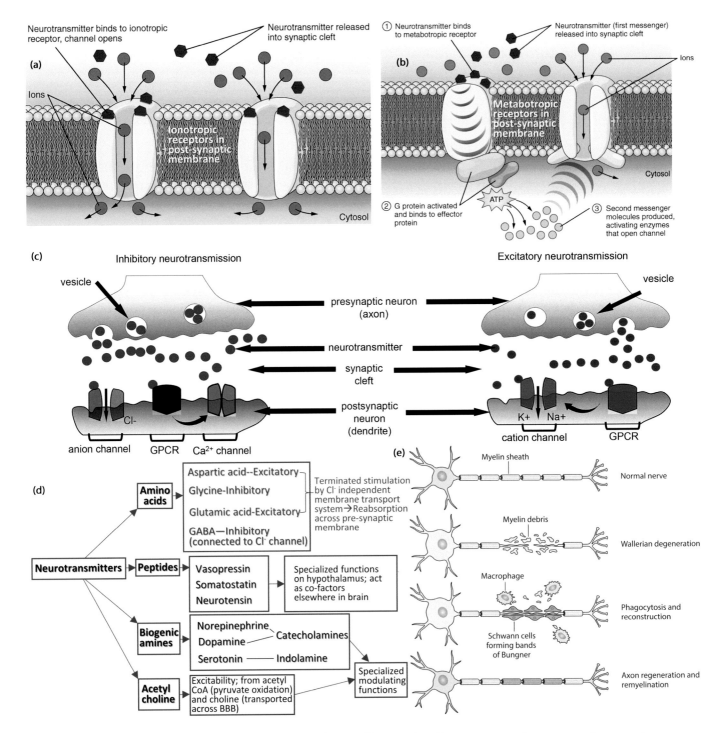

Figure 9.7 Synapse biology and nerve regeneration. (a, b) *Functionally, two types of synapses—ionotropic vs. metabotropic—allow the rapid influx of ions vs. a slower ionic influx that first requires upstream signaling, such as that involving G-proteins. (**c, d**) Depending on the particular type of neurotransmitter (NT) and post-synaptic NT receptor, synapses can be inhibitory by opening Cl- channels to hyperpolarize the post-synaptic neuron vs. excitatory by opening cation channels to depolarize the post-synaptic neuron. (**e**) Damaged peripheral nerves initially swell their soma and lose staining capacity (=chromatolysis), whereas downstream to the wound site, myelin debris is phagocytosed, and Schwann cells reorganize into bands of Bungner that help re-establish proper connections to target cells. ([a, b]: From OpenStax College (2013) Anatomy and Physiology. OpenStax. http://cnx.org/content/col11496/latest reproduced under a CC BY 4.0 creative commons license; [c]: From Leopold AV et al. (2019) Fluorescent biosensors for neurotransmission and neuromodulation: engineering and applications. Front Cell Neurosci 13:474. doi: 10.3389/fncel.2019.00474 reproduced under a CC BY 4.0 creative commons license; [d]: From Humphries, P et al. (2008) Direct and indirect cellular effects of*

Figure 9.7 (Continued) *aspartame on the brain. Eur J Clin Nutr 62: 451–462 reproduced with publisher permission; [e] From Li, R et al. (2020) Growth factors-based therapeutic strategies and their underlying signaling mechanisms for peripheral nerve regeneration. Acta Pharmacologia Sinica 41: 1289–1300 reproduced under a CC BY 4.0 creative commons license.)*

9.8   CNS—BRAIN: OVERALL FUNCTIONAL MORPHOLOGY PLUS ITS MENINGEAL COVERINGS AND EPENDYMAL LINING

The brain comprises three main parts: (1) a large and multifaceted **cerebrum**, (2) a smaller posteriorly-positioned **cerebellum**, and (3) a stalk-like **brainstem (Figure 9.8a)**. As the most complex brain component, the **cerebrum** is involved in sensory perception, fine control of skeletal muscles, regulation of behavior and emotions, as well as other processes, such as cognition, speech, learning, and memory. Alternatively, the cerebellum is traditionally viewed as **coordinating muscle movements, posture,** and **balance**, whereas the brainstem not only **relays signals** between the spinal cord and the cerebral/cerebellar regions but also controls via cranial nerves **autonomic functions** that are either regularly monitored (e.g., breathing and heart rate) or of intermittent nature (e.g., coughing and vomiting).

The **cerebrum** is divided into **left** and **right hemispheres** that are joined together by a medial **corpus callosum (Figure 9.8b)**. At their surfaces, the cerebral hemispheres are folded into deep irregular grooves (**sulci**) and intervening ridges (**gyri**) that optimize packaging of cerebral tissues into a minimally-sized cranial cavity, thereby allowing a well-cephalized head to be accommodated during birth. Conversely, in congenital cases of **lissencephaly** ("smooth brains"), loss of proper gyrification is linked to various cognitive defects.

Histologically, the periphery of each cerebral hemisphere comprises a well-developed **cerebral cortex**, which extends to the base of surface sulci and overlies **deeper cerebral components (Figure 9.8b)**. Such **subcortical** structures occurring interior to the cortex include: (1) **basal ganglia** that normally fine-tune locomotion and behaviors, but are compromised in disorders like **Parkinson's disease, Tourette syndrome**, and **Huntington's disease**; (2) the **thalamus**, which, among other roles, can **relay ascending input** to the cortex to affect cortical processing; (3) an anteriorly positioned **olfactory bulb**, which is disproportionally enlarged in mammals that depend on a keen sense of smell; and (4) a left and a right **hippocampus** that help control such functions as **memory formation** and **spatial orientation** (Box 9.2 Hippocampi: Key Modulators of Episodic Memory Formation and Spatial Orientation).

The cerebrum, cerebellum, and brain stem are housed in the skull cavity but remain separated from surrounding bones by three protective sheaths (=**meninges**), which comprise the **dura-, arachnoid-**, and

---

CASE 9.2 ABRAHAM LINCOLN'S PUTATIVE NEUROMAS

16th U.S. President
   (1809–1865)

"If I were two-faced, would I be wearing this one?"

(A.L. responding to the claim that he was a
                                        two-faced liar)

Abraham Lincoln's peculiar face caused many people, including Lincoln himself, to remark he was one of the homeliest men alive. Such facial features and other physical traits are discussed by physician J.G. Sotos in his book *The Physical Lincoln*. Sotos concludes Lincoln did not suffer from Marfan syndrome as previously proposed but rather from MEN 2B (multiple endocrine neoplasia, type 2B) that causes marfanoid skeletal defects as well as cancers in various endocrine organs. In addition, people with MEN 2B can form abnormal neuronal growths, called neuromas. Accordingly, in

digitally enhanced photographs, subtle bumps on the lips of Lincoln, his mother, and two of his sons are consistent with the proposal that such protrusions represent MEN-2B-induced neuromas. Theoretically, DNA sequencing should be able to test this hypothesis, but in response to an attempted grave robbing, Lincoln's body was re-buried in a now inaccessible site. Moreover,  other possible sources for extracting his DNA, such as blood-stained clothing either lack indisputable provenance or necessary permission to be used. Thus, until DNA data become available, one can simply speculate that Lincoln's marfanoid features and putative neuromas were due to MEN 2B.

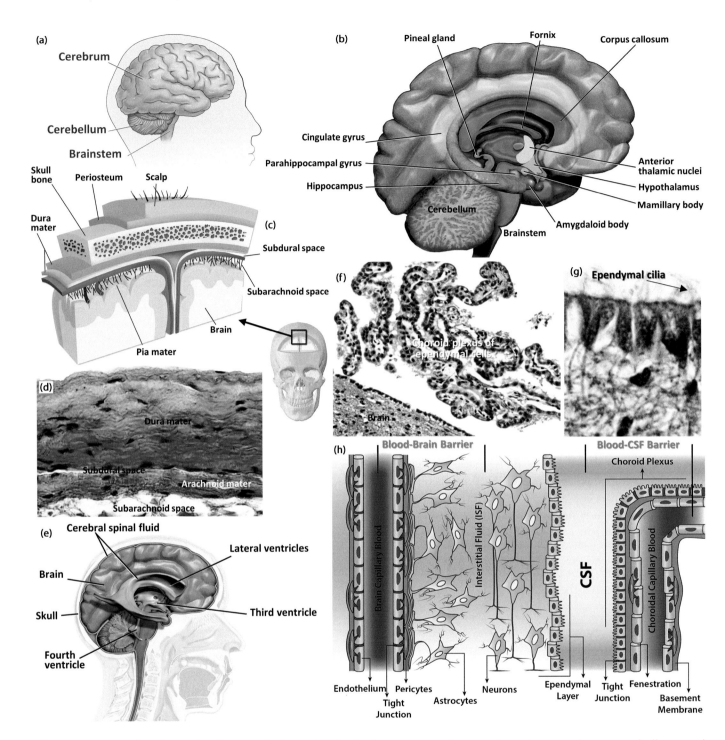

Figure 9.8 Introduction to brain morphology. *(a) The brain comprises three main regions: cerebrum, cerebellum, and brainstem. (b) Beneath the cerebral cortex are numerous deeper cerebral components, including the left and the right hippocampus. (c) In addition to the scalp, periosteum and skull bones, the brain is protected by three meningeal layers: dura-, arachnoid-, and pia mater. (d) The dura and arachnoid layers are separated by a thin subdural space, and the arachnoid layer contains a subarachnoid space that is filled with cerebral spinal fluid (CSF) (190X). (e) The brain contains a system of CSF-filled ventricles that are contiguous with the central canal of the spinal cord. (f, g) The ventricles and central canal are lined by ciliated ependymal cells. Such cells can form ventricular tufts, called choroid plexuses, which produce CSF; (f) 100X; (g) 625X. (h) Tight junctions connect both ependymal cells and endothelia of brain blood vessels, thereby helping form blood-CSF and blood-brain barriers to prevent toxins and pathogens from accessing sensitive brain tissues. ([a]: From a Netherlands Wikimedia website (https://maken.wikiwijs.nl/95908/flashmediaelement.swf#!page-3001026), which is reproduced by permission of a CC BY SA 3.0 creative commons license; [b, c, e]: From Blausen.com staff (2014)*

Figure 9.8 (Continued) *"Medical gallery of Blausen Medical 2014". WikiJournal of Medicine 1 (2): 10. doi:10.15347/ wjm/2014.010 reproduced under CC0 and CC BY SA creative commons licenses; [h]: From D'Agata, F et al. (2017) Magnetic nanoparticles in the central nervous system: targeting principles, applications and safety issues. Molecules 23: doi:10.3390/molecules23010009 reproduced under a CC BY 4.0 creative commons license.)*

BOX 9.2 HIPPOCAMPI: KEY MODULATORS OF EPISODIC MEMORY FORMATION AND SPATIAL ORIENTATION

One of the first clues that **hippocampi** could be involved in establishing memories came in the 1950s when a patient suffering from severe epileptic seizures had his two hippocampi removed during temporal lobe surgery. Although the procedure helped with his seizures, the lack of hippocampi caused **retrograde** and **anterograde amnesia** as the patient became incapable of recalling certain pre-surgery memories or making similar ones after surgery. In particular, he lost **episodic** (= "autobiographical") **memories** that depended on the specific times and places such memories were acquired during his lifetime. Conversely, non-personalized memories, such as recalling the meanings of words were less affected. Building upon the insights obtained from that surgery, numerous aspects of hippocampus-mediated episodic memory formation have since been analyzed. These studies generally show a reciprocal linkage between the hippocampus and cerebral cortex, whereby cortical input is typically relayed to the **dentate gyrus** before being transmitted to hippocampal **CA regions** and other cerebral areas for the full processing needed to establish such episodic memories (**Figure a**). In addition,

ongoing neurogenesis in the adult dentate gyrus **subgranular zone** (**SGZ**) may play important roles in this process, since such cells in humans eventually replace all hippocampal neurons that were formed before birth (Box 9.1).

The contributions of hippocampi to an overall positioning system within the brain were elucidated in Nobel-Prize-winning research by John O'Keefe and his colleagues. Such studies showed that particular **pyramidal cells** in CA regions of the hippocampus serve as **place cells** that fire when an animal encounters a specific region of its environment, called a **place field**. Alternatively, other place cells are activated as new place fields are encountered (**Figure b**). Thus, place cells help provide a **spatial reference map** in the brain that facilitate navigational capabilities. Accordingly, people like London taxicab drivers who frequently utilize this orientation system tend to have larger hippocampi compared to those in non-driver controls, and this size difference is particularly pronounced in cab drivers who have been on the job for many years. Conversely, spatial awareness capabilities are often compromised in degenerative disorders, such as Alzheimer's disease.

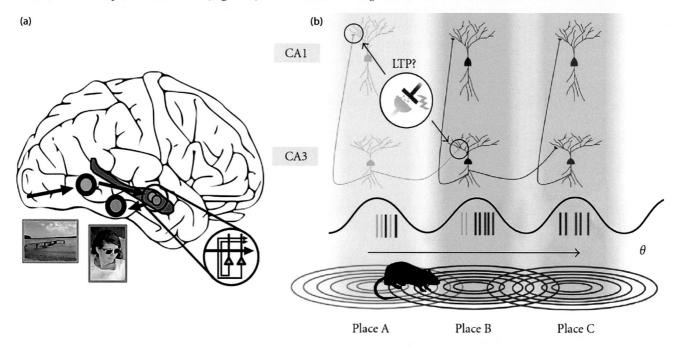

Box 9.2 (a) *Hippocampal place cells (green) and concept cells (blue) form neural circuits that allow episodic (=autobiographical) memories to be produced. (From: Horner, AJ and Doeller, CF (2017) Plasticity of hippocampal memories in humans. Curr Opin Neurobiol 43: 102–109 reproduced under a CC BY 4.0 creative commons license.) (b)* *Different place cells of the hippocampal cornu ammonis (CA) are stimulated to fire in various environs, thereby contributing to spatial awareness and tracking capabilities, which are often lost in Alzheimer's disease. (From: Sadowski, HJLP et al. (2011) Ripples make waves: Binding structured activity and plasticity in hippocampal networks. Neural Plasticity Vol 2011: doi:10.1155/2011/960389 reproduced under a CC BY 3.0 creative commons license.)*

CASE 9.3 MUHAMMAD ALI'S BRAIN

Champion Boxer
  (1942–2016)

"Float like a butterfly, sting like a bee. The hands can't hit what the eyes can't see."

(M.A. on his boxing strategy)

Muhammad Ali grew up in Kentucky as Cassius Clay Jr. and earned an Olympic gold medal in boxing at age 18. After turning professional and winning the heavyweight title in a stunning upset, Ali converted to Islam, and with his newly adopted Muslim name became an outspoken critic of racism. As champion, Ali refused to fight in the Viet Nam War and was stripped of his boxing license for three years. Upon returning to the ring from 1970 to 1981, Ali cemented his legacy of

greatness while also becoming one of the world's most widely recognized icons. However, Ali was noticeably slower as he lost his last fights and was diagnosed at age 42 with Parkinson's disease (PD), which degrades locomotory function due to inadequate dopamine production in such cerebral regions as the substantia nigra basal ganglia. The numerous blows Ali absorbed in the ring may well have exacerbated pathogenesis, and although he was able to fight the disease for 30 years, PD-induced brain pathogenesis ended the famous boxer's life in 2016.

pia-mater. The outermost menix is a relatively thick and dense **dura mater** that occurs next to periosteum of skull bones, whereas the innermost meningeal layer, called the **pia mater**, is a delicate sheet directly covering the cerebral surface (**Figure 9.8c, d**). Sandwiched between these meninges, the **arachnoid mater** underlies a thin subdural space and projects spider-like processes toward the pia, thereby forming a network of arachnoid cavities. Normally, the arachnoid layer remains closely juxtaposed to the dura, unless, for example, blood leaks into the subdural space as part of a trauma-induced **subdural hematoma.**

Contiguous with the subarachnoid space, the brain has a system of interconnected cavities, called **ventricles** (**Figure 9.8e**). Such chambers are filled with **cerebral spinal fluid** (**CSF**) that also extends into the **central canal** of the spinal cord. CSF helps protects CNS tissue from physical trauma and serves to clear neurotoxic waste via a surrounding network, called the **glymphatic system.** As part of this system, the lining of CNS cavities contains ciliated glial cells, called **ependymal cells**, which move materials via ciliary beating and form scattered aggregations (=**choroid plexuses**) that secrete CSF (**Figure 9.8f, g**).

Nervous tissue proper of the brain is penetrated by blood vessels, whose continuous endothelial lining has numerous **tight junctions** that produce a stout **blood-brain barrier (BBB)** (**Figure 9.8h**). Along with a CSF-blood barrier formed by ependymal cells, the BBB is beneficial in helping to prevent pathogens and other potentially toxic substances from reaching sensitive neural tissues. However, the sealing effects of the BBB can also complicate drug delivery within the CNS. For example, to restore depleted **dopamine** levels in patients with **Parkinson's disease** (**PD**) (**Case 9.3 Muhammad Ali's Brain**), the more permeable dopamine precursor **levodopa** (**L-DOPA**) is administered, since dopamine itself does not effectively penetrate the BBB.

Associated with blood vessel endothelia are multipotent **pericytes** that help maintain the BBB and are often degraded in pathologies, such as **amyotrophic lateral sclerosis** (=Lou Gehrig's disease) (**Case 2.5**). In addition, **end foot** processes of **astrocyte glial cells** communicate with vascular cells via **gap junctions** to help regulate vascular properties and fluid levels in the brain (**Figure 9.8h**).

The next four sections summarize the functional histology of four major brain components: (i) the **cerebral cortex** and **hippocampal** regions of the **cerebrum**, (ii) **cerebellum**, and (iii) **brainstem**. Thereafter, the **spinal cord** is considered before discussing the common neurodegenerative disorder **Alzheimer's disease.**

9.9   CNS—HISTOLOGY OF FOUR BRAIN REGIONS: CEREBRAL CORTEX, HIPPOCAMPI, CEREBELLUM, AND BRAINSTEM

Cerebral cortex: multilayered gray matter with pyramidal neurons: The peripheral-most 1–5 mm of the cerebrum comprises a highly stratified **cerebral cortex**. Histologically, the cortex consists of outermost **gray matter**, which overlies **white matter** that extends into the subcortical cerebral core (**Figure 9.9a, b**). The most common neuron in the middle to deep portions of cortical gray matter is a **pyramidal cell**, which is characterized by its triangular shape in sections (**Figure 9.9c–g**), whereas alternative neuronal types (e.g., **Cajal-Retzius cells**, **granule cells**, and **Betz cells**) predominate in other cortical layers (**Figure 9.9h**). Based on their neurons and glial cells, six gray matter

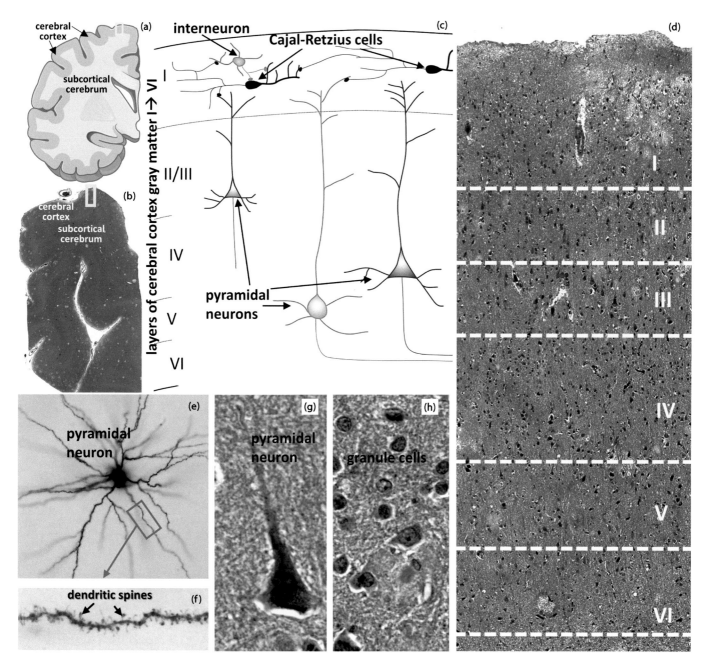

Figure 9.9 Cerebral cortex. (a, b) *The cerebral cortex constitutes the outer several millimeters of the cerebrum and along with deeper subcortical components is critical for various cognitive functions of the brain. (b) 5X. (**c, d**) Six layers of the cerebral cortex (I–VI) are assembled during fetal development in part due to reelin proteins secreted by tangentially oriented Cajal-Retzius cells located in the superficial cortex. The mid to deep cortical layers are characterized by abundant pyramidal cells that form both intra- and extracortical circuits required for proper neural functioning. (d) 125X. (**e–h**) An end-on view of a cortical pyramidal neuron (e, 400X) shows numerous dendrites with their densely arranged dendritic spines (f, 700X) that allow for highly plastic synaptic connections. Along with pyramidal cells (g, 850X), different layers of the cerebral cortex have additional types of neurons, such as granule cells (h: 1,000X). ([a]: From Harris, TC et al. (2019) The shrinking brain: cerebral atrophy following traumatic brain injury. Ann Biomed Engineer 47: 1941–1959 reproduced under a CC BY 4.0 creative commons license; [c]: From Gil, V et al. (2014) Historical first descriptions of Cajal–Retzius cells: from pioneer studies to current knowledge. Front Neuroanat 8: https://doi.org/10.3389/fnana.2014.00032 reproduced under a CC BY 3.0 creative commons license; [e, f]: From Elston, GN and Fujita, I (2014) Pyramidal cell development: postnatal spinogenesis, dendritic growth, axon growth, and electrophysiology. Front Neurosci 8: doi: 10.3389/fnana.2014.00078 reproduced under a CC BY 3.0 creative commons license.)*

layers (**I–VI**) are distinguished from the outer (**plial**) to inner (**white matter**) boundaries (**Figure 9.8c, d**). Neurons within these layers can form wholly intracortical circuits as well as pathways that extend beyond the cortex to other CNS regions, collectively allowing the cerebral cortex to act as a key initiator and regulator of neural functions.

To establish the complex layering and connections of the adult cerebral cortex, neurons begin to be generated ~40 days post-fertilization deep within a **ventricular zone** (**VZ**) that eventually forms the internal ventricles of the brain. From this origin site, newly produced neurons migrate outward, with younger neurons leap frogging over previously generated ones and thus becoming positioned more superficial to older neurons. Such patterning requires a gradient of the chemoattractant **reelin** that is secreted by peripheral **Cajal-Retzius cells** to ensure that migrating neurons find their correct destinations (**Figure 9.9c**). Neurogenesis in humans is initially quite rapid, as nearly all fetal neurons are formed by mid-gestation, and even though CNS maturation continues until ~25 years of age, the brain has already reached ~90% of its adult size by age 6. Accordingly, given the speed and complexity of cortical patterning, it is not surprising that various **neuronal migration defects** can underlie congenital brain disorders.

Hippocampi: Horseshoe-shaped cerebral components with multiple functions

A **hippocampus** develops within the left and the right cerebral hemispheres. Each hippocampus has multiple subregions and associated structures, which are variably designated in different classification schemes. However, most accounts recognize an overall **hippocampal formation** on each cerebral side as comprising two interlocking horseshoe-shaped structures—(i) the **hippocampus proper**, or cornu ammonis (CA), which comprises four distinct zones (CA1–CA4) that collectively interdigitate with (ii) the **dentate gyrus** (**Figure 9.10a–c**). The hippocampus proper contains large **pyramidal cell** neurons, whereas the dentate gyrus possesses smaller **granule cell** neurons plus unmyelinated axons (=**mossy fibers**), interneurons, and **mossy cells**. Moreover, directly beneath the dentate gyrus granular layer is a **subgranular zone** (**SGZ**), which is one of the few areas of the adult brain that continues to form new neurons throughout life (Box 9.1).

Via its anteriorly and posteriorly directed neurons, the hippocampus formation connects both with the cerebral cortex to influence cognition and with other parts of the cerebral **limbic system** to shape emotional responses. Such circuitry also allows normally functioning hippocampi to help establish **episodic** (=autobiographical) kinds of memories and to provide proper **spatial orientation** to surrounding environs (Box 9.2). Conversely, hippocampal dysfunction is linked to various pathologies, including post-traumatic stress disorder, schizophrenia, and **temporal lobe epilepsy** (**TLE**), where it is hypothesized that defective **mossy cells** and **fibers** of the dentate gyrus cause hyperactive firing that promotes epileptic seizures.

Cerebellum: More than just a brain region for coordinating muscular activities

Situated beneath the posterior end of the cerebrum, the **cerebellum** (= "little brain") exhibits regularly arranged fine grooves that differ from the coarse and irregularly folded sulci of cerebral hemispheres. Within the left and right sides of the cerebellum, each hemisphere comprises a peripheral **cortex** that is composed of **trilayered gray matter** overlying a **medullary region** composed mainly of white matter and a few embedded nuclei (**Figure 9.10d, e**). The outermost **molecular layer** of the trilaminar cerebellar cortex is mostly devoid of neuronal cell bodies, whereas the innermost **granular layer** contains numerous densely packed somata of **granule cell** neurons. Between these two main layers is a thin **Purkinje layer** with large **Purkinje cell** neurons (**Figure 9.10f**), collectively allowing the cerebellum to establish both wholly intra-cerebellar circuits and those that link the cerebellum with the cerebrum and brainstem.

As some of the largest neurons in the adult brain, Purkinje cells direct their axons away from the molecular layer while projecting their tree-like **dendritic arborizations** into the molecular layer (**Figure 9.10f**). Conversely, unlike massive Purkinje cells, cerebellar **granule cells** are among the smallest neurons in the brain and by far the most common, comprising up to 75% of the entire brain's total neurons. Granule cells send axons into the molecular layer and thereby form the bulk of the molecular layer, where they can synapse with Purkinje cell dendrites. In addition, axons from neuronal somata in both the brainstem and spinal cord extend into the cerebellum to synapse with granule cell dendrites, thereby further increasing the complexity of cerebellar wiring patterns. Via such circuits, the cerebellum helps control balance, locomotion, and posture. In addition to such well-documented functions, the cerebellum can also regulate **cognitive capabilities**. For example, genetic markers of cerebellar dysfunction have been linked to a wide range of neurologic defects that constitute **autism spectrum disorders** (**ASDs**).

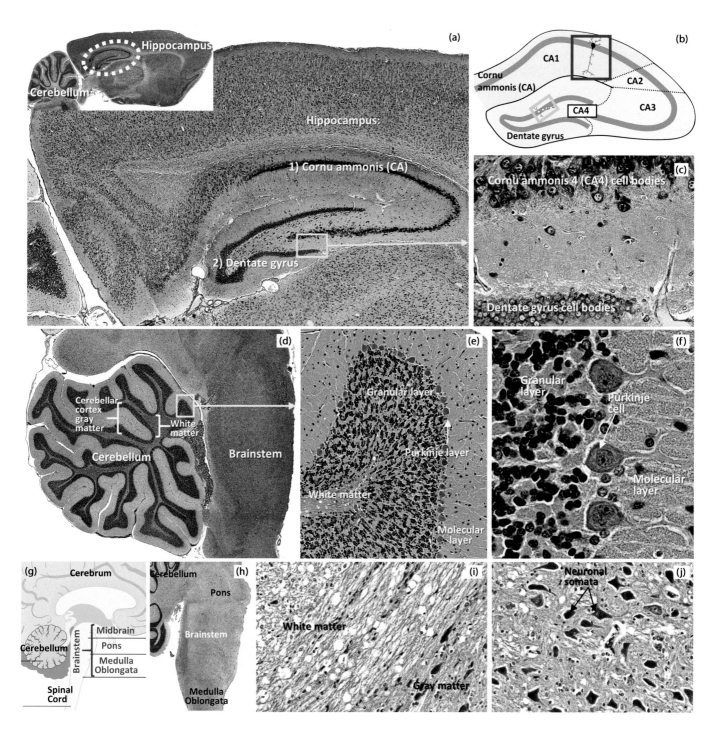

Figure 9.10 Brain components: hippocampus, cerebellum, and brainstem. (a–c) *The hippocampus, which is shown here in a low-power longitudinal section of an entire mouse brain (a, inset, 3X) and at higher magnifications (a–c) ((a) 20X; (c) 160X), comprises two interlocking horseshoe-shaped units, called the cornu ammonis (CA) and dentate gyrus, each of which contains several kinds of neurons (red and yellow boxes, figure b). With their multifaceted connections to other cerebral regions, the left and right hippocampi help shape: (i) emotional responses, (ii) spatial orientations, and (iii) certain kinds of memories. (**d–f**) The cerebellum comprises a trilayered cortex of gray matter (molecular layer, Purkinje layer, and granular layer) that covers an inner white matter core, which collectively help coordinate muscular activities while also contributing to cognitive functions that are compromised in diseases like autism spectrum disorders; (d) 10X; (e) 40X; (f) 200X. (**g–j**) As a relay station between the cerebrum, cerebellum, and spinal cord, the brainstem comprises midbrain, pons, and medulla oblongata subregions. Such brainstem components contain both white matter and gray matter with neuronal nuclei, which help integrate autonomic functioning within the CNS; (h) 7X; (i) 100X; (j) 125X. ([b]: From Sheppard, PAS et al. (2019) Structural*

Figure 9.10 (Continued) *plasticity of the hippocampus in response to estrogens in female rodents. Mol Brain 12: https://doi.org/10.1186/s13041-019-0442-7. Reproduced under a CC BY 4.0 creative common license; [g]: From a Wikimedia.org file (Cancer Research UK, CRUK_294.svg) reproduced under a CC BY SA 4.0 creative commons license.)*

Brainstem: A key connecting nexus

The brainstem consists of a superior **midbrain** region, a more inferior **pons**, and an inferior-most **medulla oblongata** that unites with the spinal cord (**Figure 9.10g, h**). The midbrain of the brainstem connects to cerebrum, whereas the pons is linked to both the cerebrum and cerebellum. In addition, ten pairs of cranial nerves arise from the brainstem to modulate various functions. Throughout its subregions, the brainstem comprises white matter plus gray matter with nuclei containing aggregated somata (**Figure 9.10i, j**). Collectively, these brainstem tissues and their connections within the CNS allow brainstem control over essential functions, such as breathing. Accordingly, in comatose patients on life support, **brainstem death** can be determined by a list of criteria, including the inability to breathe independently.

9.10   CNS—SPINAL CORD: A RECIPROCAL CNS RELAY STATION WITH GRAY AND WHITE MATTER PLUS A CENTRAL CANAL

Attached to the distal end of the brainstem is the **spinal cord** that serves as a reciprocal link between the brain and PNS. In addition to being encased by vertebrae, the spinal cord is surrounded by dura mater plus its peripherally located **epidural space** with blood vessels that allow the epidural delivery of anesthetics and other drugs. In addition, arachnoid and pia mater layers are positioned medial to the cord's dura mater.

Spinal cord tissue resembles that of the brain in comprising both **gray** and **white matter**. However, opposite to the pattern in the cerebrum and cerebellum, the spinal cord has internally positioned **gray matter** surrounded by outer **white matter**. In transverse sections, the peripheral gray matter on each side of the cord forms a **dorsal horn** and a **ventral horn**, which in the thorax are separated from each other by an intervening **lateral horn** (**Figure 9.11a–c**). The two lateral halves of the cord are connected by a central band of gray matter, thereby conferring a distinctive **butterfly** shape to the gray matter. Near the middle of the butterfly is a small **central canal** (**Figure 9.11b**) containing CSF, which also fills surrounding meningeal cavities. As is the case in the brain, spinal CSF serves protective functions and may also provide biomarkers of various disorders. Thus, CSF from the subarachnoid cistern of the lumbar region is often aspirated during **lumbar punctures** (spinal taps) for diagnostic tests, including those assaying for **meningitis** infections affecting CNS meninges.

9.11   FUNCTIONAL ANATOMY OF PNS COMPONENTS AND SOMATIC VS. AUTONOMIC REFLEXES THAT CAN BYPASS CEREBRAL PROCESSING

Although the somatic system of the PNS constitutes a single network for **voluntarily controlling** skeletal muscle, the **involuntarily controlled autonomic system** of the PNS comprises **parasympathetic** and **sympathetic** subdivisions, which differ in their positions relative to the CNS and in their general effects. Thus, parasympathetic efferent neurons project from more superior and inferior regions of the CNS, thereby constituting a **craniosacral outflow** pathway that tends to decrease activities in their targets. Conversely, sympathetic efferents of the **thoracolumbar outflow** arise from more central CNS regions and typically elevate target tissue activities.

For rapid responses, parts of the PNS are organized into **reflex arcs** that can bypass processing in the cerebrum and simply relay APs to and from the spinal cord and brainstem regions of CNS. **Somatic** reflexes link externally derived stimuli with skeletal muscle contractions, whereas **autonomic** reflexes control the actions of visceral organs, glands, and other tissues. In a typical **somatic reflex arc**, a stimulus such as pain that is sensed in the skin initiates an AP in a **pseudounipolar** afferent neuron, whose soma is located in a **dorsal root ganglion** flanking the spinal cord (**Figure 9.11d**). After being propagated into the spinal cord's **dorsal horn** (**Figure 9.11d, e**) and transmitted across a synapse with an **interneuron**, the AP travels to a motor neuron axon that exits the **ventral horn** and reaches a skeletal muscle target without an intervening synapse (**Figure 9.11f**). Once the AP arrives at the neuromuscular junction, **acetylcholine** is released to initiate skeletal muscle contractions for rapid reactions, such as withdrawal from the painful stimulus.

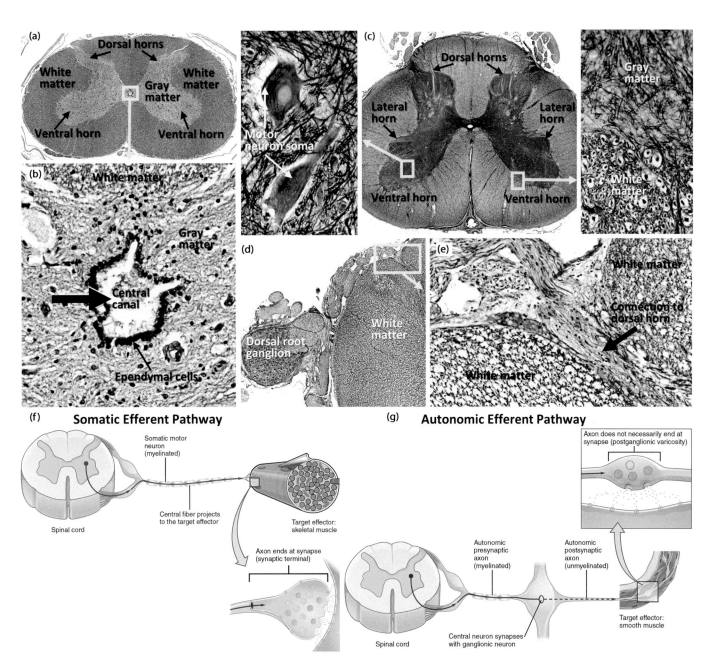

Figure 9.11 Spinal cord. *(a–c) As shown in transverse sections stained by H&E ((a) 10X; (b) 130X) or the Golgi method ((c) 15X; insets: 300X), the spinal cord has outer white matter and inner gray matter. The butterfly shaped gray matter comprises mostly paired dorsal and ventral horns with a central canal lined by ependymal cells in the connecting isthmus. Ventral horns contain motor neuron somata and in the thoracic region of the spinal cord occur anterior to lateral horns (c). (d, e) Sensory input from pseudounipolar cells in dorsal root ganglia enters the spinal cord via the dorsal horn. (d) 15X; (e) 140X. (f, g) Conversely, motor output exits the spinal cord by way of the ventral horns. In the case of skeletal muscle, each motor neuron axon extends to its neuromuscular junction without an intervening synapse (f). Conversely, autonomic efferent pathways involve a synapse between a preganglionic nerve fiber exiting the ventral side of the spinal cord and a post-ganglionic fiber, whose soma resides within an intervening efferent ganglion (g). Sympathetic ganglia occur near the spinal cord, whereas parasympathetic ganglia are located farther away. Differences in postganglionic neurotransmitters (adrenergic in sympathetic pathways vs. cholinergic in parasympathetic) contribute to the generally opposing responses elicited by these two autonomic subdivisions. ([f, g]: From OpenStax College (2013) Anatomy and Physiology. OpenStax. http://cnx.org/content/col11496/latest reproduced under a CC BY 4.0 creative commons license.)*

As with somatic reflexes, **autonomic reflex arcs** involving the spinal cord transmit sensory input dorsally and motor output ventrally to modulate bodily functions. However, unlike the single efferent neuron of somatic reflexes, each autonomic motor pathway involves two neurons that synapse within a ganglion before innervating target cells (**Figure 9.11g**). Because of this circuitry, the initial motor axon projecting from the spinal cord in an autonomic reflex arc is referred to as the **preganglionic fiber**. Conversely, the second efferent axon whose soma resides within the autonomic ganglion is called the **postganglionic fiber**. Positioning of the efferent ganglion differs in **sympathetic** vs. **parasympathetic** reflexes, as **sympathetic ganglia** occur near the vertebral column, whereas **parasympathetic ganglia** occur relatively far from vertebrae and thus closer to, or even embedded within, the organs they innervate. Because of these differences, **preganglionic fibers** are relatively short in sympathetic arcs vs. long in parasympathetic arcs, and the converse is true for **postganglionic fibers**. In addition, preganglionic fibers of both parasympathetic and sympathetic arcs secrete **acetylcholine**, as do parasympathetic postganglionic fibers. Conversely, sympathetic postganglionic fibers typically secrete **noradrenaline** or **adrenaline**. Such NT differences account for the generally antagonistic effects of parasympathetic vs. sympathetic innervations, with sympathetic input tending to stimulate **fight-or-flight** responses, and parasympathetic input usually accompanying **rest-and-digest** functions.

9.12   ALZHEIMER'S DISEASE (AD): A COMMON NEURODEGENERATIVE DISORDER WITH WELL CHARACTERIZED HISTOPATHOLOGY

Today in the United States, **Alzheimer's disease** (**AD**) afflicts ~10% of adults 65+ years-old and is the **sixth leading cause of all adult deaths**. In its typical form, initial symptoms include forgetfulness, disorientation, confusion, and/ or behavioral changes. Subsequently, AD patients progress to an incapacitated state and undergo earlier-onset deaths compared to non-affected cohorts. Previously, people with such symptoms were often diagnosed as simply exhibiting accelerated aging or a variety of psychiatric disorders (**Case 9.4 Marilyn Monroe's Mental Health**). However, live imaging analyses of patients with AD have revealed various examples of disease-specific pathogenesis, including shrinkage of brain volume, reduction of cerebral folding, and general downregulation of activity patterns that reflect a marked loss of neurons and synaptic connectivity. Particularly hard-hit areas include the **cerebral cortex** and **hippocampi**, which in turn explain why AD patients often lose both their memory and their ability to orient properly to the environment (Box 9.2). Diagnosis of AD can sometimes be achieved with targeted cognitive testing and/or several disease-linked biomarkers. However, definitive assessments that rule out other kinds of dementia may require post-mortem analyses of brain sections to identify two well-documented histological correlates of AD—**amyloid beta (Aβ) plaques** and **neurofibrillary tangles (NFTs)**. Aβ oligomers are believed to cause **hyperphosphorylation** of the microtubule-associated protein **tau**, thereby helping to generate **NFTs**, which, along with other pathological formations, can block intraneuronal transport and trigger neuronal death. Therapies aimed at these and other suspected drivers of pathology have yet to find substantial success, and without clinical breakthroughs, the numbers of Alzheimer's patients will undoubtedly rise with continued demographics of population aging.

CASE 9.4   MARILYN MONROE'S MENTAL HEALTH

Actress
(1926–1962)

"I think that when you are famous every weakness is exaggerated."

(M.M. on the burdens of fame)

Marilyn Monroe was born Norma Jean Mortensen in California to a single mother, who was committed to a mental hospital when Monroe was a child. As a result, Monroe had to persevere through stays in an orphanage and foster homes, where it was later revealed she had been sexually abused. In spite of such trauma, Monroe forged a highly successful acting career, starring in nearly a dozen movies over a ten-year stretch. Unlike the carefree characters she often played, Monroe was burdened with severe anxiety, which was sometimes manifested as panic attacks or a stutter

so incapacitating it could require dozens of re-takes for her to master a three-word line. As Monroe's mental health deteriorated, one of the psychoanalysts she regularly saw committed her to a psychiatric ward, and during her last years, Monroe abused sedatives and other prescription medications before eventually dying of a drug overdose at age 36. Even though  her short stardom occurred decades ago, Monroe remains one of Hollywood's most famous icons whose nearly mythical status can be traced back not only to her unique screen persona, beauty, and talent but also perhaps to widespread empathy for the mental health issues she had to battle.

9.13   SUMMARY—NERVOUS SYSTEM

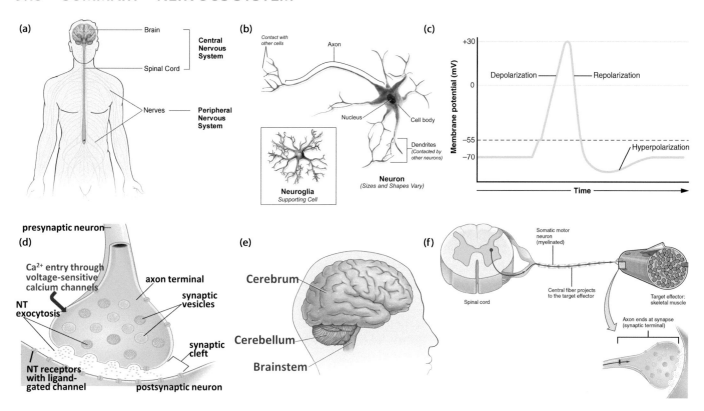

Pictorial Summary Figure Caption: *(**a**): see figure 9.1a; (**b**): see figure 9.1b; (**c**): see figure 9.5f; (**d**): see figure 9.6c; (**e**): see figure 9.8a; (**f**): see figure 9.11f.*

The **nervous system** comprises the central nervous system (**CNS**) of the **brain** and **spinal cord** plus the peripheral nervous system (**PNS**) outside the CNS. Such components **develop** from embryonic ectoderm that is internalized during a **neurulation** process while producing **nervous tissue** containing **neurons** and **glial cells** (**Figure a, b**). **Neurons** have a cell body (=**soma**) with **neurite** projections (**dendrites** and **axon**) that propagate rapid membrane potential changes, called **action potentials** (**APs**). Most neurons have multiple neurites and hence are **multipolar**, with the majority of multipolar neurons comprising CNS **interneurons**. Such neurons serve as a connecting bridge between: (i) **afferent neurons** of the PNS that bring APs to the CNS and (ii) **efferent** PNS neurons that transmit APs away from the CNS. **Glial cells** (e.g., Schwann and oligodendrocyte types) typically do not propagate APs but **play supportive roles**, such as forming myelin sheaths around axons. There is an ~ **1:1** glia-neuron ratio in the brain.

PNS nerves comprise **nerve fibers**, each consisting of an axon (with or without a myelin sheath) plus glial cells and a surrounding **endoneurium**. Nerve fibers can project from a **ganglion** where neuronal **somata** are aggregated. Myelinated axons are enveloped by a concentrically arranged **myelin sheath** made by **Schwann cells** in the PNS or **oligodendrocytes** in the CNS. Gaps in the myelin sheath (=**nodes of Ranvier**) allow **saltatory conduction** for more rapid impulse propagation, as APs hop from node to node. In the CNS, regions with dense myelination form **white matter**, whereas relatively unmyelinated areas constitute **gray matter.**

The **resting membrane potential** of unstimulated neurons is approximately **–70 mV** due to unequal ion distributions that are actively established by **Na⁺/K⁺ pumps**. Once a neuron is stimulated, **Na⁺ channels open** allowing Na⁺ influx that raises the membrane potential during the **depolarization** phase. Depolarization opens **voltage-gated Na⁺ channels** to spread the AP down the axon. At peak positive potentials, Na⁺ channels close, and **K⁺ channels open** allowing K⁺ efflux for **repolarization** (**Figure c**).

Synapses transmit APs between neurons (**Figure d**). **Excitatory** synapses use **neurotransmitters** (**NTs**) like **acetylcholine** to **depolarize** post-synaptic neurons by opening Na⁺ channels, whereas **inhibitory** NTs like **GABA hyperpolarize** post-synaptic neurons by opening Cl⁻ channels. In **ionotropic (fast) synapses**, NTs directly open ion channels, as opposed to **metabotropic (slow) synapses** that use intermediate signaling pathways between NT binding and channel gating.

The **brain** comprises: (1) a large **cerebrum**; (2) a small posterior **cerebellum**; and (3) an inferior **brainstem** with a **medulla oblongata** that connects to the spinal cord (**Figure e**). The **cerebral cortex** has multilayered gray matter with **pyramidal cells** used in **cognition**, and the subcortical cerebrum includes **hippocampi** for **memory formation** and **spatial orientation**. Within the cerebellum, a trilayered gray matter with **Purkinje cells** occurs in the cortex and helps **coordinate locomotion**. The **brainstem** controls **autonomic functions** like breathing. Three **meninges**—**dura**, **arachnoid**, and **pia maters**—help protect the brain, and brain cavities lined by ciliated **ependymal cells** contain **cerebral spinal fluid** that cushions brain tissues and helps clear wastes. Tight junctions in cerebral blood vessels form a **blood-brain barrier** to protect sensitive nervous tissue from pathogens and toxins.

The **spinal cord** has outer **white matter** and inner **gray matter** that forms **dorsal** and **ventral horns**. **Sensory input** enters **dorsal horns**, whereas **motor output** exits **ventral horns**. For rapid responses, **somatic reflex arcs** bring sensory input to the dorsal horn via **pseudounipolar** neurons whose cell bodies reside in a **dorsal root ganglion**, and after the AP is transmitted by an efferent axon from the ventral horn, the reflex induces skeletal muscle contraction via acetylcholine (**ACh**) release (**Figure f**). **Autonomic reflexes** have an efferent ganglion in which a **preganglionic** fiber synapses with a **postganglionic fiber** before innervating target cells (e.g., smooth muscles and glands). **Sympathetic** postganglionic neurons use **noradrenaline** for **fight-or-flight** responses, whereas **parasympathetic** postganglionic neurons use ACh for **rest and digest**.

During nerve regeneration, distal ends of damaged PNS nerves undergo **Wallerian degeneration**, phagocytose debris, and reorganize Schwann cells into **bands of Bungner** that guide new axonal outgrowths back to their targets; **Alzheimer's disease (AD)** causes neuronal degeneration involving **amyloid beta (Aβ) plaques** and **tau neurofibrillary tangles**. AD particularly affects the **cerebral cortex** and **hippocampi**, yielding memory loss and spatial disorientation.

SELF-STUDY QUESTIONS

1. Which specific cell type myelinates CNS axons?
2. T/F: Sensory input enters the ventral horn of the spinal cord.
3. Which is the innermost of the three meningeal layers?
4. Which of the following is an example of an inhibitory neurotransmitter?
   A. Acetylcholine
   B. Noradrenaline
   C. GABA
5. Which of the following would you expect to find in patients with Alzheimer's disease?
   A. Decreased hippocampal function
   B. Neurofibrillary tangles
   C. Aβ plaques
   D. ALL of the above
   E. NONE of the above
6. Which of the following is specifically associated with myelinated fibers?
   A. Nodes of Ranvier
   B. Gray matter
   C. Choroid plexuses
   D. ALL of the above
   E. NONE of the above
7. T/F The glial cell to neuron ratio in the brain is ~1000:1.
8. What are the two subdivisions of the autonomic nervous system?
9. Which of the following normally occurs in axons?
   A. K+ channels opening at the onset of depolarization
   B. Na+ channels closing at the onset of depolarization
   C. K+ channels opening at the onset of repolarization
   D. Na+ pumped in and K+ pumped out by Na+/K+ antiporters
   E. NONE of the above

10. What kind of intercellular junction contributes to the blood-brain barrier?
11. What fills cavities of the brain?
12. Which of the following is NOT normally a part of the adult hippocampus?
    A. Dentate gyrus
    B. Subgranular zone of neurogenesis
    C. Cornu ammonis
    D. Medulla oblongata
13. T/F Proper formation of bands of Bungner facilitates PNS nerve regeneration.

"EXTRA CREDIT" For each term on the left, provide the BEST MATCH on the right (answers A-E can be used more than once)

14. Hippocampi _____
15. Purkinje cells _____
16. Trilayered Cortex _____
17. Ganglia _____
18. Epidural space for injecting drugs _____

A. Cerebrum
B. Cerebellum
C. Brainstem
D. Spinal Cord
E. PNS

19. Compare and contrast the motor pathway in a somatic vs. sympathetic reflex arc.
20. Describe the process of action potential generation and propagation in a myelinated neuron.

ANSWERS

1) oligodendrocyte; 2) F; 3) pia mater; 4) C; 5) D; 6) A; 7) F; 8) parasympathetic and sympathetic; 9) C; 10) tight junctions; 11) cerebral spinal fluid; 12) D; 13) T; 14) A; 15) B; 16) B; 17) E; 18) D; 19) Somatic has a single motor neuron that extends to skeletal muscle and secretes ACh; sympathetic has a short pre- and a long postganglionic fiber that synapse in a paravertebral ganglion; postganglionic fiber secretes noradrenaline to activate non-skeletal muscle targets; 20) A full answer would include and define most or all the following terms: resting membrane potential, stimulus, depolarization, repolarization, hyperpolarization, refractory period, Na^+/K^+ antiporter; Na^+ channels; K^+ channels; voltage-gated Na^+ channels, initial segment, myelin sheath, nodes of Ranvier, saltatory conduction

Skin

10.1 GENERAL FUNCTIONS AND OVERALL ANATOMY OF SKIN

The **skin** (=integument or cutis) covers the body and is the largest organ in humans, comprising ~1.5–2 m² of surface area. As a first line of defense, skin serves as a **physical barrier** that helps exclude pathogens while minimizing desiccation. Skin also functions in **thermoregulation** via the insulating properties of integumental fat and hair as well as via the evaporative cooling mediated by sweat production. For its **metabolic** roles, skin is involved in synthesizing vitamin D, balancing electrolyte content, excreting water-soluble wastes, and launching immune responses. In addition, **sensory** components of skin can perceive various external cues. Conversely, skin communicates **signals** via its coloration, hair patterns, and odors.

Based on its embryological derivation and histology, skin is a bipartite organ composed of an upper **epidermis** and a lower **dermis** (**Figure 10.1a**). The epidermis (**Case 10.1 Michael Jackson's Epidermis**) develops from **ectoderm** to become a **stratified squamous epithelium**, whereas the **mesodermally derived** dermis comprises **dense irregular connective tissue**. The innermost edge of the dermis marks the bottom of skin proper, which in turn is underlain by an adipose-rich **hypodermis** (=subcutaneous layer) (**Figure 10.1a**).

The epidermal layers covering the ventral surfaces of the hands and feet optimize tactile sensitivity and gripping action via outwardly directed **ridges** and **valleys** that generate the unique patterns of **finger-** and **toeprints**. In addition, throughout the body, the base of the epidermis forms inwardly directed **epidermal pegs** that fit within indented **dermal sockets**. Such sockets in turn alternate with upwardly directed **dermal papillae** that directly surround epidermal pegs (**Figure 10.1b**). The peg-and-socket arrangement normally provides a stable interdigitation for holding the two integumental layers tightly together. However, in response to repetitive abrasion or several kinds of toxins, lymph accumulated at the epidermal-dermal border can separate the two layers by forming **blisters**.

DOI: 10.1201/9780429353307-12

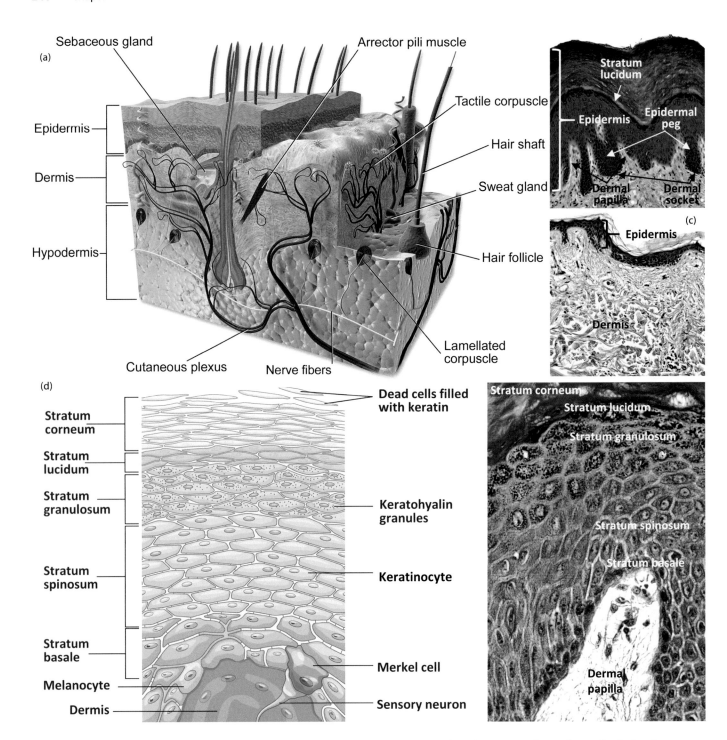

Figure 10.1 Introduction to skin. *(a)* The skin (=integument or cutaneous layer) consists of ectodermally derived epidermis overlying mesodermally derived dermis, which in turn rests on subcutaneous hypodermis. *(b, c)* Thick skin of soles and palms has a comparatively thick epidermis with a discrete stratum lucidum, whereas thin skin found everywhere else in the body has a thinner epidermis that lacks a stratum lucidum. (b) 80X; (c) 70X. *(d, e)* The stratified squamous epidermis comprises from its base to its apex the following compartments: (i) stratum basale, (ii) stratum spinosum, (iii) stratum granulosum, (iv) stratum lucidum (in thick skin), and (v) stratum corneum of dead cells. The stratum basale is the thin lowest-most layer of mainly mitotic stem cells, whereas suprabasal strata contain multiple layers of cells. Most epidermal cells are keratinocytes, although scattered melanocytes, Merkel cells, and Langerhans cells are also present. (e) 250X. ([a]: From Blausen.com staff (2014) Medical gallery of Blausen Medical 2014. Wiki J Med 1 (2): 10. doi:10.15347/wjm/2014.010 reproduced under CC0 and CC BY SA creative commons licenses; [b]: From OpenStax College (2013) Anatomy and physiology. OpenStax. http://cnx.org/content/col11496/latest reproduced under a CC BY 4.0 creative commons license.)

CASE 10.1 MICHAEL JACKSON'S EPIDERMIS

Singer
(1958–2009)

"I have a skin disorder that destroys the pigmentation of my skin."

(M.J. denying that the lightening of his skin was due to bleaching treatments)

Born to African-American parents in Indiana, Michael Jackson began singing at age 5 in a pop music group with his brothers before becoming a highly successful solo performer. During his decades as an entertainer, Jackson grew noticeably lighter-skinned, triggering rumors he had undergone skin bleaching treatments. However, Jackson denied such allegations and ascribed his condition to vitiligo, an autoimmune disease that attacks melanin-producing melanocytes in the epidermis to cause patchy

regions of depigmentation. In addition, Jackson also suffered severe scalp burns caused by a fireworks accident during the filming of a commercial. Along with ongoing stresses of super-stardom and a highly publicized trial on child molestation charges, chronic discomfort from his burns contributed to Jackson's heavy use of pain medication and sedatives, which in turn contributed to his death by cardiac arrest at age 50. In his autopsy, Jackson was reported to have vitiligo, although it remains possible that the disease was not the sole cause of his widespread epidermal lightening.

Skin ranges in overall thickness from less than 0.25 mm in eyelids to over 5 mm in parts of the back that have a particularly expansive dermis. However, historically, two types of skin—**thick** and **thin skin**—have been distinguished, based solely on the thickness of their epidermal component (**Figure 10.1b, c**). Thus, in thick skin, which occurs exclusively on **palms** and **soles**, the epidermis comprises up to several dozen cell layers, and together the thickness of these epidermal layers can exceed 1 mm. Alternatively, thin skin, which is found everywhere else in the body, has comparatively few epidermal cell layers that are collectively less than 0.5 mm thick. Moreover, the two skin types are distinguished by the presence or absence of a **stratum lucidum** (see next section), which occurs in the epidermis of thick skin but is lacking in thin skin (**Figure 10.1b, c**).

10.2 INTRODUCTION TO EPIDERMAL CELLS AND LAYERS

About 60–70% of epidermal cells are **keratinocytes**, which are so-named because they make cysteine-rich proteins, called **keratins**. In addition, the epidermis contains scattered **melanocytes**, **Langerhans cells**, and **Merkel cells**, each of which accounts for ~5–15% of all epidermal cells. Along with these four epidermal cell types, branched termini of unmyelinated sensory neurons project across the dermis and ramify as **free nerve endings** throughout the basal half of the epidermis. As **high-threshold mechanoreceptors** (**HTMRs**) that require relatively strong stimuli to be activated, such nerve endings respond to various cues including pressure, noxious input, and temperature by transmitting action potentials toward the CNS for further processing. A brief introduction to epidermal cells and layers is presented directly below, and further discussion of key epidermal processes is provided in the next section.

Keratinocytes

By undergoing a **keratinization** process that packs their cytoplasm with keratin, keratinocytes harden the epidermis and help reduce water loss. In addition, keratinocytes further **waterproof** their plasmalemmata via externally secreted lipids plus an internal envelope of proteins lining the membrane's inner surface. Keratinocytes also produce antimicrobial **defensins** and store **melanin pigment** generated by neighboring melanocytes. Moreover, keratinocytes utilize ultraviolet (UV) rays to convert 7-dehydrocholesterol into pro-vitamin D3, which is further processed in the liver and kidneys to yield active **vitamin D** (=**calcitriol**) for facilitating calcium uptake in the gut.

Keratinocytes are continually moved toward the epidermal apex in an escalator-like fashion via the basal addition of new cells, thereby resulting in the oldest keratinocytes being sloughed from the apical surface. Based on their particular properties during apical transport, keratinocytes form several epidermal compartments, or **strata**, each of which comprises a stack of cellular layers (**Figure 10.1d**). The basal-most compartment that directly overlies the basement membrane is a thin **stratum basale**, which comprises mainly cuboidal stem cells. Divisions of these cells yield either additional stem cells that undergo further mitoses or post-mitotic progeny that are transported into the overlying **stratum spinosum** to differentiate as keratinocytes.

The stratum spinosum comprises several layers of cuboidal to squamous keratinocytes that attach to their neighbors by abundant **desmosomes**. Such desmosomes and their associated intermediate filaments generate the characteristic prickly morphology of spinosum keratinocytes (**Figure 10.1d**). Directly above the stratum spinosum is a **stratum granulosum** comprising one to three layers of more squamous keratinocytes. Granular cells in this stratum not only contain small **lamellar bodies** that are destined to be exocytosed, but such cells also possess large **keratohyalin granules** that remain intracellular. While undergoing apical transport, granular keratinocytes begin to die, thereby eventually leaving only dead squamous keratinocytes packed with keratin in the superficial epidermis. In thick skin, the stratum granulosum is overlain by a translucent band of dead keratinocytes, called the **stratum lucidum**, which contains a clear keratin precursor, historically called eleidin. Above the stratum lucidum of thick skin or throughout the uppermost stratum of thin skin, a **stratum corneum** comprises up to two dozen layers of opaque dead keratinocytes (=corneocytes) filled with fully mature keratin (**Figure 10.1d, e**).

Melanocytes

Arising from neural crest cells that migrate to the epidermis at about 6 weeks of embryonic development, **melanocytes** produce **melanin** pigment granules (**Case 10.2 Bob Marley's Melanocytes**). In adults, melanocyte cell bodies occur in the epidermal stratum basale (**Figure 10.1d**), and each melanocyte extends cellular processes into the stratum spinosum to interact with about three dozen keratinocytes. Such interactions establish **epidermal-melanin units**, which allow the transfer of melanocyte-produced melanin granules into keratinocytes.

Langerhans cells

Langerhans cells are monocyte-derived **antigen-presenting cells** (**APCs**) (**Figure 10.2a–c**). After encountering and engulfing pathogens, Langerhans cells exit the epidermis and migrate to lymph nodes (**Figure 10.2a–c**), where they deliver their antigenic cargo to T lymphocytes for further pathogen processing. Intra-epidermal Langerhans cells possess paddle-shaped **Birbeck granules** (**Figure 10.2c, inset**), which contain **langerin**, a calcium-dependent type of carbohydrate-binding protein (=**lectin**) used in pathogen processing.

Merkel cells

Merkel cells are mechanoreceptors in the stratum basale (**Figure 10.1d**). Prevalent in touch-sensitive areas, such as fingertips and lips, Merkel cells connect to keratinocytes by desmosomes while also synapsing with underlying afferent neurons. Such neurons are **low-threshold mechanoreceptor (LTMR)** sensory fibers, which, unlike intra-epidermal free nerve endings that require strong stimuli to be activated, can depolarize in response to gentle pressure.

CASE 10.2 BOB MARLEY'S MELANOCYTES

Musician
    (1945–1981)

"Get up, stand up, stand up for your rights"

(From B.M.'s 1973 protest song "Get up, stand up")

Bob Marley was born in Jamaica and rose from poverty to become an international star who helped popularize reggae music in the 1970s. As a follower of the Rastafari religion that denounced materialism and believed in the spiritual use of marijuana as a sacrament ordained by the Old Testament, Marley also abided by other decrees, including a prohibition against the cutting of hair or skin. Such a belief may have hastened Marley's demise, when a growth under his toenail was diagnosed as acral lentiginous melanoma, an aggressive cancer of melanocytes that typically develops in dark-skinned people. Although doctors recommended amputation of the toe, Marley declined surgery and sought alternative therapies. However, the cancer eventually spread to his brain, and while flying back to Jamaica from treatments in Germany, Marley had to be hospitalized in Florida, where he died at age 36 from metastatic cancer arising from his malignant melanocytes.

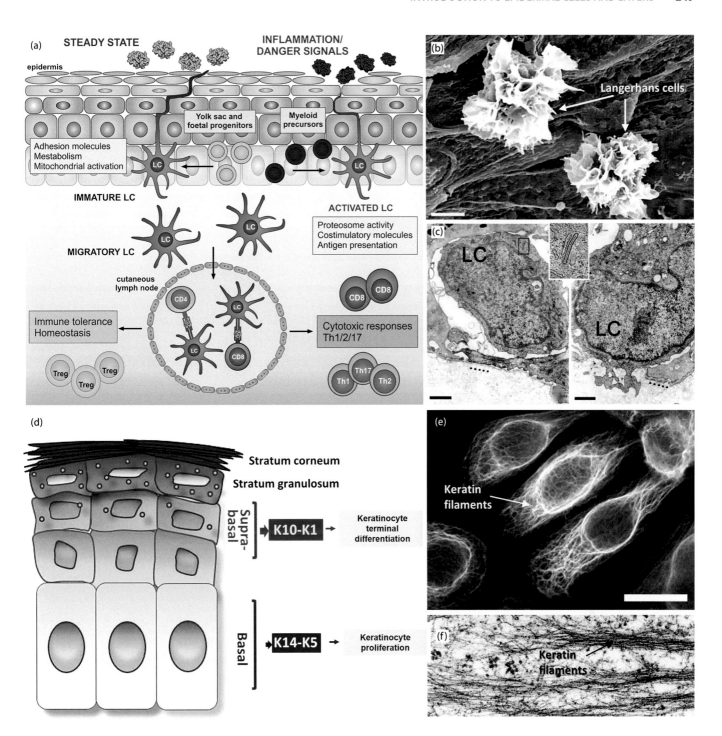

Figure 10.2 Langerhans cells (LCs) and keratinocyte keratin filaments. (*a–c*) *Langerhans cells are dendritic antigen-presenting cells that can escape the epidermis and carry antigens to nearby lymph nodes for immune processing. In (b) (SEM, scale bar = 5 μm) and (c) (TEMs, scale bars = 1 μm), Langerhans cells are fixed in the process of escaping from the epidermal base; dots mark a breach in the basement membrane, and the inset shows LC Birbeck granules; (**d**) Keratinocytes express different keratin types after moving from the stratum basale into suprabasal strata. (**e, f**) Prominent keratin filaments are shown in immunofluorescence (e, scale bar = 25 μm) and TEM (f, 10,000X) images of cultured human keratinocytes. ([a]: From Clayton, K et al. (2017) Langerhans cells—programmed by the epidermis. Front Immunol 8: doi: 10.3389/fimmu.2017.01676 reproduced under a CC BY 4.0 creative commons license; [b, c]: From Stoitzner, P et al. (2002) A close-up view of migrating Langerhans cells in the skin. J Invest Dermatol 118: 117–125 reproduced with publisher permission; [d] From Zhang, X et al. (2019) Keratin 6, 16 and 17—critical barrier alarmin molecules in skin wounds and psoriasis. Cells 8: doi:10.3390/cells8080807 reproduced under a CC BY 4.0 creative commons license; [e, f]: From Fudge, D et al. (2008) The intermediate filament network in cultured human keratinocytes is remarkably extensible and resilient. PLoS ONE 3: doi:10.1371/journal.pone.0002327 reproduced under a creative common license.)*

10.3   MAJOR CELLULAR PROCESSES OCCURRING WITHIN THE EPIDERMIS: INTRACELLULAR KERATINIZATION, CELL ENVELOPE PRODUCTION, SECRETION OF INTERCELLULAR LIPIDS FOR WATER-PROOFING, DESQUAMATION, AND MELANOGENESIS

Intracellular keratinization and cell envelope production plus secretion of intercellular lipids

While in the stratum basale, keratinocyte precursors produce **type 5** and **14 keratins** that are characteristic of non-differentiating stem cells undergoing mitotic proliferation (**Figure 10.2d**). However, after moving apically into the **suprabasal epidermis**, keratinocytes differentiate while synthesizing **type 1** and **10 keratins** that are assembled into 10–12 nm-thick intermediate filaments, called **tonofilaments** (**Figure 10.2e, f**). Such filaments and associated microtubules project inwardly from prickly cell desmosomes to form spiny cellular processes (**Figure 10.3a–c**).

Spinosum keratinocytes start making **keratohyalin granules** (**KHGs**) that eventually measure several micrometers long while also containing non-keratin proteins like **filaggrin** (**Figure 10.3d**). KHGs enlarge and become more conspicuous in the lower stratum granulosum before fusing with tonofilaments to form a large mass of cytoplasmic **keratin** (**Figure 10.3d**). In addition to making KHGs and keratin, granulosum keratinocytes also attach proteins, such as **involucrin** and **loricrin** to the inner surface of their plasmalemma, thereby generating a thickened **cell envelope** that encases the central keratin mass (**Figure 10.3e**).

Along with producing keratin and a cell envelope, keratinocytes assemble a complex mixture of lipids within oblong **lamellar bodies** that typically measure 100–400 nm long (**Figure 10.3d, e**). Lamellar bodies are produced in the spinosum and granulosum strata, and unlike KHGs or envelope components that remain within the keratinocyte, lamellar bodies are exocytosed to form a waxy coat of substances like **ceramide** lipids (**Figure 10.3e, f**). Lipids derived from lamellar bodies waterproof epidermal cells and thereby augment the anti-desiccation properties of intracellular envelopes and keratin masses.

Before being moved apically into the stratum corneum, the keratin- and envelope-containing keratinocytes of the granulosum layer lose their nuclei and organelles while undergoing an **apoptotic** form of cell death. Thus, the stratum corneum comprises only dead, keratin-packed **corneocytes** that are eventually shed at the apical surface (**Figure 10.3e**).

Desquamation

Newly added corneocytes in the stratum corneum are initially connected by desmosomes and modified types of desmosomes, called **corneodesmosomes**, which contain **corneodesmosin** plus other proteins within a conspicuous intercellular plaque (**Figure 10.4a, b**). However, during apical transport into progressively more acidic regions, the reduced pH activates **kallikrein-related peptidases** (**KLKs**) that help dissolve some of these connections, thereby allowing clumps of dead cells (=**squames**) to be disengaged from the epidermal surface (**Figure 10.4c, d**). This process of **desquamation** offsets new cell additions at the epidermal base and normally results in each keratinocyte being shed ~**1–2 months** after it had entered the suprabasal epidermis.

Melanogenesis

Melanocytes are more abundant in some regions of the human body (e.g., nipples) than in others (e.g., inner thighs). However, there is no consistent statistical difference in the overall concentrations of melanocytes that have evolved in dark- vs. fair-skinned people (**Figure 10.4e–i**). Instead, variations in pigmentation among humans are mainly due to the number of **melanin granules**, with dark-skinned individuals tending to have more melanin per unit area than do fair-skinned cohorts (**Figure 10.4g, i**). Increased granule density arose as an adaptation to protect against the greater UV exposure prevalent in Africa, where humans first evolved. Conversely, after migrating to higher latitudes with reduced UV doses, fair-skinned people with decreased melanin granule numbers gained a selective advantage by optimizing their UV-mediated synthesis of vitamin D. In addition to the effects of melanin granules, skin color is also influenced to a lesser degree by dermal vascularization, dietary intake, and endogenous metabolites.

During melanogenesis, melanocytic Golgi bodies and associated endosomal compartments form oblong **pre-melanosomes** containing a **melanofilament** scaffolding upon which melanin can be deposited (**Figure 10.5a**). In normal pre-melanosomes, melanin synthesis begins with **tyrosine** being oxidized to

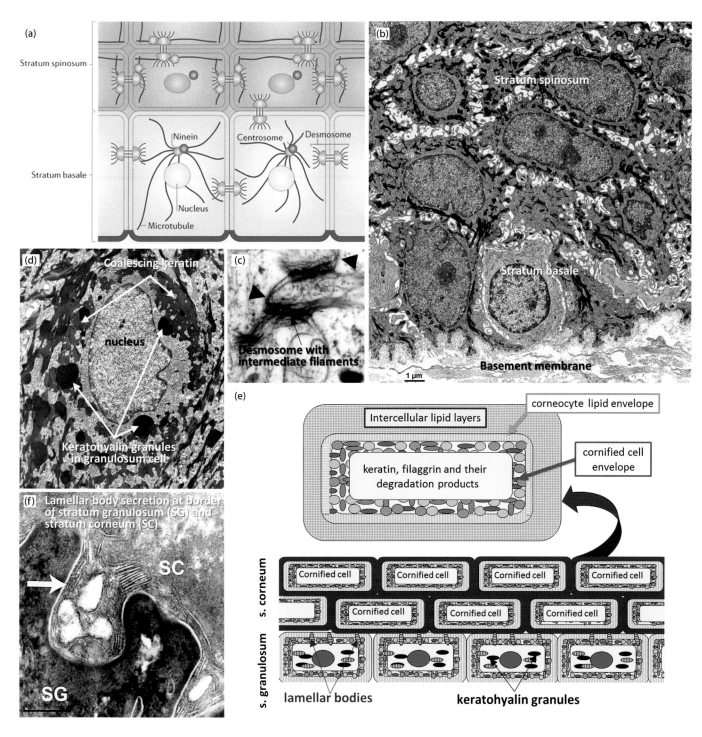

Figure 10.3 Keratinocytes in the stratum spinosum and stratum granulosum. *(a–c) After entering the stratum spinosum from the stratum basale, keratinocytes become interconnected by abundant desmosomes with attached keratin filaments and microtubules. (c) 39,000X. (**d**) Keratohyalin granules begin to form in the stratum spinosum and become more conspicuous in the stratum granulosum before eventually fusing to form masses of keratin and other proteins that fill granulosum cells. (**e, f**) Keratinocytes also coat the inside of their plasmalemma with proteins to generate a peripheral intracellular envelope (e). Keratinocytes exocytose lipid-rich lamellar bodies at the border between the granulosum and corneum strata to form a waterproofing intercellular mortar around dead cornified cells (f). Scale bar = 200 nm. ([a]: From Simpson, DL et al. (2011) Deconstructing the skin: cytoarchitectural determinants of epidermal morphogenesis. Nat Rev Mol Cell Biol 12: 565–580 reproduced with publisher permission; [b, d]: Images courtesy of Dr. H. Jastrow (http://www.drjastrow.de) reproduced with author's permission; [c]: From Ishiko, J et al. (2003) Immunomolecular mapping of adherens junction and desmosomal components in normal human epidermis. Exp Dermatol 12: 747–754 reproduced with publisher permission; [e]: From Akiyama, M (2017) Corneocyte lipid envelope (CLE), the key structure for*

Figure 10.3 (Continued) *skin barrier function and ichthyosis pathogenesis. J Dermatol Sci 88: 3–9 reproduced with publisher permission; [f]: From Zhang, L et al. (2016) Defects in stratum corneum desquamation are the predominant effect of impaired ABCA12 function in a novel mouse model of harlequin ichthyosis. PLoS ONE 11: doi:10.1371/journal. pone.0161465 reproduced under a CC BY 4.0 creative commons license.)*

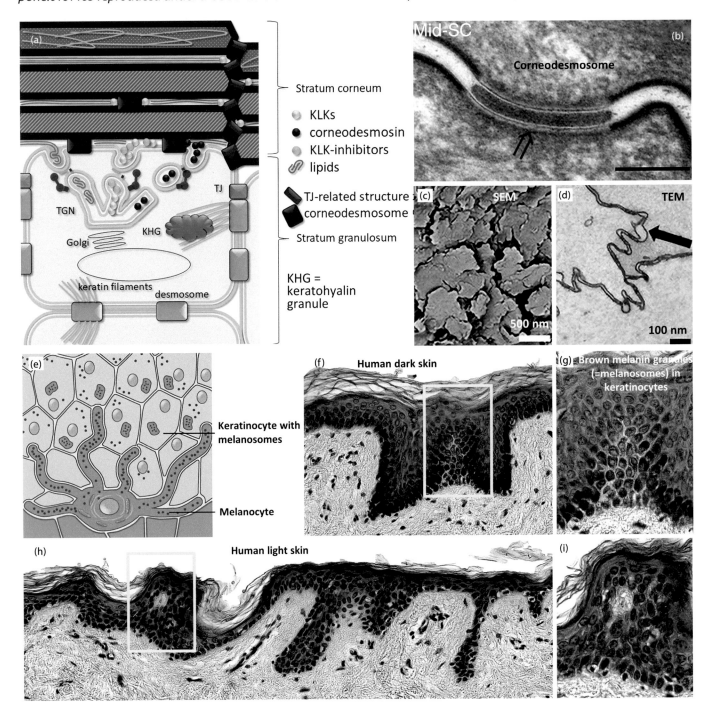

Figure 10.4 Desquamation and an introduction to melanogenesis. *(a, b) Stratum corneum keratinocytes are initially interconnected by modified desmosomes containing in their intracellular space such proteins as corneodesmosin (b, arrow). (c, d) However, after being transported more apically, kallikrein-related peptidases (KLKs) become activated and dissolve such connections so that clumps of corneal cells, called squames (c), are sloughed from the epidermal surface (d, arrow; scale bar = 100 nm). (e–i) Dark skin contains more numerous melanin granules than in light skin. (f) 10X; (g) 290X; (h) 180X; (i) 310X. ([a]: From Ishida-Yamamoto, A (2018) Molecular basis of the skin barrier structures revealed by electron microscopy. Exp Dermatol 27: 841–846 reproduced with publisher permission; [b]: From Lin, T-K et al.*

Figure 10.4 (Continued) *(2012) Cellular changes that accompany shedding of human corneocytes. J Invest Derm 132: 2430–2439 reproduced with publisher permission; [c, d]: From Vela-Romero, A et al. (2019) Characterization of the human ridged and non-ridged skin: a comprehensive histological, histochemical and immunohistochemical analysis. Histochem Cell Biol 151: 57–73 reproduced a CC BY 4.0 creative commons license; [e]: From OpenStax College (2013) Anatomy and Physiology. OpenStax. http://cnx.org/content/col11496/latest reproduced under a CC BY 4.0 creative commons license.)*

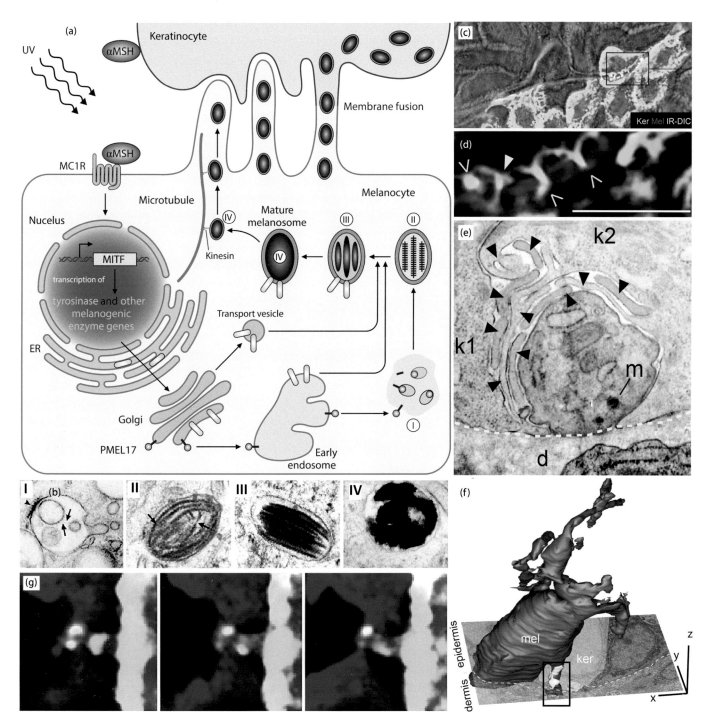

Figure 10.5 *Melanin production in melanocytes and transfer of melanin granules to keratinocytes. (a,b) UV-mediated secretion of α-melanocyte stimulating hormone (αMSH) allows αMSH binding to melanocortin 1 receptor (MCR1) on melanocytes, which in turn initiates expression of tyrosinase and other melanogenic components driven by the microphthalmia-associated transcription factor (MITF). After packaging in endosomes, PMEL17 (premelanosome protein 17) helps convert stage I melanosomes into melanofilament-containing stage II melanosomes. Melanin accumulates on*

Figure 10.5 (Continued) these filaments within stage III melanosomes before eventually filling stage IV melanosomes, which represent mature melanin granules. Such granules are transported along microtubules by kinesins to reach the cell periphery, where they can be directly engulfed by nearby keratinocytes in a type of delivery, termed cytocrine transfer. (b): 50,000X. (c–g) Fluorescence microscopy (c, d, g) and FIB-SEM (e, f) of co-cultured melanocytes (magenta) and keratinocytes (green) show contacts between keratinocyte processes and melanocyte dendrites that help mediate transfer of mature melanosomes. (c) 600X; (d) scale bar = µm; (e) 30,000X; (g) 6,200X. ([a]: From Wasmeier, C et al. (2008) Melanosomes at a glance. J Cell Sci 121: 3995–3999; [b]: From Raposo, G and Marks, MS (2007) Melanosomes – dark organelles enlighten endosomal membrane transport. Nat Rev Mol Cell Biol. 8 (10): 786–797 reproduced with publisher permission; [c–g]: From Belote, J and Simon, SM (2020) Ca²⁺ transients in melanocyte dendrites and dendritic spine-like structures evoked by cell-to-cell signaling. J Cell Biol 219: https://doi.org/10.1083/jcb.201902014 reproduced with publisher permission.)

3,4-dihydroxyphenylalanine (**DOPA**) by the enzyme **tyrosinase**. Conversely, such enzymatic activity is reduced or absent in the poorly pigmented skins of people with **albinisms**. The DOPA generated from tyrosine can be metabolized into two kinds of epidermal melanin. The most common is the brown or black **eumelanin** that is characteristic of dark-haired individuals, whereas DOPA is alternatively processed into the rarer pinkish-orange **pheomelanin** that is associated with red hair (**Figure 10.5a**). With additional melanin deposition, discrete melanofilaments become obscured, and the pre-melanosome is transformed into an electron-dense **melanosome** that further matures into a **melanin granule** (**Figure 10.5b**). Following maturation, melanin granules are transported from the melanocyte soma in a kinesin-dependent manner into cellular processes for incorporation into nearby keratinocytes. Such exchanges can occur via **cytocrine transfer** in which melanocyte and keratinocyte membranes fuse to allow direct transmission of granules (**Figure 10.5a**). Based on correlative imaging analyses, cytocrine transfers are facilitated by signaling pathways involving acetylcholine stimulation and melanocytic Ca²⁺ transients (**Figure 10.5c–g**). After being transferred into keratinocytes, melanin granules are subsequently concentrated via dynein-mediated movements over keratinocyte nuclei, thereby optimally placing such intracellular sunscreen equivalents in the path of mutagenic UV rays.

Tanning induced by moderate amounts of UV light typically leads to a two-step widespread darkening of exposed skin (Box 10.1 Ultraviolet Rays, Tanning, and Cancers). Similarly, UV rays may also deepen the coloration of **freckles** that are particularly prevalent in red-headed individuals whose alternative **melanocortin 1 receptors** (**MC1Rs**) favor the production of pheo- over eumelanin. Alternatively, excessive UV absorption leads to a pronounced reddening of skin (=**erythema**) as part of inflammatory responses associated with **sunburns**.

10.4   HARD KERATIN PRODUCTS FORMED BY THE EPIDERMIS: HAIR AND NAILS

Unlike soft keratins made by surface keratinocytes, some keratinocytes that have sunken below the epidermal surface produce **hard keratins** with a higher cysteine content for increased crosslinking and rigidity. Along with the horns, hooves, and claws of other terrestrial mammals, hard keratins form **hair** and **nails** in humans.

Hair

The human fetus initially forms a type of thin hair termed **lanugo**, which is replaced a few weeks before birth by slightly thicker hair, called **vellus.** Vellus hairs (=peach fuzz) continue to grow in most regions of a child's skin except for a few areas that lack hair, which include (i) thick skin, and (ii) normally **glabrous** (=hairless) regions of thin skin, such as lips and parts of genitalia. Alternatively, a relatively thick and pigmented hair, termed **terminal hair**, is produced in the scalp, nostrils, eyebrows, and eyelids of children. Following stimulation by extra androgen production during puberty (Chapter 15), terminal hair can also replace vellus hair in several other body regions, such as the face, axillae, and groin. Compared to other primates, the ratio of terminal to vellus hair remains low in adults, presumably as an adaptation that has helped humans reduce their ectoparasite loads and colonize hotter environments.

Human hairs are attached via elastic fibers to smooth muscle bundles, called **arrector pili** (**Figure 10.6a**). In response to cold or other sympathetic stimulation, arrector pili contraction elevates the surrounding skin into "goosebumps" and causes hair to "stand on end". For furry mammals, such contractions trap air to aid insulation while also making the body appear larger to potential foes, whereas comparable functions of arrector pili contractions have become vestigial in humans.

BOX 10.1 ULTRAVIOLET RAYS, TANNING, AND CANCERS

Although tanning of skin can occur due to either pathologies like Addison's disease or to topical applications of various sunless darkening agents, most tans are generated by **ultraviolet** (**UV**) radiation in natural sunlight or tanning beds. Nearly all of sunlight's UV that reaches the skin comprises the **UVA** class of relatively long-wavelength and deeply penetrating rays that traverse the epidermis and upper dermis, whereas the small remaining fraction of **UVB** rays has shorter wavelengths and less penetrating power (**Figure a**). UVA rays are the most effective in darkening the skin during tanning. Conversely, UVB rays contribute more to a reddening of skin (**erythema**) resulting in sunburns.

During the two phases of tanning, an **immediate pigment darkening (IPD)** induced by UV wavelengths occurs within minutes after exposure to sunlight and lasts up to a few hours. Such IPD involves both the rapid redistribution of melanosomes and a photo-oxidized darkening of melanin already present in existing melanosomes. Within 3–5 days after sunlight exposure, new melanin begins to be produced as part of a **delayed tanning (DT)** process that supplies new melanin granules to keratinocytes. The added pigment then remains in the skin for up to a few weeks before the darkened keratinocytes eventually become sloughed from the epidermal surface.

According to one model, new melanogenesis during delayed tanning results from UV-induced damage in keratinocytes activating the **p53 tumor suppressor** (**Figure b**). Enhanced p53 levels can then upregulate production of **pro-opiomelanocortin** (**POMC**), which is processed into active subunits, including **α-melanocyte stimulating hormone (α-MSH)**. Following secretion from keratinocytes, α-MSH can bind to **melanocortin 1 receptors (Mc1Rs)** on melanocytes to trigger a cAMP-mediated upregulation of the **microphthalmia associated transcription factor** (**MITF**) that drives **tyrosinase** production, which in turn initiates the conversion of tyrosine into melanin during delayed tanning.

UV absorption during tanning elevates vitamin D production and may thus help prevent certain cancers. However, dietary supplementation remains a far less risky way to raise vitamin D levels than repeatedly exposing skin to UV radiation. In fact, even moderate use of tanning beds can more than double the likelihood of developing skin cancer. Accordingly, a recent upsurge in melanoma diagnoses among younger people has been linked to increased tanning bed usage.

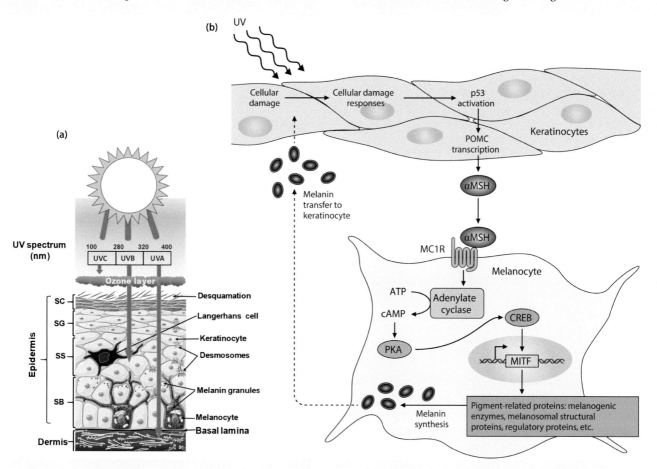

Box 10.1 (a, b) *Cell biology of tanning in response to UV in sunlight. ((a) from Camcheu, JC et al. (2019) Role and therapeutic targeting of the PI3K/Akt/mTOR signaling pathway in skin cancer: a review of current status and future trends on natural and synthetic agents therapy. Cells 8: doi:10.3390/cells8080803 reproduced under a CC BY 4.0 creative commons license; (b) modified from D'Orazio, J. et al. (2013) UV radiation and the skin. Int J Mol Sci 14: 12222–12246 reproduced under a CC BY 4.0 creative commons license.)*

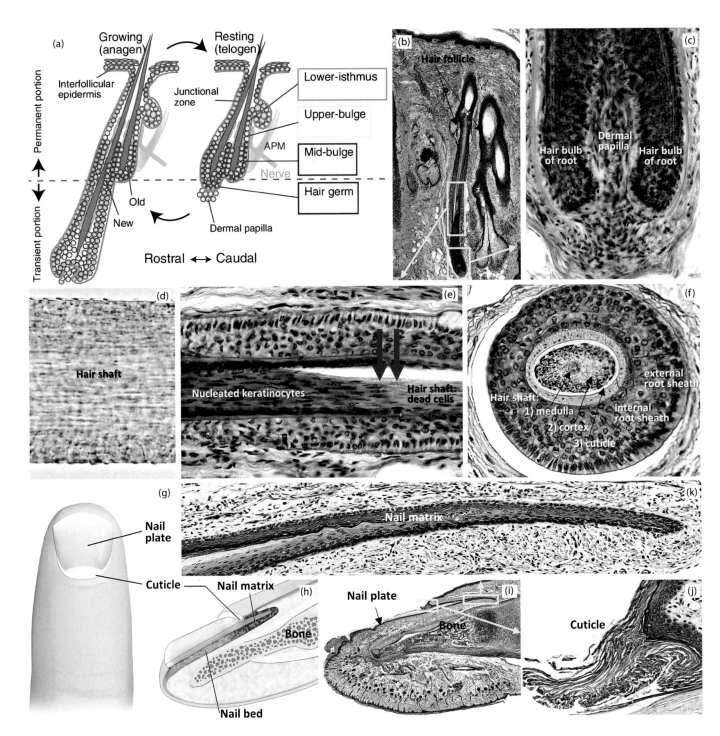

Figure 10.6 Hard keratin products of the epidermis: hair and nails. (*a*) *Developing hair follicles have several discrete niches of stem cells in bulge regions near where arrector pili muscles (APM), nerves, and sebaceous glands converge. (b, c) Stem cells migrate from bulge regions to the dermal papilla at the base of the follicle where they undergo mitoses to generate the hair shaft and its surrounding protective root sheaths. (b) 30X; (c) 200X. (d–f) Distally, the hair shaft comprises exclusively dead keratinocytes filled with hard keratin ((d), 800X), whereas closer to the dermal papilla, nucleated keratinocytes undergo apoptosis to become keratin-filled dead cells ((e), red arrows, 200X) surrounded by root sheaths ((f), 300X). (g–k) Nails resemble hairs in containing hard keratin but arise from a wide nail matrix rather than a conical dermal papilla to generate a flattened nail plate. (i) 12X; (j) 190X; (k) 150X. ([a]: From Cheng, C-C et al. (2018) Hair follicle epidermal stem cells define a niche for tactile sensation. eLIFE: e38883. doi:https://doi.org/10.7554/eLife.38883 reproduced under a CC BY 4.0 creative commons license; [g, h]: From Blausen.com staff (2014) "Medical gallery of Blausen Medical 2014". WikiJournal of Medicine 1 (2): 10. doi:10.15347/wjm/2014.010 reproduced under CC0 and CC BY SA creative commons licenses.)*

CASE 10.3 MARIE ANTOINETTE'S HAIR

Queen of France
(1755–1793)

"... in a single night, it had turned white as that of a seventy-year-old woman"

(M.A.'s servant on the rapid whitening of M.A.'s hair while imprisoned during the French Revolution)

Born in Austria, Marie Antoinette married at age 14 the future King of France Louis XVI before becoming the last Queen of France in 1774. During a time of famine catalyzed by the previous eruption of an Icelandic volcano, Marie Antoinette's lavish spending drew widespread wrath, thereby helping to trigger the French Revolution that dethroned the king and queen and led to their death by guillotining. While awaiting her fate in prison, Marie Antoinette's hair supposedly turned white overnight, and today such extremely rapid loss of hair color, or canities subita, is often referred to as the "Marie Antoinette syndrome". Many of these cases occur in response to trauma, and although typically not as rapid as overnight, whitening can be completed in as short as a few weeks. Some instances of canities subita involve another type of scalp disease, called alopecia areata, where autoimmune-mediated destruction of pigmented hair follicles leaves behind only white hair. Alternatively, accelerated color loss may proceed without noticeable alopecia areata via as-of-yet unexplained mechanisms, which in turn matches reports for the former queen, as she did not apparently experience marked hair loss before being beheaded.

Scalp hairs can exceed a meter in length and tend to grow fastest in teenagers and young adults during warmer months of the year. However, unlike surface keratinocytes that undergo continuous keratinization, hair growth is triphasic. Thus, each hair starts with an **anagen** phase of active growth that lasts several years. Anagen is then followed by an approximately two-week-long **catagen** phase of growth arrest, which in turn precedes a **telogen** phase of a few months as the hair-growing unit atrophies. At the end of a typical telogenic period of atrophy, the old hair is shed so that another anagenic phase of growth can be initiated (**Figure 10.6a**).

Histologically, each anagenic hair grows within an invaginated epidermal unit, called the **hair follicle** (**Figure 10.6a–c**). Such follicles extend into the dermis and comprise **external** and **internal root sheaths** of squamous to cuboidal cells that envelop and protect the keratinized **hair shaft** (**Figure 10.6d–f**). Approximately midway to its dermal base, the follicle connects to one or more **sebaceous glands**, and near such connections is a **bulge** region that contains several discrete populations of **stem cells** (**Figure 10.6a**). Below the bulge region, the basal **root** of each follicle is dilated to yield a **hair bulb**, into which dermal tissues invaginate, thereby forming a vascularized **dermal papilla** that is surrounded by the hair bulb (**Figure 10.6c**).

After migrating from the bulge region and infiltrating the apex of the dermal papilla, **stem cells** associate with Merkel cells, melanocytes, and Langerhans cells to constitute the functional equivalent of a stratum basale. Mitoses of follicular stem cells cause differentiating daughter cells to be transported toward the epidermal surface. During this transport, new cells added in lateral regions of the papilla become root sheath cells, whereas those emanating from a medial location differentiate into keratinocytes that form the bulk of the hair shaft. Similarly, medially situated melanocytes impart the particular color of hair, which depends on how much and what proportion of eu- vs. pheomelanins are mixed in with keratinocytes. Accordingly, the graying of hair during normal aging involves a gradual reduction in hair melanocytes, whereas diseases, such as **canities subita** cause rapid color loss (**Case 10.3 Marie Antoinette's Hair**).

As the growing hair extends from the bulb, keratinocytes in the shaft become progressively packed with hard keratins, but such keratinizing cells do not undergo desquamation as occurs in the surface stratum corneum. Instead, hair keratinocytes fuse together, and the aggregated products of keratinization plus melanocyte-derived pigments end up filling the hair shaft (**Figure 10.6d, e**). In transverse hair sections where such fusion has occurred, a thin **cuticle** delimits the outer edge of the hair, with the hair shaft consisting mostly of a peripheral **cortex** plus a smaller central **medulla** (**Figure 10.6f**). Following the telogenic phase when old hair is shed, a new hair follicle normally starts another round of anagen next to the old follicle (**Figure 10.6a**). However, either naturally or due to physiological insults, such as chemotherapy or autoimmune reactions, follicular papillae can become incapable of initiating additional hair growth, thereby leading to a balding process of hair loss, which is also referred to as **alopecia** when hair loss is excessive and/or rapid.

Nails

The distal digit of fingers and toes is covered dorsally by a **nail** that contains hard keratins (**Figure 10.6g**). To optimize attachment, the ventral side of the nail forms ridges that interdigitate with grooves in the **nail bed** region of underlying skin (**Figure 10.6h**). Each fully formed nail arises from a germinative area in the proximal nail bed, called the **nail matrix** (**Figure 10.6h, i, k**), which, like hair follicle papillae, contains stem cells, melanocytes, Langerhans cells, and Merkel cells. However, unlike the conical dermal papillae of hair, the nail matrix is broad and thin. Thus, as keratinizing cells are moved distally by cell divisions, the nail develops into a flattened **nail plate** rather than a cylindrical hair (**Figure 10.6g-i**).

The proximal nail plate is covered by an overlying fold of modified skin that is referred to as the **eponychium** with its overlying **cuticle** of dead cells (**Figure 10.6g, j**). Similarly, an underlying **hyponychium** seals the distal nail edge. Such skin flaps help protect the nail, and if disrupted, can lead to invasive fungal infections, called **onychomycoses**. Directly beyond the cuticle, the nail often exhibits a semicircular **lunula** that marks the beginning of the growing nail plate, whereas the distal edge of the nail terminates in an opaque band that is subjected to periodic trimming.

10.5 GLANDULAR ELABORATIONS OF THE EPIDERMIS: SEBACEOUS AND SWEAT GLANDS

Sebaceous glands

Throughout thin skin, the epidermis invaginates into the dermis to form **sebaceous glands** (**Figure 10.7a**). In body regions with hairy skin, several sebaceous glands typically connect to each hair follicle to form a **pilosebaceous unit** (**Figure 10.7b**). Within each of these units, sebaceous secretions are typically discharged through ducts into the lumen of the hair follicle before being spread over the epidermal surface.

Sebaceous glands are branched acinar glands that have secretory cells (=**sebocytes**) distributed within clustered acini (**Figure 10.7c**). Sebocytes fill each acinus and make a complex oily product, called **sebum**, which comprises triglycerides, wax esters, and antimicrobial agents. Sebocytes exocytose sebum via a **holocrine** process that breaks apart each secretory cell, thereby necessitating continual replacement by basally located **stem cells** derived from follicle bulges.

Sebum discharged from sebaceous glands normally lubricates the stratum corneum and hair shafts to reduce desiccation while also participating in antimicrobial defenses. However, in response to increased androgen production during puberty, excess sebum production can plug hair follicles facilitating the growth of bacteria, such as *Propionibacterium acnes*, which can trigger an inflammatory response resulting in the pimples and reddened skin of **acne**.

Sweat glands: eccrine vs. apocrine

Two types of sweat glands—**eccrine** and **apocrine**—develop as simple coiled tubular glands that invaginate from the epidermal surface into the dermis (**Figure 10.7a, d–h**). As the most common sweat gland in humans, **eccrine glands are present throughout the body** and can produce up to 10 liters of sweat per day for evaporative cooling and excretion.

Each coiled secretory tubule of eccrine glands is lined by a stratified cuboidal to pseudostratified epithelium that surrounds a relatively narrow lumen (**Figure 10.7d**). Within the epithelial lining are: (i) **stem cells**; (ii) basally positioned **myoepithelial cells** that help squeeze out secreted fluids (**Figure 10.7d**); (iii) short **secretory cells** that contribute a watery mixture of electrolytes; and (iv) taller **secretory cells** that overhang basal cells and release mainly glycoproteins. The production of eccrine sweat is normally triggered by elevated temperature, which activates **cholinergic neurons** innervating these glands. Alternatively, some sympathetic fibers regulating eccrine glands are **adrenergic** and can cause a nervous (flop sweat) particularly in the palms and forehead under anxiety-provoking conditions. During heat-induced perspiration, sweat produced by eccrine glands travels to the epidermal surface via a duct lined by a **stratified cuboidal epithelium** that normally resorbs much of the salt in sweat to conserve electrolyte loss. However, people with **cystic fibrosis** (**Case 2.2**) produce an overly salty sweat due to defective **cystic fibrosis transmembrane conductance regulators** (**CFTRs**).

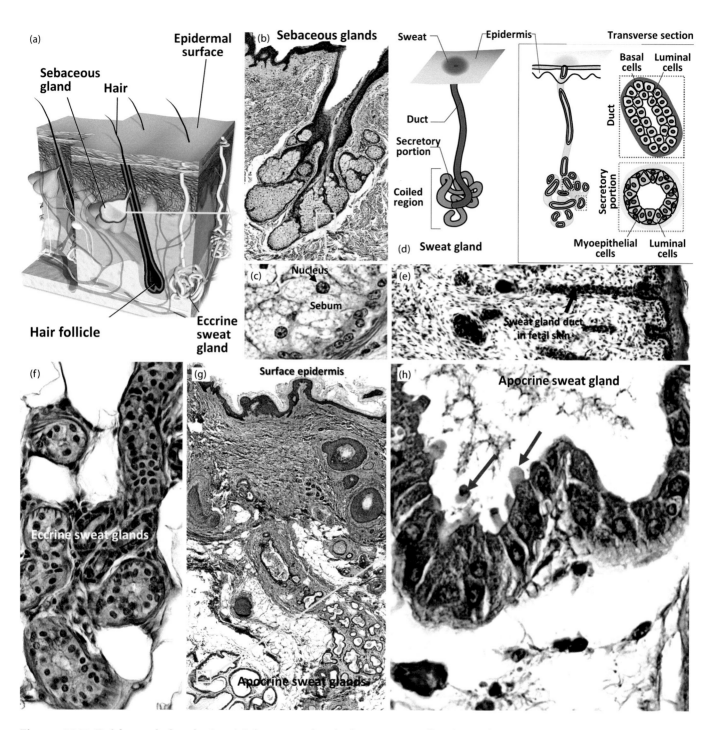

Figure 10.7 Epidermal glands. *(a–c) Sebaceous glands characteristically release their oily sebum into hair follicles via a holocrine mode of secretion, during which each glandular cell disintegrates during exocytosis. (b) 20X; (c) 280X. (d) Sweat glands are simple coiled tubular glands, whose unbranched stratified cuboidal duct connects the coiled secretory region with the epidermal surface. Myoepithelial cells surrounding secretory cells help squeeze sweat toward the epidermal apex. (e, f) Eccrine glands are widely distributed throughout the body where they undergo a merocrine mode of secretion. Their highly coiled secretory regions have small lumens (f) that exocytose a watery sweat for evaporative cooling and excretion. (e) 250X. (g, h) Conversely, apocrine sweat glands are restricted to a few sites in the body (e.g., axillae and groin) and have a comparatively wide lumen that delivers its more viscous secretions into hair follicles. Based on poorly fixed material, it was erroneously believed that apocrine glands normal lose their apical cytoplasm during exocytosis (h, arrows). In fact, these cells also undergo merocrine secretion. (g) 25X; (h) 600X. ([a]: From Blausen. com staff (2014) "Medical gallery of Blausen Medical 2014". Wiki J Med 1 (2): 10. doi:10.15347/wjm/2014.010 reproduced under CC0 and CC BY SA creative commons licenses; [d]: From Kurata, R et al. (2017) Three-dimensional cell shapes and*

Figure 10.7 (Continued) *arrangements in human sweat glands as revealed by whole-mount immunostaining. PLoS ONE 12(6): e0178709 https://doi.org/10.1371/journal.pone.0178709 reproduced under a CC BY 4.0 creative commons license.)*

Apocrine glands are so named because initial histological analyses suggested that their exocytosing cells lose substantial amounts of apical cytoplasm during an **apocrine mode** of secretion. However, current data obtained using optimally fixed specimens indicate that both apocrine and eccrine glands utilize **merocrine** secretion involving minimal loss of cytoplasm (**Figure 10.7g, h**). Nevertheless, apocrine glands differ from eccrine glands in key ways, including: (i) a **sparser distribution** that is restricted to such areas as the axillae, groin, areolae, and perianal regions; (ii) a **comparatively larger lumen** that typically empties its secretions into hair follicles; and (iii) a **single secretory cell type**, which only becomes fully functional at puberty. Post-adolescent apocrine glands are triggered by thermal stress or other adrenergic stimulation to produce a more viscous and proteinaceous sweat than the perspiration made by eccrine glands. Because of their restricted distribution, apocrine glands do not contribute substantially to evaporative cooling. Instead, apocrine secretions can emit important signals like conspecific-attracting **pheromones** or less enticing body odors, particularly when apocrine products are allowed to be fermented by bacteria.

10.6 DERMIS: GENERAL HISTOLOGY AND MECHANOSENSORY ORGANS

Dermis is generally classified as a **dense irregular connective tissue**, with finer fiber bundles tending to occur in the upper **papillary layer** than in the lower **reticular layer** of dermis (**Case 10.4 Benjamin Harrison's Dermis**) (**Figure 10.8a**). Ramifying throughout dermis are blood vessels, nerves, and adipocytes as well as stromal cells, such as macrophages, mast cells, leukocytes plus numerous **fibroblasts** that secrete dermal ECM fibers. Most of these fibers consist of type I collagen, but scattered type III reticular fibers and a few other minor collagen types are also present.

An additional dermal cell, called a **telocyte**, has been identified based on its distinctive molecular signature and its elongated cellular processes, called **telopodes** (**Figure 10.8b**). Such cells also occur in many other organs throughout the body but are particularly abundant in dermal compartments surrounding blood vessels and stem-cell-containing regions of hair follicles (**Figure 10.9c**), where telocytes not only help maintain normal homeostatic functions but also play roles in wound healing (**Case 10.5 Frida Kahlo's Skin Sores**) (Box 10.2 Cell Biology of Cutaneous Wound Healing).

Along with sensory neurons associated with either intra-epidermal Merkel cells or hair follicles, the dermis contains encapsulated mechanosensory organs referred to as **Meissner's corpuscles**, **Ruffini endings**, and **Pacinian corpuscles** (**Figure 10.8d**). Each of these mechanosensors possesses an LTMR type of afferent neuron that can respond to relatively weak mechanical stimulation. Differences in the precise functional properties of

---

CASE 10.4 BENJAMIN HARRISON'S DERMIS

23rd U.S. President
(1833–1901)

"I know my business. Benjamin Harrison had the crowd red-hot. I did not want him to freeze it out of them with his hand-shaking."

(An aide to B.H. responding to criticism he made the train leave too soon after Harrison had just finished an inspiring campaign speech)

The grandson of the ninth U.S. president William Harrison, Benjamin Harrison could move audiences with his rousing orations. However, one on one, Harrison was often uninspiring, particularly when shaking hands, which was like "holding a wilted petunia" according to one observer. Nevertheless, Harrison won the presidency in 1888 before

losing his re-election bid. Throughout his life, Harrison suffered from skin rashes and to deal with such dermatitis, frequently wore gloves, earning him the unflattering nickname of "Kid Gloves Harrison". In his youth, Harrison would bristle when compared to his famous grandfather, leading him once to exclaim: "I want it understood, I am the grandson of nobody".

However, in spite of such protests, Harrison, with his sensitive dermis, ended up following closely in his grandfather's footsteps not only by becoming president but also by dying of pneumonia, as William had done 60 years earlier.

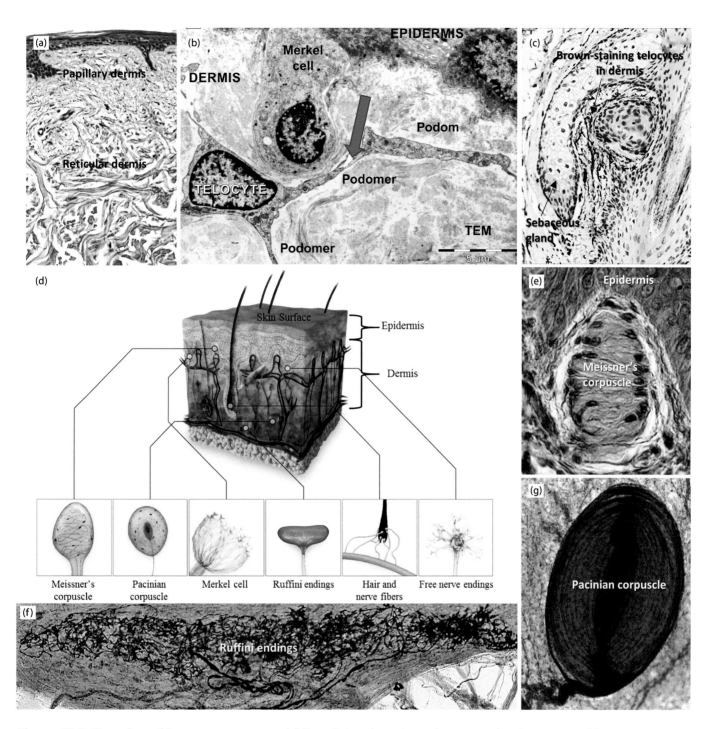

Figure 10.8 Dermis and its sensory organs. (*a*) *Dermis is a dense irregular connective tissue, comprising an upper papillary layer and a lower reticular layer. (a) 150X. (**b, c**) In addition to abundant fibroblasts and transitory cells, dermis contains telocytes with elongated telopode processes (b, blue arrow), which comprise both thin regions (podomers) and dilated termini (podoms). Such cells often occur near stem cell niches, such as bulge regions of hair follicles (c). (c) 175X. (**d–g**) Dermis also contains multiple types of sensory organs, which include from the more superficial to deeper dermal regions: Meissner's corpuscles (e), Ruffini endings (f), and Pacinian corpuscles (g). (e) 400X; (f) 75X; (g) 65X. ([b]: From Cretoiu, D et al. (2015) FIB-SEM tomography of human skin telocytes and their extracellular vesicles. J Cell Mol Med 19: 714–722 reproduced under a CC BY creative commons license; [c]: From Wang, L et al. (2020) Ultrastructural and immunohistochemical characteristics of telocytes in human scalp tissue. Sci Rep 10: 1693 https://doi.org/10.1038/s41598-020-58628-w reproduced under a CC BY4.0 creative commons license; [f]: Public domain image downloaded from https:// commons.wikimedia.org/wiki/File:Ruffini_Corpuscle_by_Angelo_Ruffini.jpg and reproduced under a CC BY SA license.)*

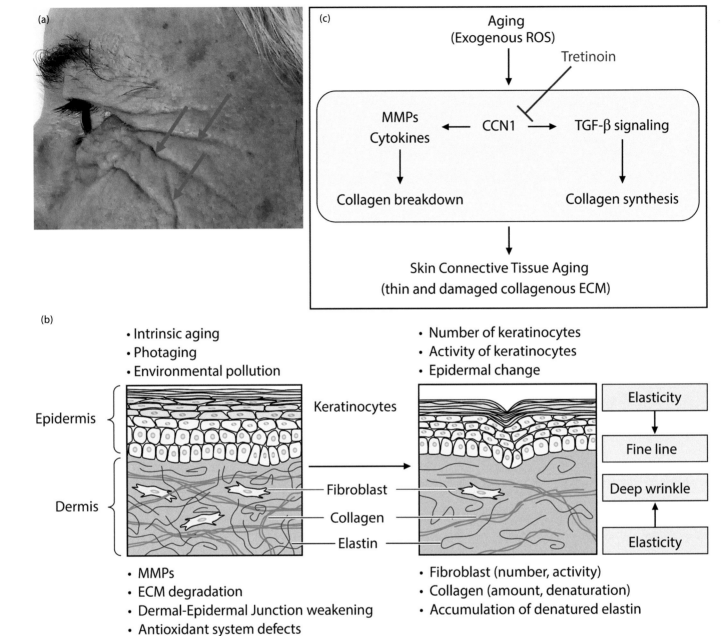

Figure 10.9 Cell biology of skin aging. (a, b) *As part of an intrinsic aging process, skin normally becomes less elastic, as the epidermis thins and develops deeper wrinkles (a, arrows) while dermal fibers decrease in number. (**c**) In addition, exogenous stressors, such as excessive sunlight exposure can increase reactive oxygen species (ROS) production in fibroblasts. Such ROS elevations may upregulate a key protein, called CCN-1, which promotes both a breakdown of existing collagen fibers and a decrease in new fiber synthesis. Anti-aging creams with a vitamin A derivative, called tretinoin, may retard the aging process by reducing CCN-1's deleterious effects. ([b]: Modified from Shin, K-O and Park, K-S (2019) Antiaging cosmeceuticals in Korea and open innovation in the era of the 4th industrial revolution: from research to business. Sustainability 11: doi:10.3390/su11030898 reproduced under a CC BY 4.0 creative commons license; [c]: Modified from Quan, T and Fisher, GJ (2015) Role of age-associated alterations of the dermal extracellular matrix microenvironment in human skin aging: a mini-review. Gerontology 61: 427–434 reproduced with publisher permission.)*

CASE 10.5 FRIDA KAHLO'S SKIN SORES

Artist
  (1907–1954)

"My painting carries with it the message of pain."

(F.K. on the impact of pain on her painting)

Essentially from the day celebrated artist Frida Kahlo was born in Mexico, pain dominated her life. Due to a congenital malformation of her neural tube, Kahlo suffered from spina bifida, which led to neural dysfunction plus bleeding sores in her right foot and leg. After contracting poliomyelitis at age 6, the effects of spina bifida were exacerbated, forcing Kahlo in later life to undergo several operations that failed to cure her condition. Instead, these surgeries generated trophic skin ulcerations and ultimately caused gangrene that required her

leg to be amputated. In addition, Kahlo's spine, pelvis, and right leg were severely injured in 1925, when the bus she was riding collided with a trolley car, necessitating her to be bed-ridden for months and never fully healing. While recuperating from the accident, Kahlo began painting self-portraits that gained critical acclaim. Via her portraits, Kahlo provided glimpses of the pain beneath her damaged skin until she eventually succumbed in 1954 to what was officially listed as a pulmonary embolism but interpreted by others as suicide by drug overdose.

BOX 10.2 CELL BIOLOGY OF CUTANEOUS WOUND HEALING

Human skin repairs wounds via multi-faceted processes that are summarized in **Figure a**. During the initial response, which starts immediately after wounding and usually lasts about 2 days, platelets quickly form a **plug** and **fibrin clot** to stem blood flow. Soon after bleeding is stopped, **neutrophils** plus monocytes that differentiate into **macrophages** enter the wound to kill and phagocytose pathogens. Over the next 2–10 days, new tissue is generated, starting with **re-epithelialization**, in which keratinocytes spread across the wound and eventually stitch together the cut edges of epidermis. Some of the spreading keratinocytes represent existing cells that were dislodged from the epidermis, but most of these repair cells arise *de novo* from mitoses of stem cells either within the stratum basale or recruited from other sites like hair follicle bulges.

While the epidermis is being rebuilt, an overlying **scab** of dried blood and encased pathogens develops above the wound, and underneath the scab, the dermis forms a **granulation tissue** that accumulates **fibroblasts**. The last stage of healing involves **tissue remodeling**, which begins 2–3 weeks post-injury and lasts a year or more, depending on wound severity. During this phase, the shed scab is replaced by a thickened epidermis, and the underlying granulation tissue is contracted by **myofibroblasts**. Eventually, apoptosis removes accumulated neutrophils, macrophages, and myofibroblasts, thereby restoring much of the pre-wound properties of the skin, although **scars** from deeper wounds can continue to display altered appearance and reduced flexibility.

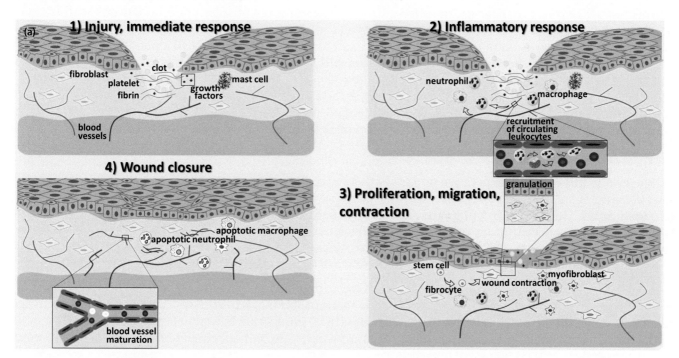

Box 10.2 (a) *Major stages in cutaneous wound healing. (From Carvalho, AR et al. (2018) Use of some Asteraceae plants for the treatment of wounds: from ethnopharmacological studies to scientific evidences. Front Pharmacol 9: doi: 10.3389/fphar.2018.00784 reproduced under a CC BY 4.0 creative commons license.)*

TABLE 10.1   FOUR MAIN MECHANORECEPTORS IN HUMAN SKIN

	Meissner corpuscle	Pacinian corpuscle	Merkel cell	Ruffini endings
Classification	(Rapid adapting) RA-I	RA-II	(Slow adapting) SA-I	SA-II
Adaptation rate	Fast	Fast	Slow	Slow
Location	Shallow	Deep	Shallow	Deep
Stimuli frequency (Hz)	10–200	70–1000	0.4–100	0.4–100
Density (units/cm^2)	140	20	70	10
Spatial resolution (mm)	3–4	10+	0.5	7+
Functions	Object slip, light touch, texture	High-frequency vibrations	Static forces with high resolution	Tension deep in the skin and fascia
Receptive field (RF)	Small and sharp, 3–5 mm	Very large and diffuse, >20 mm	Small and sharp, 2–3 mm	Large and diffuse, 10–15 mm

Source: From Park, M et al. (2018) Recent advances in tactile sensing technology. Micromachines 9: doi:10.3390/mi9070321 reproduced under a CC BY 4.0 creative commons license.

these tactile organs (Table 10.1) collectively allow dermal sensory organs to discriminate, for example, between the effects of an impinging breeze versus those of a crawling bug. Alternatively, sensory input can also aid in the gripping of objects so that the hand neither crushes nor drops what it holds.

Meissner's corpuscles are subcylindrical units that are prevalent in dermal papillae of touch-sensitive areas of the skin, such as fingers and lips. The sensory neuron in the corpuscle is surrounded by spiraling Schwann cell layers that give the appearance of a loosely wound skein of yarn (**Figure 10.8e**). Located deeper in the dermis than Meissner's corpuscles, spindle-shaped **Ruffini endings** possess afferent neurons with branched dendrites that can sense the stretching of skin (**Figure 10.8f**). **Pacinian corpuscles** possess several dozen layers of fibroblasts that wrap around a centrally located axon to give the appearance of a paddle-shaped onion-like structure (**Figure 10.8g**). Such sensory organs are typically found in the deep dermis to superficial hypodermis and are particularly abundant in fingertips, where they detect higher frequency vibrations than those tracked by Meissner's corpuscles. According to some accounts, an additional encapsulated sensory organ called a **Krause end bulb** occurs in such areas as genital skin, the tongue, and recto-anal region. However, whether this organ constitutes a distinct sensory organ or simply a subtype of Meissner's corpuscle remains controversial.

10.7   CELL BIOLOGY OF SKIN AGING

Skin ages not only due to degenerative processes that naturally occur during decades-long **intrinsic aging** but also because of superimposed stresses, such as excessive sunlight exposure that drive what is referred to as **extrinsic aging**. Compared to young skin, skin subjected to normal intrinsic aging tends to be more fragile and transparent, owing to its thinner epidermis and reduced dermal component that contains fewer fibers than in young dermis. Conversely, UV damage during extrinsic aging can markedly darken the epidermis as well as thicken it by increasing the stratum corneum to produce a leathery texture.

Older skin also contains more prominent **wrinkling**, owing in large part to a reduction and remodeling of its collagen fiber content (**Figure 10.9a, b**). This is because dermal fibroblasts in aged skin tend to secrete less collagen, often in response to increased **reactive oxygen species** (**ROS**) production brought about by intrinsic aging as well as UV-mediated photoaging. In addition, higher ROS levels can cause enhanced secretion of cytokines and **matrix metalloproteases** (**MMPs**) that serve to reorganize and cleave dermal ECM fibers (**Figure 10.9b**).

According to one model of skin aging, elevated ROS levels in fibroblasts of aged skin increases the production of a cysteine-rich protein, called **CCN-1**, which both upregulates existing collagen breakdown and reduces additional collagen production by downregulating pathways involving **transforming growth-beta** (**TGFβ**) (**Figure 10.9c**). Thus, anti-wrinkling creams containing vitamin A derivatives, such as all-trans retinoic acid (**tretinoin**) may help block CCN-1 activity to retard the aging process.

10.8   SOME SKIN DISORDERS: PSORIASIS AND SKIN CANCERS

The most common form of **psoriasis**, called psoriasis vulgaris, causes red and scaly **psoriatic plaques** that usually develop on skin of the elbows, knees, scalp, and back. Such reddening is due to enhanced vascularization and immune responses, whereas scales involve an accelerated transport of keratinocytes that can be sloughed in as little as 3–5 days, as opposed to the 1–2-month residence time in normal skin. This rapid cycling is in turn triggered by pro-inflammatory **cytokines** like **interleukin 17A**.

With ~1 in 5 adults developing **non-melanoma skin cancer** (**NMSC**) over their lifetimes, such cancers are the most common malignancies in the United States. The two most frequently diagnosed NMSCs, **basal cell carcinomas** (**BCCs**) and **squamous cell carcinomas** (**SCCs**), tend to originate in the stratum basale and stratum corneum, respectively. BCCs are less likely to spread beyond the epidermis than are SCCs, and following early diagnoses, these NMSCs can often be cured by surgery. Conversely, compared to NMSCs, **melanomas** arising from unregulated melanocyte mitoses are relatively rare, but also more aggressive, skin cancers that have an increased likelihood of metastasizing to other organs, thereby driving higher mortality rates (**Case 10.2**).

10.9   SUMMARY—SKIN

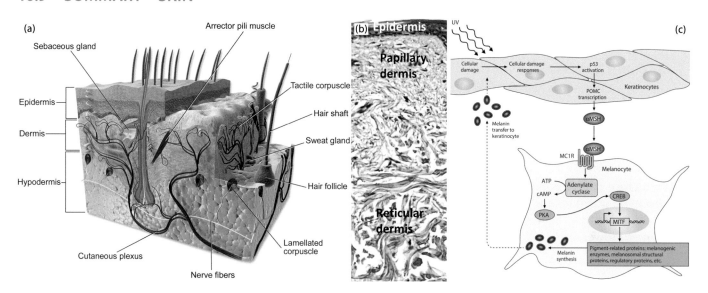

Pictorial Summary Figure Caption: (a)*: see figure 10.1a; **(b)**: see figure 10.8a; **(c)**: see box 10.1b.*

As the largest organ of the body, **skin** (=integument) **functions** in: (i) protection, (ii) thermoregulation, (iii) metabolism (e.g., excretion and vitamin D synthesis), (iv) sensation, and (v) signaling (odors, coloration, and hair distribution). Skin comprises ectodermally derived **epidermis** and mesodermally derived **dermis**, with **hypodermis** occurring below skin proper (**Figure a, b**). **Thick skin** of the palms and soles has a relatively thick epidermis, whereas **thin skin** found everywhere else has a thinner epidermis.

Epidermis has as its main cell type **keratinocytes** that produce **keratins** during a **keratinization** process. Epidermis also contains **melanocytes** (**melanin** pigment producers), **Merkel cells** (sensory), and **Langerhans cells** (antigen-presenting cells). The epidermis is a **stratified squamous epithelium** organized into layers (strata), which from base to apex are: **stratum basale**, **stratum spinosum**, **stratum granulosum**, **stratum lucidum** (in thick skin), and **stratum corneum** of dead cells.

The process of **keratinization** continually occurs as mitoses in the stratum basale add new keratinocytes to the suprabasal epidermis, and older keratinocytes are thus moved toward the epidermal apex. During apical movement, keratinization changes the structure and function of each keratinocyte by such processes as—(1) keratin filaments attaching to desmosomes, thereby giving the stratum spinosum cells a spiny appearance; (2) stratum granulosum cells forming large keratohyalin granules; and (3) granulosum cells destined for the stratum corneum becoming filled with fused filaments and keratohyalin granules to form a keratin mass. Keratinocytes in the stratum granulosum also: (i) line the inside of their plasmalemma with proteins like **loricrin** to form a peripheral **cell envelope** and (ii) secrete **lamellar bodies** to coat their outer membrane with waxy lipids. Before entering the stratum corneum, keratinocytes undergo apoptosis and eventually detach from the epidermal apex by dissolving corneodesmosomes via KLK enzymes. Most keratinocytes are sloughed ~1–2 months after entering suprabasal layers.

During **melanin formation**, melanocytes convert **tyrosine** via **tyrosinase**-requiring processes into **melanins**—the common brown eumelanin and rarer pinkish pheomelanin. Accordingly, the unpigmented skin of albinism can arise from the lack of tyrosinase activity. Melanin granules formed in melanocytes are moved into keratinocytes, often via **cytocrine transfer**, thereby protecting keratinocyte nuclei from the mutagenic effects of UV light. Dark- vs light-skinned people have approximately equal melanocyte numbers but differ in melanin granule amounts.

Hairs and **nails** are made from epidermal cells, which contain **hard keratins** and are not continuously shed like surface epidermal cells that are filled with soft keratins. Hair follicles have a **bulge region** with **stem cells** that drive hair formation and help with **wound healing.**

Sebaceous glands are typically associated with hair follicles and exocytose via **holocrine secretion** an oily **sebum** to protect hair and skin. **Sweat glands** of the body are mostly **eccrine glands** that make watery sweat for cooling. **Apocrine** sweat glands occur in restricted sites (e.g., armpits) and secrete a more viscous sweat that does not cool as efficiently as eccrine-produced sweat.

Dermis has an upper **papillary layer vs.** lower **reticular layer** with mainly type I collagen. **Arrector pili** muscles that are attached to hair follicles can cause "goose-bumps", which represent evolutionary relics of adaptive responses used by furry mammals.

Integumental sensory structures that sense pressure and/or vibration include: (1) **free nerve endings in epidermis**, which are high-threshold mechanoreceptors (**HTMRs**) sensing pressure, pain, and noxious cues; (2) other sensory units (e.g., intraepidermal **Merkel cell units** plus **Meissner's, Ruffini, Pacinian corpuscles** in the dermis), which have low-threshold mechanoreceptors (**LTMRs**).

Wrinkling from excessive UV exposure results in reduced collagen fiber production plus enhanced fiber breakdown. Wrinkles may be exacerbated by reactive oxygen species (**ROS**) upregulating **CCN-1** protein in dermal fibroblasts of aged skin. **Tanning** is a **two-step process** comprising an initial rapid darkening of existing melanin, followed after several days by an increase in melanin granules (**Figure c**).

Some skin disorders: psoriasis involves **interleukin 17A**-associated increases in cycling times so that cells are shed in as little as 3–5 days, thereby causing patches of red, scaly skin; **skin cancers** usually arise as non-melanoma skin cancers (NMSCs) and are typically treatable if caught early on, whereas **melanomas** are less common but can be more lethal.

SELF-STUDY QUESTIONS

1. What is the name of the superficial dermal layer that has a finer bundle of CT fibers than deeper in the dermis?
2. T/F: Lamellar bodies of keratinocytes are the same as keratohyalin granules.
3. What is the most common sweat gland throughout the human body?
4. Which of the following undergo stem cell divisions to aid cutaneous wound healing?
    A. Stratum basale
    B. Hair follicle bulge cells
    C. BOTH of the above
5. Which layer is characterized by desmosomes and keratin-containing tonofilaments that form prickly cells?
    A. Stratum basale
    B. Stratum spinosum
    C. Stratum granulosum
    D. Stratum lucidum
    E. NONE of the above

6. Where is thick skin found?
    A. Throughout the body
    B. Eyelids
    C. Palms of the hands and soles of the feet
    D. Scalp
    E. NONE of the above
7. Approximately how long do keratinocytes normally remain in the epidermis before being shed?
8. What is the major collagen type in dermis?
9. Which of the following is a type of high threshold mechanoreceptor (HTMR)?
    A. Intraepidermal free nerve endings
    B. Meissner's corpuscles
    C. Pacinian corpuscles
    D. Ruffini corpuscles
10. What is the name of smooth muscle bundles attached to hair follicles that can contract to form "goosebumps"?
11. Which of the following statements regarding tanning is correct?
    A. Most of sun-induced tanning is due to UVA rays
    B. Tanning reduces the risk of developing skin cancer
    C. Melanocytes increase in number within minutes after sun exposure
    D. ALL of the above
    E. NONE of the above
12. Which of the following statements regarding skin wrinkles is correct?
    A. Wrinkles are reduced by UV light exposure
    B. Reactive oxygen species increase skin wrinkles
    C. CCN-1 tends to block wrinkles
    D. ALL of the above
    E. NONE of the above
13. Which epidermal layer is found only in thick skin?

"EXTRA CREDIT" For each term on the left, provide the BEST MATCH on the right (answers A-G can be used more than once)

14. Loricrin _____
15. Interleukin 17A _____
16. KLKs _____
17. Bulge region _____
18. Lack of tyrosinase _____

    A. Tanning
    B. Wrinkle formation
    C. Stem cells
    D. Psoriasis
    E. Cell envelope
    F. Albinism
    G. Detachment of corneum cells

19. If dark-skinned people tend to develop skin cancer less frequently than fair-skinned people, why are not all people dark-skinned?
20. Describe keratinization.

ANSWERS

1) papillary; 2) F; 3) eccrine; 4) C; 5) B; 6) C; 7) 1-2 months; 8) Type I; 9) A; 10) arrector pili; 11) A; 12) B; 13) stratum lucidum; 14) E; 15) D; 16) G; 17) C; 18) F; 19) Fair skin optimizes vitamin D production in regions with low sunlight; also, if dark skin's adaptive advantage was solely to reduce skin cancer, there would be little selective pressure for dark skin to evolve widely, given that most skin cancers arise in older, post-reproductive individuals. 20) A full answer would include and define most or all the following terms: strata basale, spinosum, granulosum, lucidum, corneum; stem cells, keratins, tonofilaments, tonofibrils, keratohyalin granules, apoptosis, enucleation, KLKs, squames, desquamation; normal cycling vs. psoriasis

ORGAN HISTOLOGY— DIGESTIVE, RESPIRATORY, AND ENDOCRINE SYSTEMS

PART THREE

Digestive System I: Oral Cavity

11.1 INTRODUCTION TO THE OVERALL STRUCTURE AND FUNCTION OF THE DIGESTIVE SYSTEM

The **digestive system** comprises three **accessory organs**—the **liver, gallbladder,** and **pancreas**—plus the main **gastrointestinal (GI) tract** (=alimentary canal, digestive tract, or gut) that extends from the mouth to the anus (**Figure 11.1a**). The **upper GI tract** consists of the **oral cavity** (=mouth) (**Figure 11.1b, c**), pharynx, esophagus, stomach, and the initial portion of the small intestine (=duodenum), whereas the **lower GI tract** contains the post-duodenal small intestine, large intestine, and anal canal. Collectively, the digestive system functions in (1) food **ingestion**; (2) mechanical and chemical **digestion**; (3) nutrient **absorption**; (4) **waste removal** via solid materials in feces and solubilized byproducts in bile; (5) production of **hormones** and other bioactive molecules; and (6) **protection** against invasive pathogens via non-encapsulated lymphoid tissues in the GI lining. This chapter describes oral cavity structures, whereas the rest of the digestive tract and its associated organs are covered in the next two chapters.

DOI: 10.1201/9780429353307-14

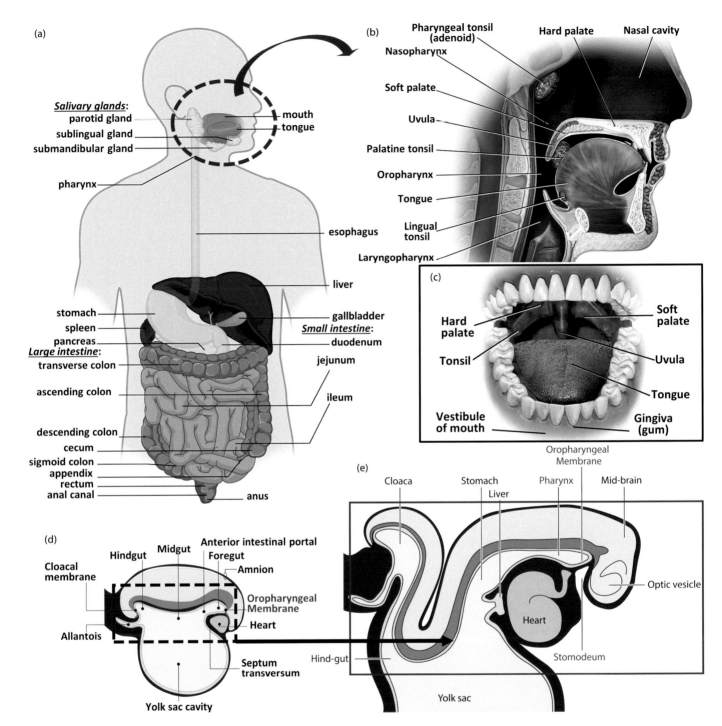

Figure 11.1 Digestive system, oral cavity, and development of the mouth. (a–c) *Figure (a) summarizes the major components of the digestive system, whereas figures (b) and (c) show key features of the oral cavity (=mouth).* **(d, e)** *Two successive developmental stages diagram the formation of a stomodeal invagination that is separated from the pharynx by the oropharyngeal membrane. After the stomodeum and pharynx fuse, the oral cavity begins to differentiate at the anterior end of the fused chambers. ([a]: From OpenStax College (2013) Anatomy and Physiology. OpenStax. http://cnx .org/content/col11496/latest, reproduced under CC BY 4.0 creative commons license. [b, c]: From Blausen.com staff (2014) "Medical gallery of Blausen Medical 2014". Wiki J Med 1 (2): 10, doi: 10.15347/wjm/2014.01010, reproduced under CC0 and CC-SA creative commons licenses; [d]: From DeSesso, JM (2016) Vascular ontogeny within selected thoracoabdominal organs and the limbs. ReprodToxicol. http://dx.doi.org/10.1016/j.reprotox.2016.10.007, reproduced with publisher permission.)*

11.2   THE ORAL CAVITY: GENERAL DEVELOPMENT, ANATOMY, AND HISTOLOGICAL ORGANIZATION OF ITS MUCOSA

Development of the human oral cavity begins ~4 weeks post-fertilization with an ectodermal invagination, called the **stomodeum**, extending toward endodermal **pharyngeal tissue** of the developing foregut (**Figure 11.1d, e**). After these two primordia break through an intervening **oropharyngeal membrane** and fuse, the prospective mouth that develops anterior to the fusion site begins to form an **oral mucosa** (**Case 11.1 Ada Lovelace's Oral Mucosa**). Initially, the inner lining of the mucosa is a simple squamous epithelium that subsequently becomes a stratified squamous epithelium by ~10 weeks (**Figure 11.2a, b**). All of the oral cavity components have begun to develop by 2 months (**Figure 11.2a–e**), and by 6 months, the oral mucosa generates a discrete subepithelial connective compartment, called the **lamina propria**. Eventually, the mucosal lining epithelium diversifies into keratinized vs. non-keratinized regions that are maintained throughout life.

The general features of the adult oral cavity are illustrated in **Figure 11.1b, c**, with the posterior-most components (lingual root, tonsils, and posterior soft palate) sometimes being assigned in other classification schemes to the pharynx. Except for teeth that have erupted through the mucosa, oral structures are covered by three main types of stratified squamous epithelia: (i) a keratinizing kind localized to the tough **masticatory mucosae** of the hard palate and teeth-anchoring gums (=gingivae); (ii) widely distributed non-keratinizing forms in **flexible mucosae** of the lips, cheeks, soft palate, and most of the rest of the mouth; and (iii) a mix of keratinizing and non-keratinizing types in the **specialized mucosa** covering the dorsal surface of the tongue.

11.3   FUNCTIONAL HISTOLOGY OF LIPS

Lips not only seal the mouth during swallowing and chewing but also play important roles in forming sounds during phonation and in contributing to informative facial expressions such as smiles. In addition, via their shape, coloring, and sensory capabilities, lips function in sexual attraction and intimacy. Accordingly, the upper lip's profile is often referred to as **cupid's bow** (**Figure 11.3a**).

The outer surface of each lip is covered by a **highly keratinized stratified squamous epithelium** with hair follicles, glands, and other features that resemble those of cutaneous epidermis in contiguous regions of the face (**Figure 11.3b, c**). In addition, a central depression in the upper lip, called the philtrum, extends from the nasal septum to the median dip in cupid's bow, thereby providing a reserve of covering epithelium for muscle-mediated stretching of the lips. As opposed to the outer lip surface, a glabrous **non-keratinized stratified squamous epithelium** lines the inside of each lip and is penetrated by ducts of lubricating **labial salivary glands** that are often associated with lymphocyte aggregates (**Figure 11.3d**).

Connecting the outer and inner lip mucosae is a well-vascularized transitional region, called the **vermilion zone**, which is where lipstick is applied (**Figure 11.3a–c**). Epithelial keratinization in this zone is either reduced or

---

CASE 11.1   ADA LOVELACE'S ORAL MUCOSA

Mathematician/Pioneering Computer Scientist (1815–1852)

"The Analytical Engine weaves algebraical patterns just as the Jacquard-loom weaves flowers and leaves."

(A.L.'s visionary view of the potential held by the "Analytical Engine", a proposed prototype for a computing machine).

Ada Lovelace was a mathematical prodigy, who was mentored by Charles Babbage, the inventor of a computing machine, called the Analytical Engine. Babbage was so impressed with Lovelace's skills that he asked her to add supplementary notes to an article describing his Analytical Engine prototype. Lovelace's contribution included various groundbreaking insights including a way of calculating the sequence of

Bernoulli numbers in what is considered by some to be the first computer program. While growing up, Lovelace contracted a debilitating case of measles, whose widespread rash is often preceded by bumps on the mucosal lining of the mouth. Several decades before Dr. H. Koplik formally described these oral bumps, which are now referred to as Koplik spots, Lovelace's promising career was cut short, when the pioneering mathematician died at age 36 due to uterine cancer.

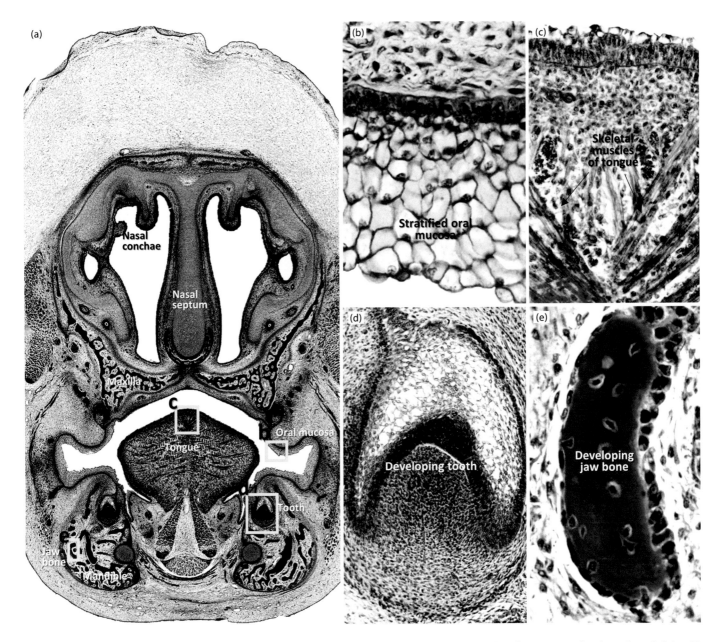

Figure 11.2 Fetal oral cavity. *(a–e) Key components of the oral cavity are shown developing in a fetal pig head. (a) 15X; (b) 230X; (c) 170X; (d) 120X; (e) 290X.*

lacking, as some sites exhibit **parakeratinization** with nucleated apical cells that contain reduced keratin levels. Because of its diminished keratinization and exposed positioning, the vermilion zone is prone to chapping.

11.4   DEVELOPMENT AND ANATOMY OF THE TONGUE: MUCOSAL PAPILLAE OVERLYING A LINGUAL CORE CONTAINING SKELETAL MUSCLES, LINGUAL GLANDS, AND NERVES

By 5 weeks, the primordium of the **tongue** (**Case 11.2 Gerald Ford's Tongue**) begins to develop from progenitor cells arising within and between embryonic pharyngeal arches. The paddle-shaped tongue (**Figure 11.3e, f**) that subsequently forms from such anlagen consists of: (i) a posterior **root** region situated closest to the pharynx; (ii) a **body** portion comprising the anterior two-thirds of the tongue; and (iii) a **terminal sulcus** that forms the narrow V-shaped border between the root and body (**Figure 11.4a**). A lingual **tonsil** involved in anti-pathogenic defenses

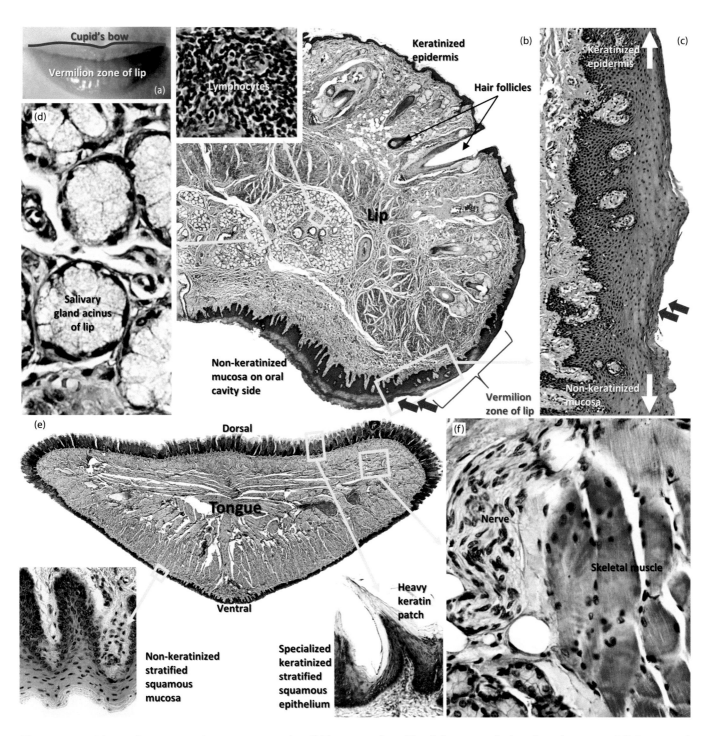

Figure 11.3 Lip and tongue microanatomy. (*a–d*) *The curved profile of the upper lip is referred to as cupid's bow, and the vermilion zone of lips where lipstick is applied represents a transitional region between highly keratinized skin and non-keratinized oral mucosa. In addition to labial salivary glands (d), the lip often has lymphocyte aggregates (b, inset, 250X). (b) 15X; (c) 80X; (d) 300X. (**e, f**) The core of the tongue is filled with crisscrossing skeletal muscle fibers and nerves allowing voluntary control over tongue movements for mastication and phonation. The ventral tongue surface is covered by a non-keratinizing mucosa, whereas the dorsal surface has a specialized stratified squamous epithelium, which in animals like rodents, such as shown here, can be heavily keratinized. (e) 15X; (e, left inset) 175X; (e, right inset) 100X; (f) 200X.*

CASE 11.2 GERALD FORD'S TONGUE

38th U.S. President
  (1913–2006)

"My fellow Americans, our long national nightmare is over."

(From G.F.'s inaugural address signaling the close of the Watergate scandal)

Gerald Ford was the only U.S. president not to receive any electoral votes, having been selected to replace Vice President Spiro Agnew and soon thereafter becoming president when Richard Nixon resigned due to the Watergate scandal. While campaigning to win another term, Ford committed a famous slip-of-the-tongue in stating that Eastern Europe was not under Soviet domination—a faux pax which, along with a deteriorating economy, contributed to his defeat in the 1976 presidential race. During his short stint in the White House,

Ford had no notable health problems and was able to maintain an athletic vigor that dated back to his college years as an All-American football lineman, who turned down contracts to play professionally. However, long after leaving Washington, Ford suffered a stroke that caused slurred speech, and late in life, he also underwent surgery for an infected tongue. Thereafter, Ford's health steadily declined, and in 2006, the appointed president with multiple tongue-related problems died at the age of 93.

develops at the base of the lingual root and is underlain by a mucosa that lacks substantial keratinization. Conversely, the dorsal mucosa of the tongue forms a mixture of keratinized and non-keratinized regions containing elevated clusters, called **lingual papillae** (**Figure 11.4a**). Such mucosal protrusions provide a roughened surface to facilitate mastication and can also form **taste buds** that mediate the sense of taste (=gustation). Four types of lingual papillae are present—**filiform** papillae, which lack taste buds, plus three taste-bud-containing gustatory papillae, termed **circumvallate**, **foliate**, and **fungiform** papillae (**Figure 11.4a**).

As the smallest and most abundant type of papilla, **filiform papillae** are densely packed throughout the dorsal mucosa. Such papillae are covered by a keratinized stratified squamous epithelium that extends an apical cluster of thread-like keratinized cells, called **fila**. Filiform papillae of humans are relatively poorly keratinized, and in some sites, the covering epithelium may be parakeratinized (**Figure 11.4a**). Conversely, in animals such as rodents and cats, filiform papillae are highly keratinized to form large plaques that confer a sandpaper-like texture for use in processes like grooming (**Figure 11.3e**). Given their lack of taste buds, human filiform papillae do not function in gustation but instead increase the tongue's gripping ability, while also playing roles in pressure reception.

Circumvallate papillae are the largest and least common papillae. Approximately 10 of these papillae form a V-shaped array at the caudal end of the lingual body, directly anterior to the terminal sulcus (**Figure 11.4a**). Each circumvallate papilla is separated from a surrounding collar of elevated mucosa by an annular, saliva-containing trench, into which ~250 basally positioned taste buds open. Summed across all circumvallate papillae, such taste buds equal about half of the ~4,600 total in a typical adult tongue, with foliate and fungiform papillae each containing roughly another quarter of all lingual taste buds.

Foliate papillae occur on a series of ~10–20 ridges (=folia) that run along the left and right posterolateral regions of the lingual body near the molars (**Figure 11.4a**). As in circumvallate papillae, taste buds are restricted to the basal half of each folium and are oriented toward the interfolial groove, which accumulates saliva produced by subjacent lingual glands.

Fungiform papillae comprise mushroom-shaped elevations scattered throughout the dorsal lingual body (**Figure 11.4a**). Unlike the other two gustatory papillae, each fungiform papilla possesses only a few apically situated taste buds and often appears as a red spot on the tongue, owing to its well-vascularized core.

Between the dorsal and ventral mucosae, most of the tongue is filled with crisscrossing bundles of **skeletal muscle** that allow flexible tongue movements (**Figure 11.3f**), and interspersed among these muscles are lingual salivary glands that supplement the saliva produced by other oral glands. In addition, the lingual core also contains nerve fibers, including efferents controlling muscle and gland functioning plus afferents that collect sensory input from the tongue (**Figure 11.3f**).

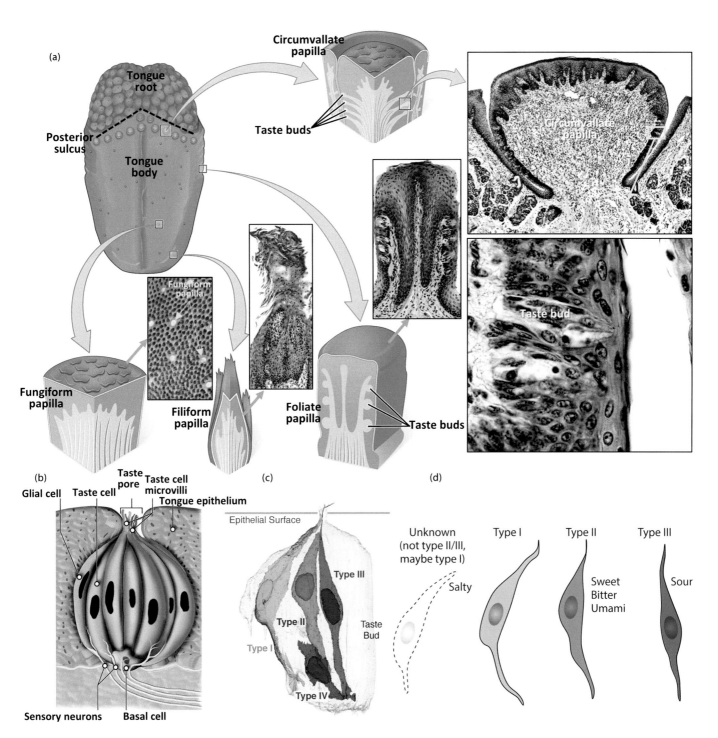

Figure 11.4 Lingual papillae and taste buds. (a) *Dorsally, the tongue contains four kinds of papillae. The most common are filiform, which lack taste buds and instead aid in mastication. The other three kinds—circumvallate, foliate, and fungiform—are gustatory papillae, which contain taste buds; (fungiform inset) 4X; (filiform inset) 50X; (foliate inset) 40X; (circumvallate, upper inset) 25X; (circumvallate, lower inset) 400X.* **(b, c)** *Each taste bud is an ovoid collection of cells with an apical pore that allows tastant molecules to reach microvilli of type I, II, and III sensory cells. Shortened type IV at the base of the bud serve as stem cells, and underlying afferent neurons transmit action potentials to the brain for taste perception.* **(d)** *Type I cells act as glial supporting cells and may also perceive certain types of salty tastes. Type II cells sense sweet, bitter, and umami tastes, whereas type III cells react to sour tastes. ([a]: Line drawings from OpenStax College (2013) Anatomy and Physiology. OpenStax. http://cnx.org/content/col11496/latest, reproduced under CC BY 4.0 creative commons license; [a]: inset of fungiform papillae from Hsu, J-C et al. (2018) Unilateral nasal obstruction induces degeneration of fungiform and circumvallate papillae in rats. J Formosan Med Assoc 117: 220–226, reproduced*

11.5   FUNCTIONAL HISTOLOGY OF TASTE BUDS IN GUSTATORY PAPILLAE THAT CAN SENSE SALTY, BITTER, SWEET, UMAMI, AND SOUR TASTES

Although scattered **taste buds** and solitary taste-sensing cells occur in the palate, pharynx, esophagus, and respiratory tract, taste buds are mainly localized to lingual papillae, where they mediate the perception of five basic tastes—**salty**, **bitter**, **sweet**, **umami**, and **sour** (Box 11.1 Umami and the Cell Biology of Taste). Lingual taste buds promote the intake of desirable foods rich in salty, sweet, and umami properties that signal high electrolyte, carbohydrate, and protein content, respectively. Conversely, taste buds can help avoid sour or bitter tastes that may be associated with putrid or noxious substances. For these functions, each lingual taste bud is typically able to detect all five basic tastes.

Taste buds start to differentiate in the tongue by 8 weeks of gestation and by 14 weeks acquire an apical **pore** indicating the onset of functional maturation. When fully developed, each adult taste bud is filled with ~50–100 elongated cells plus a few basally restricted **type IV** cells that represent stem cells (**Figure 11.4b, c**). Collectively, these components form an ~50 μm-long ovoid unit, whose apical pore leads into a shallow depression where elongated taste bud cells extend receptor-bearing microvilli for sensing **tastant** molecules in saliva.

Elongated taste bud cells were originally classified into dark vs. light types corresponding to supporting vs. receptor cells, respectively. However, further molecular and physiological studies have revealed a more complicated mode of gustation involving three types of elongated cells—I, II, and III (**Figure 11.4c, d**) (Box 11.1). Type I dark cells are glial cells that support type II and III cells, while also perhaps sensing some salty tastes. Type II light cells comprise three subpopulations, each expressing a single receptor type that can sense bitter, sweet, or umami tastes. Conversely, type III cells of intermediate density sense sour tastes and are the only receptor cells that form traditional synapses with gustatory afferent neurons. Moreover, type III cells can depolarize not only after binding sour tastants but also in response to **ATP** that tastant-stimulated type II cells secrete, thereby relaying bitter-, sweet-, or umami-triggered output from type II cells (Box 11.1).

In addition to taste-bud-mediated gustation, sensory neurons in the dorsal lingual mucosa can express **transient receptor potential (TRP) channels** to perceive **cool** vs. **hot** tastes associated with such ingested foods as mints vs. chiles. For example, lingual TRP vanilloid 1 (TRPV1) channels bind to **capsaicin**, the active ingredient of spicy peppers and are thereby able to convey the sensation of heat, even though no thermal energy is transferred.

11.6   TEETH: STAGES IN DEVELOPMENT AND THE MINERALIZATION OF ENAMEL AND DENTIN LAYERS THAT SURROUND THE CENTRAL PULP

Humans produce two sets of teeth—**primary** (=deciduous, baby) **teeth** vs. **secondary** (=permanent, adult) **teeth** (**Figure 11.5a, b**) (Box 11.2 Repairing and Regrowing Teeth in Humans). Based on differing morphologies exhibited by the **crown** of the tooth that occurs above the gumline as well as the variations that occur in the numbers and shapes of dental **roots** below the gumline, teeth can be classified into four main types: **incisors**, **canines**, **premolars**, or **molars**. Adults produce all four types, whereas babies form only incisors, canines, and molars (**Figure 11.5a–f**).

Deciduous tooth formation begins during the 6th week of gestation, with the lower **central incisors** typically being the first teeth to develop. A midline thickening of oral ectoderm in the future lower and upper jaw bones (**mandible** vs. **maxilla**, respectively) signals the onset of tooth formation, and at discrete sites along this thickening, each tooth primordium progresses through a morphogenetic sequence that can be broadly divided into the **bud**, **cap**, **bell**, and **maturation** stages (**Figure 11.6a**). During the **bud stage**, oral ectodermal cells and

BOX 11.1 UMAMI AND THE CELL BIOLOGY OF TASTE

As the last of the five basic tastes to be recognized, **umami** was discovered in 1908 by Japanese chemistry professor Kikunae Ikeda (**Figure a**), who determined that the distinctive flavor of his wife's seaweed soup was due to **monosodium glutamate (MSG)** (**Figure b**). Ikeda called such glutamate-mediated flavoring "**umami**", which is variously translated into English as savory, meaty, pleasant, or delicious. Although Ikeda was able to help launch what is now a multibillion-dollar MSG industry, acceptance of umami as a bone fide taste of its own required decades of further research until: (i) taste bud receptors composed of **T1R1/T1R3** heterodimers (Taste receptor type 1, members 1 and 3) (**Figure b**) were shown to be specifically activated by glutamate, and (ii) nerve fibers in the tongue were discovered that routinely transmit action potentials in response

to MSG but only rarely respond following sodium stimulation. Today, umami tastants are viewed as key indicators of protein content in various food items and are believed to trigger sensory output typically by binding to umami receptors on type II cells in taste buds. According to one model, such binding activates a G-protein-mediated cascade within the cell that leads to Ca^{2+} release and the activation of TRPM5 (transient receptor potential cation channel subfamily M member 5) on type II plasma membranes, thereby depolarizing these cells. Following depolarization, type II cells may secrete ATP to bind ATP receptors on type III cells. This in turn can trigger neurotransmitter release to activate post-synaptic afferent fibers whose action potentials are then transmitted to gustatory regions of the brain (**Figure c**).

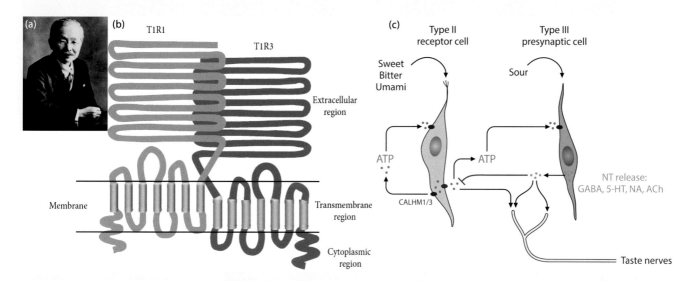

Box 11.1 (a) *Prof. K Ikeda (1864–1936) discovered umami taste by analyzing the chemistry of seaweed soup.* **(b)** *Umami was subsequently shown to be sensed by its own particular receptor on type II taste bud cells.* **(c)** *Scheme of how umami taste can be relayed to the brain via type III receptors. ([a]: From https://commons.wikimedia. org/wiki/File:Kikunae_Ikeda.jpg; public domain, unknown author; [b]: From Kurihara, K (2015) Umami the fifth basic taste: history of studies on receptor mechanisms and role as a food flavor. BioMed Res Int 2015, http//dx.doi. org/10.1155/2015/189402, reproduced under a CC BY 3.0 creative commons license; [c]: Modified from Von Molitor, E et al. (2020) Sensing senses: optical biosensors to study gustation. Sensors 20, 1811, doi: 10.3390/s20071811, reproduced under a CC BY 4.0 creative commons license.)*

neural-crest-derived mesenchyme assemble a clustered **placode** with a centrally located organizer, called the **primary enamel knot**, in which morphogenetic regulators like Wnt, β-catenin, and **fibroblast growth factor** are concentrated (**Figure 11.6a, b**). Subsequently, the placode and nearby cells sink into jaw mesenchyme to form an ovoid **bud** that remains connected to the oral cavity via a stalk-like **dental lamina** (**Figure 11.6c**). In the **cap stage**, the enamel-knot-containing primordium folds back toward the oral cavity to overlie a **dental papilla** and underlie a network of star-shaped cells, called the **stellate reticulum**, which secretes glycosaminoglycans to cushion the tooth (**Figure 11.6c, d**). Further morphogenesis of the folded cap generates the **bell stage** that comprises a bilayered **enamel organ** with four main regions: (1) an **outer enamel epithelium** of ectodermal cells that occur nearest the oral cavity; (2) an **inner enamel epithelium** of differentiating **ameloblast** cells that cover the dental papilla; (3) recurved **cervical loops** at the enamel organ periphery that connect the outer and inner enamel epithelia; and (4) the intervening **stellate reticulum** between the inner and outer epithelial layers (**Figure 11.6d**). Beneath the inner enamel epithelium, most cells within the dental papilla go on to form the tooth core (=**dental**

BOX 11.2 REPAIRING AND REGROWING TEETH IN HUMANS

As opposed to the polyphyodontic pattern of tooth formation in lower vertebrates like sharks that can continually produce teeth throughout their lives, the diphyodontic dentition mode of humans generates only two sets of teeth. Thus, when adult teeth are badly damaged or lost, no replacements are formed, and instead, artificial surrogates are supplied as overlying caps, removable dentures, or permanently anchored implants. However, various treatments are being tested that may allow new versions of healthy living teeth to be generated. These therapies include: (1) treatments that promote **tooth re-mineralization**, and (2) **stem-cell-mediated growth** of entire new teeth:

1. *Re-mineralization therapies:* As the most widely used method to re-build the apatite mineral phase of adult teeth, topical applications of fluoride can help mend minor cracks and thinning of adult teeth. However, such treatments are unable to reverse substantial mineral loss and may also present health risks. Thus, novel re-mineralization therapies are currently being tested. For example, stimulation with a low-power pulsed laser (LPL) causes dental pulp stem cells to upregulate key mediators of enamel and dentin production that are

linked to increased re-mineralization. In addition, dentin production in damaged murine teeth can be increased by **glycogen synthase kinase 3 (GSK3)** antagonists that promote the Wnt/β-catenin pathway mediating tooth mineralization. Similarly, use of a small peptide fragment of amelogenin on extracted adult human teeth increases the amount of enamel rod production and the overall hardness of treated teeth compared to either untreated controls or fluoride administration.

2. *Growing new teeth with stem cells:* Over a decade ago, researchers were able to form new teeth in mice by culturing autologous non-dental stem cells in renal capsules before transplanting the resultant tooth germ in the mouth. Such results indicated "*...teeth could be produced 'to order' by the use of autologous adult non-dental cells to create tooth primordia in vitro...for replacement in humans*" (Ohazama, A et al. (2004) Stem-cell-based tissue engineering of murine teeth. J Dent Res 83: 518–522). Figure (a) outlines some of the basic strategies that are being tested with the aim of making human tooth replacement a reality in the near future.

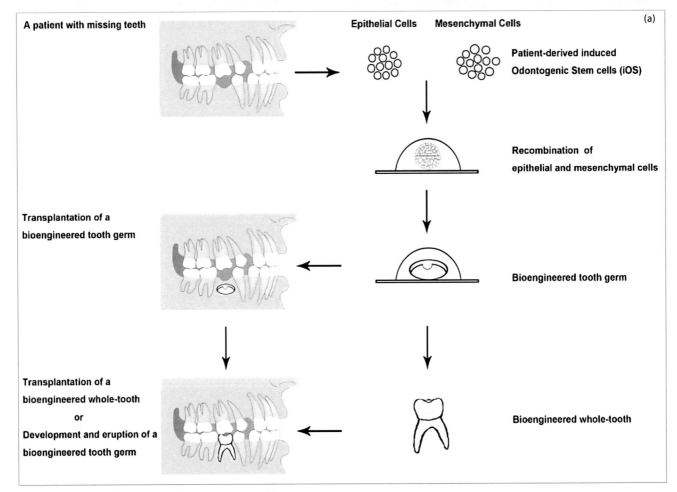

Box 11.2 (a) *Diagram of how one day stem cell culturing may allow adult human teeth to be regrown. (From: Volpini et al. (2018) Tooth repair and regeneration. Current Oral Health Rep 5: 295–303 published under a CC BY 4.0 creative commons license.)*

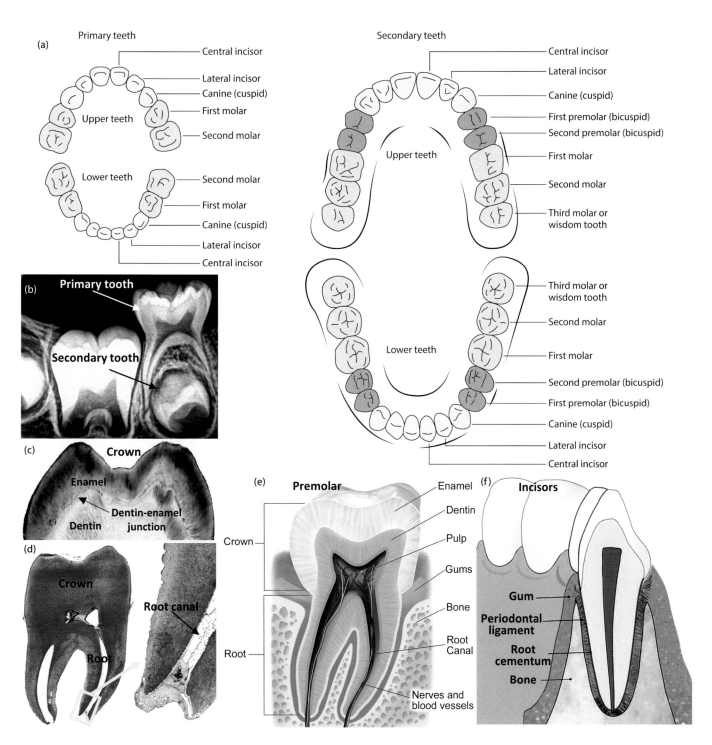

Figure 11.5 Introduction to teeth. (a, b) *Humans form two sets of teeth: (1) primary (=baby, deciduous), and (2) secondary (=adult, permanent). (**b**) An X-ray shows a secondary tooth underlying a primary tooth before it is shed; (**c, d**) ground sections of whole teeth allow the apatite mineral phases of enamel and dentin to be observed (c), whereas decalcified sections, such as shown in (d), remove the mineral phase while retaining the organic matrix and cells of teeth. (**e, f**) As shown for premolars vs. incisors, the tooth crown occurs above the gumline and contains an enamel outer coat overlying dentin, whereas in the sub-gumline root, cementum covers dentin and is anchored to bone by a periodontal ligament. Nerve and blood vessels enter the central pulp of teeth via one or more root canals. ([a]:From recolored version of a line drawing from OpenStax College (2013) Anatomy and Physiology. OpenStax. http://cnx.org/content/col11496/latest, reproduced under CC BY 4.0 creative commons license; [b]: From Shah, N, reproduced under a CC SA creative commons license, and downloaded from https://commons.wikimedia.org/wiki/*

Figure 11.5 (Continued) *File:Intraoral_Periapical_Radiograph_(IOPA)_showing_Deciduous(Milky_or_Primary)_ Tooth_75_and_developing_crown_of_Permanent_or_Secondary_Teeth_35,_36_and_37.jpg; [e]: From Blausen.com staff (2014) Medical gallery of Blausen medical 2014. Wiki J Med 1(2):10, doi:10.15347/wjm/2014.01010, reproduced under CC0 and CC SA creative commons licenses; [f]: From Muñoz-Carrillo, JL et al. (2019) doi: 10.5772/intechopen.86548, reproduced under a CC-BY 3.0 creative commons license.)*

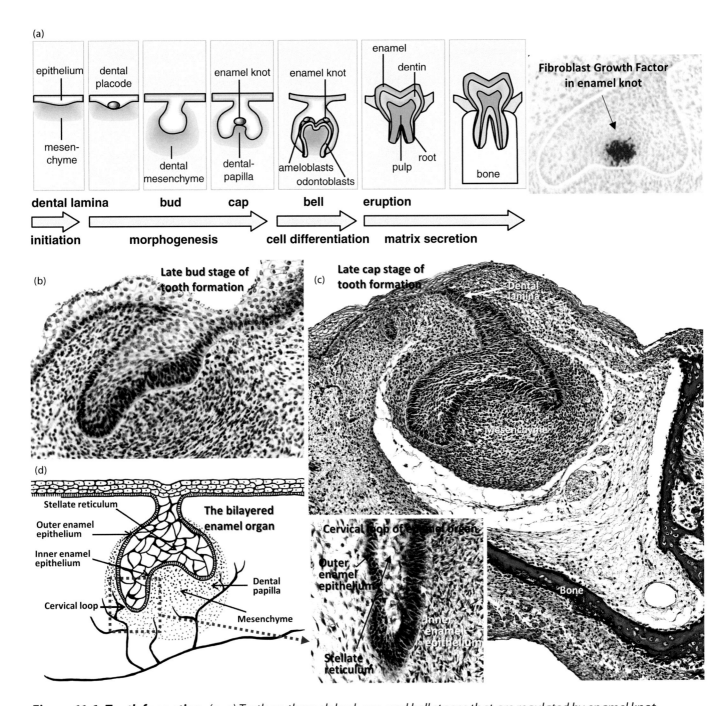

Figure 11.6 Tooth formation. *(**a–c**) Teeth go through bud, cap, and bell stages that are regulated by enamel knot morphogenetic centers. (a) 75X; (b) 150X; (c) 75X; (c, inset) 175X. (**d**) During the late cap stage, a bilayered enamel organ forms with odontoblasts subjacent to the inner enamel epithelium laying down dentin, and ameloblasts in the inner enamel epithelium depositing enamel. ([a]: From Thesleff, I (2014) Current understanding of the process of tooth formation: transfer from the laboratory to the clinic. Austral Dental J 59: 48–54, reproduced with publisher permission; [d]: From AL Shami, A et al. (2019) Tooth morphology overview. doi: 10.5772/intechopen.87153, reproduced under a CC BY 3.0 creative commons license.)*

pulp), whereas neural-crest-derived **odontoblasts** assemble a layer directly beneath the inner enamel epithelium. Such odontoblasts eventually produce a mineralized tooth layer, called **dentin**, whereas ameloblasts in the inner enamel epithelium form the outer mineralized **enamel** layer that encases dentin. As morphogenesis proceeds, the dental lamina of the tooth primordium disconnects from the oral cavity, and the tooth differentiates further while encased within the developing jaw. In addition, the primary enamel knot apoptoses and is replaced by one or more **secondary enamel knots** that regulate morphological traits, such as the number of major surface protrusions (=**cusps**) that a tooth will form.

Starting at around 14–16 weeks of gestation and continuing through the **maturation stage** of tooth development, late bell-stage primordia initiate a mineralization process within their dentin layer by depositing **apatite** crystals of calcium phosphate. When fully mineralized, dentin resembles bone in consisting of ~60% apatite and ~40% organic matrix and water. As a prelude to such mineralization, odontoblasts secrete a non-mineralized **pre-dentin** of organic matrix comprising mainly type I collagen plus other proteins that collectively help mediate apatite deposition on collagen fibrils (**Figure 11.7a, b**). During this sequence of dentinogenesis, apical processes arising from odontoblasts are eventually surrounded by deposited dentin. Retraction of these odontoblastic extensions creates narrow channels, called **dentinal tubules**, which confer the characteristic striations of mature dentin (**Figure 11.7c**). Such channels normally play beneficial roles by helping to maintain both **primary dentin** that starts forming *in utero* and **secondary dentin** that begins being made after tooth eruption and continues throughout life (**Figure 11.7d**). However, if the overlying enamel layer is breached, dentinal tubules create easier access for pathogens that can trigger heightened sensitivities and infections in the underlying dental pulp.

After the onset of dentinogenesis, **ameloblasts** begin **enamel** production by reorganizing their former basal compartments into sawtooth-shaped **Tomes' processes** that face the developing dentin layer (**Figure 11.7e, f**). Tomes' processes exocytose enamel matrix proteins like amelogenins and enamelins, upon which calcium and phosphate ions are seeded into apatite crystals. This produces mineralized **enamel rods** directly over Tomes' processes and **interrod enamel** in between rods, with these two enamel components possessing differing orientations of apatite crystals that help further strengthen enamel (**Figure 11.7e, f**). After the secretory phase is completed, Tomes' processes degenerate, and ameloblasts transition to a maturation stage as they take up excess water and matrix. Such maturation allows enamel to contain **>95% apatite** and thus become the hardest tissue in the body. Post-maturation ameloblasts are eventually dislodged from the tooth once it erupts through the gums. Thus, unlike dentin, which is continuously re-generated, enamel lost after birth is not replaced by ameloblasts. Instead, enamel of adult teeth can be re-mineralized via non-ameloblast-mediated mechanisms, such as the seeding of calcium and phosphate in saliva by **fluoride ions** that are routinely introduced via toothpastes, oral rinses, or fluoridated water supplies (Box 11.2).

By ~20 weeks, fetuses begin to develop secondary teeth next to the primary set. Secondary tooth formation occurs via morphogenetic stages resembling those used by primary teeth, albeit over longer time spans that generate thicker mineralized layers. Both sets of teeth coexist under the gumline until after birth (**Figure 11.5b**), when primary tooth eruption begins and erupted primary teeth are eventually replaced by secondary teeth. One of the lower incisors is typically the first tooth to erupt by 6 months post-birth, and the entire primary set usually emerges by ~2.5 years. A mandibular incisor also normally initiates secondary tooth emergence at ~6–7 years, and all permanent teeth including third molars (wisdom teeth) tend to erupt by age 22 (**Case 11.3 Gordie Howe's Teeth**).

11.7   THE PERIODONTAL LIGAMENT AND CEMENTUM LAYER THAT HELP ANCHOR TOOTH ROOTS IN JAW BONES

Each secondary tooth remains firmly attached to the jaw bone via surrounding tissues, collectively called the **periodontium**. Most of the periodontium comprises a tough **periodontal ligament** (**PDL**) (**Figure 11.5f**) containing bundles of collagen, called **Sharpey's fibers**, which help anchor the PDL to tooth and jaw bone surfaces. In addition, at its interface with the tooth, the periodontium contains **cementoblasts** that produce a bilayered adhesive **cementum** of ~50% apatite and ~50% organic matrix plus water, thereby ensheathing sub-gumline dentin while helping to hold teeth in place (**Figure 11.5f**). The outer layer of cementum farthest from dentin has lacunar spaces with entrapped **cementocytes**, which can also re-model cellular cementum to compensate for minor movements of teeth within jaws.

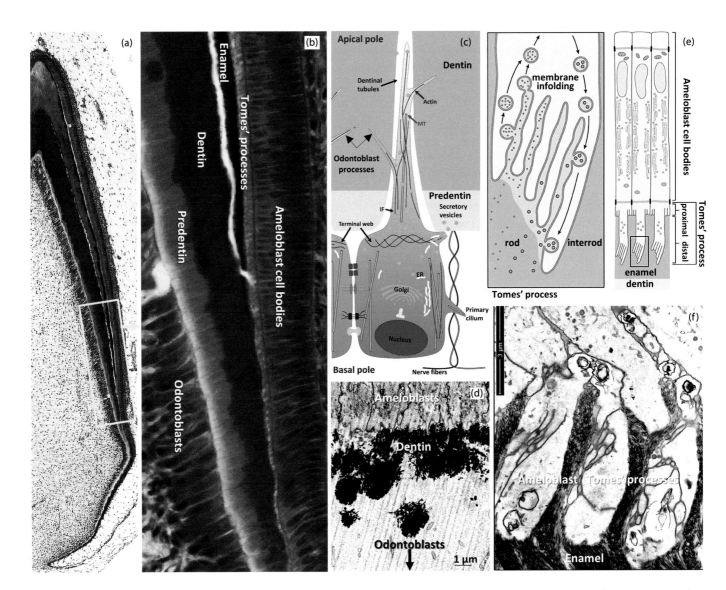

Figure 11.7 Tooth mineralization: odontoblasts and ameloblasts. (a, b) Low- and higher-magnification views of an enamel organ periphery show the positions of dentin-producing odontoblasts vs. overlying ameloblasts with their Tomes' processes that deposit enamel. **(c, d)** Odontoblasts extend slender processes within dentinal tubules that allow unmineralized predentin to be secreted before it becomes mineralized with apatite to form dentin. **(e, f)** Ameloblasts generate saw-tooth-shaped surfaces, called Tomes' processes, which allow enamel to be secreted as a rod directly over each Tomes' process and as an interrod in the spaces between neighboring rods. ([a, b]: images courtesy of Dr. A. Hand; [c]: From Chang, B et al. (2019) Cell polarization: from epithelial cells to odontoblasts. Eur J Cell Biol. https://doi.org/10.1016/j.ejcb.2018.11.003, reproduced with publisher permission; [d, f]: From Smith, DA et al. (2016) Ultrastructure of early amelogenesis in wild-type, Amelx⁻/⁻, and Enam⁻/⁻ mice: enamel ribbon initiation on dentin mineral and ribbon orientation by ameloblasts. Mol Gen Genom Med, doi: 10.1002/mgg3.253, reproduced under a CC BY 4.0 creative commons license; (e): From Pham, C-D. et al. (2017) Endocytosis and enamel formation. Front Physiol 8: doi: 10.3389/fphys.2017.00529, reproduced under a CC BY 4.0 creative commons license.)

With its enamel, dentin, and cementum kinds of mineralized layers, each emerged tooth comprises three main regions (**Figure 11.5e**): (i) an **anatomical crown** of enamel and dentin, which is visible within the oral cavity and contrasts with a **clinical crown** that may also have some exposed cementum, owing to receding gums, (ii) an anatomical **root**, which occurs beneath the anatomical crown and consists of dentin covered by cementum, and (iii) a central **pulp** compartment of living tissues with blood vessels, nerves, and stem cells. Depending on the particular tooth, a fully differentiated crown may overlie one to three roots, each of which is penetrated by a **canal** that allows nerves and blood vessels to maintain tooth viability (**Figure 11.5d**). Based on

CASE 11.3 GORDIE HOWE'S TEETH

Hockey Player
  (1928–2016)

"To me, hockey was always tremendous fun. That's what kept me going for so long. I simply love to play hockey."

  (G.H. describing his love of hockey)

As one of hockey's greatest and most durable players, "Mr. Hockey" Gordie Howe played professionally from the 1940s to 1980s. Growing up on a farm in Canada, Howe struggled in school due to dyslexia but excelled at hockey and in 1944 was signed by the Detroit Red Wings of the National Hockey League (NHL). Howe flourished for 25 years in Detroit, before playing with his sons in the World Hockey Association and ending his NHL career in Hartford. Howe scored 801

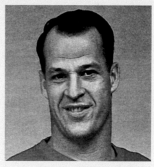

NHL goals and was also a fearlessly tough guy, as evidenced by the three teeth he lost in his very first game as a Red Wing. After retiring, Howe suffered from arthritis and dementia, no doubt exacerbated by his lengthy hockey career, and in 2016, he died at age 88 of undisclosed causes. Following his death, Howe's casket was brought into the Detroit arena, where thousands payed their respects to a superstar, who certainly sacrificed more than just his teeth to the sport he long played.

crown and root shape, each left and right half of the upper and lower sets of fully erupted secondary teeth (i.e., each **tooth quadrant**) comprises two incisors: one canine: two premolars: three molars. Such 2:1:2:3 dentition patterns in turn yield a total of 32 teeth distributed across all four quadrants in each adult (**Figure 11.5a**).

11.8 THE HARD AND SOFT PALATE IN THE ROOF OF MOUTH USED IN MASTICATION AND LUBRICATION

At ~6–8 weeks of gestation, a **palate** begins to form the roof of the mouth, and when fully differentiated, the anterior 2/3 of the definitive palate is termed the **hard palate**, due to its overlying maxillary and palatine bones. Conversely, the posterior 1/3 of the palate located near the pharynx lacks bone and hence is called the **soft palate** (**Figure 11.1b**) (**Case 11.4 Grover Cleveland's Palate**).

The mucosa covering the inferior surface of the **hard palate** contains a rugose and highly keratinized stratified squamous epithelium (**Figure 11.8a, b**). This masticatory mucosa rests upon dense irregular connective tissue plus bone to provide a tough surface for chewing. Conversely, the superior side of the hard palate underlying the nasal cavity is covered by a pseudostratified ciliated columnar epithelium (PCCE) type of respiratory epithelium.

Compared to the hard palate, the non-keratinized squamous epithelium of the **soft palate** is less ridged and can contain scattered **taste buds** that help monitor the desirability (=palatability) of ingested food. Between its

CASE 11.4 GROVER CLEVELAND'S PALATE

22nd and 24th U.S. President
  (1837–1908)

"Four good reasons for electing Cleveland: 1. He is honest. 2. He is honest. 3. He is honest. 4. He is honest."

  (An endorsement of G.C. before it was known he hid from the public critical information regarding his health)

As the only U.S. president to serve two non-consecutive terms, Grover Cleveland developed palatine cancer after his second election, presumably due to his longtime habit of smoking cigars. To avoid fueling public angst, Cleveland chose not to disclose his illness, and under the ruse of taking a fishing trip, Cleveland embarked on a friend's yacht that had secretly boarded several doctors. Once the boat was in Long Island Sound where it could not be observed from shore, the physicians strapped the president to a chair and gave him mild anesthesia

before removing two teeth and much of his palate. After the yacht docked, Cleveland recuperated in a cottage for what was reported by the White House to be dental surgery, and eventually he was also fitted with an artificial palate to help restore his ability to speak. Incredibly, the undisclosed cancer and its well-orchestrated cover-up remained for many years nothing more than discredited gossip until after Cleveland's death, when the story regarding the president and his diseased palate was eventually revealed.

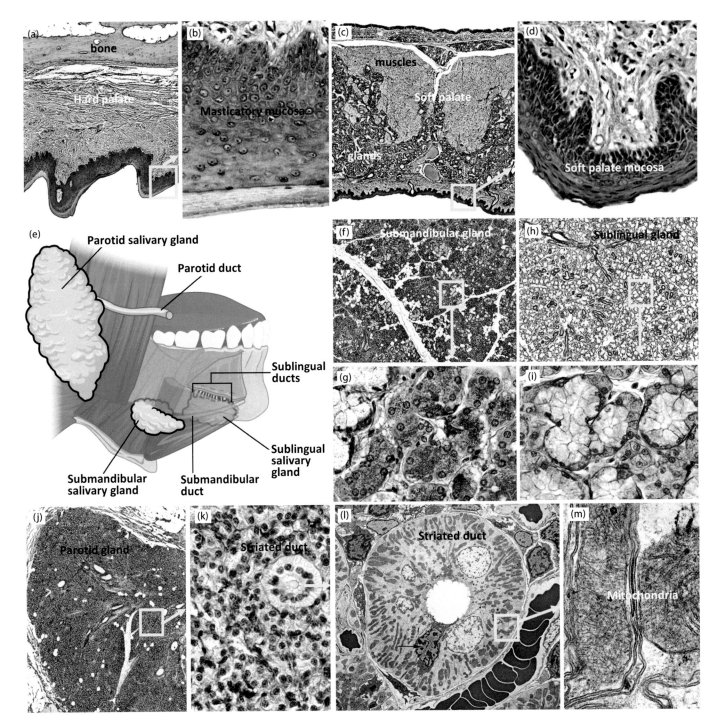

Figure 11.8 Palate and salivary glands. (a–d) *The hard palate at the front of the oral roof is overlain by bone and contains a ridged masticatory mucosa (a, b). Conversely, the more posterior soft palate lacks bone or masticatory mucosa and instead contains muscles plus glands for flexibility and lubrication (c, d). (a) 30X; (b) 210X; (c) 20X; (d) 230X. (e) Along with secretions from minor salivary glands, the oral cavity receives saliva produced by three major salivary glands: submandibular, sublingual, and parotid. (e–g) The submandibular gland comprises mainly darkly staining serous cells with relatively few lightly staining mucous cells. (f) 20X; (g) 190X. (h–k) Conversely, mucous cells predominate in the sublingual gland (h, i), and the parotid gland (j, k) contains exclusively serous cells. (h) 20X; (i) 180X; (j) 20X; (k) 250X. (l, m) Striated duct cells contain numerous basal infoldings and mitochondria, generating a striated appearance in sections. Such ducts actively pump ions to help convert pre-saliva into mature saliva. (l) 1,000X; (m) 20,000X. ((e) from OpenStax College (2013) Anatomy and Physiology. OpenStax. http://cnx.org/content/col11496/latest, reproduced under CC BY 4.0 creative commons license; (l, m) from Tandler, B et al. (2001) Secretion by striated ducts of mammalian major salivary glands: review from an ultrastructural, functional, and evolutionary perspective. Anat Rec 264:121–145, reproduced with publisher permission.)*

oral mucosa and respiratory epithelium lining the nasal cavity, the soft palate is filled with abundant **glands** as well as **skeletal muscles** that lubricate and reshape the malleable soft palate in order to facilitate swallowing, breathing, and phonation (**Figure 11.8c, d**).

11.9   MAJOR SALIVARY GLANDS (PAROTID, SUBLINGUAL, SUBMANDIBULAR): SEROUS VS. MUCOUS CELLS AND PRIMARY SALIVA CONVERSION INTO SALIVA

Along with smaller clusters of salivary glands spread throughout the mouth, three pairs of major salivary glands—**parotid**, **submandibular**, and **sublingual**—collectively produce ~0.5–1.5 l of **saliva** per day. Such fluid helps maintain normal oral functioning by: (i) lubricating the mouth, (ii) initiating carbohydrate and lipid digestion, (iii) facilitating gustation, bolus production, and swallowing; (iv) neutralizing acids that can damage teeth, and (v) warding off oral pathogens (**Figure 11.8e**) (**Case 11.5 Andrew Jackson's Salivary Glands**).

Morphologically, the major salivary glands are compound branched tubuloalveolar glands, whose secretory acini can contain darkly staining **serous** cells and lighter **mucous** cells (**Figure 11.8e–k**). Serous cells produce a watery fluid rich in proteins, such as **α-amylase** that breaks down carbohydrates, whereas mucous cells produce a more viscous solution containing various mucopolysaccharides that aid lubrication. **Parotid glands** are composed exclusively of **serous cells** (**Figure 11.8j, k**), whereas submandibular and sublingual glands are **mixed salivary glands** with both serous and mucous cells. Submandibular glands possess mostly serous cells (**Figure 11.8f, g**). Conversely, sublingual glands contain mainly mucous cells (**Figure 11.8h, i**). In conventionally prepared material, serous cells can form half-moon-shaped caps (=serous demilunes) over groups of mucous cells. However, after freeze-substitution processing, discrete demilunes are generally lacking, suggesting that these serous caps may represent artifacts of aqueous-based fixations.

During saliva production, protein secretion by acinar cells is mainly triggered by noradrenaline derived from nearby sympathetic neurons. Conversely, mucin exocytosis is predominantly stimulated by parasympathetic neurons releasing acetylcholine. During secretion of an initial product, called **primary saliva**, secretory cell Cl^- channels allow chloride efflux into the acinar lumen, and along with Na^+ and water derived from the extracellular matrix (ECM), such ions establish NaCl concentrations essentially **isotonic** to that of blood (Chapter 16). The newly-formed salivary fluid is then moved through ducts by contractile myoepithelial cells surrounding duct lining cells, and while in transit, primary saliva is modified into a more alkaline and less salty saliva by ATP-requiring active transport. Many of these modifications involve cystic fibrosis transmembrane conductance receptor (**CFTR**) **channels** and other membrane proteins localized to **striated duct** cells, whose basal infoldings and abundant mitochondria confer a striped appearance that is visible by light microscopy (**Figure 11.8k–m**).

CASE 11.5 ANDREW JACKSON'S SALIVARY GLANDS

7th U.S. President
    (1767–1845)

"I am swollen from my toes to the crown of my head and in bandages to my lips."

(A. J. in a letter describing his deteriorating health late in life)

Before being elected the seventh U.S. President, Andrew Jackson was a rugged and combative military man, but soon after leaving the White House, the ex-president became a frail invalid with numerous ailments. Many of Jackson's medical problems were due to advanced kidney dysfunction, chronic dysentery, and respiratory disease. However, Jackson also suffered from regularly ingesting calomel, a mercury-containing compound that was widely used for various maladies before it became clear that mercury not only wreaked havoc on the central

nervous system but also served to overstimulate salivary glands. Such hypersalivation was pronounced in miners who excavated quicksilver forms of elemental mercury, and similarly, Jackson noted that calomel caused him to salivate excessively. In fact, his comment toward the end of his life that he was "blubber from head to foot" presumably referenced not

only the general swelling triggered by his various illnesses but also the excessive drooling caused by a mercury-induced overactivation of his salivary glands.

11.10 TONSILS: POSITIONS WITHIN HEAD, B-CELL FOLLICLES, AND INVAGINATED CRYPTS LINED BY A MODIFIED LYMPHOEPITHELIUM

To counteract a barrage of pathogens entering the body through the nose and mouth, the lamina propria of fetal nasal and oral mucosae develop secondary lymphoid organs, called **tonsils**. Such structures lack a connective tissue capsule and comprise: (i) a nasopharyngeal (=adenoid) tonsil in the nasopharynx, (ii) a tubal tonsil near each Eustachian tube connecting the nasopharynx to a middle ear, (iii) two **palatine tonsils** that sit laterally near the posterior soft palate, and (iv) a **lingual** tonsil within the lingual base (**Figure 11.1a**).

When fully developed, the palatine and lingual tonsils associated with the oral cavity resemble the spleen and lymph nodes in possessing numerous **B-cell follicles**, interfollicular **T cells**, and a rich blood supply that collectively help fight invasive pathogens (**Figure 11.9a–c**). However, unlike other secondary lymphoid organs, the surface mucosa of tonsils folds inward at one or more sites to form invaginated **crypts** that are lined by unusual epithelia, called **lymphoepithelia** (=reticulated epithelia) (**Figure 11.9a**). Lymphoepithelia overlie a discontinuous basement membrane and contain various leukocytes, reticular cells, and blood vessels, including high endothelial venules (**Figure 11.9b–d**). Located beneath the lymphoepithelium are B-cell follicles, in which appropriately stimulated B cells can mature within germinal centers to become antibody-secreting plasma cells. In addition, each follicle abuts: (i) **interfollicular tissue** that is enriched in helper T cells, and (ii) an overlying **mantle zone** between the follicle and lymphoepithelium, in which B cells accumulate (**Figure 11.9b**). In spite of such defenses, the thin crypts of **palatine tonsils** can trap food debris and thereby help sustain pathogenic microbes. Accordingly, chronically infected palatine tonsils may in some cases prompt surgical removals (=tonsillectomies) (**Case 11.6 Elvis Presley's Tonsils**).

As opposed to palatine tonsils, the **lingual tonsil** contains a single wide crypt that divides the organ into left and right lobes. The lingual crypt is lined by a lymphoepithelium that is not as highly modified as in palatine tonsils, and because of its wider lumen it does not become chronically infected as frequently as do palatine tonsils with their narrower crypts (**Figure 11.9e, f**).

11.11 SOME ORAL CAVITY DISORDERS: SJÖGREN'S SYNDROME AND CLEFT LIP/PALATE

Sjögren's syndrome (SjS) is an autoimmune disease involving dysfunction of ocular lacrimal glands and oral salivary glands, thereby causing dry eyes and mouth as part of a sicca (=dryness) syndrome. In addition, salivary glands are often infiltrated by aggregations of lymphocytes, as B and T cells react to auto-antigens released from the apoptosis of epithelial cells. SjS is currently managed by treatment of sicca-related symptoms with eyedrops and fluoride for ameliorating eye and tooth problems. In addition, medications that upregulate secretory capabilities or suppress immune responses may also be prescribed.

CASE 11.6 ELVIS PRESLEY'S TONSILS

Singer/Actor
(1935–1977)

"Rock and roll music, if you like it, if you feel it, you can't help but move to it."

(E.P. on of rock and roll)

Born in Mississippi, Elvis Presley released his first record at age 19 and soon thereafter launched a meteoric rise to worldwide fame that was fueled by numerous No. 1 hit songs, movie roles, and live performances, which often caused frenzied fans to storm the stage. Over his career, Presley sold 600 million records, which is more than any other solo act, and at the young age of 36 received a Grammy Lifetime Achievement Award. Of the various maladies Presley suffered early on, the most frightening may have been a severe case tonsillitis for which he was hospitalized in 1959. Doctors reportedly avoided a tonsillectomy for fear it could ruin Presley's voice. Eventually, penicillin cured his infected tonsils and allowed him to continue his singing career until his death at age 42 caused by a lethal combination of various pharmaceuticals.

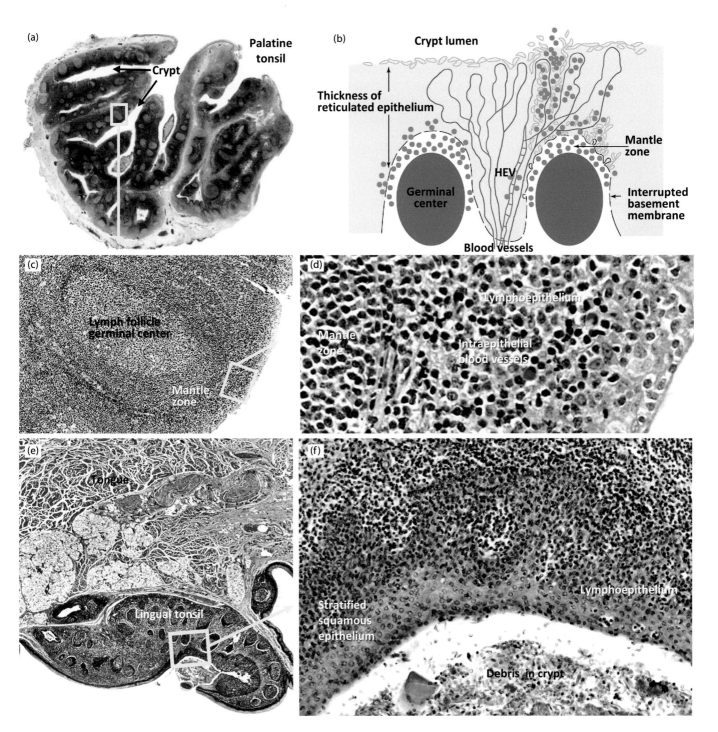

Figure 11.9 Tonsils. (a–c) *The palatine and lingual tonsils of the oral cavity are characterized by: (i) numerous lymph follicles with germinal centers, (ii) one to multiple invaginated crypts, and (iii) a specialized lymphoepithelium (=reticulated epithelium) lining crypts. (a) 6X; (c) 70X. (**d**) The lymphoepithelium has: (i) intraepithelial blood vessels, including permeable high endothelial venules (HEVs), (ii) infiltrating lymphocytes (orange circles in figure b) from a mantle zone near the germinal center of each follicle, and (iii) a discontinuous basement membrane. Such adaptations allow tonsils to launch immune responses against pathogens that gain access to the oral cavity during food ingestion (250X). (**e, f**) The lingual tonsil has but a single wide crypt that does not become infected as easily as in palatine tonsils and thus is not commonly removed in tonsillectomies. (e) 7X; (f) 125X. ((b) from Perry, M and Whyte, A (1998) Trends Immunol 19: 414–421 reproduced with publisher permission.)*

Cleft lip and palate (**CL/P**) is a common craniofacial malformation that has a worldwide incidence of ~1/700 live births. Owing to various genetic mutations associated with CL/P, insufficient fusion of maxillary and medial nasal processes of the developing head generates fissures near the philtrum of the upper lip (=cleft lip) and/or medial gaps in the palate (=cleft palate). Collectively, such defects often cause numerous problems for affected infants, including an inability to suckle properly. In most cases, CL/P can be at least partially corrected by surgery. However, children with severe clefts may still encounter speech, hearing, or other physical and emotional problems post-surgery.

11.12 SUMMARY—ORAL CAVITY

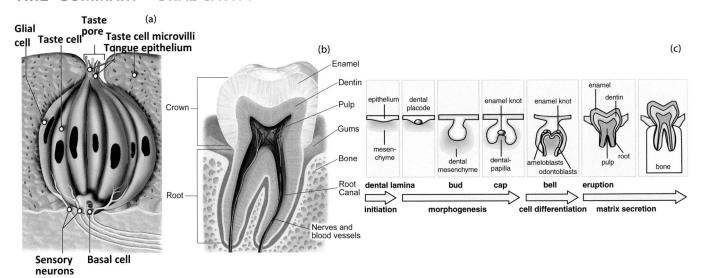

Pictorial Summary Figure Caption: (a): *see figure 11.4b;* ***(b):*** *see figure 11.5e;* ***(c):*** *see figure 11.6a*

The **oral cavity** (=mouth) has: (i) **keratinized** stratified squamous epithelium (**sse**) in oral mucosae covering the hard palate and the gingiva of teeth, (ii) **non-keratinized** oral sse throughout much of the mouth, or (iii) **mixed keratinized** and **non-keratinized** sse on the dorsal surface of the tongue. **Lips** contain (i) keratinized, hairy sse on their outer surface, (ii) non-keratinized, glabrous sse linings, and (iii) a transitional region in between, called the **vermilion zone**.

The **tongue** contains: (i) crisscrossed skeletal muscles for flexible positioning during chewing and speaking, (ii) **lingual glands** for lubrication, and (iii) a **lingual tonsil** for pathogen defense. The dorsal tongue surface has four **lingual papillae** types—**filiform** (the smallest, most common papillae, which lack taste buds and enhance the gripping action of the tongue) plus three gustatory types with taste buds consisting of **fungiform papillae** (mushroom-shaped papillae with ~25% of total lingual taste buds), **foliate papillae** (laterally-restricted papillae possessing ~25% of the tongue's taste buds), and **circumvallate papillae** (the largest papillae, which occur in a V-shaped array near the tongue base and contain ~50% of all taste buds). Each **taste bud** has ~50–100 elongated cells (type I, II, III cells) plus basal stem cells (type IV cells) and can perceive five basic tastes: salty, sweet, bitter, umami, and sour (**Figure a**). Tastant molecules enter taste bud apical pores, bind receptors on microvilli of elongated cells, and trigger depolarizations for output via gustatory afferent nerve fibers.

Major salivary glands—**parotid**, **submandibular**, **sublingual**—are compound branched tubuloalveolar glands distinguished by their proportions of two types of gland cells—**serous** (darker cells that produce a watery, protein-rich fluid) vs. **mucous** (lighter cells that secrete a viscous, carbohydrate-rich product). The parotid gland is all serous, whereas the submandibular is mostly serous, and the sublingual gland is mostly mucous. During transport down ducts, the initial **primary saliva** is converted into saliva by ion exchanges, particularly in mitochondria-rich **striated ducts**.

Humans form two sets of **teeth**—**primary** (=baby) vs. **secondary** (=adult) **teeth**. Each tooth has a central **pulp** of living tissues, which are surrounded by mineralized layers containing the calcium-phosphate crystal **apatite**. Apatite occurs in both the overlying **enamel layer** and underlying **dentin layer** of teeth. In addition,

cementoblasts in the periodontium secrete a calcified cementum (~50% apatite), which along with a **periodontal ligament** anchor the tooth base into the jaw. After eruption through the mucosa, the tooth **crown** above the gumline has enamel and dentin, whereas the sub-gumline **root** below possesses dentin and cementum. Roots have **canals** so that blood vessels and nerves can enter the pulp (**Figure b**).

Each tooth begins to develop from an ectodermal **placode** with a central signaling center, called the **primary enamel knot**. During the **bud stage**, the placode invaginates into the jaw and develops into a **cap**-like primordium. Folding of the cap at its **cervical loops** during the cap and **bell stages** produces a bilayered **enamel organ** with **outer enamel epithelium**, **inner enamel epithelium**, and intervening cushion of **stellate reticulum cells**. Neural crest cells differentiate as **odontoblasts** on the inner surface of the inner enamel epithelium to secrete **dentin**, which resembles bone in containing ~60% apatite and thus resembling bone. Oral-ectoderm-derived **ameloblasts** in the inner enamel epithelium produce **enamel**, which is >95% apatite and hence forms the hardest tissue in the body. Dentin production starts during the bell stage as odontoblasts initially produce unmineralized **pre-dentin** before dentin mineralization. Subsequently, ameloblasts with **Tomes' processes** form **enamel rods** and interrod enamel.

The **palate** in the roof of oral cavity comprises a **hard palate** with bone and keratinized sse for mastication plus a **soft palate** containing glands and a flexible non-keratinized sse for lubrication and swallowing.

Tonsils are non-encapsulated lymphoid tissue with specialized **lymphoepithelia** that help ward off invasive pathogens. **Palatine tonsils** near the oropharyngeal opening comprise **B-cell follicles** and narrow **lymphoepithelium-lined crypts** that can become chronically infected, thereby requiring a tonsillectomy. In addition, a **lingual tonsil** within the tongue base has B-cell follicles and a single wide lymphoepithelium-lined crypt.

Some oral cavity disorders: **Sjögren's syndrome**—an autoimmune disease involving focal infiltrations of lymphocytes in salivary glands associated with hyposecretions. **Cleft lip/palate** (**CL/P**): a common abnormality of development leaving gaps in the newborn's lip and/or palate.

SELF-STUDY QUESTIONS

1. Which calcified layer occurs at the interface between root dentin and the periodontal ligament?
2. T/F: Approximately 95% of enamel is composed of apatite.
3. What is the name of salivary gland ducts with numerous mitochondria and basal infoldings?
4. Which of the following salivary glands has exclusively serous secretory cells?
   A. Parotid
   B. Submandibular
   C. Sublingual
   D. ALL of the above
   E. NONE of the above
5. What kind of epithelium lines the inner surfaces of lips?
   A. A typical keratinized stratified squamous epithelium
   B. A mixture of typical keratinized and non-keratinized
   C. Lymphoepithelium
   D. A simple squamous epithelium
   E. A non-keratinized stratified squamous epithelium
6. Which is the hardest tissue in the body?
   A. Dentin
   B. Cementum
   C. Bone
   D. Enamel
   E. Cartilage
7. Which is the largest lingual papilla?
8. T/F Cleft lip/palate is a relatively rare malformation that occurs in only 1/70,000,000 births.
9. T/F The hard palate contains bone.
10. Along with salty, sweet, bitter, and sour, what the fifth basic taste?

11. Which secrete dentin?
    A. Odontoblasts
    B. Ameloblasts
    C. Cementoblasts
    D. Stellate reticulum cells
    E. Oral epithelium derivatives
12. Which lingual papillae lack taste buds?
    A. Filiform
    B. Fungiform
    C. Foliate
    D. Circumvallate
13. T/F. The root of the tooth contains enamel, dentin, and cementum.
    "EXTRA CREDIT" For each term on the left, provide the BEST MATCH on the right (answers A-F can be used more than once)

14. Tomes' process ___   A. Tonsil
15. Amylase____   B. Odontoblast
16. Crypt lymphoepithelium ____   C. Soft palate
17. Neural crest derived___   D. Lip
18. Vermilion border ____   E. Ameloblast
                                                    F. Salivary gland

19. Describe the functional histology of a taste bud.
20. Describe tooth formation *in utero*.

ANSWERS

(1) cementum; (2) T; (3) striated ducts; (4) A; (5) E; (6) D; (7) circumvallate; (8) F; (9) T; (10) umami; (11) A; (12) A; (13) F; (14) E; (15) F; (16) A; (17) B; (18) D; (19) The answer should cover along with other pertinent topics: lingual gustatory papillae, light vs. dark cells; Type I-IV cells; apical pore, tastants for five basic tastes; microvillar receptors, depolarization, ATP secretion; type III activation, post-synaptic afferent neurons; (20) The answer should include and clearly define such terms as baby v. adult teeth; dental lamina; placode-, bud-, cap-, bell-, maturation stages; enamel organ; stellate reticulum; dentin and enamel organic matrices; apatite; dentinal tubules; Tomes' processes; enamel rods; enamel matrix resorption; cementum, periodontal ligament, crown vs. root vs. dental pulp; root canal

Digestive System II: Pharynx to Anus

12.1 THE DIGESTIVE TRACT: OVERVIEW OF DEVELOPMENT AND GENERAL ORGANIZATION OF ITS WALL

During the 4th week of gestation, the endodermally derived primordium of the digestive tract becomes organized into future **fore-**, **mid-**, and **hindgut** regions (**Figure 11.1d**). The **foregut** gives rise to the lining of anterior gut components, including the pharynx, esophagus, stomach, upper duodenum, and accessory digestive organs. Alternatively, the **midgut** forms the lining of the lower duodenum, small intestine, and proximal large intestine, whereas the **hindgut** produces the lining of the distal large intestine. Developmental milestones and the approximate post-fertilization timings of digestive tract differentiation include: (i) growth of the intestines as they herniate into the umbilical cord cavity, weeks 5–6; (ii) retraction of the intestines back into the abdominal cavity, weeks 10–12; (iii) formation of gastric-acid-secreting cells, pancreatic islets, and intestinal enzymes, week 12; (iv) onset of swallowing, weeks 16–17; and (v) maturation of gut motility, week 36.

DOI: 10.1201/9780429353307-15

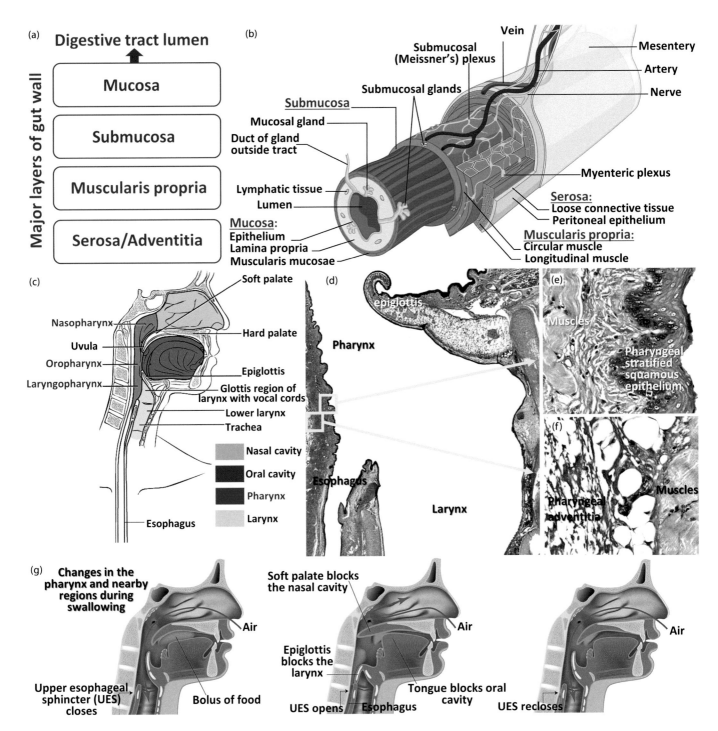

Figure 12.1 Introduction to the digestive tract wall and functional microanatomy of the pharynx. (**a, b**) *The wall of the digestive tract is typically composed of four main layers: (1) mucosa, (2) submucosa, (3) muscularis propria, and (4) serosa or adventitia. (**c**) The pharynx comprises three regions—nasopharynx, oropharynx, and laryngopharynx. (**d–g**) A stratified squamous epithelium lines the oro- and laryngopharynx. Striated muscles in the pharyngeal wall allow the pharynx, in coordination with other nearby structures, such as the tongue, soft palate, and epiglottis, either to deliver food into the esophagus or to send air into the larynx; (d) 6X; (e) 175X; (f) 100X. ([b, c]: From OpenStax College (2013) Anatomy and Physiology. OpenStax. http://cnx.org/content/col11496/latest, reproduced under CC BY 4.0 creative commons license; [g]: From Fujiso, Y et al. (2018) Swall-E: A robotic in-vitro simulation of human swallowing. PLoS ONE 13(12): e0208193. https://doi.org/10.1371/journal.pone.0208193, reproduced under a CC BY 4.0 creative commons license.)*

When fully developed, the wall of the digestive tract typically comprises: (i) a **mucosa** with a lining **epithelium**, subepithelial **lamina propria** of loose connective tissue, and smooth muscle component (=**muscularis mucosae**); (ii) a **submucosa** of dense connective tissue; (iii) multiple smooth muscle layers of the **muscularis propria** (=muscularis externa); and (iv) an outermost layer of mainly loose connective tissue that is termed either the **serosa** or **adventitia** depending on the particular gut component under analysis (**Figure 12.1a, b**). For components suspended in the abdominal cavity (e.g., distal-most esophagus, stomach, small intestine, and parts of the large intestine), the gut wall is bounded by a lubricating **serosa** that possesses at its periphery a **peritoneal** epithelium. Conversely, in attached areas (e.g., pharynx, upper esophagus, rectum, and anal canal), a connective tissue **adventitia** constitutes the wall periphery.

12.2 THE PHARYNGEAL CONNECTION BETWEEN THE ORAL CAVITY AND ESOPHAGUS

The adult **pharynx** comprises three regions: (i) a superior **nasopharynx** that connects with the nasal cavity; (ii) a middle **oropharynx**; and (iii) an inferior **laryngopharynx** that joins the oropharynx with the larynx and esophagus (**Figure 12.1c**). Thus, the pharynx can either deliver air into the larynx or direct food into the esophagus (**Figure 12.1d–g**). Throughout much of the oro- and laryngopharynx, the mucosal lining comprises a non-keratinized **stratified squamous epithelium**, and sandwiched between this lining and the peripheral adventitia are striated muscles used in swallowing (**Figure 12.1e, f**).

12.3 THE MUSCULAR ESOPHAGUS THAT JOINS THE PHARYNX TO THE STOMACH

The **esophagus** is a muscular tube that transports food from the pharynx to the stomach. From its proximal connection with the laryngopharynx to its distal junction with the stomach, the esophagus comprises three main regions: (i) an uppermost **cervical** portion; (ii) a **thoracic** component that lies within the thoracic cavity; and (iii) a short **abdominal** segment between where the esophagus penetrates the diaphragm and connects distally to the stomach (**Figure 12.2a**).

The cervical esophagus is guarded by an **upper esophageal sphincter** (**UES**) of circularly arranged skeletal muscle. Normally, the UES remains contracted during breathing, thereby allowing air to be deflected from the esophagus and delivered to the larynx (**Figure 12.1g**). However, during swallowing, the UES relaxes while the epiglottis folds over the larynx (Chapter 14) so that food can move into the esophagus (**Figure 12.1g**). Similarly, vomiting reflexes open the UES in preparation for retrograde regurgitations.

To help protect against abrasion during food transport, the esophageal mucosa contains a **non-keratinized stratified squamous epithelium** and characteristically forms deep folds to accommodate stretching by large food boluses (**Figure 12.12b, c**). At the edge of the mucosa is a well-developed muscularis mucosae, and beneath the submucosa, an inner circular and outer longitudinal muscle layer constitute the muscularis propria (**Figure 12.2b**). In the upper esophagus, skeletal muscles in the muscularis propria allow voluntarily controlled initiations of swallowing. Conversely, mixed striated and smooth muscles occur in the middle esophagus (**Figure 12.2d**), whereas the distal third of the esophagus contains only smooth muscles. Such muscle distributions mean that once swallowing has started, food transported down the esophagus cannot be voluntarily halted before reaching the stomach.

At the connection between the esophagus and stomach, the stratified squamous epithelium of the esophagus abruptly transitions into a **simple columnar epithelium** lining the stomach, thereby yielding a distinct **esophagogastric junction** (**EGJ**), with the term **gastric** referring to stomach (**Figure 12.2a, e–g**). Flow of materials through the EGJ is controlled by an intrinsic **lower esophageal sphincter** (**LES**) plus an associated extrinsic **diaphragmatic sphincter**. During swallowing or regurgitation, relaxation of these sphincters opens the EGJ, whereas in between such events, tonic contractions of sphincter muscles normally block the reflux of stomach contents into the esophagus. However, with bulimia (**Case 12.1 Princess Diana's Esophagus**) and **gastroesophageal reflux disease** (**GERD**), frequent regurgitations and/or incomplete EGJ closures promote retrograde transport of gastric fluids, thereby damaging the esophageal mucosa and leading to **heartburn** and other disease states.

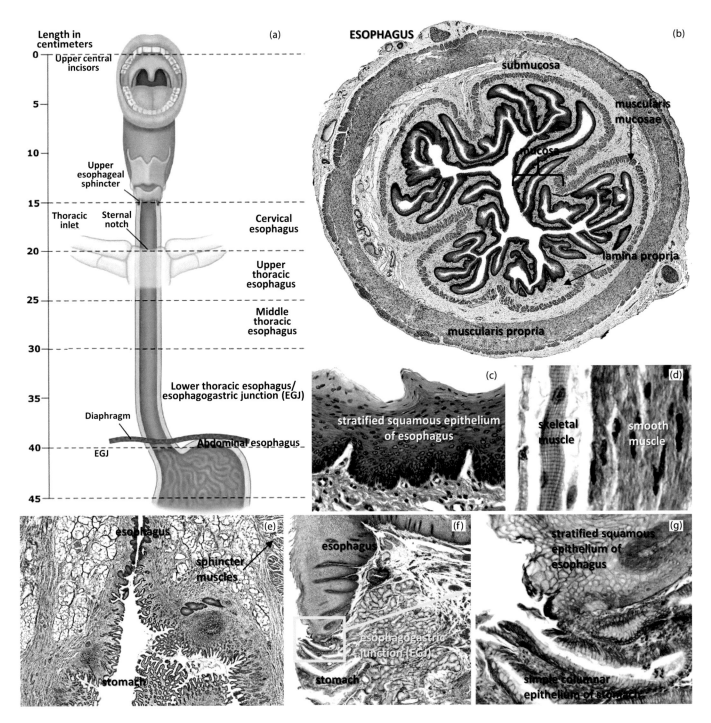

Figure 12.2 Esophagus. *(a) Extending from the pharynx to the stomach, the esophagus comprises upper, middle, and lower subregions. (**b–d**) Transverse sections of the esophagus show its: (i) stratified squamous lining epithelium (c: 200X), (ii) muscular wall with mixed skeletal and smooth muscle in the middle esophagus (d: 400X), and (iii) deeply folded mucosa (b: 15X). (**e–g**) The esophagus joins the stomach at the esophagogastric junction (EGJ), which is guarded by sphincter muscles to prevent the backflow of acidic stomach contents into the esophagus. (f) 25X; (g) 125X. ([a]: From Ferhatoglu, MF and Kıvılcım, T (2017) Anatomy of esophagus. doi: 10.5772/intechopen.69583, reproduced under a CC BY 3.0 creative commons license.)*

CASE 12.1 PRINCESS DIANA'S ESOPHAGUS

Princess of Wales, Humanitarian
  (1961–1997)

"You inflict it upon yourself because your self-esteem is at a low ebb...You fill your stomach up four or five times a day ...and it gives you a feeling of comfort...Then you're disgusted at the bloatedness of your stomach, and then you bring it all up again."

(P.D. on her battles with the disease bulimia nervosa)

Born in Norfolk, England, Diana was married at the age of 20 to Prince Charles, the heir to the British throne, before eventually divorcing him and dying at age 36 in a car crash. During her short life, Princess Di gained worldwide fame owing to her movie-star-like persona and laudable humanitarian efforts, such as garnering support for cancer and AIDS research. While a member of the royal family, Diana suffered from bulimia nervosa, a disease that is characterized by eating binges

and self-inflicted cycles of vomiting to prevent weight gain. The bulimic episodes lasted several years until she found professional help to cure the disorder. Over the long term, repeated regurgitations of acidic materials from the stomach can cause significant esophageal damage, with such harm ranging from a chronic cough and hoarseness to bleeding abrasions and even cancer. Thus, although the histopathology that Diana sustained from her bulimia has not been detailed, there can be little doubt that her esophagus was degraded by the debilitating effects of the disease.

12.4   INTRODUCTION TO STOMACH ANATOMY AND THE GENERAL CONTROL OF ACID SECRETION

As the most dilated region of the adult digestive tract, the J-shaped **stomach** can contain up to a liter of food at any one time. The major role of the stomach is to initiate protein catabolism and to generate a liquefied product, called **chyme**, which can be further processed for nutrient absorption and feces formation in the intestines. Such gastric chyme is formed by the mechanical churning of stomach contents combined with the secretion of **enzyme** progenitors and copious amounts of **acid**. The resulting low pH in the gastric lumen not only denatures proteins but also kills pathogens and facilitates the uptake of metals from ingested food.

The stomach is traditionally viewed as comprising four regions: (i) a short **cardia** component near the heart where the esophagus joins the stomach; (ii) an upwardly directed **fundus** superior to the cardia; (iii) the major **corpus** (=body) of the organ; and (iv) a **pylorus** with a proximal **antrum** and distal **canal** connecting the stomach to the duodenum of the small intestine (**Figure 12.3a**). Alternatively, other classification schemes distinguish only two gastric regions—corpus and pylorus—with the fundus being viewed as a sub-compartment of the corpus, and the cardia considered to be a modified esophageal component.

The stomach wall exhibits a typical quadripartite organization except that the muscularis propria has three layers of smooth muscle arranged as inner **oblique**, middle **circular**, and outer **longitudinal** fibers (**Figure 12.3a, b**). In addition, irregular folding of the wall forms macroscopic ridges, called **rugae**, which extend into the lumen, thereby increasing surface area while also facilitating the mixing of chyme (**Figure 12.3a**).

In and among gastric rugae, the **columnar epithelial cells** lining the lumen not only act as a physical barrier but also as producers of some of the bicarbonate ions (HCO_3^-) and mucus that protect the mucosa from luminal acid. Scattered throughout the surface lining are several million pores. Each of these invaginates toward the lamina propria as a **gastric pit** (=**foveola**) that in turn connects basally with a few tubular **gastric glands** (**Figure 12.3c**). Collectively, the lining epithelium of the foveolar region and subjacent glands possesses at least three cell types: (i) **mucous cells** that produce copious amounts of mucus; (ii) **enteroendocrine cells** that secrete hormones into underlying blood vessels; and (iii) **stem cells** that continually replenish the epithelium. In addition, gastric glands in the corpus also contain: (i) basophilic **chief** (=zymogenic) cells that produce enzyme precursors, called **zymogens**; and (ii) acidophilic **parietal** (=oxyntic) cells that secrete acid and a glycoprotein, called **intrinsic factor** (**Figure 12.3c**). Moreover, beneath the mucosal epithelium, the lamina propria contains several types of cells, including **enterochromaffin-like** (**ECL**) cells that secrete **histamine** (**Figure 12.3d**).

As a key function of the stomach, **acid secretion** is controlled by three main stimuli: (i) **acetylcholine** released from parasympathetic neurons in the gastric wall; (ii) **gastrin** produced by G-type **enteroendocrine cells** that are located within mucosal glands of the gastric antrum; and (iii) **histamine** secreted by lamina propria **ECL cells**

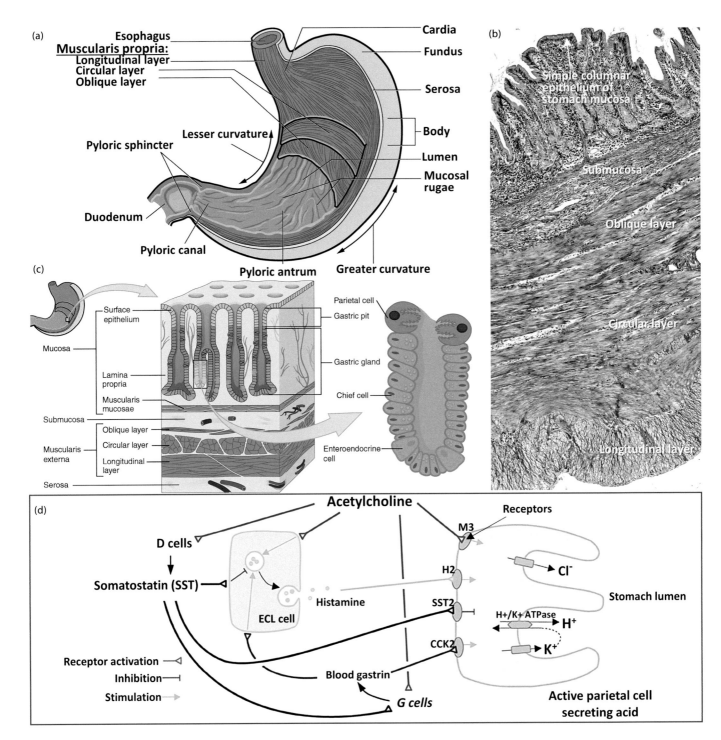

Figure 12.3 Introduction to the stomach and control of acid secretion. (a–c) *The stomach is characterized by: (i) a mucosa with a simple columnar epithelial lining; (ii) numerous gastric pits (=foveolae) that lead into mucosal glands with several cell types, including parietal, chief, and enteroendocrine cells; and (iii) three smooth muscle layers (oblique, circular, and longitudinal) in the muscularis propria. (b) 50X. (**d**) Acid secretion by parietal cells in gastric glands is stimulated by acetylcholine, gastrin, and histamine. Histamine is secreted by enterochromaffin-like (ECL) cells in the lamina propria, and acid production is downregulated by somatostatin. ([d]: From Arin, RM et al. (2017) Adenosine: direct and indirect actions on gastric acid secretion. Front Physiol 8: doi: 10.3389/fphys.2017.00737, reproduced under a CC BY 4.0 creative commons license.)*

(**Figure 12.3d**). Conversely, acid production is downregulated by inhibitory molecules such as **somatostatin**, which is secreted by D-type enteroendocrine cells in stomach glands (**Figure 12.3d**).

In response to gastric expansion triggered by food ingestion, **vagovagal reflexes** can cause efferent fibers of the **vagus nerve** to release acetylcholine, which diffuses toward nearby gastric glands. Acetylcholine increases acid production by: (i) binding to M3-type muscarinic receptors on parietal cells (**Figure 12.3d**), (ii) upregulating G-cell secretion of the acid stimulator gastrin, and (iii) downregulating D cell release of the acid inhibitor somatostatin. Accordingly, surgical incisions of the vagus nerve (=vagotomies) were once standard practice for treating gastric disease due to excessive acid production before either effective anti-acid medications became available or research clearly demonstrated the roles of *Helicobacter pylori* endosymbionts in triggering pathogenesis (Box 12.1 The Discovery of Helicobacter pylori as a Causative Agent of Peptic Ulcers).

In addition to acetylcholine-mediated G-cell stimulation, food reaching the antrum can also cause antral G cells to **release gastrin** into nearby blood vessels. Blood-delivered gastrin can then bind particular cholecystokinin receptors (=CCK2Rs) occurring on parietal and ECL cells. Gastrin bound to these receptors not only stimulates acid production by parietal cells but also causes **histamine** release by ECL cells. Histamine secreted from ECL cells can then diffuse to nearby parietal cells, where it binds **H2 histamine receptors** and further promotes acid production. Accordingly, this pathway is the target of drugs that reduce acid secretions by inhibiting H2 receptor functioning.

12.5 FUNCTIONAL MICROANATOMY OF GASTRIC GLANDS: CHIEF CELLS AND ENZYME SECRETION VS. PARIETAL CELLS AND ACID PRODUCTION

Within the gastric body, mucosal gland complexes usually consist of a relatively short foveola connected to long glands. Each gland comprises an upper **isthmus**, middle **neck**, and lower **base**, with parietal and chief cells occurring in the lower two parts of the gland (**Figure 12.4a**). Conversely, the pylorus is characterized by long foveolae and short glands. Gastric glands in the orad 70–80% of the stomach (i.e., corpus, cardia and fundus) typically contain abundant **parietal cells**, whereas the pylorus characteristically lacks large numbers

---

BOX 12.1 THE DISCOVERY OF HELICOBACTER PYLORI AS A CAUSATIVE AGENT OF PEPTIC ULCERS

Peptic ulcer disease (**PUD**) is characterized by substantial breaks in the mucosa, called **ulcers**, which extend into the submucosa and can trigger life-threatening blood loss in the case of bleeding ulcers. Such ulceration typically occurs in the stomach or proximal duodenum and for decades was believed to be caused by improper diets and emotional stress. However, today, it is well established that many gastric and duodenal ulcers result from pathogenesis initiated by the bacterium *Helicobacter pylori* (**Figure a**) This paradigm-shift in the field of gastroenterology is in large part due to studies conducted by two Australians, **Barry Marshall** and **Robin Warren**, who were awarded the 2005 Nobel Prize for Medicine in honor of their ground-breaking work (**Figure b, c**). The road to their Nobel Prize began in 1979, when Warren first noticed curved bacteria later to be identified as *H. pylori* growing in a gastric biopsy. After recognizing that the bacteria were associated with gastric diseases like PUD but absent in normal biopsies, Warren shared his findings with Marshall, who was a physician trainee looking for a research project during his gastroenterology rotation. Over the next few years, Marshall gathered further data while presenting his findings. However, most physicians at that time believed that either bacteria could not survive in the harsh stomach environment or that any found there were transient commensals unrelated to disease progression. Marshall and Warren went on to show that after successful treatment of *H. pylori* infections, PUD rarely re-occurred, whereas without treatment, ulcers persisted. Nevertheless, their work continued to be met with skepticism, particularly since no appropriate model system was available to confirm a direct link between *H. pylori* colonization and PUD onset. Thus, to demonstrate cause-and-effect, Marshall initially documented his lack of gastritis before infecting himself with an *H. pylori* culture from a diseased patient. After quickly developing severe gastritis, Marshall was able to achieve a rapid cure with antibiotics. Along with such anecdotal evidence, the efficacy of antibiotics in treating PUD was eventually documented in large datasets, and in 1994, a consensus meeting held by the National Institutes of Health issued its conclusion that the key to treating PUD was detecting and eradicating *H. pylori*. Subsequently, various aspects of *H. pylori* biology have been elucidated, including how the bacterium can survive luminal acid by secreting **urease** to convert **urea** into ammonia and CO_2, thereby elevating the surrounding pH and facilitating colonization. Today, the widespread availabilities of simple diagnostic tools, such as the **urease breath test** and effective treatments like **triple-antibiotics-plus-bismuth cocktails** have helped drive a general decline in both *H. pylori*-triggered peptic ulcers and overall PUD cases, while at the same time increased usage of non-steroidal anti-inflammatory drugs (**NSAIDs**) has led to an uptick in NSAID-induced PUD (**Figure d**).

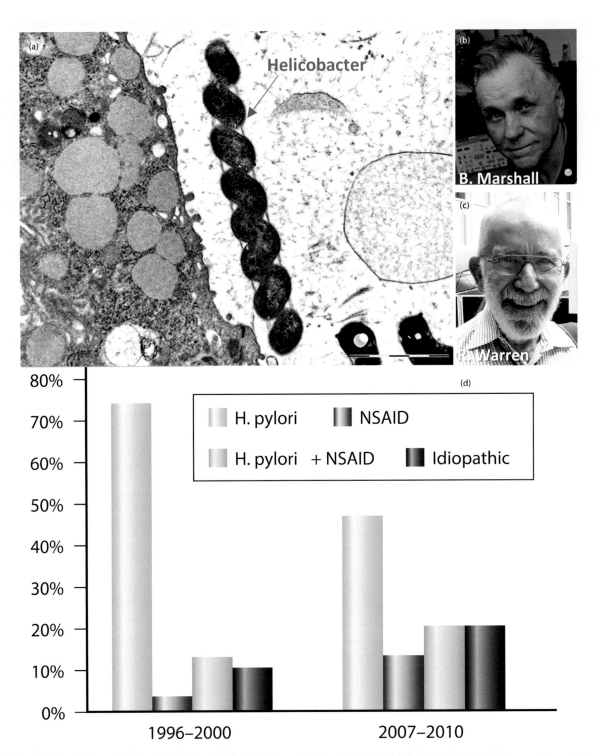

Box 12.1 (**a**) *Transmission electron micrograph (TEM) of Helicobacter bacterium in a dog's stomach. ([a]: From Lanzoni, A et al. (2011) Localization of Helicobacter spp. in the fundic mucosa of laboratory Beagle dogs: an ultrastructural study. Veterinary Res 42: 42. http://www.veterinaryresearch.org/content/42/1/42, reproduced with publisher permission; scale bar = 2 μm; image courtesy of Dr. C. Recordati; (**b**) 2005 Nobel Laureate Barry Marshall (b. 1951) downloaded from https://commons.wikimedia.org/wiki/File:Dr_Barry_Marshall_-_Nobel_Laureate.jpg and reproduced under a CC SA 4.0 creative commons license; (**c**) 2005 Nobel Laureate Robin Warren (b. 1937) downloaded from https://commons.wikimedia.org/wiki/File:Robin_Warren.jpg and reproduced under a CC SA 3.0 creative commons license; (**d**) drawn based on data presented in De Carli, DM et al. (2015) Peptic ulcer frequency differences related to H. pylori or AINES. Arq Gastroenterol. doi: 10.1590/s0004-280320150001000010.)*

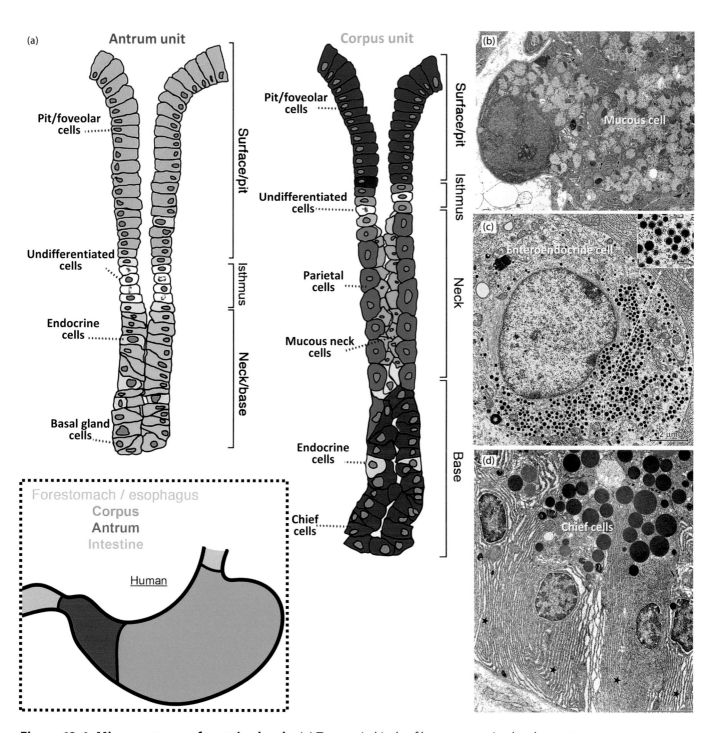

Figure 12.4 Microanatomy of gastric glands. *(a) Two main kinds of human gastric glands—antrum vs. corpus— occur in the pyloric vs. body regions of the stomach, respectively. Such glands differ in the relative sizes of their foveolar and glandular regions as well as in the particular types of cells they contain. For example, parietal and chief cells occur mainly in corpus glands. (**b–d**) Transmission electron micrographs (TEMs) show mucous, enteroendocrine (with hormone granules at higher magnification in the inset), and chief cells from a mammalian stomach. (b) 1,700X; (c) 6,700X; (d) 3,500X. ([a]: From Willet, SG, and Mills, JC (2016) Stomach organ and cell lineage differentiation: from embryogenesis to adult homeostasis. Cell Mol Gastroenterol Hepatol 2(5): 546–559, reproduced with publisher permission; [b–d]: From Ziolkowska, N et al. (2014) Light and electron microscopy of the European beaver (Castor fiber) stomach reveal unique morphological features with possible general biological significance. PLoS ONE 9(4): e94590. doi:10.1371/journal. pone.0094590, reproduced under a CC BY 4.0 creative commons license.)*

of such cells. Thus, human gastric glands have historically been divided into **corpus** (=oxyntic) vs. **antrum** (=pyloric) subtypes, based on the presence vs. absence of parietal cells.

As part of a typical functional unit within the **gastric corpus**, numerous **mucous cells** are present both in the surface epithelium and within foveolae. In addition, **mucous neck cells** occur in the glandular neck region and often protrude their apices farther into the glandular lumen than do nearby parietal cells (**Figure 12.4b**). Collectively, **mucin** glycoproteins in the mucus produced by these cells help protect the stomach wall from acidic chyme, enzymatic digestion, and pathogen-mediated assaults. Such mucus typically forms two layers—an innermost dense stratum that is firmly connected to the mucosal lining plus a more loosely arranged overlying coat. Under most conditions, outer gastric mucus contains numerous commensal bacteria, but the inner layer usually does not, owing to its tightly meshed network of mucin polymers. However, the bacterium ***H. pylori*** is capable of penetrating both mucous layers and damaging the lining epithelium, thereby causing breaches in the mucosa, termed **ulcers** (Box. 12.1).

The isthmus region of gastric glands possesses self-renewing **stem cells** that can differentiate into various cell types, including **enteroendocrine** cells (**Figure 12.4c**) that secrete such hormones as the appetite regulator **ghrelin** (Table 12.1). In addition, stem cells generate mucous neck cells, a subpopulation of which can transdifferentiate into **chief cells** (**Figure 12.4d**).

Zymogen-producing chief cells are basophilic due to their ribosome-studded sheets of endoplasmic reticulum that help synthesize inactive enzyme precursors like **pepsinogen**. Following exocytosis into the stomach lumen, chief-cell-derived zymogens are activated by, and adapted to function in, the reduced pH of the stomach lumen.

TABLE 12.1 SUMMARY OF SOME ENTEROENDOCRINE CELLS, LOCALIZATIONS, HORMONE SECRETIONS, AND FUNCTIONS

Cell	Localization	Peptide	Function
G cell	Pylorus, antral part	Gastrin	Regulation of acid secretion
X or A-like cells	Stomach	Ghrelin	Food intake stimulation
K cells	Proximal intestine	GIP, Glucose-dependent insulinotropic peptide	Enhancement of insulin secretion and gastric acid secretion and reduction of LPL activity in adipose tissue
I cell	Proximal intestine	CCK, cholecystokinin	Gallbladder contraction, gastric motility reduction, stimulation of pancreatic enzyme secretion, inhibition of food intake
S cell	Proximal intestine	Secretin	Stimulation of bicarbonate secretion and inhibition of gastric acid secretion, colonic contraction, and motility
M cell	Proximal intestine	Motilin	Gut motility
N cell	Distal intestine	Neurotensin	Gastric acid secretion, biliary secretion, and intestinal mucosal growth
L cell	Distal intestine and colon	PYY, peptide tyrosine tyrosine; GLP-1/2, glucagon-like peptide 1/2; glicentin	Inhibition of gastric acid secretion and gastric emptying and enhancement of insulin secretion

Source: From Moran-Ramos, S et al. (2012) Diet: friend or foe of enteroendocrine cells: how it interacts with enteroendocrine cells. Adv Nutr 3: 8–20, reproduced with publisher permission.

CASE 12.2 DOREEN KARTINYERI'S STOMACH

Activist for Aboriginal Rights
  (1935–2007)

"A fiery girl who talked too much."

*(D.K. in a 2007 interview when asked
to characterize herself as a child)*

Doreen Kartinyeri was an indigenous Australian born within the Point MacLeay Aboriginal Reserve of South Australia. After moving to the nearby city of Adelaide, Kartinyeri often acted up in school and was expelled when she was caught fighting with white classmates who had been bullying a student. As an adult,

Kartinyeri used her considerable energy and feisty personality to help gain justice for aboriginal peoples and became famous in the 1990s while protesting a proposed bridge that would link Goolwa, South Australia with Hindmarsh Island. Kartinyeri argued that the bridge and the added development it could bring would ruin a sacred site of Ngarrindjeri tribes to which she belonged. However, after various battles in the press and courtroom, the Hindmarsh Bridge was eventually built. The prolonged fight took a toll on Kartinyeri's health, as she not only suffered gastric ulcers but also had to have surgery to remove a tumorous growth from her stomach before dying of stomach cancer in 2007.

Via such processes, chief cells normally play beneficial roles in the stomach. However, transformed types of chief cells can also contribute to the metastasis of aggressive forms of stomach cancers (**Case 12.2 Doreen Kartinyeri's Stomach**).

As opposed to basophilic chief cells, **parietal cells** possess abundant acidophilic mitochondria that produce the ATPs needed for pumping H$^+$ ions into an **apical membrane vacuole** (=canaliculus) during acid production (**Figure 12.5a–c**). Accordingly, resting parietal cells contain an elaborate **tubulovesicular system** and relatively short actin-containing microvilli that extend into the apical vacuole. Following stimulation, however, the tubulovesicular system fuses with the apical vacuole thereby helping to elongate apical microvilli while also delivering proton pumps to the vacuolar membrane (**Figure 12.5d–j**).

Stimulated parietal cells ramp up their secretion of hydrogen and chloride ions into the gastric lumen, thereby delivering on average ~1–2 liters of HCl per day and reducing luminal pH to ~1–2. To produce acid, the enzyme **carbonic anhydrase** within the parietal cell cytoplasm catalyzes the efficient coupling of H$_2$O with CO$_2$ to form the transient reaction product **carbonic acid** (H$_2$CO$_3$), which in turn rapidly dissociates into bicarbonate (HCO$_3^-$) and H$^+$ ions. The resulting bicarbonate ions are mostly pumped into underlying blood vessels creating an **alkaline tide** within the bloodstream following meals. Conversely, H$^+$ ions are delivered to the apical canaliculus, where they are actively extruded into the stomach lumen, mainly by **H$^+$/K$^+$ ATPase** pumps (**Figure 12.5j**). Prior to parietal cell stimulation, these ATP-dependent pumps are sequestered within sub-membrane vesicles, but after stimulation, H$^+$/K$^+$ ATPase pumps are rapidly recruited to the plasmalemma for extruding H$^+$ in exchange for K$^+$ influx. Given their key downstream positioning in acid-secretory pathways, proton pumps are targeted by **proton pump inhibitor** (**PPI**) drugs for effective treatments of excessive acid secretion linked to GERD, heartburn, and other disorders.

Along with their roles in acid production, parietal cells also secrete **intrinsic factor** to facilitate the absorption of **vitamin B$_{12}$** within the small intestine. Since vitamin B$_{12}$ is required for efficient erythropoiesis and is not produced by human tissues, inadequate vitamin B$_{12}$ uptake due to parietal cell dysfunction can lead to a type of reduced erythropoiesis, called **pernicious anemia**. Although such pathogenesis may arise from certain parasitic infections and gastric cancers, autoimmune-mediated attacks on parietal cells are the more common causes of pernicious anemia, which in turn is typically treated with vitamin B$_{12}$ injections.

12.6 INTRODUCTION TO THE SMALL INTESTINE: INCREASED SURFACE AREA FOR DIGESTION AND ABSORPTION PLUS THE ROLES OF CHOLECYSTOKININ (CCK) IN DUODENAL FUNCTIONING

Designated the **small intestine** because it is about half as wide as the large intestine, this anterior bowel region spans ~3–6 m in various adults and thus constitutes the longest component of the digestive tract. The small intestine begins posterior to the gastric pylorus as the ~20–25 cm-long **duodenum** that is followed in order by the **jejunum** and **ileum** (**Figure 12.6a**). Each of these two posterior regions usually extends ~2–3 m, with the jejunum typically being the shorter of the two by ~30% (**Case 12.3 Maurice Gibb's Small Intestine**).

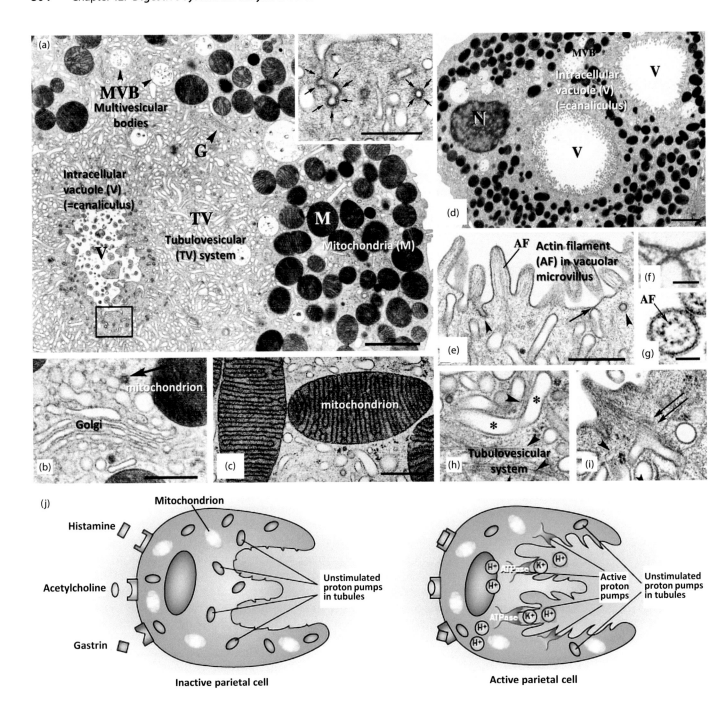

Figure 12.5 Parietal cell microanatomy and acid secretion. (a–i) *As illustrated in these transmission electron micrographs (TEMs) of high-pressure-frozen parietal cells from a rabbit, the acidophilic parietal cells of gastric glands contain large apical vacuoles plus numerous mitochondria, Golgi bodies, and multivesicular bodies. During acid secretion, an elaborate tubulovesicular system helps deliver H+/K+ ATPases (=proton pumps) to actin-containing microvilli that extend into the apical vacuole. (j) Following stimulation by acetylcholine, gastrin, and/or histamine, enlarging vacuoles form numerous long microvilli as the tubulovesicular system transports proton pumps to the vacuolar membrane for active extrusion of H+ into the lumen. Scale bars = (a, d) 2 μm; (a) inset; (b, c, e) 500 nm; (f, g) 50 nm; (h, i) 40,000X. ([a–h]: From Sawaguchi, A et al. (2002) A new approach for high-pressure freezing of primary culture cells: the fine structure and stimulation-associated transformation of cultured rabbit gastric parietal cells. J Microscopy 208: 158–166, reproduced with publisher permission; [j]: from DeVault, KR and Talley, NJ (2009) Insights into the future of gastric acid suppression. Nat Rev Gastroenterol Hepatol 6: 524–532, reproduced with publisher permission.)*

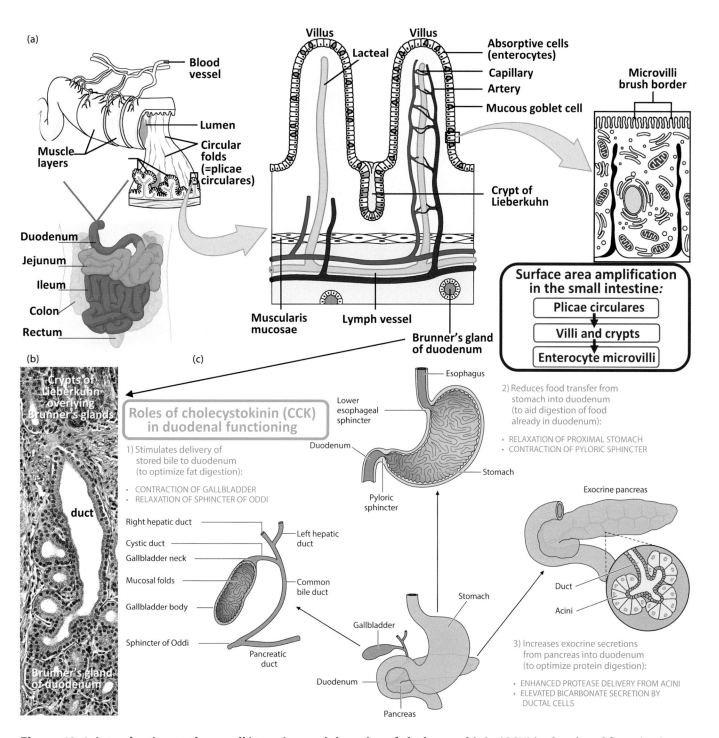

Figure 12.6 Introduction to the small intestine and the roles of cholecystokinin (CCK) in duodenal functioning.
*(**a, b**) The small intestine comprises the duodenum, jejunum, and ileum, all of which have adaptations for increasing surface area, thereby assisting secretory and absorptive processes during digestion. Such adaptations include large plicae circulares, villi, crypts of Lieberkuhn, duodenal Brunner's glands (see fig. b), and microvillar brush borders on enterocytes. (b) 100X. (**c**) CCK secreted by I-type enteroendocrine cells facilitates digestion within the duodenum by: (1) stimulating bile delivery from the gallbladder to assist fat digestion; (2) reducing gastric motility for optimizing digestion of food already present in the duodenum; and (3) releasing pancreatic enzymes and bicarbonate to aid protein digestion while also neutralizing acidic chyme from the stomach. ([a]: From OpenStax College (2013) Anatomy and physiology. OpenStax. http:// cnx.org/content/col11496/latest, reproduced under CC BY 4.0 creative commons license; [c] redrawn, based on Chandra, R and Liddle, RA (2018) Cholecystokinin. Pancreapedia: Exocrine Pancreas Knowledge Base, doi: 10.3998/panc.2018.18.)*

CASE 12.3 MAURICE GIBB'S SMALL INTESTINE

Musician
(1949–2003)

"Life goin' nowhere, somebody help me, yeah, somebody help me, I'm staying alive…"

(From the 1977 song Staying Alive, co-written by M.G. and performed by the Bee Gees)

At age 8, Maurice Gibb moved with his family from England to Australia where he and his brothers formed a highly successful rock band, called the Bee Gees. Gibb was an accomplished musician and the technical force behind the Bee Gees achieving worldwide fame during the heyday of disco music in the 1970s. Throughout those years, the Bee Gees recorded numerous hit songs, such as "Staying Alive" on the Saturday Night Fever soundtrack and ranked behind only the Beatles

and Supremes for number-one songs by musical groups on the Billboard top 100 charts. Years after the band's popularity had peaked, Gibb collapsed from severe abdominal pain and underwent emergency surgery to correct a congenitally twisted digestive tract that over the years triggered intestinal ischemia. However, the procedure was unsuccessful, and Gibb died at age 53, with post-mortem analyses indicating extensive necrosis caused by an inadequate blood supply in his intestinal wall.

Throughout its length, the small intestine has evolved several features that increase **surface area** for **digestion** and **absorption** (**Figure 12.6a, b**). From a coarse to fine scale, such adaptations include: (i) large folds of the mucosa and underlying submucosa, called **plicae circulares** (=valves of Kerckring); (ii) finger-like mucosal **villi** that project from plicae circulares; (iii) intestinal glands (=**crypts of Lieberkuhn**) that extend from the bases of villi toward the submucosa; (iv) submucosal Brunner's glands that connect via ducts to the duodenal lumen; and (v) regularly arranged microvilli constituting the apical **brush border** of **enterocytes**, which are the most common cells in the **simple columnar epithelium** lining the mucosa. Collectively, such features are estimated to amplify surface area ~130-fold over that of a smooth cylinder.

As discussed further in the next chapter, the **duodenum** serves as an important mixing chamber where intestinal digestion is initiated. Thus, acidic chyme entering the duodenum from the stomach combines not only with **bile** delivered from the **gallbladder** but also with **enzymes** plus **bicarbonate ions** secreted by the **pancreas**. These processes are primarily controlled by the hormone **CCK** that is released into the bloodstream by I-type enteroendocrine cells situated within the duodenal and jejunal mucosal lining. In response to fat- and protein-rich chyme reaching the duodenum, increased CCK secretion triggers: (1) **downregulation of gastric emptying** via reduced stomach contractions coordinated with enhanced constriction of the pyloric sphincter to prevent further influx of chyme; (2) waves of **gallbladder wall contractions** coupled with relaxation of the sphincter of Oddi surrounding the common bile duct, thereby allowing bile delivery for facilitating, in particular, fat digestion; and (3) **stimulation of enzyme and bicarbonate secretion from the pancreas** through the main pancreatic duct and a relaxed sphincter of Oddi to help digest chyme while also buffering its acidity (**Figure 12.6c**).

12.7   GENERAL MICROANATOMY OF THE SMALL INTESTINE AND DISTINGUISHING HISTOLOGICAL FEATURES OF ITS THREE REGIONS

The brush border of **enterocytes** is coated with a thick **glycocalyx** that contains enterocyte-specific mucins (**Figure 12.7a–e**). Within this glycocalyx as well as among inter-microvillar spaces of the brush border, enterocytes secrete various proteases, carbohydrases, and lipases, which along with pancreatic enzymes delivered to the duodenum help mediate digestion (Chapter 13). In addition, enterocytes extrude HCO_3^- through apically situated cystic fibrosis transmembrane conductance regulator (CFTR) channels to help buffer chyme. Following catabolism of luminal contents, enterocytes absorb nutrients, vitamins, and water before residual components of undigested chyme are transported to the large intestine for forming feces.

Scattered among enterocytes throughout the small intestine are **mucous goblet cells** that constitute ~10–15% of villar and crypt lining cells (**Figure 12.7f–i**). Such goblet cells tend to be more abundant in the distal small intestine, thereby providing increased lubrication for solidifying materials, as water is further resorbed from unutilized chyme.

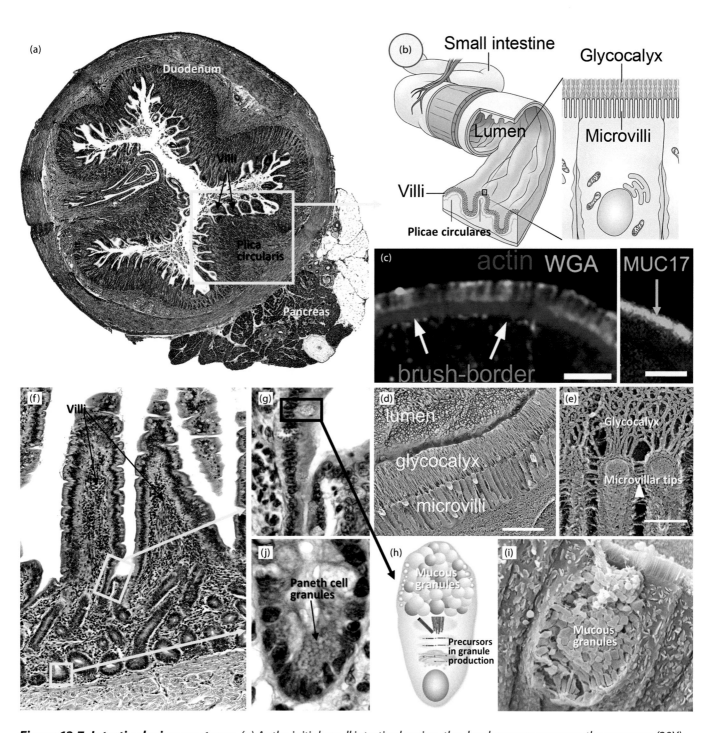

Figure 12.7 Intestinal microanatomy. (a) *As the initial small intestinal region, the duodenum occurs near the pancreas (20X).* ***(b–e)*** *Actin-rich microvilli in the brush border of enterocytes are overlain by a thick glycocalyx containing various carbohydrates, such as mucin 17 (MUC17), which can be stained with fluorophore-conjugated wheat germ agglutinin (WGA) or anti-mucin antibodies (scale bars = 2 μm [b, c], 1 μm (d), 50 nm [e]).* ***(f–i)*** *Interspersed among enterocytes are mucous goblet cells that secrete a protective layer of mucus. (f) 90X; (g) 300X; (i) 2,400X.* ***(j)*** *Paneth cells in intestinal crypts contain large granules with anti-microbial components, such as lysozyme (450X). ([b–e]: From Sun, WW et al. (2020) Nanoarchitecture and dynamics of the mouse enteric glycocalyx examined by freeze-etching electron tomography and intravital microscopy. Commun Biol 3:5. https:// doi.org/10.1038/s42003-019-0735-5, reproduced under a CC BY 4.0 creative commons license; [h]: From Birchenough, GMH (2015) New developments in goblet cell mucus secretion and function. Mucosal Immunol 8, doi:10.1038/mi.2015.32, reproduced with publisher permission; [i]: From Bohórquez, DV et al. (2011) Ultrastructural development of the small intestinal mucosa in the embryo and turkey poult: A light and electron microscopy study. Poultry Sci 90: 842–855, reproduced with publisher permission; image courtesy of Dr. D. Bohórquez.)*

At the base of crypts in the duodenum, jejunum, and ileum, **stem cells** frequently divide to replenish senescent cells that are continually sloughed from villar tips. Such processes result in the entire epithelium being replenished every 2–6 days, which represents one of the highest turnover rates of tissues in the body. Adjacent to these basally located progenitors are intracrypt **Paneth cells** with large granules (**Figure 12.7j**). Such granules contain the enzyme **lysozyme** for degrading bacterial cell walls plus several types of **antimicrobial peptides** (**AMPs**), enabling Paneth cells to serve as key mediators of intestinal innate immunity.

Peripheral to the mucosa and submucosa, each small intestine region contains a well-developed muscularis propria that is arranged as **inner circular** and **outer longitudinal** smooth muscle fibers sandwiched around the **myenteric plexus** (of Auerbach) (**Figure 12.8a, b**). Along with a **submucosal plexus** (of Meissner), the myenteric plexus forms part of the **enteric nervous system** (**ENS**) that occurs throughout the digestive tract and interacts with the CNS to control digestive functions (Box 12.2 Functional Morphology of the Enteric Nervous System). For example, the ENS helps coordinate activities within muscularis layers to generate **peristaltic waves** of contractions and relaxations in the gut wall that propel materials in an aboral direction (i.e., away from the mouth and toward the anus) (**Figure 12.8c**). Conversely, ENS-controlled **segmentation** processes downregulate peristalsis and associated aboral translocations, thereby facilitating churning actions that break up luminal contents prior to absorption.

Although such features are generally observed throughout the small intestine, a few distinctive characteristics help to distinguish the duodenum, jejunum, and ileum. For example, wide-field views of the duodenum often include portions of the nearby pancreas (**Figure 12.7a**). Moreover, bicarbonate- and mucus-secreting **Brunner's glands** occur exclusively in the duodenal submucosa (**Figure 12.6b**), where they respond to S-cell-derived **secretin** by exocytosing their contents to help buffer acidic chyme. Alternatively, the submucosa of the **ileum** contains aggregations of lymph follicles, called **Peyer's patches**, which extend to the mucosal lining and underlie a modified **follicle-associated epithelium** (**FAE**) containing **M-cells** (=microfold cells) (**Figure 12.8d–f**). Such M cells lack a brush border on their thin dome-like apex and represent sites where underlying lymphoid cells can aggregate near the lumen. Collectively, these features facilitate the transcytosis of luminal pathogenic antigens across M cells to underlying lymphocytes and antigen-presenting cells. Conversely, the **jejunum** is devoid of both Brunner's glands and Peyer's patches, thereby helping to identify this region via its absence of distinctive duodenal and ileal features (**Figure 12.8g**).

12.8 ANATOMY OF THE LARGE INTESTINE AND FUNCTIONAL HISTOLOGY OF THE APPENDIX

In adults, the **large intestine** measures 1–2 m long by ~0.5 cm wide and comprises several distinct regions that terminate distally at the anus (**Figure 12.9a**). From its junction with the ileum, the proximal large intestine extends downward as a short pouch, called the **cecum**, which gives rise to a thin blind-ended diverticulum, termed the **appendix** (**Figure 12.9a, b**). Superior to the cecum, the main part of the large intestine, or **colon**, comprises ascending, transverse, and descending segments. The colon then forms a looped sigmoid part that

---

BOX 12.2 FUNCTIONAL MICROANATOMY OF THE ENTERIC NERVOUS SYSTEM

Unlike other organs, the digestive tract possesses a well-developed intrinsic network of nerves, called the **enteric nervous system** (**ENS**), which normally interacts with the central nervous system (CNS) but can also function autonomously (**Figure a**). In humans, such ENS circuits are estimated to comprise 400–600 million neurons, which is about equal to the total neurons in the spinal cord. Much of the ENS within the gut wall is organized into two interconnected networks: (i) a **myenteric plexus** of relatively large ganglia that extend efferent axons into the muscularis, and (ii) a **submucosal plexus** of smaller ganglia that project fibers toward the mucosa. Via these plexuses and their associated

neurons emanating from the CNS, the ENS controls various key functions, including: (i) **movement of materials** along the digestive tract; (ii) **secretion of gastric acid**; (iii) regulation of both **transmucosal fluid movement** and **local blood flow patterns**; and (iv) **interactions** with **immune** and **endocrine** tissues of the digestive tract. Accordingly, in the absence of a fully functional ENS, such activities are severely compromised. For example, in congenital Hirschsprung's disease, ganglia fail to develop in the distal bowel, thereby preventing defecations and requiring surgical removal of the aganglionic region to avoid necrosis.

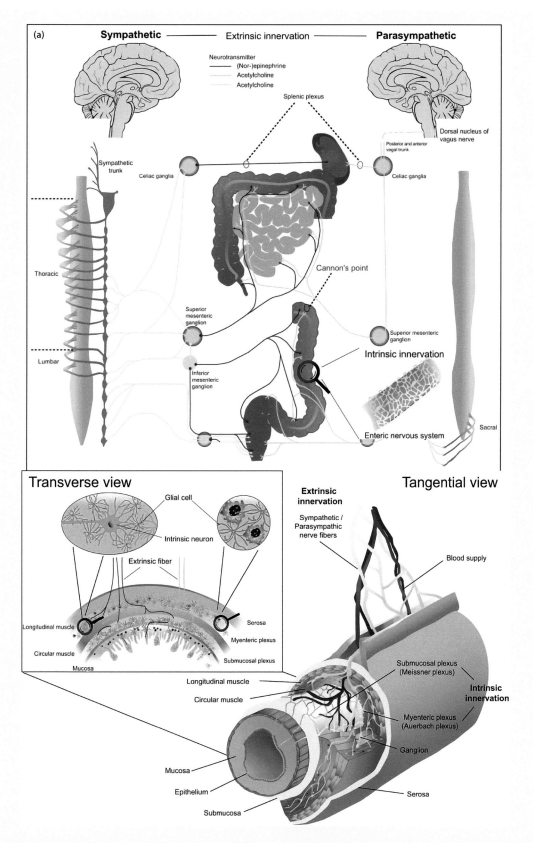

Box 12.2 (a) *The enteric nervous system. (From Jakob, MO et al. (2020) Neuro-immune circuits regulate immune responses in tissues and organ homeostasis. Front Immunol 11, doi: 10.3389/fimmu.2020.00308, reproduced under a CC BY 4.0 creative commons license.)*

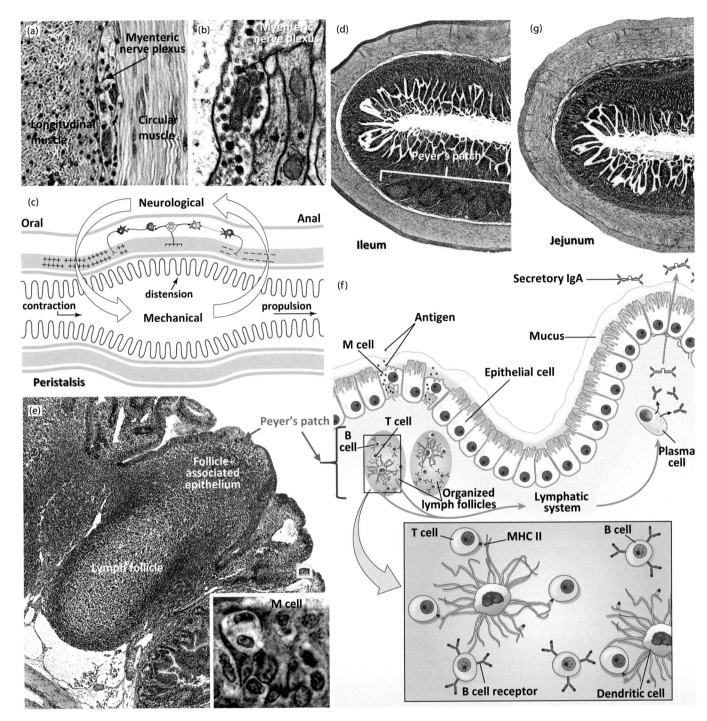

Figure 12.8 Peristalsis, Peyer's patches, and distinguishing features of small intestine regions. (a–c) *The myenteric nerve plexus located between circular and longitudinal layers of the muscularis propria constitute part of the enteric nervous system (ENS), which along with the CNS and mechanical forces can help coordinate complex processes, such as the aboral propulsion of luminal contents via peristalsis. (a) 330X; (b) 5,000X. **(c–f)** The ileum contains organized collections of lymph follicles, collectively termed Peyer's patches, which utilize a modified follicle-associated epithelium with microfold (M) cells to process pathogenic antigens. (c, d) 10X; (e) 40X. **(g)** The jejunum can be distinguished by its lack of either submucosal glands or Peyer's patches. ([b]: TEM image courtesy of Dr. M. Morroni; [c]: from Spencer, NJ et al. (2016) Insights into the mechanisms underlying colonic motor patterns. J Physiol 594: 4099–4116, reproduced with publisher permission; [f]: from OpenStax College (2013) Anatomy and Physiology. OpenStax. http://cnx.org/content/col11496/latest, reproduced under CC BY 4.0 creative commons license.)*

CASE 12.4 EDWARD PLUNKETT'S APPENDIX

Writer
  (1878–1957)

"But, logic, like whiskey, loses its beneficial effect when taken in too large quantities."

(E.P., a writer of fantasy stories, on the
need to suspend logical thought at times)

Edward Plunkett was prolific writer, whose best-known novels belonged to a new fantasy genre that he helped create. Born to one of the most famous of Irish families, Plunkett was the 18th Baron of Dunsany and for a time lived at the Dunsany Castle, which is one of the oldest continuously inhabited abodes in Ireland. Decades before J.R.R. Tolkien published his popular Lord of the Rings series, Plunkett wrote under the penname of Lord Dunsany a series of stories set in the mythical world

of Pegana, which had its own gods, history, and geographical features. Over the years, Plunkett published 90 books and often socialized with other literary giants of Ireland and England. At one such dinner in 1957, Plunkett took ill with what was initially thought to be food poisoning. Instead, his appendix had become infected, and before he could receive proper treatment, the accomplished writer succumbed to complications of a burst appendix.

is contiguous with a **rectum** region connecting the sigmoid colon to the **anal canal**, which in turn ends at the anal opening.

As the large intestine region with the greatest pathlength to the anus, the worm-shaped **appendix** is lined by a simple columnar epithelium that possesses sparsely distributed crypts of Lieberkuhn overlying numerous **lymph follicles** (**Figure 12.9c, d**). In spite of its well-endowed lymphoid tissue, this narrow diverticulum is prone to inflammation (=**appendicitis**), which can progress to life-threatening ruptures (**Case 12.4 Edward Plunkett's Appendix**). Because the appendix was once considered a vestigial organ without useful properties, removal of a healthy appendix was sometimes prescribed before trips to isolated locales in order to avoid possible sepsis triggered by appendicular rupture. However, prophylactic appendectomies are no longer regularly advised, based on new evidence that the appendix and cecum have evolved to incubate beneficial members of the gut microbiome that can help optimize digestive functions.

12.9 THE COLON WITH ITS MUCOSA THAT LACKS VILLI BUT POSSESSES TALL CRYPTS OF LIEBERKUHN AND NUMEROUS MUCOUS GOBLET CELLS

In early fetal life, the mucosal lining of the developing colon resembles that of the small intestine in possessing villi. However, by 30 weeks, colonic villi are lost, thereby yielding a more uniform apical profile than the irregular outline of small intestinal villi (**Figure 12.9e, f**). In adults, a simple columnar epithelium lines the colon lumen and connects to tall **crypts of Liebekuhn** that contain mainly enterocytes, stem cells, Paneth cells plus numerous mucous goblet cells, which account for ~50% of the surface and crypt lining (**Figure 12.9g**). In addition to their greater density compared to the small intestine, colonic goblet cells secrete a thicker mucus that provides extra protection and lubrication as solidifying materials are transported toward the anal canal. Moreover, the muscularis propria of the colon is distinguished by three bundles of longitudinal muscle, historically termed the **tenia coli**, owing to the mistaken belief that they corresponded to tapeworm-like parasites of the colon (**Figure 12.9h, i**). In actuality, tenia coli are normal components of the colonic wall that can undergo tonic contractions to constrict the colon into broad saclike subdivisions, called **haustra.** Such sub-compartments facilitate colonic transport and the processing of materials as they are moved from one haustrum to its neighbor. Eventually, luminal contents enter the rectum, whose mucosal lining resembles that of the colon (**Figure 12.9j, k**).

As opposed to the small intestine where both the digestion and absorption of multiple kinds of molecules occur, the colon functions in the absorption of water, electrolytes, and vitamins while forming **feces** (=stools) (**Figure 12.9j**). Feces typically comprise ~75% water and ~25% solid material, with most of the solids consisting of: (i) bacteria, (ii) residual components of ingested food, and (iii) added wastes, such as the **bilirubin** breakdown

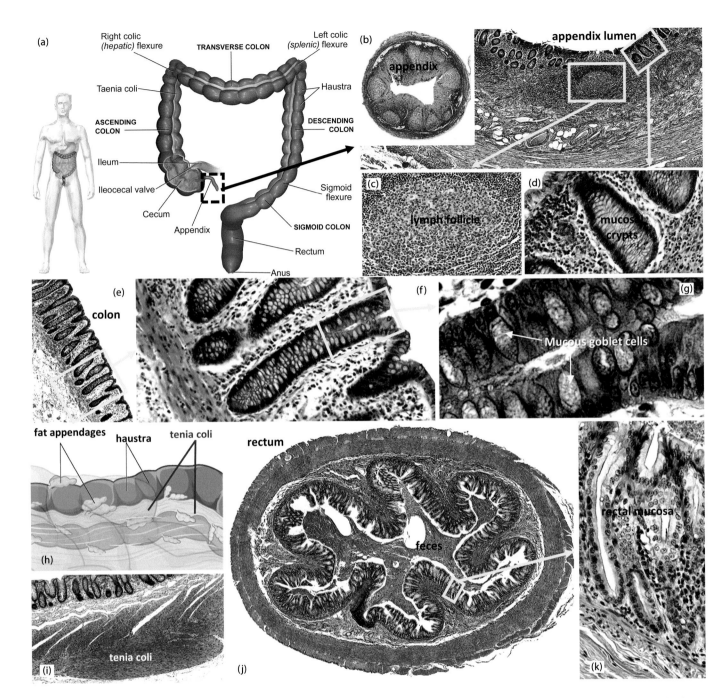

Figure 12.9 Large intestine. *(a) The multipartite large intestine mostly comprises a wide colon that connects with the anal canal via the rectum. (b–d) The appendix has numerous lymph follicles and sparse crypts of Lieberkuhn. Owing to its narrow lumen, the appendix is prone to infections, which, if not treated, can lead to fatal ruptures. (b) 25X; (b, inset) 10X; (c, d) 100X. (e–i) The colon is characterized by: (i) partial segments (haustra); (ii) longitudinal bands of smooth muscle (tenia coli); (iii) numerous mucous goblet cells to protect against abrasion; (iv) tall crypts of Lieberkuhn; and (v) a lack of villi which provides a smoother luminal edge than in the villus-containing small intestine. (e) 15X; (f) 75X; (g) 350X; (i) 10X. (j, k) The rectal mucosa resembles that of the colon and may temporarily store fecal material before passage into the anal canal. (j) 20X; (k) 175X. ([a]: From Blausen.com staff (2014) Medical gallery of Blausen medical 2014. Wiki J Med 1 (2): 10, doi:10.15347/wjm/2014.01010, reproduced under CC0 and CC-SA creative commons licenses; [h]: from OpenStax College (2013) Anatomy and Physiology. OpenStax. http://cnx.org/content/col11496/latest, reproduced under CC BY 4.0 creative commons license.)*

CASE 12.5 CORAZON AQUINO'S COLON

11th President of the Philippines
(1933–2009)

"I don't like politics. I was only involved because of my husband."

(*C.A. on how she was reluctantly drawn into politics after the assassination of her husband*)

From 1954 to 1983, Corazon Aquino remained a self-described "plain housewife" who was married to Senator Benigno Aquino Jr., a leading opponent of the long-time dictatorial president of the Philippines, Ferdinand Marcos. However, that all changed in 1983, when Sen. Aquino was assassinated. Even though she had never been directly involved in politics, Aquino took up her husband's cause in response to a petition with a million signatures urging her to oppose Marcos in the 1986 presidential election. After the votes were counted, Marcos

was declared the winner. However, allegations of rampant voter fraud on Marcos's behalf triggered widespread protests until Marcos stepped aside, enabling Aquino to become the first female elected president not only in the Philippines but also in all of Asia. In 2008, 16 years after her presidential term had ended, Aquino was diagnosed with cancer of the colon and underwent bouts of chemotherapy before eventually dying at the age 76 due to complications from her diseased colon.

product of recycled erythrocytes that contributes to the brownish coloration of stools. Small amounts of feces made in the colon may be temporarily stored in the rectum, whereas large boluses typically trigger a recto-anal reflex that relaxes anal sphincter muscles and converts the rectum into a transmitting conduit during defecation (**Case 12.5 Corazon Aquino's Colon**).

12.10 THE ANAL CANAL: STRATIFIED SQUAMOUS EPITHELIAL LINING PLUS HAIR FOLLICLES, GLANDS, VASCULARIZATION, AND MUSCLES

The simple columnar epithelium of the rectum and proximal portion of the anal canal transitions distally into a **stratified squamous epithelium** (**Figure 12.10a**). Near this junction, the stratified lining epithelium is non-keratinized. However, closer to the anal opening, keratinization increases, and both hair and apocrine glands are present. The submucosa of the anal canal possesses a well-developed bed of blood vessels that in response to over-straining during defections can bulge outward as dilated sacs, called **hemorrhoids** (**Figure 12.10b–d**). In addition, the sub-adventitial muscularis contains muscle bundles that form dual sphincters, with the outer sphincter containing striated muscle under voluntary control.

12.11 SOME DIGESTIVE TRACT DISORDERS: BARRETT'S ESOPHAGUS AND DIVERTICULAR DISEASE

Barrett's esophagus (**BE**) is characterized by pathological changes in the distal esophageal mucosa that are typically triggered by chronic GERD. In particular, parts of the stratified squamous epithelium in the esophagus of BE patients tend to undergo a metaplastic transdifferentiation into a **simple columnar epithelium** with associated glands that help counteract acid-mediated damage to the esophageal wall. Such glands can resemble colonic crypts in containing abundant mucous goblet cells as well as scattered enterocytes and Paneth cells. To prevent a potential progression into esophageal cancer, management of BE normally includes aggressive anti-reflux treatments.

Diverticular disease involves the formation of herniated outpocketings (=**diverticula**) at weakened sites along the large intestinal wall. In western countries, such **diverticulosis** occurs in about half of adults 60 years and older. Conversely, much lower rates of diverticulosis occur in undeveloped countries where fiber-rich diets predominate, suggesting that diverticular formation may be promoted by low-fiber western diets. Approximately 10–25% of diverticulosis cases develop into **diverticulitis**, where hardened feces within diverticular sacs can cause inflammation and rupturing. To prevent such life-threatening complications, treatments of diverticulitis can involve antibiotics, probiotics, increased physical activity, and/or surgeries.

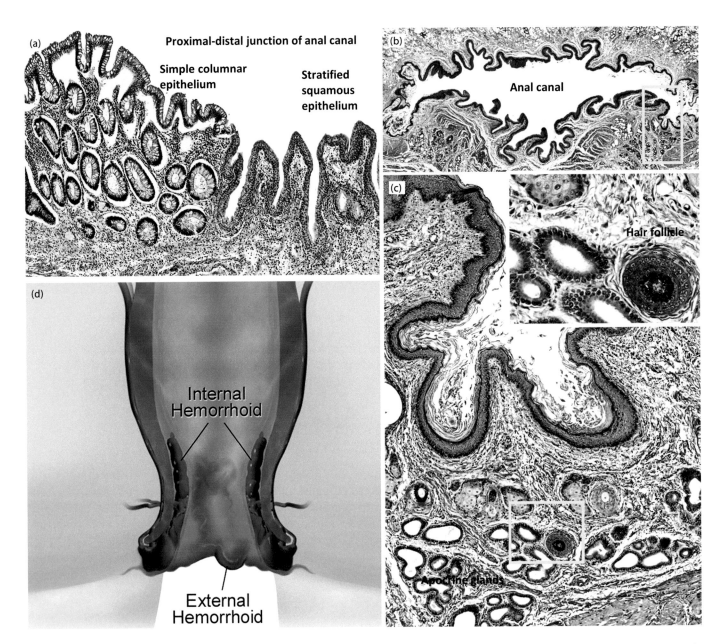

Figure 12.10 Anal canal. (a–c) *The distal portion of the anal canal can be distinguished from the rectum and proximal anal canal by its stratified squamous epithelium as well as by its keratinization, hair follicles, and apocrine glands. (a) 25X; (b) 8X; (c) 50X; (c, inset) 125X. (**d**) Blood vessels in the anal wall can become dilated to form hemorrhoids. ([d]: From Blausen.com staff (2014) "Medical gallery of Blausen Medical 2014". Wiki J Med 1 (2): 10, doi:10.15347/wjm/2014.01010, reproduced under CC0 and CC-SA creative commons licenses.)*

12.12 SUMMARY—DIGESTIVE TRACT II: PHARYNX TO ANUS

Pictorial Summary Figure Caption: (a)*: see figure 12.1a; (****b****): see figure 12.3c; (****c****): see figure 12.6a; (****d****): see figure 12.7b; (****e****): see Figure 12.9a.*

The **post-oral digestive tract** consists of the pharynx, esophagus, small intestine, and large intestine that terminates at the anus. Such regions typically possess a **quadripartite wall** comprising a: (i) **mucosa**, (ii) **submucosa**, (iii) **muscularis**, and (iv) either a peripheral-most **adventitia** in gut components attached to other tissues or an outer **serosa** in regions suspended within the abdominal cavity (**Figure a**).

The **pharynx** conveys food to the esophagus and is mostly lined by a **stratified squamous epithelium**. The **esophagus** connects the pharynx to the stomach and is lined by a **stratified squamous epithelium**. **Striated** muscles in upper esophagus can be voluntarily controlled to initiate swallowing, whereas involuntary **smooth** muscle predominate distally. The distal end of the esophagus has **sphincters** whose ability to block stomach acid backflow is compromised in gastroesophageal reflux disease (**GERD**).

The **stomach** secretes **acid** and **enzyme precursors** to convert food into a liquefied product, called **chyme**. The gastric wall has macroscopic folds (**rugae**) and contains three smooth muscle layers—oblique, circular, and longitudinal. The mucosal lining is a **simple columnar epithelium** perforated by pores. Each pore leads into a **foveola** (=gastric pit) that connects with underlying **glands** (**Figure b**). Gastric glands in the main **body** region have: (i) **mucous cells** that form a protective coat of mucus; (ii) **stem cells** that replenish senescent cells; (iii) **enteroendocrine** cells that secrete hormones (e.g., ghrelin); (iv) **chief cells** that produce enzyme precursors (e.g., **pepsinogen**); and (v) **parietal** (=oxyntic) cells that secrete **acid** and **intrinsic factor**—a glycoprotein that normally facilitates vitamin B_{12} uptake for erythropoiesis and thus when released at insufficient levels can result in **pernicious anemia**. Acid secretion by parietal cells in the stomach body requires: (i) **carbonic anhydrase** to generate H^+, and (ii) H^+/K^+ pumps on **apical canaliculi** to extrude H^+. Acid secretion is inhibited by **somatostatin** and stimulated by: (1) **acetylcholine** from cholinergic neurons, (2) **gastrin** from G-type enteroendocrine cells of the pylorus, and (3) **histamine** from **enterochromaffin-like cells** (ECLs) in the gastric lamina propria.

The **small intestine** between the stomach and large intestine comprises the **duodenum**, **jejunum**, and **ileum**. **Surface area** in these regions is increased for optimal digestion and absorption via: (i) **plicae circulares** (mucosa and submucosa folds); (ii) finger-like mucosal **villi** extending from plicae circulares; (iii) invaginated mucosal **crypts of Lieberkuhn**; and (iv) the brush border of **microvilli** on **enterocytes** in the **simple columnar epithelial**

lining of the mucosa (**Figures c, d**). Scattered among enterocytes are **mucous goblet cells** that secrete protective mucus. Crypts have: (i) active **stem cells** that accommodate a rapid epithelial turnover rate, and (ii) **Paneth cells** with large granules containing anti-microbial agents. In comparing the duodenal and ileal regions, the submucosa of the duodenum has **Brunner's glands** for producing bicarbonate-rich mucus, whereas **Peyer's patches** of the ileum possess lymph follicles with overlying **M-cells** for immune responses. The jejunum lacks both Brunner's glands and Peyer's patches. Small intestinal activities, such as waves of **peristaltic contractions** for propelling food boluses toward the anus are controlled by an enteric nervous system (ENS) containing **submucosal** and **myenteric** plexuses in the intestinal wall. Duodenal digestion is mainly regulated by the **cholecystokinin (CCK)** hormone that is secreted by I-type enteroendocrine cells. CCK: (i) **inhibits gastric emptying**; (ii) **contracts the gallbladder** to deliver stored **bile**; and (iii) **causes pancreatic enzyme and bicarbonate secretion**.

The **large intestine** comprises a: (i) proximal pouch (**cecum**) with a worm-shaped **appendix**, which when inflamed can require removal; (ii) main part, called the **colon**; (iii) **rectum** that connects the colon with the anal canal; and (iv) **anal canal** that terminates at the anus (**Figure e**). The colonic **mucosa** has a **simple columnar lining** that (i) lacks villi; (ii) has numerous **mucous goblet cells** for extra lubrication; (iii) forms tall crypts of Lieberkuhn; and (iv) is underlain by **tenia coli** muscle bands. The colon absorbs water, ions, and vitamins while forming **feces** from residual chyme and from added wastes, such as **bilirubin** breakdown products of recycled erythrocytes that contribute to the brown color of feces. The distal **anal canal** has a **stratified squamous epithelium** plus hairs, glands, vascular beds, and surrounding sphincters.

Some digestive tract disorders: **ulcers**—breaks in the mucosa of the stomach or proximal duodenum that are often caused by *Helicobacter pylori* bacteria; **Barrett's esophagus**—metaplastic conversion of the esophageal lining into a simple columnar epithelium due to chronic GERD; **diverticulosis**—the formation of colonic outpocketings (**diverticula**) that can become inflamed during **diverticulitis** to cause bowel obstructions and hemorrhages.

SELF-STUDY QUESTIONS

1.   What is the name of the bacterium that can cause ulcers?
2.   T/F Chief cells secrete pepsinogen.
3.   Which gastric cells secrete acid?
4.   Which one best characterizes the jejunum?
     A.   Brunner's glands
     B.   Foveolae
     C.   Peyer's patches
     D.   An absence of Brunner's glands or Peyer's patches
     E.   Tenia coli
5.   What kind of epithelium lines crypts of Lieberkuhn?
     A.   Non-keratinized stratified squamous epithelium
     B.   Simple columnar
     C.   Lymphoepithelium
     D.   Simple squamous
     E.   NONE of the above
6.   Which does NOT stimulate acid secretion?
     A.   Acetylcholine
     B.   Somatostatin
     C.   Histamine
     D.   Gastrin
     E.   ALL of the above stimulate acid secretion
7.   Which disease is characterized by a GERD-mediated metaplastic transformation of the mucosal lining into a simple columnar epithelium?
8.   T/F The colon has plicae circulares and villi.
9.   What is the name of the most proximal region of the small intestine?

10. T/F Striated muscles occur in the upper esophagus.
11. Where are rugae found?
    A. Duodenum
    B. Colon
    C. Appendix
    D. Esophagus
    E. Stomach
12. Which is NOT a direct effect of CCK?
    A. Inhibits gastric emptying
    B. Closes the pyloric sphincter
    C. Stimulates acid secretion
    D. Causes gall bladder contraction
    E. Stimulates pancreatic secretions
13. T/F The pouch-like region of the large intestine adjacent to the ileum is called the cecum.

"EXTRA CREDIT" For each term on the left, provide the BEST MATCH on the right (answers A-F can be used more than once)

14. Generates H^+ ___   A. Crypt of Lieberkuhn
15. Paneth cells____   B. Overlying Peyer's patch
16. M cells ____   C. Tenia coli
17. I cells___   D. Carbonic anhydrase
18. Intrinsic factor ____   E. CCK secretion
                               F. Prevents pernicious anemia

19. Describe structural adaptions and hormonal control of digestion and absorption in the duodenum.
20. Describe acid production in the stomach, including as many pertinent molecules, cellular features, and contributing cell types as possible.

ANSWERS

(1) *Helicobacter pylori*; (2) T; (3) parietal (=oxyntic) cells; 4) D; (5) B; (6) B; (7) Barrett's esophagus; (8) F; (9) duodenum; (10) T; (11) E; (12) C; (13) T; (14) D; (15) A; (16) B; (17) E; (18) F; (19) The answer should discuss: surface area adaptations—plicae circulares, villi, crypts of Lieberkuhn, enterocytes and their brush border of microvilli—mucous goblet cells, and CCK plus its effects on the stomach, gall bladder, pancreas, and gut wall; (20) The answer should discuss: parietal cells, mitochondria, intracellular canaliculus, carbonic anhydrase, bicarbonate secretion, proton pumps, Cl^- K^+ channels, acetylcholine, gastrin, G cells, histamine, ECL cells, H2 receptors, somatostatin

Digestive System III: Liver, Gallbladder, and Pancreas

13.1 AN OVERVIEW OF LIVER, GALLBLADDER, AND PANCREAS DEVELOPMENT

With their ducts that attach to the duodenum, the **liver, gallbladder**, and **pancreas** deliver key enzymes and other secretory products needed for proper intestinal functioning (**Figure 13.1a**). After summarizing how these accessory digestive organs develop, their microanatomy in adults is described while focusing mainly on their roles in digestion. However, given the dual nature of the pancreas as both an exocrine and endocrine organ, this chapter also discusses pancreatic hormones secreted by the **islets of Langerhans**.

The three accessory digestive organs initially arise from two outgrowths of foregut endoderm that subsequently undergo a complex morphogenesis to generate a duodenal-connected liver, gallbladder, and pancreas. Organ development begins during the 4th week of gestation, when the first foregut evagination, called the **hepatic bud**, arises ventrally to begin producing the liver (**Figure 13.1b, c**). As the nascent liver enlarges to fill much of the embryonic abdomen, two types of hepatic epithelial cells develop—(1) **hepatocytes** that form most of the liver's chief secretory product, called **bile**, and (2) duct-lining **cholangiocytes** that secrete and modify bile during its transport through the liver.

DOI: 10.1201/9780429353307-16

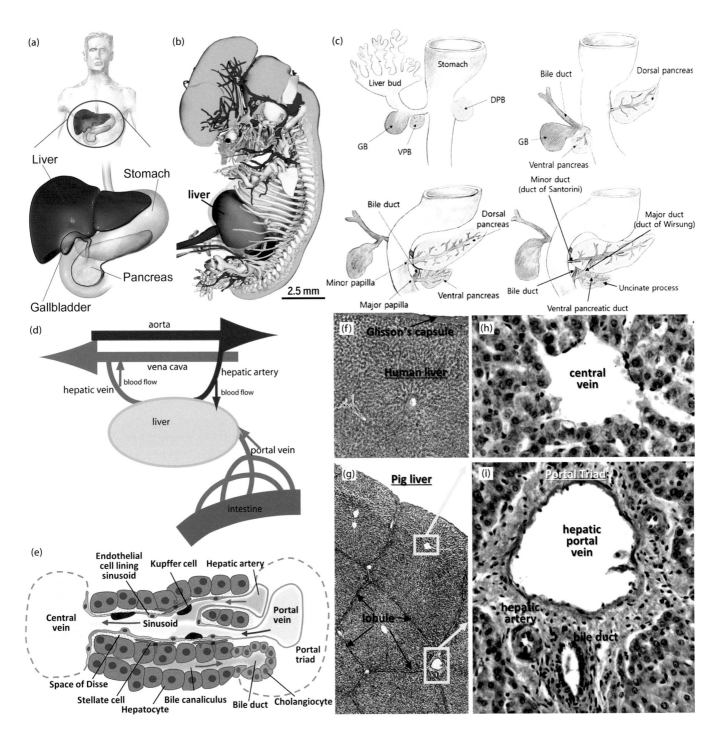

Figure 13.1 Introduction to accessory digestive organs and liver anatomy. (*a*) The duodenum is connected to three accessory digestive glands—the liver, gallbladder, and pancreas—which facilitate proper digestion and absorption of nutrients. (*b*) As shown in this 3D reconstruction of a 7.5-week-old human embryo, the developing liver fills much of the abdomen. (*c*) The liver, gallbladder, and pancreas arise from a ventral and a dorsal bud that extend from the future duodenal region of the gut. The ventral bud initially forms the liver and later produces a gallbladder bud (GB) and a ventral pancreatic bud (VPB). With the liver removed from the last three drawings for clarity, the ventral pancreatic bud is shown rotating around the duodenum and fusing with the dorsal pancreatic bud to form a minor portion of the adult pancreas, whereas the gallbladder bud forms the gallbladder that stores bile. (*d, e*) The liver receives most of its vascular supply in the form of nutrient-rich but oxygen-poor blood via the hepatic portal vein that drains the intestines and spleen. In addition, the hepatic artery supplies nutrient-poor and oxygen-rich blood. Throughout the interior packing tissue of the liver (=hepatic parenchyma), both types of blood mix within sinusoidal capillaries and are drained via numerous

Figure 13.1 **(Continued)** *central veins, eventually returning to the heart via the inferior vena cava. (**f–i**) Parenchyma of the human liver (f) lacks obvious sub-structuring, as opposed to livers of pigs and a few other mammals, where the parenchyma is divided by connective tissue sheaths into hexagonally shaped lobules (g–i). Each of these classic lobules has a medial central vein surrounded by up to six peripheral portal triads, which comprise branches of the hepatic portal vein, hepatic artery, and bile duct. (f) 35X; (g) 20X; (h) 200X; (i) 200X. ([a]: From Blausen.com staff (2014) Medical gallery of Blausen Medical 2014. Wiki J Med 1(2):10. doi:10.15347/wjm/2014.01010, reproduced under CC0 and CC SA creative commons licenses; [b]: From De Bakker, BS et al. (2016) An interactive three-dimensional digital atlas and quantitative database of human development. Science 354:10.1126/science.aag0053, reproduced with publisher permission; [c]: From Kim, SS et al. (2019) Various congenital abnormalities and anatomic variants of the pancreas: a pictorial review. J Belgian Soc Radiol 103(1):39, 1–9. doi: https://doi.org/10.5334/jbsr.1780, reproduced under a CC BY 4.0 creative commons license; [d]: From Aitsebaomo, J et al. (2008) Molecular insights into arterial–venous heterogeneity. Circulation Res 103:929–939, reproduced with publisher permission; [e]: From Gordillo, M et al. (2015) Orchestrating liver development. Development 142:2094–2018, reproduced under a CC BY 4.0 creative commons license.)*

With continued hepatic growth, the proximal part of the hepatic bud forms two additional primordia: (i) a **cystic bud** that gives rise to the **gallbladder**, and (ii) the **ventral pancreatic bud** that produces a minor portion of the adult pancreas (**Figure 13.1c**). A new foregut outgrowth also evaginates from the dorsal side of the developing gut to generate the **dorsal pancreatic bud** (**Figure 13.1c**). The dorsal anlage grows rapidly and eventually forms the bulk of the adult pancreas, whereas the ventral pancreatic bud rotates around the prospective duodenum to fuse with the dorsal anlage and generate parts of the proximal pancreas (**Figure 13.1c**). Following such fusion, a **main pancreatic duct** (of Wirsung) ends up draining most of the pancreas, while an **accessory pancreatic duct** (of Santorini) can provide supplemental drainage in some adults (**Figure 13.1c**).

13.2 LIVER ANATOMY AND HISTOLOGY: DUAL BLOOD SOURCES SUPPLYING A PARENCHYMA OF HEPATOCYTIC PLATES SEPARATED BY SINUSOIDS

As the most massive accessory digestive organ, the adult **liver** weighs on average ~1.5 kg and is traditionally viewed as comprising four anatomical lobes. Beneath an overlying sheath of connective tissue (=**Glisson's capsule**), internal hepatic tissues are highly vascularized with ~800–1,200 ml of blood percolating through the liver each minute. Blood entering the liver is delivered from two sources (**Figure 13.1d**). Approximately 75% of the blood supply comes from the **hepatic portal vein** that provides **nutrient-rich** but **oxygen-poor** blood from the digestive tract and spleen for multifaceted processing by hepatocytes. This vessel is a portal vein because it connects initial capillaries in the digestive tract and spleen with a second plexus of hepatic **sinusoidal** capillaries. To maintain proper oxygen levels, the remaining ~25% of the liver's vascular supply is contributed by the **hepatic artery**, which delivers **oxygen-rich** but **nutrient-poor** blood that is pumped from the heart to reach sinusoidal capillaries (**Figure 13.1e**). Once inside a sinusoid, arterial blood combines with venous blood from the hepatic portal vein, and after completing its intrahepatic circulation, the mixed blood is eventually returned to the heart.

The densely packed epithelial tissue supplied by these two blood sources is called **parenchyma** and mostly comprises branched plates of **hepatocytes** that account for ~70% of resident hepatic cells. Interspersed among these plates are sinusoidal capillaries, which are separated from neighboring hepatocytes by a thin perisinusoidal **space of Disse** (**Figure 13.1e**). In addition to hepatocytes and sinusoid-associated cells, parenchyma also contains **cholangiocytes** in the epithelial lining of bile-transporting ducts, which comprise ~3% of resident hepatic cells (**Figure 13.1e**).

When sectioned and examined at low magnification (**Figure 13.1f**), human hepatic parenchyma lacks an obvious sub-structure, such as exhibited by pigs and some other mammals, where connective tissue sheaths divide the parenchyma into distinct units, termed **classic lobules** (**Figure 13.1g**). In each of these hexagonally shaped lobules, a mesially positioned **central vein** (**Figure 13.1h**) is surrounded by radially arranged plates of hepatocytes plus interspersed sinusoids that span across the lobule to the peripheral connective tissue sheath. Such sheaths contain up to six **portal triads**, which comprise branches of the **hepatic portal vein**, **hepatic artery**, and **bile duct** (**Figure 13.1i**). Between portal triads and the central vein, classic lobules generate oppositely directed flows of bile and blood. Thus, hepatocyte-secreted **bile** moves centrifugally from central vein surroundings toward portal bile ducts, whereas **blood** flows centripetally within sinusoids from portal vessels to the central vein (**Figure 13.1e**).

Although discrete connective tissue sheaths are lacking, hepatic lobules of human livers exhibit regularly spaced portal triads and central veins, and among the several models of lobular organization that have been proposed for human livers (Box 13.1 Differing Views of Lobular Organization in Human Livers), a classic-lobule-like arrangement is the most widely accepted view. Regardless of the precise lobular organization, blood delivered from central veins eventually exits the liver in **hepatic veins** and returns to the heart via the **inferior vena cava** (**Figure 13.1d**). Conversely, bile from portal bile ducts coalesces into progressively larger intrahepatic bile ducts and exits the liver to travel down the **common hepatic duct**. In addition to these two fluids, lymph derived from sinusoidal plasma is carried by hepatic **lymph vessels** that tend to originate in periportal spaces (of Mall) surrounding portal triads.

13.3   FUNCTIONAL MICROANATOMY OF PARENCHYMAL HEPATOCYTES AND CHOLANGIOCYTES DURING BILE FORMATION AND TRANSPORT

Based on its vascularization pattern, classic lobule parenchyma can be functionally divided into three concentric zones (**Figure 13.2a**), with the highest oxygen and nutrient levels occurring near the portal triads in zone I. Conversely, the lowest oxygen and nutrient levels surround the central vein in zone III, and intermediate values are established in middle zone II that is situated between zones I and III (**Figure 13.2a, b**).

Within these zones, **hepatocytes** process nutrients and toxins derived from the digestive tract in key ways that help maintain homeostasis throughout the body. For example, hepatocytes: (1) **reduce blood glucose** after meals by converting glucose into glycogen (=glycogenesis); (2) **increase blood glucose** during fasting both by breaking down glycogen into glucose (=glycogenolysis) and by forming new glucose from building-blocks like glycerol, lactate, and glucogenic amino acids (=gluconeogenesis); (3) **synthesize** the majority of **proteins circulating in blood** (e.g., albumins, fibrinogen, and globulins); (4) **contribute to fat stored in adipose tissue** by producing and secreting into the bloodstream **very low density lipoproteins** (**VLDLs**); (5) **catabolize dietary proteins** and **convert** the resulting harmful **ammonia** byproduct into **urea** for excretion by the kidneys; (6) **help rid the body** of ingested toxins (e.g., alcohol and pharmaceuticals); and (7) **aid antimicrobial defenses**.

To perform these functions, hepatocytes possess abundant endoplasmic reticulum arrays, Golgi bodies, mitochondria, peroxisomes, and glycogen particles (**Figure 13.2c–e**). In addition, up to 40% of hepatocytes are **polyploid** with supernumerary amounts of DNA beyond normal diploid (2n) levels. The extra DNA is sometimes distributed in two nuclei (**Figure 13.2f**) and is thought to provide backup genes for these active cells that are often subjected to genotoxic stress.

BOX 13.1 DIFFERING VIEWS OF LOBULAR ORGANIZATION IN HUMAN LIVERS

Three main models have been proposed to explain how lobules are organized in human livers (**Figure a**). The **central-vein-oriented** view of a classic lobule has portal triads at the lobular periphery sending blood to a vein in the center of each lobule. Conversely, according to the **portal-triad-centered** view, each lobule is arranged around a central portal triad that along with other triads transports blood to collecting veins at the lobular periphery. Similarly, the **hepatic acinar** model postulates that blood from a pair of centrally located portal triads diverges into two collecting veins at opposite ends of lobule, with these veins also receiving blood from other nearby triads.

Regardless of their details, all three models maintain that mixed arterial and venous blood delivered to sinusoids flows from portal triads toward collecting veins before being drained from the liver. Similarly, the countercurrent transport of bile from the vicinity of collecting veins toward portal triads is also conserved across lobule models.

However, the precise positioning of the collecting veins relative to their associated portal triads within each lobule affects various parameters of lobular function. For example, the size and shape of parenchymal zones I, II, and III, which represent differing oxygen and nutrient levels, are fundamentally altered, depending on how lobules are organized. In addition, the particular way that parenchymal plates and sinusoids are distributed within each lobule can affect fluid flow rates, with so-called "axial" regions of each lobule that are centered around portal triads (see **Figure 13.2a**) being predicted to have faster flow rates than those in interspersed "facial" regions between axial components (**Figure b**). Various empirical and computational data validate certain aspects of all models, and although no single model is fully supported, the central-vein-centered version is generally viewed as the reigning paradigm in the field of hepatology.

(a)

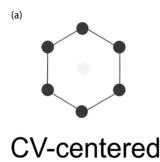

CV-centered hepatic lobule

PT-centered hepatic lobule

Hepatic acinus

(b)

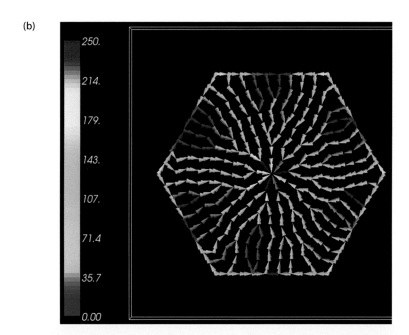

Box 13.1 *(**a**) Three main models for hepatic lobular organization. (**b**) Computer modeling suggests that flow rates differ within various regions of a classic lobule (high velocities = red; low velocities = blue). ([a, b]: From Fu, X et al. (2018) Modeling of xenobiotic transport and metabolism in virtual hepatic lobule models. PLoS ONE 13(9):e0198060, reproduced under a CC BY 4.0 creative commons license.)*

In addition to such adaptations, hepatocytes form unusual apical-basal polarities. Thus, the sides of the hepatocyte overlying nearby sinusoids represent its basal surfaces, whereas opposite to these surfaces, the hepatocyte generates multiple apices, with each apex forming an indented gutter along the length of the cell (**Figure 13.3g**). In conjunction with a similar indentation in a neighboring hepatocyte, such grooves generate narrow intercellular channels, called **bile canaliculi**, which form the initial conduits of the bile-transporting system. To facilitate bile secretion, hepatocytes seal off these spaces with tight junctions and extend microvilli into each canaliculus (**Figure 13.2g–j**).

Within parenchyma surrounding central veins, individual bile canaliculi coalesce with similar units before eventually connecting with larger transporting conduits, called **canals of Hering** (**Figure 13.2k**). Such canals are distinguished from bile canaliculi in being lined by a mixture of hepatocytes and **cholangiocytes**. In addition, canals of Hering contain hepatic **stem cells** that in response to trauma can differentiate into new hepatocytes and cholangiocytes, thereby providing considerable regenerative capabilities for damaged livers. At their peripheral

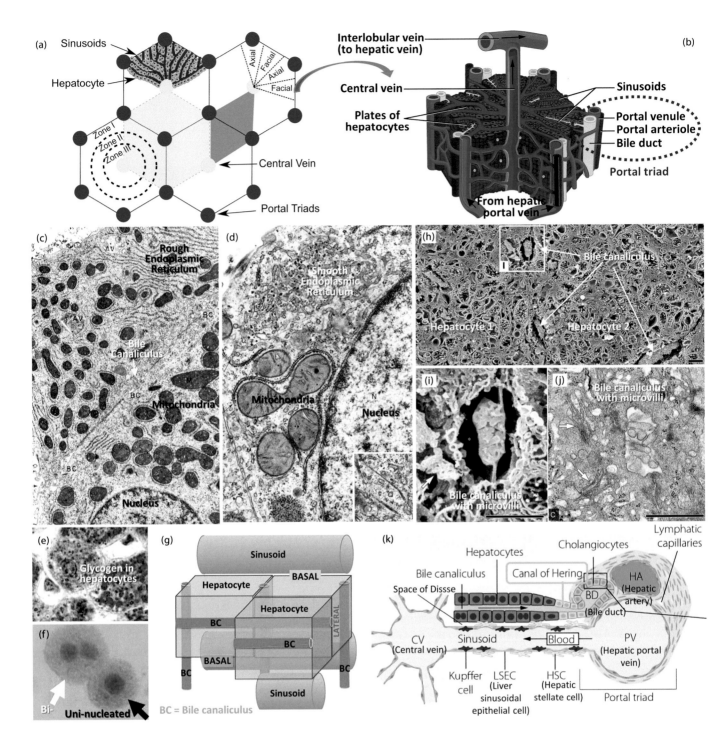

Figure 13.2 Lobule organization and hepatocyte microanatomy. (*a*) Human liver parenchyma is generally viewed as being organized into classic lobules with three zones (I, II, III) of decreasing nutrient and oxygen levels. (*b*) Interspersed among plates of hepatocytes are sinusoids that bring mixed blood from the portal triad to the lobular central vein. Conversely, bile produced by hepatocytes in the vicinity of the central vein moves toward bile ducts in portal triads. (*c, d*) As metabolically active cells, hepatocytes have numerous mitochondria and ER arrays, while also forming narrow channels, called bile canaliculi. (c) 5,000X; (d) 12,000X. (*e*) Hepatocytes contain abundant glycogen particles ((e) 370X). (*f*) Up to 40% of hepatocytes are polyploid and in some of these cases contain two nuclei ((f) 350X). (*g*) Each hepatocyte can border: (i) a thin perisinusoidal space of Disse that surrounds sinusoids; (ii) lateral membranes of neighboring hepatocytes; and (iii) indented gutter-like bile canaliculi occurring between neighboring hepatocytes (BCs); (*h–j*) As shown in scanning electron micrographs (SEMs) (h, i) and a TEM (j) of sectioned pig livers, hepatocytes extend microvilli into bile canaliculi, thereby increasing surface area to optimize bile secretion; scale bars = 1 μm. (*k*) After traversing bile

Figure 13.2 (Continued) *canaliculi that are lined exclusively by hepatocytes, bile enters canals of Hering, whose lining comprises both hepatocytes and cholangiocytes, which in turn modify and supplement bile produced by hepatocytes. Such canals connect with bile ducts that are lined exclusively by cholangiocytes. ([a]: From Fu, X et al. (2018) Modeling of xenobiotic transport and metabolism in virtual hepatic lobule models. PLoS ONE 13(9):e0198060. https://doi. org/10.1371/journal.pone.0198060, reproduced under a CC BY 4.0 creative commons license; [b]: From OpenStax College (2013) Anatomy and Physiology. OpenStax. http://cnx.org/content/col11496/latest, reproduced under CC BY 4.0 creative commons license; [c, d]: From Chapman, GS et al. (1973) Parenchymal cells from adult rat liver in non-proliferating monolayer culture: II. Ultrastructural studies. J Cell Biol 59:635–747, reproduced with publisher permission; [f]: From Baratta, JL et al. (2009) Cellular organization of normal mouse liver: a histological, quantitative immunocytochemical, and fine structural analysis. Histochem Cell Biol 131:713–726, reproduced with publisher permission; [g]: From Gissen, P and Arias, IM (2015) Structural and functional hepatocyte polarity and liver disease. J Hepatol 63:1023–1037, reproduced under a CC BY 4.0 creative commons license; [h–j]: From Ishihara Y et al. (2020) The ultrastructural characteristics of bile canaliculus in porcine liver donated after cardiac death and machine perfusion preservation. PLoS ONE 15(5):e0233917. https://doi.org/10.1371/journal.pone.0233917, reproduced under a CC BY 4.0 creative commons license; [k]: From Tam, PKH et al. (2018) Cholangiopathies – towards a molecular understanding. EBioMedicine 35:381–393, reproduced with publisher permission.)*

ends, canals of Hering connect to bile ducts in portal triads, and distal to these junctions, the conducting network is lined solely by cholangiocytes.

Each day, the adult liver secretes ~1 liter of yellowish-green bile. About 75% of this fluid is **primary bile** that is constitutively exocytosed into canaliculi by hepatocytes. The remainder of bile is formed by canal- and duct-residing cholangiocytes, which secrete a watery mixture of Cl^- and HCO_3^- ions while also resorbing some of the components present in primary bile. Such bile modifications are mainly stimulated by **secretin** released from intestinal S-cells in response to food ingestion. In addition, each cholangiocyte has projecting from its apical surface a **primary cilium** with receptors and channels that help regulate secretory and absorptive processes.

On a molar basis, bile's three major components comprise bile acids and salts, phospholipids, and cholesterol. These and other constituents allow bile to: (i) **aid lipid and water mixing** (=**fat emulsification**) within the small intestine; (ii) **sequester toxic compounds**, such as endogenously formed compounds like **bilirubin** from hemoglobin recycling; (iii) **eliminate excess cholesterol**; and (iv) help reduce enteric infections via **antimicrobial properties of bile salts**.

13.4 OTHER RESIDENT CELLS: HEPATIC STELLATE CELLS IN SPACES OF DISSE PLUS ENDOTHELIAL, KUPFFER, AND PIT CELLS OF SINUSOIDS

Situated within the **space of Disse** between hepatocytes and sinusoidal endothelial cells are flattened **hepatic stellate cells (HSCs)** (=fat-storing cells of Ito) (**Figure 13.3a, b**). HSCs constitute ~5–10% of resident hepatic cells and play key roles in both healthy and diseased livers. For example, HSCs normally store most of the body's vitamin A supply within clusters of cytoplasmic lipid droplets. Alternatively, during chronic liver disease, such as caused by alcohol, drugs, or viruses, HSCs transdifferentiate into **myofibroblast-like** cells that secrete type I collagen fibers, thereby contributing to connective tissue scarring (=**fibrosis**) and reduced hepatic functioning. In some cases, removal of the damaging stimulus and/or use of myofibroblast-targeting therapies can lead to substantial recoveries. However, if left unchecked, fibrosis may progress to end-stage **cirrhosis** associated with an irreversible loss of hepatic functions requiring either a liver transplant or emerging stem cell therapies in order to prolong life (**Case 13.1 Billie Holiday's Liver**).

As opposed to other blood cells that transiently circulate in sinusoids and normally pass through the liver, phagocytic **Kupffer cells** constitute ~5–15% of resident hepatic cells and represent key mediators of hepatic innate immunity (**Figure 13.3c**). During embryogenesis, Kupffer cells derived from bone-marrow-generated monocytes differentiate into sinusoidal phagocytes with two main phenotypes—M1 vs. M2 subtypes—which in turn correspond to predominantly pro- vs. anti-inflammatory forms, respectively.

Liver sinusoidal endothelial cells (LSECs) are widely distributed throughout hepatic parenchyma, thereby constituting ~5–10% of resident liver cells (**Figure 13.3d–f**). Unlike in most other capillaries, LSECs normally

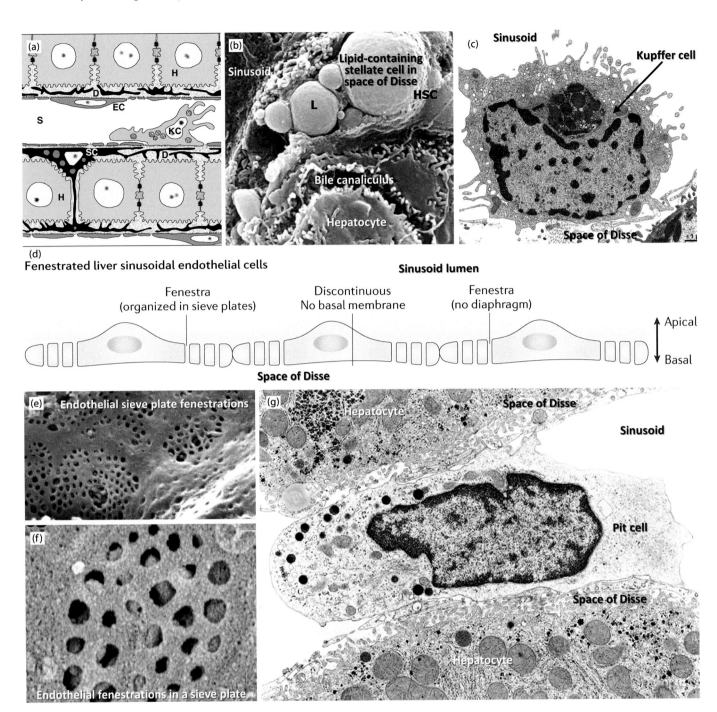

Figure 13.3 Non-hepatocyte resident cells of the liver. (*a, b*) *Hepatic stellate cells (SC) occurring within the space of Disse (D) contain lipid droplets for storing vitamin A and other hydrophobic substances. Such cells can transdifferentiate into fiber-producing myofibroblasts during pathogenesis. (b) 4,000X.* (*c*) *Phagocytic Kupffer cells occur in sinusoids overlying the space of Disse (4,500X).* (*d–f*) *Liver sinusoidal endothelial cells (LSECs) constitute a discontinuous epithelium typically lacking a basement membrane. LSEC fenestrations form sieve plates that facilitate trans-epithelial molecular flow while retaining cells in the lumen. (e) 12,000X; (f) 46,000X.* (*g*) *Pit cells are natural killer types of innate immune cells that occur in hepatic sinusoids (8,300X). ([a]: From Frevert, U et al. (2005) Intravital observation of Plasmodium berghei sporozoite infection of the liver. PLoS Biol 3(6):e192, reproduced under a CC BY creative commons license; [b]: From Le Couteur, DG et al. (2008) Old age and the hepatic sinusoid. Anat Rec 291:672–683, reproduced with publisher permission; [c]: TEM image courtesy of Drs. H. Jastrow and H. Wartenberg, reproduced with permission from Jastrow's Electron Microscopic Atlas, htttp://www.drjastrow.de; [d]: From Shetty, S et al. (2018) Liver sinusoidal endothelial cells— gatekeepers of hepatic immunity. Nat Rev Gastroenterol Hepatol 15:555–567, reproduced with publisher permission;*

Figure 13.3 (Continued) *[e, f]: From Warren, A et al. (2006) T lymphocytes interact with hepatocytes through fenestrations in murine liver sinusoidal endothelial cells. Hepatology 44:1182–1190, reproduced with publisher permission; [g]: From Peng, H et al. (2016) Liver natural killer cells: subsets and roles in liver immunity. Cell Mol Immunol 13: 328–336, reproduced with publisher permission.)*

CASE 13.1 BILLIE HOLIDAY'S LIVER

Singer
(1915–1959)

"Them that's got shall get, them that's not shall lose. So the Bible says, but it still is news."

(B.H. on life's inequalities and injustices, from the "God Bless the Child" song that she co-wrote)

Born in Philadelphia, Eleanora Fagan became known by her stage name of Billie Holiday while gaining praise as a groundbreaking jazz and blues singer. After a tough childhood that involved neglect and abuse, Holiday began singing as a teenager at various jazz clubs in New York City. Following those early gigs, Holiday landed recording contracts and honed her distinctive style while performing with prestigious bands. However, in 1947, Holiday was

arrested for narcotics possession and sentenced to prison. Following her release and an initial successful comeback that included sold-out shows, she was sent back to prison for narcotics use. With her health and voice deteriorating, Holiday continued to be targeted by authorities and was even handcuffed in her hospital bed as she lay dying from liver failure. In the end, the 44-year-old acclaimed singer succumbed to hepatic cirrhosis.

lack a basement membrane and are perforated by numerous **fenestrations**. Clusters of these perforations serve to retain blood cells within the sinusoidal lumen while providing selective **sieve plates** for transepithelial movement of metabolites, small lipid droplets, and signaling molecules (**Figure 13.3e, f**). For example, in heathy livers, LSEC fenestrations allow delivery of **nitric oxide** (**NO**) from the endothelium into the space of Disse to help keep HSCs in a quiescent state. Conversely, during chronic liver disease, LSECs lose fenestrations and form a basement membrane, thereby reducing NO delivery and promoting the conversion of HSCs into fibrosis-generating myofibroblasts.

Intra-sinusoidal **pit cells** occur next to about 1 in 10 Kupffer cells and are so-named for their cytoplasmic granules that resemble grape pits (=seeds) (**Figure 13.3g**). Pit cells can play several roles but predominantly serve as hepatic-specific **natural killer** (**NK**) **cells** that work synergistically with Kupffer cells to destroy pathogens.

13.5 THE GALLBLADDER: A LIVER-ASSOCIATED SAC FOR STORING AND MODIFYING BILE

In humans and many other vertebrates (Box 13.2 The Gallbladder of Mice and Men), bile can be stored and modified in a **gallbladder** that is located near the liver (**Figure 13.4a, b**). The adult human gallbladder is a pear-shaped sac with unusual histological features. For example, the gallbladder lining increases surface area via numerous interconnected mucosal folds that generate a honeycomb-like pattern of polygonal crypts when viewed *en face* in bisected organs (**Figure 13.4c**). Each of these mucosal folds is lined by a **simple columnar epithelium** plus underlying lamina propria, and beneath the lamina propria, both a muscularis mucosae and a submucosa are lacking. Instead, subjacent smooth muscle fibers constitute a **muscularis propria** without clearly defined circular vs. longitudinal layers (**Figure 13.4c–f**). Peripheral to the muscularis propria is a loose connective tissue compartment that comprises an adventitia where the gallbladder abuts the liver vs. a peritoneal-covered serosa in portions facing the abdominal cavity (**Figure 13.4g**).

The mucosal lining of the gallbladder consists mainly of columnar **principal cells** that are characterized by: (i) apical **mucous granules**, which help protect against bile acids (**Figure 13.4h**), and (ii) **basolateral spaces** into which ions are actively pumped so as to draw in water osmotically from the gallbladder lumen, thereby concentrating bile solutes (**Figure 13.4i**). In addition to such mucifying and concentrating roles, principal cells also contain apical microvilli that facilitate secretory and absorptive processes. For example, bicarbonate secretion by principal cells helps buffer bile acids, whereas cholesterol resorption reduces the likelihood of

BOX 13.2 THE GALLBLADDERS OF MICE AND MEN

In many vertebrates, bile that is produced by the liver can be delivered not only to the gut but also to a saclike **gallbladder**. In general, gallbladders have evolved for accumulating and modifying hepatic bile and are believed to be particularly adaptive in animals, such as carnivores and primates with high-fat diets and/or infrequent feeding modes. Conversely, herbivores that graze continuously have often lost their gallbladder. Contrary to such general trends, comparisons of two closely related murine omnivores reveal that mice have a large gallbladder that develops under the influence of the transcription factor Sox17, whereas rats lack Sox17 expression near the differentiating liver and fail to form a gallbladder (**Figure a**). Thus, the underlying selective pressures for either

the formation or loss of a gallbladder across various lineages have yet to be fully clarified for all vertebrate groups.

In humans, the gallbladder was traditionally viewed as a non-essential organ, since patients who underwent surgical removal of their gallbladder for various reasons, such as excessive gallstone formation often survived for many years without this organ. However, more recently, gallbladder removal has been shown to be linked to increased rates of obesity and metabolic syndrome disease. Such pathologies could result from the normally intermittent release of bile by the gallbladder being replaced by a steady supply from the liver, thereby altering systemic metabolism via as-of-yet poorly defined pathways that tend to promote metabolic syndrome onset (**Figure b**).

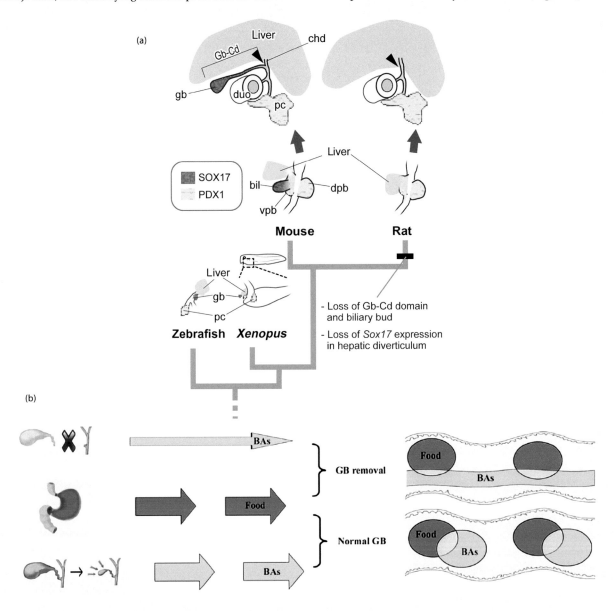

Box 13.2 (a) *Gallbladder evolution. (**b**) Intermittent bile release from an intact gallbladder compared to continual release from the liver following gallbladder removal. ([a]: From Higashiyama, H et al. (2018) Anatomy and development of the extrahepatic biliary system in mouse and rat: a perspective on the evolutionary loss of the gallbladder. J Anat 232:134–145, reproduced with publisher permission; [b]: From Qi, L et al. (2019) Gall bladder: the metabolic orchestrator. Diabetes Metab Res Rev, doi: 10.1002/dmrr.3140, reproduced with publisher permission.)*

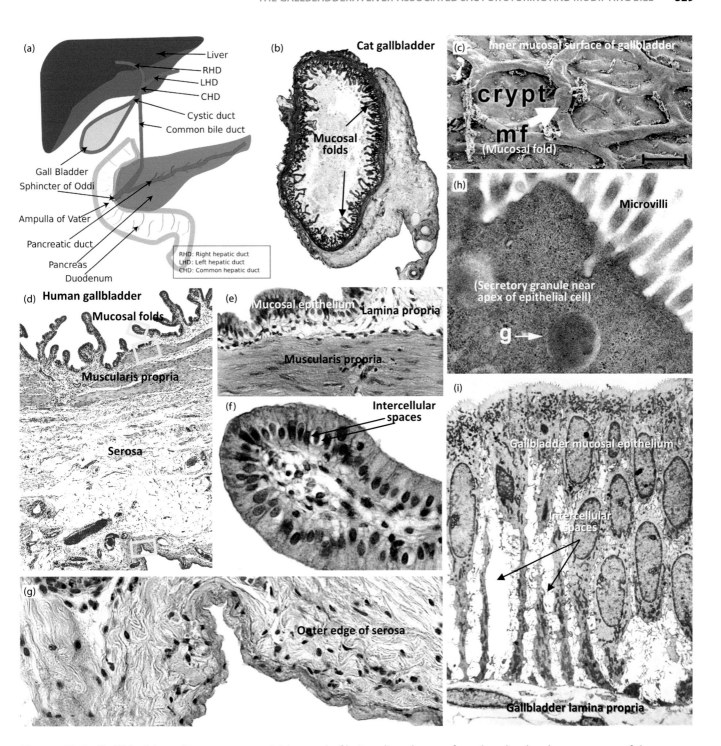

Figure 13.4 Gallbladder microanatomy. *(a) Instead of being directly transferred to the duodenum, some of the bile produced in the liver can be diverted via the cystic duct for storage and modification in the gallbladder. (b, c) The pear-shaped gallbladder contains mucosal folds that form a honeycomb-like pattern of surface crypts when viewed en face, such as in this SEM of a dog gallbladder (c, scale bar = 250 μm). (d–g) The gallbladder mucosa comprises a simple columnar lining epithelium and subjacent lamina propria. A muscularis mucosae and submucosa are lacking, and instead the mucosa overlies a muscularis propria, which in freely suspended regions of the organ is surrounded by serosa. (d) 15X; (e) 120X; (f) 250X; (g) 170X. (h, i) Principal cells of the lining epithelium secrete mucous granules to protect the mucosa. Such cells also possess large intercellular spaces into which ions are pumped in order to draw in luminal water and thereby concentrate bile. (h) 28,000X; (i) 700X. ([a]: From https://commons.wikimedia.org/wiki/File:Biliary_system_new.svg, a public domain image by vishnu2011; [c, h]: From Kesimer, M et al. (2015) Excess secretion of gel-forming mucins and associated innate defense proteins with defective mucin un-packaging underpin gallbladder mucocele formation in*

Figure 13.4 (Continued) *dogs. PLoS ONE 10(9):e0138988, doi:10.1371, reproduced under a CC BY 4.0 creative commons license; [i]: From Kaye, JL et al. (1976) Fluid transport in the rabbit gallbladder. A combined physiological and electron microscopic study. J Cell Biol 30:237–268, reproduced with publisher permission.)*

CASE 13.2 MARY MALLON'S GALLBLADDER

"Typhoid Mary", Transmitter of Typhoid Fever Bacteria (1869–1938)

"Never has there been an instance, as the present, where a woman who never had typhoid fever should prove a veritable germ factory."

(From a 1907 newspaper article describing the case of M.M., who became known as "Typhoid Mary")

After emigrating from Ireland to the United States in 1883, Mary Mallon worked as a seemingly healthy cook until she was identified as an asymptomatic transmitter of typhoid fever. Mallon earned her nickname of "Typhoid Mary" in 1906, when several people staying at a vacation home contracted the disease. To investigate the outbreak, the homeowner hired a sanitary engineer, who was able to link the illnesses and other cases to Mallon. Upon being apprehended, Mallon argued she was perfectly healthy. However, tests confirmed she had the strain of Salmonella bacteria that causes typhoid fever, and

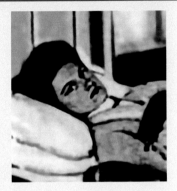

authorities offered Mallon a cholecystectomy operation in order to eliminate the pathogen reserves that were harbored in her gallbladder. After refusing to believe the diagnosis and rejecting the surgery, Mallon was isolated for three years at an island hospital for contagious diseases until she was released after having promised not to work as a cook. A few years later, though, Mallon was found cooking under another name and once again spreading typhoid fever. Thus, she was re-isolated for 23 more years, before eventually dying of pneumonia, thereby finally inactivating a gallbladder that had served as a persistent germ factory known to infect at least 50 and kill 3.

generating cholesterol-rich forms of gallstones (**Figure 13.5a–d**). Such stones not only enhance the risk of sepsis due to obstructed outflow from the gallbladder, but, along with epithelial lining cells, can also serve as reservoir sites for typhoid-fever-inducing bacteria, allowing apparently healthy carriers to transmit the disease for years (**Case 13.2 Mary Mallon's Gallbladder**).

13.6   FILLING AND EMPTYING OF THE GALLBLADDER AND THE ROLE OF BILE SALTS IN FAT EMULSIFICATION

Between meals, bile secreted by the liver into the **common hepatic duct** is either directly delivered into the duodenum via the **common bile duct** or shunted into the **cystic duct** supplying the gallbladder (**Figure 13.4a**). Flow into the gallbladder involves molecules like **vasoactive intestinal peptide** (**VIP**) that relaxes the gallbladder muscularis propria to promote filling. Conversely, after food ingestion, VIP levels drop, and **cholecystokinin** (**CCK**) secreted by duodenal I cells causes gallbladder contractions that empty upper parts of the sac while typically retaining fluid in the lowest region for additional processing. Bile ejected from the gallbladder can then be delivered to the duodenum in coordination with a CCK-induced relaxation of the **sphincter of Oddi** that surrounds the distal end of the **common bile duct** (**Figure 13.4a**).

To help digest lipids, hepatocytes utilize cholesterol to form **bile acids** like cholic acid. These are then typically conjugated to taurine or glycine during conversion into sodium and potassium **bile salts** (**Figure 13.5e–g**). Within the intestinal lumen, bile salts facilitate **fat emulsification**, whereby lipids from ingested meals are more readily mixed with water for optimal functioning of water-soluble lipases. To achieve such mixing, bile salts are amphipathic molecules with hydrophobic and hydrophilic sites that attach both to lipid droplets and to water molecules, thereby aiding mechanical churning within the duodenum as it breaks up bile-salt-coated lipids into smaller **micelles** for efficient lipase-mediated digestion (**Figure 13.5e–g**).

13.7   THE EXOCRINE PANCREAS: SECRETORY ACINI, CENTROACINAR CELLS, AND PANCREATIC DUCTS

The adult human pancreas is an ~10–15 cm-long organ that extends laterally from the duodenum across the abdomen (**Figure 13.6a, b**) (**Case 13.3 Sally Ride's Pancreas**). From its attachment site, the pancreas comprises a proximal head and neck, middle body, and distal tail of mixed exocrine and endocrine tissues. Approximately

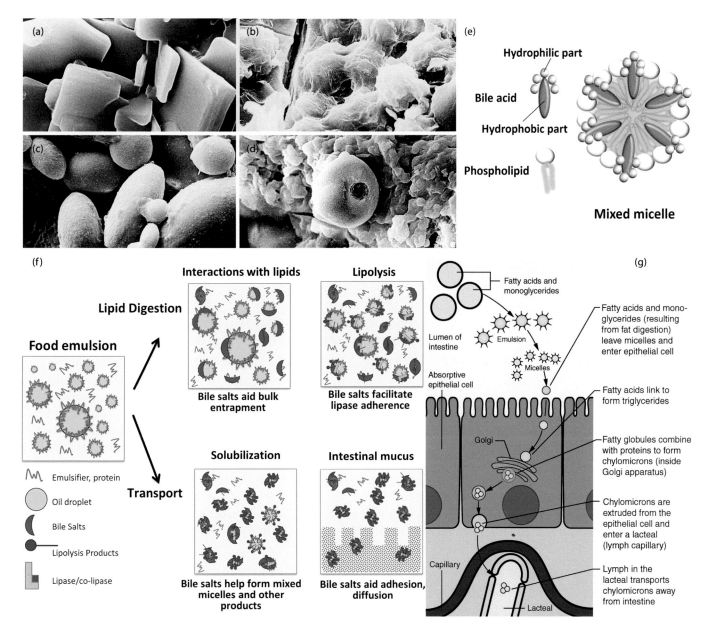

***Figure 13.5 Gallbladder stones and the roles of bile salts in digestion. (a–d)** The three most common types of gallbladder stones are shown in scanning electron micrographs (SEMs) (a = cholesterol; b = pigment; c, d = calcium carbonate) (a–d) 2,000X. **(e–g)** Cholesterol derivatives with both hydrophobic and hydrophilic regions are conjugated to sodium and potassium to form bile salts that help solubilize fats in the intestines. ([a–d]: From Qiao, T et al. (2013). The systematic classification of gallbladder stones. PLoS ONE 8:e74887, doi:10.1371/journal.pone.0074887 creative commons doi:10.1371/journal.pone.0094590, reproduced under a CC BY 4.0 creative commons license; [e]: From Pavlovic, N et al. (2018) Bile acids and their derivatives as potential modifiers of drug release and pharmacokinetic profiles. Front Pharmacol 9, doi: 10.3389/fphar.2018.01283, reproduced under a CC BY 4.0 creative commons license; [f]: From Macierzanka, A et al. (2019) Historical perspective. Bile salts in digestion and transport of lipids. Adv Colloid Interface Sci 274, https://doi.org/10.1016/j.cis.2019.102045, reproduced with publisher permission; [g]: From OpenStax College (2013) Anatomy and Physiology. OpenStax. http://cnx.org/content/col11496/latest, reproduced under CC-BY-4.0 creative commons license.)*

CASE 13.3   SALLY RIDE'S PANCREAS

Physicist and Astronaut
(1951–2012)

"Yes, I did feel a special responsibility to be the first American woman in space."

(S.R. on obligations she felt being a role model
as the first American female in space)

Making good use of her athleticism and expertise in Physics, Sally Ride became the first female from America in space. To reach such lofty heights, Ride decided before completing her Ph.D. in Physics to apply to NASA's space program and in 1978 was one of 35 accepted applicants from a pool of 8,000. Ride made two successful flights on Challenger space shuttle missions in the early 1980s. However, in 1986, the Challenger spacecraft exploded soon after launching, and Ride was later

tasked with investigating both the Challenger disaster and the subsequent explosion of the Columbia spacecraft. After leaving NASA, Ride became a Physics professor at the University of California, San Diego and co-founded Sally Ride Science to promote science education, which she guided with her life-partner, another female professor. Ride was subsequently diagnosed with pancreatic cancer and eventually succumbed to complications from her diseased pancreas at age 61.

85% of the pancreas consists of exocrine glandular cells (**Figure 13.6c–f**), whereas <2% comprises endocrine islets of Langerhans, with the remainder of the organ corresponding to exocrine ducts, blood vessels, ECM, and neural tissue.

The main role of the exocrine pancreas is to secrete **digestive enzymes** in a **bicarbonate-rich fluid**. For these functions, secretory regions form grape-like acini of basophilic cells. Such **acinar cells** contain abundant arrays of ribosome-studded endoplasmic reticulum, which in conjunction with Golgi bodies produce apically situated **zymogen granules** containing inactive forms of enzymes, such as **trypsinogen** and **chymotrypsinogen** (**Figure 13.6g, h**). In response to **CCK** that is secreted by I cells following the ingestion of fatty foods, acinar cells exocytose their zymogen granules for delivery to the duodenum. In addition, each acinus has lightly staining **centroacinar cells** that constitute the proximal duct lining within the acinar core (**Figure 13.6d**). Such cells release bicarbonate ions to buffer acidic gastric chyme, thereby providing a more optimal pH for pancreatic enzymes. The enzyme- and bicarbonate-rich fluids derived from acini are transported in progressively larger ducts (**Figure 13.6f**) before being delivered to the duodenum, predominantly via the main duct of Wirsung.

In addition to these features, the exocrine pancreas also contains scattered **Pacinian corpuscles**, which have been hypothesized to perform functions ranging from perceiving intraorgan pressure to regulating blood and lymph flow. However, the actual adaptive value, if any, of these pancreatic components remains enigmatic.

13.8   THE ENDOCRINE PANCREAS: HORMONE-SECRETING ISLETS OF LANGERHANS FOR REGULATING BLOOD GLUCOSE VIA INSULIN AND GLUCAGON

Islets of Langerhans aggregate into functional units soon after birth and eventually constitute several million sub-spherical clusters scattered among acini of the adult pancreas (**Figure 13.7a, b**). Islets normally measure 50–300 μm wide and contain three principal endocrine cells, called **β** (**Figure 13.7c, d**), **α** (**Figure 13.7e, f**), and **δ** cells. The proportions of these cells can vary substantially in different regions of the human pancreas but are generally found at organ-wide averages of about 60–80% β cells, 10–30% α cells, and 1–5% δ cells. In addition, islets contain a few other minor types of hormone-producing cells, such as those that secrete pancreatic polypeptide (PP), with PP cells occurring at substantially higher numbers in certain regions derived from the ventral pancreatic primordium (Table 13.1).

Pancreatic islets function primarily in regulating blood glucose levels, thereby helping to modulate energy sources for various cell types. Glucose regulation is accomplished via the opposing effects of two polypeptide hormones, called **insulin** and **glucagon**. In response to high blood glucose levels (=**hyperglycemia**) that can occur for example after eating, **β cells** secrete **insulin** to reduce blood glucose by triggering glucose uptake into various target cells throughout the body (**Figure 13.8a**). Conversely, **α cells** are stimulated by periods of fasting and concomitant low blood glucose levels (=**hypoglycemia**) to secrete **glucagon**, which elevates the amounts of

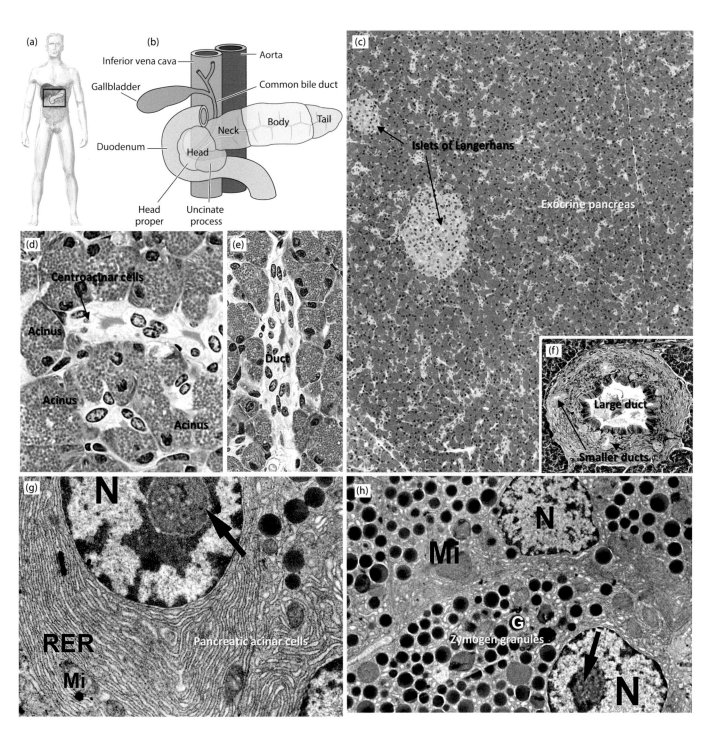

Figure 13.6 Exocrine pancreas. *(**a, b**) The pancreas extends laterally across the abdominal cavity from its attachment to the duodenum and comprises three main regions: head (=head proper + uncinate process), neck, and tail. (b) 300X. (**c**) Except for scattered endocrine islets of Langerhans, the pancreas is filled with exocrine tissue containing blood vessels plus glandular acini and their associated ducts. (c) 85X. (**d–f**) Pancreatic acini secrete their enzyme and bicarbonate products into a duct system that begins as centroacinar cells and terminates as large ducts, most of which connect with the duodenum via the main pancreatic duct of Wirsung. (d) 500X; (e) 350X; (f) 50X. (**g, h**) Acinar cells are characterized by nuclei (N) with prominent nucleoli (arrow) plus numerous mitochondria (Mi), Golgi bodies, and rough ER (RER) that help package inactive proenzymes into zymogen granules (G). (g) 8,500X; (h) 5,000X. ([a]: From Blausen.com staff (2014) Medical gallery of Blausen Medical 2014. Wiki J Med 1(2):10, doi:10.15347/wjm/2014.01010, reproduced under CC0 and CC-SA creative commons licenses; [g, h]: From Jin, Y et al. (2016) Reduced pancreatic exocrine function and organellar disarray in a canine model of acute pancreatitis. PLoS ONE 11(2):e0148458, doi:10.1371/journal.pone.0148458, reproduced under CC-BY-4.0 creative commons license.)*

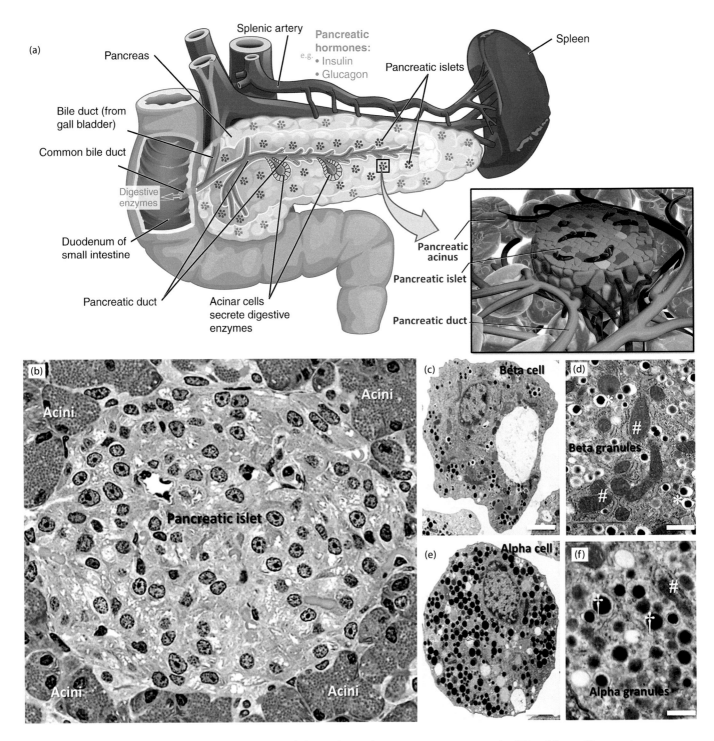

Figure 13.7 Endocrine pancreas. *(**a**) Scattered throughout the pancreas are several million islets of Langerhans.*
*(**b**) Each islet comprises three main types of cells: (1) insulin-secreting beta cells; (2) glucagon-secreting alpha cells;*
*and (3) somatostatin-secreting delta cells. (b) 460X; (**c–f**) Beta cells (c, d) tend to have smaller hormone granules than*
those produced by alpha cells (e, f); scale bars (c, e) = 2 μm; (d, f) = 0.5 μm. ([a, main figure]: From OpenStax College
(2013) Anatomy and Physiology. OpenStax. http://cnx.org/content/col11496/latest, reproduced under CC-BY-4.0 creative
commons license; [a, inset]: From Blausen.com staff (2014) Medical gallery of Blausen Medical 2014. Wiki J. Med 1(2):10,
doi:10.15347/wjm/2014.01010, reproduced under CC0 and CC-SA creative commons licenses; [c–f]: From Basford, CL
et al. (2012) The functional and molecular characterisation of human embryonic stem cell-derived insulin-positive cells
compared with adult pancreatic beta cells. Diabetologia 55:358–371, reproduced with publisher permission.)

TABLE 13.1 SECRETORY PRODUCTS OF CELLS IN THE ISLETS OF LANGERHANS AND THEIR ENDOCRINE EFFECTS

Cell type	Frequency of cell type by region (~% vol) Post. head; Ant. head, body, tail	Hormone released	Action
α cell	Very low (<1%); moderate (15%)	Glucagon	Stimulates breakdown of stored hepatic glycogen during fasting
β cell	Moderate (20%); high (80%)	Insulin	Promotes storage of nutrients in liver, muscle, and adipose tissue; paracrine inhibition of α cells
δ cell	Very low (<1%); low (5%)	Somatostatin	Slowing of nutrient absorption from intestinal tract; paracrine inhibition of glucagon and insulin
PP cell	High (80%); very low (<1%)	PP, pancreatic polypeptide	Potentiates the effect of insulin on liver

Source: From Wynne K et al. (2019) Diabetes of the exocrine pancreas. J Gastroenterol Hepatol 34: 346–354, reproduced with publisher permission (in addition, <1% of islet cells comprise minor constituents, such as ε-cells that secrete ghrelin).

circulating glucose mainly by its effects on the liver (**Figure 13.8a**). **Somatostatin** production by **δ cells**, on the other hand, can inhibit secretions by both β cells and α cells.

These three major hormones are not only controlled by blood glucose levels but also by autonomic stimulation and paracrine factors diffusing within islets (**Figure 13.8b**). Thus, according to one model of islet regulation, β cells respond to **high blood glucose** and amplifying molecules like **acetylcholine** by secreting insulin. Insulin that remains within the islet ECM can act as a paracrine factor that **inhibits glucagon secretion** by α cells, thereby further promoting a decrease in blood glucose. Conversely, **low blood glucose levels** cause the adrenal medulla and sympathetic axons terminating in the islet cortex to release **catecholamines**, which downregulate the insulin signal from β cells and allow **glucagon secretion** by α cells to increase blood glucose levels.

After being released into the bloodstream, insulin can bind insulin receptors on various target cells such as **myofibers**, **adipocytes**, and **hepatocytes** to increase both **glucose uptake** from blood and the conversion of **glucose into glycogen** (**Figure 13.8c**). In addition, insulin also **decreases glucose production** (=gluconeogenesis) in hepatocytes. Conversely, glucagon elevates blood glucose predominantly by acting on the liver to increase both **gluconeogenesis** and the **breakdown of glycogen into glucose** (=glycogenolysis).

13.9 TYPE 1 VS. TYPE 2 DIABETES MELLITUS

As a common pathology of insulin dysfunction, **diabetes mellitus** (**DM**) is characterized by elevated blood glucose levels, excessive urination (=**polyuria**), and complications, such as blindness, cardiovascular disease, and renal failure (**Case 13.4 Sejong the Great's Pancreatic Islets**). In addition, high blood glucose can spill over into urine instead of being resorbed back into blood as occurs either in normal patients or in those with excessive urination due to diabetes insipidus (Chapter 15). In fact, these two types of diabetic polyurias were once distinguished by the sweet taste of urine in DM that was conversely lacking in insipidus patients, with "mellitus" referring to "honeyed".

DM comprises two subtypes: **Type 1 diabetes** (=**T1D**) vs. **Type 2 diabetes** (=**T2D**). In people with T1D, which accounts for only ~10% of DM diagnoses, autoimmune-mediated destruction of β cells can cause insufficient insulin secretion and hence high blood glucose levels. Such hyperglycemia is typically treated with exogenous insulin supplementation while also focusing on increased exercise and balanced diets. Conversely, in T2D, insulin levels can be normal or even elevated. Nevertheless, the cumulative effects of overeating and sedentary habits result in hyperglycemia, owing to **insulin resistance**, in which target cells respond ineffectively to circulating insulin and thus do not adequately take up glucose. Such insulin resistance can involve insufficient

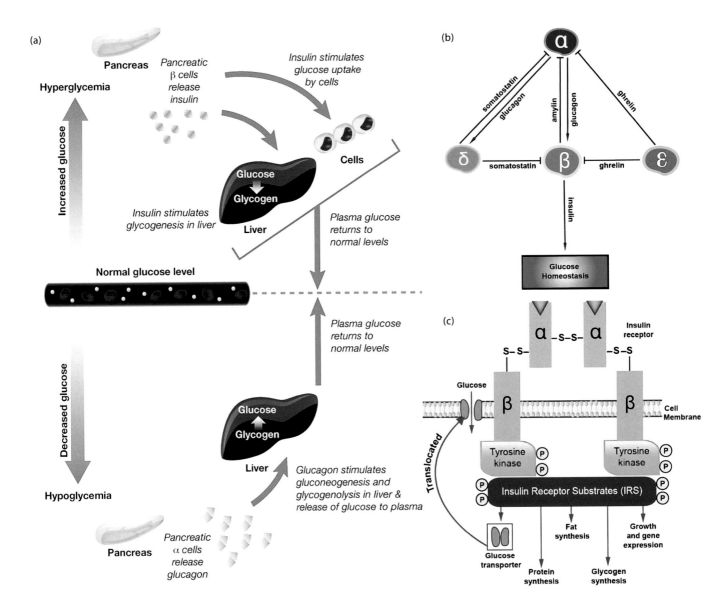

Figure 13.8 Blood glucose regulation by insulin and glucagon. *(**a**) Insulin lowers blood glucose levels, whereas glucagon has the opposite effect. (**b**) Somatostatin plus other factors, such as ghrelin and amylin help modulate alpha- and beta-cell secretions during regulation of glucose homeostasis. (**c**) Insulin binds to heterodimeric insulin receptors on cells to trigger glucose uptake. Downstream to such binding, various pathways are normally stimulated in target cells. However, in Type 2 diabetes mellitus, elevated blood glucose levels persist in the presence of normal or even elevated insulin levels, owing to abnormal reactions within target cells to insulin, collectively constituting insulin resistance. ([a]: From Haedersdal, S et al. (2018) The role of glucagon in the pathophysiology and treatment of Type 2 diabetes. Mayo Clin Proc 93:217–239, reproduced with publisher permission; [b, c]: From Sayed, SA and Mukherjee, S (2016) Statpearls Treasure Island (FL): StatPearls Publishing; 2020 Jan, https://www.ncbi.nlm.nih.gov/books/NBK430685/, reproduced under a CC BY 4.0 creative commons license.)*

expression of normal insulin receptors and/or defects in insulin-stimulated pathways downstream to receptor functioning (**Figure 13.8c**). Accordingly, instead of relying on exogenous insulin to control disease progression, T2D therapies often supplement exercise and healthier diets with alternative **hypoglycemic** agents, such as **metformin**. Such drugs are able to reduce insulin resistance via various mechanisms, including by upregulating the incorporation of glucose transporters into target cell plasmalemmata for more effective glucose uptake.

CASE 13.4 SEJONG THE GREAT'S PANCREATIC ISLETS

Korean King and Inventor
(1397–1450)

"The people are the roots of a nation, and the roots should be strong so as to create a peaceful nation."

(Quote attributed to Korean ruler Sejong the Great)

Considered one of the most influential Korean monarchs, Sejong the Great was born in present-day South Korea and assumed the throne of the Joseon dynasty in 1418. Sejong was a progressive ruler who guided his nation into a Golden Age that included not only military victories but also cultural innovations. For example, Sejong spearheaded improvements to printing press technologies and also helped design the first rain gauge and highly accurate water clock. However, by far his most

significant achievement was his invention of a phonetic alphabet to supplant the Chinese version that had been in use, thereby making it easier for Koreans to learn a writing system. Toward the end of his three-decade rule, Sejong's health deteriorated due to diabetes-related dysfunction of his pancreatic islets of Langerhans. Thus, Sejong went blind at age 50, and not long thereafter, as his pancreatic islets could no longer sustain him, the Korean ruler and inventor died due to diabetes.

13.10 SOME DISORDERS OF THE LIVER, GALLBLADDER, AND PANCREAS: HEPATIC STEATOSIS AND PANCREATIC CANCERS

Steatosis, or the abnormal buildup of lipids, can occur in various organs, but most commonly affects the liver. First identified in heavy drinkers, fatty livers have since been diagnosed in non-alcoholics with a type of steatosis termed **non-alcoholic fatty liver disease** (**NAFLD**). NAFLD is linked to obesity and underactivity and thus is also associated with other metabolic syndrome diseases, such as diabetes. In addition to excess lipid storage, fatty livers are often fibrotic and may eventually progress to end-stage cirrhosis with full liver failure. Thus, treatment typically relies on early detection coupled with dietary and lifestyle changes.

With 55,000 new cases reported in the United States during 2018, **pancreatic cancer** (**PAC**) currently constitutes only a minor fraction of all cancer diagnoses. However, given that this cancer is often detected after it has progressed substantially, PAC has a low survival rate and accounts for >40,000 deaths per year in the United States. Nearly all PACs are adenocarcinomas affecting exocrine tissues, whereas relatively few arise as islet tumors. Current therapies are not very effective, and without major advances in detecting and treating PAC, it is estimated that by 2030, pancreatic cancer will become the second leading cause of cancer-induced mortality, behind only lung cancer.

13.11 SUMMARY—LIVER, GALLBLADDER, AND PANCREAS

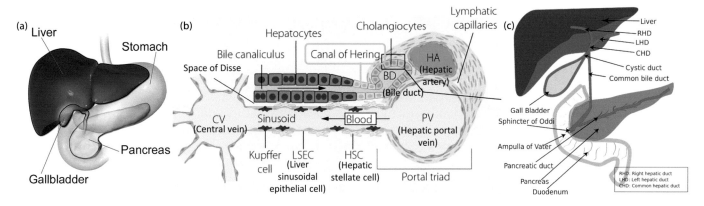

Pictorial Summary Figure Caption: (a): *see figure 13.1a;* ***(b):*** *see figure 13.2k;* ***(c):*** *see figure 13.4a.*

The accessory digestive organs—**liver**, **gallbladder**, **and pancreas** (**Figure a**)—develop from **foregut evaginations**, with the pancreas initially arising from two separate primordia that later fuse to form a single adult pancreas.

The **bile-producing liver** has a dual blood supply: (i) ~75% is nutrient-rich, oxygen-poor blood that arrives from the gut and spleen carried by the **hepatic portal vein**, and (ii) ~25% is oxygen-rich, nutrient-poor blood from the heart that is delivered by the **hepatic artery**. Both blood types mix together within hepatic **sinusoidal** capillaries (**Figure b**), supply hepatic tissues, and are drained by **hepatic veins** before returning to the heart. Liver packing tissue (=**parenchyma**) contains a few cell types like **cholangiocytes** in bile duct linings but is mainly composed of **hepatocytes** in branched plates that are separated from sinusoids by the **space of Disse**. Hepatic parenchyma is organized into functional units (=**lobules**). According to the "classic" model, each lobule has: (i) peripheral **portal triads** with branches of a **hepatic portal vein**, **hepatic artery**, and **bile duct**, (ii) a **central vein**, and (iii) numerous hepatocytic plates interspersed with sinusoids. Blood flows from portal triads into sinusoids and then to central veins, whereas **bile** flows from the areas surrounding central veins to bile ducts within portal triads. **Hepatic stellate cells** in the space of Disse normally store vitamin A but can be converted into **myofibroblasts** that secrete collagen during hepatic fibrosis. Liver sinusoidal endothelial cells (**LSECs**) are fenestrated and normally lack a basal lamina to aid molecular exchanges. Such LSECs surround **Kupffer cell** macrophages and **pit cell** types of natural killer cells occurring within sinusoids that help provide innate immunoprotection. **Hepatocytes** are often polyploid. Each cell is multifunctional (e.g., glucose formation/storage, plasma protein synthesis, fat metabolism, and detoxification) but mainly serves to produce **bile** that it secretes into apical **bile canaliculi**. The many components dissolved in bile fluid (e..g. cholesterol, bile salts, and bilirubin breakdown products of hemoglobin recycling) are transported from canaliculi into **canals of Hering** that are surrounded by hepatocytes plus **cholangiocytes** before reaching portal triad bile ducts lined solely by cholangiocytes. After exiting the liver and entering the **common hepatic duct**, bile is either directly transported to the duodenum by the **common bile duct** or shunted via the **cystic duct** into the gallbladder (**Figure c**). To optimize lipase functioning, **bile salts** aid the mixing of lipids with water (=**fat emulsification**).

The **gallbladder** stores and modifies hepatic bile before duodenal delivery and has an unusual histological organization (e.g., honey-combed-shaped mucosal folds plus no muscularis mucosae or submucosa).

The **pancreas** is a dual exocrine/endocrine gland that produces: (i) digestive enzymes (e.g., trypsinogen secreted by **acinar** glandular cells) in a bicarbonate-rich fluid (secreted by **centroacinar** cells), plus (ii) hormones from scattered **islets of Langerhans**. Such islets are clusters of hormone-secreting cells, the most common being β and α cells. β cells secrete **insulin** to lower blood glucose, whereas α cells secrete **glucagon** to raise blood glucose levels. **Diabetes mellitus** (**DM**) is characterized by **high blood glucose** and excessive urination. **Type 1 DM** involves low insulin levels due to **autoimmune destruction of β cells** and is typically treated with insulin shots. Approximately 90% of DM cases are **Type 2 DM** resulting from **insulin resistance**, which involves defects in target cells' insulin receptors and/or post-receptor signaling pathways. Thus, non-insulin drugs are used to reduce blood glucose levels in T2D.

Some disorders of the liver, gallbladder, and pancreas—hepatic steatosis: excess lipid accumulation in hepatocytes typically due to obesity or alcohol abuse; **gallstones**: hardened concretions in the gallbladder often containing **cholesterol** and arising from metabolic syndrome diseases; **pancreatic cancers**: mostly adenocarcinomas of the exocrine pancreas that are projected to be the 2nd leading cause of cancer mortality within a decade.

SELF-STUDY QUESTIONS

1. Which accessory digestive organ arises from two embryonic primordia?
2. T/F Liver sinusoidal endothelial cells typically lack a basement membrane.
3. What is the name of the vessel that supplies most of the blood to the liver?
4. Which of the following has a lumen that is surrounded entirely by hepatocytes?
   A. Bile canaliculus
   B. Canal of Hering
   C. Bile duct
   D. ALL of the above
   E. NONE of the above

5. What kind of layer directly underlies the gallbladder mucosa?
   A. Striated muscles
   B. Muscularis mucosae
   C. Muscularis propria
   D. Serosa with adrenergic neurons
   E. NONE of the above

6. Which of the following contains exclusively oxygen-rich, nutrient-poor blood?
   A. Central veins
   B. Hepatic portal vein
   C. Sinusoids
   D. Hepatic artery

7. T/F Islets of Langerhans are typically 50–300 µm wide.

8. T/F Kupffer cells reside mainly in the space of Disse.

9. T/F Liver sinusoidal endothelial cells are fenestrated.

10. What is the name of the duct that delivers bile into the gallbladder?

11. Which cells store vitamin A?
    A. Cholangiocytes
    B. Hepatocytes
    C. β-Cells
    D. Hepatic stellate cells
    E. Centroacinar cells

12. Which of the following directly aids fat emulsification?
    A. Bilirubin
    B. Glucagon
    C. Bile salts
    D. ALL of the above
    E. NONE of the above

13. T/F. Compared to most other kinds of cancers, pancreatic cancers have a high survival rate.

"EXTRA CREDIT" For each term on the left, provide the BEST MATCH on the right (answers A-F can be used more than once)

14. Natural killer cell ___	A. Centroacinar
15. Bicarbonate-secreting cell____	B. δ-Cell
16. Macrophage____	C. Pit cell
17. Secretes trypsinogen___	D. Kupffer cell
18. Can transdifferentiate into myofibroblast ____	E. Acinar cell
	F. Hepatic stellate cell

19. Describe bile composition, production, transport, and function.

20. Compare and contrast type I vs type II diabetes mellitus.

ANSWERS

(1) pancreas; (2) T; (3) hepatic portal vein; (4) A; (5) C; (6) D; (7) T; (8) F; (9) T; (10) cystic; (11) D; (12) C; (13) F; (14) C; (15) A; (16) D; (17) E; (18) F; (19) The answer should include among other pertinent topics: hepatocyte ultrastructure/polarity; bile canaliculi, canals of Hering, bile ducts, common hepatic duct, cystic duct, gallbladder, common bile duct; duodenum, bile salts, fat emulsification; (20) The answer should include such terms as islets of Langerhans, beta, alpha, delta cells; insulin, glucagon, glycogenesis, glycogenolysis, gluconeogenesis, polyuria, diabetes insipidus vs. diabetes mellitus (DM), prevalence of DMI vs. DMII, autoimmune, obesity, insulin resistance, treatment options

Respiratory System

14.1 INTRODUCTION TO THE RESPIRATORY SYSTEM WITH ITS CONDUCTING AND RESPIRATORY COMPONENTS

The primary role of the respiratory system is to **supply oxygen** while **removing carbon dioxide** and **volatile wastes**. In addition, the nasal cavity contains an **olfactory epithelium** involved in the sense of **smell**, and air moved through **vocal cords** of the **larynx** can generate various sounds.

To carry out such functions, the respiratory system comprises two main parts: (i) an initial **conducting component**, which not only optimizes inhaled air quality but also transports air to and from the lungs, and (ii) a **respiratory component** of numerous **pulmonary alveoli** (=air sacs) where gases are exchanged during respiration in the lungs (**Figure 14.1a**). In addition, a **ventilating** system consisting of the **ribcage, diaphragm,** and lung-encasing **pleural membranes** enables mechanical movements of air during cycles of inhalation and exhalation.

DOI: 10.1201/9780429353307-17

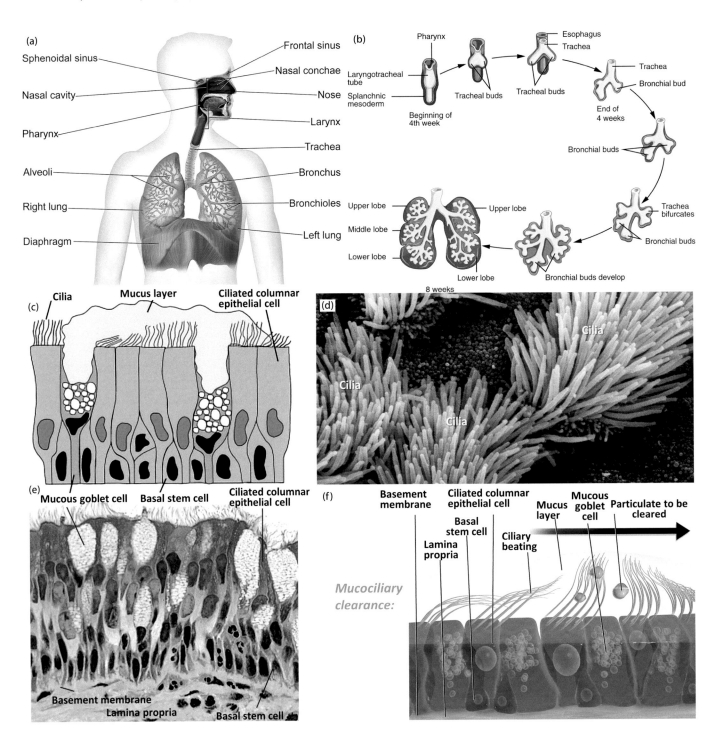

Figure 14.1 The respiratory system and its pseudostratified ciliated columnar epithelium (PCCE). (a) *The respiratory system consists of an air-transporting conducting component (nasal cavity, pharynx, larynx, trachea, bronchi, and bronchioles ending in terminal bronchioles) plus a respiratory component in the lungs comprising alveoli where gas exchanges occur. (**b**) The conducting system develops as a foregut out-pocketing that undergoes repeated branching and eventually forms alveoli within lungs. (**c–e**) Much of the conducting system is lined by a PCCE with staggered nuclei giving the appearance of stratification, when in fact PCCE is a simple epithelium, whose cells all contact the basement membrane. (**d**) 8,000X; (**e**) 650X. (**f**) As key constituents of the PCCE, ciliated cells and mucous goblet cells help move mucus toward the oral cavity during mucociliary clearance so that particulate wastes can be swallowed or expectorated. ([a, f]: From Blausen.com staff (2014) Medical gallery of Blausen Medical 2014. Wiki J Med 1(2):10, doi:10.15347/wjm/2014.01010, reproduced under CC0 and CC-SA creative commons licenses; [b]: From OpenStax College (2013) Anatomy and Physiology. OpenStax. http://cnx.org/content/col11496/latest, reproduced under CC-BY-4.0*

The nasal cavity and upper portions of the respiratory tract start differentiating at ~4 weeks post-fertilization, and about week later, the remainder of the respiratory system begins to develop from a single foregut evagination that later bifurcates and continues to branch on each side of the bifurcation (**Figure 14.1b**). The trachea forms proximal to the initial bifurcation, whereas bronchi and bronchioles arise at more distal branching sites (**Figure 14.1b**). During the second trimester, lung tissues initially assume a densely packed **pseudoglandular** morphology before gradually transforming into numerous hollow alveoli that become capable of postnatal functioning near the end of the third trimester. At birth, fewer than 10% of the ~100–400 million alveoli in each adult lung have formed, but by the time children reach 4 years old, ~80–90% of adult alveolar totals are already present.

14.2 GENERAL ORGANIZATION OF THE CONDUCTING SYSTEM MUCOSA: PSEUDOSTRATIFIED CILIATED COLUMNAR EPITHELIUM (PCCE) AND ASSOCIATED LYMPHATIC TISSUE

Much of the conducting system's **mucosa** is lined by a **pseudostratified ciliated columnar epithelium** (**PCCE**) containing short cells interspersed among taller ones. Although its staggered arrangement of nuclei suggests cellular stratification, PCCE is in fact a simple epithelium with all of its cells contacting the basement membrane (**Figure 14.1c–e**). The major cell type of PCCE is a **ciliated columnar cell** that helps transport pathogens and other foreign objects in a mucous sheath for removal by expectoration, nasal discharge, or swallowing. For such **mucociliary clearance,** mucus is secreted by subepithelial glands as well as by the second most common PCCE cells, called **mucous goblet cells** (**Figure 14.1f**). The PCCE also possesses sensory **brush cells** containing stubby microvilli. Some of these cells react to bitter cues by triggering pathways that decrease breathing rates in order to reduce inhalations of potentially noxious substances. In addition, two types of basally restricted short cells occur in the epithelium: (1) **stem cells** that replenish lost or damaged epithelial cells, and (2) **enteroendocrine cells** with small hormone-containing granules that can regulate PCCE secretions.

Although PCCE lines much of the conducting system, **stratified squamous epithelium** normally occurs in scattered patches of the nasal cavity and larynx in order to protect against damage due to high-velocity air flow. Similarly, chronic coughing induced by long-term smoking can cause a **metaplastic transition** of PCCE into stratified squamous epithelium. The metaplasia-induced loss of cilia triggered by such damage hampers mucus removal, thereby generating a positive feedback loop that intensifies the coughing.

In normal respiratory tracts, PCCEs transition distally into simple ciliated columnar epithelia. These in turn gradually reduce their height and ciliary content before becoming the non-ciliated simple squamous epithelia that line pulmonary alveoli.

To defend against invasive pathogens, the respiratory mucosa possesses **non-encapsulated lymphatic tissue** comprising: (i) scattered subepithelial leukocytes, (ii) discrete subepithelial lymph follicles, and (iii) leukocytes that infiltrate the lining epithelium. The earliest of these tissues to develop in humans is **nasal associated lymphatic tissue** (**NALT**) of the nasal cavity that begins to differentiate before birth, whereas more distal regions of the respiratory tract develop similar lymphatic tissues after birth (**Case 14.1 Amelia Earhart's Nasal Cavity**).

14.3 FUNCTIONAL MICROANATOMY OF THE NASAL CAVITY: TURBINATE BONES, OLFACTORY EPITHELIUM, AND VOMERONASAL ORGANS

Split into left and right halves by a **median septum**, the nasal cavity receives outside air through two anteriorly positioned **nostrils** (=external nares) (**Figure 14.2a, b**). Toward its outer edge, each nostril has a distal vestibule with stratified squamous epithelium containing thickened hairs, called **vibrissae**, which can trap large particles in inspired air. Proximal to the vestibule, the rest of the nasal cavity is lined mainly by PCCE and

CASE 14.1 AMELIA EARHART'S NASAL CAVITY

Aviator

(1897–1937/*1939) (last documented as alive/*legally declared dead)

"I have often said that the lure of flying is the lure of beauty."

(A.E. on her attraction to flying)

Born in Kansas, Amelia Earhart knew during her first 10-min ride in an airplane that she wanted to dedicate her life to flying. After obtaining a pilot's license, Earhart honed her skills, and in 1928, she became the first woman to fly solo across the Atlantic Ocean. In 1937, Earhart embarked with a navigator on what would have been at the time the longest flight around the world. However, after completing roughly three-quarters of

the 29,000-mile journey, the plane vanished while over the Pacific Ocean. Prior to her disappearance, Earhart had to overcome chronic sinusitis that caused debilitating pressure headaches and copious mucus production. In seeking relief, Earhart underwent several surgeries and often had to use a sinus drainage tube that she covered with a bandage. Given these maladies, it is even more amazing that Earhart was able to set so many flight records while dealing with her balky nasal cavity and sinuses.

serves to **condition** inhaled air by optimizing its temperature and water content. As part of this conditioning process, the cavity has three flattened bones—the **superior**, **middle**, and **inferior turbinate bones**—which are also called **conchae**, based on their resemblance to turbinate-like snail shells (**Figure 14.2a–d**). These shelf-like bones extend horizontally and divide the nasal cavity into restricted channels (=**meatuses**) that generate mini-vortices during air transport. Because such turbulent air flow prolongs retention time within the nasal cavity compared to a more streamline (=laminar) transport through the cavity, inspired air has a greater opportunity to reach optimal temperature and water content before being relayed toward the lungs. As a further adaptation for air conditioning, parts of the nasal cavity lamina propria also contain **erectile tissue** comprising abundant **veins** and **venules** (**Figure 14.2c–e**). Such vessels typically fill with blood during several-hour-long **nasal cycles** that periodically alternate from one side of the nose to the other. This in turn leaves half of the nasal cavity less congested at any single time point, while helping to optimize air conditioning in the engorged half.

Following conditioning, most of the air in the nasal cavity is sent toward the lungs via two posterior openings, called the **nasal choanae**. The nasal cavity also connects to each middle ear by a **Eustachian tube** that equalizes pressures across the ear drum. Moreover, small perforations, called **ostia**, link the nasal cavity with the large spaces in skull bones, called (para-) **nasal sinuses**, which reduce skull weight while also aiding air conditioning (**Figure 14.2a, b**).

Along with such roles in conditioning air, the top of the **nasal cavity** detects smell-encoding **odorant** molecules via a main **olfactory epithelium** that extends over parts of the nasal septum and upper turbinates (**Figure 14.3a–c**). This pseudostratified epithelium contains short basal **stem cells** for replenishing epithelial components plus three columnar cell types: (i) glia-like **sustentacular cells** for maintaining normal tissue functioning, (ii) **brush cells** that interact with afferent neurons for generalized sensory perception, and (iii) **bipolar** types of **olfactory sensory neurons (OSNs)** that function in olfaction via their non-motile cilia extending from an apical **dendritic knob** into the nasal cavity (**Figure 14.3d–g**). In addition, below the olfactory epithelium are **Bowman's glands** that secrete protective mucus with **odorant-binding proteins** for concentrating odorants (**Figure 14.3b**).

During olfaction, **olfactory receptors (ORs)** on ciliary membranes of OSNs bind volatile odorant ligands brought into the nasal cavity via inspired air. Several hundred functional OR genes have been identified in humans, with each OSN expressing only a single type of OR. Once bound to its ligand, the OR triggers a **G-protein-mediated signaling cascade** to depolarize the OSN (**Figure 14.3e**). Odorant-induced action potentials are then transmitted via OSN axons that extend through holes in the **cribriform plate** of the **ethmoid bone** separating the nasal cavity from the overlying **olfactory bulb** region of the brain (**Figure 14.3b**). After reaching the olfactory bulb, bipolar axons synapse in spherical clusters of neurons and glia, called **glomeruli**, which organize this input and transmit action potentials to nearby **mitral cells**. From there, action potentials are relayed via an **olfactory tract** to other parts of the brain to elicit various reactions, including avoidance behaviors and modulations of perceived tastes (**Box 14.1 Anosmia in the Elderly as a Potential Harbinger of Imminent Death**).

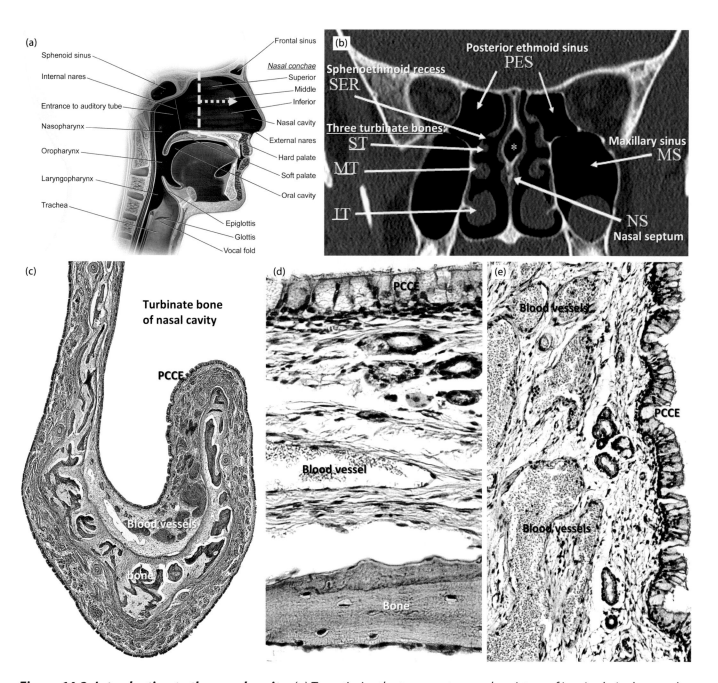

Figure 14.2 Introduction to the nasal cavity. *(**a**) To optimize the temperature and moisture of inspired air, the nasal cavity contains shelf-like turbinate (=conchal) bones that slow down air transport and thus allow maximal conditioning of air before it is sent toward the lungs. (**b**) As shown in this frontal X-ray, the three conchal bones–superior (ST), middle (MT), and inferior (IT) turbinates–possess curved distal ends that help produce non-laminar air flow. (**c–e**) To aid in air conditioning, parts of the nasal cavity also contain a well-vascularized lamina propria. (c) 25X; (d) 250X; (e) 65X. ([a]: From Blausen.com staff (2014) Medical gallery of Blausen Medical 2014. Wiki J Med 1(2):10, doi:10.15347/wjm/2014.01010, reproduced under CC0 and CC-SA creative commons licenses. [b]: From Abdulmalik, SA (2017) Paranasal sinus anatomy: what the surgeon needs to know, paranasal sinuses, Balwant Singh Gendeh, IntechOpen, doi: 10.5772/intechopen.69089. IntechOpen, doi: 10.5772/intechopen.69089. Available from https://www.intechopen.com/books/paranasal-sinuses/paranasal-sinus-anatomy-what-the-surgeon-needs-to-know, reproduced under a CC BY 3.0 creative commons license.)*

In addition to the main olfactory epithelium, most vertebrates have **vomeronasal organs (VNOs)** for detecting **pheromones**, which are externally released substances mediating chemical communication between conspecific individuals. Whether or not adult humans possess a functional VNO that supplements the main olfactory epithelium remains controversial (Box 14.2 Vomeronasal Organs (VNOs) and Pheromone Detection).

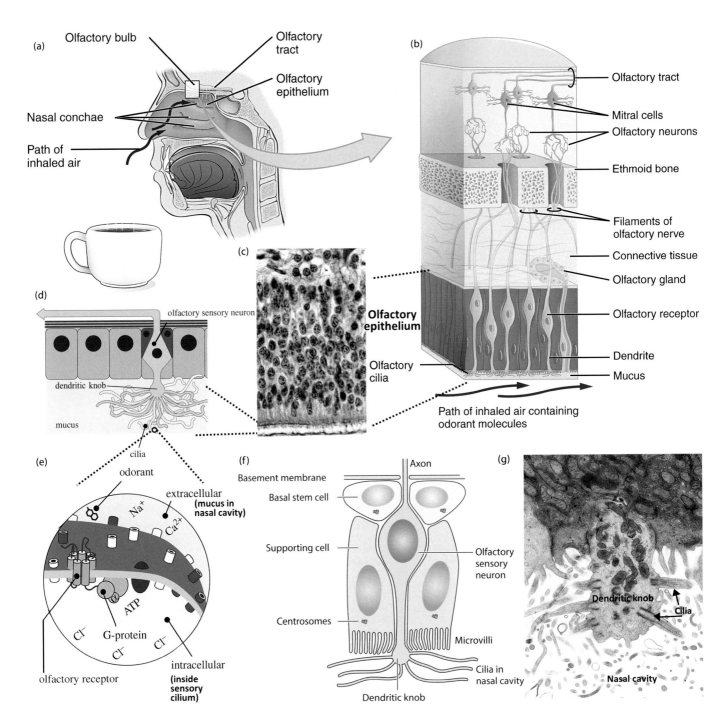

Figure 14.3 Olfactory epithelium. (a–c) *The upper nasal cavity contains an olfactory epithelium that senses odorant molecules and transmits action potentials via axons that project through a perforated ethmoid bone to reach the olfactory bulb region of the brain. (c) 300X. (**d–g**) Bipolar olfactory sensory neurons (OSNs) contain odorant receptors on elongated immotile cilia that extend into the nasal cavity from a dendritic knob region of each OSN. Odorant binding triggers a G-protein-mediated signaling cascade that results in OSN depolarization and action potential transmission to the olfactory bulb. (g) 14,000X. ([a, b]: From OpenStax College (2013) Anatomy and Physiology. OpenStax. http://cnx.org/ content/col11496/latest, reproduced under CC-BY-4.0 creative commons license; [d, e]: From Brookes, JC (2010) Science is perception: what can our sense of smell tell us about ourselves and the world around us? Phil Trans R Soc A368:3491– 3502, doi: 10.1098/rsta.2010.0117, reproduced under a CC BY creative commons license; [f, g]: From Falk, N et al. (2015) Specialized cilia in mammalian sensory systems. Cells 4:500–519, doi: 10.3390/cells, reproduced under CC BY 4.0 creative commons license; drawing in [f] modified from that in the original article; TEM in (g) courtesy of Dr. A. Giessl.)*

BOX 14.1 ANOSMIA IN THE ELDERLY AS A POTENTIAL HARBINGER OF IMMINENT DEATH

The olfactory system allows people with a normal sense of smell (=**normosmia**) to discriminate various odors and thereby orient to, or avoid, potentially beneficial or harmful cues, respectively. Conversely, in conjunction with such disorders as **Parkinson's disease, Covid-19,** and **Alzheimer's disease (AD)**, the capacity for sensing smells can become reduced (=**hyposmia**) or essentially lacking (=**anosmia**). In addition, without being linked to a discernable medical condition, elderly patients may suffer from idiopathic anosmia. Collectively, such degradation of olfactory capabilities often leads to diminished taste sensations and an associated loss of appetite (=**anorexia**).

To quantify olfactory defects, recent studies have used **standardized scratch-and-sniff cards** with distinctive odors that are readily discernable by normally functioning olfactory systems. Somewhat surprisingly, results obtained using these cards have shown that elderly patients who scored well on these tests tended to have a reduced risk for mortality in the 5 years following the test, whereas those who made a higher number of errors in identifying the odors were statistically more likely to die over the 5-year timeframe (**Figure a, b**)

Currently, there is no clear explanation for the correlation between anosmia and imminent death in the elderly. In some cases, such as failing to detect smoke from a fire or lacking proper sustenance due to anosmia-induced anorexia, the inability to detect smells could certainly jeopardize survival. Alternatively, anosmia itself may not be particularly maladaptive and instead simply arises as a byproduct of more widespread health problems. For example, given that the olfactory epithelium requires continual renewal from stem cells, anosmia could represent the most readily apparent defect in stem cell capabilities that generally decline throughout the body during aging. In this way, anosmia could be the **"canary in the coal mine"** signaling a more imminent onset of death that is due to systemic degenerative processes, including an inability to perceive smells.

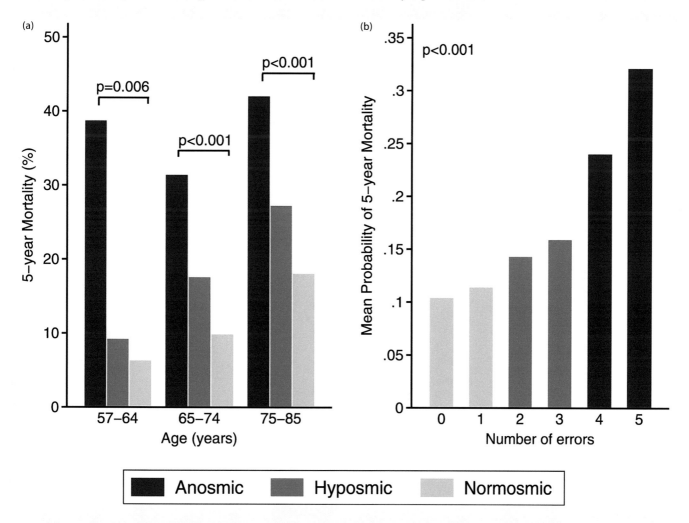

Box 14.1 (a, b) *Increased mortality rates correlated with loss of olfaction (ansomia) in the elderly as quantified by the number of errors on scratch-and-sniff tests. (From: Pinto, J M et al. (2014). Olfactory dysfunction predicts 5-year mortality in older adults. PLoS ONE 9, doi: 10.1371/journal.pone.0107541, reproduced under a CC BY 4.0 creative commons license.)*

BOX 14.2 VOMERONASAL ORGANS (VNOS) AND PHEROMONE DETECTION

First analyzed in detail by L. Jacobson in the 19th century, **vomeronasal organs** (=**VNOs** or **Jacobson's organs**) have been identified near the **vomer bone** within the floor of the nasal cavity in primitive fish, amphibians, reptiles, and most mammals (**Figure a**). Conversely, such organs are generally lacking in bony fish, birds, most bats, and various primates, such as Old World monkeys. In humans, VNOs form during gestation, and immunostaining analyses of fetuses reveal receptor-type neuronal cells in developing VNOs. However, human VNOs morphologically regress after birth to become simple indented **pits** within the nasal cavity floor rather than full-fledged organs, such as found in other adult vertebrates (**Figures b–d**).

As components of the **accessory olfactory system (AOS)** that augments the **main olfactory epithelium** in the roof of the nasal cavity, VNOs detect **pheromone** odors released from one individual to affect the behavior or physiology of another conspecific individual. Two main classes of G-protein-coupled **vomeronasal receptors (VRs)**, called type 1 (**V1Rs**) and type 2 (**V2Rs**) VRs, are present on microvilli of **knob cells** in the VNO sensory epithelium, with V1Rs generally detecting air-bone pheromones, and V2Rs usually sensing water-soluble ligands. Of the hundreds of VNO receptor genes that have been identified across the animal kingdom, each VNO sensory cell expresses a single type of VR gene,

and after binding to their cognate pheromones, activated receptors can trigger action potentials. Such electrical signals are then sent to an **accessory olfactory bulb** at the anterior end of the brain before being relayed to other brain regions, where the perceived pheromones can help shape behavioral or physiological responses (**Figure e, f**).

In humans, electrode recordings of adult VNO pits have revealed electrical activity in response to femtomolar doses of naturally occurring pheromones from human skin. However, whether or not bone fide action potentials are actually transmitted from VNO pits to the brain remains unclear, especially since neither a functional accessory olfactory bulb nor a distinct VNO-to-brain connection has been identified. Moreover, many human genes that are involved in pheromone sensing have undergone a **pseudogenization** process that ablates functionality. Thus, although humans may respond to pheromones, the mechanisms by which such responses occur remain unclear. Given the vestigial morphology of VNO pits as well as the absence of an associated neurocircuitry and lack of key gene functionality, it seems unlikely human VNOs represent the same kind of robust pheromone sensors as found in other mammals. Instead, other potential receptive organs, or the even the main olfactory epithelium itself should be further assessed viz. pheromone detection in humans.

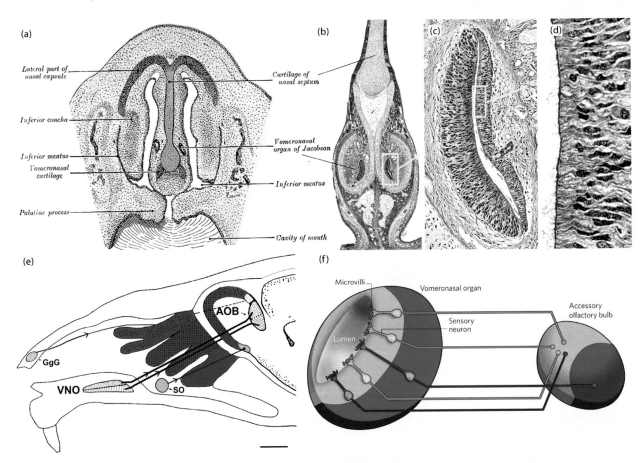

Box 14.2 *(a)* Vomeronasal organs. (b) 20X; (c) 100X; (d) 350 ([a]: From Gray, H (1918) Anatomy of the Human Body. Lea and Febiger, Philadelphia, public domain document; [e] Salazar, I and Quintero, PS (2011) The risk of extrapolation in neuroanatomy: the case of the mammalian vomeronasal system. Front Neuroanat 3: doi: 10.3389/neuro.05.022.2009, reproduced under an CC BY creative commons license); [f]: Vomeronasal connection to accessory olfactory organ from Munger, SD (2009) Noses within noses. Nature 459: 521–522, reproduced with publisher permission.)*

14.4 THE POST-PHARYNGEAL EXTRA-PULMONARY CONDUCTING SYSTEM: EPIGLOTTIS, LARYNX, AND TRACHEA

Inferior to the nasal cavity, the pharynx connects with both the esophagus of the digestive tract and the **larynx** of the respiratory system. As an adaptation for this dual connection, the larynx is guarded by an overlying flap of tissue, called the **epiglottis**, which is displaced during swallowing over the laryngeal opening (=glottis) to ensure food is sent into the esophagus rather than "down the wrong pipe" of the respiratory tract (**Figure 14.4a–c**). The superior epiglottal surface is covered by **stratified squamous epithelium**, which continues over the distal

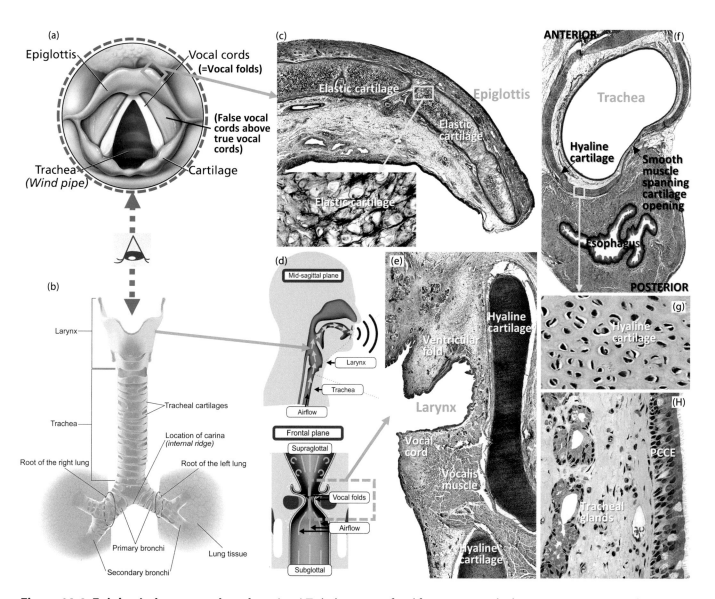

Figure 14.4 Epiglottis, larynx, and trachea. *(a–c) To help prevent food from entering the lower respiratory tract, the larynx is guarded by an overlying epiglottis with an elastic cartilage core. (c) 25X; (c, inset) 100X. (d, e) Air expelled past the laryngeal vocal cords can make different sounds based on: (i) the size and shape of the laryngeal and nasal cavities, and (ii) contributions from the oral cavity (e.g., tongue and lip positioning). (e) 20X. (f–h) Inferior to the larynx, the trachea is protected anteriorly by hyaline cartilage and is lined by PCCE with underlying tracheal glands. (f) 25X; (g) 250X; (h) 200X. ([a]: From A. Hoofring, illustrator, the National Cancer Institute of the National Institutes of Health, downloaded from https://commons.wikimedia.org/wiki/File:Larynx_(top_view).jpg and reproduced as a public domain document; [b]: From OpenStax College (2013) Anatomy and Physiology. OpenStax. http://cnx.org/content/col11496/latest, reproduced under CC-BY-4.0 creative commons license; [d]: From Thornton, F et al. (2019) Impact of subharmonic and aperiodic laryngeal dynamics on the phonatory process analyzed in ex vivo rabbit models. Appl. Sci. 9, 1963, doi: 10.3390/app9091963, reproduced under a CC BY 4.0 creative commons license.)*

CASE 14.2 GEORGE WASHINGTON'S EPIGLOTTIS

1st President of the United States
(1732–1799)

"Doctor, I die hard; but I am not afraid to go; I believed from my first attack that I should not survive it; my breath cannot last long."

(Attributed to a G.W. on his deathbed)

After completing two terms as America's first president, George Washington would often set out on horseback to supervise day-to-day operations of his large estate in Mount Vernon, Virginia. Following a particularly long ride in inclement weather, Washington developed breathing difficulties, prompting a team of doctors to be summoned. As was common in those days, the doctors treated Washington's ailment by removing large quantities of blood in order to drain "malevolent humors". Noting that the treatment was not helping, one of the doctors

proposed an alternative plan involving tracheotomy surgery. However, his suggestion was rejected, and Washington soon thereafter died of suffocation. Based on historical accounts, it has been concluded that an acute infection of Washington's epiglottis caused marked swelling and airway obstruction. Without supplemental antibiotics, the proposed tracheotomy would not have cured Washington's ailment, but, unlike bloodletting, might at least have given him the chance to fight the effects of his swollen epiglottis.

inferior surface before transitioning into a PCCE that also lines the upper larynx. Sandwiched between these two epiglottal epithelia is a connective tissue compartment containing an internal skeleton composed of **elastic cartilage** (**Figure 14.4c**). This cartilage normally provides a flexible scaffold for properly sealing the laryngeal aperture, although pathogen-induced **epiglottitis** can adversely affect the size and shape of the protective flap to obstruct breathing (**Case 14.2 George Washington's Epiglottis**).

Within the **larynx** (**Case 14.3 Mary Wells's Larynx and Trachea**), much of the mucosal lining contains **PCCE**, which abuts a connective tissue compartment with anteriorly positioned cartilage for protection (**Figure 14.4d, e**). In addition, the laryngeal wall contains scattered patches of **larynx associated lymphatic tissue** (**LALT**) that remains present throughout life to help ward off infections that can lead to **laryngitis**. In the upper larynx, **ventricular folds** (=false vocal cords) that are covered mainly by PCCE protrude into the laryngeal cavity and augment the epiglottis in closing off the glottis during swallowing (**Figure 14.4a, e**). Situated inferior to the ventricular folds are the **true vocal cords**, which are covered by a **stratified squamous epithelium** to help protect against damage from forceful airflows during repetitive shouting and loud singing. The vocal cord mucosa is underlain by striated **vocalis muscles** that are used to change vocal cord positions and thus laryngeal shape, thereby helping to alter the pitch of emitted sounds (**Figure 14.4d, e**). Sound production initiated in the larynx is further modified by tongue positioning, mouth and lip shape, as well as nasal cavity morphology.

Distally, the larynx connects with the **trachea**, which is protected anteriorly and laterally by horseshoe-shaped plates of **hyaline cartilage** (**Figure 14.4f, g**). The open posterior ends of the plates connect with each

CASE 14.3 MARY WELLS'S LARYNX AND TRACHEA

Singer
(1943–1992)

"I'm here today to urge you to keep the faith. I can't cheer you on with all my voice, but I can encourage, and I pray to motivate you with all my heart and soul and whispers."

(M.W. testifying in 1991 at a congressional hearing regarding funding for cancer research)

As the first real star for the Motown label of rhythm and blues music, Mary Wells signed a recording contract at age 17 and went on to sing several hits like "My Guy" during the early 1960s. However, as Motown began promoting rising stars like the Supremes, Wells decided to sign with a new record label. Never able to replicate the success she had achieved with Motown, Wells struggled with poor health and drug

addiction for the next two decades until she began losing her voice while relegated to undercard acts on 1980s nostalgia tours. In 1990, Wells was diagnosed with laryngeal cancer and had a tube fitted into her trachea to avoid further damage to her larynx as she underwent chemoradiation therapies. While dealing with laryngeal cancer and the tracheal tube she had to use, Wells helped lobby for increased cancer research funding, before dying of cancer complications at age 49.

other via **smooth muscle** for added flexibility during deep breathing, and internally, the tracheal lumen is lined by PCCE, whose goblet cells, along with sub-epithelial tracheal **glands,** produce mucus to protect the trachea (**Figure 14.4h**). The distal end of the trachea bifurcates into **left** and **right primary bronchi** that lead into the lungs and thereby initiate the intrapulmonary conducting system (**Figure 14.1a, b**).

14.5 BRONCHI VS. BRONCHIOLES AND THE END OF THE CONDUCTING SYSTEM IN TERMINAL BRONCHIOLES

After entering the lungs, the left and right main **bronchi** branch asymmetrically since a left-side indentation for accommodating the heart causes the smaller left lung to form only two sub-compartments (=**lobes**) vs. the three lobes of the right lung (**Figure 14.1b**). Larger bronchi produced from such bifurcations are nearly fully surrounded by plates of **hyaline cartilage** and are lined by a **PCCE**. Subjacent to the lining epithelium is a **bronchus associated lymphatic tissue** (**BALT**), which, unlike persistent LALT of the larynx, tends to degenerate during aging, thereby increasing the chances of the elderly contracting bronchial infections (**Case 14.4 Pocahontas's Bronchi**). In addition, bronchi contain submucosal glands and **smooth muscle fibers** that can become overly constricted during bouts of **asthma**. Along with minimizing exposure to environmental triggers, asthmatic patients often use short- or long-acting **broncho-dilating** drugs that relax smooth muscles to facilitate breathing. Smaller bronchi are surrounded by fewer cartilaginous plates (**Figure 14.5a**) as their PCCE transitions into a simple ciliated sub-columnar epithelium with scattered goblet cells. Eventually, the smallest bronchi connect with **bronchioles** that differ from bronchi both in **lacking cartilage** (**Figure 14.5b**) and in possessing more squamous simple epithelia with fewer ciliated and goblet cells. Bronchioles that lack alveoli belong to the conducting system, with the distal-most of these conducting bronchioles constituting **terminal bronchioles** (**Figure 14.5c**).

14.6 THE CONDUCTING-RESPIRATORY TRANSITION: RESPIRATORY BRONCHIOLES WITH CLUB CELLS AND ALVEOLAR OUT-POCKETINGS

Distal to terminal bronchioles, **respiratory bronchioles** have scattered alveoli incorporated into their walls and hence correspond to the initial part of the respiratory component where gas exchange occurs. The inter-alveolar epithelium of these bronchioles contains non-ciliated domed cells, which were originally coined Clara cells but have since been renamed **club cells** (**CCs**) (**Figure 14.5d, e**). Such cells, which are also present in terminal bronchioles and other scattered respiratory tract sites, secrete several beneficial products, including the **CC16** (=uteroglobin) protein for protection against viral infections and oxidative stress.

Alveolar out-pocketings in respiratory bronchioles (**Figure 14.6a–c**) become more densely packed distally until eventually the bronchioles terminate in **chambers of interconnected alveoli** that make up the bulk of

---

CASE 14.4 POCAHONTAS'S BRONCHI

Native American Celebrity
(˜1596–1617)

"She hazarded the beating out of her own brains to save mine."

(John Smith on how Pocahontas supposedly risked her life to save him from execution)

Pocahontas was the daughter of chief Powhatan who led several Native American tribes in what is now is now Virginia, USA. Over the years, descriptions of Pocahontas's short life have often morphed into fairy-tale-like legends without solid foundation in fact. For example, it is unlikely she helped save an English settler named John Smith, as he later embellished in his memoirs. However, Pocahontas did marry another Englishman, John Rolfe, and had a son with him, thereby promoting a period of peace between the English settlers and native tribes. In 1616, the Rolfe family visited England, where Pocahontas was presented as a Princess in order to sell the notion that the New World was a civilized destination, ready for further colonization. After completing the publicity tour, Pocahontas suddenly took ill, and she died soon thereafter in England. Although various causes of death have been proposed, Pocahontas's rapid demise may have involved a bronchopneumonia affecting bronchi and alveoli in her lungs.

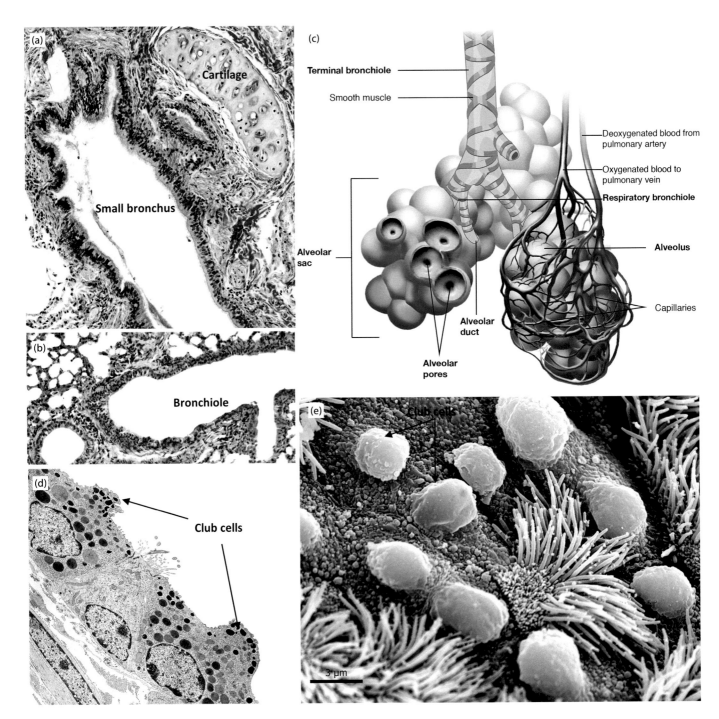

Figure 14.5 Bronchi, bronchioles, and club cells. (**a, b**) Larger bronchi are surrounded by cartilage plates and contain a PCCE lining, whereas bronchioles lack cartilage and gradually transition into simple epithelia that are contiguous with the simple squamous epithelium of alveoli. (a) 150X; (b) 90X. (**c**) Terminal bronchioles lack alveoli and are thus the last of the conducting system before entering alveoli-containing regions of the lungs, where gas exchange occurs. (**d, e**) Bronchiolar club cells contain a dilated, non-ciliated apex and produce beneficial compounds, such as the protein CC-16. (d) 3,600X; (e) 4,600X. ([c]: From OpenStax College (2013) Anatomy and Physiology. OpenStax. http://cnx.org/content/col11496/latest, reproduced under CC-BY-4.0 creative commons license; [d, e]: From Alessandrini, F et al. (2010) Effects of ultrafine particles-induced oxidative stress on Clara cells in allergic lung inflammation. Particle Fibre Toxicol **7**:11. http://www.particleandfibretoxicology.com/content/7/1/11, reproduced under a CC BY 2.0 creative commons license; images courtesy of Dr. F. Alessandrini.)

CASE 14.5 INDIRA GANDHI'S LUNGS

Prime Minister of India
  (1917–1984)

"I do not care whether I live or die… I shall continue to serve until my last breath."

(From I.G.'s last speech, a day before she was assassinated)

As the first female Prime Minister of India, Indira Gandhi was recognized as "Woman of the Millennium" by a British poll for the key roles she played in helping to modernize India. However, during her 15 years of leadership, Gandhi also made some fervent foes, including her own bodyguards, who ended up assassinating Gandhi in retribution for the way she had handled an uprising of Sikh separatists. As a teenager, Gandhi cared for her mother, who had contracted tuberculosis

(TB), a once common and often deadly pulmonary infection caused by the bacterium *Mycobacterium tuberculosis* that generates granuloma tubercles in the lungs. After her mother's death from TB, Gandhi was also treated for the disease, and although her lungs eventually recovered, they and other internal organs could not survive the assassins' bullets that riddled her body just a day after she had given a foreboding speech perhaps portending her imminent demise.

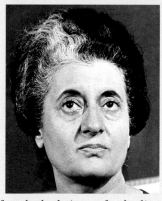

the lungs (**Case 14.5 Indira Gandhi's Lungs**). Thus, along with scattered profiles of blood vessels, bronchi, and bronchioles, alveoli predominate in lung sections.

14.7 MICROANATOMY OF ALVEOLAR WALLS: CAPILLARIES ABUTTING TYPES I AND II EPITHELIAL CELLS PLUS THE GENERAL ROLES OF SURFACTANT

The central cavity (=**air space**) of each inflated alveolus is ~100–200 μm wide and is separated from neighboring alveoli by a thin wall, called the **alveolar septum** (**Figure 14.6d**). This septum comprises: (1) a simple squamous **respiratory epithelium** that directly abuts air spaces, and (2) an underlying **interstitial** space that contains scattered cells, abundant **elastic fibers**, and **pulmonary capillaries** (**Figure 14.6e**). These septal regions are normally kept quite narrow, with the entire distance between the air space and capillary lumen measuring as little as a micrometer in some sites, thereby facilitating the diffusion of O_2 into capillary erythrocytes and the countercurrent flow of CO_2 from blood into the air space (**Figure 14.6e**).

The alveolar epithelium comprises two major kinds of lining cells (**Figure 14.7a**). Type I alveolar epithelial cells (=**AE1** cells, or Squamous Alveolar Cells) are extremely thin and distributed throughout the epithelium. Conversely, type 2 cells (=**AE2** cells, or Great Alveolar Cells) are much taller cells that bulge into the air space at scattered sites and thus contact only a small portion of the air space compared to that lined by AE1 cells (**Figure 14.7a, b**).

AE2 cells carry out wide-ranging functions, including transdifferentiating into AE1 cells and interacting with fibroblasts and white blood cells to help repair alveolar damage. However, the primary role of AE2 cells is to produce a complex mixture of 90% lipids and 10% proteins, called **surfactant**, which is stored in **lamellar bodies** containing onion-like arrays of densely packed surfactant molecules (**Figure 14.7b–d**). In response to stimuli, such as alveolar stretching, lamellar bodies are exocytosed into the watery layer overlying the well-developed **glycocalyx** of the alveolar epithelium (**Figure 14.7e**). Hydrophilic proteins in surfactant bind surface molecules on invasive pathogens to trigger immune-mediated clearance of microbes. Conversely, hydrophobic proteins, in conjunction with surfactant lipids, ensure proper alveolar functioning by **facilitating the inflation of all alveoli** and by **stabilizing smaller alveoli**, in accordance with principles delineated by **LaPlace's law**.

14.8 LAPLACE'S LAW (P = 2T/R) AND HOW SURFACTANT AIDS ALVEOLAR FUNCTIONING

LaPlace's law (P = 2T/R) states that the inwardly directed **pressure** (P) of a gas-filled sac is: (i) directly proportional to twice the **surface tension** (T) on the inside of the sac wall, and (ii) inversely related to the sac's **radius** (R) (**Figure 14.7f**). In the case of alveoli, surface tension at the air-water interface overlying septa arises from strong intermolecular bonding of water that causes surface fluids to contract. By interacting with water to reduce this

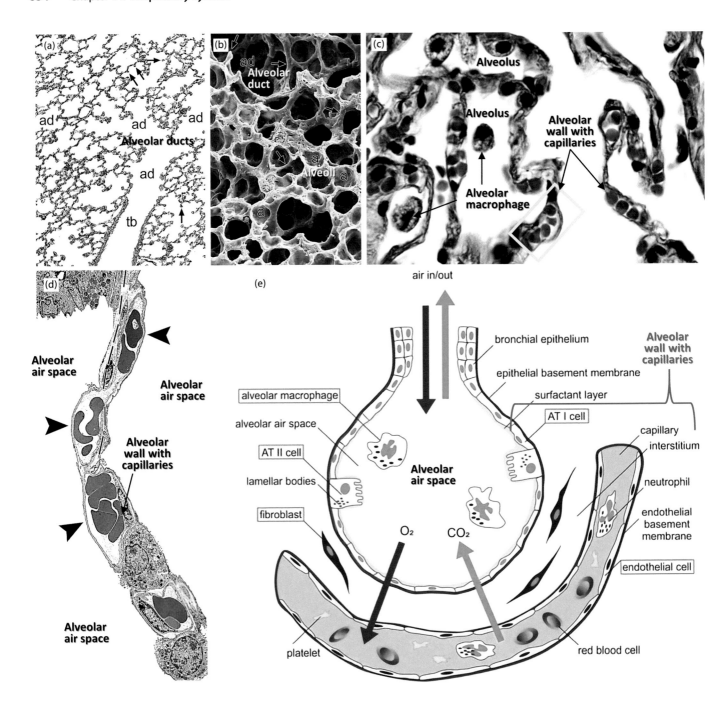

Figure 14.6 Alveolar microanatomy. (a, b) Intrapulmonary channels, called alveolar ducts, are connected to numerous alveoli. (a) 20X; (b) 275X. **(c–e)** The wall of each alveolus is lined by a simple squamous epithelium, which is underlain by capillaries that offload CO_2 into, and absorb O_2 from, the alveolar air space. (c): 425X; (d) 1,100X. ([a, b, d]: From Burri, PH (2006) Structural aspects of postnatal lung development – alveolar formation and growth. Biol Neonate 89:313–322, reproduced with publisher permission; [e]: From Nova, Z et al. (2019) Alveolar-capillary membrane-related pulmonary cells as a target in endotoxin-induced acute lung injury. Int J Mol Sci 20, 831, doi: 10.3390/ijms20040831, reproduced by permission of a CC BY 4.0 creative commons license.)

contractility, surfactant lowers the surface tension of pure water from ~70 dynes/cm to <10 dynes/cm. This in turn reduces inward pressure so that each alveolus can more readily initiate inflation. As a way of demonstrating principles of LaPlace's law, repeatedly stretching a balloon mechanically decreases surface tension and hence reduces inwardly directed pressure, thereby making it easier to start inflation. Similarly, the inverse relationship between sac radius and inward pressure is evident when blowing up the balloon, as increased radii during

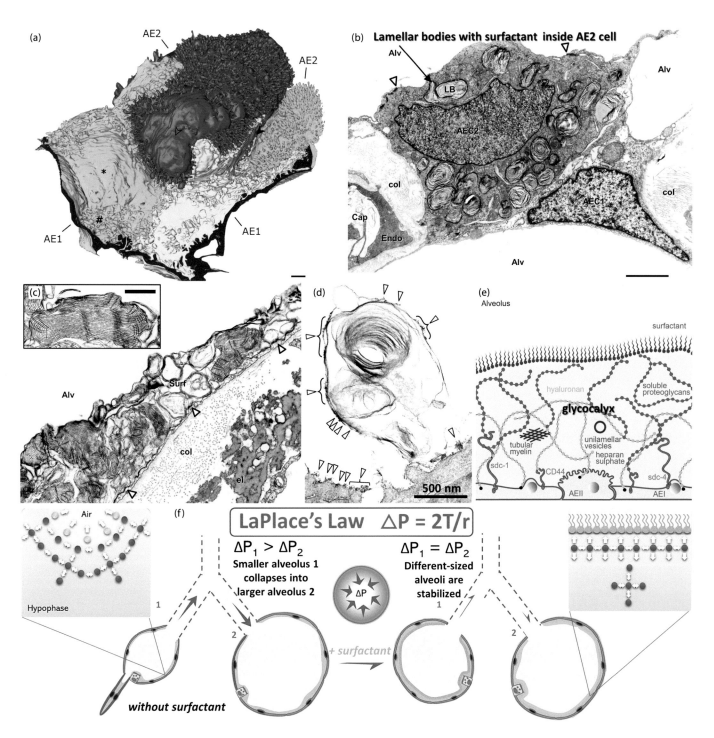

Figure 14.7 Type I vs. type 2 alveolar epithelial cells, surfactant, and LaPlace's law. (*a*) As illustrated in this 3D reconstruction of serial FIB-SEM sections, the alveolar epithelium comprises squamous type 1 alveolar epithelial (AE1) cells with taller type 2 cells (AE2) interspersed among AE1 cells. Scale bar = 1 µm. (*b, c*) AE2 cells contain lamellar bodies filled with a complex lipoprotein mixture, called surfactant. (b) Scale bar = 2 µm; (c, inset) scale bar = 0.5 µm. (*d, e*) After being exocytosed, surfactant coats the thick proteoglycan-containing glycocalyx that covers the apices of alveolar epithelial cells. (*f*) LaPlace's law defines the inwardly directed pressure of a gas-filled sac (P) as being directly proportional to surface tension (T) and inversely related to sac radius (r). Thus, smaller sacs without surfactant would tend to collapse under greater P and cause larger sacs to enlarge, thereby reducing surface:volume ratios and negatively affecting gas exchanges. However, the overlying surfactant layer reduces surface tension more effectively in smaller vs. larger alveoli. This in turn balances inwardly directed pressures in interconnected alveoli of differing sizes and stabilizes smaller sacs for more effective gas exchange. ([*a*]: From Schneider, JP et al. (2020) The three-dimensional ultrastructure of the human

Figure 14.7 (Continued) *alveolar epithelium revealed by focused ion beam electron microscopy. Int J Mol Sci 21, 1089, doi: 10.3390/ijms21031089, reproduced under a CC BY 4.0 creative commons license; [b, c]: From Knudsen, L and Ochs, M (2012) The micromechanics of lung alveoli: structure and function of surfactant and tissue components. Histochem Cell Biology 150:661–676, reproduced with publisher permission; images courtesy of Dr. M. Ochs; [d, e]: From Ochs, M et al. (2020) On top of the alveolar epithelium: surfactant and the glycocalyx. Int J Mol Sci 21, 3075, doi: 10.3390/ijms21093075, reproduced under a CC BY 4.0 creative commons license; [f]: Sehlmeyer, K et al. (2020) Alveolar dynamics and beyond – the importance of surfactant protein C and cholesterol in lung homeostasis and fibrosis. Front Physiol 11, doi: 10.3389/fphys.2020.00386, reproduced under a CC BY 4.0 creative commons license.)*

expansion provide progressively lower inward pressures and thus require less effort to complete the inflation process.

Comparing variously sized alveoli interconnected with each other, LaPlace's law predicts that smaller alveoli with greater inward pressure would tend to collapse and send their air into larger neighboring alveoli (**Figure 14.7f**). Without some way of equalizing this pressure imbalance, fewer and fewer small alveoli would remain open, while larger alveoli would expand their volumes, thereby **reducing the collective surface area** available for gas exchange. However, owing to synergistic effects that occur when surfactant molecules become densely packed in small alveoli, surfactant causes a greater reduction in surface tension in smaller alveoli than in larger alveoli with sparser surfactant distributions. This in turn **reduces inward pressures to a greater extent in small vs. large alveoli** to counteract potential pressure imbalances, thereby maximizing the number of functional alveoli available in lungs.

The key roles that surfactant normally plays in respiration are underscored by two interrelated findings: (1) fetal AE2 cells only start making surfactant around weeks 24–34 of gestation, and (2) highly premature babies can suffer **neonatal respiratory distress syndrome** (**NRDS**), because their alveoli have inadequate surfactant supplies and are thus unable to fill properly with air. Previously, such problems resulted in high mortality rates, but today premature babies can be treated with surfactant replacement therapies (**SRTs**), thereby enabling survival until endogenous surfactant production ramps up sufficiently.

14.9 ALVEOLAR MACROPHAGES (DUST CELLS): THEIR ROLES IN NORMAL LUNGS AND DURING EMPHYSEMA

Within the lumen of alveoli are numerous monocyte-derived phagocytic cells, called dust cells or **alveolar macrophages (AMs)**, which engulf particulate matter and are thus particularly abundant in sections of smoker's lungs (**Figure 14.8a–c**). In addition to phagocytosing debris and pathogens, AMs also catabolize surfactant, ensuring that a fresh supply coats the apical surface of alveolar epithelia.

A common pathology involving lung macrophages is a type of **chronic obstructive pulmonary disease** (**COPD**), called **emphysema**, which typically develops following long-term smoking. Emphysema reduces the natural elasticity of alveolar walls and causes neighboring alveoli to fuse together, thereby forming large **blebs** with reduced surface area and diminished elasticity (**Figure 14.8d, e**). Collectively, such processes hamper breathing and eventually cause patients to use oxygen tanks to counteract abnormally low oxygen concentrations (=**hypoxia**).

At a cellular level, emphysema pathogenesis involves toxins in inhaled smoke activating AMs and thereby causing them to attract neutrophils to alveoli (**Figure 14.8f**). Along with lymphocytes and interstitial macrophages within alveolar septa, activated alveolar macrophages and neutrophils initiate inflammatory responses that include the exocytosis of neutrophilic **elastase** that breaks down elastic fibers in the alveolar wall. In healthy lungs, elastic fiber disintegration is normally kept in check by the elastase inhibitor **alpha-1 anti-trypsin** (**A1AT**), which is produced by the liver and circulated in the bloodstream to reach pulmonary capillaries. However, in smokers, not only is elastase secretion elevated, but **matrix metalloproteinases** (**MMPs**) produced during inflammation reduce A1AT levels, thereby further promoting the widespread elastic fiber degradation of emphysema (**Figure 14.8f**).

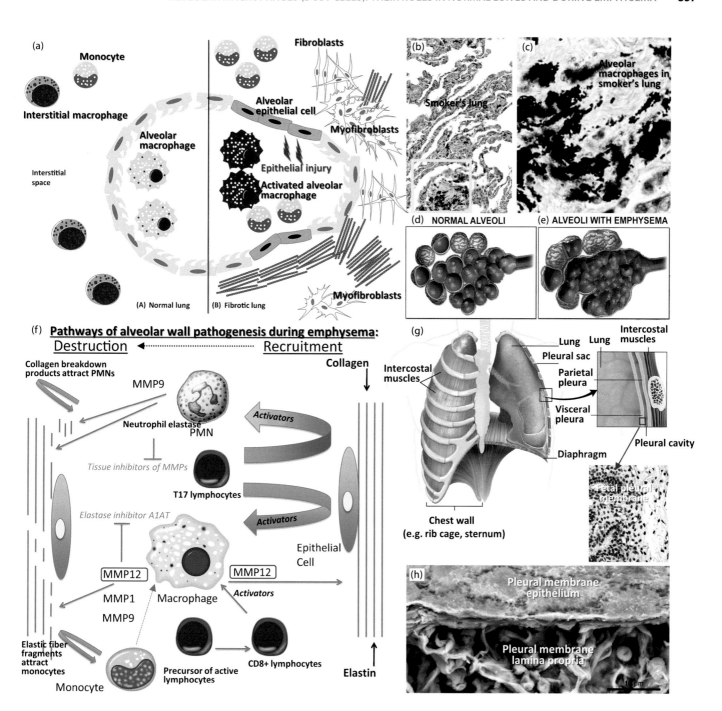

***Figure 14.8 Alveolar macrophages, emphysema, and pleural membranes.** (**a**) Alveolar macrophages (=dust cells) normally help maintain pulmonary homeostasis, but activated macrophages can contribute to various forms of pathogenesis, including emphysema. (**b, c**) Alveolar macrophages are particularly abundant in smoker's lungs, where they phagocytose toxic particulates. (b) 45X; (c) 300X. (**d, e**) During emphysema, alveoli become less elastic and tend to fuse forming larger sacs that cannot exchange gases as efficiently. (**f**) Much of the pathogenesis of emphysema involves breakdown of alveolar elastic fibers owing to increased amounts of elastase coupled with decreased levels of the elastase inhibitor, alpha-1 anti-trypsin (A1AT), triggered by elevated amounts of matrix metalloproteases (MMPs). (**g**) The lungs reside within a pleural sac bounded externally and internally by the parietal vs. visceral pleural membranes, respectively. (g, micrograph) 125X. (**h**) As shown in this paraffin section that was subsequently subjected to SEM imaging, the pleural membrane comprises a simple squamous epithelium overlying a lamina propria. ([a]: From Byrne, A et al. (2016) Pulmonary macrophages: a new therapeutic pathway in fibrosing lung disease? Trends Mol Med 22:303–322, reproduced with publisher permission; [d, e]: From Blausen.com staff (2014) Medical gallery of Blausen Medical 2014. Wiki J Med 1(2): 10, doi: 10.15347/wjm/2014.01010, reproduced under CC0 and CC-SA creative commons licenses; [f]: from Houghton, AM*

Figure 14.8 (Continued) *(2018) Matrix metalloproteinases in destructive lung disease. Matrix Biol 44–46:167–174, reproduced with publisher permission; [g]: From OpenStax College (2013) Anatomy and Physiology. OpenStax; [h]: From Sawaguchi, A et al. (2018) Informative three-dimensional survey of cell/tissue architectures in thick paraffin sections by simple low-vacuum scanning electron microscopy. Sci Rep 8:7479, doi: 10.1038/s41598-018-25840-8, reproduced under a CC BY 4.0 creative commons license.)*

14.10 THE PLEURAL MEMBRANES IN HEALTH AND DISEASE

Each lung is surrounded by a thin **pleural cavity** that is bounded on its inner and outer sides by **pleurae** consisting of an inner **visceral-** and an outer **parietal pleural membrane**, respectively (**Figure 14.8g**). The simple squamous lining layers of pleural membranes are examples of **mesothelia** (**Figure 14.8h**) that can be major origin sites for asbestos-induced cancers, called **mesotheliomas**. Subjacent to each lining epithelium is a well-vascularized connective tissue compartment, and within the pleural cavity, serous fluid helps lubricate the two pleural membranes to reduce friction during inhalations and exhalations.

Pressure within the pleural cavity is normally kept lower than in the lungs in order to facilitate each ventilation cycle. However, due to lung disease or trauma to the chest, introduced air pressure within the pleural space can cause a collapsed lung (=**pneumothorax**), which triggers breathing difficulties and potentially could become fatal if left uncorrected. Typically more benign than a pneumothorax are inflammations of pleural membranes caused by viral, bacterial, or fungal infections as well as by autoimmune reactions. Such pathogenesis can result in a condition, called **pleurisy**, which is characterized by sharp pains during breathing (**Case 14.6 Catherine the Great's Pleural Membranes**).

14.11 SOME RESPIRATORY PATHOLOGIES: LUNG CANCER AND CYSTIC FIBROSIS

As the leading cause of cancer-induced mortality, **lung cancer** accounted for **1.6 million deaths** worldwide in 2015. Approximately 85% of lung cancers occur in smokers vs. ~15% that develop in non-smokers, most commonly in response to various air pollutants, including second-hand smoke. Of the primary cancers that begin in the lungs, two main types—**small cell lung carcinomas (SCLCs)** and **non-small cell lung carcinomas (NSCLCs)**—can be distinguished by their predominant cell morphologies. SCLCs are relatively rare lung cancer cases involving abundant small cells displaying high nuclear-to-cytoplasmic ratios, whereas NSCLCs account for the vast majority of lung cancers overall and share in common a general lack of small cells with high nuclear-to-cytoplasmic ratios. Long-term prognosis for lung cancers is often poor, although newer targeted therapies can extend longevity.

CASE 14.6 CATHERINE THE GREAT'S PLEURAL MEMBRANES

Russian Empress
(1729–1796)

"Nothing is so bad as to have a child for a husband."

(C.t.G. on her husband and unhappy marriage)

Born in what is now Poland, Empress Catherine II became known as Catherine the Great while ruling Russia during its Golden Age of prosperity. Soon after arriving in Russia for an arranged marriage with the future Emperor Peter III, Catherine developed pleurisy, an inflammation of the pleural membranes surrounding her lungs. Although currently a relatively benign condition that is typically cured using antibiotics or bedrest, Catherine's pleurisy was treated with bloodlettings that nearly proved fatal until she eventually recovered and was wed at age 16. From its onset, Catherine's marriage was an unhappy one, as the Emperor-to-be would often play with his toy soldiers when he was not drinking heavily or abusing his wife. Soon after Peter III began his unpopular and short-lived reign, Catherine led a coup to oust her husband and was subsequently crowned Empress in 1762. The acclaimed ruler was able to hold on to power for 34 years before dying in bed due to a stroke, several decades after surviving the botched treatment of her pleural membranes.

Particularly common in Caucasians with northern European ancestry, **cystic fibrosis (CF)** can arise from over a thousand types of mutations in the **cystic fibrosis transmembrane conductance regulator (CFTR)** gene, which encodes an essential anion-specific channel of epithelial cells throughout the body. Thus, CF causes such problems as overly salty sweat, abnormal pancreatic secretions, and **widespread dysfunction in respiratory airways**. In particular, impaired CFTR functioning triggers **viscous mucous secretions** in the respiratory tract, owing to decreased fluid content and overabundant mucus crosslinking, thereby resulting in compromised mucociliary clearance and increased chronic infections. CF patients typically undergo physical therapy to dislodge built-up phlegm and are often prescribed medications to kill infective microbes and dilate bronchi. In addition, drugs that target CFTRs to increase their open probability have shown some promise in reducing CF-associated hospitalizations.

14.12 SUMMARY—RESPIRATORY SYSTEM

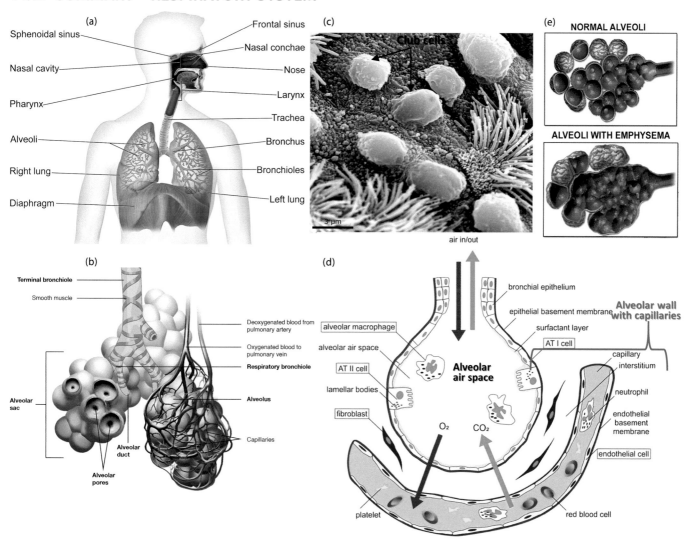

Pictorial Summary Figure Caption: (a): *see figure 14.1a; (b): see figure 14.5c; (c): see figure 14.5e; (d): see Figure 14-6e; (e): see Figure 14-8d, e.*

The **respiratory system serves to**—(1) supply O_2 while removing CO_2 and volatile wastes, (2) sense smells, (3) defend against pathogens, and (4) produce sounds. The system develops mainly from a **foregut evagination** beginning ~5 weeks post-fertilization in humans and functionally comprises a **conducting portion** (nasal cavity, nasopharynx, larynx, trachea, bronchi, and proximal bronchioles) that transports air to and from **alveoli** in the **respiratory portion**, where gas exchange occurs (**Figure a**).

Much of the respiratory system's conducting portion is lined by **pseudostratified ciliated columnar epithelium** (**PCCE**)—a stratified-appearing simple epithelium with ciliated cells and mucous goblet cells that function in **mucociliary clearance** of pathogens and debris. For immune responses, PCEE is underlain by **mucosa-associated lymphatic tissue** (e.g., NALT of the nasal cavity and BALT of bronchi). The roof of the nasal cavity has an **olfactory epithelium** with olfactory receptors on cilia of **bipolar neurons**. The binding of odorants to these receptors causes action potentials to be sent to the **olfactory bulb** of the brain for subsequent perception of smells. In addition to the main olfactory system, many vertebrates have well-developed **vomeronasal organs** near the nasal cavity floor that utilize vomeronasal receptors (VRs) for perceiving secreted **pheromones** during intraspecific communication. Within the nasal cavity, **turbinate bones** create non-laminar air flow that retains inspired air longer, thereby aiding **air conditioning** so that the temperature and humidity of air is optimized before transport toward the lungs. The **larynx** contains **vocal cords** for sound production and is guarded by an overlying **epiglottis** with elastic cartilage. Hyaline cartilage surrounds the **trachea** and **bronchi** but is lacking around **bronchioles**. The last bronchioles in the conducting system are **terminal bronchioles**, whereas alveoli-containing **respiratory bronchioles** are the first part of the respiratory component where gases are exchanged (**Figure b**). Bronchioles can contain **club cells** that secrete beneficial compounds, such as the 16-kD protein **CC16** (**Figure c**).

Alveolar walls comprise capillaries, scattered CT cells (e.g., interstitial macrophages), elastic fibers, plus two types of a̲lveolar e̲pithelial lining cells—squamous **AE1** cells and taller **AE2** cells (**Figure d**). AE2 cells secrete a mixture of lipids and proteins, called **surfactant**, which maintains proper alveolar functioning by reducing **surface tension** (**T**) in all alveoli and by lowering T to a greater degree in smaller ones. This ensures that alveoli with small radii (**R**) do not collapse due to their relatively high inwardly directed pressures (**P**), as defined by **LaPlace's law: P = 2T/R.**

Some common respiratory pathologies—**Neonatal respiratory distress syndrome** involves impaired breathing in highly premature babies due to insufficient surfactant production. **Smoking-induced emphysema** causes: (i) **macrophage activations**, (ii) excessive **elastase** secretions, and (iii) reduced availability of the elastase inhibitor **A1AT**. Such processes degrade alveolar elastic fibers, reduce pulmonary elasticity, and trigger alveolar fusion into larger blebs (**Figure e**), thereby making breathing difficult. **Non-small cell lung cancers** (**NSCLCs**) are the most common type of lung cancers. **Cystic fibrosis** arises from mutations of the cystic fibrosis transmembrane conductance regulator (**CFTR**) gene that cause abnormal anion transport. In the respiratory system, CF is associated with **viscous mucus production** that make patients prone to infections.

SELF-STUDY QUESTIONS

1. T/F The PCCE lining much of the conducting system is a stratified epithelium.
2. T/F Anosmic elderly have higher mortality rates that normosmic cohorts who have not lost their sense of smell.
3. Which accessory olfactory organs are used to detect pheromones in many mammals?
4. Which of the following would you expect to be associated with neonatal respiratory distress syndrome?
   A. Births that are delayed beyond normal gestation
   B. Elevated surfactant levels
   C. Highly premature births
   D. ALL of the above
   E. NONE of the above
5. Which of the following would normally be found in terminal bronchioles?
   A. Surrounding cartilage
   B. AE1 cells
   C. Club cells
   D. ALL of the above
   E. NONE of the above
6. Where is the olfactory bulb located?
   A. Within the main olfactory epithelium
   B. On the floor of the nasal cavity
   C. In the brain
   D. Next to the vocal cords
   E. NONE of the above

7. What is the general name of the process by which PCCE transdifferentiates into stratified squamous epithelium due to heavy coughing?
8. Which organ in the respiratory system has an internal skeleton made of elastic cartilage?
9. Which of the following is NOT found in the nasal cavity?
   A. NALT
   B. Turbinate bones
   C. AE1 cells
   D. PCCE
   E. ALL of the above are found in the nasal cavity
10. Where is BALT located?
   A. Choanae of nasal cavity
   B. Alveoli
   C. Bulb region of olfactory system
   D. Bronchi
   E. NONE of the above
11. Which of the following statements regarding alveolar macrophages is correct?
   A. Also referred to as dust cells
   B. Secrete a chemoattractant for neutrophils
   C. Serve to clear old surfactant secretions
   D. ALL of the above
   E. NONE of the above
12. Which statement regarding emphysema is correct?
   A. Involves abnormally high A1AT levels
   B. Results from excess elastic fibers in alveolar walls
   C. Involves abnormally low elastase levels
   D. ALL of the above
   E. NONE of the above
13. What is the name of bronchioles that contain alveoli?

   "EXTRA CREDIT" For each term on the left, provide the BEST MATCH on the right (answers A-F can be used more than once)

14. NSCLC_____   A. Surfactant source
15. CC16 _____   B. Main olfactory bulb
16. AE2 _____   C. Larynx
17. VR _____   D. Lung cancer
18. Ventricular   E. Pheromone detector
    folds_____   F. Club cells

19. Describe the cause, common respiratory-related symptoms, and general treatments of cystic fibrosis.
20. Describe LaPlace's law and how surfactant helps mediate normal alveolar function.

ANSWERS

(1) F; (2) T; (3) vomeronasal organs; (4) C; (5) C; (6) C; (7) metaplasia; (8) epiglottis; (9) C; (10) D; (11) D; (12) E; (13) respiratory; (14) D; (15) F; (16) A; (17) E; (18) C; (19) Caused by mutations in cystic fibrosis transmembrane conductance regulator (CFTR) gene that alter anion transport, which leads to viscous mucous secretions and chronic infections. Treated with antibiotics, mucus-thinning agents, bronchodilators, direct CFTR potentiators, and physical therapy. (20) A full answer would include $P=2T/R$, define all of the terms of the equation, describe pressure imbalance between small and large alveoli, state what surfactant is, where it is made, and how it not only aids the functioning of all alveoli, but also how it stabilizes small alveoli so that they do not readily collapse

Endocrine System

15.1 INTRODUCTION TO ENDOCRINE ORGANS

The body contains **endocrine organs** that release into the bloodstream bioactive molecules, called **hormones** (**Figure 15.1a**). Unlike ducted exocrine glands, endocrine organs lack ducts after having lost their epithelial connections and becoming vascularized during development (**Figure 15.1b**). By distributing hormones via the bloodstream, endocrine organs can elicit more wide-ranging effects than are achieved by either: (i) **juxtacrine** molecules that affect only directly contacted target cells, or (ii) **diffusible autocrine** and **paracrine** factors that modulate just the secretory cell itself or nearby tissues, respectively (**Figure 15.1c**).

To ensure that their levels in the bloodstream remain at optimal concentrations, most hormones trigger a **negative feedback** response from their target cells. Thus, when excessive blood hormone levels are reached, hormone-stimulated target cells can produce their own hormones to inhibit further production of the original

DOI: 10.1201/9780429353307-18

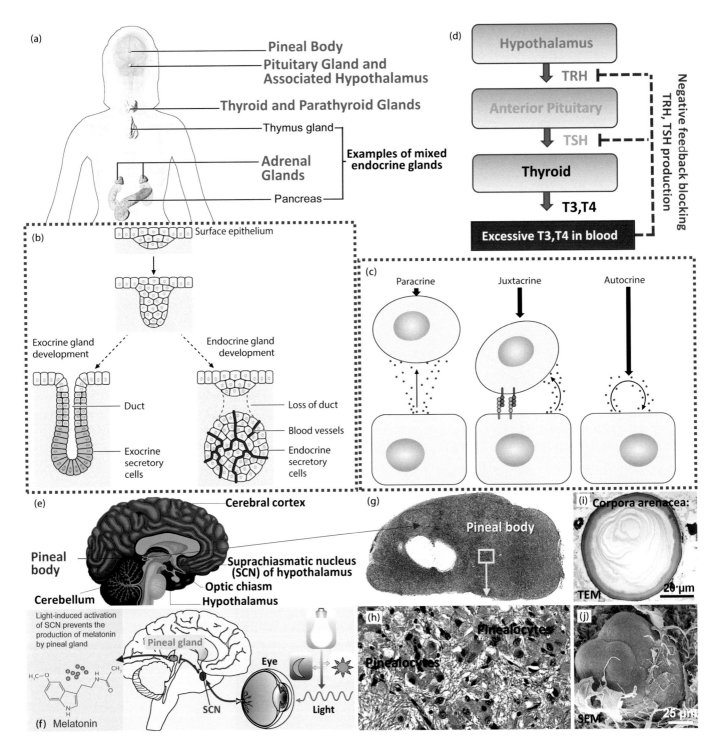

Figure 15.1 Introduction to the endocrine system and the pineal body. (*a*) *Along with mixed endocrine organs like the pancreas, the endocrine system comprises predominantly endocrine organs (pineal body, pituitary gland and its associated hypothalamus, thyroid and parathyroid glands, and adrenal glands) plus a diffuse system of individual endocrine cells.* (***b***) *As opposed to duct-containing exocrine glands that secrete their products onto the body surface or into non-vascular internal compartments, endocrine glands lose their ducts during development and release their products, called hormones, into the bloodstream.* (***c***) *Blood-borne hormones of endocrine organs and tissues can have wide-ranging effects throughout the body compared to more localized modes of para-, juxta-, and autocrine signaling, where cells use secreted or attached compounds (red dots and cell membrane components in diagram) to modulate nearby cells, juxtaposed cells, or the signaling cells themselves, respectively.* (***d***) *As illustrated by this example of thyroid hormones T3 and T4, most hormones maintain optimal blood concentrations via negative feedback loops, in which*

Figure 15.1 (Continued) *excessive amounts of circulating hormones trigger inhibition of upstream regulators of hormone production. (**e–h**) The pineal body (=pineal gland) occurs in the posterior midline of the brain and comprises numerous pinealocytes which primarily secrete high levels of the hormone melatonin at night. Melatonin is released nocturnally, because light perception by the eye stimulates the suprachiasmatic nucleus (SCN) of the hypothalamus to inhibit melatonin production during the day. Note: the brain and pineal gland shown here are oriented with anterior to the right, whereas subsequent views of the pituitary and hypothalamus have anterior to the left. (g) 10X; (h) 100X. (**i, j**) Aging pineal bodies are characterized by the accumulation of calcified corpora arenacea (=brain sand), whose functions, if any, remain unclear. ([a]: From Blausen.com staff (2014) Medical gallery of Blausen Medical 2014. Wiki J Med 1(2):10, doi: 10.15347/wjm/2014.01010, reproduced under CC0 and CC-SA creative commons licenses; [e] downloaded from Wikipedia site: https://commons.wikimedia.org/wiki/File:Suprachiasmatic_Nucleus.jpg and reproduced under a CC SA 3.0 creative commons license; [f]: From Ma, Z et al. (2016) Melatonin as a potential anticarcinogen for non-small-cell lung cancer. Oncotarget 7:46768–46784, reproduced under a CC BY 3.0 creative commons license; [i, j]: Kim, J et al. (2012) Growth patterns for acervuli in human pineal gland. Sci Rep 2:984, doi: 10.1038/srep00984, reproduced with publisher permission.)*

stimulatory hormone (**Figure 15.1d**). Alternatively, in a few cases, such as when a surge in luteinizing hormone is elicited during the menstrual cycle (Chapter 18), a **positive feedback** loop develops so that the two interacting hormones co-stimulate each other and reciprocally cause rapid increases in hormone levels.

Organs and tissues that primarily produce hormones constitute the **endocrine system**, which includes the: (i) **pineal body**, (ii) **pituitary gland** plus associated **hypothalamus**, (iii) **thyroid gland**, (iv) **parathyroid glands**, and (v) **adrenal glands**. In addition, hormones are secreted by various organs and tissues that play other non-endocrine roles (e.g., pancreas, kidneys, liver, thymus, gonads, and fat) as well as by individual cells, particularly in digestive and respiratory mucosae. Such scattered cells form a **diffuse endocrine system** (**DES**) and along with mixed endocrine organs are covered in other chapters of this book.

15.2 THE PINEAL BODY AND ITS MAIN SECRETORY PRODUCT, MELATONIN

A small **pineal body** (=pineal gland) occurs along the posterior midline of the forebrain in humans and most other vertebrates (**Figure 15.1e–g**). Similar in shape to a miniature pine cone, the pineal body evolved in association with a light-sensitive **parietal** ("third") eye, which was lost in mammals. Due to its secluded positioning and former connection to the parietal eye, the human pineal body is sometimes ascribed lofty metaphysical powers, even though it is scarcely larger than a **grain of rice**.

Beneath its connective tissue capsule, most of the pineal body comprises secretory **pinealocytes**, with the remainder consisting mainly of supportive **glial** cells (**Figure 15.1h**). In addition, during aging, the pineal body accumulates calcified concretions, called **corpora arenacea** (= brain sand, acervuli), which often appear as mulberry-shaped inclusions that can fill a substantial portion of pineal volume in older adults (**Figure 15.1i, j**). Currently, corpora arenacea have not been definitively linked to any particular function or pathology, although higher amounts have been reported in some disorders, such as Alzheimer's disease.

The major hormone produced by pinealocytes is **melatonin**, which is made via enzymatically catalyzed steps that convert **tryptophan** into **serotonin** and then serotonin into melatonin (**Figure 15.1f**). Following the discovery that constant illumination causes pineal bodies to atrophy, it was shown that melatonin production is normally inhibited by daylight and stimulated by darkness, owing to the markedly increased activity of melatonin-generating enzymes at night (**Figure 15.1f**). Such nocturnal production of melatonin is controlled by an endogenous clock system that sends photoperiod-encoded cues to the pineal body on both a 24-hour (**circadian**) and **seasonal** basis. This system involves light—particularly in blue wavelengths—being perceived by **melanopsin** photopigment in specialized ganglion cells of the **retina**. Light-induced melanopsin signaling from the retina in turn stimulates the **suprachiasmatic nucleus** (**SCN**) in the hypothalamus, thereby inhibiting melatonin production during the day. Conversely, a nocturnally inactive SCN allows melatonin secretion at night (**Figure 15.1f**).

CASE 15.1 ZELDA FITZGERALD'S PINEAL BODY

Socialite and Writer
  (1900–1948)

*"Why should all life be work, when we all can borrow?
Let's think only of today, and not worry about tomorrow."*

(From an inscription under Z.F.'s
picture in her high school yearbook)

Zelda Fitzgerald was a writer and an eccentric icon of the Roaring Twenties in America. Born in Alabama as Zelda Sayre, Fitzgerald married aspiring writer F. Scott Fitzgerald at age 19. Early on, Fitzgerald served as an inspirational muse for her husband, who fashioned characters in his novels after her. Buoyed by the success of her husband's early novels, the Fitzgeralds became popular partygoers, widely known for their hard drinking, constant reveling, and wild deeds until Fitzgerald was committed at age 29 to a sanatorium where she was diagnosed with schizophrenia. Fitzgerald wrote an autobiographically based novel in between being shuttled

into psychiatric institutions before she eventually died in a fire while locked in her hospital room at age 47. Today, schizophrenia is recognized as a complex set of illnesses linked to various causative factors, including in some cases pineal body dysfunction. Thus, although numerous other well-established drugs can help

manage disease progression, an augmentation of the pineal's major product, melatonin, might also serve to reduce some symptoms of schizophrenia, begging the question of whether or not Fitzgerald could have benefited from therapy targeting her pineal body.

Of the various roles that have been proposed for nightly melatonin production, most relate to **promoting sleep** or **entraining other circadian rhythms**. Accordingly, exogenous melatonin intake may aid sleep and mitigate time-shift maladies like jet lag. Along with such functions, melatonin also helps **block precocious puberty** by reducing hypothalamic secretions, with melatonin levels normally dropping in the months before puberty onset. In addition, beneficial **antioxidant properties** of melatonin are believed to reduce cardiovascular disease, inflammation, and certain cancers. Conversely, abnormally low melatonin levels have been linked to such disorders as Alzheimer's disease, autism, and schizophrenia (**Case 15.1 Zelda Fitzgerald's Pineal Body**).

15.3   OVERVIEW OF THE PITUITARY (=HYPOPHYSIS): DEVELOPMENT, GENERAL ORGANIZATION, AND HORMONES OF ITS ANTERIOR LOBE (=ADENOHYPOPHYSIS) VS. POSTERIOR LOBE (=NEUROHYPOPHYSIS)

The pea-sized **pituitary** (=**hypophysis**) lies directly beneath the brain and is often referred to as the body's master gland. However, instead of being a single, fully autonomous controller, the pituitary actually comprises a distinct glandular and neural lobe, which in turn are governed by the overlying **hypothalamus** region of the brain.

The bipartite nature of the pituitary is evident in 4–5 week-old human embryos, when oral ectoderm evaginates a sac, called **Rathke's pouch**, toward the brain, and the brain in turn directs a stalked downgrowth toward the evagination (**Figure 15.2a, b**). Eventually, Rathke's pouch separates from the oral ectoderm and adheres anteriorly to the neural downgrowth while partially surrounding it (**Figure 15.2b, c**). Because of their relative positions and histological features, the more glandular part of the pituitary arising from oral ectoderm is called the **anterior pituitary** (=**adenohypophysis**), whereas the neural derivative of the brain is termed the **posterior pituitary** (=**neurohypophysis**) (**Figure 15.2b, c**). When fully developed, the anterior pituitary comprises three parts: (i) a large anterior **pars distalis**; (ii) a smaller **pars tuberalis** that partially wraps around the posterior pituitary's stalk; and (iii) a slender **pars intermedia**, which occurs between the pars distalis and posterior pituitary in various vertebrates but is mostly vestigial in adult humans. Conversely, the posterior pituitary consists of a posterior **pars nervosa** that connects to the hypothalamus by an **infundibular** stalk (**Figure 15.2b, c**).

As described further in the next section, the hypothalamus mediates hormone secretions from the anterior and posterior lobes, thereby eliciting wide-ranging effects throughout the body (**Figure 15.3a–d**). In response to hypothalamic signals, seven major adenohypophyseal hormones are exocytosed: (i) **somatotropin** (=**growth hormone [GH]**), (ii) **prolactin (PRL)**, (iii) **luteinizing hormone (LH)**, (iv) **follicle-stimulating hormone (FSH)**, (v) **thyroid stimulating hormone (TSH)**, (vi) **adrenocorticotropic hormone (ACTH)**, and (vii) **α-melanocyte stimulating hormone (α-MSH)**. Conversely, the pars nervosa of the neurohypophysis releases two

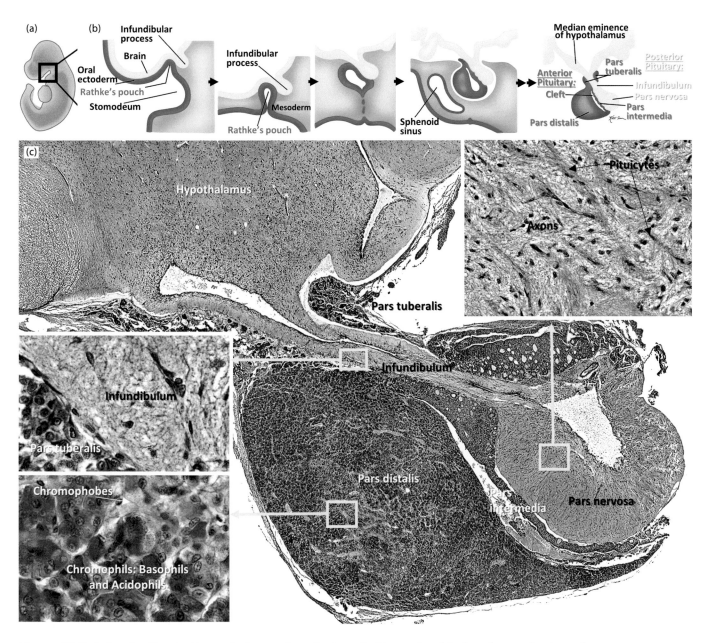

Figure 15.2 Introduction to the anterior and posterior pituitary gland (adenohypophysis and neurohypophysis). *(a, b)* *The anterior and posterior lobes of the pituitary gland arise from separate primordia. The anterior pituitary (=adenohypophysis) develops from an evagination of oral ectoderm, called Rathke's pouch, whereas the posterior pituitary (=neurohypophysis) forms as a downgrowth of the developing brain. (c) When fully differentiated, the anterior pituitary comprises mainly a par distalis and pars tuberalis, with a vestigial pars intermedia occurring in adult humans. The posterior pituitary has a pars nervosa of axons and pituicytes that is connected to the hypothalamus by an infundibular stalk, around which the pars tuberalis is wrapped. (c) 50X; (c, right inset) 275X; (c, left upper inset) 300X; (c, left lower inset) 330X. ([a]: From Rizzoti, K (2015) Genetic regulation of murine pituitary development. J Mol Endocrinol 54, R55–R73, reproduced under a CC BY 3.0 creative commons license; [b]: From Shields, R et al. (2015) Magnetic resonance imaging of sellar and juxtasellar abnormalities in the paediatric population: an imaging review. Insights Imaging 6:241–260, reproduced under a CC BY creative commons license.)*

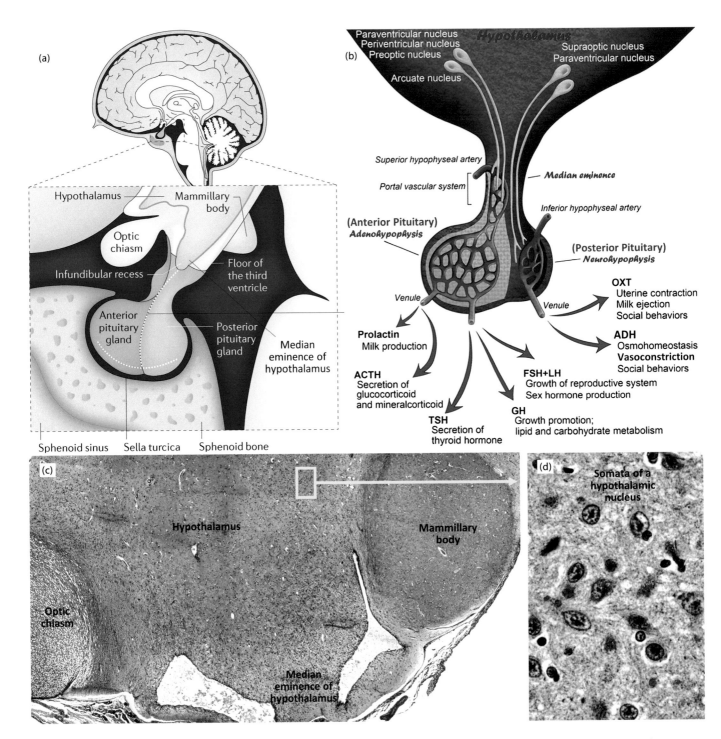

Figure 15.3 Hypothalamus. (a–d) *The hypothalamus sends releasing- or inhibiting hormones through a median eminence into the anterior pituitary in order to trigger or block adenohypophyseal hormone secretion. Axons arising from hypothalamic nuclei also extend into the posterior pituitary where they release their two hormone products. (c) 50X; (d) 380X. ([a]: From Mueller, HL et al. (2017) Craniopharyngioma. Nat Rev Dis Primer 5: article 75, https://doi.org/10.1038/ s41572-019-0125-9, reproduced with publisher permission; [b]: From Miyata, S (2017) Advances in understanding of structural reorganization in the hypothalamic neurosecretory system. Front Endocrinol 8, doi: 10.3389/fendo.2017.00275, reproduced under a CC BY 4.0 creative commons license.)*

TABLE 15.1 HORMONES OF THE ADULT PITUITARY GLAND AND HYPOTHALAMUS

Pituitary cell types and the hormones they release	Hypothalamic releasing/ inhibiting hormones (+/–)	Major target organs
Anterior pituitary basophils		
FSH	GnRH (+)	Gonads
LH	GnRH (+)	Gonads
ACTH	CRH (+)	Adrenal cortex
TSH	TRH (+)	Thyroid
Anterior pituitary acidophils		
Prolactin	PRH (+), dopamine (–)	Mammary gland
GH (=somatotropin)	GHRH (+), somatostatin (–)	Tissues throughout body
Posterior pituitary axons (from hypothalamic nuclei)		
ADH		Kidneys
Oxytocin		Uterus, mammary gland

Source: Based on data presented in Rawindraraj, AD et al. (2020) Physiology, Anterior Pituitary. Stat Pearls downloaded from https://www.ncbi.nlm.nih.gov/books/NBK499898/, reproduced under a CC BY 4.0 creative commons license.

hormones—**antidiuretic hormone (ADH)** and **oxytocin (OT)**—which are synthesized by clustered neuronal somata in **hypothalamic nuclei** before being transported to the posterior pituitary for storage and release into the bloodstream (**Figure 15.4a**) (Table 15.1).

15.4 HYPOTHALAMIC INTERACTIONS WITH THE ANTERIOR PITUITARY: PORTAL BLOOD SUPPLY, HYPOTHALAMIC NUCLEI, AND HORMONES THE HYPOTHALAMUS PRODUCES TO REGULATE ANTERIOR PITUITARY FUNCTION

To provide increased surface area and compartmentalized hormone delivery, a **hypothalamic-adenohypophyseal portal system** forms two fenestrated capillary beds (=plexuses)—a superior **primary capillary plexus** and an inferior **secondary capillary plexus** (**Figure 15.4b**). Above this portal system, several nuclei (e.g., arcuate, paraventricular, periventricular, and preoptic) are the primary sources of hypothalamic hormones, which either stimulate or inhibit adenohypophyseal hormone release and hence are referred to as **hypothalamic releasing-** or **inhibiting hormones** (Table 15.1) (**Case 15.2 Barbara Bush's Hypothalamus**). Hypothalamic hormones are transported down axons for secretion into the **primary capillary plexus** and from there can be sent to the **secondary capillary plexus** for regulating adenohypophseal secretions (**Figure 15.4b**). In response to releasing hormones for LH, FSH, TSH, and ACTH, adenohypophyseal cells become tightly associated with the secondary plexus (**Figure 15.4c, d**) and generate cell-type-specific secretory niches for delivering their hormones throughout the body (**Figure 15.4e–h**).

Alternatively, release of both **GH** and **prolactin** is dually regulated by hypothalamic hormones with opposing effects. **Somatostatin** is a hypothalamic-inhibiting hormone that stops GH secretion, whereas growth hormone releasing hormone (GHRH) triggers GH release. Similarly, the hypothalamus inhibits vs. stimulates **prolactin** secretion via dopamine vs. prolactin-releasing hormone, respectively (Table 15.1).

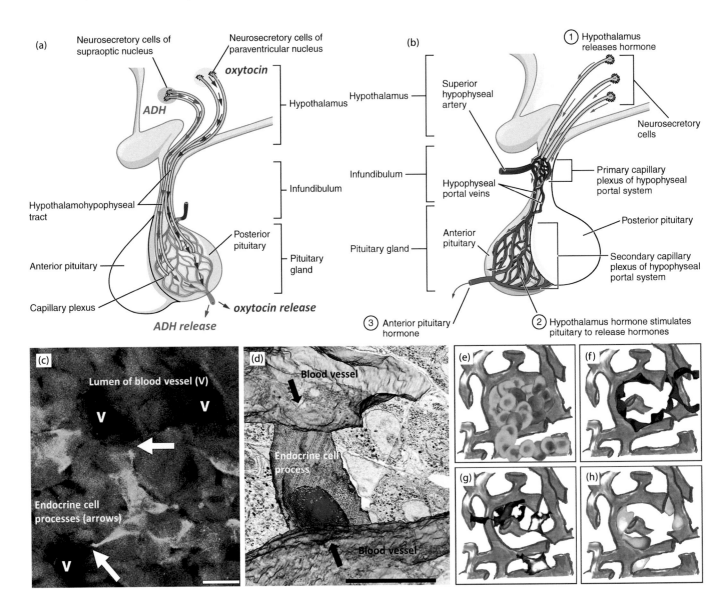

Figure 15.4 Vascular organization of the hypothalamic-hypophyseal axis. (*a*) The hormones ADH and oxytocin are produced in hypothalamic nuclei and released into the bloodstream via a posterior pituitary capillary plexus. (*b*) Conversely, releasing- and inhibiting-hormones made by neurosecretory cells of the hypothalamus are initially secreted in a primary capillary plexus before being distributed throughout the anterior pituitary. In response to such hypothalamic factors, adenohypophyseal endocrine cells secrete their hormones into the secondary capillary plexus for distribution throughout the body. (*c, d*) As shown in this immunofluorescence (c) and 3D-reconstruction of FIB-SEM datasets (d), processes of pars distalis cells (arrows) extend toward blood vessels (V) to deliver hormones into the bloodstream, scale bars = 10 μm. (*e–h*) The precise relationships between adenohypophyseal endocrine cells and blood vasculature differ, thereby providing cell-type-specific niches for hormone secretion. A schematic diagram shows sinusoidal capillaries in gray vs. endocrine cells that secrete growth hormone (green, e), prolactin (red, f), ACTH (magenta, g) or LH (cyan, h). ([a, b]: From OpenStax College (2013) Anatomy and Physiology. OpenStax. http://cnx.org/content/col11496/latest, reproduced under CC-BY-4.0 creative commons license; [c, d]: From Yoshitomi, M (2016) Three-dimensional ultrastructural analyses of anterior pituitary gland expose spatial relationships between endocrine cell secretory granule localization and capillary distribution. Sci Rep 6:36019, doi: 10.1038/srep36019, reproduced under a CC BY 4.0 creative commons license; [e–h]: From Le Tissier, PR et al. (2012) Anterior pituitary cell networks. Front Neuroendocrinol 33:252–266, reproduced under a CC BY 3.0 creative commons license.)

CASE 15.2 BARBARA BUSH'S HYPOTHALAMUS

First Lady of the United States
(1925–2018)

"I married the first man I ever kissed. When I tell this to my children, they just about throw up."

(B.B. on her marriage to George H. Bush)

As a popular U.S. First Lady, Barbara Bush earned high praise for her charitable acts, including the sizable donations she made from the sales of a book she had written from her dog Millie's point of view. After her husband George H. Bush won the 1988 presidential election, Bush was diagnosed with Graves' disease, an autoimmune disorder associated with elevated thyroid hormone production. Via negative feedback mechanisms, such hyperthyroidism can cause the hypothalamus to reduce its production of thyrotropin-releasing hormone (TRH), and even though such reductions decrease thyroid stimulating hormone (TSH) levels, thyroid hormones remain elevated due to the TSH-mimicking effects of circulating autoantibodies. Not long

after her diagnosis, President Bush also developed Graves' disease, and even the dog Millie was afflicted with an autoimmune disorder. The low probability of these non-contagious diseases occurring in genetically unrelated individuals suggested the possibility of environmental toxins playing a role. However, no potential candidates in the water supplies of the White House or Bush's previous homes were identified. In any case, after managing her hyperthyroidism plus its wide-ranging effects like altered hypothalamic functions, Bush eventually succumbed to complications of Graves' disease at age 92.

15.5   ANTERIOR PITUITARY CELL TYPES: CHROMOPHOBES VS. HORMONE-SECRETING CHROMOPHILS, WHICH COMPRISE ACIDOPHILS THAT MAKE GH AND PROLACTIN PLUS BASOPHILS THAT PRODUCE LH, FSH, TSH, ACTH, AND α-MSH

Based on their staining properties, two major morphotypes of anterior pituitary cells have been distinguished: (i) **chromophobes**, which typically lack hormones and are poorly stained by commonly used dyes vs. (ii) **chromophils**, which secrete hormones and readily absorb stains (**Figure 15.2c**). Chromophobes include: (i) **stem cells**, (ii) **degenerative stages** of aged chromophils, and (iii) **folliculostellate cells** (**FSCs**). Such star-shaped FSCs can aggregate to form follicle-like structures within the anterior pituitary (**Figure 15.5a, b**) and are capable of both phagocytosis and regulating adenohypophyseal functions via signaling molecules like **annexins** (**Figure 15.5b**).

Chromophils can be distinguished from chromophobes by their denser cytoplasm and hormone-containing granules (**Figure 15.5c**). Among chromophils, **acidophils** stain with acidic dyes, whereas **basophils** bind basic dyes (**Figure 15.2c**). Two populations of **acidophils** secrete **GH** or **prolactin**, whereas **basophils** release **LH, FSH, TSH, ACTH**, or **α-MSH**, with the latter two being formed from enzymatic cleavages of a large precursor protein, called **proopiomelanocortin** (**POMC**) (**Figure 15.5d**). Based on their functional properties, acidophils can be subdivided into **somatotropes** vs. **lactotropes**, which secrete GH (=somatotropin) vs. prolactin. Similarly, among basophils, **thyrotropes**, **corticotropes**, and **melanotropes** produce TSH, ACTH, and α-MSH, respectively, whereas **gonadotrope** basophils produce LH and FSH, which both target gonads and hence are termed **gonadotropin** hormones. The majority of chromophils occupy cord-like clusters distributed throughout the **pars distalis**, and each of these clusters generally comprises a mixture of several chromophil subtypes, although certain kinds of basophils or acidophils can also predominate within distinct regions of the adenohypophysis. For example, lactotropes tend to be most concentrated in the posterior portion of the pars distalis, whereas somatotrope levels are often highest laterally.

In human fetuses, the **pars intermedia** contains functional melanotropes that have the necessary enzymes to produce α-MSH from ACTH, thereby supplementing various extra-pituitary α-MSH sources, including the hypothalamus. However, soon after birth, production of α-MSH by the pituitary essentially ceases, as the pars intermedia eventually becomes reduced to a thin vestige with cyst-like **clefts** representing leftover luminal portions of Rathke's pouch. In adults, α-MSH from non-adenohypophyseal sources not only promotes melanocytic differentiation and tanning within the epidermis (Box 10.1) but also downregulates food intake and increases metabolism.

The **pars tuberalis** (**PT**) lacks acidophils but contains numerous FSCs plus scattered basophils that mostly consist of gonadotrope-, corticotrope-, and thyrotrope-like cells. Potential functions of the human PT have not

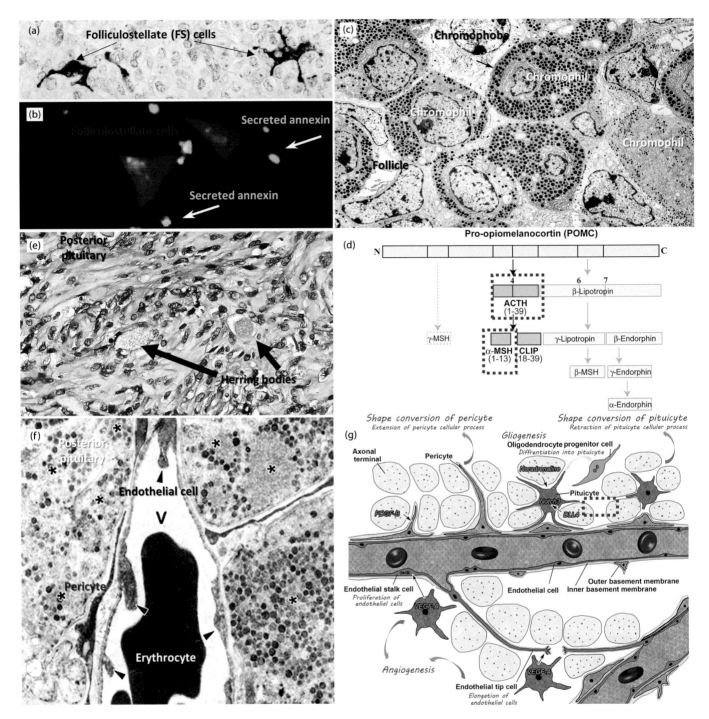

Figure 15.5 Chromophobes vs. chromophils of the adenohypophysis and functional microanatomy of the neurohypophysis. *(a, b)* *Star-shaped folliculostellate types of chromophobe cells form follicles in the anterior pituitary and can modulate hormone secretion patterns by physical means and by secretory products, such as annexins. (a) 250X; (b) 400X. (c) Compared to chromophobes, chromophils have a denser cytoplasm plus hormone-containing granules of various shapes and sizes (1,800X). (d) Of the hormones secreted by chromophobes, ACTH and α-melanocyte stimulating hormone (α-MSH) (dotted boxes) are generated from the cleavage of a large precursor molecule, called POMC. (e) Neurohypophyseal axons of the pars nervosa can either release ADH or oxytocin directly into the bloodstream or temporarily store these products in dilated axonal regions, called Herring bodies (300X). (f) Release of neurohypophyseal hormones stored in electron-dense granules (asterisks) can be regulated by pericyte cells (colorized red in this 25,000X TEM), which associate with the endothelial cells (arrowheads) of pars nervosa blood vessels (V). (g) By interacting with axons and capillary endothelia, pituicytes can modulate rates of hormone release into the posterior pituitary plexus.*

Figure 15.5 (Continued) *([a]: From Pires, M and Tortosa, F (2016) Update on pituitary folliculo-stellate cells. Int Arch Endocrinol Clin Res 2, doi: 10.23937/2572-407X.1510006, reproduced under a CC BY creative commons license; [b]: From Chapman, L et al. (2002) Externalization of annexin I from a folliculo-stellate-like cell line. Endocrinol 143(11):4330–4338, reproduced with publisher permission; (c) Topilko, P et al. (2002) Multiple pituitary and ovarian defects in Krox-24 (NGFI-A, Egr-1)-targeted mice. Mol Endocrinol 12: 107–122 reproduced with publisher permission; [d]: From Ross, AP et al. (2013) Multiple sclerosis, relapses, and the mechanism of action of adrenocorticotropic hormone. Front Neurol 4, doi: 10.3389/fneur.2013.0002, reproduced under a CC BY 3.0 creative commons license; [e]: From Cao, Y et al. (2020) A rare case report of pituicytoma with biphasic pattern and admixed with scattered Herring bodies. World J Surg Oncol 18:108. https://doi.org/10.1186/s12957-020-01889-6, reproduced under a CC BY 4.0 creative commons license; [f, g]: From Miyata, S (2017) Advances in understanding of structural reorganization in the hypothalamic neurosecretory system. Front Endocrinol 8, doi: 10.3389/fendo.2017.00275, reproduced under a CC BY 4.0 creative commons license; figure [f] is a colorized version of a TEM originally published by Nishikawa K et al. (2017) Structural reconstruction of the perivascular space in the adult mouse neurohypophysis during an osmotic stimulation. J Neuroendocrinol 29, doi: 10.1111/jne.12456 and reproduced with publisher permission.)*

been fully elucidated. However, in various mammals whose reproductive cycles are sensitive to photoperiod changes, the PT plays key roles in regulating seasonally induced behavioral and physiological responses (Box 15.1 The Central Role of the Pars Tuberalis (PT) in Regulating Circannual Clocks in Mammals).

As might be expected given the numerous targets of adenohypophyseal hormones (Table 15.1), wide-ranging maladies can arise from abnormally low or high levels of hormone production by the anterior pituitary. Examples of adenohypophyseal dysfunction include: (i) **gigantism** and **acromegaly** (Case 15.3 Pio Pico's Anterior Pituitary) that result from excessive GH production in children and post-adolescents, respectively; (ii) hypopituitarism-induced **short stature** due to insufficient GH secretion; (iii) pituitary-mediated forms of **Cushing's disease** that trigger adrenal cortex hypertrophy due to excessive ACTH production (see also pg. 384); and (iv) **Sheehan's syndrome** (=postpartum hypopituitarism), which typically follows uterine hemorrhaging during childbirth and concomitant ischemia-mediated deficiencies in multiple hormone secretions.

15.6 THE POSTERIOR PITUITARY AND THE TWO HORMONES IT RELEASES: ANTIDIURETIC HORMONE (ADH) AND OXYTOCIN (OT)

Unlike the anterior pituitary's glandular cords, the posterior pituitary contains thousands of axons that originate in the **supraoptic** and **paraventricular nuclei** of the hypothalamus and extend through the **infundibulum** to terminate in the **pars nervosa**. In addition, the pars nervosa has its own capillary bed, and scattered among pars

---

BOX 15.1 THE CENTRAL ROLE OF THE PARS TUBERALIS (PT) IN REGULATING CIRCANNUAL CLOCKS IN MAMMALS

For long-lived mammals living at temperate latitudes, photoperiod length is the major cue that synchronizes seasonally modulated functions, such as reproduction, energy usage, hibernation, and migration. Photoperiod-induced changes in physiology and behavior are mediated by an endogenous **circannual clock** that tracks and responds to **short-photoperiod (SP)** days of winter vs. **long-photoperiod (LP)** days of summer. In domesticated animals where such adaptations have been extensively analyzed, seasonal clock functioning involves melatonin stimulation of the anterior pituitary's **pars tuberalis (PT)** region (**Figure a**).

PT cells coordinate the circannual clock by expressing **melatonin receptors (MTs)** that bind pineal-derived melatonin. During long nights of **SP** winter days, prolonged melatonin production inactivates tuberalis cells that normally produce the **β subunit of thyroid stimulating hormone (TSHβ)** (**Figure a**). Conversely, with reduced melatonin production during **LP** days of summer, TSHβ-generating cells increase

their hormone production, which can stimulate both increased T3/T4 production from the thyroid and **GnRH release** from the hypothalamus to affect reproductive functions.

Previously, it was thought that the human PT lacks photoperiod-mediated seasonal adaptations, but as various components of signaling pathways used by other mammals become identified in humans, similar roles played by the PT in humans now appear more likely. For example, increased inflammation and behavioral disorders, such as seasonal affective disease (SAD) are associated with SP days of winter. Accordingly, endocannabinoid production by the human PT has been demonstrated to rise during LP days of summer, and such factors bind pars distalis corticotropes to block ACTH production. Conversely, a decrease in PT-released endocannabinoid during the SP of winter could elevate adrenal production of cortisol, as shown for other mammals, and thereby increase inflammatory responses and mood disorders like SAD.

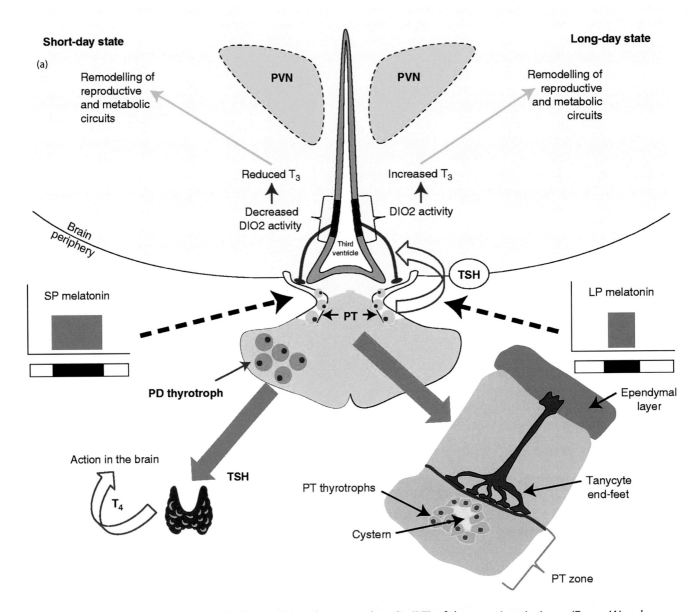

Box 15.1 (a) *Circannual seasonal clocks involving the pars tuberalis (PT) of the anterior pituitary. (From: Wood, S, and Loudon, A (2018) The pars tuberalis: the site of the circannual clock? Gen Comp Endocrinol 258: 222–235, reproduced under a CC BY 4.0 creative commons license.)*

nervosa neurons and blood vessels are numerous **pituicyte** glial cells that can be distinguished by their bundles of intermediate filaments containing **glial fibrillary associated protein (GFAP)**.

Neurohypophyseal axons are unmyelinated and contain the small peptide hormones **anti-diuretic hormone (ADH)** and **oxytocin (OT)**, which are made mainly in the supraoptic and paraventricular nuclei of the hypothalamus, respectively (**Figure 15.4a**). After being transported down axons, ADH- or OT-containing vesicles of hypophyseal neurons are occasionally secreted directly into the pars nervosa capillary bed. Alternatively, these hormones can be temporarily stored bound to proteins called neurophysins, thereby forming dilations in axonal termini, called **Herring bodies (Figure 15.5e)**. In either case, pituicytes and pericytes surrounding pars nervosa blood vessels modulate hormonal secretion by either facilitating or obstructing access to neurohypophyseal capillaries (**Figure 15.5f, g**). Pituicytes also secrete growth factors such as, VEGF (vascular endothelial growth factor) that affect blood vessel proliferation (=angiogenesis) and hence vascularization patterns within the pars nervosa (**Figure 15.5f, g**).

CASE 15.3 PIO PICO'S ANTERIOR PITUITARY

Last Governor of Mexican California
(1801–1894)

"...an uglier man than Pio Pico rarely had entered this world."

(A description of P.P.'s homeliness caused by acromegaly)

A 19th century rancher and politician, Pio Pico was the last Governor of Mexican California before it was taken over by the United States following the Mexican-American war of 1846–1848. In addition to his role in early California politics, Pico was noteworthy not only for having acromegaly due to excessive growth hormone (GH) secretion by the anterior pituitary, but also for showing a marked regression of acromegalic pathologies before he died at age 93 of undisclosed causes. Based on photographs from 1847–1858, Pico began exhibiting the coarsening facial features of acromegaly in his

40s and 50s presumably due to elevated GH levels caused by hyperplasia of his anterior pituitary. Unlike most people with untreated acromegaly who die within a decade after the disease reaches its peak pathology, Pico survived another several decades. In fact, he showed a remarkable loss of acromegalic features as he aged, indicating a spontaneous case of pituitary apoplexy, in which his hyperplasia regressed while still maintaining normal operations in the rest of the gland. Thus, the politician who witnessed California's major changeover from Mexican to U.S. rule also saw radical transitions in the functioning of his anterior pituitary over the course of his life.

Once released into the bloodstream, ADH and OT have wide-ranging effects (**Figure 15.3b**). Although sometimes referred to as **vasopressin** owing to its vasoconstrictive properties, ADH primarily targets the kidneys to reduce water loss during urine formation, thereby helping to prevent a process termed **diuresis** that results in frequent urination (polyuria) (Chapter 16). Accordingly, certain neurohypophyseal tumors trigger the diuresis that is characteristic of **diabetes insipidus** (Chapter 13) by downregulating ADH release from the posterior pituitary. As opposed to ADH's functions, **OT** is mainly associated with: (i) **uterine muscle contractions** during childbirth, and (ii) the **milk ejection response** of nursing that involves **myoepithelial cells** squeezing out secretions from mammary glands (Chapter 18).

15.7 DEVELOPMENT AND FUNCTIONAL HISTOLOGY OF THE THYROID

During human embryogenesis, the thyroid arises near the base of the developing tongue as an invagination of **endoderm** that separates from the digestive tract by the 8th week (**Figure 15.6a**). Subsequently, the initial thyroid anlage fuses with a second endodermal thyroid primordium containing **parafollicular** (or **C-**) **cells** that will produce a minor thyroid hormone, called **calcitonin**. By the 10th week of gestation, the fused thyroid primordia start forming spherical **follicles** that fill much of the organ's interior, except for thin compartments of the interfollicular interstitium containing blood vessels, fibroblasts, and few other stromal cells. Each of these developing follicles is lined by a simple squamous to cuboidal epithelium of secretory **thyrocytes**. Thyrocyte apices border the follicular lumen, whereas basally, such cells overlie blood vessels, into which thyrocytes secrete two kinds of **iodine-containing thyroid hormones**, called **T3** (**tri-iodothyronine**) and **T4** (**tetra-iodothyronine**, or **thyroxine**).

When fully developed, the adult **thyroid** occurs near the laryngeal-tracheal junction and is often shaped like a butterfly with two **lateral lobes** that are interconnected by an **isthmus** region (**Figure 15.6b**). The numerous spherical **follicles** filling thyroid lobes store within their lumens abundant **colloid** material containing **thyroglobulin** proteins with thyroid hormone precursors (**Figure 15.6c, d**). The production of colloid and the subsequent delivery of hormones into the bloodstream are both stimulated by adenohypophysis-derived **thyroid stimulating hormone** (**TSH**) that in turn is secreted in response to **cold** or other demands for **raising basal metabolism**. Production and secretion of thyroid hormones by thyroid follicles can be viewed as involving two oppositely directed pathways—an **exocytosis** leg oriented toward the follicular lumen vs. an **endocytosis** sequence moving away from the lumen. Such processes are summarized in **Figure 15.6e** and described further below in terms of four major stages:

1. **Iodine uptake:** iodine atoms obtained from food or vitamin supplements are readily ionized within the digestive tract to yield **iodide**, which is absorbed by the small intestine and delivered via the bloodstream to the thyroid. In thyroid follicles, iodide and sodium ions are actively co-transported into thyrocytes by **sodium-iodide symporters** (**NISs**) on thyrocyte basal surfaces, with iodide subsequently being released

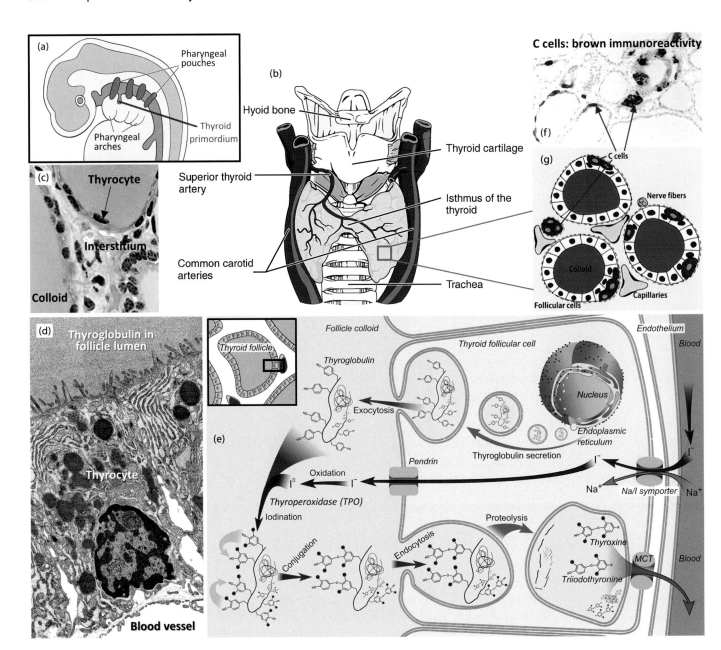

Figure 15.6 Thyroid gland. (a, b) *The thyroid gland develops from foregut endoderm and when fully formed typically constitutes a butterfly-shaped organ at the base of neck. (**c, d**) Unlike other endocrine glands, the thyroid is filled with spherical follicles of thyrocytes that surround a central lumen. Each follicle stores hormone precursors in a colloid mixture containing iodinated thyroglobulin protein. (c) 300X; (d) 2,300X. (**e**) Stages in the formation and secretion of the thyroid hormones T3 (tri-iodothyronine) and T4 (tetra iodothyronine, thyroxine) involve: (i) an exocytosis pathway in which iodine is conjugated to tyrosine residues on thyroglobulin within luminal colloid, and (ii) an endocytosis pathway in which iodinated thyroglobulin engulfed by thyrocytes is cleaved into T3 and T4 before being secreted basally into the bloodstream. Both pathways are stimulated by the binding of thyroid stimulating hormone (TSH) to its receptors on thyrocytes. (**f, g**) In addition to the thyroid's main hormones T3 and T4, C cells both within follicles and in the interfollicular interstitium produce a hormone, called calcitonin, which plays a key role in reducing excessive calcium in circulating blood of non-human vertebrates. (f) 75X. ([a]: From Policeni, BA et al. (2012) Anatomy and embryology of the thyroid and parathyroid glands. Sem Ultrasound CT MRI 33(2):104–114, reproduced with publisher permission; [b]: From OpenStax College (2013) Anatomy and Physiology. OpenStax. http://cnx.org/content/col11496/latest, reproduced under CC-BY-4.0 creative commons license; [d]: Image courtesy of Dr. H. Jastrow and Dr. H. Wartenberg reproduced with permission from Dr. Jastrow's Electron Microscopic Atlas. http://www.drjastrow.de; [e]: From Häggström, Mikael (2014) Medical gallery of Mikael Häggström 2014. Wiki J Med 1(2), doi: 10.15347/wjm/2014.008, ISSN2002-4436, reproduced under a creative*

Figure 15.6 (Continued) commons CC0 1.0 Universal Public Domain Dedication license; [f, g]: From Fernandez-Santos, JM et al. (2012) Paracrine regulation of thyroid-hormone synthesis by C cells. Intech Open, doi: 10.5772/46178, reproduced under a CC BY 3.0 creative commons license.)

into the follicular lumen via anion transporters, called **pendrin**. Without adequate iodine intake, low thyroid hormone production (=**hypothyroidism**) can lead to an enlarged thyroid (=**goiter**). Such hypertrophy arises from insufficient negative feedback of thyroid hormones on TSH release, which in turn allows elevated TSH levels to expand the thyroid (**Figure 15.1d**) (Box 15.2 Iodine, Goiters, and the Evolution of Thyroid Hormones).

2. **Thyroglobulin production:** thyroglobulin is a large ~660 kD homodimer glycoprotein that is synthesized in thyrocytes ER arrays and glycosylated in Golgi bodies before being exocytosed into the lumen for subsequent iodination. Within the lumen, thyroglobulin dimers become crosslinked to form aggregates that constitute the vast majority of the colloid, which also contains minor amounts of processing enzymes and other accessory constituents.

3. **Iodination of thyroglobulin:** intraluminal **iodide** is oxidized into **iodine** via **thyroperoxidase** (**TPO**) enzymes and **hydrogen peroxide** derived from thyrocytes. TPO also catalyzes the attachment of one or two iodine atoms on thyroglobulin tyrosines to form hormone intermediates, called mono-iodotyrosine (MIT) and di-iodotyrosine (DIT), respectively. The covalent coupling of MIT to DIT generates tri-iodothyronine (T3) regions on thyroglobulin, whereas two DITs linked together produce tetra-iodothyronine (T4) sites (**Figure 15.6e**).

4. **Endocytosis and cleavage of iodinated thyroglobulin**: colloid endocytosed by thyrocytes is cleaved by **cathepsin** enzymes to yield T3 and T4 molecules. These two thyroid hormones are then moved basally and delivered into nearby blood vessels via transporters in thyrocyte basal membranes (**Figure 15.6e**).

After delivery into the bloodstream, nearly all T4 is converted into the more active, but also more ephemeral, **T3** form within peripheral tissues, such as the **liver**. T3 acts as the main thyroid hormone regulator of **essentially all cells** in the body. Once T3 is incorporated within a target cell, T3 can bind to receptor molecules and enter nuclei

BOX 15.2 IODINE, GOITERS, AND THE EVOLUTION OF THYROID HORMONES

Iodine is required for production of the **thyroid hormones** T3 and T4. To meet such needs, the recommended minimum daily intake of ~150–250 µg of iodine can be obtained from vitamin supplements and/or from a balanced diet that includes iodine-rich foods (e.g., milk, yogurt, potatoes, marine fish, and seaweed). In addition, since the 1920s, iodine has been added to table salt in the United States. However, not all countries have ready access to iodized salt or iodine-rich food, and thus inadequate intake of this trace mineral poses a health risk for 2 billion people worldwide.

Severe iodine deficiency can lead to **hypothyroidism**, which if left untreated produces an enlarged thyroid, called a **goiter** (**Figure a**). Such goiters arise from the accumulation of insufficiently iodinated thyroglobulin in thyroid follicles, which in turn results from the anterior pituitary secreting excessive TSH owing to insufficient negative feedback from circulating T3/T4. Similarly, even with adequate iodine intake, hypothyroidism due to **Hashimoto's thyroiditis** can trigger goiter formation via an autoimmune-mediated reduction in T3/T4 production, thereby allowing excessive TSH to enlarge the thyroid. As the most common form of hypothyroidism, Hashimoto's disease affects ~5% of Caucasian adults and typically involves autoantibodies against **thyroperoxidase** (**TPO**) or **thyroglobulin** proteins used in making T3/T4. Conversely, a goiter can also form during **Graves hyperthyroidism**, with TSH-mimicking

autoantibodies causing excessive colloid and hormone secretion (pg. 384).

To understand T3/T4 signaling in humans, insights can be gained from tracing the evolutionary origin of thyroid hormones. Several kinds of marine algae produce T3/T4 and other iodinated compounds, which collectively can constitute up to 1% of the alga's weight. Moreover, signaling pathways mediated by T3/T4 are widely used in extant animal groups to regulate energy usage, growth, and development. Such findings led to the proposal that thyroid hormones were originally obtained from food sources before organs evolved to produce such hormones endogenously. Organs capable of secreting thyroid hormones initially arose as simple endostyle structures within the feeding apparatus of protochordates followed by the production of separate follicle-based thyroids in vertebrates (**Figure b**).

With their ability to concentrate iodine and secrete T3/T4, vertebrate thyroids control various functions, both via non-genomic pathways, such as utilized by invertebrates and via the coupling of thyroid hormones to nuclear thyroid hormone receptors to exert transcriptional effects. Accordingly, it has been hypothesized that non-genomic and genomic pathways using thyroid hormones helped establish cardinal vertebrate characteristics, including the maintenance of core body temperatures within a narrow range that allowed **endothermy** to evolve.

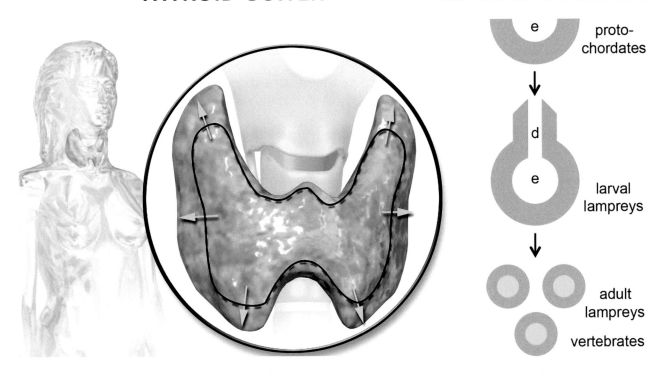

(a) **THYROID GOITER**

(b) **THYROID EVOLUTION**

Box 15.2 (**a**) *An enlarged thyroid (goiter). (From Blausen.com staff (2014) "Medical gallery of Blausen Medical 2014". Wiki J Med 1 (2): 10. doi:10.15347/wjm/2014.01010, reproduced under CC0 and CC-SA creative commons licenses.)* (**b**) *Diagram showing the evolution of follicle-based thyroids in adult vertebrates from a simplified endostyle (e) organ without or with a duct (d) in protochordates and larval lampreys. (From Liang, S et al. (2017) A branching morphogenesis program governs embryonic growth of the thyroid gland. Development 145: doi:10.1242/dev.146829, reproduced under a CC BY 3.0 creative commons license.)*

to modulate T3-sensitive gene expression. Such effects are particularly evident during neural development, where congenital forms of hypothyroidism, including **cretinism**, diminish intellectual capabilities in the absence of proper T3 signaling. Alternatively, T3 can also catalyze non-genomic processes, the net effect of which is typically to **increase energy consumption and cellular metabolism**, thereby raising basal metabolic rates.

In the case of excessive T3/T4 production (=**hyperthyroidism**), several **anti-thyroid drugs (ATDs)** have been developed as alternatives to surgical procedures and **radioactive iodine** treatments that were once commonly used to dampen thyroid output (**Case 15.4 Vera Schmidt's Thyroid**). For example, **sodium-iodide symporter** blockers downregulate iodide transport into thyrocytes, whereas **TPO inhibitors** prevent **thyroglobulin iodination**. Conversely, hypothyroidism is typically treated by supplementing thyroid output with a synthetic form of T4 (levothyroxine) (Box 15.2).

In addition to T3 and T4, the thyroid also secretes the 3400-MW polypeptide **calcitonin (CT)**, which is made by shorter and more lightly staining **parafollicular cells** occurring next to thyrocytes, either within the follicular epithelium or outside it in interstitial spaces (**Figure 15.6f, g**). Parafollicular cells release calcitonin into the bloodstream in response to excess blood calcium levels. In other vertebrates, **calcitonin** (CT) upregulates osteoblastic osteogenesis to reduce blood calcium levels. However, it remains controversial if CT also plays

---

CASE 15.4 VERA SCHMIDT'S THYROID

Psychoanalyst and Educator
   (1889–1937)

"...the major part of the children are children of the Party executives who give all their time to their work and are not able to rear their children..."

(A Russian historian describing the clientele of the Detski Dom home for children run by V.S.)

Vera Schmidt was a Russian psychologist and educator who ran a school in Moscow for children of elite communist party members. Born Vera Yanitskaia, Schmidt received

psychoanalytical training at the Kiev Women's Educational Institute before marrying Otto Schmidt, who went on to become a well-connected member of the communist party after the 1917 Russian Revolution. In 1921, Schmidt began overseeing the Detski Dom school that permanently boarded young children from families of various communist party executives and groomed these wards to become new party leaders. However, after only a few years, a lack of consistent funding caused the Detski Dom project to be shut down. Schmidt continued conducting research on child learning until her health declined due to Graves' disease, and she died at age 48 while undergoing surgery on her diseased thyroid.

---

crucial roles in modulating blood calcium in humans, given that no marked phenotype is typically observed with either CT insufficiency or excess.

15.8 PARATHYROID GLANDS AND REGULATING BLOOD CALCIUM HOMEOSTASIS VIA PTH

In humans, **parathyroid glands** produce **parathyroid hormone** (**PTH**), which maintains **blood calcium levels** within the precise range required for normal body functions. The parathyroids of most adults total four in number and are distributed as paired sets of **superior** and **inferior** glands on the posterior side of each lateral thyroid lobe (**Figure 15.7a**), although alternative numbers and positioning can also occur. Each lentil-shaped parathyroid contains a cord-like parenchyma that is easily distinguished from nearby thyroid follicles (**Figure 15.7b**). Along with scattered adipocytes and blood vessels, the parathyroid parenchyma contains two cell types—(i) PTH-secreting **principal** (=**chief**) **cells**, which constitute most of the gland, and (ii) mitochondria-rich **oxyphil cells** that increase in numbers during aging and play enigmatic roles (**Figure 15.7c, d**).

When blood calcium levels **fall below minimal levels**, principal cells secrete increased amounts of PTH to maintain skeletomuscular, cardiovascular, and neurological functioning. PTH raises blood calcium by indirectly **increasing osteoclastic bone resorption**, thereby releasing Ca^{2+} into the bloodstream. Such resorption involves PTH binding to its receptors on osteocytes and osteoblasts, which in turn activate osteoclasts via RANK/RANKL/osteoprotegerin signaling (Box 4.2). In addition, PTH upregulates **active vitamin D** (calcitriol) production for optimal intestinal uptake of calcium, while also increasing renal phosphate excretion into urine in order to prevent ectopic calcium phosphate mineralization (**Figure 15.7e**).

In other vertebrates, **calcitonin** (**CT**) antagonizes PTH's effects to balance blood calcium levels, whereas in humans, blood calcium regulation appears to be handled primarily by PTH- and vitamin-D-mediated processes without major CT-mediated input. The key role of PTH in maintaining calcium homeostasis is underscored by the fact that low blood calcium levels (=**hypocalcemia**) due to either parathyroid-disabling tumors or surgeries can trigger seizures, irregular heartbeats, and severe muscle cramping (**tetany**). Conversely, overly elevated blood calcium levels (=**hypercalcemia**) caused by hyperactive parathyroids are linked to cardiovascular maladies (**Case 15.5 Garry Shandling's Parathyroid Glands**), mental disorders, and osteoporosis. Counterintuitively, recombinant PTH (rPTH) is used as a treatment for osteoporosis, because instead of triggering bone resorption at physiological levels that normally circulate in the bloodstream, intermittent injections of low amounts of rPTH act as anabolic agents to stimulate bone formation (Box 4.2).

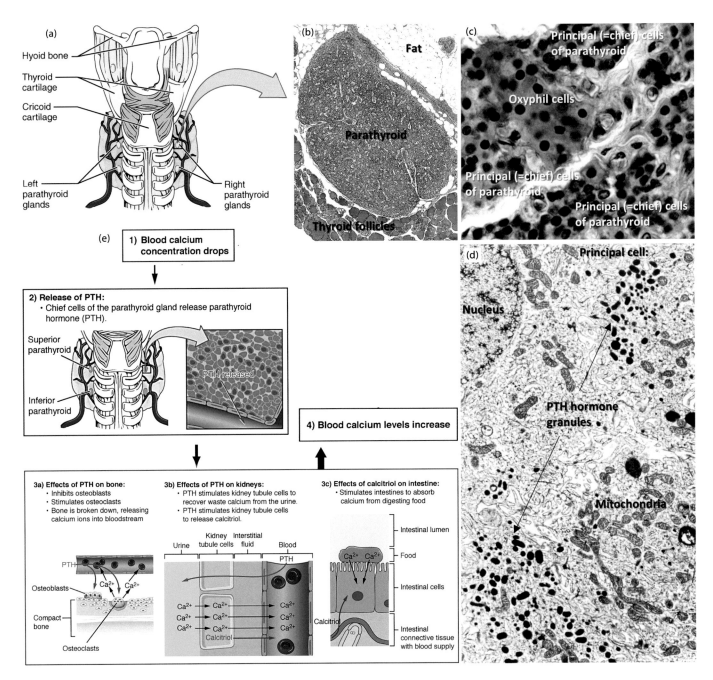

Figure 15.7 Parathyroid glands. *(a) Adults typically possess four parathyroid glands that occur on the posterior surface of the thyroid. (b–d) Each parathyroid contains cords of principal (=chief) cells that secrete parathyroid hormone (PTH). In addition, aged glands exhibit increasing numbers of larger oxyphil cells, which may represent senescent chief cells or a functional cell type whose roles have yet to be fully clarified. (b) 20X; (c) 300X; (d) 3,300X. (e) In response to low blood calcium levels, PTH is released by principal cells to trigger osteoclastic bone resorption and increased uptake of dietary calcium in the intestines. ([a, e]: From OpenStax College (2013) Anatomy and Physiology. OpenStax. http://cnx. org/content/col11496/latest reproduced under CC-BY-4.0 creative commons license; (d) Marti, R et al. (1987) Parathyroid ultrastructure after aldehyde fixation, high-pressure freezing, or microwave irradiation. J Histochem Cytochem 35: 1415–1424, reproduced with publisher permission.)*

CASE 15.5 GARRY SHANDLING'S PARATHYROID GLANDS

Comedian
  (1949–2016)

"So, turns out I had a hyperparathyroid gland that was undiagnosed because the symptoms mirror the exact same symptoms an older Jewish man would have."

(G.S. commenting about his hyperparathyroidism on a show called Comedians in Cars Getting Coffee)

Garry Shandling was a comedian, whose death raised awareness regarding the key roles that parathyroid glands play in maintaining human health. As a young man, Shandling received a marketing degree from the University of Arizona and later moved to Los Angeles where he wrote for TV comedy shows until he was seriously injured in an automobile accident. After recovering, Shandling pursued his dream of being a stand-up comedian before eventually starring in highly praised TV shows and serving as host of Grammy and Emmy telecasts. Shandling tried to keep

physically fit by boxing in a gym and regularly playing basketball. However, in his later years, Shandling's health began to deteriorate for reasons that were not well understood until he was finally diagnosed with hyperparathyroidism. Normally, the parathyroid glands maintain proper blood calcium levels, but if overactive, these glands can allow excess blood calcium to trigger cardiovascular and pulmonary pathogenesis. Accordingly, at age 66, Shandling died of pulmonary thrombosis, which may have resulted from his dysfunctional parathyroid glands.

15.9 ADRENAL GLANDS: CORTEX WITH THREE REGIONS (ZONA GLOMERULOSA, FASCICULATA, RETICULARIS) AND THEIR HORMONES (MINERALCORTICOID, GLUCOCORTICOIDS, ANDROGENS) PLUS THE ADRENAL MEDULLA AND CATECHOLAMINE PRODUCTION

As mediators of systemic responses to stress, two **adrenal glands** occurring on the superior poles of the kidneys secrete hormones and sympathetic neurotransmitters to regulate wide-ranging processes (**Figure 15.8a, b**). Each adrenal comprises two regions—an outer **cortex** and an inner **medulla**—which develop from different embryonic sources. During the 4th week of gestation, **mesodermally-derived** progenitor cells of the **cortex** start to form both a large **fetal cortical zone** and a thin overlying prospective **adult** (=definitive) **cortical zone** (**Figure 15.8c, d**), and a few weeks later, **neural crest ectodermal cells** generate a central **medulla** beneath the fetal cortex within each developing adrenal gland. During the second half of gestation, the fetal cortex helps maintain a full-term pregnancy by supplying pro-estrogen precursors before eventually degenerating by a year after birth as the adult cortex continues to grow and encase the subjacent medulla (**Figure 15.8c, d**).

When fully developed, adrenal cortical tissues constitute ~70% of the adult gland and comprise three regions: (i) an outermost **zona glomerulosa**, (ii) an intermediate **zona fasciculata**, and (iii) an innermost **zona reticularis** that abuts the medulla (**Figure 15.8d**). All three cortical regions produce **steroid** hormones via the differential processing of **cholesterol** within hormone-synthesizing cells. Proper levels of such corticosteroids are in turn required for maintaining systemic homeostasis, as evidenced by various maladies associated with either adrenal insufficiency (**Case 15.6 John Kennedy's Adrenal Glands**) or excessive adrenal cortical output (see pg. 384).

Corticosteroids of the **zona glomerulosa** regulate sodium and water retention in vertebrate kidneys and hence are termed **mineralcorticoids**. As the mineralcorticoid of humans, **aldosterone** is secreted in response to **angiotensin II** (Chapter 16) that is circulated to adrenal glands for binding to **angiotensin II receptors** on glomerulosa cells. Secreted aldosterone can then bind to **mineralcorticoid receptors** on target cells in the kidneys to help concentrate urine (Chapter 16).

Zona fasciculata cells in vertebrate adrenals secrete a group of closely related hormones that include **corticosterone, cortisone,** and **cortisol**. Such secretions are stimulated by **ACTH** binding to its receptors on fasciculata cells. Fasciculata corticosteroids activate **glucocorticoid receptors** (GRs) on virtually all cells in the body to modulate carbohydrate metabolism and various reactions to stress, including downregulating immune responses during inflammation. Because of their roles in modulating glucose levels in cells, these corticosteroid

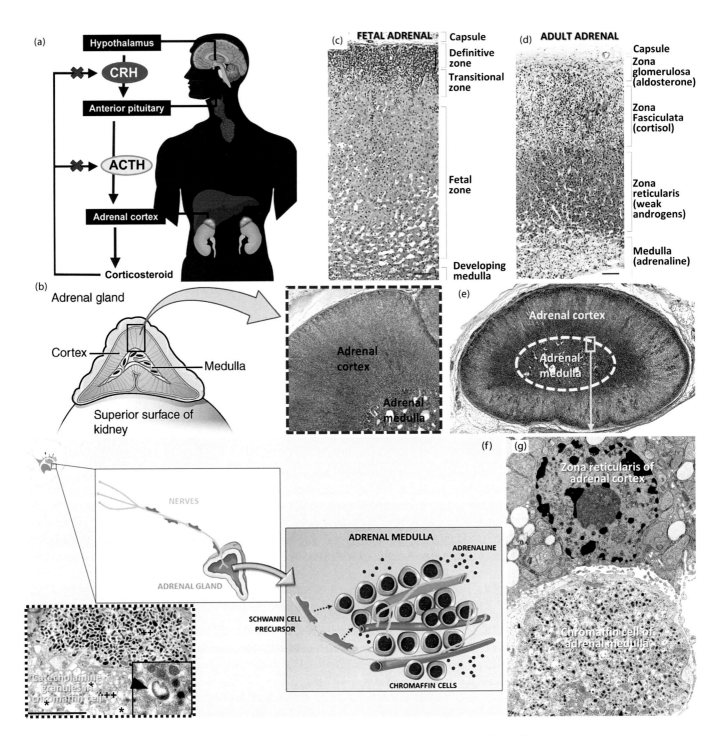

Figure 15.8 Adrenal glands. *(a, b) Adrenal glands occur on the superior poles of the kidneys and secrete several kinds of hormones (e.g., aldosterone, the corticosteroid cortisol, and weak androgens) plus catecholamine types of neurotransmitters like adrenaline. (c, d) The fetal adrenal gland contains: (i) a peripheral zone that will become the definitive adult cortex, (ii) a subjacent fetal cortical zone, and (iii) a central medulla. After birth, the fetal cortex degenerates and is replaced by the definitive cortex that comprises three layers: zona glomerulosa, zona fasciculata, and zona reticularis, which secrete aldosterone, cortisol, and weak androgens, respectively. Scale bars = 100 μm. (b, dotted inset) 25X. (e–g) Secretory cells of the adrenal medulla arise during development from neural-crest-derived Schwann cell precursors, and when fully functional store catecholamine neurotransmitters like adrenaline in electron-dense granules (f, inset). Since catecholamines form a grayish-brown product when reacted with chromium-containing stains, adrenal medullary cells are called chromaffin cells. (e) 10X; (f) scale bar = 5 μm; (g) 2,000X. ([a]: From Ross, AP et al. (2013) Multiple sclerosis, relapses, and the mechanism of action of adrenocorticotropic hormone. Front Neurol 4, doi: 10.3389/*

hormones are termed **glucocorticoids**. Corticosterone is the primary glucocorticoid of non-primate vertebrates. Conversely, **cortisol** serves such functions in various primates, including humans, and as an ingredient in medications is often referred to as **hydrocortisone**.

The last of the three cortical regions to develop fully is the **zona reticularis**, which begins its major differentiation in 4–8-year-old children during a process, called **adrenarche**. After acquiring full functionality, the reticularis zone can be stimulated by ACTH to secrete weak male sex hormones (=**androgens**) that mostly affect hair formation and apocrine gland development.

The **adrenal medulla** occurs internal to the cortex and constitutes ~30% of the entire gland. Along with blood vessels and scattered stromal cells, the medulla comprises modified neural-crest-derived sympathetic neurons that arise from Schwann cell precursors (**Figure 15.8e, f**). Such medullary cells convert tyrosine into three kinds of **catecholamine neurotransmitters—epinephrine** (=adrenaline), **norepinephrine** (=noradrenaline), and to a lesser extent, **dopamine**. These substances are normally stored as granules within medullary cells (**Figure 15.8f, g**) and are oxidized by chromic-acid-containing stains to form a grayish-brown product, called **chromaffin**. Hence, medullary secretory cells are referred to as **chromaffin cells**.

In the rest of the body, catecholamines secreted by sympathetic neurons affect nearby targets like involuntary muscles, glands, neurons, and adipocytes. However, because chromaffin cells have lost their axons and are instead associated with intramedullary capillaries, blood-borne catecholamines serve as widely dispersed hormones during **fight-or-flight** situations. In this way, the adrenal medulla coordinates with **cortisol secretions** from the zona fasciculata for multi-pronged reactions to stress by increasing energy usage while readying the body for action.

CASE 15.6 JOHN KENNEDY'S ADRENAL GLANDS

35th U.S. President
(1917–1963)

"John F. Kennedy does not now nor has he ever had an ailment described classically as Addison's disease, which is a tuberculose destruction of the adrenal gland."

(A press release that sidestepped directly addressing whether or not J.K. suffered from Addison's disease by narrowly defining classical Addison's as being caused by tuberculosis—a disease which J.K. had not contracted)

More so than any other U.S. President, John Kennedy hid from the public serious health problems that plagued him until his assassination at age 46. As a scrawny and sickly boy, Kennedy suffered from a nearly endless list of medical maladies, and during World War II, he contracted malaria while also further injuring his already chronically weak back. After the war, Kennedy became seriously ill on a trip in England. Doctors there diagnosed his various problems as resulting from Addison's disease, an atrophying of the adrenal gland cortex requiring regular cortisol treatments to replace endogenous products

of healthy adrenals. Although Kennedy was so incapacitated on the return voyage home that he was given his last rites, he eventually recuperated, and with the aid of cortisol therapy plus major back surgery, he was elected to the U.S. Senate. During his run for the Democratic presidential nomination in 1960, Kennedy was accused of suffering from Addison's disease, which at the time was a serious, and sometimes fatal, illness. However, a carefully worded press release was able to defuse the issue by simply stating Kennedy did not have classical Addison's disease induced by tuberculosis. This in turn helped Kennedy win the election without ever having to acknowledge his defective adrenal glands, which were verified in a post-assassination autopsy to display Addison-induced atrophy.

15.10 SOME ENDOCRINE DISORDERS: GRAVES' HYPERTHYROIDISM PLUS CUSHING'S AND CONN'S HYPERADRENALISM VS. ADDISON'S ADRENAL INSUFFICIENCY

As the most common form of excessive thyroid hormone production (=hyperthyroidism), **Graves' disease (GD)** is an autoimmune disorder that is far more prevalent in women than in men. Although the underlying causes of GD remain unknown, its effects are mediated by **thyroid-stimulating immunoglobulin (TSI)** antibodies that bind TSH receptors on thyrocytes and thereby mimic endogenous TSH in triggering thyroid hormone (T3/T4) production. TSI can thus cause thyroid hypertrophy (=**goiter**), **bulging** of the **eyes** (exophthalmia), and **systemic over-stimulation** resulting in weight loss, increased heart rate, and anxiety. Treatment of GD may involve surgical ablation, radio-iodine administration, and/or anti-thyroid drugs (ATDs) that target T3/T4 production.

Conn's and **Cushing's syndrome** are the two most common forms of **hyperadrenalism**, where the adrenal cortex secretes excessive amounts of corticosteroids. In **Conn's syndrome**, the zona glomerulosa region overproduces **aldosterone**. Because of aldosterone's role in raising blood pressure (Chapter 16), Conn's syndrome is characterized by hypertension and is often treated with surgery and/or aldosterone antagonists. In **Cushing's syndrome**, the zona fasciculata secretes too much **cortisol**. Symptoms of Cushing's syndrome include: (i) excess weight gain, (ii) hypertension; (iii) increased urination, and (iv) fragile skin, with treatment typically involving surgery and/or anti-glucocorticoid medications.

As opposed to Conn's and Cushing's syndromes, **Addison's disease** results in the **atrophying** of the adrenal cortex and thus triggers corticosteroid underproduction. Most Addison's cases are caused by an **autoimmune-mediated** destruction of the adrenal cortex that triggers such symptoms as fatigue, low blood pressure, and dizziness/fainting upon standing (orthostatic hypotension). In addition, without negative feedback from corticosteroids, upregulated ACTH production can cause **hyperpigmentation**, owing to ACTH cleavage generating the skin-darkening agent α-MSH. Currently, Addison's disease is treated mainly via daily intake of corticosteroid replacements and/or high salt diets.

15.11 SUMMARY—ENDOCRINE SYSTEM

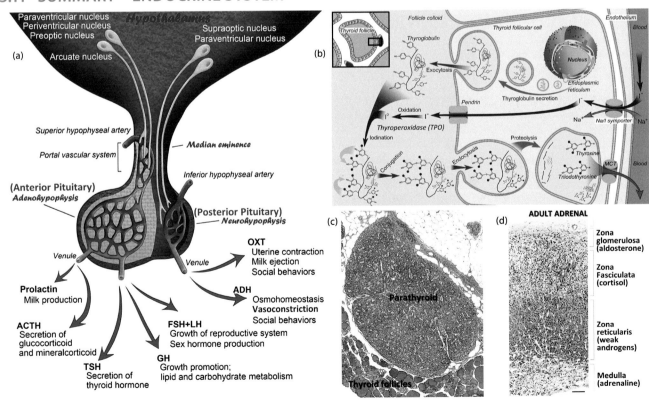

***Pictorial Summary Figure Caption: (a)**: see figure 15.3b; **(b)**: see figure 15.6e; **(c)**: see Figure 15.7b; **(d)**: see Figure 15.8d*

Endocrine organs lack ducts and secrete **hormones** into the bloodstream for controlling systemic **homeostasis**. Such organs include: the **pineal body** and **pituitary** plus associated **hypothalamus**, as well as the **thyroid, parathyroids**, and **adrenal glands.**

The **pineal body** of the brain secretes **melatonin** at night to regulate sleep and other **circadian rhythms**.

The **pituitary** arises from two embryonic primordia—(1) **oral ectoderm** (=Rathke's pouch) that forms the glandular **anterior lobe**, and (2) a **downgrowth of the brain** that gives rise to the neural **posterior lobe**. The anterior pituitary lobe comprises a large **pars distalis** (**pd**) and smaller **pars tuberalis** (**pt**) plus a **pars intermedia** (**pi**) in fetuses. The pd has **chromophobes** (e.g., **folliculostellate cells**) and **chromophils** divided into **acido-** vs. **basophils**. Chromophils secrete hormones in response to **releasing/inhibiting hormones** made in the hypothalamus. Hypothalamic hormones are delivered to the **primary capillary plexus** of the **hypothalamic-hypophyseal portal system** to regulate pituitary secretions into a **secondary capillary plexus** (**Figure a**). Pd **acidophils** and (*their hormones*) plus [targets] = (1) **somatotropes** (*growth hormone*) [epiphyseal plates, muscles, etc.], and (2) **lactotropes** (*prolactin*) [mammary glands]. Pd **basophils** and (*their hormones*) plus [targets] = (1) **gonadotropes** (*LH, FSH*) [gonads]; (2) **thyrotropes** (*TSH*) [thyroid]; and (3) **corticotropes** (*ACTH*) [adrenal cortex]. The pt coordinates with the pineal body to regulate endogenous **seasonal** clocks, whereas the pi is vestigial in adults but can make **melanocyte-stimulating hormone** from a **POMC** precursor in fetuses. The posterior pituitary lobe comprises **axons** from the hypothalamus plus supportive glial cells (**pituicytes**) and makes no hormones but can store in intra-axonal Herring bodies and subsequently release two hypothalamus-synthesized hormones: (1) **ADH** for urine processing, and (2) **oxytocin** to contract uterine muscles and to cause milk flow during lactation.

In order to produce iodinated **thyroid hormones (T3, T4)** in response to **TSH** from the anterior pituitary, **thyrocytes of thyroid follicles** take up iodide and eventually attach iodine to tyrosines of **thyroglobulin** protein occurring within follicular lumens. Iodinated tyrosines are then connected together to form **T3** and **T4** precursors, which after thyroglobulin endocytosis are cleaved into T3/T4 by **cathepsin** enzymes. T3/T4 released into the bloodstream can increase **basal metabolic rates** (**Figure b**). **Goiters** (=hypertrophied thyroids) develop via several pathways, including: (i) insufficient iodine intake causing low T3/T4 levels (=**hypothyroidism**), which in turn triggers high TSH levels and excessive colloid production, or (ii) from **Graves' hyperthyroidism**, where T3/T4 levels are high and TSH levels are low, but auto-antibodies mimic TSH's effects to overstimulate colloid secretion. In addition to thyrocyte-produced T3/T4, **parafollicular cells** in the thyroid make **calcitonin**, which in some animals plays a key role in decreasing blood calcium.

Parathyroids typically comprise two pairs of small glands next to the thyroid with principal cells secreting **PTH** to **increase blood calcium** by elevating osteoclastic bone resorption (**Figure c**).

An **adrenal gland** overlying each kidney comprises a mesodermally derived **cortex** surrounding a central **medulla** that contains modified sympathetic neurons arising from **neural crest cells**. The adrenal cortex has three zones—**glomerulosa, fasciculata**, and **reticularis**—that produce steroid hormones (**Figure d**). In response to angiotensin II, glomerulosa cells secrete **aldosterone** to reduce water loss in urine. ACTH causes fasciculata and reticularis cells to secrete **cortisol** and **weak androgens**, respectively. Cortisol regulates carbohydrate metabolism and inhibits immune responses. Weak androgens supplement gonadal testosterone. The adrenal medulla has **chromaffin cells** that secrete **catecholamines** (e.g., noradrenaline) in response to stress for inducing **fight-or-flight** actions.

Some endocrine disorders—**acromegaly**: too much GH secreted by the anterior pituitary after puberty; **Sheehan's syndrome**: post-partum hypopituitarism (low hormone secretions from the anterior pituitary); **diabetes insipidus**: excessive urination due to low ADH release from the posterior pituitary; **Hashimoto's thyroiditis**: autoimmune-mediated hypothyroidism (low T3 and T4); **Cushing's/Conn's syndromes**: excessive cortisol/aldosterone secretion from the adrenal cortex; **Addison's disease**: adrenal cortex atrophy, typically due to autoimmune reactions.

SELF-STUDY QUESTIONS

1. Which region of the adrenal cortex produces mainly cortisol?
2. T/F The posterior lobe of the pituitary synthesizes ADH.

3. What is the name of the oral ectoderm evagination that will form the anterior pituitary?
4. Of the following modes of cell signaling, which can have the most wide-ranging effects?
   A. Endocrine
   B. Merocrine
   C. Paracrine
   D. Juxtacrine
5. Which of the following is NOT made in the anterior pituitary?
   A. ACTH
   B. Calcitonin
   C. Prolactin
   D. TSH
   E. FSH
6. Which of the following is made in the adrenal cortex?
   A. PTH
   B. Glucagon
   C. Aldosterone
   D. T3
   E. GH
7. Which hormone triggers uterine smooth muscle contraction and milk ejection?
8. T/F Adrenaline and noradrenaline are catecholamines.
9. Which of the following is an example of a pituitary chromophobe?
   A. Chief cell
   B. Folliculostellate cell
   C. Oxyphil cell
   D. Chromaffin cell
   E. Parafollicular cell
10. T/F The outermost layer of secretory cells in the adrenal cortex is called the zona glomerulosa.
11. Which of the following is involved in controlling circadian rhythms?
    A. Suprachiasmatic nucleus (SCN)
    B. Pineal gland
    C. Melatonin
    D. ALL of the above
    E. NONE of the above
12. Which of the following is characterized by hypothyroidism?
    A. Diabetes mellitus
    B. Acromegaly
    C. Graves' disease
    D. Hashimoto's disease
    E. Diabetes insipidus
13. T/F PTH normally elevates blood calcium levels via an increase in osteoclastic bone resorption.

"EXTRA CREDIT" For each term on the left, provide the BEST MATCH on the right (answers A-F can be used more than once)

14. Chromaffin cell ___   A. Parathyroid gland
15. MSH precursor ____   B. POMC
16. Herring bodies ____   C. Pendrin
17. Iodide transporter ___   D. Adrenal medulla
18. Seasonal rhythm ____   E. Posterior pituitary
    regulator   F. Pars tuberalis

19. Describe thyroid hormone production.
20. Describe hypothalamic-pituitary processes involved in secretion of basophil-synthesized hormones as well as the major targets and functions of such hormones.

ANSWERS

(1) zona fasciculata; (2) F; (3) Rathke's pouch; (4) A; (5) B; (6) C; (7) oxytocin; (8) T; (9) B; (10) T; (11) D; (12) D; (13) T; (14) D; (15) B; (16) E; (17) C; (18) F; (19) The answer should expand on: iodide transport, thyroglobulin iodination, colloid endocytosis, T3/T4 cleavage and secretion, while defining such terms as NIS, pendrin, TOP, hydrogen peroxide, iodide oxidation into iodine, monoiodotyrosine, diiodotyrosine, cathepsin, T4 to T3 conversion, etc.; (20) The following should be addressed: hypothalamic nuclei, hypothalamic-hypophyseal portal system, chromphil vs. chromophobe; acidophil vs. basophil, basophil hormone types and targets

ORGAN HISTOLOGY—URINARY SYSTEM, REPRODUCTIVE SYSTEMS, AND SENSORY ORGANS

PART FOUR

Urinary System

16.1   INTRODUCTION TO THE URINARY SYSTEM

The **urinary system** comprises two **kidneys**, two **ureters**, a **bladder**, and a **urethra**. Each kidney connects via a ureter to the bladder, which in turn is drained exteriorly by the urethra (**Figure 16.1a**). The primary function of these organs is to filter blood plasma during the production of **urine**, thereby regulating **water** and **electrolyte ion levels** throughout the body. Urine also eliminates **soluble waste** via the process of **excretion**. Moreover, the urinary system removes water that could contribute to vascular volume and thereby raise **blood pressure**. Accordingly, the kidneys constitute a key part of the **renin-angiotensin system** (**RAS**) for modulating blood pressure. Along with their urine-related functions, the kidneys help maintain homeostasis by producing an active vitamin D metabolite, called **calcitriol** (1,25-dihydroxycholecalciferol), which increases intestinal uptake of calcium. In addition, low oxygen levels trigger interstitial kidney cells to secrete the hormone **erythropoietin** (**Epo**) for stimulating erythropoiesis.

DOI: 10.1201/9780429353307-20

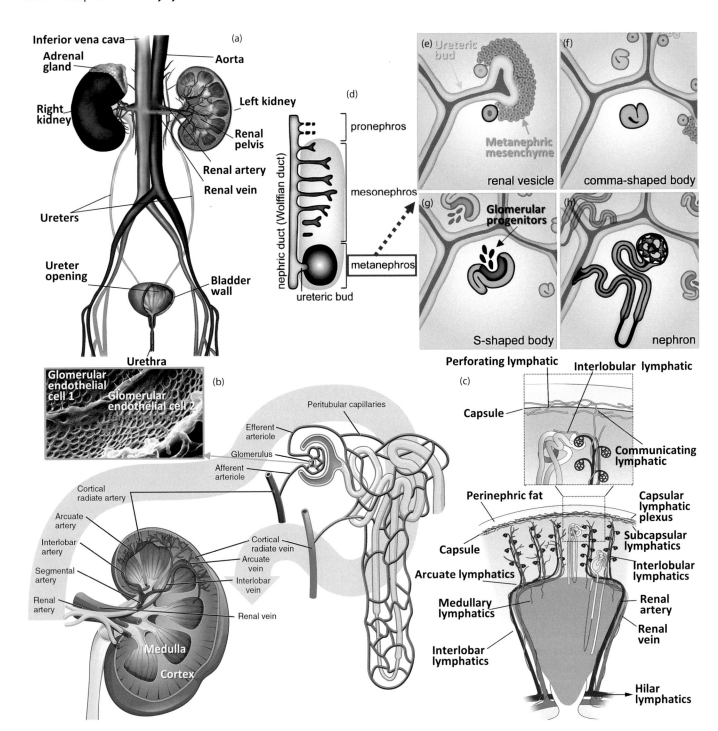

Figure 16.1 Introduction to the urinary system, renal circulation, and nephron development. (*a*) *The urinary system consists of two kidneys, two ureters, a bladder, and a urethra. (**b**) In order to form urine by filtering blood, a rich vascular supply permeates the kidney's cortex and medulla. Each urine-producing unit of the kidney, which is termed a nephron, receives blood into its glomerular capillary via an afferent arteriole. Some of the glomerular plasma is initially filtered through endothelial cell fenestrations (scanning electron micrograph [SEM], inset: 27,000X), and the remaining blood exits the glomerulus via an efferent arteriole to be drained from the kidneys by renal veins. (**c**) The kidneys also possess a well-developed network of lymphatic vessels that help maintain renal functioning. (**d-g**) During the second trimester, each nephron develops from a cluster of metanephric mesenchyme cells and transitions from a hollow vesicle through comma- and S-shaped units. The nephron then differentiates its subregions and attaches via its distal end to a developing collecting duct that is derived from ureteric bud cells. ([a]: Downloaded from https://nroer.gov. in/55ab34ff81fccb4f1d806025/file/56a87a7c81fccb6cb2edd0f9 of the National Repository of Open Educational Resources (NROER) of India and reproduced under a CC BY SA creative commons license; [b, c]: From OpenStax College (2013)*

Figure 16.1 (Continued) *Anatomy and Physiology. OpenStax. http://cnx.org/content/col11496/latest, reproduced under CC BY 4.0 creative commons license; [b, inset]: SEM inset from Rice, WL et al. (2013) High resolution helium ion scanning microscopy of the rat kidney. PLoS ONE 8, doi: 10.1371/jounal.pone.0057051, reproduced under a CC BY creative commons license; [d–h]: From Schell C et al. (2014). Glomerular development—shaping the multi-cellular filtration unit. Sem Cell Dev Biol 36:39–49, reproduced with publisher permission.)*

16.2 KIDNEY ANATOMY: CORTEX VS. MEDULLA, BLOOD VESSELS, AND NEPHRON DEVELOPMENT

Situated posteriorly in the upper abdominal cavity, each bean-shaped kidney in an adult human is about the size of a fist (**Figure 16.1a**) (**Case 16.1 Idi Amin's Kidneys**). The indented **hilum** of the kidney faces medially and is the site where: (i) blood vessels enter and exit the kidney, and (ii) the proximal end of the **ureter** is located (**Figure 16.1a, b**). Directly surrounding the kidney is a tough connective tissue capsule that is overlain at its superior pole by an adrenal gland. Subjacent to the capsule, kidney tissues are organized into an outer **cortex** and an underlying **medulla**, which in turn possesses conical subregions, called **renal pyramids** (**Figure 16.1a, b**).

As the site where blood is filtered to form urine, the kidneys receive a disproportionately plentiful vascular supply amounting to ~20–25% of total cardiac output. Such blood reaches renal tissues via the left or right **renal artery**, each of which branches into several **segmental arteries** near the hilum (**Figure 16.1b**). Arising from segmental arteries are **interlobar arteries** that connect with a curved **arcuate artery** running parallel to the kidney's convex surface at the cortico-medullary border (**Figure 16.1b**). The arcuate artery gives rise to numerous **interlobular arteries** that extend perpendicularly into the cortex and branch into **afferent arterioles**, each of which connects to a blind-ended tuft of capillaries, called a **glomerulus**. Some of the blood is filtered within the glomerulus, whereas residual unfiltered blood is drained from the glomerulus by an **efferent arteriole** that connects with peritubular **capillaries** of a **renal portal system** before eventually returning blood to the heart via the **renal vein** (**Figure 16.1b**). Along with their blood vascular system, the kidneys also contain numerous lymphatic vessels that help maintain proper renal functioning (**Figure 16.1c**).

Each adult kidney has on average about a million **urine-forming units**, called **nephrons**, which are assembled from **metanephric mesenchyme** within the **metanephros** primordium that will form the adult **kidney**. The metanephric kidney replaces two transitory versions produced earlier in embryogenesis, called the pro- and mesonephros, which represent evolutionary vestiges of kidneys formed by other vertebrate groups (**Figure 16.1d**).

In 3- to 5-month-old fetuses, presumptive adult nephrons begin to assemble as **hollow vesicles** of metanephric mesenchyme near **ureteric buds** that will form the collecting duct system for transporting urine to the ureters (**Figure 16.1d, e**). By ~4–6 months, each mesenchymal vesicle transitions through a **comma-like** stage prior to assuming an **S-shaped** configuration that incorporates glomerular progenitor cells (**Figure 16.1f, g**). The developing nephron then differentiates its major anatomical regions and fuses with a collecting duct by 6–7 months of gestation (**Figure 16.1h**). Such morphogenesis is completed by ~36 weeks of gestation, but each nephron undergoes considerable maturation for a year after birth before attaining full functionality.

CASE 16-1 IDI AMIN'S KIDNEYS

3rd President of Uganda
(1925?–2003)

"His Excellency, President for Life, Field Marshal Al Hadji Doctor Idi Amin Dada, VC, DSO, MC, CBE, Lord of All the Beasts of the Earth and Fishes of the Seas and Conqueror of the British Empire in Africa in General and Uganda in Particular"

(I.A.'s self-bestowed title after seizing the presidency of Uganda)

After leading a coup to take over the presidency of Uganda in 1971, Idi Amin suspended constitutional rights and expanded his authority during a brutal eight-year reign. In addition to expelling all Asians who had been living in Uganda, Amin used his military to kill hundreds of thousands of foes and

other targets, including various ethnic groups and journalists, while continually enriching himself and his polygamous family of at least 40 children and 6 wives. In 1979, a failing economy and waning support for Uganda's war with Tanzania triggered an uprising that prompted Amin's hasty escape. While exiled in Saudi Arabia, Amin's health deteriorated, and as large man with a voracious appetite, he developed various signs of metabolic syndrome disease, including chronic kidney dysfunction. Thus, in 2003, Amin lapsed into a coma in a Saudi hospital and was taken off life support to die from complications of kidney failure.

16.3 KIDNEY HISTOLOGY: SUBREGIONS OF THE KIDNEY, RENAL PELVIS OF THE URETER, AND NEPHRON COMPARTMENTS

When hemisected and viewed by an unaided eye, the **cortex** and **medulla** of the adult kidney can be seen overlying a **renal pelvis**, which represents the dilated proximal end of the ureter into which the medulla drips urine (**Figure 16.2a**). Not only does the more vascularized cortex look redder than the medulla in living samples, but when examined with a magnifying glass, the cortex appears more granular due to small inclusions, called **renal** (=Malpighian) **corpuscles**, which are lacking in the medulla (**Figure 16.2a**). Beneath the cortex, the medulla comprises an **inner** and **outer medulla**, with the outer medulla being further subdivided into a peripheral **outer stripe** plus an **inner stripe** that borders the inner medulla (**Figure 16.2b, c**).

Adjacent to the inner medulla, the funnel-shaped renal pelvis of the ureter is elaborated into large and small cup-like extensions, called the **major** and **minor calyces**, respectively, with each minor calyx surrounding the pointed tip of a **renal pyramid** (**Figure 16.2a**). The human kidney contains on the order of a dozen pyramids, and each pyramid plus its overlying cortical tissue constitutes a **renal lobe**. Thus, humans have **multilobar** kidneys as opposed to the **unilobar** type of rodents, in which the entire kidney contains but a single pyramid (**Figure 16.2a, b**). The cortex of renal lobes is further subdivided by interlobular arteries into numerous microscopic units, called **renal lobules**. Each cortical lobule in turn contains within its center a tract of mostly straight renal tubules forming an intracortical **medullary ray** that extends downward into the medulla (**Figure 16.2c**).

After differentiating from metanephric mesenchyme, nephrons are composed of four main anatomical regions, which based on nomenclature proposed in 1988 by a committee of renal scientists comprise the: (1) **renal corpuscle**, (2) **proximal tubule**, (3) **intermediate tubule** (which is often referred to as the descending and ascending thin limbs of the U-shaped **loop of Henle**), and (4) **distal tubule** (**Figure 16.2c**). In addition, distal tubules fuse via **connecting tubules** to **collecting ducts** of the **urine-transporting** system within each kidney (**Figure 16.2c**).

16.4 GENERAL MORPHOLOGY OF A RENAL CORPUSCLE (=BOWMAN'S CAPSULE + GLOMERULUS + INTERVENING URINARY SPACE)

In fully formed human kidneys, each **renal corpuscle** measures ~200 μm wide and comprises **Bowman's capsule** surrounding a **glomerular capillary** (**Figure 16.3a**). The end of the corpuscle where the afferent arteriole enters and the efferent arteriole exits is referred to as the **vascular pole**, whereas opposite to the vascular pole is the **urinary pole** that connects Bowman's capsule to the proximal tubule (**Figure 16.3a, b**).

Owing to the way glomerular capillaries form within differentiating nephric vesicles during embryogenesis, Bowman's capsule comprises two concentrically arranged layers. A simple cuboidal epithelium, called the **visceral layer** of Bowman's capsule, directly covers fenestrated endothelial cells of the glomerular capillary, whereas a simple squamous **parietal layer** forms the capsule periphery and is separated from the visceral layer by the capsule's **urinary space** (**Figure 16.3b**).

16.5 RENAL CORPUSCLE PODOCYTES, FILTRATION SLITS, AND MESANGIAL CELLS

Cells in the visceral layer of each renal corpuscle are termed **podocytes** in reference to their slender termini (=foot processes, or **pedicels**) that arise from thicker proximal branches of the podocyte soma (**Figure 16.3c–i**). Each pedicel interdigitates with nearby pedicels to form ~30–40 nm-wide **filtration slits**, through which filtered plasma from the glomerulus reaches the urinary space of Bowman's capsule. Such slits possess near their base a unique 5–10 nm-thick structure, called a slit **diaphragm**, which contains **nephrin** proteins capable of interacting with an intra-pedicle **actin** network (**Figure 16.3i**) as well as with ECM proteins. Although various parameters influence filtration dynamics, molecules less than ~3–4 nm in size normally pass through slits. In doing so, each filtered particle traverses: (1) a **fenestration** in the capillary endothelium, (2) the **glomerular basement membrane** which comprises the fused basement membranes of podocyte and endothelial epithelia, and (3) a filtration **slit diaphragm** (**Figure 16.3i**).

Renal corpuscles also contain **mesangial cells** (**MCs**) that do not contact the urinary space but instead lie within an interstitial compartment (=**mesangium**) bounded by glomerular capillaries and the glomerular basement membrane (**Figure 16.4a**). Most MCs are smooth-muscle-like cells that secrete ECM fibers to maintain structural integrity of the mesangium while also using **phago**- and **pinocytosis** to remove obstructive debris generated during blood filtration.

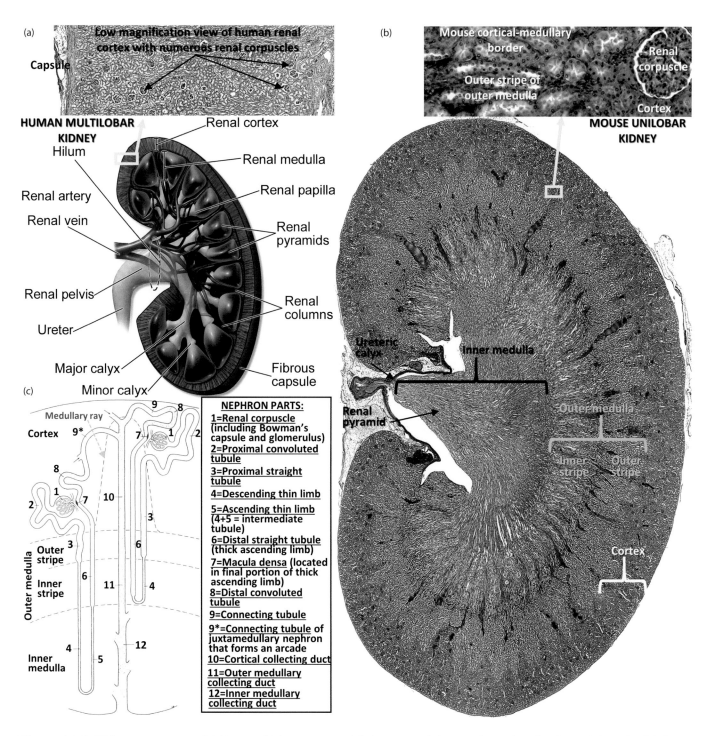

Figure 16.2 Kidney anatomy. (a) *Human kidneys are multilobar, containing on the order of a dozen lobes. Each lobe has a central renal pyramid that drains urine into a cuplike minor calyx at the dilated head of the ureter (=renal pelvis) (15X).* **(b, c)** *Unilobar rodent kidneys have only a single renal pyramid within the inner medulla, and the outer medulla comprises an outer and inner stripe, which possess differing arrays of nephron subregions. Peripheral to the medulla is the cortex, which is where all renal corpuscles are situated. (b, top image) 100X; (b, bottom image) 6X. ([a]: From Blausen.com staff (2014) Medical gallery of Blausen Medical 2014. Wiki J Med 1(2):10, doi: 10.15347/wjm/2014.01010, reproduced under CC0 and CC-SA creative commons licenses; [c]: From Kriz, W and Bankir, L (1988) A standard nomenclature for structures of the kidney. Kidney Int 33: 1–7, reproduced with publisher permission.)*

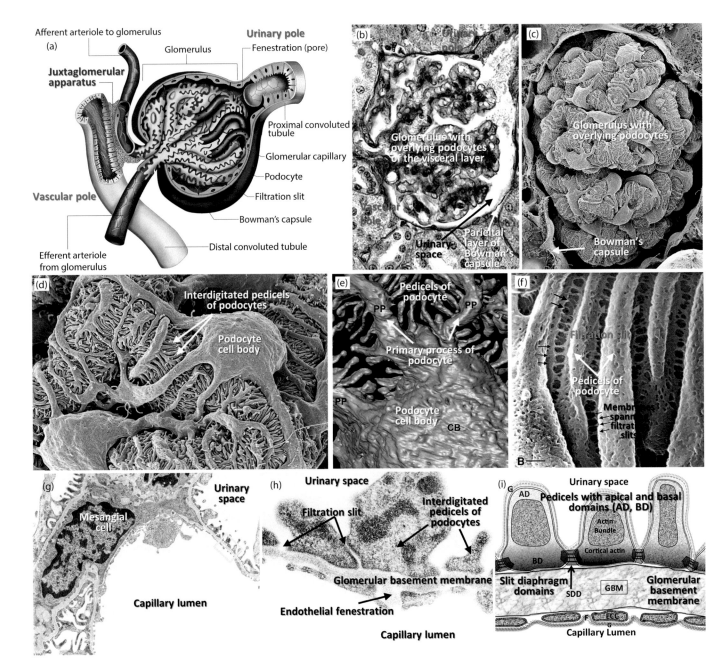

Figure 16.3 Functional microanatomy of renal corpuscles. *(a–c) Each renal corpuscle comprises Bowman's capsule, a glomerulus with overlying podocytes, and an intervening urinary space into which glomerular plasma is filtered to generate pre-urine. The vascular pole is where: (i) the afferent arteriole enters and the efferent arteriole exits each corpuscle, and (ii) a juxtaglomerular apparatus is present to help regulate blood pressure. Opposite to the vascular pole is the urinary pole, where Bowman's capsule connects with the proximal tubule. (b) 280X; (c) 430X. (d–h) As shown in SEMs (d, f), a 3D reconstruction of a FIB-SEM dataset (e), and TEMs (g, h), podocytes in each renal corpuscle have numerous branched processes that terminate in slender interdigitating pedicels. (d) 2,000X; (e) 11,000X; (f) scale bar=100 nm; (g) 12,000X; (h) 43,000X. (i) Adjacent pedicels with actin bundles form narrow inter-pedicel spaces, called filtration slits, whose slit diaphragm domains are the last structures through which blood is filtered before entering the urinary space. ([a]: Downloaded from https://nroer.gov.in/55ab34ff81fccb4f1d806025/file/56a87adb81fccb6cb2edd34f of the National Repository of Open Educational Resources (NROER) of India and reproduced under a CC BY SA creative commons license; [c–e]: From SEM and 3D FIB SEM reconstruction, Miyaki, T et al. (2020) Three-dimensional imaging of podocyte ultrastructure using FE-SEM and FIB-SEM tomography. Cell Tissue Res 379:245–254, reproduced under a CC BY 4.0 creative commons license; [f]: From Rice, WL et al. (2013) High resolution helium ion scanning microscopy of the rat kidney. PLoS ONE 8, doi: 10.1371/jounal.pone.0057051, reproduced under a CC BY creative commons license; [g, h]: From Takahashi-Iwanaga, H (2015) Three-dimensional microanatomy of the pericapillary mesangial tissues in the renal glomerulus:*

Figure 16.3 (Continued) *comparative observations in four vertebrate classes. Biomed Res 36:331–341, reproduced with publisher permission; images courtesy of Drs. Takahashi-Iwanaga and Iwanaga; [i]: From Neal, CR (2015) Podocytes… what's under yours? (Podocytes and foot processes and how they change in nephropathy) Front Endocrinol 6, doi: 10.3389.fendo.2015.00009, reproduced under a CC BY 4.0 creative commons license.)*

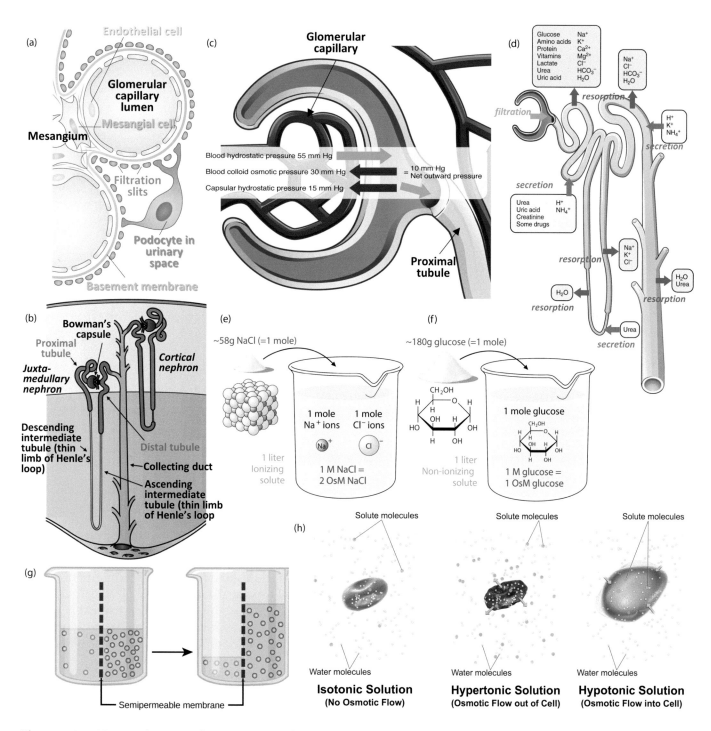

Figure 16.4 Mesangium, nephron types, and osmolarity-related definitions. *(a) Occurring within the mesangium compartment of the glomerular capillary, mesangial cells help clean the glomerular basement membrane and maintain structural integrity of the glomerulus. (b) Juxtamedullary nephrons have long intermediate tubules that extend into the inner medulla, whereas cortical nephrons have a shorter intermediate tubule that reaches only the outer medulla. (c) The filtration process of excretion occurs exclusively in renal corpuscles and uses the higher hydrostatic pressure of blood compared to opposing osmotic and capsular forces to generate pre-urine for delivery into the proximal tubule. (d) During transport in post-corpuscular regions of the nephron, pre-urine is converted into urine by: (1) the removal of various*

Figure 16.4 (Continued) *pre-urine components (e.g., glucose, Na+, and water) during the process of resorption, and (2) the addition of solutes to urine (e.g., urea and certain drugs) during the process of secretion. (**e, f**) As opposed to molarity which measures the number of moles of a solute in a liter of solution, osmolarity measures the total number of dissolved particles formed through ionizations in a liter of solution (note: osmolality is the number of dissolved particles in a kg of solution, which for water is essentially the same as osmolarity). Thus, a 1 M NaCl solution is 2 OsM due to the production of 1 osmole of Na+ ions plus 1 osmole of Cl- ions, whereas a 1 M solution of non-ionizing solutes like glucose is 1 OsM. (**g, h**) Osmosis is the movement of water down its concentration gradient as it passes through a semi-permeable membrane, such as the plasmalemma. Tonicity considers the concentration of non-penetrating particles, such as sodium ions that do not readily traverse cell membranes and thereby contribute to non-equivalent distributions across the membrane that affect osmosis. Isotonic solutions have the same concentration of non-penetrating particles as in a reference solution (e.g., intracellular milieu) and thus do not experience a net osmotic flow of water either into or out of the reference solution. Conversely, cells placed in hypertonic solutions will shrink as water moves by osmosis from the cell, whereas cells in hypotonic solutions will swell as water moves into the cell. ([a]: From Milner, JH (2020) Type IV collagen and diabetic kidney disease. Nat Rev Nephrol 16, https://doi.org/10.1038/s41581-019-0229-1, reproduced with publisher permission; [b]: Downloaded from https://commons.wikimedia.org/wiki/File:Kidney_Nephron.png; author H Fischer, reproduced under a CC BY 3.0 creative commons license; [c, d, g]: From OpenStax College (2013) Anatomy and Physiology. OpenStax. http://cnx.org/content/col11496/latest, reproduced under CC BY 4.0 creative commons license; [h]: Blausen.com staff (2014) Medical gallery of Blausen Medical 2014. Wiki J Med 1(2):10, doi: 10.15347/wjm/2014.01010, reproduced under CC0 and CC-SA creative commons licenses.)*

16.6   NEPHRON SUBTYPES AND MEDULLARY SUBREGIONS

Although all nephrons begin in the cortex as a renal corpuscle that extends proximal, intermediate, and distal tubular regions into the medulla, two main nephric types can be recognized: (1) **juxtamedullary nephrons** with corpuscles occurring near the cortico-medullary border that eventually connect to relatively long intermediate tubules extending into the **inner medulla** vs. (2) **cortical nephrons** containing not only corpuscles that are situated farther from the medulla but also shorter intermediate tubules that protrude only into the **outer medulla** (**Figure 16.4b**). These varying nephron morphologies and distributions across the medulla in turn allow juxtamedullary nephrons to play the driving roles in establishing standing salt gradients used in concentrating urine, as described further below.

16.7   FILTRATION, SECRETION, AND RESORPTION PROCESSES OF EXCRETION DURING URINE FORMATION AND DEFINITIONS OF RELATED TERMS—MOLARITY VS. OSMOLARITY AND ISO- VS. HYPER- VS. HYPOTONIC SOLUTIONS

Excretion of soluble wastes during urine formation is accomplished via three processes: **filtration**, **secretion**, and **resorption**. **Filtration** occurs exclusively in renal corpuscles and involves the movement of glomerular plasma across filtration slits into the urinary space of Bowman's capsule (**Figure 16.4c, d**). The amount of plasma that traverses all filtration slits in both kidneys per minute is the **glomerular filtration rate** (**GFR**), which under normal conditions averages **~125 ml/min**. Glomerular-derived fluid filtered into the urinary space is termed ultrafiltrate, or **pre-urine**, and after being substantially modified during its transport through post-corpuscle regions of the nephron is eventually delivered as **urine** to the ureter. Such modifications can involve **secretion**, whereby components that were absent in the original pre-urine are added to maturing urine as it moves toward the ureter (**Figure 16.4d**). Conversely, the process of removing pre-urine components and delivering them to the ECM is variously referred to as absorption, reabsorption, or **resorption** (**Figure 16.4d**). For example, based on a GFR of 125 ml/min that produces 180 liters of pre-urine per day, **~99% of the water** in pre-urine is resorbed to yield average daily urine outputs of just 1–2 liters.

As water is resorbed from pre-urine, the concentration of its dissolved molecules (=**solutes**) increases. Further descriptions of solute changes in solutions like pre-urine rely on several key terms. For example, **molarity** measures the number of moles of a solute dissolved in a liter of solution, where a one-molar (1 M) solution contains one mole of solute per liter. Similarly, **molality** measures the number of moles of solute per kg of solution, which for water-based solutions is essentially the same as molarity.

However, a more pertinent term is **osmolarity**, which tracks the number of **osmoles** per liter of solution, with osmoles equaling the total moles of **dissolved particles** that are generated in a solution. Thus, a one osmolar (1 OsM) solution contains one osmole of dissolved particles in a total liter of solution. Since most body fluids are less than 1 OsM in concentration, osmolarity is typically expressed in **milliosmolar** (mOsM) units, where 1000 mOsM = 1 OsM. For example, the osmolarity of human blood is ~275–300 mOsM (= 0.275–0.3 OsM). It is important to emphasize that **the concentration of water is inversely related to its solute content**. Thus, the concentration of water is **highest** in its pure state without any dissolved solutes. Conversely, adding solutes to water progressively **decreases the concentration of water**, as each dissolved solute reduces the number of water molecules that are present per unit volume of solution.

To contrast molarity vs. osmolarity, two cases can be considered—an ionizing solute like NaCl (molecular weight [MW] ~58 g) vs. a solute like glucose (MW ~180 g) that does not form ions when solubilized. Dissolving ~58 g of NaCl in a total volume of 1 liter of water generates a 1 M solution (**Figure 16.4e**). However, because each mole of NaCl ionizes into 1 mole of Na^+ cations and 1 mole of Cl^- anions, a 1 M NaCl solution contains 2 osmoles of dissolved particles and is thus 2 OsM (=2000 mOsM). Conversely, dissolving ~180 g of glucose in total of 1 liter of water produces a 1 M solution that is just 1 OsM, since only a single osmole of the non-ionized glucose molecules is present in the liter (**Figure 16.4f**).

A term related to osmolarity is **osmosis**, which is the **diffusion** of water down its concentration gradient while traversing a **semi-permeable** membrane like a plasmalemma (**Figure 16.4g**). Water movement during osmosis is a facilitated form of diffusion, since as a **hydrophilic** molecule, water does not readily traverse **hydrophobic** cell membranes except through water-specific intramembrane channels, called **aquaporins**. Accordingly, in various tissues, cell membranes have evolved aquaporin channels that allow rapid osmosis across plasmalemmata (Box 16.1 Aquaporins and Cell Volume Regulation Channels for Mediating Rapid Transcellular Water Movement).

The direction of water flow through aquaporins depends on the concentration of water within the cell vs. that in the surrounding ECM. These concentrations are in turn directly affected by the amounts of **non-penetrating particles**, such as Na^+ and Cl^-, which do not traverse the plasmalemma without facilitation. Conversely, **penetrating particles** like **urea** can more readily cross the cell membrane and equilibrate to the same concentration inside and outside the cell, thereby often canceling out their effects on osmosis. Accordingly, unlike osmolarity which measures the osmoles of both penetrating and non-penetrating particles, **tonicity** considers only the concentrations of non-penetrating particles while ignoring penetrating particles and is generally reported in relative terms. Thus, in **isotonic** cases, where concentrations of non-penetrating particles inside and outside the cell are the same, intra- vs. extracellular concentrations of water are also equal, and there is no net osmotic-driven flow of water (**Figure 16.4h**). Alternatively, in **hypertonic** situations where the ECM has a higher concentration of non-penetrating particles than is present in the cellular milieu, **water is more concentrated inside the cell** than outside, and thus osmosis drives intracellular **water outward** to shrink the cell (**Figure 16.4h**). Conversely, in **hypotonic** solutions, more dissolved non-penetrating particles, but **less water molecules**, occur per unit volume **inside** vs. outside the cell. Hence, osmosis will force **water into the cell**, causing it to swell (**Figure 16.4h**).

16.8 FUNCTIONAL HISTOLOGY OF POST-CORPUSCLE NEPHRON REGIONS DURING URINE FORMATION

Each post-corpuscle portion of the nephron possesses key microanatomical adaptations for optimizing urine production (**Figure 16.5a–j**). For example, the **proximal tubule** is lined by a **simple cuboidal epithelium** and comprises an initial **convoluted** region, which, after winding around in the cortex, connects with a **straight part**. The straight proximal tubule terminates in the outer stripe of the outer medulla and is also referred to as the **thick descending limb** of the loop of Henle.

Proximal tubules modify pre-urine by resorbing a wide array of solutes, including most of the sodium and water that was in Bowman's capsule, while also **secreting** a few molecules. Examples of secreted components include protons as well as excess pharmaceuticals and other **xenobiotic** toxins from the environment. In order to increase surface area for resorption and secretion, proximal tubule cells form an apical **brush border** of long microvilli that in turn reduce the size of the tubule lumen (**Figure 16.5a, b, e–h**). In addition, the basal plasmalemma folds inward and is associated with numerous mitochondria that supply the ATPs needed for energy-requiring processes, such as **active Na^+ resorption**.

BOX 16.1   AQUAPORINS AND CELL VOLUME REGULATION CHANNELS FOR MEDIATING RAPID TRANSCELLULAR WATER MOVEMENT

Many human tissues are capable of rapidly transporting water across cells. Such water transport occurs via two major routes—(1) a **paracellular pathway**, whereby water flows across intercellular junctions and spaces that separate neighboring cells, and (2) a **transcellular pathway** that involves water traversing cell membranes and cytoplasmic domains. Given the hydrophilic properties of water vs. the hydrophobic nature of cell membranes, water movement through cells would be extremely slow without mechanisms to facilitate transport. Thus, cells have evolved plasmalemmal channels composed of different **aquaporin (AQP)** proteins to accelerate transcellular water flow.

AQPs are integral membrane proteins ~24–30 kD in monomeric size that assemble as **homo-tetrameric units** to form water-specific channels in the plasmalemma (**Figure a**).

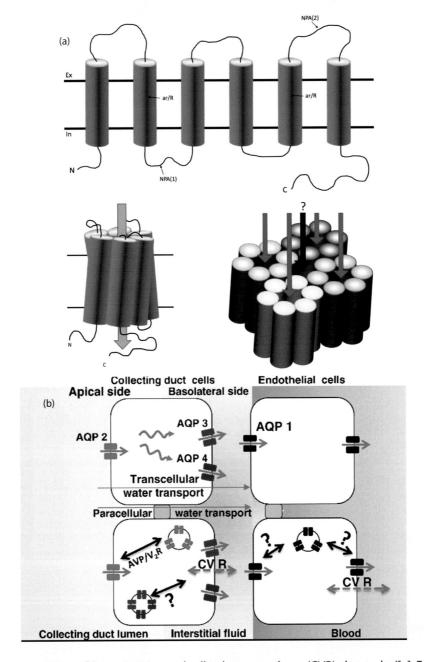

Box 16.1 (a) *Aquaporin structure.* **(b)** *Aquaporins and cell volume regulator (CVR) channels. ([a]: From Halsey, AM et al. (2018) Aquaporins and their regulation after spinal cord injury. Cells 7: doi:10.3390/cells7100174 reproduced under a CC BY 4.0 creative commons license; [b]: From Day, RE et al. (2014) Human aquaporins: regulators of transcellular water flow. Biochim Biophys Acta 1840: 1492–1506 reproduced under a CC BY 3.0 creative commons license.)*

Humans have evolved at least 13 AQPs, which are expressed in various body regions, and within renal tissues, multiple AQPs have been identified (**Figure a**). AQP tetramers allow water molecules to move in a single file through each tetrameric channel, while at the same time selectively blocking passage of other molecules. Once water molecules are situated near a channel opening, they are moved through the membrane by osmotically driven rapid **diffusion** without requiring either energy input or channel modifications.

Along with AQPs, cells can use additional channels for rapid **cell volume regulation** (**CVR**) following osmotic-induced cellular shape changes (**Figure b**). Such channels coordinate with AQPs by passively or actively transporting ions and other osmotic modulators (=**osmolytes**) like the organic compound **taurine**. Thus, for example, after osmotic-driven water gain in hypotonic environments, swollen cells can utilize CVRs along with other ion channels to regain normal cellular shape and turgidity.

Beyond the proximal tubule, the **intermediate tubule** comprises a U-shaped, relatively thin region that is also referred to as the descending and ascending **thin limbs of the loop of Henle**. The proximal **descending limb** of the intermediate tubule extends toward the medullary tip, whereas the distal **ascending limb** returns toward the cortex. At its distal end within the medulla (**Figure 16.5a, c, i**), the ascending intermediate tubule connects with the initial straight portion of the distal tubule, which is also called the **thick ascending limb** (**TAL**) of the loop of Henle.

Throughout its length, the intermediate tubule is lined by a simple squamous to sub-cuboidal epithelium that possesses sparse microvilli (**Figure 16.5c, i**). The major role of the intermediate tubule is to **concentrate urine** by establishing within the interstitial ECM a continually present **gradient of osmolarity** that normally ranges from ~300 mOsM in the cortex (i.e., essentially isotonic to blood) to hypertonic levels of ~1,200 mOsM in the inner medulla. This standing salt gradient is established by a **countercurrent exchange mechanism**, which in turn relies on two key properties of intermediate tubules: (1) **oppositely directed fluid flow** in descending vs. ascending limbs, and (2) **differing permeability and ion-transporting capabilities** of the two limbs. In particular, **the descending limb** has AQP1 aquaporins and is thus **permeable to water** but **impermeable to sodium**. Conversely, the aquaporin-deficient ascending limb is **impermeable to water** but can **actively transport sodium** and other ions into the ECM. In addition, the selective movement of **urea** also contributes to the standing osmolarity gradient in the case of long loops of juxtamedullary nephrons that extend into the inner medulla.

The initial TAL of the **distal tubule** contains basal infoldings with interspersed **mitochondria** that provide the energy needed for ion pumping (**Figure 16.5d, j**). This TAL region eventually terminates within the cortex as a group of lining cells with densely packed nuclei, called the **macula densa**, which helps regulate blood pressure (see upcoming section on the Renin-Angiotensin System). Beyond the macula densa, the distal tubule is folded in the cortex to form a **distal convoluted tubule (DCT)**, which connects via a straight **connecting tubule** to a collecting duct. Distal tubules are generally lined by a simple cuboidal epithelium, and exhibit a prominent central lumen, owing to the absence of a brush border (**Figure 16.5j**).

The DCT lacks aquaporins, making this region impermeable to water. However, in response to **aldosterone** stimulation, DCT cells can decrease intra-tubular fluids to hypotonic levels by actively resorbing Na^+ via **sodium chloride cotransporter** (**NCC**) pumps that are sensitive to **thiazide** types of diuretic agents (i.e., drugs that cause dilute urine production by blocking sodium resorption) (Box 16.2 Anti-Hypertensive Therapies: From Snake Venom Derivatives to Alternative RAS Modulators). Conversely, the DCT secretes not only **K^+ ions** to prevent excessive potassium build-up (**hyperkalemia**) but also secretes **H^+ ions** to help control pH levels. Distal to the DCT, **connecting tubule** cells resemble those of collecting ducts in expressing aquaporins, and thus these two regions are sometimes grouped as a functional unit, called the **distal nephron**.

Collectively, the variations in ECM osmolarities and nephron functional properties described above help move water via osmosis into the interstitial ECM during transport of nascent urine down descending limbs of intermediate tubules. Such osmotic processes in turn yield **hypertonic intratubular fluid** at the recurved portions of intermediate tubules that are located farthest away from the cortex (**Figure 16.6a–d**). Conversely, owing to active ion pumping and water impermeability, nascent urine within ascending intermediate tubules and distal tubules returns toward an **isotonic state** as it travels toward the superficial cortex. This decrease in osmolarity provides flexibility in urine output, because subsequent processing during transport through collecting ducts can generate either dilute urine to help flush out toxic substances or the more concentrated urine commonly used in maintaining homeostasis, as discussed further below.

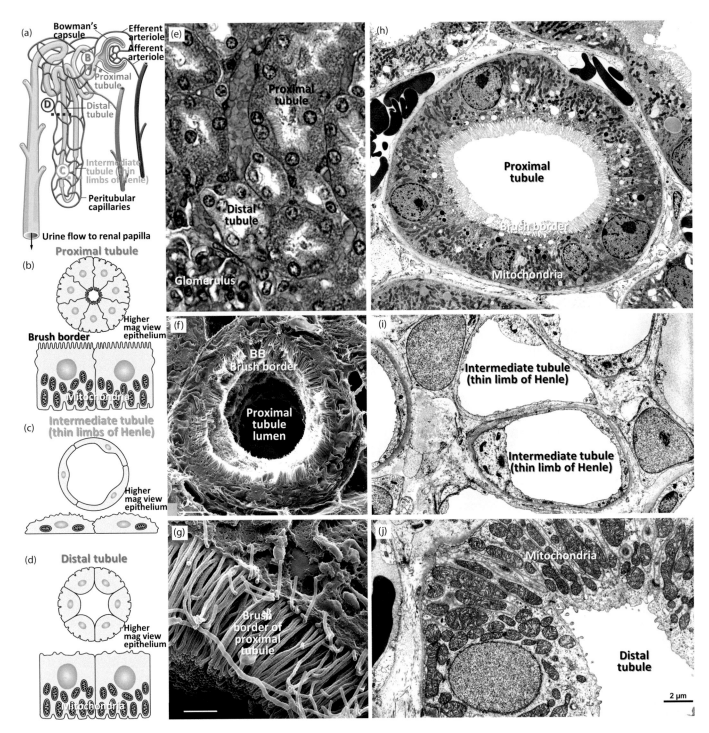

Figure 16.5 Microanatomy of nephron components. (a–j) The three main post-corpuscle regions of the nephron can be distinguished by: (1) the brush border, basal mitochondria, and more occluded lumen of proximal tubules; (2) the thinner diameter and squamous epithelium of intermediate tubules; and (3) the mitochondria, lack of a brush border, and more patent lumen of distal tubules. (e) 380X; (f) scale bar = 5 μm; (g) scale bar = 0.5 μm; (h) 1,700X; (i) 1,800X; (j) 6,000X. ([a]: From OpenStax College (2013) Anatomy and Physiology. OpenStax. http://cnx.org/content/col11496/ latest, reproduced under CC BY 4.0 creative commons license; [f, g]: From Rice, WL et al. (2013) High resolution helium ion scanning microscopy of the rat kidney. PLoS ONE 8, doi: 10.1371/jounal.pone.0057051, reproduced under a CC BY creative commons license; [h–j]: TEM images courtesy of Drs. H. Jastrow and H. Wartenberg reproduced with permission from Dr. Jastrow's electron microscopic atlas, htttp://www.drjastrow.de.)

BOX 16.2 ANTI-HYPERTENSIVE THERAPIES: FROM SNAKE VENOM DERIVATIVES TO ALTERNATIVE RAS MODULATORS

As the first oral medication for treating high blood pressure (**BP**), a thiazide diuretic was introduced in the 1950s. Such diuretics inactivate **sodium chloride cotransporters**

(**NCCs**) in the distal tubule and thus counteract elevated BP (=**hypertension**) by downregulating sodium resorption and the concomitant water transfer from urine to blood. Soon

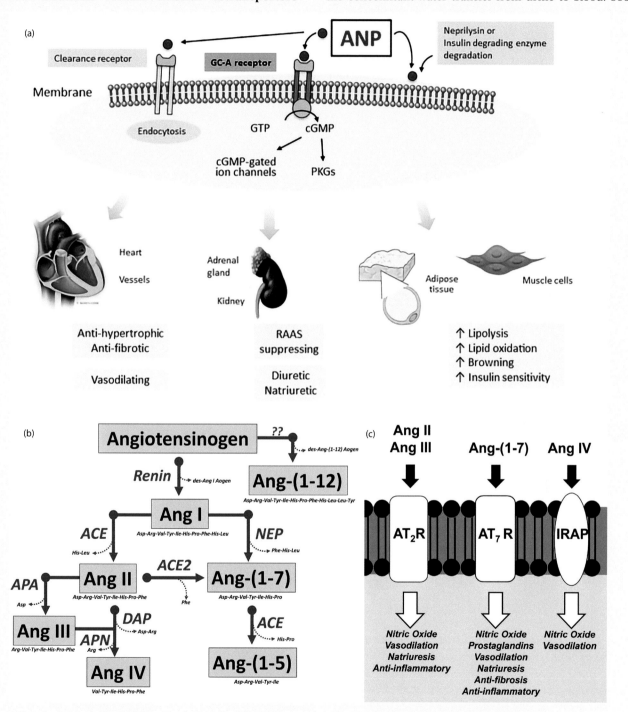

Box 16.2 (a–c) *Antihypertensive therapies involving atrial natriuretic peptide (ANP) and non-classical RAS signaling. ([a]: From Cannone, V et al. (2019) Atrial natriuretic peptide: a molecular target of novel therapeutic approaches to cardio-metabolic disease. Int J Mol Sci 20: doi: 10.3390/ijms20133265, reproduced under a CC BY 4.0 creative commons license; (b, c): From Chappell, MC (2014) The non-classical renin-angiotensin system and renal function. Compr Physiol 2:2733–2752, reproduced with publisher permission.)*

thereafter, β-blockers and calcium channel antagonists were developed for lowering BP.

However, a key innovation in anti-hypertensive therapies came with the introduction of an **angiotensin converting enzyme inhibitor** (**ACE inhibitor**) that was generated based on studies of snake venom that immobilizes prey by inducing a rapid BP drop. Such venom-induced BP decreases are achieved by preventing the prey's ACE enzymes from converting **angiotensin I** (**Ang I**) into its more active form **angiotensin II** (**Ang II**), which normally serves to raise BP. Accordingly, in the 1970s, an analog of the snake toxin, called **captopril**, began being marketed, and today, improved versions of ACE inhibitors are commonly prescribed to treat hypertension.

In addition to such drugs, **natriuretic peptides** (**NPs**) produced by various tissues, including atrial cells of the heart, counteract the hypertensive effects of the classical **RAS** pathway by stimulating vasodilation, diuresis, and increased sodium excretion in urine (=**natriuresis**) (**Figure a**). Similarly, non-classical cleavage products of angiotensin also have hypotensive properties (**Figure b, c**). Thus, along with well-established drugs targeting classical RAS signaling (e.g., β-blockers, direct renin inhibitors, ACE inhibitors, angiotensin receptor blockers [ARBs], and aldosterone antagonists), NPs and non-classical RAS effectors are also being investigated for anti-hypertensive therapies that include, for example, either **Ang-(1-7) analogs** or the inhibition of **neprilysin** enzymes that degrade NPs (**Figure a**).

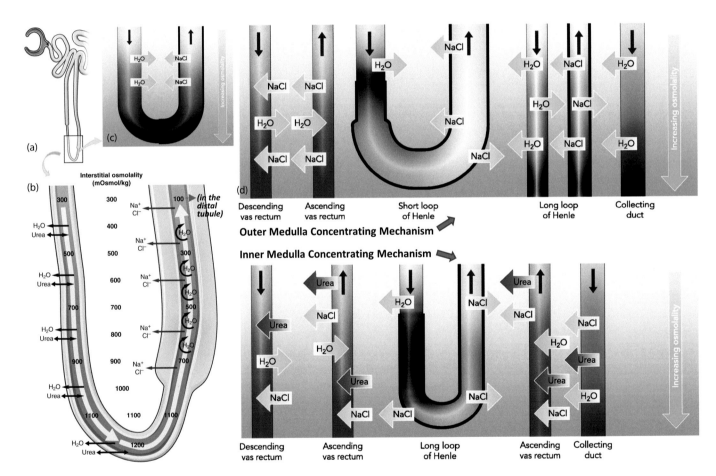

Figure 16.6 **Counter-current multiplier mechanism for concentrating urine.** *(a–d) To produce a concentrated urine for minimal water loss from the body, the U-shaped intermediate tubule (=thin limbs of the loop of Henle) has nascent urine flowing toward the medullary tip in the descending limb vs. returning toward the cortex in the ascending limb. Such countercurrent flow combines with differing permeability properties—a water-permeable/sodium-impermeable descending limb vs. a water-impermeable ascending limb that actively pumps out sodium—to generate a standing gradient of salt in the renal ECM. Thus, the cortex is isotonic compared to blood osmolarity but progressively more hypertonic toward the medullary tip. In addition to sodium ion pumping, selective urea movement also contributes to the standing osmolarity gradient in the inner medulla. ([a, b]: From OpenStax College (2013) Anatomy and Physiology. OpenStax. http://cnx.org/content/col11496/latest, reproduced under CC BY 4.0 creative commons license; [c, d]: From Layton, AT et al. (2015) The mammalian urine concentrating mechanism: hypotheses and uncertainties. Physiology 24:250–256, reproduced with publisher permission.)*

After differentiating from ureteric buds and fusing with nephrons, **collecting ducts** extend from the cortex to fill much of the medullary pyramid (**Figure 16.7a, b**). Within the simple cuboidal epithelium lining each collecting duct are both **principal cells** (**PCs**) and **intercalated cells** (**ICs**) (**Figure 16.7a**). Apically, PCs possess stubby microvilli plus a prominent primary cilium (**Figure 16.7c**), and their relatively light cytoplasm contains few mitochondria. Conversely, ICs lack a primary cilium and instead form curved apical ridges, called **microplicae**, which overlie a darker cytoplasm with abundant mitochondria (**Figure 16.7c**). PCs are the major renal cells controlling the final water and Na$^+$ content of urine, whereas ICs play alternative roles, such as modulating acid-base balance.

Collecting ducts can restore hypertonicity both by resorbing sodium to increase ECM levels and by allowing intra-duct water to move through aquaporins into the ECM (**Figure 16.7d**). Sodium resorption is mainly stimulated by the binding of **aldosterone** to PC receptors, thereby triggering sodium efflux through **epithelial sodium channels** (**ENaCs**). Such efflux in turn involves signaling pathways mediated by primary cilia that help modulate the overall size and functioning of PCs.

Aquaporin activities, on the other hand, are chiefly mediated by **antidiuretic hormone** (**ADH**) from the posterior pituitary. Without enough ADH, aquaporin-deficient collecting ducts are impermeable to water and thus produce dilute urine that requires frequent urination (=polyuria). Such polyuria can be triggered not only by diseases like diabetes (Chapter 15) but also by ingestion of diuretics like **caffeine** and **alcohol**, which downregulate ADH secretion and hence aquaporin expression. Alternatively, with sufficient ADH binding to receptors on PCs, such cells become permeable to water via: (1) rapid **recruitment** to the plasmalemma of previously synthesized aquaporins that had been **stored in apical vesicles**, and (2) **neosynthesis** of additional aquaporins (**Figure 16.7d**).

16.9 HISTOLOGY OF TRANSPORTING AND STORAGE COMPONENTS: URETERS, BLADDER, AND URETHRA

At the distal end of each medullary pyramid, large **papillary ducts** derived from coalesced collecting ducts drip their contents through a perforated plate (=area cribrosa) in the pyramidal tip and thereby deliver urine into a **minor calyx** of the renal pelvis (**Figure 16.8a**). From there, urine is transported through the rest of the **ureter** to an extensible **bladder** and subsequently drained exteriorly by the **urethra** (**Figure 16.8b**). The ureters, bladder, and urethra are lined by a stratified **urothelium** that is also called a **transitional epithelium**, because its cells change morphology depending on luminal fluid levels (**Figure 16.8c–g**).

Urothelial structure and function have been most intensively analyzed in the bladder, where the apical layer of the urothelium contains **umbrella cells** that shift from cuboidal to squamous in shape as the lumen becomes filled with urine (**Case 16.2 Christine Jorgensen's Bladder**) (**Figure 16.8e**). Such cells not only must remain pliable for marked changes in bladder volume but also need to provide a stable barrier that preserves urine composition and prevents toxins or pathogens from crossing the bladder lining. Thus, as in other protective epithelia, apical umbrella cells connect to each other via **tight junctions** and cover their apices with a thick **glycocalyx**.

In addition, stretching of the bladder causes umbrella cells to intercalate numerous **uroplakin vesicles** into their apical membranes in order to increase surface area and modify membrane properties (**Figure 16.8h, l**). As a result of such uroplakin incorporations, umbrella cells form a unique **asymmetric unit membrane** (**AUM**), in which the thickened outer plasmalemmal leaflet possesses ~0.3–1-μm-wide **membrane plaques**, with each of these plaques comprising numerous ~16-nm **hexagonal arrays of uroplakins**. Such uroplakin-containing plaques are separated from each other by flexible **hinge regions** and collectively enhance urothelial barrier properties while also strengthening the apical membrane during periodic shape changes (**Figure 16.8h–k**).

16.10 THE JUXTAGLOMERULAR APPARATUS AND RENIN-ANGIOTENSIN SYSTEM (RAS): A NETWORK FOR INCREASING BLOOD PRESSURE

Numerous **juxtaglomerular** (**JG**) **apparatuses** are spread throughout the renal cortex, where they contribute to a multi-organ **RAS** that serves to **raise blood pressure** (**BP**) (**Figure 16.9a–c**). Each JG apparatus occurs at the vascular pole of a renal corpuscle and comprises: (1) the **macula densa** at the distal end of a TAL near the renal corpuscle, (2) modified smooth muscle cells in the afferent arteriole, called **juxtaglomerular** (**JG**) **cells**, and (3) a cluster of **extraglomerular mesangial cells** (=Lacis cells), which help regulate blood flow through the glomerulus (**Figure 16.9a–c**).

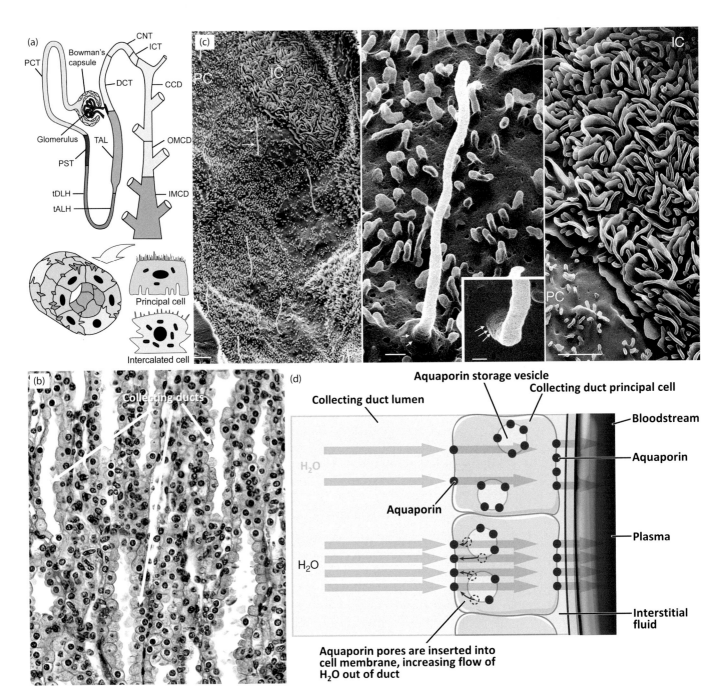

Figure 16.7 Collecting ducts. (a–c) *After traversing distal convoluted tubule (DCT), connecting tubule (CNT), and initial collecting tubule (ICT) regions, urine enters collecting ducts (CDs) that extend from cortical collecting ducts (CCDs) through outer medulla collecting ducts (OMCDs) before reaching inner medulla collecting ducts (IMCDs). IMCDs then coalesce to form large papillary ducts, which in turn drip urine into the ureter. The collecting duct lining epithelium comprises (i) principal cells (PCs) with a long primary cilium rising above short microvilli, and (ii) intercalated cells (ICs) that lack cilia or microvilli and instead have unusual apical folds, called microplicae. (b) 200X; (c) scale bars: left = 2 μm; middle = 200 nm; middle, inset = 100 nm; right = 2 μm. (**d**) In response to ADH, principal cell plasma membranes incorporate aquaporins to increase water permeability for concentrating urine. ([a]: From Staruschenko, A (2012) Regulation of transport in the connecting tubule and cortical collecting duct. Compr Physiol. 2, doi: 10.1002/cphy.c110052, reproduced with publisher permission; [c]: From Rice, WL et al. (2013) High resolution helium ion scanning microscopy of the rat kidney. PLoS ONE 8, doi: 10.1371/joural.pone.0057051, reproduced under a CC BY creative commons license; [d]: From OpenStax College (2013) Anatomy and Physiology. OpenStax. http://cnx.org/content/col11496/latest, reproduced under CC BY 4.0 creative commons license.)*

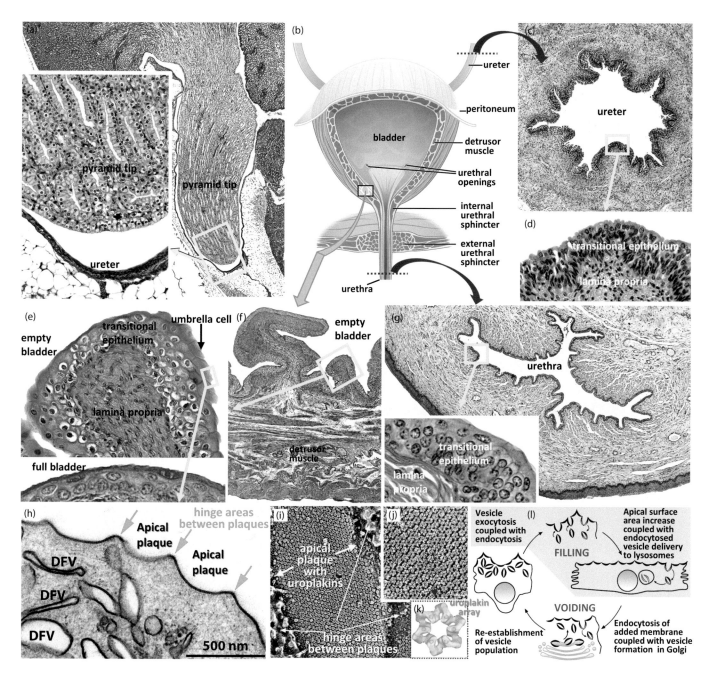

Figure 16.8 Ureter, bladder, and urethra microanatomy. (a–g) *After dripping into the ureter from medullary pyramids, urine is stored in the bladder before being drained via the urethra. The ureter, bladder, and urethra are lined by a transitional stratified urothelium, whose apical-most umbrella cells change shape depending on urine content (see figure e). (a) 13X; (a, inset) 75X; (c) 20X; (d) 125X; (e, top) 200X; (e, bottom) 375X; (Fp 35X). (g) 20X; (g, inset) 300X.* **(h–l)** *During stretching as the bladder fills with urine, the apical membranes of umbrella cells incorporate uroplakin-containing discoidal/fusiform vesicles (DFVs), thereby increasing surface area and re-enforcing the apex with ~0.3–1-μm-wide plaques. Plaques are composed of ~1000 hexagonal arrays of uroplakins, with each array measuring ~16 nm in diameter (k). (h) 46,000X; (i) 60,000X; (i, inset) 140,000X. ([b]: From OpenStax College (2013) Anatomy and Physiology. OpenStax. http://cnx.org/content/col11496/latest, reproduced under CC BY 4.0 creative commons license; [h]: From Truschel, ST et al. (2018) Age-related endolysosome dysfunction in the rat urothelium. PLoS ONE 13, e0198817. https://doi.org/10.1371/journal.pone.0198817, reproduced under a CC BY 4.0 creative commons license; [i, j, l]: From Apodaca, G (2004) The urothelium: not just a passive barrier. Traffic 5: 117–125 reproduced with publisher permission; [k]: From Wu, X-R et al. (2009) Uroplakins in urothelial biology, function, and disease. Kidney Int 75:1153–1165, reproduced with publisher permission.)*

CASE 16.2 CHRISTINE JORGENSEN'S BLADDER

Transgender Celebrity
(1926–1989)

"We may not have started it, but we gave it a good swift kick in the pants."

(C.J. in a 1988 interview on what role she and her cohorts had played in launching a sexual revolution)

Born George Jorgensen Jr. in New York City, Christine Jorgensen gained fame as the first U.S. citizen to undergo sex reassignment surgery and later became an unabashed advocate for transgender rights. After growing up as a boy and serving in the U.S. Army, Jorgensen traveled to Denmark in 1951 to undergo hormone replacement therapies and numerous operations. When Jorgensen returned to the United States in 1952, newspapers ran front-page stories with quips like "Christine Jorgensen went abroad and came back a broad". Jorgensen went on to become a popular talk show guest, who openly discussed various sex-related topics. In addition to such appearances, Jorgensen wrote several accounts of her life and worked as both an actress and singer. As she grew older, Jorgensen developed bladder cancer, and after all the surgeries in Denmark that had altered her urogenital systems, it was ultimately her diseased bladder that caused her death at age 62.

The **macula densa** (**MD**) is the sensory component of the JG apparatus consisting of closely positioned nuclei that constitute a **dense spot** (= "macula densa") (**Figure 16.9b, c**). Such cells are able to monitor BP-associated parameters, and following BP drops, can release paracrine factors like **prostaglandins** and **nitric oxide** that stimulate nearby JG cells to secrete into the bloodstream an enzyme, called **renin**.

The 37-kD **renin** protease targets a precursor protein, called **angiotensinogen** (**Agt**), which continually circulates in the bloodstream following its secretion primarily by **liver hepatocytes** (**Figure 16.9d, e**). Via a renin-mediated cleavage of its N-terminus, Agt yields the decapeptide **angiotensin I** (**Ang I**) that is transported to pulmonary capillaries (**Figure 16.9d, e**). In the presence of **angiotensin-converting enzyme** (**ACE**) produced by alveolar endothelial cells, Ang I is converted into the more active octapeptide **angiotensin II** (**Ang II**), which is the main effector of the RAS network (**Figure 16.9d, e**). By binding to receptors on various target cells, Ang II **serves to raise BP** by (i) **triggering peripheral vasoconstriction**; (ii) **elevating ADH secretion** from the **posterior pituitary** so that water can exit urine for increasing blood volume; and (iii) **stimulating aldosterone secretion** by the **adrenal cortex** in order to facilitate Na^+ resorption by the distal nephron, thereby also contributing to blood volume increases.

Use of these **classical** RAS signals can counteract life-threatening ischemia due to severe hypotension during hemorrhaging. However, chronic RAS stimulation is maladaptive, because it causes hypertension and various related pathologies (**Case 16.3 Franklin Roosevelt's Renin-Angiotensin System**). Accordingly, a **non-classical RAS** pathway has evolved to balance the hypertensive effects of classical RAS signals (Box 16.2). Similarly, various body tissues, including atria of the heart (Chapter 7), produce **natriuretic peptides** (**NPs**) that can antagonize classical RAS-mediated hypertension before such peptides are eventually degraded by NP-targeting **neprilysin** enzymes (Box 16.2).

16.11 SOME DISORDERS OF THE URINARY SYSTEM: KIDNEY STONES AND GLOMERULONEPHRITIS

Kidney stones (=renal calculi) formed during a process, called nephrolithiasis, are often readily passed in urine without symptoms. However, large calculi can cause ureteric damage, obstruction, and intense pain, thereby requiring dissolving treatments or surgical intervention. Aside from a few non-calcified calculi, such as those containing uric acid, most kidney stones are **calcified**, with ~80% of such stones comprising **calcium oxalate** vs. ~20% consisting of **calcium phosphate**.

Glomerulonephritis (**GN**) is a heterogeneous collection of predominantly autoimmune-induced glomerular diseases characterized by such symptoms as hypertension, excess protein excretion (proteinuria), and reduced urine output (oliguria). In **non-proliferative GNs**, the number of glomerular cells does not change substantially. Conversely, **proliferative GNs** involve increased cell numbers within affected glomeruli, often exhibiting **crescentic infillings** of new cells from the glomerular periphery into Bowman's space. Depending on the particular type and severity of the GN, treatment can involve pathogen eradication and/or mitigation of auto-immune pathways.

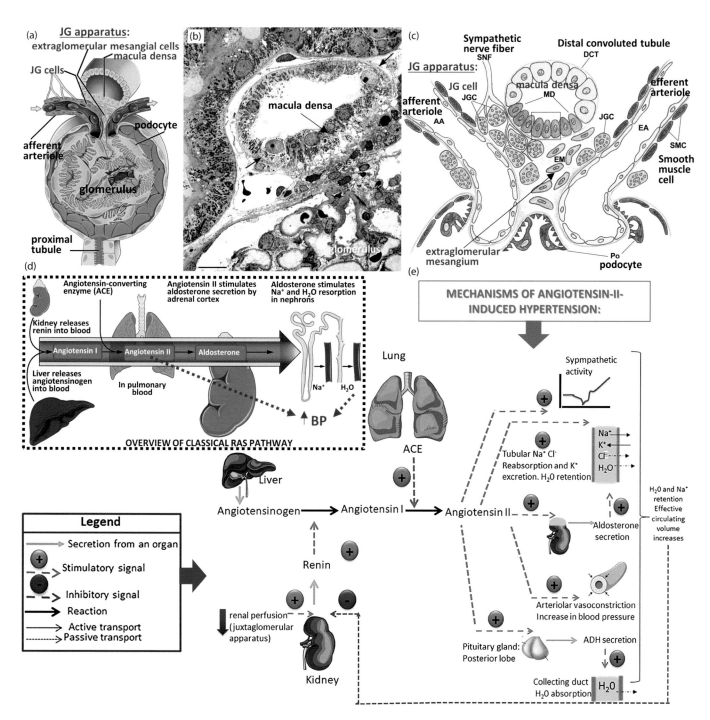

Figure 16.9 Renin-angiotensin system (RAS) for raising blood pressure. *(a–c) As a key component of the RAS system, each nephron has at its vascular pole a juxtaglomerular (JG) apparatus. The JG apparatus utilizes macula densa (MD) cells at the end of the distal tubule's thick ascending limb to sense parameters of low blood pressure (BP) during events, such as hemorrhaging. Macula densa cells can stimulate JG cells in the afferent arteriole wall to release the enzyme renin, thereby initiating a physiological cascade that ultimately raises BP. (b) 460X. (**d, e**) Renin from JG cells cleave a liver-derived circulating protein, called angiotensinogen, to form the angiotensin I (AGI) peptide. AGI is converted into the more active angiotensin II form by angiotensin-converting enzyme (ACE) secreted in large amounts by vessels in the lungs. AGII raises BP via multiple mechanisms, including by triggering: (1) vasoconstriction; (2) ADH release from the posterior pituitary; and (3) aldosterone secretion by the adrenal cortex to stimulate sodium resorption by the distal nephron. The combined effects of ADH and sodium resorption extract water from urine to increase vascular volume, thereby elevating BP. (a, d) from OpenStax College (2013) Anatomy and Physiology. OpenStax. http://cnx.org/content/col11496/latest, reproduced under a CC BY 4.0 creative commons license; (b, c) from Cangiotti, AM et al. (2018) Polarized ends of human*

Figure 16.9 (Continued) *macula densa cells: ultrastructural investigations and morphofunctional correlations. Anat Rec 301: 922–931 reproduced with publisher permission; images courtesy of Dr. M. Morroni; (e) de Almeida LF and Coimbra TM (2019) When less or more isn't enough: renal maldevelopment arising from disequilibrium in the renin-angiotensin system. Front Pediatr 7:296. doi: 10.3389/fped.2019.00296 reproduced under a CC BY 4.0 creative commons license.)*

CASE 16.3 FRANKLIN ROOSEVELT'S RENIN-ANGIOTENSIN SYSTEM

32nd U.S. President
  (1882–1945)

"The President looked old, and thin, and drawn...he sat looking straight ahead with his mouth open as if he were not taking in things. Everyone was shocked by his appearance..."

(Winston Churchill's physician on F.R.'s gravely ill appearance just before the 1945 Yalta conference)

As the only U.S. President to serve more than two terms, Franklin Delano Roosevelt, aka FDR, became paralyzed from the waist down at age 39 when he developed either polio or perhaps Guillain-Barre syndrome. Later in life, FDR also had severe hypertension, which his White House physician ignored until finally diagnosing nascent heart failure during FDR's third term. By then, FDR's demise in 1945 due to a cerebral hemorrhage could not have been prevented, not only

because of the advanced nature of his disease but also because of a lack of effective treatment options in the 1940s. At that time, the connection had not yet been made between hypertension and the RAS network involving juxtaglomerular apparatuses in the kidneys. Thus, after traveling to Yalta, Russia in 1945 to discuss the post-war fate of Europe, FDR was so incapacitated that, according to some historians, he allowed Russia to take over Eastern Europe. Exactly how much FDR's maladies affected outcomes of the Yalta conference remains debatable. However, there is little doubt that his cardiovascular heath would have greatly benefited from current treatments that counteract the hypertensive effects of the classical RAS network.

16.12 SUMMARY—URINARY SYSTEM

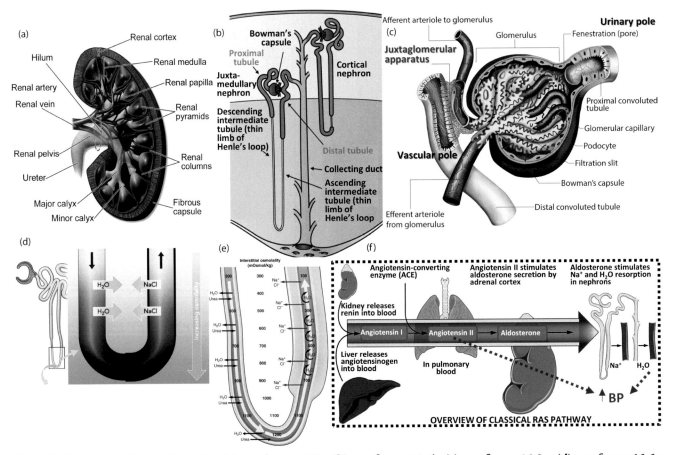

Pictorial Summary Figure Caption *(a): see figure 16.2a; (b): see figure 16.4b; (c): see figure 16.3a; (d): see figure 16.6a; (e): see figure 16.6b; (f): see figure 16.9d.*

The **urinary system** comprises two **kidneys,** two **ureters**, a **bladder**, plus a **urethra**, which collectively serve to generate **urine**, thereby helping to **excrete water-soluble wastes** and to **maintain water and electrolyte balance** throughout the body. Kidneys also regulate **blood pressure** and produce **bioactive compounds** (e.g., erythropoietin).

Each bean-shaped kidney contains an outer **cortex** plus an inner **medulla**, which is indented medially to form a concave **hilum** where the expanded ureteric head (=**renal pelvis**) is located. The renal **cortex** has ~1 million **renal corpuscles**, which serve as: (i) the initial components of urine-generating units, called **nephrons**, and (ii) the exclusive sites where blood in **glomerular** capillaries is filtered to form **pre-urine**. The **medulla** lacks corpuscles and drains urine via **collecting ducts** into **calyces** of the renal pelvis (**Figure a**). Each kidney has multiple **medullary pyramids** plus their associated cortical tissues, thereby forming about a dozen **renal lobes** that are highly vascularized for efficient blood filtering. **Nephrons** arise from **metanephric mesenchyme** and possess four major regions: **renal corpuscle**, **proximal tubule**, **intermediate tubule**, and **distal tubule**. Late in gestation, nephrons fuse with **collecting ducts** that are derived from **ureteric bud** cells. Two nephric morphotypes are **juxtamedullary nephrons** (with a corpuscle near the cortico-medulla border and a long intermediate tubule that extends into the inner medulla) vs. **cortical nephrons** (with a corpuscle occurring farther from the medulla and a short intermediate tubule that extends only into the outer medulla) (**Figure b**).

Each renal corpuscle has (i) a bilayered **Bowman's capsule** with an inner cuboidal **visceral** layer of **podocytes** and an outer squamous **parietal** layer at the capsule periphery; (ii) a **glomerulus** that is covered by podocytes, and (iii) a **urinary space** occurring between the visceral and parietal layers. The corpuscle's **vascular pole** is where an **afferent arteriole** enters and an **efferent arteriole** exits the glomerulus, whereas the **urinary pole** occurs opposite to the vascular pole and joins Bowman's capsule to the **proximal tubule** (**Figure c**).

Urinary excretion involves: (i) **filtration** (blood plasma filtered into the urinary space as pre-urine); (ii) **secretion** (addition of components not in pre-urine); and (iii) **resorption** (removal of pre-urine components— e.g., by proximal tubule cells with a **brush border** that helps resorb most Na$^+$, glucose, and water). Resorption and secretion processes convert pre-urine into urine. All filtration occurs in renal corpuscles through 30–40 nm-wide **filtration slits** between podocyte **foot processes**. Each **slit** has a **diaphragm** with **nephrins** that normally allows only fluid and small molecules into the urinary space. Corpuscles have phagocytic **mesangial cells** to clean the filtering apparatus.

Urine is a **solution** (**sln**) with particles (=**solutes**), such as ions that are dissolved in a water-based **solvent**. **Molarity (M)** = **moles** of **solute/l** of **sln; osmolarity (OsM)** = total **osmoles** of all dissolved particles (**including ions** from ionizing solutes)/liter. Thus, 1 mole of NaCl/l is **1 M** and **2 OsM (=2,000 mOsM)** since each NaCl mole forms 1 Na$^+$ osmole and 1 Cl$^-$ osmole. **Osmosis** is the diffusion of water down its concentration gradient across a semi-permeable membrane. **Tonicity** is the concentration of **non-penetrating particles** (like **Na$^+$**) that do not readily cross semi-permeable membranes. A sln is **hypertonic** if it has more non-penetrating particles than a reference sln, such as ~**300 mOsM** blood; hypertonic solns gain water via osmosis from **hypotonic** slns that possess less non-penetrating particles. No net osmotic water flow occurs between **isotonic** slns with equal concentrations of non-penetrating particles.

Countercurrent exchange during urine concentration forms a **standing gradient** of ECM osmolarity (low in the cortex vs. high at the medullary pyramid tip) by means of: (1) opposite fluid flow in **intermediate tubules**, and (2) differing properties of **descending** limbs (water permeable but salt impermeable) vs. **ascending** limbs (water impermeable with active salt transport) (**Figure d, e**). The ECM gradient **concentrates urine** by coupling **aldosterone**-induced Na$^+$ resorption in the distal nephron with osmotic-driven water outflow from collecting ducts when the posterior pituitary releases sufficient **ADH** to increase duct **aquaporin** channels. Dilute urine can be formed by diuretics (e.g., alcohol, caffeine) that block ADH release.

Urine transporter/storage components (ureter, bladder, and urethra) are lined by a **urothelium** with apical membrane **uroplakins** for barrier strengthening.

The classical **renin-angiotensin system (RAS) raises blood pressure (BP)** via a **juxtaglomerular (JG) apparatus** at each corpuscle's vascular pole. For such functions, **macula densa** cells in the distal tubule sense low BP to trigger **renin** secretion by **JG cells** in the afferent arteriole. Renin cleaves the liver-derived protein

angiotensinogen (Agt), thereby forming **angiotensin I (AngI)**. AngI is converted into the more active **angiotensin II (AngII)** by angiotensin-converting enzyme (**ACE**) in the lungs. Ang II raises BP by: (i) **vasoconstriction**; (ii) **ADH** release; and (iii) **aldosterone** secretion for Na^+ resorption (**Figure f**). Classical RAS signaling is counteracted by: (i) **natriuretic peptides** (**NPs**) that are degraded by **neprilysin** enzymes; (ii) **non-classical RAS** signals (e.g., **Ang1-7**), and/or (iii) various **anti-hypertensive drugs** (e.g., **ACE inhibitors**).

Some renal disorders: **kidney stones** are typically composed of **calcium oxalate**; **glomerulonephritis** involves corpuscle damage that is usually caused by autoimmune attacks.

SELF-STUDY QUESTIONS

1. Which region of the kidney contains renal corpuscles?
2. T/F Collecting ducts are derived from metanephric mesenchyme.
3. Which part of the nephron normally resorbs most of the Na^+, water, and glucose in pre-urine?
4. Which of the following would tend to raise blood pressure?
    A. Angiotensin converting enzyme inhibitors
    B. Renin
    C. Natriuretic peptides
    D. ALL of the above
    E. NONE of the above
5. Which of the following solutions would lose water via osmosis to mammalian blood?
    A. 100 mOsM NaCl
    B. 275–300 mOsM glucose
    C. 1,200 mOsM NaCl
    D. ALL of the above
    E. NONE of the above
6. Which part of the urinary system contains minor calyces?
    A. Renal cortex
    B. Renal medulla
    C. Ureter
    D. Bladder
    E. Urethra
7. What is the term for a solution that has a higher concentration of dissolved non-penetrating solutes than occurs in a reference solution such as blood?
8. T/F Podocytes constitute a transitional epithelium.
9. Which cells phagocytose debris within a renal corpuscle?
    A. Endothelial cells
    B. Podocytes
    C. JG cells
    D. Parietal cells
    E. Mesangial cells
10. About how wide are typical renal corpuscles?
11. What is the name of the blood vessel that supplies blood to the glomerulus?
    A. Arcuate
    B. Afferent arteriole
    C. Interlobar
    D. Renal
    E. Efferent arteriole
12. Which of the following does NOT normally express aquaporins?
    A. Collecting ducts
    B. Connecting tubules
    C. Thick ascending limb

13. T/F Each human kidney contains a single lobe.

"EXTRA CREDIT" For each term on the left, provide the BEST MATCH on the right (answers A-F can be used more than once)

14. Degrades natriuretic peptides ___

15. Renal calculi ____

16. Slit diaphragm ____

17. ACE inhibitor ___

18. Strengthens ____ urothelium

A. Calcium oxalate
B. Uroplakin
C. Captopril
D. Neprilysin
E. Metanephros
F. Nephrin

19. Describe the various components of a renal corpuscle and their major functions.

20. Describe how the classical RAS pathway raises blood pressure.

ANSWERS

(1) cortex; (2) F; (3) proximal tubule; (4) B; (5) A; (6) C; (7) hypertonic; (8) F; (9) E; (10) 200 μm; (11) B; (12) C; (13) F; (14) D; (15) A; (16) F; (17) C; (18) B; (19) The answer should expand on: vascular vs. urinary poles, afferent vs. efferent arterioles, Bowman's capsule parietal vs. visceral layers, urinary space, glomerulus, podocytes, filtration, GFR, filtration slits, diaphragm, basal laminae, endothelial cells, mesangial cells, JG apparatus, etc. (20) Consider: the JG apparatus, macula densa, JG cells, renin, angiotensinogen, Ang I, ACE, Ang II, vasoconstriction, ADH, aldosterone, hemorrhage, roles of non-classical pathway/NPs.

Male Reproductive System

17.1 INTRODUCTION TO MALE REPRODUCTIVE SYSTEM

Along with two **seminal vesicles**, a **prostate gland**, and two **bulbourethral glands** that produce **seminal fluid**, the male reproductive system includes: (i) paired **testes** where gametes termed spermatozoa (=**sperm**) are formed; (ii) sperm-transporting conduits comprising paired **epididymi** (singular = epididymis), **vasa deferentia** (singular = vas deferens), and **ejaculatory ducts**, plus (iii) a vascularized **penis** (**Figure 17.1a**). Each testis occurs within a **scrotal sac** and delivers sperm via transporting conduits to the penile **urethra** for releasing **semen** that contains sperm suspended in seminal fluid (**Figure 17.1a**). Via such components, the male reproductive system functions in **gamete production** and **sexual intercourse**, while also producing male sex hormones (=**androgens**) that can have wide-ranging effects throughout the body.

17.2 DEVELOPMENT OF THE MALE REPRODUCTIVE SYSTEM AND FUNCTIONAL HISTOLOGY OF THE SCROTUM

Human sexual characteristics are determined by genes located on the **X** and **Y sex chromosomes**, which together with 22 pairs of autosomal non-sex chromosomes constitute the 46-chromosome array of diploid cells (**Figure 17.1b, c**). During normal development, fertilized zygotes having an **XX** configuration of sex chromosomes become females, whereas **XY** zygotes develop into males. As such differentiation proceeds, each 5–7 week-old embryo initially passes through a **sexually indifferent stage** with: (i) two **bipotential gonadal primordia**, (ii) two pairs of tubes, called the **Wolffian** and **Mullerian ducts**, and (iii) prospective **genitalia** that appear similar in males and

DOI: 10.1201/9780429353307-21

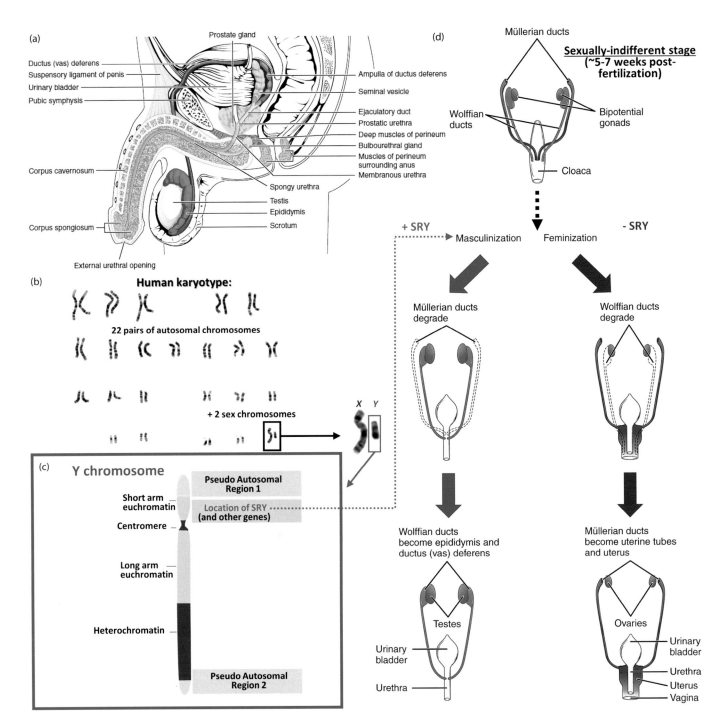

Figure 17.1 Introduction to the male reproductive system and its development. (**a**) Major components of the male reproductive system include: (i) a scrotal sac containing two sperm-producing testes, with each connecting to an epididymis; (ii) a pair of sperm-transporting vasa deferentia and ejaculatory ducts; (iii) accessory glands for seminal fluid production (two seminal vesicles, a prostate, and two bulbourethral glands); and (iv) a penis. (**b, c**) Of the 46 chromosomes in a diploid human cell, males possess a Y sex chromosome, whose SRY (sex-determining region of Y) gene plays a central role in male development. (**d**) After embryos reach a sexually indifferent stage by ~5–7 weeks post-fertilization, SRY expression triggers degradation of Mullerian ducts, differentiation of Wolffian ducts into vasa deferentia, and formation of testes from bipotential precursor gonads. Without SRY expression, females degrade their Wolffian ducts, while retaining the Mullerian ducts as uterine tubes (oviducts) that become situated near developing ovaries. ([a, d]: From OpenStax College (2013) Anatomy and Physiology. OpenStax http://cnx.org/content/col11496/latest, reproduced under a CC BY 4.0 creative commons license; [b]: Downloaded from https://commons.wikimedia.org/wiki/File:NHGRI_human_male_karyotype.png and reproduced as a public domain image that had been generated by the National Institutes of Health as part of the federal government; [c]: From Colaco, S, and Modi, D (2018) Genetics of the human Y chromosome

Figure 17.1 (Continued) and its association with male infertility. Reprod Biol Endocrinol 16, https://doi.org/10.1186/ s12958-018-0330-5, reproduced under a CC BY 4.0 creative commons license.)

CASE 17.1 ADOLF HITLER'S TESTES

Dictator
  (1889–1945)

  "Hitler has only got one ball"

  (From a British song used as war propaganda)

Arguably the most widely reviled man in history, Adolf Hitler sought world domination during World War II while ordering the execution of millions of non-combatants he regarded as "life unworthy of life". Although convinced of the superiority of his race, Hitler himself was hardly a picture of health. For example, amphetamine and cocaine usage degraded his mental stability, while a collection of medical disorders, such as throat polyps, chronic indigestion, skin lesions, apparent Parkinsonism, and tertiary syphilis would have otherwise

hastened his demise, had he not committed suicide in 1945. However, Hitler's most famous affliction was his supposed possession of a single testis within his scrotum. This widely cited rumor was recently substantiated by newly discovered medical records that noted Hitler had right-side cryptorchidism, or failure of his right testis to descend fully from the abdomen. Consistent with this finding, Hitler did not have any officially recognized offspring, suggesting that his undescended testis may have impaired his fertility.

---

females (**Figure 17.1d**). Over the next three weeks, sex-specific gene expression driven mainly by the **SRY** (<u>s</u>ex-determining <u>r</u>egion of <u>Y</u>) gene on the Y chromosome (**Figure 17.1b**) causes differentiating testes to secrete **anti-Mullerian hormone** (**AMH**), which triggers the apoptotic degradation of Mullerian ducts and the retention of the Wolffian ducts as vasa deferentia (**Figure 17.1d**). With further development, increased **testosterone** production by the developing testes yields male-specific genitalia.

Beginning at ~3 months, abdominally situated fetal testes begin to translocate into the developing scrotal sac outside the body proper (**Figure 17.2a**). Testicular descent is necessary for full fertility, as optimal sperm formation (=**spermatogenesis**) fails to occur at core body temperatures and instead requires the ~2–3 °C cooler environment of the scrotum. Movement to the scrotum is usually accomplished before birth but in a few cases may take up to several years during early post-natal life to complete. In **cryptorchidism** (**Case 17.1 Adolf Hitler's Testes**), testes permanently fail to reach their normal landing sites in the scrotum (**Figure 17.2a**), and without surgical corrections, fertility may be reduced.

When fully developed, a median septum in the scrotum forms a left and right chamber, with each containing a testis (**Figure 17.2b, c**). Beneath its sub-integumental dartos and cremaster muscle layers, the scrotum contains a bilayered **tunica vaginalis capsule** with an intervening thin cavity (**Figure 17.2b**). Scrotal muscles contract in response to cold, anxiety, or sexual arousal, thereby pulling the scrotum toward the body to protect and warm the testes, whereas the tunica vaginalis serves to reduce friction between the testis and scrotum.

17.3 TESTICULAR MORPHOLOGY: SEMINIFEROUS TUBULES AND INTERSTITIAL LEYDIG CELLS

Medial to the scrotal wall, each testis is covered by a tough capsule, called the **tunica albuginea** (**Figure 17.2c**), which extends numerous connective tissue **septa** inward to subdivide the testis into a few hundred **lobules**. Within each lobule, one to a few convoluted **seminiferous tubules** fill much of the intralobular space and represent the sites where spermatogenesis occurs in post-adolescent males (**Figure 17.2c, d**). If stretched out and laid end-to-end, seminiferous tubules in each testis would reach several hundred meters in length.

Connective tissue compartments adjacent to seminiferous tubules contain blood vessels and nerves plus clusters of **Leydig cells** (**Figure 17.2e**). Such cells remain quiescent until puberty, when luteinizing hormone from the anterior pituitary stimulates their secretion of **testosterone** to drive adult sexual characteristics and sperm production. Activated Leydig cells become filled with hormone-containing vesicles plus rod-shaped inclusions, called **Reinke crystals**, which are dissolved during conventional aqueous fixations but can be preserved by specialized protocols (**Figure 17.2f**). Immunostaining analyses indicate that Reinke crystals contain **3β-hydroxysteroid dehydrogenase** (**3β-HSD**), which is a key enzyme needed for steroid hormone synthesis. Along with Leydig cells, each seminiferous tubule is enveloped by flattened **myoid cells** whose contractions help move sperm along seminiferous tubules (**Figure 17.3a**).

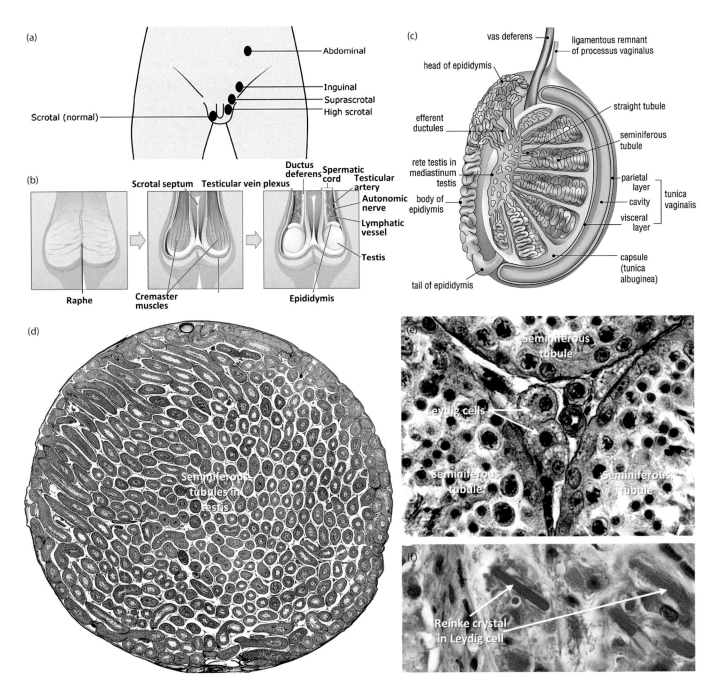

Figure 17.2 Scrotal sac and testis microanatomy. (a) *Testes begin development within the abdomen and normally descend into a scrotal sac, whose cooler microenvironment optimizes spermatogenesis. However, in cryptorchidism, incomplete testicular descent leaves testes at abnormal sites that can compromise sperm production.* **(b)** *The scrotum is divided into two halves by a median septum, and cremaster muscles surround the testes.* **(c)** *The testicular capsule (=tunica albuginea) forms connective tissue septa that divide each testis into numerous lobules. Testicular lobules are filled by one or a few coiled seminiferous tubules, which coalesce distally within a rete testis to form multiple vasa efferentia that connect with the epididymis.* **(d, e)** *As the sites where spermatogenesis occurs, coils of seminiferous tubules are separated from each other by an interstitial CT compartment containing blood vessels and Leydig cells that produce testosterone. (d) 15X; (e) 450X.* **(f)** *Leydig cells normally contain Reinke crystals that comprise enzymes required for testosterone production (375X). ([a]: From Rodprasert, W et al. (2020) Hypogonadism and cryptorchidism. Front Endocrinol 10, doi: 10.3389/fendo.2019.00906, reproduced under a CC BY 4.0 creative commons license; [b]: From OpenStax College (2013) Anatomy and Physiology. OpenStax. http://cnx.org/content/col11496/latest, reproduced under aCC BY 4.0 creative commons license; [c]: From Biddle, C (2015) Trichilemmal cysts of the scrotal wall. Sonography 2: 39–42, reproduced with publisher permission.)*

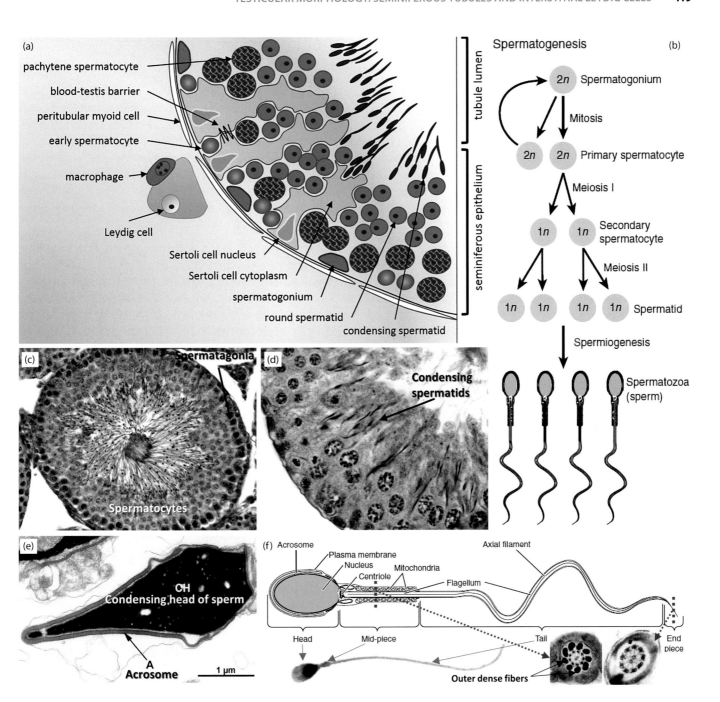

Figure 17.3 Introduction to spermatogenesis. (a–d) *Spermatogenesis begins with spermatogonia undergoing mitotic divisions at the base of the seminiferous epithelium. After moving more apically toward the blood-testis barrier formed by Sertoli cell tight junctions, spermatogonia commit to meiosis and thereby become primary spermatocytes. Two meiotic divisions of spermatocytes yield haploid spermatids that undergo the process of spermiogenesis (acrosome formation, tail production, cytoplasmic sloughing, and nuclear condensation) before being discharged into the tubular lumen. (c) 125X; (d) 325X. (**e, f**) Mature sperm have a head, midpiece, and tail with a flagellar axoneme containing 9+2 microtubules surrounded by outer dense fibers in the proximal part of the flagellum. (f, bottom left LM) 1,800Z; (f, bottom middle TEM) 35,000X; (f, bottom right TEM) 50,000X. ([a]: From Loveland, KL et al. (2017) Cytokines in male fertility and reproductive pathologies: immunoregulation and beyond. Front Endocrinol 8:307, doi: 10.3389/fendo.2017.00307, reproduced under a CC BY 4.0 creative commons license; [b, f, drawing]: From OpenStax College (2013) Anatomy and Physiology. OpenStax. http://cnx.org/content/col11496/latest, reproduced under a CC BY 4.0 creative commons license; [e]: From Bragina, IE et al. (2017) Ultrastructure of spermatozoa from infertility patients, Spermatozoa – facts and perspectives, doi: 10.5772/intechopen.71596, reproduced under a CC BY 3.0 creative commons license; [f, middle and right inset TEMs]: From Zhao, W et al. (2018) Outer dense fibers stabilize the axoneme to maintain sperm motility. J Cell Mol Med 22:1755–1768, reproduced under a CC BY 4.0 creative commons license.)*

17.4   SPERMATOGENESIS: FROM SPERMATOGONIAL DIFFERENTIATION TO SPERM RELEASE INTO THE TUBULE LUMEN

Spermatogenesis within the lining epithelia of **seminiferous tubules** involves two distinct cell types: (i) mitotically dividing somatic cells, called **Sertoli cells**, and (ii) germline-derived **spermatogenic cells**, which undergo meiosis to become sperm (**Figure 17.3a**). Characterized by their pale cytoplasm and large nucleus, Sertoli cells connect to each other by junctional complexes that form a **blood-testis barrier** between the apical portion of the tubule and more basal compartments. This barrier helps minimize autoimmune responses against haploid sperm that form after puberty and could be sensed as non-autologous invaders. In addition, Sertoli cells directly abut, and continuously support, developing spermatogenic cells. Among their supportive roles, Sertoli cells secrete **sex hormone binding glycoprotein (SHBG)** (=**androgen-binding protein**) that binds testosterone, thereby concentrating androgens within the seminiferous tubule to promote spermatogenesis.

Of the various spermatogenic stages that are associated with Sertoli cells, four basic cell types can be discerned: (1) **spermatogonia** that divide mitotically at the base of the epithelium during spermatogenesis onset; (2) **spermatocytes** that undergo meiotic divisions while translocating toward the epithelial surface; (3) **spermatids** that are meiotically generated haploid cells, whose post-meiotic maturation into sperm is termed **spermiogenesis**; and (4) **sperm** that are spermiogenesis-produced gametes released into the tubular lumen during a process, called **spermiation** (**Figure 17.3b**). Accordingly, such spermatogenic stages can be viewed as undergoing four key developmental steps during the overall sequence of sperm formation:

1. *Spermatagonial proliferation and differentiation*—Initially, **A-type** spermatogonia undergo mitotic divisions near the base of the seminiferous epithelium to increase spermatogonial numbers. Some of these progeny mature into **B-spermatogonia**, which: (1) detach from the basement membrane and move closer to the blood–testis barrier in response to paracrine cues, such as Sertoli-derived **retinoic acid**, and (2) eventually transition into meiosis-committed **primary spermatocytes**.

2. *Spermatocytic meiosis*—Each **primary spermatocyte** begins meiosis by duplicating its DNA and undergoing crossing-over and independent assortment events that increase genetic diversity (Chapter 2). The primary spermatocyte completes its first meiotic division to generate two **secondary spermatocytes**, and without an intervening phase of DNA synthesis, each secondary spermatocyte undergoes the second meiotic division to yield a pair of haploid **spermatids** (**Figure 17.3b–d**). To complete these meiotic divisions, spermatocytes require an adequate supply of **follicle-stimulating hormone** (**FSH**) from the anterior pituitary as well as sufficient **testosterone** produced by Leydig cells in post-adolescent testes. However, throughout spermatogenesis, spermatogenic cells lack receptors for both FSH and testosterone, and instead, these two essential stimuli bind **receptors on Sertoli cells**, which then relay appropriate signals to spermatocytes.

3. *Spermiogenesis*—After being generated by the second meiotic division, spermatids initially resemble secondary spermatocytes with their circular profile, spherical nucleus, and abundant cytoplasm. Subsequently, spermatids undergo a multi-step **spermiogenesis** process over ~16 days to become streamlined sperm. During spermiogenesis, each spermatid: (i) forms at its anterior end an **acrosome** that comprises a Golgi-derived sac with enzymes required for fertilization (Box 17.1 Cell Biology of Sperm Functioning in Preparation for, and during, Fertilization); (ii) develops a **flagellum** (=**tail**) at the posterior end for sperm motility; (iii) undergoes **nuclear condensation** via arginine-rich DNA-binding proteins, called **protamines**, which allow tighter packing of chromatin within the sperm **head**; (iv) generates a mitochondria-containing **midpiece** between the head and tail to provide the ATPs needed for tail motility; and (v) sloughs **excess cytoplasm** to optimize hydrodynamic properties for locomotion in the female reproductive tract.

4. *Release of sperm into the tubular lumen during spermiation*—Following spermiogenesis, sperm are detached from the seminiferous epithelium to reside in the tubular lumen. This process of **spermiation** relies heavily on adjacent Sertoli cells, which extend cellular processes that not only help slough excess spermatid cytoplasm but also aid in detaching sperm from the epithelium.

Post-spermiation sperm measure ~**50 μm** long, with a head, midpiece, and tail that average about 4, 2, and 44 μm, respectively. Anteriorly, the **acrosome** forms a light crescentic cap over the denser nuclear region of the head (**Figure 17.3e, f**), and posterior to the head, the midpiece contains ~50–75 spirally arranged mitochondria. To help stabilize the tail, midpiece mitochondria surround a longitudinally oriented group of nine stabilizing **outer dense fibers (ODFs)** that extend from the midpiece through the initial 3/4 of the tail comprising the flagellum's principal piece (**Figure 17.3f**). The ODFs in turn surround nine pairs of microtubules (MTs), which

BOX 17.1 CELL BIOLOGY OF SPERM FUNCTIONING IN PREPARATION FOR, AND DURING, FERTILIZATION

After coitus, sperm undergo an ~5-6 hour-long process of **capacitation** in the uterus, which convert sperm into functional gametes capable of fertilizing eggs. In addition to **hyperactivating** sperm so that they can swim more effectively, capacitation removes inhibitory substances on sperm surfaces to expose molecules that mediate sperm-egg interactions. Capacitated sperm also have enhanced capacities both for orienting to oviductal fluid flow (**rheotaxis**)

and for following gradients of temperature (**thermotaxis**) and chemoattractants (**chemotaxis**) that collectively allow sperm to approach the egg and its surrounding **cumulus cells** (Chapter 18). According to some studies, capacitated mammalian sperm undergo an **acrosome reaction** (**AR**) while penetrating the cumulus layer by producing an **SP8** protein that causes **progesterone** release from cumulus cells, thereby triggering the AR (**Figure a**). Alternatively, other data suggest sperm

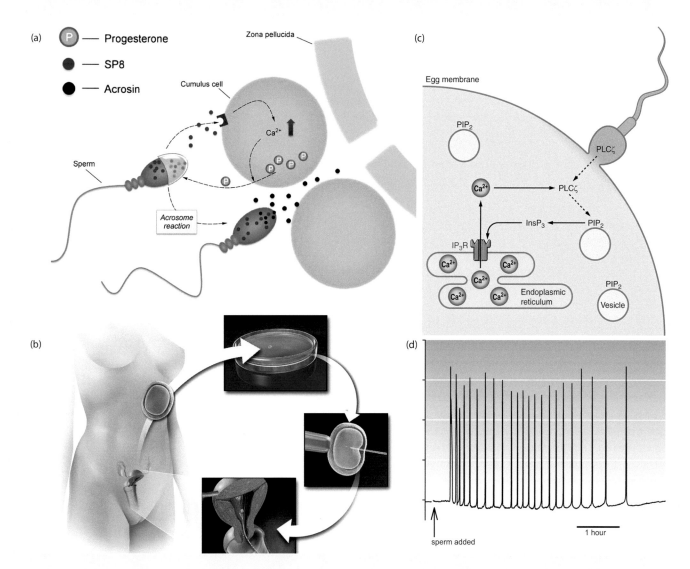

Box 17.1 *(**a**) A model for the secretion of acrosin enzymes during the acrosome reaction driven by progesterone and SP8 protein signaling. (From Sun, TT et al. (2011) Acrosome reaction in the cumulus oophorus revisited: involvement of a novel sperm-released factor NYD-SP8. Protein Cell 2: 92–98 reproduced with publisher permission.) (**b**) ICSI (intracytoplasmic sperm injection) and embryonic transfer as a form of assisted reproductive technology (ART). (From Blausen.com staff (2014) "Medical gallery of Blausen Medical 2014". Wiki J Med 1 (2): 10. doi:10.15347/wjm/2014.01010, reproduced under CC0 and CC-SA creative commons licenses.) (**c, d**) The delivery of a soluble sperm factor PLC-zeta to release calcium via IP3 receptors on the egg's ER membrane, thereby triggering key calcium oscillations over a several-hour-long period during fertilization. (From Swann, K and Lai, FA (2015) Egg activation at fertilization by a soluble sperm protein. Physiol Rev 96: 127-149 reproduced with publisher permission.)*

initiate an AR only after contacting **ZP3** proteins in the **zona pellucida** coat surrounding the egg. In either case, the AR process releases enzymes like **acrosin** to form holes in the egg's zona (**Figure b**).

After capacitated sperm breach the zona, compatibility-related proteins ensure proper sperm-egg interactions (**Figure c**). For example, **Izumo** proteins in the equatorial region of acrosome-reacted sperm heads attach to **Juno** receptors on the oolemma to aid fusion. Without Izumo, penetration of sperm through the zona can occur normally, but sperm-egg fusion is inhibited, thereby halting further steps in fertilization and subsequent development.

Cytoplasmic continuity established during gamete fusion allows the fertilizing sperm to deliver a soluble factor into the ooplasm in order to trigger a proper calcium response needed for successful fertilization. Accordingly, introducing sperm into the egg cytoplasm via **intracytoplasmic sperm injection (ICSI)** (**Figure b**) generates substantially higher pregnancy rates than are achieved by simply mixing sperm and eggs together in a dish. The sperm factor introduced during fertilization is most likely **PLCzeta**, which can hydrolyze **PIP$_2$** lipids to yield **IP$_3$**. IP$_3$ then releases Ca^{2+} from the egg's endoplasmic reticulum to produce repeated **Ca^{2+} oscillations** that drive proper fertilization and embryogenesis (**Figure c, d**).

encircle a central microtubular pair within the flagellar shaft (=**axoneme**) to yield a 9 + 2 configuration of MTs. Distal to the principal piece, ODFs are lacking, and instead only MTs plus their associated structures are present (**Figure 17.3f**). In particular, two arm-like projections of dynein ATPase (=**dynein arms**) normally extend from each peripheral MT pair to hydrolyze ATP, thereby providing the energy need for tail beating. However, in men suffering from **Kartagener's syndrome**, dynein arms are characteristically lacking or highly defective, causing sperm dysmotility and hence infertility.

17.5 CELLULAR ASSOCIATIONS AND CYCLES OF THE SEMINIFEROUS EPITHELIUM: STAGGERED SPERMATOGENESIS THAT PROVIDES A CONTINUAL SUPPLY OF FRESHLY MADE SPERM

Within each mammalian seminiferous tubule, **the onset of spermatogenesis is staggered** along the length of the tubule so that no single transverse section contains all spermatogenic stages from early spermatogonia through released sperm (**Figure 17.4a–d**). A key consequence of such staggering is that any given time, new sperm are released from discrete sites along the tubule rather than being produced simultaneously throughout its entire length. This in turn avoids boom-and-bust cycles of a single pan-tubular mode of sperm formation and thus provides a continuous supply of freshly made sperm.

During this punctuated spermatogenesis, several stages of spermatogenesis consistently co-occur with each other as part of discrete clusters, called **cellular associations** (**Figure 17.4b-d**). Such associations are repeated at regularly spaced sites along the length of the seminiferous tubule, because the orientations of spindles that form during meiotic divisions cause more advanced spermatogenic stages to be obliquely translocated relative to the luminal surface (**Figure 17.4b**). As a result of these oblique translocations, each cellular association is cyclically repeated by a new iteration of that association further down the tubule during what is referred to as a **cycle of the seminiferous epithelium**, which in turn takes about 8.6 days in mice vs. 16 days in humans (**Figure 17.4e**). Every individual sperm undergoes ~4.6 seminiferous cycles as it carries out the entire sequence of spermatogenesis spanning from spermatogonial differentiation to intraluminal sperm release. Thus, human spermatogenesis is completed in ~4.6 × 16 = ~**74 days**.

Currently, it remains unclear exactly how such cycles are coordinated. However, localized pulses of **retinoic acid** produced by Sertoli cells at staggered positions are believed to play a key role. Regardless of the underlying mechanisms, a healthy young man can generate 1,000 sperm per second, with such rates being subject to both a normal aging-related decline and to various pathological factors that can dramatically reduce sperm production (**Box 17.2 Are Sperm Counts Declining in Western Countries?**).

17.6 SPERM-TRANSPORTING CONDUITS: RETE TESTIS, VAS EFFERENS, EPIDIDYMIS, VAS DEFERENS, AND EJACULATORY DUCT

Toward the superior pole of each testis, seminiferous tubules join a network of irregularly shaped chambers, called the **rete testis**, which in turn connects to short tubes, referred to as the **vasa** (or ductuli) **efferentia**. (**Figure 17.2c**). The simple columnar epithelium lining each vas efferens contains both ciliated cells that move luminal contents and non-ciliated absorptive cells that take up water to help concentrate fluids surrounding newly generated sperm. After exiting the testes via efferent ducts, sperm plus surrounding seminal fluid are eventually delivered to the urethra by several kinds of transporting ducts and their associated glands, with

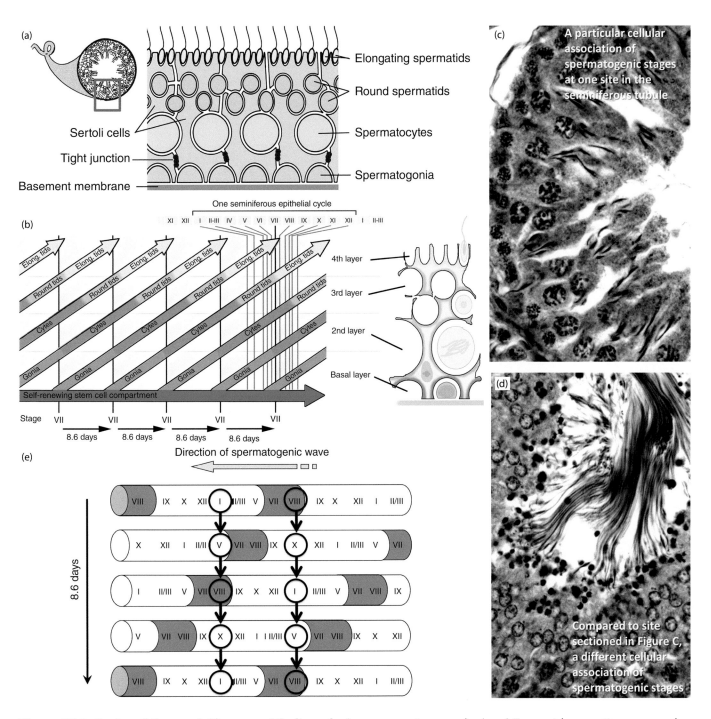

Figure 17.4 Cycles of the seminiferous epithelium during spermatogenesis. (a–e) *To provide a continuous supply of freshly made sperm, seminiferous tubules have staggered onsets of spermatogenesis so that several discrete stages of sperm formation (e.g., spermatogonia, spermatocytes, round spermatids, and elongated spermatids) co-occur at regularly arranged sites along the tubule. Identical cellular associations are separated from each other due to the oblique orientations of spindles moving more advanced spermatogenic stages both apically and down the tubule rather than directly perpendicular toward the epithelial surface. Each cellular association is repeated every 8.6 days in mice vs. 16 days in humans during what is referred to as a cycle of the seminiferous epithelium. Every sperm undergoes ~4.6 cycles during its development within the seminiferous epithelium so that in humans the entire process of spermatogenesis from spermatogonial differentiation to sperm release into the tubular lumen takes* $\sim 4.6 \times 16 = \sim 74$ *days. (c) 700X; (d) 450X. ([a, b, e]: From Yoshida, S (2015) From cyst to tubule: innovations in vertebrate spermatogenesis. Wiley Interdiscip Rev Dev Biol 5(1), doi: 10.1002/wdev.204, reproduced with publisher permission.)*

BOX 17.2 ARE SPERM COUNTS DECLINING IN WESTERN COUNTRIES?

In 1992, an article analyzing 61 previously conducted investigations of human sperm counts was published in the British Medical Journal. The report concluded that over the preceding 50 years, **the concentration of sperm in semen decreased** from 119 million/ml to 66 million/ml, and the average ejaculate volume declined from 3.4 to 2.75 ml. Thus, **the total sperm count in an average ejaculate fell ~55%** from ~400 million to ~180 million. The article garnered substantial publicity, particularly in light of the fact that continuing drops in sperm counts could soon fall below the World Health Organization's estimates of **20 million sperm/ml** and **40 million sperm/ejaculate** as the lower cut-off values required for normal male fertility. The report also sparked speculation as to what might cause this decline, with proposed explanations ranging from increased cancers and congenital malformations to various forms of pathogenesis triggered by **endocrine-disrupting chemicals** (**EDCs**). In particular, **xenoestrogen** types of EDCs, such as **Bisphenol-A** (**BPA**) derived from certain plastics were postulated to mimic estrogen's effects and thereby reduce sperm counts.

After the article appeared, other researchers raised doubts about the validity of the study's findings, generally criticizing the fact that as a retrospective study which did not follow the same group of subjects throughout the decades-long timeframe, improper comparisons were made. However, in 2017, a new team of scientists published a far more comprehensive analysis using meta-regression models to assess reliable datasets from 244 studies of sperm counts obtained from >42,000 men during 1973-2011. Unlike the 1992 metanalysis, the newer study excluded potentially confounding results from studies that analyzed: (i) sperm counts from fertility clinics; (ii) subjects with known fertility risks (e.g., due to smoking, cancers, or occupational hazards); and (iii) small datasets or those based on non-standard collection methods. The 2017 study then binned data from **Western countries** (North America, Europe, Australia, and New Zealand) vs. **Non-western countries** (South America, Africa, and Asia). Based on such analyses, the paper corroborated the 1992 study by finding a **50-60% decline in sperm concentration/ml** and **total sperm count per ejaculate** between 1973 and 2011 for men in western countries (**Figure a, b**). Conversely, men from non-western countries did not exhibit significant decreases in these parameters. Such basic trends were also observed in analyses that included only data from demonstrably fertile men who had successfully contributed to a pregnancy, thereby ruling out a skewing of the data due entirely to infertile subjects. Although these metanalyses remain controversial, additional studies are certainly warranted, particularly in order to ascertain if elevated xenoestrogens and/or other factors enhanced in western countries mediate the observed declines in sperm counts.

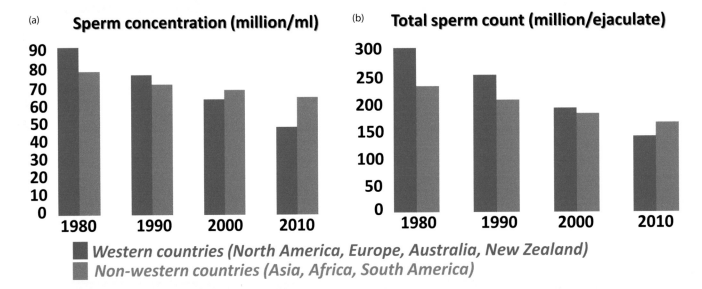

Box 17.2 *Meta-analysis of literature reports showing a decline in: (**a**) sperm concentrations (millions of sperm/ml) and (**b**) total sperm counts (millions of sperm/ejaculate) from 1980 to 2010 for men in western countries (North America, Europe, Australia, New Zealand) vs. no significant declines for men in non-western countries (South America, Africa, and Asia). (Graph bars are estimated based on data presented by Levine, H et al. (2017) Temporal trends in sperm count: a systematic review and meta-regression analysis. Hum Reprod Update 23:646–659.)*

many of these reproductive tract components being susceptible to persistent infections by various pathogens, including, for example, Zika virus (**Figure 17.5a**).

The first wholly extra-testicular conducting tube is a coiled conduit, called the **epididymis** (**Figure 17.5b–d**), which from its proximal to distal ends comprises a head, body, and tail region. The epididymal lumen is lined by pseudostratified columnar epithelium containing predominantly **principal cells**. Such cells possess long

CASE 17.2 JOHN TYLER'S EPIDIDYMI AND VASA DEFERENTIA

10th U.S. President
  (1790–1862)

"I am the president and...I shall be pleased to accept your counsel and advice. But I can never consent to being dictated to...When you think otherwise, your resignations will be accepted."

(J.T. laying down the law to his cabinet after transitioning from Vice President to President)

When William Harrison became the first U.S. president to die in office, Vice President John Tyler began an uncharted journey governing the nation. Some referred to Tyler as "His Accidency" and argued he was obligated to conduct a committee-run presidency. However, Tyler rejected such views and proceeded to assume all presidential powers. In so doing, Tyler set a standard for subsequent vice presidents who had to take over owing to presidential deaths by natural causes (M. Fillmore, C. Coolidge, and H. Truman) or assassination (A. Johnson,

C. Arthur, T. Roosevelt, and L. Johnson). As an adult, Tyler suffered from several debilitating conditions, such as chronic gastroenteritis and temporary idiopathic paralysis. However, in spite of these maladies, Tyler fathered 15 children, with his last being born soon before he died. Such offspring not only set a record for U.S. presidents, but in an era preceding the advent of assisted reproductive technologies, also indicated Tyler possessed a fully functioning reproductive system throughout his adult life. Thus, along with active sperm, Tyler must have also had functional epididymi, vasa deferentia, and other conducting components needed for procreation, thereby allowing the precedent-setting role model for other ascending vice-presidents to become a father many times over.

microfilament-containing **stereovilli** that increase surface area for aiding cellular reorganizations that maturing sperm undergo during their several-day-long transit within the epididymis (**Figure 17.5d**). For example, epididymal sperm flagella are modified to enhance subsequent motility, and **P34H** protein is added to the sperm surface to enable normal fertilization.

The tail of the epididymis connects with the **vas deferens**, which extends from the scrotum into the abdomen and can be cut during a **vasectomy** for male contraception. Lining the lumen of the vas deferens is a **pseudostratified columnar epithelium** with abundant **stereovilli**, and surrounding this mucosal lining are several layers of **smooth muscles** (**Figure 17.5e**) (**Case 17.2 John Tyler's Epididymi and Vasa Deferentia**).

Distally, the vas deferens combines with a duct emanating from the seminal vesicle to form an **ejaculatory duct** (**Figures 17.5a** and **17.6a**), which in turn extends through the prostate gland to reach the urethra. The lining epithelium of ejaculatory ducts is typically **pseudostratified columnar**, and unlike the vas deferens, little surrounding smooth muscle is present. Thus, ejaculatory ducts serve mainly as passive conduits for transmitting semen to the urethra.

17.7 ACCESSORY GLANDS THAT MAKE SEMINAL FLUID: SEMINAL VESICLES, PROSTATE, AND BULBOURETHRAL GLANDS

The seminal fluid bathing sperm is made predominantly by paired **seminal vesicles**, a large **prostate gland**, and two small **bulbourethral glands** (= Cowper's glands) (**Figures 17.1a** and **Figure 17.6a–h**), with ~65–75% of the fluid being produced by the seminal vesicles. Each almond-shaped **seminal vesicle** is a simple tubuloalveolar gland that is located superior to the prostate gland. The vesicle lumen is highly folded to form chambers that are lined by a **pseudostratified columnar epithelium** (**Figure 17.6a–c**). To optimize sperm functioning, seminal vesicles secrete: (i) **fructose** that is used as an energy source to fuel sperm metabolism, and (ii) **prostaglandins**, **prolactin**, and **bicarbonate** ions that enhance sperm flagellar beating.

As a walnut-sized organ connected to the urethra, the **prostate** is a compound tubuloalveolar gland comprising several distinct subregions (**Figure 17.6a, d–f**) (**Case 17.3 Pablo Neruda's Prostate**). Glandular portions of the prostate are lined by a **pseudostratified columnar epithelium** that can undergo both merocrine and apocrine secretion to contribute ~20–30% of total seminal fluids in an ejaculate (**Figure 17.6f–h**). During ejaculation, smooth muscle fibers in **fibromuscular** regions help expel prostatic secretions, which are rich in zinc, glucose, and kallikrein enzymes. In middle-aged and elderly men, the prostate typically undergoes a gradual swelling, called **benign prostatic hyperplasia**, which can cause: (i) difficulties in voiding the bladder, and (ii) increased incidences of lower urinary tract infections. In addition, older prostate glands often develop small **corpora amylacea** concretions containing abundant lamellae of glycoproteins that typically do not generate significant health problems (**Figure 17.6e**). However, larger and more heavily calcified formations, called prostatic **calculi** (=stones),

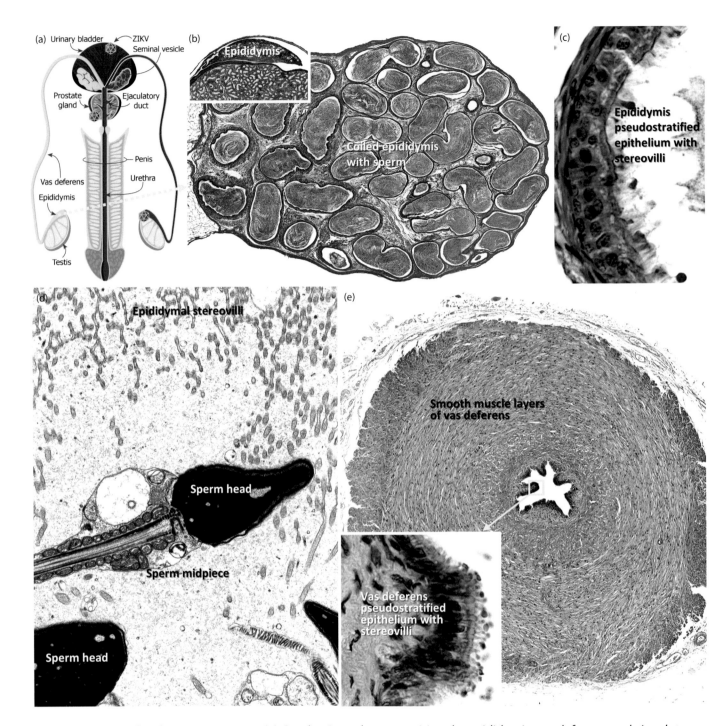

Figure 17.5 Conducting components. (a) *Conducting tubes comprising the epididymis, vas deferens, and ejaculatory duct deliver sperm from each testis to the urethra and are subject to chronic infections like those involving Zika virus (ZIKV). (**b–e**) Both the epididymis (b–d) and vas deferens (e) are lined by a pseudostratified epithelium containing apical stereovilli, with the vas deferens being distinguished by its non-coiled morphology and highly muscular wall. (b) 30X; (b, inset) 6X; (c) 350X; (d) 3,300X; (e) 70X; (e, inset) 275X. ([a]: From Stassen, L et al. (2018) Zika virus in the male reproductive tract. Viruses 10, 198, doi: 10.3390/v10040198, reproduced under a CC BY 4.0 creative commons license; [d]: TEM by Drs. H. Wartenberg and H. Jastrow, reproduced with permission from Dr. H. Jastrow's electron microscopic atlas, http//www.drjastrow.de.)*

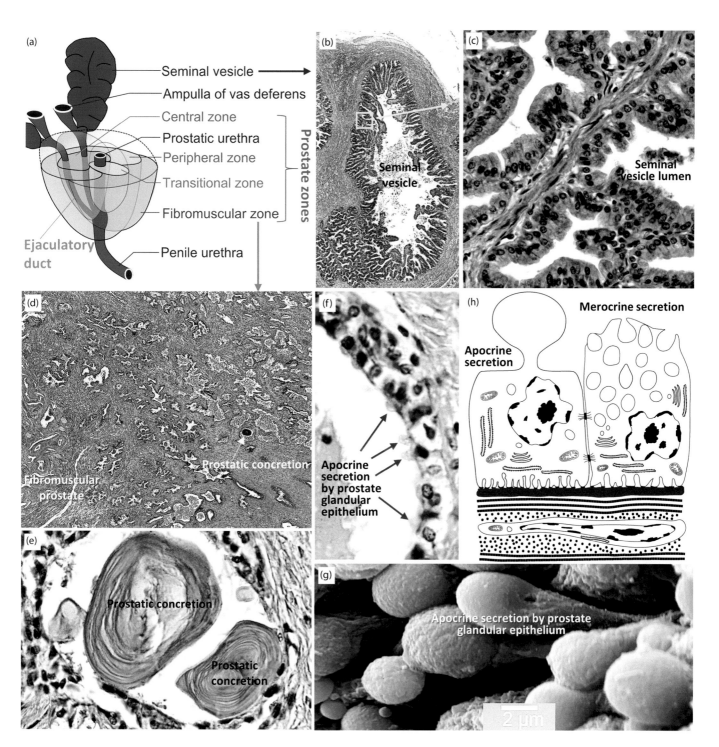

Figure 17.6 Accessory glands. *(a–h) The vast majority of seminal fluid is made by a pair of seminal vesicles (a–c) plus a prostate gland (d–h), with the seminal vesicles producing ~70% of the fluid and the prostate supplying ~25%. The prostate has multiple discrete histological regions and can be distinguished from seminal vesicles by its (i) irregular luminal profiles (d), (ii) intraluminal concretions (e), and (iii) apocrine as well as merocrine secretory modes (f–h); (b) 20X; C: 200X; D: 20X; E: 270X; F: 325X. ([a]: Downloaded from https://commons.wikimedia.org/wiki/File:Prostate_zones.png; Häggström, M (2014) Medical gallery of Mikael Häggström2014. Wiki J Med 1(2), doi: 10.15347/wjm/2014.008, ISSN 2002-4436 and reproduced under a CC 0 1.0 universal public domain license; [g, h]: From Fullwood, NJ et al. (2019) An analysis of benign human prostate offers insights into the mechanism of apocrine secretion and the origin of prostasomes. Sci Rep 4582. https://doi.org/10.1038/s41598-019-40820-2, reproduced under a CC BY 4.0 creative commons license.)*

CASE 17.3 PABLO NERUDA'S PROSTATE

Poet
    (1904–1973)

"Look around—there's only one thing of danger for you here—poetry."

(*P.N. to armed government soldiers searching his house for evidence of criminal activity*)

Born in Chile, Ricardo Eliécer Neftalí Reyes Basoalto began writing poems at age 10 and in order to avoid parental disapproval used a penname of Pablo Neruda, which he later legally adopted. At 19, Neruda published *Veinte poemas de amor y una canción desesperada* (*Twenty Love Poems and A Desperate Song*), which remains the best-selling Spanish book of poetry, and for this and other acclaimed works, Neruda was awarded the 1971 Nobel Prize for Literature. In addition to his writings, Neruda was active in the left wing of Chilean politics, serving as a senator in the communist party and an advisor to the socialist president of Chile, Salvadore Allende.

When Allende's government was overthrown in a coup led by General Augusto Pinochet, Neruda was undergoing treatment in a hospital for prostate cancer. While hospitalized, Neruda phoned his wife asking to be retrieved, fearing he had been given a toxic injection by an agent of Pinochet's who had posed as a doctor. Soon after arriving home, Neruda died, and even though he was corpulent at the time of his death, his official cause of death was listed as cachexia (severe withering) due to prostatic cancer. Recently, however, Neruda's remains were tested. No chemical poison was found, but the presence of a suspicious bacterial strain cast further doubt on a cachexic death driven by prostatic neoplasia and instead left open the possibility of murder via an injected bio-weapon.

can cause pain and obstructions often in association with chronic inflammation (=**prostatitis**) that may require medical intervention.

Bulbourethral glands are two pea-sized glands located inferior to the prostate, which, along with the testis, epididymis, and vas deferens, produce the remaining ~5% of seminal fluid not secreted by the seminal vesicles or prostate gland. The left and right bulbourethral glands connect to the proximal end of the penile urethra and secrete a thickened mucus that lubricates the urethra while also potentially protecting sperm from the acidic pH of any urine that might mix with semen.

17.8 THE PENIS: HISTOLOGICAL ORGANIZATION AND PHYSIOLOGICAL BASIS OF AN ERECTION

Functioning both in urination and in discharging sperm, the **penis** is a cylindrical organ that houses the urethra, through which semen can be expelled. The penile urethra is surrounded by a single, relatively small vascular bed, called the **corpus spongiosum**, and superior to such tissue are two large plexuses of blood vessels, termed the **corpora cavernosa** (**Figure 17.7a–d**). Also referred to as **erectile tissues**, each of the three vascular beds comprises numerous **sinusoids** that are lined by endothelia and fed by terminal **helicine** arteries whose walls contain endothelial and smooth muscle cells. Normally, such arteries are tonically contracted and thereby serve as a shunt that essentially bypasses the sinusoidal beds and instead sends blood to penile veins. However, during an erection, terminal arteries supplying erectile tissues relax, allowing blood to flow into sinusoids so that the penis can be used as an intromittent organ for delivering sperm during intercourse (**Figure 17.7a**).

The physiological mechanisms of an erection can involve various sensory or mechanical stimuli initially triggering the secretion of **acetylcholine** from parasympathetic nerves in penile tissues. The released acetylcholine causes nitric oxide synthase (NOS) in nearby neurons and endothelial cells to synthesize **nitric oxide** (**NO**), a substance which was once termed **endothelial-derived relaxation factor** (**EDRF**), when it was first isolated from endothelial cells and shown to relax smooth muscle (**Figure 17.7e**). Neuronal- and endothelial-produced NO can then diffuse into smooth muscle cells in terminal arteries and thereby activate the enzyme **guanylate cyclase** (**GC**), which raises **cGMP** levels within the muscle cell cytoplasm (**Figure 17.7e**). Elevated cGMP reduces calcium ion concentrations by closing plasmalemmal calcium channels and promoting calcium re-uptake into the sarcoplasmic reticulum, thereby preventing muscular contraction (**Figure 17.7e**). In this relaxed state, arterial blood rushes into the sinusoids of the three erectile tissues, with engorgement of spongiosum tissues serving to keep the urethra from collapsing so that semen can be expelled.

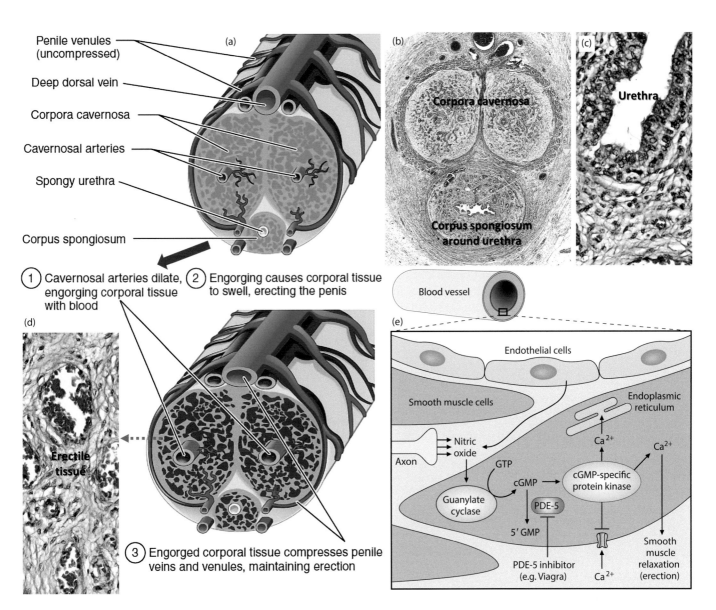

***Figure 17.7 Functional microanatomy of the penis.** (**a–d**) The penis contains well-vascularized erectile tissue distributed as two superior corpora cavernosa plus a single inferior corpus spongiosum surrounding the urethra. Normally, contracted muscular shunts cause blood to bypass erectile tissues. However, in response to cholinergic neural stimulation, blood can be directed into the erectile sinusoids to initiate an erection. (b) 10X; (c) 200X; (d) 180X. (**e**) Erectile tissues become engorged with blood following relaxation of smooth muscle in arteries supplying the three vascular beds. Relaxation involves neurons and endothelial cells releasing nitric oxide (NO), which diffuses into vascular smooth muscle cells to increase cGMP production. This reduces intracellular calcium levels via the blockage of external calcium influx and the sequestration of calcium into the ER. Reduced intracellular calcium prevents muscle contraction thereby promoting vessel engorgement during erections. The PDE-5 inhibitor Viagra promotes and maintains erections by keeping cGMP levels elevated, thereby blocking smooth muscle contraction. ([a]: From OpenStax College (2013) Anatomy and Physiology. OpenStax. http://cnx.org/content/col11496/latest, reproduced under CC-BY-4.0 creative commons license.)*

At climax, **adrenergic neurons** trigger muscular contractions needed for ejaculation, while also helping to undo the effects of the original cholinergic stimulation at the onset of the erection. As a key part of this reversal, **type V phosphodiesterase (PDE5)** in smooth muscle cells reduces cGMP levels, which in turn elevates calcium ion levels to promote muscular contraction and thereby returns the penis to a flaccid state by shunting blood from erectile tissues. Accordingly, drugs like **Viagra** treat **erectile dysfunction** by serving as **PDE5 inhibitors**, which relax arterial smooth muscle to promote penile engorgement (**Figure 17.7e**).

17.9 EVOLUTION OF FERTILIZATION PROCESSES WITH A FOCUS ON SPERM FUNCTIONING

During intercourse, each man can deposit many tens of millions of sperm to ensure that a sufficient number may reach an ovulated egg for fertilization. Directly after entering the female reproductive tract, human sperm are unable to fertilize eggs and must first undergo a maturation process, called **capacitation**. A key event in capacitation is activation of a <u>cat</u>ion channel of <u>sper</u>m (**CatSper**) in tail plasmalemmata, which allows calcium ion influx to optimize sperm motility. Hypermobilized sperm can then reach the oviductal site of fertilization within ~1–12 hours and remain viable there for ~3–5 days.

Capacitation also enhances tracking capabilities of sperm. For example, capacitated sperm can utilize **rheotaxis** and **thermotaxis** for orienting to gradients of currents and temperature, respectively. To guide sperm during their final leg of migration, short-range **chemotaxis** is employed, as sperm follow gradients of **attractants** like **non-steroidal molecules** and **progesterone**, which are released by the egg and its surrounding follicle cells.

To approach an ovulated egg, sperm must first traverse the layer of surrounding follicle cells (=**cumulus cells**) and their thickened ECM that is enriched in **hyaluronic acid**. Such **cumulus penetration** involves a hyaluronidase enzyme on the sperm surface, thereby allowing sperm to reach the extracellular coat (=**zona pellucida**) encasing the egg's plasmalemma (=**oolemma**). During such movements or directly thereafter (Box 17.1), sperm undergo an **acrosome reaction** (**AR**) that exocytoses enzymes like **acrosin** to digest the zona and thereby allow sperm-oolemma binding. After binding and fusing with the oolemma, the sperm then triggers a series of calcium ion elevations (=**calcium oscillations**) within the egg that are required for monospermic fertilization and subsequent development (Box 17.1).

17.10 SOME MALE REPRODUCTIVE SYSTEM DISORDERS: MALE INFERTILITY AND PROSTATE CANCER

Male infertility can result from **erectile dysfunction**, a blockage in sperm transport, or **abnormal semen** containing reduced numbers of functional sperm. Diminished semen quality can in turn be due to congenital aberrations or other maladies that specifically target male reproduction (e.g., cryptorchidism, Kartagener's syndrome, or an enlarged venous plexus in the scrotum [=**varicocele**]). Alternatively, poor health, for example due to excessive alcohol intake, medications, and **endocrine-disrupting chemicals** (**EDCs**) derived from food or environmental pollutants (Box 17.2) can also cause infertility. As long as some normal sperm are produced, infertility can potentially be treated by **assisted reproductive technology** (**ART**), with the most successful ART mode conducted in fertility clinics typically being **intracytoplasmic sperm injection** (**ICSI**). In such cases, an isolated sperm is injected into an egg to achieve *in vitro* fertilization, and the resultant embryo can then be implanted in the woman's endometrium to be carried to term (Box 17.1).

Approximately one in in seven U.S. men develop **prostate cancer** during their lifetimes, making it the most common male-specific cancer. If detected early, surgical removal of cancerous tissues and/ or radiation- and chemo-therapies can often cure the disease. However, at more advanced stages, particularly in African-American men, such cancer is difficult to contain within the prostate and once it has metastasized yields 5-year survival rates of <30%. Since 1986, a blood test for **prostate-specific antigen** (**PSA**) corresponding to a kallikrein enzyme has been used to detect early stages of prostate cancer and thereby help reduce mortality rates.

17.11 SUMMARY—MALE REPRODUCTIVE SYSTEM

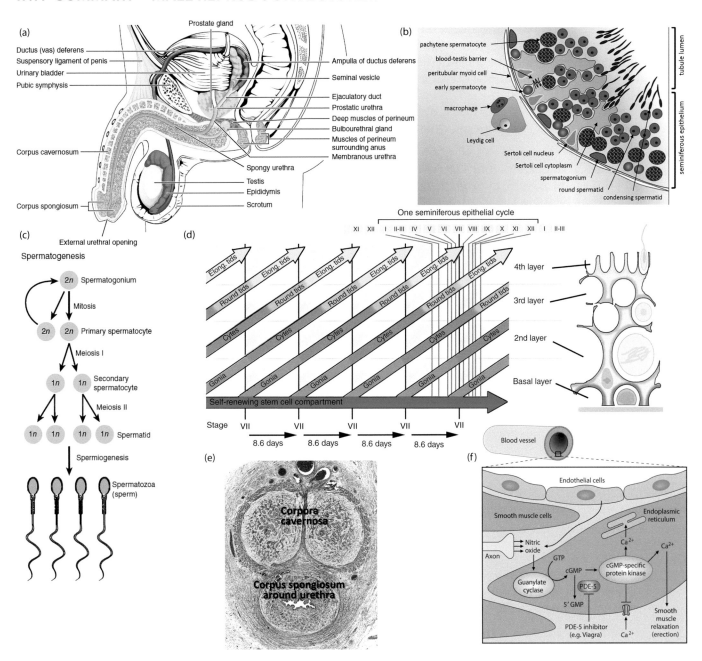

Pictorial Summary Figure Caption: *(**a**): see figure 17-1a; (**b**): figure 17-3a; (**c**): see figure 17.3b; (**d**): see figure 17.4b; (**e**): see figure 17.7B; (**f**): see figure 17.7e.*

Two **testes** occur in the **scrotum** (**Figure a**), where lower temperature optimizes sperm formation (=**spermatogenesis**). Each testis connects via intermediary structures to an **epididymis** with long immotile **stereovilli** on its pseudostratified epithelial lining. The epididymis joins a **vas deferens** (the tube cut in vasectomies) that exits the scrotum and connects with the seminal vesicle to form an **ejaculatory duct**. Ejaculatory ducts deliver sperm to the **urethra** that extends through the **penis**. The reproductive tract also has **three sets of accessory glands** (two **seminal vesicles**, a **prostate** with its characteristic **corpora amylacea,** and two **bulbourethral glands**) that secrete alkaline and fructose-rich **seminal fluid** to maintain sperm, with most of this fluid being made by the seminal vesicles.

Each testis has **seminiferous tubules** lined by a stratified epithelium where spermatogenesis occurs. The **seminiferous epithelium** comprises: (i) somatic **Sertoli cells** that nurture developing **spermatozoa** (=**sperm**) and form a **blood-testis barrier** to ward off pathogens and autoimmune reactions; and (ii) germline-derived **spermatogenic cell**s that undergo meiosis to form sperm. **Leydig cells** in ECM compartments surrounding tubules produce sufficient **testosterone** after puberty to enable spermatogenesis, which involves: (i) **Type A spermatogonia** undergoing mitotic divisions before differentiating into **B spermatogonia**, (ii) B spermatogonia committing to meiosis to become **primary spermatocytes**, (iii) the first meiotic division forming **secondary spermatocytes**, (iv) the second meiotic division forming **spermatids**, and (v) haploid spermatids maturing during **spermiogenesis** to form sperm, which are released into tubule lumen during **spermiation** (**Figure b, c**).

No single transverse section of seminiferous tubules exhibits all spermatogenic stages, because the onset of sperm formation is staggered along the tubule to provide a continuous supply of freshly made sperm. Within the lining epithelium of seminiferous tubules, discrete sub-stages of spermatogenesis co-occur with each other to constitute **cellular associations** (**Figure d**). The time it takes for a cellular association to be repeated in an adjacent tubule region is 16 days in humans and corresponds to a **cycle of the seminiferous epithelium**. Each developing sperm goes through ~4.6 seminiferous cycles to complete spermatogenesis in ~**74 days**, with staggered **retinoic acid** release by Sertoli cells potentially coordinating such cycles.

Spermiogenesis converts each spermatid into a 50 μm-long sperm with a **head, midpiece, tail**. This 16-day-long process involves: (i) **acrosome** production to provide enzymes needed for fertilization; (ii) posterior tail (=**flagellum**) formation with **9+2 microtubules** (**MTs**) plus **dynein arms** in the shaft (=**axoneme**), although functional **dynein arms** are lacking in the sperm of **Kartagener's** patients; (iii) head condensation via **protamine** types of DNA-binding proteins; (iv) mitochondria-rich **midpiece** production, providing energy for tail beating; and (v) **excess cytoplasm sloughing** for better sperm locomotion.

The penis has three vascular beds (=**erectile tissues**) (**Figure e**). Under parasympathetic stimulation, release of **acetylcholine** causes **nitric oxide** (**NO**) secretion from endothelial cells and neurons. NO diffuses into smooth muscle cells of terminal arteries to increase **cGMP**. High cGMP levels reduce Ca^{2+} concentration to relax the arterial wall and allow blood flow into erectile tissues for an erection. Adrenergic stimulation causes ejaculation and reverses the erection by raising **PDE5** activity so that: (i) cGMP levels fall, (ii) Ca^{2+} rises, and (iii) smooth muscle contracts. **Viagra** promotes erections by blocking PDE5 (**Figure f**).

Sperm cannot fertilize eggs without first undergoing **capacitation** and an **acrosome reaction** (**AR**) in the female tract. During capacitation, **CatSper** channels on the sperm tail are stimulated, thereby enabling **hyperactivated** swimming. Capacitated sperm also orient better to gradients for finding eggs. An AR occurs in the egg vicinity so that sperm can release acrosomal enzymes like **acrosin** to digest the zona pellucida and reach the egg membrane. After fusing with the egg during fertilization, the sperm delivers a soluble factor (probably **PLCzeta**) to trigger Ca^{2+} **oscillations** needed for development, and such sperm-factor-mediated modes of egg activation are supported by findings that treatments with **intracytoplasmic sperm injection** (**ICSI**) in fertility clinics enhance pregnancy rates.

Some male reproductive system disorders: Low sperm counts (**oligospermia**) have been documented for western countries, perhaps due to toxins that act like estrogen (=**xenoestrogens**). Worldwide, **prostate cancer** is the most common male-specific cancer.

SELF-STUDY QUESTIONS

1. What is the name of a spermatogenic cell that has just completed the first meiotic division?
2. T/F Most of the seminal fluid is made by the prostate.
3. Which cells form the "blood-testis barrier"?
4. Which of the following occurs exclusively outside the scrotum?
   A. Epididymis
   B. Seminiferous tubules
   C. Bulbourethral glands
   D. Rete testis
   E. Vasa efferentia

5. Which of the following would promote an erection?
    A. PDE5 activation
    B. cGMP hydrolysis (breakdown)
    C. Adrenergic stimulation
    D. Helicine artery contraction
    E. Nitric oxide production
6. Which sperm component mediates hyperactivation during capacitation in the female reproductive tract?
    A. Retinoic acid
    B. CatSper
    C. PSA
7. In which region of the sperm do you find dynein arms?
8. T/F Epididymal lining cells have motile cilia.
9. Which disease is characterized by the lack of dynein arms?
    A. Prostate cancer
    B. Gonorrhea
    C. Kartagener syndrome
    D. Cryptorchidism
    E. NONE of the above
10. About how long are fully formed human sperm (head plus tail)?
11. Which statement regarding human spermatogenesis is correct?
    A. The full sequence takes about 74 days
    B. Its onset sites are staggered along each seminiferous tubule
    C. It involves repeated cycles of cellular associations
    D. ALL of the above
    E. NONE of the above
12. Of the following processes involving sperm, which occurs last?
    A. Sperm-egg fusion
    B. Zona penetration by the sperm
    C. Capacitation
    D. Acrosome reaction
    E. PLCzeta delivery
13. T/F Fertilization-induced calcium responses involve a soluble sperm factor.

"EXTRA CREDIT" For each term on the left, provide BEST MATCH on the right (answers A-F can be used more than once).

14. Reinke crystals _____	A.	Sertoli cell
15. PDE5 inhibitor _____	B.	Leydig cell
16. Retinoic acid _____	C.	Prostate
17. Corpora amylacea _____	D.	Viagra
18. Protamines _____	E.	Seminal vesicle
	F.	Sperm

19. Describe the stages and their characteristic features that occur during human spermatogenesis.
20. Describe the physiological steps involved in an erection.

ANSWERS

(1) secondary spermatocyte; (2) F; (3) Sertoli cells; (4) C; (5) E; (6) B; (7) tail (axoneme); (8) F; (9) C; (10) 50 μm; (11) D; (12) E; (13) T; (14) B; (15) D; (16) A; (17) C; (18) F; (19) The answer should expand on: Type A spermatogonium (basal and differentiating mitotic cell)→B spermatogonium (more apical mitotic cell)→primary spermatocyte (commits to meiosis)→secondary spermatocyte (after first meiotic division)→spermatid (haploid sperm precursor)→sperm (after all the events of spermiogenesis are completed); (20) A full answer would include and define most or all the following terms: the three erectile tissues and blood flow patterns during the flaccid state; acetylcholine, NO (EDRF), cGMP and calcium signaling during smooth muscle relaxation; adrenergic innervation, PDE5

Female Reproductive System

18.1 INTRODUCTION TO THE FEMALE REPRODUCTIVE SYSTEM

Organs of the female reproductive system include: (i) paired **ovaries** and **oviducts**, (ii) a single **uterus** and **vagina**, and (iii) **external genitalia** (**Figure 18.1a, b**). In addition, uterine tissues contribute to a **placenta** that nourishes the fetus during intra-uterine development (=**gestation**), and **mammary glands** in the breasts provide babies with milk following birth. Collectively, the various components of the female reproductive system function primarily in **gamete production**, **sexual intercourse**, and **procreation** but can also play ancillary roles in regulating homeostasis throughout the body via their secretion of hormones.

Situated laterally within the pelvic cavity, each **ovary** occurs nearby, but not attached to, a neighboring **oviduct** (= fallopian tube or uterine tube) that extends from the superior end of the **uterus** (**Figure 18.1a**). From its oviductal connection, the expanded body of the uterus continues toward a narrowed **cervix**, which joins with the **vagina**, an organ that allows coitus and childbirth. Unlike most other animals that reproduce during a distinct breeding season or at irregular times during the year, women have evolved a hormonally regulated **menstrual cycle** that occurs every 25–30 days (Box 18.1 Why Has Menstruation and Menopause Evolved in Humans?). Accordingly, the microanatomical features of female organs described below are subject to considerable variation depending on the particular stage of the menstrual cycle that is examined.

DOI: 10.1201/9780429353307-22

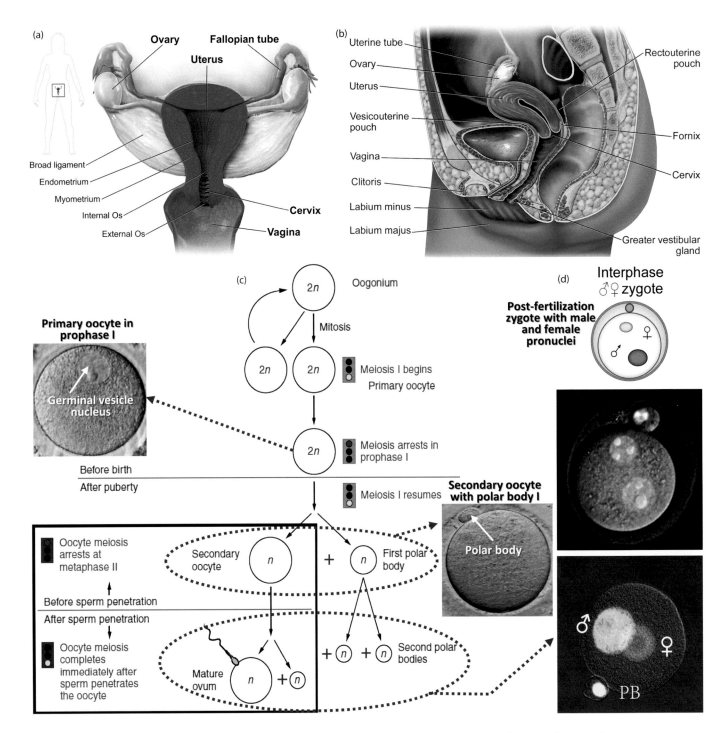

Figure 18.1 Introduction to the female reproductive system and oogenesis. (a, b) *Female reproductive organs include: (i) two ovaries, each with an associated oviduct (=fallopian tube or uterine tube); (ii) a uterus with a narrow cervical region; (iii) a vagina; and (iv) external genitalia. (**c, d**) Ovaries produce haploid eggs via two meiotic divisions during the process of oogenesis. Such divisions begin with a primary oocyte that remains arrested in prophase I until puberty and thus has a large nucleus, called the germinal vesicle. In response to appropriate hormonal cues, the primary oocyte proceeds through the first meiotic division that yields a secondary oocyte with a small polar body. Secondary oocytes can mature into fertilizable eggs, which, after incorporating sperm and completing the second meiotic division, become zygotes with a male and a female pronucleus containing genes from the sperm and egg, respectively (d). (c, left image) 330X; (c, right image) 300X; (d: 310X). ([a, b]: From Blausen.com staff (2014) Medical gallery of Blausen Medical 2014. Wiki J Med 1(2):10, doi:10.15347/wjm/2014.01010, reproduced under CC0 and CC-SA creative commons licenses; [c]: From OpenStax College (2013) Anatomy and Physiology. OpenStax. http://cnx.org/content/col11496/latest, reproduced*

BOX 18.1 WHY HAS MENSTRUATION AND MENOPAUSE EVOLVED IN HUMANS?

In women undergoing **menstrual cycles**, a hormone-driven process, called **spontaneous decidualization**, remodels the uterine endometrium in preparation for embryonic implantation. Without a pregnancy, the decidualized endometrium is sloughed as another cycle is initiated. Such menstruation also occurs in various non-human primates, several bats, an elephant shrew, and even a rodent (**Figure a**). Conversely, in non-menstruating species that comprise the vast majority of mammals, preparation for pregnancy occurs post-implantation in response to embryonic cues, thereby begging the question: **why has menstruation evolved?**

Previously, menstruation was postulated to provide a selective advantage either by **eliminating sperm-borne pathogens** or by **reducing energetic costs** associated with maintaining a continually receptive state. However, neither proposal is well supported by empirical evidence. Nor do these explanations address the lack of menstruation in most mammals.

Alternatively, menstruation can be viewed as a byproduct of uterine evolution driven by multifaceted immune responses related to potential placenta formation. Thus, in humans and other menstruating mammals with highly invasive placentas, signaling cues can be viewed as having become **genetically assimilated** into the maternal reproductive program in order to initiate spontaneous decidualization before an embryo reaches the uterus, thereby providing time to prepare for implantation and extensive infiltration by the placenta. During the luteal phase of maternally controlled menstrual cycles, the endometrium undergoes a **biphasic immune response** initially involving inflammatory reactions that may select embryos with high fitness before switching to anti-inflammatory responses that aid implantation (**Figure b**). In the absence of conception, a third phase triggers the massive inflammation of menstruation (**Figure b**). Conversely, in mammals with less invasive placentas, there is little selective pressure to evolve such mechanisms. Viewed in this way, menstruation itself is not adaptive but instead is a consequence of co-evolving immune and placentation processes.

Related to menstruation is the **cessation of menstrual cycles in menopausal women**. Unlike in most other animals where sexually mature females can reproduce throughout their lives, women, as well as females of a few cetacean species, stop reproducing well before they die, thereby yielding a prolonged **post-reproductive lifespan** (**PRLS**) after menopause. According to "helper type" hypotheses, menopause is adaptive given that older women who could suffer increased mortality risk at childbirth forego reproduction and instead invest energy into **ensuring the survival of their children and grandchildren**. Conversely, others have argued that menopause is not actually adaptive but instead is the by-product of mismatched rates of senescence in reproductive vs. somatic tissues. Accordingly, with declining mortality rates in most modern societies, a woman is simply able to survive well beyond the functionality of her more-rapidly degrading reproductive organs. Regardless of the forces that have led to menopause, women in industrialized countries appear to have

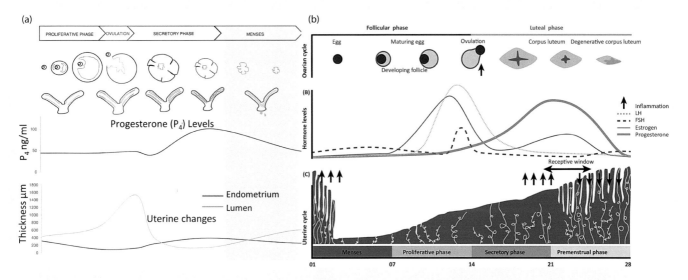

Box 18.1 (a) *Menstruation in a spiny mouse.* ***(b)*** *Inflammation as a key driver in the evolution of the spontaneous decidualization process underlying menstrual cycles. ([a]: From Bellofiore, N et al. (2017) First evidence of a menstruating rodent: the spiny mouse (Acomys cahirinus). J Obstet Gynecol 216:40.e1–11, reproduced with publisher permission, [b]: From Alvergne, A and Tabor, VH (2018) Is female health cyclical? Evolutionary perspectives on menstruation. Trends Ecol Evol 33, https://doi.org/10.1016/j.tree.2018.03.006, reproduced with publisher permission.)*

very different menstrual cycles and PRLSs than those of their ancestors, based on extrapolations from extant hunter-gatherer societies in Africa. In such societies, women start menstruating at age 20 and spend much of their reproductive lives without menstrual cycles because they are either pregnant or nursing their young after childbirth. As a consequence, such women have <80 pre-menopausal menstrual cycles and live only ~30 years beyond menopause onset. By comparison, women in western countries tend to have earlier-onset menarches, fewer pregnancies, shorter post-natal nursing cycles, and longer lifespans, resulting in ~400–500 menstrual cycles and prolonged PRLSs that can exceed 50 years.

18.2   OVARIES: OOGENESIS, HISTOLOGY, FOLLICULAR MATURATION, AND POST-OVULATORY FORMATION OF THE CORPUS LUTEUM

During embryogenesis, primordial germ cells migrate to the developing ovaries and divide mitotically to generate prospective gametes, termed **oogonia (Figure 18.1c)**. To initiate the process of **oogenesis** within ovaries, oogonia can commit to undergoing meiosis, thereby becoming **primary oocytes**. Such oocytes arrest at prophase I of meiosis and are characterized by their large nucleus, termed the **germinal vesicle (GV) (Figure 18.1c)**. Beginning at puberty, primary oocytes can be triggered by hormonal stimulation to resume meiosis and complete the first meiotic division (**Figure 18.1c**). Because their meiotic apparatus is eccentrically positioned, cytokinesis yields two unequal daughter cells—a large **secondary oocyte** plus a much smaller **polar body** attached to one pole of the oocyte (**Figure 18.1c**). The secondary oocyte is then transferred from the ovary to the oviduct via the process of **ovulation** that is discussed further in subsequent sections. The ovulated oocyte remains arrested at **metaphase** of the **second meiotic division** and matures into a fertilizable **egg** that can fuse with a sperm during fertilization (**Figure 18.1c**). Fertilization causes the egg to complete its second meiotic division by emitting another polar body and thus becoming a fertilized **ovum**. The ovum in turn transitions into a diploid **zygote** containing both a **female pronucleus** of maternally derived genes and a **male pronucleus** with genes from the incorporated sperm (**Figure 18.1d**).

Measuring ~3–5 cm long when fully grown, each almond-shaped ovary is suspended by ligaments near the oviductal opening (**Figure 18.2a**) (**Case 18.1 Coretta Scott King's Ovaries**). Histologically, ovaries are surrounded by a connective tissue **capsule**, called the **tunica albuginea**. Beneath its capsule, the ovarian **cortex** contains numerous **follicles**, each of which comprises a primary oocyte surrounded by mitotically dividing somatic cells, called **follicle cells** (**Figure 18.2b**). Medial to the cortex is a **medulla** that lacks follicles and instead is filled with a connective tissue stroma housing large blood vessels and nerves (**Figure 18.2a**).

Over time, the cortex contains decreasing numbers of follicles. Most of this follicular loss is due to the apoptotic degradation of immature follicles as they undergo the process of **atresia**, which starts during gestation and continues after birth. In addition, over the reproductive lifespan of a woman that begins at puberty and ends with **menopause** onset in ~45–55-year-olds (Box 18.1), follicular atresia is augmented by monthly releases of oocytes during **ovulations**. Thus, female newborns are estimated to possess about 0.5–1 million follicles, down from their mid-gestation peak numbers of several million due to atresia, and as a result of continued atresia plus ~500 ovulations during reproductively active years, few follicles remain at menopause onset.

The post-natal ovarian cortex contains **primordial follicles**, which have not yet begun follicular maturation. Each of these quiescent follicles measures ~30–50 µm wide and possesses a single layer of **squamous follicle cells**, called **granulosa cells**, which surround a small primary oocyte (**Figure 18.2b–d**). Primordial follicles lack a blood supply and in almost all cases undergo atresia (**Figure 18.2d**). However, once puberty is reached, some primordial follicles can avoid apoptotic atresia and undergo further follicular maturation (=**folliculogenesis**). In most menstrual cycles, a single follicle fully matures. Alternatively, in rare cases, more than one follicle can complete folliculogenesis during a menstrual cycle, an outcome which becomes more likely when fertility-enhancing drugs are taken.

During the entire process of folliculogenesis, each maturing follicle transitions through several distinct stages that are categorized here as **primary**, **secondary** (=**pre-antral**), and **tertiary** (=**antral**) **follicles** (**Figure 18.2c–e**). Primary follicles are slightly larger than primordial follicles and have a single layer of **cuboidal granulosa cells**. Conversely, secondary follicles are initially ~100–200 µm wide and possess multiple granulosa cell layers plus an external sheath of fibroblast-like **thecal cells**. At this secondary stage, the enclosed primary oocyte with its prominent **GV** is surrounded by an extracellular coat, termed the **zona pellucida**, which functions in fertilization and embryonic encapsulation (**Figure 18.2f**). In tertiary follicles, a fluid-filled cavity, termed

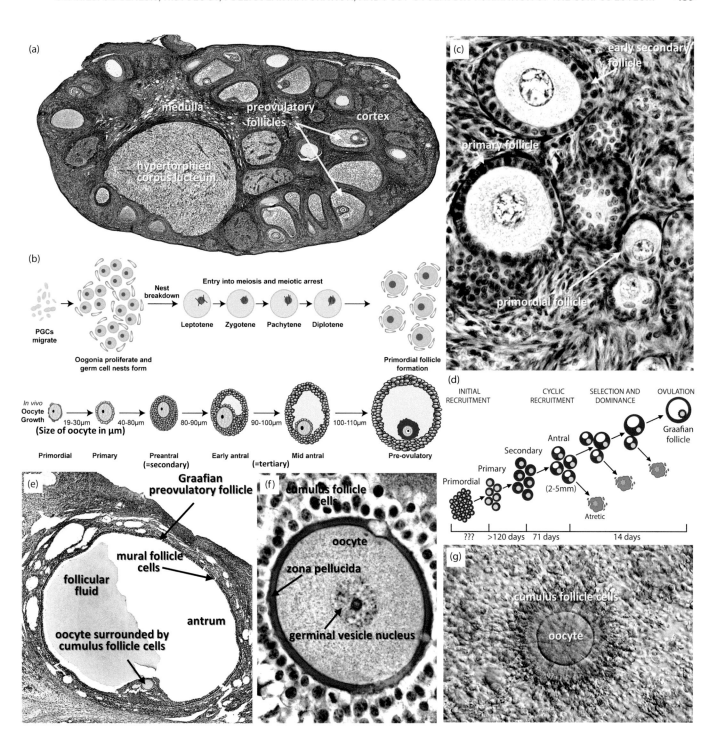

Figure 18.2 Ovarian microanatomy and folliculogenesis. (a) In this section of a cat ovary, the medulla is surrounded by a cortex, which has numerous pre-ovulatory follicles capable of undergoing multiple ovulations and subsequent fertilizations to yield a litter of kittens. Each of these follicle comprises an oocyte within a large fluid-filled antrum that is surrounded by follicle cells. (20X). **(b, c)** After proliferating mitotically, oogonia commit to meiosis to become prophase-I-arrested primary oocytes that are ensheathed by one to numerous layers of follicle cells. Follicles with a single layer of squamous follicle cells are primordial follicles. These can become primary follicles with a simple cuboidal layer of follicle cells before transitioning into secondary preantral follicles that have multiple follicle cell layers but lack an antrum. Tertiary (=antral) follicles have an enlarging antrum before maturing into a pre-ovulatory (Graafian) follicle. (c) 200X. **(d)** Following puberty, hormonal stimulation during each menstrual cycle triggers one (or a few) dominant follicle(s) to proceed to the Graafian stage of folliculogenesis, whereas other follicles undergo apoptosis during a process, called atresia. **(e, f)** The oocyte occurs eccentrically within Graafian follicles and is surrounded by follicle cells, term cumulus

Figure 18.2 (Continued) *cells, whereas follicle cells forming the wall of the follicle are called mural cells. Prior to ovulation, each oocyte has a germinal vesicle nucleus, and during later stages of folliculogenesis, an extracellular coat, called the zona pellucida, forms between cumulus cells and the oocyte plasma membrane. (e) 30X; (f) 400X. (**g**) Cumulus cells remain around the oocyte following ovulation and become spread apart during cumulus expansion. (g) 160X. ([b]: From Anderson, RA and Telfer, EE (2018) Being a good egg in the 21st century. Br Med Bull 127: 83–89 reproduced under a CC BY 4.0 creative commons license; [d]: From Broekmans, FJ (2019) Individualization of FSH doses in assisted reproduction: facts and fiction. Front Endocrinol 10, doi: 10.3389/fendo.2019.00181, reproduced with colorization under a CC BY 4.0 creative commons license; [g]: From Reinzi, L et al. (2012) The oocyte. Hum Reprod. 27(S1):i2–i21, doi:10.1093/humrep/des200, reproduced with publisher permission.)*

CASE 18.1 CORETTA SCOTT KING'S OVARIES

Activist
(1927–2006)

"Struggle is a never-ending process. Freedom is never really won. You earn it and win it in every generation."

(C.S.K.'s advice to younger activists)

Born in Alabama, Coretta Scott King was one of the first African Americans to attend Antioch College in Ohio, where she studied music and participated in the civil rights movement. After transferring to the New England Conservatory of Music in Massachusetts for further training in singing and piano playing, she married Martin Luther King Jr. The newlyweds then relocated to Alabama and began leading peaceful crusades seeking equal rights for African Americans. In response to such activism, King's home was bombed and hit

by bullets before her husband was eventually assassinated in 1968. Following his murder, King led civil rights movements around the world, including those advocating Lesbian, Gay, Bisexual, and Transgender equality, while also successfully campaigning for a holiday to honor her husband's legacy. Late in life, King's health deteriorated, as she suffered a heart attack and developed cancer of her ovaries. Such malignant growths were deemed incurable by traditional methods, and soon after seeking treatment at a holistic health center in Mexico, the life-long activist died of ovarian cancer.

the **antrum**, develops within the granulosa cell compartment. Each month, typically one, but occasionally up to a few, **dominant antral follicle(s)** continue to enlarge as the other antral follicles undergo atresia (**Figure 18.2d**). Dominant follicles expand their antrum while adding more granulosa and thecal cells until becoming fully developed **pre-ovulatory** (=**Graafian**) **follicles** that can measure ~20–25 mm wide (**Figure 18.2e**). The fully grown primary oocyte of pre-ovulatory follicles occurs eccentrically at one follicular pole, and most granulosa cells consist of **mural** cells in the follicular wall, whereas a small cloud-like cluster of granulosa cells encircles the oocyte as a **cumulus oophorus** (**Figure 18.2e–g**). Many of these cumulus cells extend slender **transzonal projections** through the zona pellucida to connect with the plasmalemma of the oocyte via **gap junctions** (**Figure 18.3a–d**), which in turn play key roles in mediating oocyte maturation (Box 18.2 Cell Biology of Mammalian Oocyte Maturation).

The full sequence from primordial follicle recruitment to pre-ovulatory follicle production takes nearly a year. Most of this folliculogenesis involves pre-antral stages that are then followed by rapid differentiation of the antral follicle into a pre-ovulatory state lasting just a few weeks (**Figure 18.2d**).

Even in fully grown Graafian follicles, each enclosed oocyte is not capable of being fertilized until it is triggered to resume meiosis and then released from the ovary into the oviduct to become a mature, fertilizable egg. According to one model of how such oocyte maturation and release proceeds, cumulus cells normally deliver **cGMP** via gap junctions into the primary oocyte and thus keep the immature oocyte in meiotic arrest. However, as discussed further in section 18.6, **LH stimulation** during each menstrual cycle stops this delivery of cGMP and thereby helps activate **maturation-promoting factor** (**MPF**), which in turn initiates meiotic maturation (Box 18.2). Soon after the maturing oocyte completes its first meiotic division, the surrounding follicle ruptures during an **ovulation** process that results in the ovulated oocyte plus surrounding cumulus cells being expelled through a protease-weakened region of the ovarian wall, called the **stigma** (**Figure 18.4a–c**). After being released through the stigma and transported into the oviduct, the oocyte further matures into an egg that remains fertilizable for about a day.

The follicular residuum remaining within the ovary after ovulation becomes an oblong yellowish organ, called the **corpus luteum** (**Figure 18.4d–f**). As elaborated in section 18.6, hormones secreted by the corpus luteum help

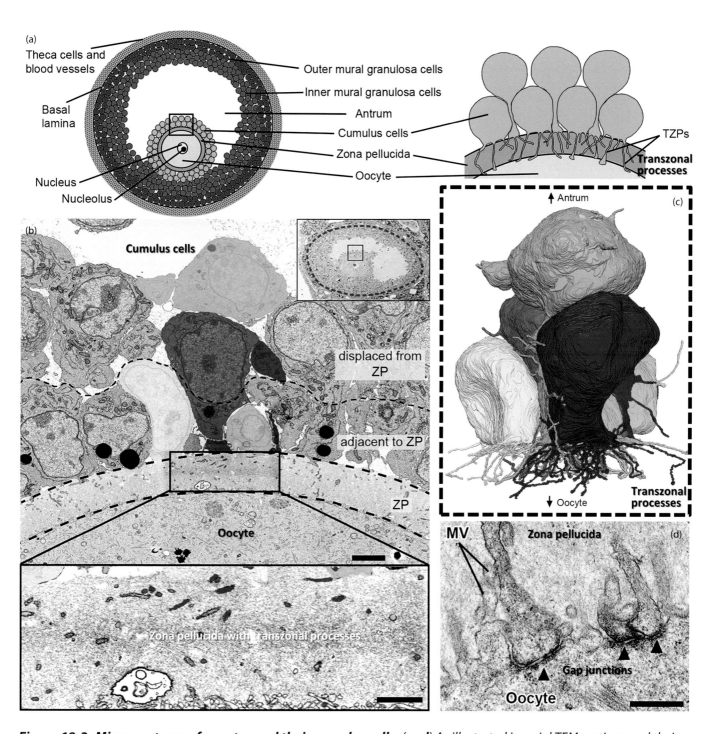

Figure 18.3 Microanatomy of oocytes and their cumulus cells. *(a–d) As illustrated in serial TEM sections and their 3D-reconstructions, cumulus cells extend slender transzonal processes through the zona pellucida to communicate with oocytes via gap junctions, which allow the transduction of key signals for stimulating oocyte maturation and ovulation. Scale bars = (b, top) 5 μm; (b, bottom) 2 μm; (d) 500 nm. (From Baena, V and Terasaki, M (2019) Three-dimensional organization of transzonal projections and other cytoplasmic extensions in the mouse ovarian follicle. Sci Rep 9: 1262 https://doi.org/10.1038/s41598-018-37766-2 reproduced under a CC BY 4.0 creative commons license; images courtesy of Drs. V. Baena and M. Terasaki.)*

BOX 18.2 CELL BIOLOGY OF MAMMALIAN OOCYTE MATURATION

Mammalian oocytes are kept in an immature state by **elevated cAMP levels** and must reduce their cAMP concentrations in order to mature into fertilizable eggs. Based on various lines of evidence, cAMP is elevated in immature oocytes, owing to two major processes: (i) the G-protein coupled receptor **GPR3** located in the oocyte membrane serves to stimulate type III adenylate cyclase (**ACIII**) and thereby increases **intraoocytic cAMP levels**, and (ii) **cGMP made by follicle cells** is delivered to immature oocytes via **gap junctions** in order to block degradation of intraoocytic cAMP by **phosphodiesterase 3A (PDE3A)** (**Figure a**).

Conversely, at ovulation, **LH binding** to follicle cell receptors can trigger **a decrease in intraoocytic cAMP levels** by reducing intraoocytic cGMP so that PDE3A is able to degrade cAMP and the oocyte can begin to mature (**Figure b**). Such LH-triggered pathways involve activation of

epidermal growth factor receptors (**EGFRs**) on follicle cells by EGFR agonists like **amphiregulin** and eventually result in: (i) **downregulation of cGMP synthesis** in follicle cells; (ii) **upregulation of PDE-mediated cGMP degradation** in follicle cells; and (iii) **alterations in gap junctions** so that cGMP flows away from, rather than into, the maturing oocyte.

In response to the drop in intraoocytic cAMP, maturing oocytes undergo a decrease cAMP-dependent protein kinase (**PKA**) activation. PKA deactivation in turn allows **maturation-promoting factor** (**MPF**) (=CDK1/cyclin B) to become activated, thereby stimulating meiotic resumption, as evidenced by germinal vesicle breakdown (**GVBD**) and polar body formation as meiosis I is completed to produce a secondary oocyte. The maturing secondary oocyte and its associated cumulus cells are then ovulated into the oviduct to become a fertilizable metaphase-II-arrested egg.

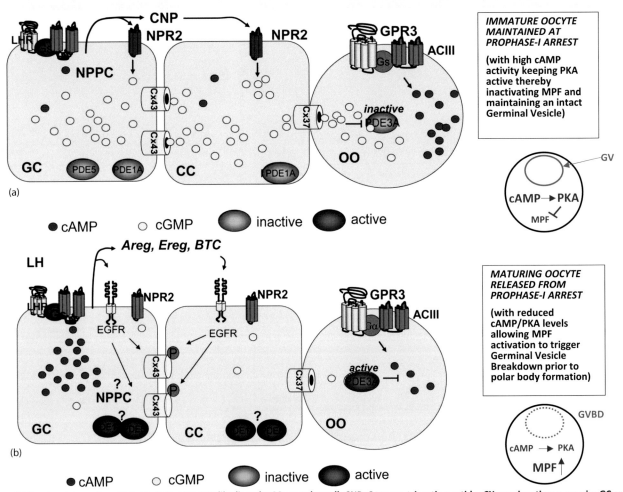

ACIII=adenylate cyclase III; Areg, Ereg, BTC=EGF-like ligands; CC=cumulus cell; CNP=C type natriuretic peptide; CX=gap junction connexin; GC granulosa cell; GPR3=G protein coupled receptor 3; GV= germinal vesicle; GVBD= germinal vesicle breakdown; MPF= maturation promoting factor; NPPC= natriuretic peptide precursor C; NPR2=natriuretic peptide receptor 2; OO=oocyte; PDE=phosphodiesterase; PKA= cAMP dependent protein kinase

Box 18.2 (a, b) *A model for the roles of cGMP, EGF-like ligands, gap junctions, and cAMP during LH-induced oocyte maturation. (Blue-green drawings on the left are from Conti, M et al. (2012) Novel signaling mechanisms in the ovary during oocyte maturation and ovulation. Mol Cell Endocrinol 356: 65–73, reproduced with publisher permission.)*

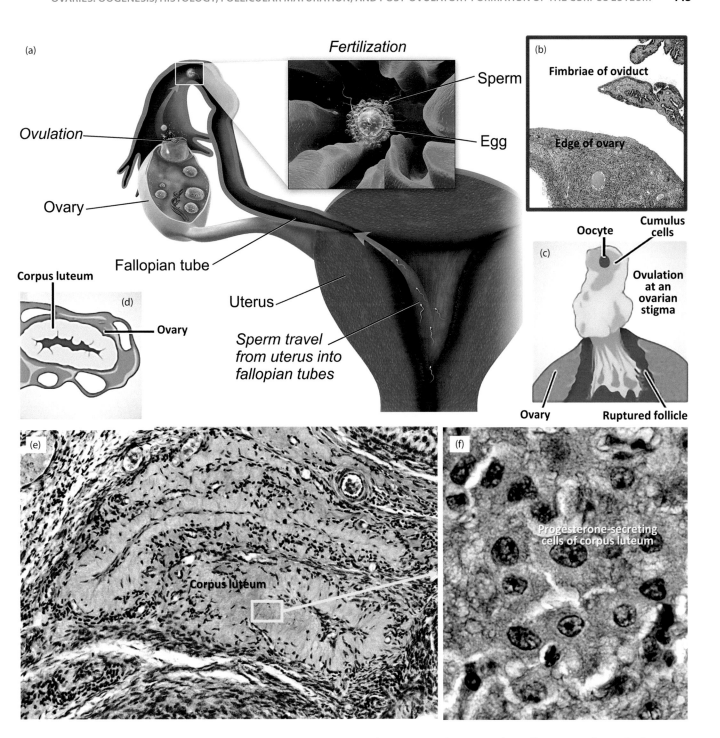

Figure 18.4 Ovulation. *(a–d) During ovulation, an oocyte and its surrounding cumulus cells rupture through the ovarian stigma region and are delivered via ciliary and muscular movements of oviductal fimbriae into the oviduct in preparation for fertilization. (b) 20X. (e, f) Intra-ovarian remnants of the ovulated follicle re-organize into a corpus luteum that secretes copious amounts of progesterone. (e) 200X; (f) 800X. ([a]: From Blausen.com staff (2014) Medical gallery of Blausen Medical 2014. Wiki J Med 1(2):10, doi: 10.15347/wjm/2014.01010, reproduced under CC0 and CC-SA creative commons licenses; [c, d]: From OpenStax College (2013) Anatomy and Physiology. OpenStax. http://cnx.org/content/col11496/latest, reproduced under CC-BY-4.0 creative commons license.)*

prepare the uterus for a possible pregnancy, but in the absence of a pregnancy, the corpus luteum degenerates into a whitish structure, called the **corpus albicans**, which is eventually resorbed as scar tissue.

18.3   OVIDUCTS: CILIATED CONDUITS FOR TRANSPORTING GAMETES AND EMBRYOS

Human **oviducts** are cylindrical organs that typically measure ~10–15 cm long (**Figure 18.5a**). At its distal opening near the ovary, each oviduct contains an **infundibular** region whose ciliated finger-like extensions (=**fimbriae**) help sweep ovulated oocytes into the oviduct (**Figure 18.5a**). More proximal to the infundibulum, the **ampullary** region connects with an **isthmus** that extends through the uterine wall as an **interstitial** portion to reach the uterine lumen.

The oviductal **ampulla** represents the site where fertilization normally occurs and is characterized by branched mucosal folds that fill much of the ampullary lumen. The core of each fold contains lamina propria connective tissue and is covered by a **simple columnar epithelium** that possesses **ciliated cells** (**Figure 18.5b–e**). Along with muscular contractions of the oviductal wall, ciliary beating helps mediate the transport of fluid, gametes, and embryos within oviducts. Scattered among ciliated oviductal cells are non-ciliated columnar **secretory cells** that exocytose various products, thereby helping to maintain oviductal viability as well as proper sperm functioning (**Figure 18.5d**). In addition, a few basally situated **peg cells** serve as stem cells for replacing lost columnar cells.

18.4   GENERAL HISTOLOGY OF THE UTERUS

The wall of the uterine body consists of: (i) an outer **perimetrium** with a thin serosal covering of mainly loose connective tissue, (ii) an intermediate **myometrium** containing multiple layers of smooth muscle fibers that contract in response to **oxytocin** during childbirth, and (iii) a complex inner **endometrium** which undergoes substantial reorganizations during menstrual cycles (**Figure 18.5f, g**). The endometrium comprises an inner **functionalis** layer, which is shed during each menstrual cycle, plus an outer **basalis** layer that is retained during menstruation for restoring lost functionalis components during the next cycle (**Figure 18.5f**).

In general, the endometrium is lined by a **simple columnar epithelium** with proliferative stem cells, ciliated cells, and secretory cells (**Figure 18.5f**). Subjacent to this surface lining are invaginated endometrial glands, blood vessels, and connective tissue, which collectively fill the stromal space between the luminal epithelium and myometrium (**Figure 18.5f, g**). During the pre-ovulatory **proliferative phase** of the menstrual cycle, the endometrium is relatively thin and has shallow unbranched glands with few ciliated cells (**Figure 18.5h**). Conversely, after ovulation, the thickened endometrium adopts a more glandular and ciliated configuration during a **secretory phase** that prepares the endometrium for a possible pregnancy (**Figure 18.5i, j**).

Between the body proper of the uterus and the vagina is a constricted connecting portion, called the uterine **cervix** (**Figure 18.6a–d**). The proximal cervix consists of a glandular **endocervix**, which is lined by a simple columnar epithelium and serves to produce copious amounts of **cervical mucus** for lubricating the vagina and preserving sperm viability (**Figure 18.6a–c**). More distally, the **ectocervix** is lined by a stratified squamous epithelium that connects with the vagina (**Figure 18.6a**). The border between these two regions comprises a **squamocolumnar junction (SCJ)** with high mitotic activity. By scraping the SCJ and nearby regions, cancerous cells can be identified based on abnormal staining and/or altered morphologies in diagnostic tests, referred to as **Pap smears** (**Case 18.2 Eva Perón's Uterus**).

As opposed to cervical cancers occurring within the uterus, **endometriosis** is the **ectopic growth** of endometrium-like tissues outside the uterus, with most of these abnormal growths developing on the outer surfaces of oviducts and ovaries. According to the **retrograde movement** theory, abnormal cellular translocations during menstruation can spread viable cells toward, rather than away from, the oviduct, thereby allowing entry into, and subsequent growth within, the pelvic cavity. Such growths typically trigger chronic pain and widespread inflammatory responses, which can contribute to reduced fertility.

18.5   MICROANATOMY OF THE VAGINA

The **vagina** is a 7–9 cm-long tubular organ that extends from the uterine cervix to the **vulval** region where external genitalia (labial folds, clitoris, and vestibular glands) surround the vaginal orifice (**Figure 18.6d–g**). Histologically, the vaginal wall comprises: (i) an inner **mucosa** with a non-keratinized **stratified squamous**

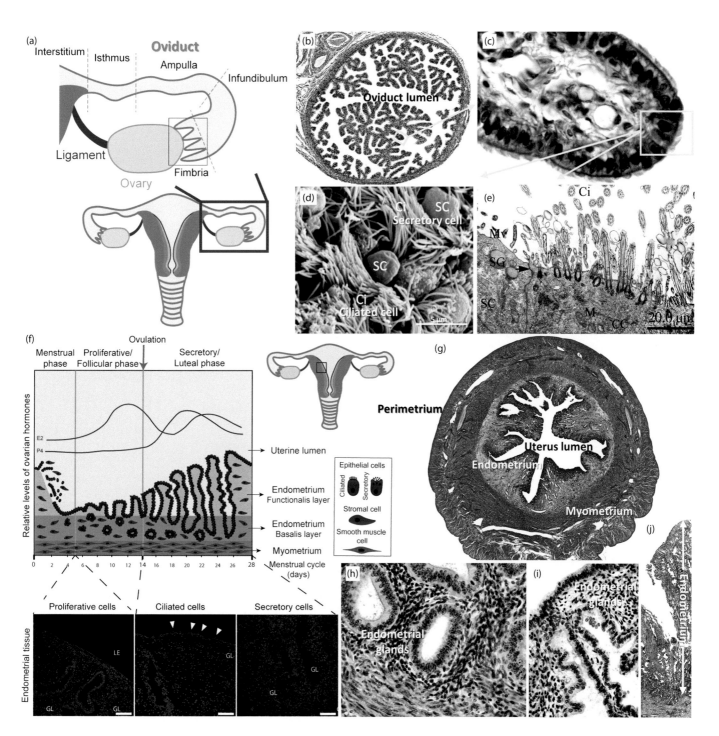

Figure 18.5 Oviductal and uterine microanatomy. *(a–e)* Oviducts extend from the uterine wall and are suspended by ligaments to lie near, but not physically connected to, ovaries. Each oviduct has several subregions, with the ampulla being the most common site for fertilization. The simple columnar epithelium lining the oviduct lumen has secretory cells interspersed among numerous ciliated cells that help move ovulated oocytes into the ampulla and transport post-fertilization embryos into the uterus. (b) 15X; (c) 250X. *(f–j)* The endometrium lining the uterine lumen contains proliferative stem cells, ciliated cells, and secretory cells plus invaginated epithelial glands that extend toward the surrounding myometrial and perimetrial layers of the uterine wall. The endometrium comprises: (i) a superficial stratum functionalis, which is shed during menstruation, and (ii) a stratum basalis that remains in the endometrium to re-constitute the functionalis layer after menstruation. The histological organization of the endometrium changes markedly from a relatively thin layer containing shallow, smooth glands during the pre-ovulatory proliferative phase (h) to a much thicker lining with deep, elaborate glands that can accommodate embryonic implantation. (g) 20X; (h) 150X; (i)

Figure 18.5 (Continued) 90X; (j) 30X. ([a, f]: From Alzamil, L et al. (2020) Organoid systems to study the human female reproductive tract and pregnancy. Cell Death Differ, https://doi.org/10.1038/s41418-020-0565-5, reproduced under a CC BY 4.0 creative commons license; [d]: From Aviles, M et al. (2015) The oviduct: a key organ for the success of early reproductive events. Animal Frontiers 5:25–31, reproduced with publisher permission; [e]: From Zhao, W et al. (2015) Levonorgestrel decreases cilia beat frequency of human fallopian tubes and rat oviducts without changing morphological structure. Clin Exper Pharmacol Physiol 42:171–178, reproduced with publisher permission.)

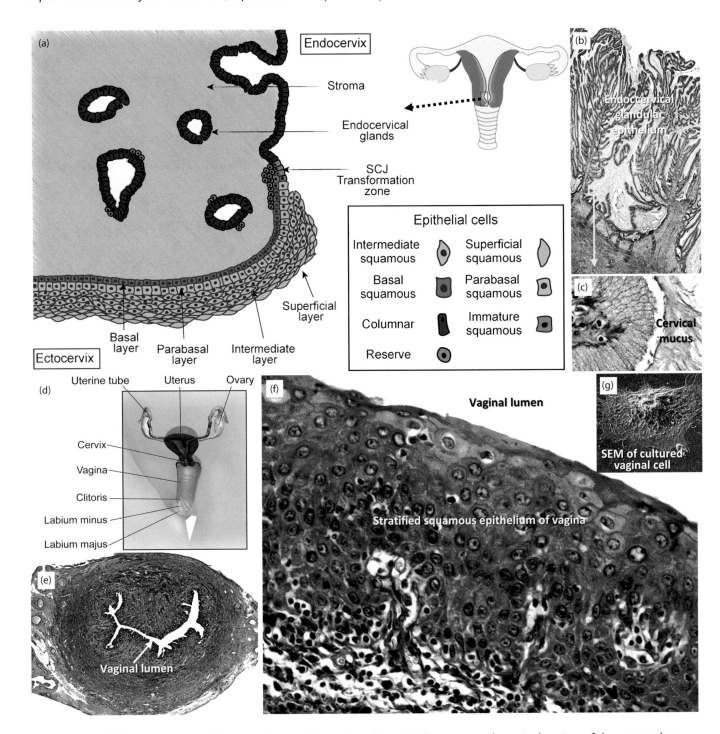

*Figure 18.6 **Microanatomy of the uterine cervix and vagina.*** (*a*) The narrowed cervical region of the uterus that connects with the vagina has an abrupt squamocolumnar junction (SCJ) between the glandular simple columnar epithelium of the endocervix and the non-glandular stratified squamous epithelium of the ectocervix. The SCJ is targeted for cell scrapings in Pap smears. (***b, c***) Cervical mucus produced by glandular cells can help lubricate the vagina and

Figure 18.6 (Continued) *maintain sperm viability following coitus. (b) 10X; (c) 240X. (**d–g**) The lumen of the vagina is lined by a stratified squamous epithelium and allows both sexual intercourse and childbirth. (e) 8X; (f) 325X. ([a]: From Alzamil, L et al. (2020) Organoid systems to study the human female reproductive tract and pregnancy. Cell Death Differ, https://doi.org/10.1038/s41418-020-0565-5, reproduced under a CC BY 4.0 creative commons license; [d]: From Blausen. com staff (2014) Medical gallery of Blausen Medical 2014. Wiki J Med 1(2):10, doi: 10.15347/wjm/2014.01010, reproduced under CC0 and CC-SA creative commons licenses; [g]: From Li, T (2016) Baofukang suppository promotes the repair of vaginal epithelial cells in response to Candida albicans. AMB Expr 6: 109, doi: 10.1186/s13568-016-0281, reproduced under a CC BY 4.0 creative commons license.)*

CASE 18.2 EVA PERÓN'S UTERUS

First Lady of Argentina
(1919–1952)

"It is our sad duty to inform the people of the Republic that Eva Perón, the Spiritual Leader of the Nation, died at 8:25 P.M."

(Translation of official announcement reporting the death of E.P., the First Lady of Argentina)

Eva Perón became First Lady Evita when her husband Juan Perón was elected President of Argentina in 1946. Five years later, Perón suffered a fainting spell and underwent an operation for what was reported to be an appendectomy but was actually surgery to deal with advanced cervical cancer. Near the end, Perón was greatly weakened by cancer, chemotherapy, and perhaps a lobotomy that she had been given in order to control her outbursts. Thus, for a public appearance during her husband's re-election campaign, Perón had to be held upright by a scaffold hidden under

her coat. After dying at age 33 in 1952, Perón's embalmed body was put on display until 1955, when Juan Perón hastily fled a military coup. Following the takeover, the new regime had Perón's corpse secretly buried in Italy in order to suppress a resurgence of Perónism. In 1971, her gravesite was revealed, and Perón's exhumed body was subsequently kept in the home of Juan Perón and his third wife in Spain. After Juan Perón regained the presidency in 1973 and died a year later, Perón's corpse was returned to Argentina and displayed next to his coffin before being entombed. Today, Perón is still revered by many, as tales of her untimely death due to uterine cancer, coupled with the bizarre saga of her well-traveled corpse, have elevated her lifestory to near mythical status.

epithelium, (ii) a middle **smooth muscle** component, and (iii) an outer **adventitial** covering that adheres to various tissues, including a striated sphincter near the external vaginal orifice. Throughout the vaginal wall, parasympathetic and sympathetic nerve fibers can help relay input during sexual intercourse and coordinate uterine contractions during childbirth. The vaginal mucosa forms large folds that accommodate stretching during coitus, and its stratified squamous epithelium lacks glands (**Figure 18.6e, f**). Thus, fluids within the vagina are derived mainly from cervical secretions plus products of vulval glands.

18.6 AN OVERVIEW OF THE ENDOCRINOLOGY AND HISTOLOGICAL REORGANIZATIONS OF MENSTRUAL CYCLES

For girls in the United States, the first menstrual cycle (=**menarche**) occurs at an average age of ~12–13 years old and signals the start of the reproductive lifespan. Menarche is triggered by increased body mass plus elevated levels of several closely related hormones, called **estrogen**, with the most common form in non-pregnant women being **estradiol** (**E2**).

From menarche on, the menstrual cycle is regulated by an integrated system of endocrine organs and their hormone products (**Figure 18.7a**). Such organs consist of the: (i) **hypothalamus**, (ii) **anterior pituitary**, (iii) **ovarian follicle** and **corpus luteum**, and, if a pregnancy occurs, (iv) the **chorion**, which in turn produce, respectively: (i) **gonadotropin-releasing hormones** (**GnRHs**), (ii) the **gonadotropins FSH** and **LH**, (iii) three types of ovarian hormones—the steroids **estrogen** and **progesterone** plus **inhibin** proteins, and (iv) **human chorionic gonadotropin** (**hCG**). Interactions among these hormones drive progression through the cycle, which is typically viewed as comprising two roughly equal halves, called the **follicular** and **luteal** phases.

The follicular phase at the onset of each cycle begins with a several-day-long **menstruation** ("the period"), which discharges uterine blood plus sloughed tissues, collectively termed **menses** (**Figure 18.7b**). Menstruation and the remainder of the follicular phase last about two weeks and involve the dramatic growth of a maturing

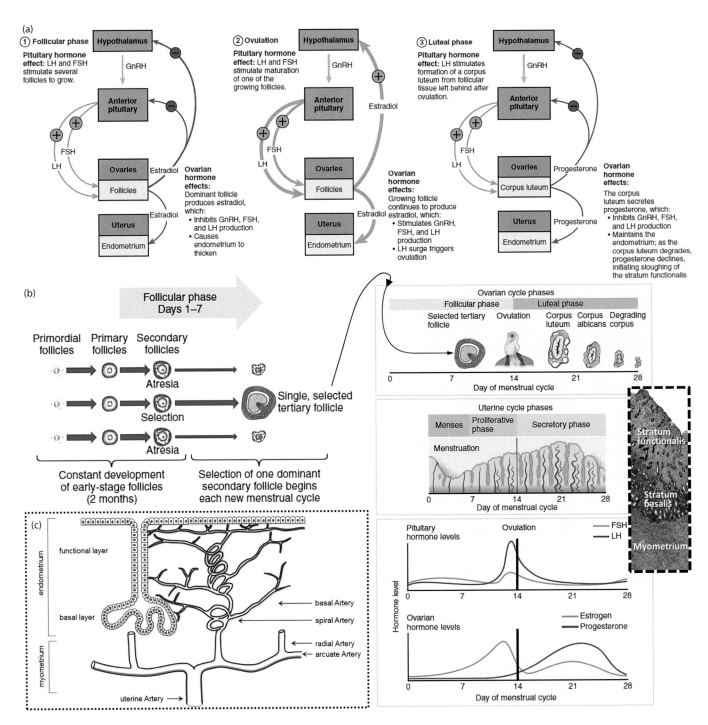

Figure 18.7 The menstrual cycle. (a, b) *Day 0 of the menstrual cycle is marked by the onset of menstruation, as the endometrial functionalis layer begins to slough. During the first half of the cycle (=follicular phase), increased GnRH production from the hypothalamus causes the anterior pituitary to secrete the gonadotropins, FSH and LH, which in turn trigger the completion of folliculogenesis by a dominant follicle within the ovary. Follicle cells of the maturing follicle secrete progesterone and the estradiol (E2) form of estrogen. Initially, estradiol feeds back negatively on the hypothalamus/anterior pituitary to reduce gonadotropin release, but after surpassing a critical threshold level in the blood, estrogen forms a positive feedback loop, thereby generating a rapid and marked rise in LH. The midcycle LH surge causes oocyte maturation and ovulation, leaving behind a follicular remnant that reorganizes into a corpus luteum. During the second half of the menstrual cycle (=luteal phase), the corpus luteum secretes progesterone to: (i) block further GnRH/LH/FSH release, and (ii) maintain the endometrium in a receptive state for possible embryonic implantation.*

Figure 18.7 (Continued) Without a pregnancy, the corpus luteum is programmed to degenerate, thereby reducing progesterone levels so that GnRH/LH/FSH levels rise again, and a new cycle is initiated. (*c*) As a key driver of menstruation, reduced progesterone levels constrict spiral arteries in the stratum functionalis. The resulting ischemia combined with protease secretions leads to endometrial sloughing. ([a, b]: From OpenStax College (2013) Anatomy and Physiology. OpenStax. http://cnx.org/content/col11496/latest, reproduced under CC BY 4.0 creative commons license; [c]: From Mizutani, S (2020) New insights into human endometrial aminopeptidases in both implantation and menstruation. BBA-Proteins Proteomics 1868, https://doi.org/10.1016/j.bbapap.2019.140332, reproduced with publisher permission.)

follicle as it completes **folliculogenesis**. At the end of the follicular phase, the mature follicle undergoes a **midcycle ovulation**, and the cycle transitions into the **luteal phase**, where post-ovulatory follicular remnants reorganize into a **corpus luteum** (**Figure 18.7b**). Various attributes of the female reproductive tract change during each menstrual cycle. However, the following summary focuses on the uterine **endometrium**, ovarian **follicle**, and **corpus luteum** during a menstrual cycle that does not lead to a pregnancy, before also outlining what occurs after a fertilization.

Toward the end of the previous menstrual cycle, decreased progesterone levels trigger constriction of coiled **spiral arteries** that extend perpendicularly into the endometrial functionalis layer (**Figure 18.7c**). Along with ischemia triggered by these constrictions, secreted **matrix metalloproteinases** (**MMPs**) cause the apical stratum functionalis to break away from the underlying stratum basalis, thereby causing the bleeding and tissue sloughing of menstruation at the onset of a new cycle. The drop in progesterone that triggered menstruation also relieves an inhibition of gonadotropin production that was established during the previous cycle. Thus, in response to a new supply of **GnRHs** from the hypothalamus, the **anterior pituitary** releases increasing amounts of **FSH** and **LH** to start another follicular phase (**Figure 18.7a, b**).

In particular, increased FSH production typically causes just one **dominant** follicle to complete the process of follicular maturation, while other stimulated follicles undergo atresia (**Figures 18.2d** and **18.7b**). In response to FSH binding its receptors on **granulosa cells**, the dominant follicle: (i) expands its antrum, (ii) adds new granulosa cells that produce **estrogen**, and (iii) produces **inhibin B** that feeds back on the anterior pituitary to downregulate FSH secretion. Along with FSH-mediated stimulation, **LH** also increases **estrogen** and **progesterone** production by follicle cells.

Toward the end of the follicular phase, circulating blood levels of estrogen rise substantially, and after surpassing a critical threshold, estrogen switches from its normally negative mode of gonadotropin regulation to a stimulatory effect on the hypothalamus and anterior pituitary. The positive feedback loop generated by this switch leads to a **midcycle LH surge** that typically peaks ~12 hours before ovulation and triggers: (1) **meiotic resumption** in the primary oocyte involving decreases in intraoocytic **cGMP** and **cAMP** levels that drive the completion of meiosis I (Box 18.2); (2) **cumulus expansion** as cumulus cells secrete gel-like **mucins** to become more spread apart, thereby facilitating post-ovulatory transport and fertilization; and (3) **ovulation** of the secondary oocyte plus surrounding cumulus cells, initially from the follicle and then, via protease-mediated rupturing of the ovarian stigma, from the ovary, as the **cumulus-oocyte complex** enters the oviduct.

Once ovulation is achieved, the **luteal phase** begins, and follicular remnants become reorganized into the **corpus luteum** (**Figure 18.7b**). With increased vascularization in the surrounding connective tissue compartment, luteal cells secrete into the bloodstream **inhibins A** and **B, estrogen**, as well as copious amounts of **progesterone**. Luteal progesterone helps downregulate gonadotropin production by the anterior pituitary, thereby preventing another ovulation from occurring while an ovulated egg is already positioned in the oviduct for possible fertilization. Such effects in turn are the mechanisms by which progesterone-containing **oral contraceptive pills** can provide birth control.

In addition to blocking ovulation, high progesterone levels during the luteal phase keep the endometrium in a well-vascularized state as part of a **spontaneous decidualization** process (Box 18.1), which prepares the endometrium for a pregnancy before embryonic implantation occurs. During decidualization, **decidual stromal cells** (**DSCs**) in the endometrial stroma both remodel endometrial tissues and help regulate inflammatory responses so that an implanted embryo can survive while also not overwhelming maternal tissues (Box 18.1).

In a menstrual cycle that **does not result in a pregnancy**, the corpus luteum degenerates via **apoptosis** about two weeks after it was first formed. This degeneration triggers a drop in circulating progesterone levels that reverses the inhibition of gonadotropin secretion and allows a new menstrual cycle to be initiated. However, if fertilization occurs during the luteal phase and the developing embryo implants in the endometrium, **chorionic tissues** (see next section) form part of the developing **placenta** and secrete into the bloodstream **human chorionic gonadotropin** (**hCG**), which is the hormone detected in over-the-counter pregnancy tests. Circulating hCG prevents degeneration of the corpus luteum so that it can continue to produce progesterone. This in turn maintains ovulation inhibition and blocks menstruation so that a "period is missed" as pregnancy begins.

18.7 EARLY STAGES OF A PREGNANCY

About a day after fertilization, the zygote undergoes its first mitotic division (=**cleavage**), and the cleaving embryo is then moved by ciliary beating and muscular contractions toward the uterus over the next few days. During transport, the embryo normally remains encased within its zona pellucida, which helps to minimize ectopic implantation in the oviduct and thereby prevent a **tubal pregnancy** that in some cases can lead to serious health risks. After reaching the uterus and forming a large internal cavity, the embryo becomes a **blastocyst**, which **hatches** from the zona pellucida ~4–6 days post-fertilization and **implants** in the endometrium ~6–7 days post-fertilization (**Figure 18.8a**).

During the week following implantation, the thin outer trophoblast cells of the implanted blastocyst divide to form the **chorionic** contribution to a disc-like **placenta** (**Case 1.3 Mumtaz Mahal's Placentas**). Initially, during this process of **placentation**, the chorion forms a syncytial mass of nuclei that are not separated by cell membranes (=**syncytiotrophoblast**) before generating a cellularized **cytotrophoblast** (**Figure 18.8b**). Eventually, these two trophoblastic derivatives form branched **chorionic villi** containing fetal blood vessels (**Figure 18.8c–f**), thereby increasing surface area to facilitate nutrient, gas, and waste exchanges between the maternal and fetal circulatory systems.

Throughout gestation, fetal and maternal blood supplies remain separate from each other, thereby maintaining an anatomical barrier that helps prevent a potential rejection of the fetus. The distinct nature of these two blood supplies is particularly evident during the first half of a pregnancy, given that maternal vessels contain enucleated erythrocytes, whereas fetal vessels exhibit nucleated red blood cells, which become much less common by birth (**Figure 18.8e, f**) and then essentially disappear in early post-natal life.

During gestation, the fetus grows away from the placenta but remains connected to it by an **umbilical cord** (**Figure 18.8g–i**) that contains large blood vessels surrounded by an unusual type of loose embryonic connective tissue, called **Wharton's jelly** (**Figure 18.8i**). After birth, the umbilical cord is severed and its healed detachment site becomes the umbilicus (=belly button).

18.8 MAMMARY GLANDS AND THE BIOLOGY OF LACTATION

As a defining feature of the class Mammalia, **mammary glands** produce **milk** during a **lactation** process that nourishes newborns. In humans, a mammary gland develops on each side of the upper thorax and drains externally through a raised **nipple** in highly pigmented skin, called the **areola** (**Figure 18.9a**). At puberty, additional connective tissues that are enriched in **fat deposits** begin to enlarge the breasts as their **compound tubuloalveolar** mammary glands undergo further development (**Case 18.4 Yao Beina's Mammary Glands**). The branched duct system of each gland connects with multiple **lobes** containing numerous secretory alveoli. Such alveoli in turn constitute functional subunits (=**lobules**) within each lobe (**Figure 18.9a**), with **myoepithelial cells** occurring at the periphery of intralobular alveoli.

In non-pregnant women, inactive mammary glands lack stored milk and contain relatively squamous secretory cells (**Figure 18.9b**). However, during a pregnancy, elevated ovarian hormones in conjunction with **prolactin** from the anterior pituitary upregulate glandular mitoses and branching, thereby increasing mammary gland size and activity. In lactating mammary glands, secretions from cuboidal to columnar glandular cells fill the

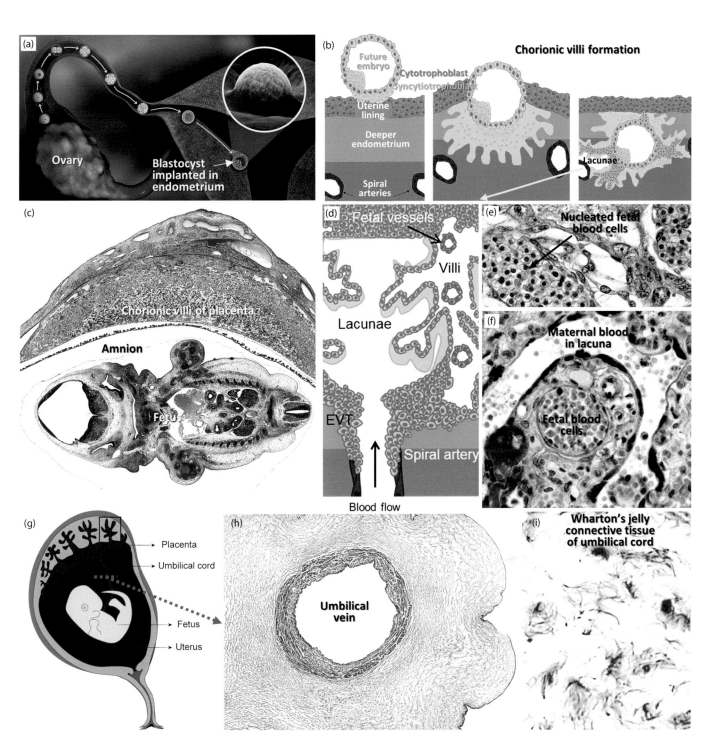

Figure 18.8 Implantation and development. *(**a**) After fertilization in the oviduct, the dividing embryo becomes a blastocyst which can implant in the endometrial lining of the uterus. (**b–f**) The outermost layer of the implanted blastocyst forms syncytiotrophoblastic and cytotrophoblastic components of a chorion that generates numerous villi and constitutes the fetal contribution to the placenta. Fetal blood cells within developing blood vessels of chorionic villi are initially nucleated before gradually becoming enucleated before birth. Throughout gestation, the fetal circulation remains separate from maternal blood located in circum-villar lacunae. (c) 25X; (e) 350X; (f) 280X. (**g–i**) With further development, the growing fetus extends into the uterine lumen but remains connected to the placenta via an umbilical cord that is filled with an unusual type of connective tissue, called Wharton's jelly. (h) 70X; (i) 350X. ([a]: Downloaded from http://www.scientificanimations.com/wiki-images and reproduced under a CC BY SA 4.0 creative commons license; [b, d]: From Apicella, C et al. (2019) The role of epigenetics in placental development and the etiology of preeclampsia. Int J Mol*

Figure 18.8 (Continued) Sci 20, doi: 10.3390/ijms20112837, reproduced under a CC BY 4.0 creative commons license; [g]: From Alzamil, L et al. (2020) Organoid systems to study the human female reproductive tract and pregnancy. Cell Death Differ, https://doi.org/10.1038/s41418-020-0565-5, reproduced under a CC BY 4.0 creative commons license.)

CASE 18.3 MUMTAZ MAHAL'S PLACENTAS

Mughal Empress Entombed in the Taj Mahal
  (1593–1631)

"...the proud passion of an emperor's love wrought in living stones."

(The poet E. Arnold on the majestic Taj Mahal built by the Mughal Emperor to commemorate his wife)

The woman known today as Mumtaz Mahal was born in what is now Agra, India. While growing up, Mumtaz Majal attracted numerous suitors before being betrothed to Shah Jahan, whose father ruled the mighty Mughal Empire encompassing much of present-day India, Kashmir, and Pakistan. When Shah Jahan became Emperor, Mumtaz Mahal became his Chief Empress and began taking an active role in advising her husband as he reigned over his vast empire. Mumtaz Mahal also proceeded to have 14 of Shah Jahan's children over a span

of less than 20 years. During her 14th pregnancy, Mumtaz Mahal was accompanying her husband on a military campaign, and while giving birth to her last child, Mumtaz Mahal died, most likely due to placental hemorrhaging. Following her death, Shah Jahan mourned for a year before ordering the construction of a great mausoleum for Mumtaz Mahal in Agra, and after 22 years of construction, the Taj Mahal was completed. Today, the building ranks as one of the seven wonders of the modern world while also acting as the final resting place for Mumtaz Mahal, who generated numerous placentas and offspring during her short life.

lumen of alveoli and ducts (**Figure 18.9c**), and during nursing, suckling action stimulates **oxytocin** release from the posterior pituitary, thereby triggering **myoepithelial cell contractions** that help move secreted milk toward the nipple during the **milk let-down** response.

The initial product that is released during the first week of nursing is a relatively viscous and fat-enriched fluid, called **colostrum**. Milk generated after colostrum production comprises a more aqueous mixture of various beneficial components, such as **IgA** that can be exocytosed via an **apocrine** mode of secretion. In the absence of breastfeeding or after its termination, withdrawal of physical and hormonal stimuli serves to end milk production as the mammary gland returns to its non-lactating state.

18.9 SOME DISORDERS OF THE FEMALE REPRODUCTIVE SYSTEM: BREAST CANCERS AND POLYCYSTIC OVARY SYNDROME

As the second most common cancer of women behind only skin cancer, **breast cancer** develops in about 1 of 8 U.S. women over the course of their lives. Such cancers comprise a heterogeneous collection of subtypes that can be distinguished based on their sites of origin. Luminal cancers tend to arise more apically within mammary glands and thus closer to glandular lumens, whereas basal-like cancers originate more basally toward enveloping myoepithelial cells. In general, **luminal cancers** tend to be slower-growing neoplasias that respond well to various therapies, thereby yielding **relatively high cure rates**. Conversely, **basal-like cancers** are often **aggressive (Figure 18.9d, e)** and can alter blood vascular and lymphatic networks making them more difficult to treat with currently available options.

About 5–10% of reproductive-age women suffer from **polycystic ovary syndrome (PCOS)**, which is characterized by elevated levels of male hormones (**androgens**), infrequent menstrual cycles, and reduced fertility, often in association with fluid-filled ovarian structures, called **cysts**. For women with PCOS, therapies may involve: (i) **birth control pills** for those not seeking to become pregnant; (ii) fertility-enhancing **treatments to boost FSH levels**, thereby facilitating ovulations and potential pregnancies; and (iii) **anti-androgen medications** to restore normal androgen levels and combat side effects, such as hirsutism and excess acne.

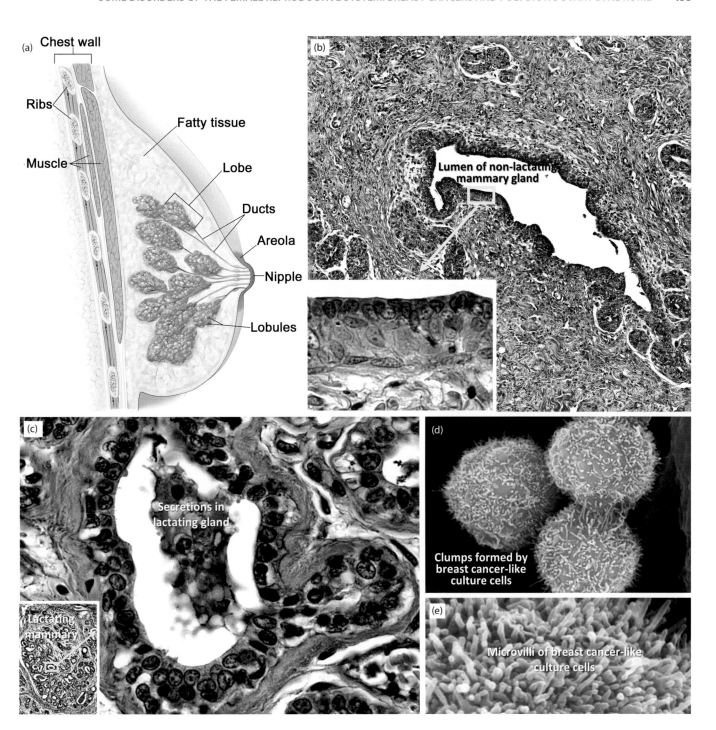

Figure 18.9 Breast microanatomy. (*a*) *Beneath their covering skin, breasts contain a well-developed hypodermis with abundant adipose tissue plus compound tubuloalveolar mammary glands whose ducts reach the skin surface at a protruded nipple. (**b**, **c**) Compared to non-lactating versions, mammary glands in nursing women have more densely packed glandular lobules with milk secretions in their lumens. (b) 100X; (b, inset) 500X; (c) 200X. (**d**, **e**) As investigated in cultured cell lines that model a type of inflammatory breast cancer (IBC), aggressive forms of breast cancer cells can often clump together and increase tissue vascularization in order to supply growing tumors with nutrients and oxygen needed for rapid growth. (d) 3,600X; (e) 8,800X. ([a]: Downloaded from https://nci-media.cancer.gov/pdq/media/images/415520. jpg and reproduced as a public domain illustration generated by the NIH National Cancer Institute of the federal government; [d, e]: From Barreno, L et al. (2019) Vasculogenic mimicry-associated ultrastructural findings in human and canine inflammatory breast cancer cell lines. BMC Cancer 19 (750), https://doi.org/10.1186/s12885-019-5955-z, reproduced under a CC BY 4.0 creative commons license.)*

CASE 18.4 YAO BEINA'S MAMMARY GLANDS

Singer and Advocate for Breast Cancer Awareness
(1981–2015)

"I hope I can bring strength to people and boost their confidence with my songs."

(Y.B. on her using her songs to help people who are affected by breast cancer)

Yao Beina was a Chinese pop singer, songwriter, and actress who began a rise to stardom in 2008, when she won her nation's Young Singer's Championship. In addition to giving concerts and making critically acclaimed albums, Yao recorded numerous songs for highly popular TV shows and films as she gained celebrity status throughout her country. In 2011, Yao was diagnosed with breast cancer and underwent a mastectomy as part of her treatment. While apparently achieving remission of the disease, Yao helped with a campaign to spread awareness about the increasing rates of breast cancers in China, which were rising by more than 3% per year around the time of Yao's diagnosis. Although Yao continued her outreach efforts, the neoplasia that had begun in her mammary glands eventually metastasized to her lungs and brain, thereby causing her death at the age of 33.

18.10 SUMMARY—FEMALE REPRODUCTIVE SYSTEM

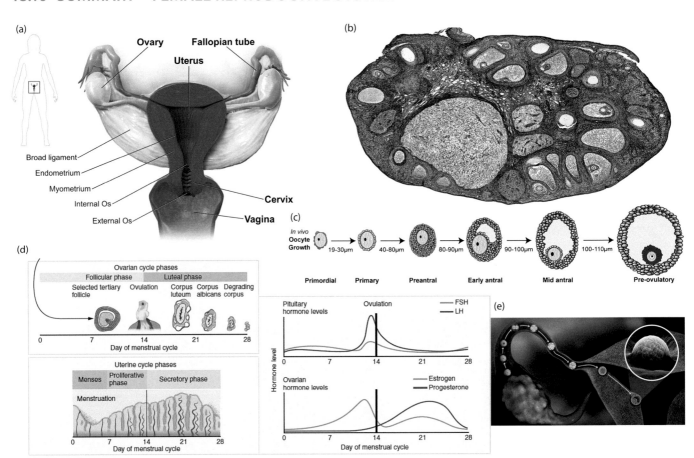

Pictorial Summary Figure Caption: (**a**): see figure 18.1a; (**b**): see figure 18.2a; (**c**): see figure 18.2b; (**d**): see figure 18.7b; (**e**): see Figure 18.8a

The female reproductive system includes: (i) two **ovaries**, each with an inner **medulla** plus an outer **cortex** that contains numerous **follicles**; (ii) two **oviducts** with a **simple columnar ciliated** type of epithelium lining their mucosae; (iii) a **uterus** containing a dynamic **endometrial** lining that changes during **menstrual cycles** plus a narrow **cervical** region that is assayed for cancer in **Pap tests**; (iv) a **vagina** with a **stratified squamous** epithelial lining; and (v) **mammary glands** that nourish newborns (**Figure a**).

Within the cortex of the **ovary**, **oogenesis** occurs in **follicles**, each of which comprises an oocyte surrounded by follicle cells (**Figure b, c**). When mitotically dividing **oogonia** commit to undergoing **meiosis**, they become **primary oocytes**. Under hormonal stimulation, the primary oocyte completes its first meiotic division to become a **secondary oocyte** plus an associated **polar body** cell. **Ovulation** sends a secondary oocyte with its polar body and follicle cells into the oviduct to become a fertilizable **egg**, leaving behind a residual follicle that forms a **corpus luteum**. The **egg** can be fertilized in the oviduct to complete the second meiotic division and become a diploid **zygote**. Peak follicle numbers occur in the fetal ovary before apoptotic degradation (=**atresia**) reduces follicular levels prior to birth and continues such reductions after birth. **Follicles** can progress through several stages: (i) **primordial follicle** = primary oocyte plus a single layer of squamous granulosa cells; (ii) **primary follicle** = primary oocyte plus a single layer of cuboidal granulosa cells; (iii) **secondary follicle** = primary oocyte plus multiple granulosa layers and peripheral **thecal** cells; and (iv) **tertiary follicle** = a follicle with a fluid-filled cavity (**antrum**), and in **pre-ovulatory tertiary follicles**, the primary oocyte is surrounded by **cumulus** cells (**Figure c**).

Menstrual cycles occur from **menarche** (first cycle at puberty) to **menopause** (end of cycling at middle age) allowing an ovulation each month. The **organs** [*and hormones they produce*] that regulate the cycle are: **hypothalamus** [*GnRHs*], **anterior pituitary** [*FSH and LH*], and ovarian **follicle** [*estrogen and progesterone*] (**Figure d**). The **follicular phase** during the first half of the cycle begins with **menstruation** (i.e., bleeding during the menstrual period as the uterine **endometrium is sloughed** due to low progesterone levels). During menstruation, FSH/LH levels begin to rise causing follicle growth and increasing **estrogen** production. At suprathreshold levels, estrogen generates **positive feedback** to cause a **midcycle LH surge**, which triggers (1) **oocyte maturation** as the primary oocyte resumes meiosis and begins to become a fertilizable egg; (2) **cumulus expansion** via **mucin** secretion; and (3) **ovulation** via **protease** secretion at the ovarian **stigma**. After ovulation occurs at the end of the follicular phase, the intraovarian follicular remnant is re-organized into a **corpus luteum** during the **luteal phase** that represents the last half of the cycle. The corpus luteum secretes progesterone to: (1) **block ovulation** (which is how birth control pills serve as contraceptives), and (2) **maintain the endometrium** for embryonic implantation. Without fertilization, the **corpus luteum degenerates**, reducing progesterone levels to trigger the next period.

If the **ovulated egg** is fertilized in the oviduct, a dividing embryo can be transported to the uterus to hatch as a **blastocyst** and **implant** in the endometrium. **Trophoblast** cells of the implanted embryo form the fetal contribution to the **placenta**, called the **chorion**, which produces **human chorionic gonadotropin (hCG)**, the hormone detected in pregnancy tests. hCG prevents corpus luteum degeneration, keeping progesterone levels high and the endometrium intact so that a "period is missed". During pregnancy, the **placenta** facilitates the exchange of nutrients, gases, and wastes while keeping fetal blood separate from maternal blood. The growing fetus remains connected to placenta by an **umbilical cord** that contains **Wharton's jelly** connective tissue. As gestation proceeds, maternal **mammary glands** produce milk in response to **prolactin** and following birth release milk for nursing via **myoepithelial cell** contractions induced by **oxytocin**.

Approximately 1 in 8 U.S. women will develop **breast cancer**. Other common disorders include (1) **endometriosis**: ectopic growth of endometrial tissues outside the uterus; and (2) **polycystic ovary syndrome**: abnormal ovulations and ovarian cysts associated with elevated male hormone levels.

SELF-STUDY QUESTIONS

1. What is the name of the small daughter cell that occurs on a secondary oocyte?
2. T/F Each primordial follicle contains an oocyte surrounded by a single layer of squamous follicle cells.
3. What do the remnants of a post-ovulatory follicle that are left over in the ovary become?
4. What is the name of the envelope surrounding an unhatched embryo?
   A. Zona pellucida
   B. Antrum
   C. Graafian
   D. Germinal vesicle
   E. Polar body
5. When do peak numbers of oocytes occur?
   A. Before birth
   B. Directly after birth
   C. At the onset of puberty
   D. At peak reproductive age (~20–30 years old)
   E. At menopause onset

6. Where does fertilization in humans normally occur?
   A. Ovary
   B. Oviduct
   C. Uterus
   D. Vagina
   E. Abdominal cavity
7. What is the name of the narrow region of the uterus that connects to the vagina?
8. T/F Cells surrounding ovulated oocytes are called cumulus cells.
9. Which disease is characterized by abnormally high levels of male sex hormones (androgens)?
   A. Endometriosis
   B. Polycystic ovary syndrome
10. In which region of the ovary are follicles found?
11. Which of the following statements regarding menstrual cycles is correct?
    A. They affect the histology of just the endometrium
    B. They last on average about 28 days in humans
    C. Humans are the only mammals with menstrual cycles
    D. ALL of the above
    E. NONE of the above
12. Which of the following would occur at the onset of the follicular phase in a menstrual cycle?
    A. Menstruation
    B. Formation of a corpus luteum
    C. Secretion of relatively high amounts of progesterone
    D. ALL of the above
    E. NONE of the above
13. T/F Pap tests analyze cervical scrapings.
    "EXTRA CREDIT" For each term on the left, provide the BEST MATCH on the right (answers A-F can be used more than once).

14. Fimbriae _____	A.	Hypothalamus
15. FSH/LH production _____	B.	Umbilical cord
16. Stigma _____	C.	Anterior Pituitary
17. Wharton's jelly _____	D.	Ovary
18. GnRH production _____	E.	Oviduct
	F.	Endometrium

19. Describe in proper order the stages and distinguishing features of follicle maturation.
20. Describe, in correct order, the stages, regulatory hormones, and major changes that occur in the ovary and endometrium during a normal menstrual cycle that does result in a pregnancy.

ANSWERS

(1) polar body; (2) T; (3) corpus luteum; (4) A; (5) A; (6) B; (7) cervix; (8) T; (9) B; (10) cortex; (11) B; (12) A; (13) T; (14) E; (15) C; (16) D; (17) B; (18) A; (19) Primordial: single layer squamous follicle cells; Primary: single layer of cuboidal fcs; Secondary: multiple layers of fcs; Tertiary: multiple fc layers and fluid-filled antrum; (20) A full answer would include and define most or all the following terms: hypothalamus, anterior pituitary, ovary, endometrium; GnRHs, gonadotropins (FSH, LH); estrogen, progesterone, inhibin; menstruation, follicular phase, midcycle ovulation; luteal phase; corpus luteum; cumulus cells, secondary oocyte, stigma

Eyes

19.1   INTRODUCTION TO SENSORY ORGANS

Humans can interact with various parts of their environment based on input received from scattered **sensory organs**, such as skin-based units and specialized regions of the **nasal** and **oral** cavities. In addition, unlike sensors occurring in organs that also play non-sensory roles, **eyes** and **ears** are devoted exclusively to sensory functions and typically serve as the primary means by which environmental cues are perceived. Compared to many other animals, humans possess inferior visual or auditory capabilities. Nevertheless, human eyes and ears have allowed wide-ranging radiations into various niches, owing in large part to key microanatomical adaptations that these organs have evolved. The development and histology of eyes are summarized below, whereas similar topics related to ears are covered in the following chapter.

19.2   DEVELOPMENT OF THE EYE

Human eyes begin to develop during the fourth week of gestation in the form of **optic vesicles** that evaginate on the left and right sides of the forebrain beneath lateral ectodermal regions, called **lens placodes** (**Figure 19.1a–c**). Each optic vesicle initially signals a lens placode to invaginate and produce a **lens pit**, which later detaches from the surface to become a **lens vesicle** for generating the lens (**Figure 19.1a–c**). Conversely, the developing lens causes the optic vesicle to become a stalked **optic cup** that will form the **retina** and part of the **optic nerve** (**Figure 19.1a–c**). By the tenth week of development, major eye components, such as the **retina, optic nerve**, and **iris** are established from neuroectoderm, whereas the **lens, cornea**, and associated structures, such as the **conjunctiva** arise from surface ectoderm. Additional eye components (e.g., **ciliary muscles, choroid**, and parts of the **vitreous body**) develop from mesodermal and neural crest precursors in head mesenchyme.

DOI: 10.1201/9780429353307-23

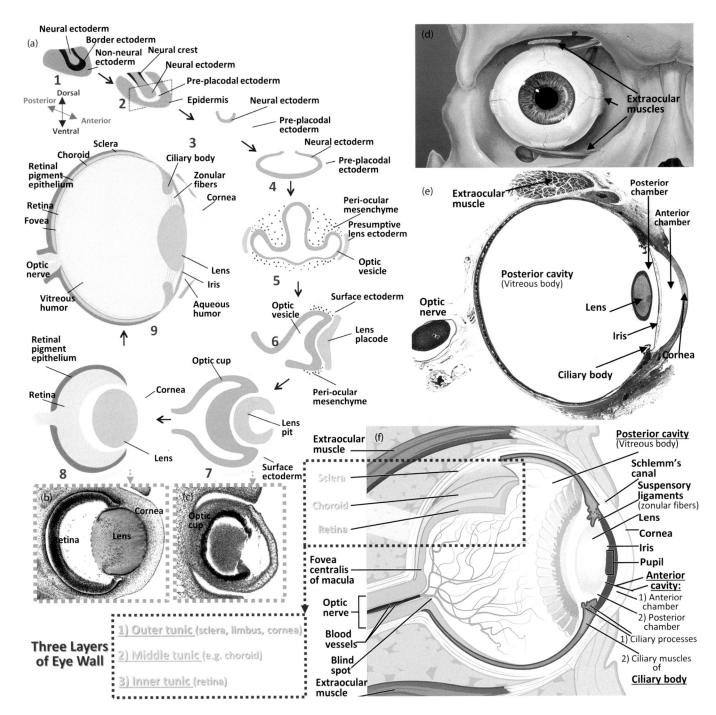

Figure 19.1 Introduction to eye development and anatomy. *(a–c) The eye develops via reciprocal inductions occurring between an ectodermal lens placode and an optic vesicle derived from a forebrain outgrowth. The optic vesicle eventually forms the photoreceptive retina of the eye and connects via an optic nerve to the brain. (b) 40X; (c) 50X. (d, e) When fully developed, the eye can be translocated by three pairs of extraocular muscles in order to keep vision properly centered on regions of interest. (e) 5X. (f) The ocular wall comprises three concentrically arranged layers: (1) an outer tunic (lens, limbus, and cornea); (2) a middle tunic (e.g., choroid); and (3) an inner tunic (retina). Other key eye components include: (i) the lens, (ii) iris with its central pupil, (iii) ciliary body, (iv) zonular fibers that attach to the lens, (v) macula region of the retina, (vi) anterior cavity with anterior and posterior chambers containing aqueous humor, and (vii) posterior cavity filled with a gel-like vitreous body. ([a]: From Dash, S et al. (2016) RNA binding proteins in eye development and disease: implication of conserved RNA granule components. Wiley Interdiscip Rev RNA 7(4):527–557, reproduced with publisher permission; [d]: Downloaded from https://commons.wikimedia.org/wiki/File:Eye_orbit_anatomy_anterior2.jpg; image by Patrick J. Lynch, medical illustrator and reproduced under a CC BY 2.5 creative common license; [f]: From OpenStax College (2013) Anatomy and Physiology. OpenStax. http://cnx.org/content/col11496/latest, reproduced under a CC BY 4.0 creative commons license.)*

19.3 EYE ANATOMY: THE LENS AND THREE OCULAR TUNICS (OUTER, MIDDLE, AND INNER) PLUS THE VITREOUS BODY AND EYELIDS

When fully grown, the adult eye averages ~25 mm in diameter and is connected to three pairs of extraocular muscles that control eye movements (**Figure 19.1d–f**). The anterior part of the eye contains a **cornea** and underlying **lens** that serve to focus light on photoreceptors within the retina (**Figure 19.1d, e**). Posterior to the lens, the ocular wall comprises three concentric layers (=**tunics**) (**Figure 19.1f**). The **outer tunic** consists of an opaque capsule, called the **sclera**, which anterior to the lens connects with a transitional **limbus** region before becoming the clear cornea at the anterior-most eye surface (**Figure 19.1f**). The **middle tunic** (=**uvea**) comprises: (i) a **choroid** layer occurring directly beneath the sclera; (ii) an anterior **ciliary body** encircling the lens; and (iii) an anterior-most annular **iris** that is arranged around a central hole, called the **pupil** (**Figure 19.1f**). Interior to the uvea, the **inner tunic** consists of the **retina** that connects posteriorly to the **optic nerve**. Enveloped by these tunics, the posterior cavity of the eye that is situated behind the lens is filled with a gel-like **vitreous body**, whereas the bipartite anterior cavity occurring in front of the vitreous body comprises a thin fluid-filled **posterior chamber** behind the iris, which in turn connects via the pupil to a larger **anterior chamber** between the iris and cornea (**Figure 19.1e, f**).

The **sclera** is an ~0.7 mm-thick fibrous sheath that is visible anteriorly as the "white" of each eye. Most of the sclera is composed of densely packed **type I collagen fibers** that attach to tendons of extraocular muscles used for eye movements (**Figure 19.1d, e**). Such muscle-mediated translocations allow rapid tracking of objects while keeping the central region of the retina with its optimal resolving capabilities oriented toward the area of interest.

As the narrow transition zone between the opaque sclera and clear cornea, the **limbus** maintains the cornea and provides an outflow pathway for anterior chamber fluid. In conventionally stained sections, the precise boundaries of the limbus can be difficult to discern, as the sclera, limbus, and cornea all comprise densely packed collagen fibers (**Figure 19.2a–e**). However, the limbus is distinguished by elastic fiber bundles in a flexible hinge region (**Figure 19.2e**) as well as by abundant **limbal stem cells** (**LSCs**) that normally move centripetally to replenish lost corneal cells (**Figure 19.2b**). Accordingly, LSC dysfunction leads to a common form of blindness, called **limbal stem cell deficiency** (**LSCD**).

Sandwiched between the sclera and retina is a highly vascularized **choroid** that noticeably thins during aging, averaging ~200 µm wide in newborns vs. only ~80 µm in the elderly. The choroid comprises the following main layers from the retinal to scleral sides: (1) **Bruch's membrane** of dense collagen fibers beneath the retina; (2) the **choriocapillaris** of fenestrated capillaries; (3) a vascular compartment subdivided into small-to-medium blood vessels vs. large blood vessels in **Sattler's** vs. **Haller's layers**, respectively; and (4) a **suprachoroid** layer abutting the sclera (**Figure 19.2f, g**). Scattered among choroid blood vessels are **melanocytes** plus other stromal cells (e.g., macrophages and lymphocytes), lymph vessels, and nerves, which collectively allow the choroid to supply **oxygen** and **nutrients** to the retina.

Anteriorly, the choroid connects with a **ciliary body**, whose macroscopic folds (=**ciliary processes**) are covered by a bilayered epithelium comprising an apical **non-pigmented** (**NP**) layer and a basal **pigmented** (**P**) layer (**Figure 19.3a–c**). Via tight junctions among NP epithelial cells, the epithelium forms a **blood-aqueous barrier** that protects the eye and helps keep light-scattering blood components from interfering with image formation. Beneath its epithelium, the ciliary body contains stromal connective tissue plus **ciliary muscles** with smooth muscle cells, whereas extending apically from ciliary processes are suspensory ligaments comprising **zonular fibers** that traverse the posterior chamber to connect with the lens periphery (**Figure 19.3a, d**). Most of these several-micrometer-wide fibers are arranged radially along the ligamental long axis, but a few form circumferentially oriented re-enforcing hoops that wrap around axial fibers (**Figure 19.3e, f**). Zonular fibers are in turn composed of numerous **fibrillin-containing microfibrils** that measure ~20 nm wide (**Figure 19.3g**).

In conjunction with ciliary muscle contractions, zonular fiber attachments to the lens mediate a process, called **accommodation**, which allows the eye to focus on nearby objects, such as during reading (**Figure 19.4a**). Thus, for distant objects more than 6 m away, relaxed ciliary muscles expand the ciliary body so that it is maximally separated from the lens, thereby generating full tension on the ciliary body's zonular fibers to flatten the lens into an optimal configuration for distance imaging (**Figure 19.4b**). Alternatively, when viewing closely positioned objects, the ciliary muscles contract to fold the ciliary body into a pleated configuration that moves ciliary processes closer to the lens. Such movement reduces zonular fiber tension and thickens the lens for better close-in focus (**Figure 19.4b**). Due to decreased flexibility of the lens as well as a weakening of ciliary muscles, accommodation is impaired during the aging-related process of **presbyopia** ("aged eye"), thereby often requiring corrective reading glasses to see closely positioned objects (**Figure 19.4a**).

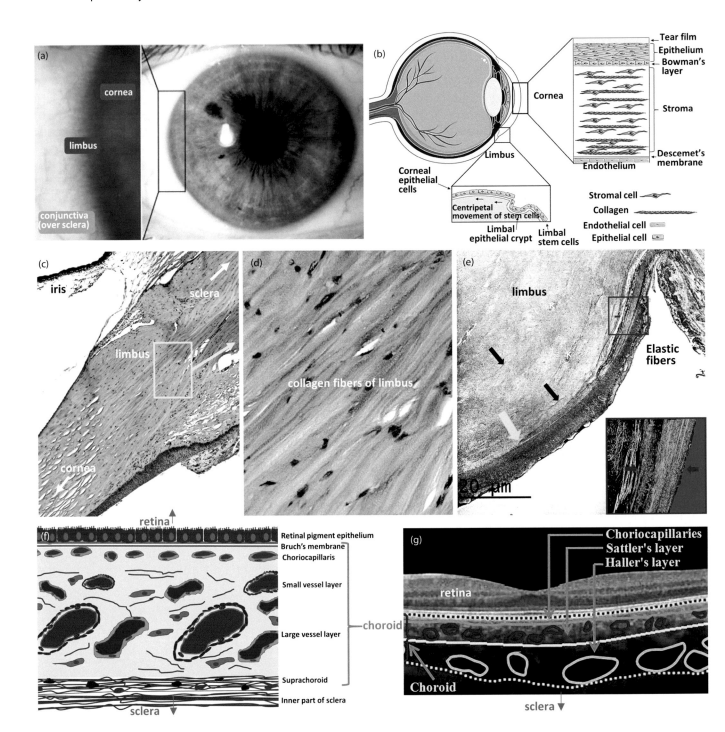

Figure 19.2 Sclera, limbus, and choroid microanatomy. (a–e) As the transition zone between the opaque sclera and translucent cornea, the collagen-rich limbus also has a few elastic fibers (figure e) for flexibility. In addition, stem cells from the limbus help maintain proper function of the cornea, while drainage of aqueous humor through Schlemm's canal in the limbus helps prevent overly elevated intraocular pressures. (c) 35X. (d) 250X. **(f, g)** The vascularized choroid of the middle tunic comprises several layers with differing blood vessel diameters. ([a]: From Casaroli-Marano, RP (2015) Potential role of induced pluripotent stem cells (IPSCs) for cell-based therapy of the ocular surface. J Clin Med 4:318–342, reproduced with publisher permission; [b]: From Masterton, S and Ahearne, M (2018) Mechanobiology of the corneal epithelium. Exp Eye Res 177:122–129, reproduced with publisher permission; [e]: From Lewis, PN et al. (2016) Three-dimensional arrangement of elastic fibers in the human corneal stroma. Exp Eye Res 146:43–53, reproduced under a CC BY 4.0 creative commons license; image courtesy of Drs. K. Meek, P. Lewis, and T. White; [f]: From Campos, A et al. (2017) Viewing the choroid: where we stand, challenges and contradictions in diabetic retinopathy and diabetic macular oedema. Acta Ophthalmol 96:446–459, reproduced with publisher permission; [g]: From Uppugunduri, SR et al. (2018)

Figure 19.2 (Continued) *Automated quantification of Haller's layer in choroid using swept-source optical coherence tomography. PLoS ONE 13:e0193324, https://doi.org/10.1371/journal.pone.0193324, reproduced under a CC BY 4.0 creative commons license.)*

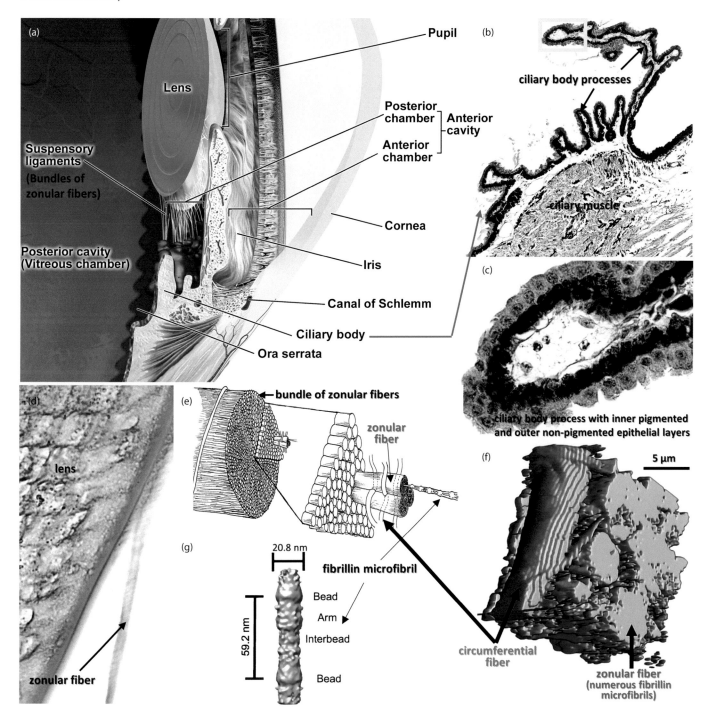

Figure 19.3 Ciliary body. *(**a–e**) Encircling the lens is a ciliary body with numerous folded processes. Basally, the ciliary body overlies smooth muscle cells of the ciliary muscle, whereas apically, ciliary processes extend ligaments comprising bundles of zonular fibers that attach to the lens periphery. (b) 50X; (c) 325X; (d) 450X. (**f, g**) Each zonular fiber is composed of numerous 20 nm-thick fibrillin microfibrils that allow the ciliary body and its associated muscle to change the shape of the lens during an accommodation process for viewing nearby objects. ([a]: From Blausen.com staff (2014) Medical gallery of Blausen Medical 2014. Wiki J Med 1(2):10, doi: 10.15347/wjm/2014.01010, reproduced under CC0 and CC-SA creative commons licenses; [e–g]: From Godwin, ARF et al. (2018) Multiscale imaging reveals the hierarchical organization of fibrillin microfibrils. J Mol Biol 430:4142–4155, reproduced with publisher permission.)*

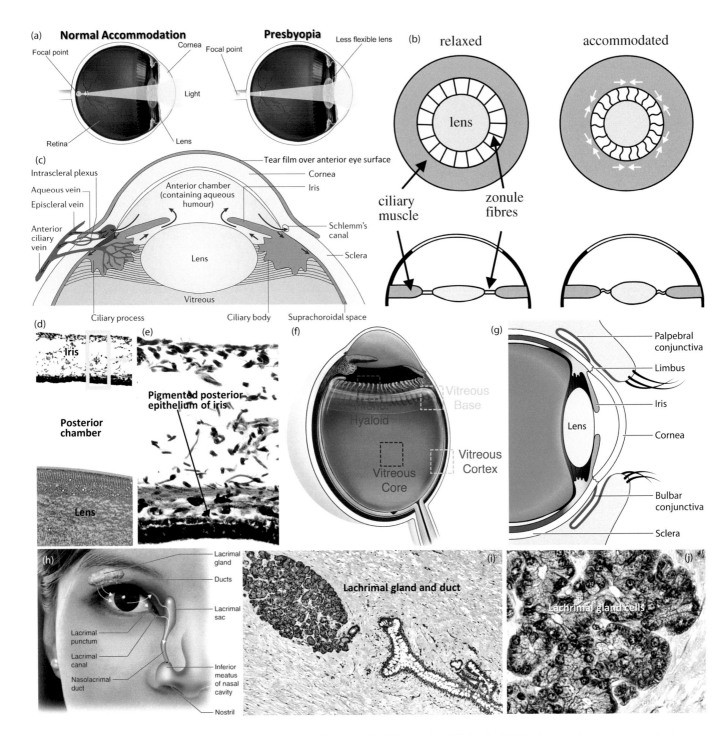

Figure 19.4 Accommodation, aqueous humor, vitreous body, and eyelids. (*a, b*) *During ocular accommodation, the ciliary muscle normally contracts. This reduces tension in zonular fibers to thicken the lens and thereby optimize imaging of nearby objects. However, aged lenses are less flexible and cannot thicken as well as young lenses. Thus, presbyopia ("aged eye") often requires reading glasses to bolster accommodation. (*c*) The ciliary body produces aqueous humor to fill the anterior and posterior chambers of the eye, thereby maintaining proper lens and corneal shapes. To avoid elevated intraocular pressures that can lead to glaucoma, excess aqueous humor is normally drained via Schlemm's canal in the limbus. (*d, e*) The posterior iris contains a highly pigmented epithelium to help eliminate stray light rays that can degrade imaging. (d) 45X; (e) 175X. (*f*) The vitreous body is a gel-like structure that helps maintain proper positioning of the lens and retina. (*g*) The inside of each eyelid is lined by a stratified epithelium, called the conjunctiva, which continues over the visible sclera and contributes to tear film production. (*h–j*) Most of the aqueous component and solutes of tear films are secreted by a superior main lachrimal gland with its associated ductal network.*

Figure 19.4 (Continued) *(i) 70X; (j) 400X. ([a]: From Blausen.com staff (2014) Medical gallery of Blausen Medical 2014. Wiki J Med 1(2):10, doi: 10.15347/wjm/2014.01010, reproduced under CC0 and CC-SA creative commons licenses; [b]: From Land, M (2015) Focusing by shape change in the lens of the eye: a commentary on Young (1801) 'On the mechanism of the eye'. Philos Trans R Soc Lond B Biol Sci 370(1666): 20140308, reproduced under a CC BY 4.0 creative commons license; [c]: From Weinreb, RN and Khaw, PT (2004) Primary open-angle glaucoma. Lancet 363(9422):1711–1720, reproduced with publisher permission; [f]: From Skeie, JM et al. (2015) Proteomic insight into the molecular function of the vitreous. PLoS ONE 10(5):e0127567, doi: 10.1371/journal.pone.0127567, reproduced under a CC BY creative commons license; [h]: Downloaded from https://courses.lumenlearning.com/austincc-ap1/chapter/special-senses-vision/; this work by Cenveo is licensed under a Creative Commons Attribution 3.0 United States, http://creativecommons.org/licenses/by/3.0/us/.)*

In addition to its roles in accommodation, the ciliary body secretes the **aqueous humor** fluid of the posterior and anterior chambers so that proper **intraocular pressure** (**IOP**) can maintain the shape and position of the cornea (**Figure 19.4c**). To produce aqueous humor, non-pigmented cells of the ciliary body epithelium actively pump ions into the posterior chamber thereby drawing in extracellular matrix (ECM) fluids. After filling the posterior chamber, aqueous humor traverses the pupil to keep the IOP of the anterior chamber slightly higher than the pressure outside the eye, and conversely, excessively high IOPs are normally avoided by constantly draining aqueous humor into limbal tissues via **Schlemm's canal** (**Figure 19.4c**).

Connected to the ciliary body is an ~12 mm-wide annular **iris** surrounding a central **pupil** that allows light to reach more posterior eye regions (**Figure 19.4c**). The iris comprises: (i) an anterior **stroma** of fibrous connective tissue and blood vessels, and (ii) a posterior **pigmented epithelium**, which not only confers eye color depending on its particular mix of eu- vs. pheo-melanins but also prevents off-axis light from reaching the lens (**Figure 19.4d, e**). To increase light delivery under low-light conditions or sympathetic stimulation, the pupil is **dilated** by contractions of radially arranged myoepithelial cells that run like spokes of a wheel toward the ciliary body. Conversely, circularly oriented smooth muscles contract under parasympathetic input to **constrict** the pupil and reduce light input. The overlap zone of radial and circular contractile elements occurs within a **collarette** region, which has individual-specific sub-features, thereby allowing the use of **iris imaging** for biometric identifications.

The **vitreous body** filling the posterior eye cavity helps maintain proper eye shape while also preventing retinal detachments from the choroid (**Figure 19.4f**). With a composition of 99% water plus such solutes as electrolytes, glucose, ascorbic acid, and hyaluronic acid, vitreous fluid forms a clear and viscous mass that is attached to the retina and lens via type II collagen fibrils. Invasive materials are typically phagocytosed by vitreous **hyalocytes**. However, suspended debris may accumulate with age, particularly as the vitreous shrinks slightly away from the retina, thereby generating clumps perceived as "floaters".

The eye is protected by a retractable upper and lower **eyelid**, which from their outer to inner surfaces comprise: (i) a keratinized **stratified squamous epithelium** with thickened **eyelash hairs** that intercept potential contaminants, (ii) a **connective tissue core** with glands and scattered stromal cells that can contribute to the red and watery eyes of hayfever or infections, and (iii) an innermost **non-keratinized epithelium** and underlying lamina propria, collectively termed the **conjunctiva**, which also extends over the visible sclera (**Figure 19.4g**). The stratified conjunctival epithelium contains abundant **mucous goblet cells** that secrete mucin carbohydrates into **tear films** coating the conjunctiva, thereby helping to prevent chronic inflammation, referred to as **conjunctivitis** (="pink eye") (Box 19.1 Meibomian Glands and the Production of Trilayered Tear Films). In addition to components of the eyelid proper, a main **lachrimal gland** is situated above each eye and secretes the aqueous layer of tear films through a network of lachrimal ducts (**Figure 19.4h-j**). Lachrimal gland cells are supplemented in their production of tears by lipid secretions from tubuloalveolar glands in the inner eyelid, called **Meibomian glands** (Box 19.1).

19.4 FUNCTIONAL HISTOLOGY OF THE LIGHT-FOCUSING APPARATUS: THE CORNEA AND LENS

As the anterior-most part of the eye, the **cornea** is a light-focusing translucent disc that measures ~0.55 mm thick at its center vs. ~12 mm wide (**Case 19.1 Aldous Huxley's Corneas**) (**Figure 19.5a–d**). The cornea provides ~2/3 of the eye's total focusing power, with the remaining ~1/3 being supplied by the lens. For its interactions with light, the cornea possesses: (1) an outer **stratified squamous epithelium**, which is continually desquamated at its apex and renewed by limbic stem cell mitoses; (2) **Bowman's membrane**, which is an ~15 μm-thick layer

BOX 19.1 MEIBOMIAN GLANDS AND THE PRODUCTION OF TRILAYERED TEAR FILMS

In addition to keeping debris and excessive light from hitting the eye, eyelids secrete complex **tear films** for lubricating and protecting the eye. Formed by multiple glands, tears are trilayered, consisting of an **inner mucin coat**, middle **aqueous component**, and an outer **lipid cap** (**Figure a**). The mucin coat is derived mainly from individual **mucous goblet cells** in the conjunctival epithelium, whereas the aqueous portion of tears is secreted by multicellular tubuloalveolar glands comprising: (i) the main **lacrimal gland** that secretes watery fluids via ducts in the upper eyelid, and (ii) accessory lacrimal organs associated with the conjunctiva. Conversely, the lipid overlay is derived mainly from sebaceous-like **Meibomian glands** located at the inner eyelid margins (**Figure b, c**) and to a lesser

degree from mixed sebaceous and sweat glands in the large pink **caruncle** occurring at the medial corner of the eye.

Along with reducing friction during eyelid closures and preventing desiccation when eyes are exposed, tear films also contain numerous antimicrobial compounds, such as secretory immunoglobulin A (sIgA), secretory phospholipase A2 (sPLA2), secretory leukocyte protease inhibitor (SLPI), and surfactant protein D (SP-D) as well as other more commonly known protectants like lysozyme and lactoferrin (**Figure d**). Most of these compounds are secreted by lacrimal glands, conjunctival cells, corneal epithelial cells, and stromal leukocytes. However, Meibomian glands also supplement the secretion into tears of an antimicrobial peptide, called lacritin.

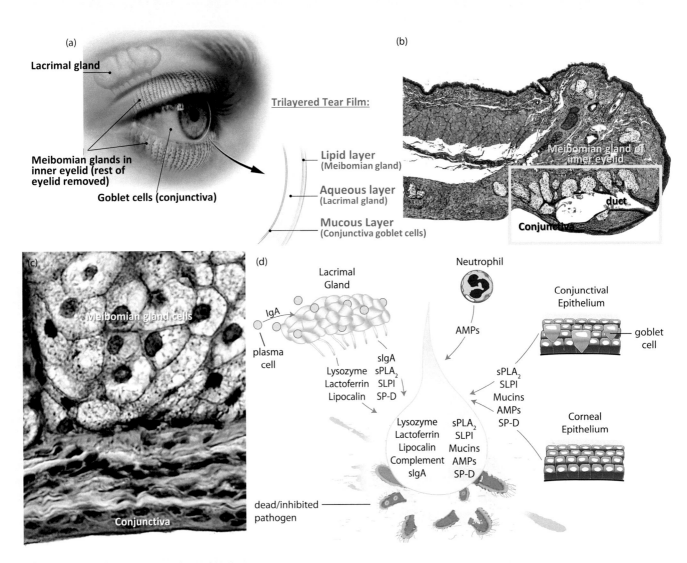

Box 19.1 *(**a–d**) Meibomian gland functional morphology and its roles in producing tear films. ([a]: Downloaded from https://medialibrary.nei.nih.gov/media/1812 and reproduced as a public domain document produced by US federal agency, https://www.usa.gov/government-works; [d]: From McDermott, AM (2013) Antimicrobial compounds in tears. Exp Eye Res 117, doi: 0.1016/j.exer.2013.07.014 and reproduced with publisher permission.)*

CASE 19.1 ALDOUS HUXLEY'S CORNEAS

Writer
(1894–1963)

"There are things known and there are things unknown, and in between are the doors of perception."

(A.H.'s metaphysical outlook on life)

Born in England to a distinguished family of scientists, Aldous Huxley ruled out pursuing a scientific career when he contracted superficial punctate keratitis, which destroyed patches of his cornea and left him nearly blind for several years. After using an unorthodox therapy to regain his sight, Huxley established himself as an acclaimed author, publishing several widely praised books, including *Brave New World* and the *Doors*

of Perception, the latter of which chronicled his mescaline-induced visions and gave rise to the name of the 1960s rock band, the Doors. Like many writers and artists of his era, Huxley smoked heavily, which no doubt contributed to the laryngeal cancer he developed. While on his deathbed, Huxley requested an injection of lysergic acid diethylamide (LSD), and soon after receiving the psychedelic drug, the influential author, who overcame his diseased corneas, died.

that helps protect underlying corneal tissues; (3) the main **stroma** which constitutes ~90% of the cornea and contains a few hundred sublayers of tightly packed and regularly arranged **type I collagen fibers**; (4) **Descemet's membrane**, an ~10–20 μm-thick basement membrane consisting of mainly type IV and III collagen fibers; and (5) an innermost simple epithelium, traditionally referred to as the **endothelium** (**Figure 19.5a, d**).

Situated posterior to the cornea, the **lens** is a biconvex disc, which in adults measures ~10 mm wide by ~4 mm thick (**Case 19.2 Marie Curie's Lenses**). The lens comprises from its periphery to center: (1) a protective clear **capsule** composed mainly of type IV collagen, laminin, and proteoglycans; (2) a generative **epithelium**, which in adult eyes is restricted to the anterior lens hemisphere; and (3) an internal mass of degenerating epithelial cells plus lens **fibers**, which represent the transformed derivatives of living epithelial cells (**Figure 19.5e, f**). During this transformation process, lens epithelial cells remain interconnected by **gap junctions** but lose their organelles while becoming filled with **crystallin proteins** that provide the transparent properties of the lens. However, in the elderly, or even earlier under pathological stress, damaged lens fibers increase the opacity of the lens during **cataract** formation, thereby reducing visual acuity and requiring artificial lens replacement.

The cornea and lens form a **light-focusing apparatus** that **refracts** incident light rays, thereby changing their paths so that they can converge at a common focal point on the retina to produce a tightly focused image. However, with mismatches between the eye's refractive powers and optical pathlength, light is either brought to a focus in front of the retina or would converge behind the retina in the case of near-sightedness (**myopia**) vs. far-sightedness (**hyperopia**), respectively (**Figure 19.5g, h**). Such focusing defects can be rectified by properly shaped lenses placed in front of the cornea (**Figure 19.5g, h**). Alternatively, surgeries can re-sculpt the cornea and thereby eliminate common vision defects without the need for additional corrective lenses (Box 19.2 Refractive Surgery to Correct Vision Defects).

19.5 RETINA HISTOLOGY AND ROD VS. CONE PHOTORECEPTORS PLUS THE SIGNAL TRANSDUCTION PATHWAYS OF LIGHT PERCEPTION

As the innermost ocular tunic posterior to the lens, the **retina** is a concave disc that measures ~0.1–0.55 mm thick. Along with its more peripherally situated regions, the retina contains near its center: (i) an **optic disc**, where blood vessels converge, and a connection to the **optic nerve** is made, thereby generating a blind spot due to the absence of photoreceptors; (ii) the **macula**, which is an ~5 mm-wide pigmented region near the eye center; and (iii) the **fovea**, which is an indentation in the central macula containing a foveolar center where densely packed photoreceptors in the thinnest part of the retina provide optimally sharp focusing (**Figure 19.6a, b**).

Functionally, the retina senses light via a multi-layered collection of supporting **glial cells** plus three classes of neurons: (1) **photoreceptor cells** that convert light into neural impulses, (2) **interneurons** (e.g., bipolar, amacrine, and horizontal cells) that relay and modulate photoreceptor-generated action potentials, and (3) **ganglion cells** near the inner retinal edge that receive input from interneurons and transmit impulses along the optic nerve to the brain.

Such retinal components constitute from the inner (closest to the vitreous body) to outer (abutting the choroid) surfaces the following main layers: (1) **inner limiting membrane** containing mainly ganglion cell axons that

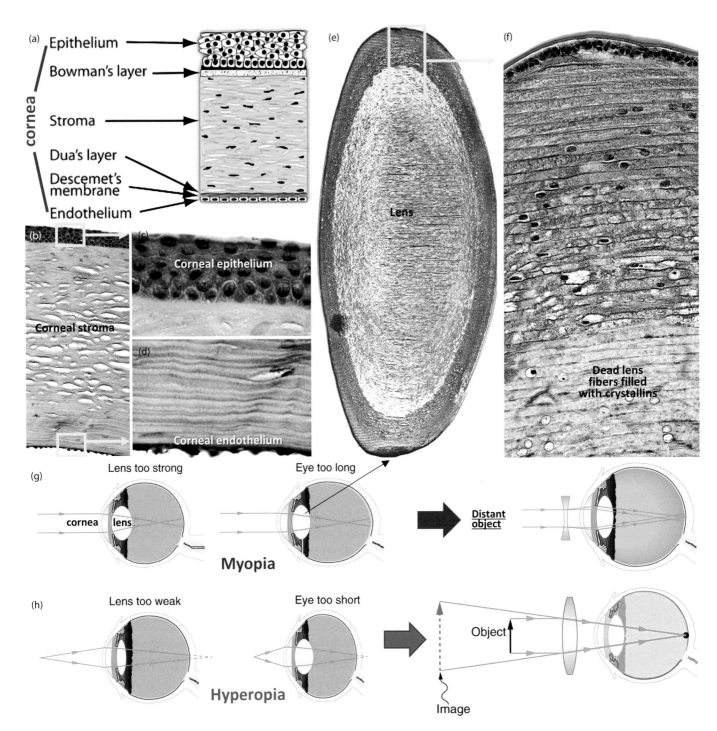

Figure 19.5 Cornea, lens, and refraction defects. *(a–f) Each translucent cornea comprises numerous distinct layers (a–d) and when functioning properly with the lens (e, f) serves to refract light so that it focuses on the retina. (b) 75X; (c) 375X; (d) 250X; (e) 40X; (f) 300X. (**g**) However, eyes with either overly strong refraction or too long of a pathlength will focus light in front of the retina and thus generate the blurry images associated with myopia (near-sightedness). (**h**) Conversely, in hyperopia (far-sightedness), insufficient refraction or too short of an eyeball would focus light at a plane behind the retina. Each of these refraction defects can usually be rectified with properly shaped eyeglasses or contact lenses. ([a]: From Navaratnam, J et al. (2015) Substrates for expansion of corneal endothelial cells towards bioengineering of human corneal endothelium. J Funct Biomater 6:917–945, reproduced under a CC BY 4.0 creative commons license; [g, h]: From OpenStax College (2013) College Physics. OpenStax. http://cnx.org/content/col11496/latest, reproduced under a CC BY 4.0 creative commons license.)*

CASE 19.2 MARIE CURIE'S LENSES

Nobel Laureate in Physics and Chemistry
  (1867–1934)

"I am among those who think that science has great beauty."

(M.C. on the beauty of scientific endeavors)

For her ground-breaking research on radioactivity, Marie Curie became the only dual winner of a Nobel Prize in two scientific fields. Curie studied Physics in France where she married another physicist, Pierre Curie. Together, the Curies discovered two new radioactive elements for which they received the 1903 Nobel Prize in Physics. After her husband was run over and killed by a horse-drawn carriage, Curie isolated pure radium and researched its applications, breakthroughs that resulted in her 1911 Nobel Prize in Chemistry.

Unfortunately, at that time, the health risks of radioactivity were not yet fully appreciated, and Curie did not handle radium with adequate protection, causing her to exhibit signs of chronic radiation sickness. For example, the particularly sensitive lens fibers in her eyes formed opaque cataracts, for which she underwent multiple surgeries. However, far more serious was her development of aplastic anemia, a condition that reduced her hematopoietic output. Thus, after her longtime work on radioactivity had damaged her lenses and compromised her ability to form blood cells, the Nobel Prize laureate died at age 66.

---

exit the eye via the optic nerve, (2) **ganglion cell layer** containing somata of ganglion cells, (3) **inner plexiform layer** where interneurons synapse with ganglion cells, (4) **inner nuclear layer** containing interneuron cell bodies (=perikarya), (5) **outer plexiform layer**, where photoreceptors synapse with interneurons, (6) **outer nuclear layer** of photoreceptor perikarya, (7) **rods and cones layer** composed of light-sensing parts of photoreceptors, and (8) **retinal pigmented epithelium**, a melanin-containing simple cuboidal epithelium with tight junctions forming a **blood-retina barrier** that blocks entry of toxic substances while allowing passage of beneficial substances from the vascularized choroid (**Figure 19.6c-e**). Normally, the outer surface of the retina remains close to the choroid, but retinal-choroid detachment compromises maintenance of the retina and can thus lead to blindness (**Case 19.3 Theodore Roosevelt's Retinas**).

The two retinal photoreceptors—**rods** and **cones**—are so-named based on the shape of their outermost region, called the **outer segment**. Each outer segment has a stack of membranous discs that invaginate from the plasmalemma, thereby increasing surface area for housing light-sensitive **photopigment** molecules (**Figure 19.7a–c**). On average, the retina has ~120 million rods and ~6 million cones, with rods predominating peripherally and cones being centrally concentrated, particularly in the foveolar region of the macula (**Figure 19.6b**). Rods can function under dim-light **scotopic** conditions to generate low-resolution monochromatic images. Conversely, cone cells utilize the brighter light in **photopic** conditions to produce higher-resolution color images.

For such functions, the stacked membranes in rod and cone outer segments contain numerous photopigments, each of which comprises: (i) a large **opsin** protein, which is a type of transmembrane **G protein coupled receptor** (**GPCR**), and (ii) **retinal**, a small vitamin A derivative that is covalently linked to opsin (**Figure 19.7d–f**). In rods, opsin plus retinal forms a **rhodopsin** type of photopigment, whereas cone cells contain red, green, or blue **photopsin** photopigments with differing sensitivities to red, green, and blue light. Photopigments are produced by synthetic organelles in the **inner segment** of each photoreceptor, which in turn joins the outer segment by a **connecting cilium** region that facilitates reciprocal trafficking of molecules between the two segments (**Figure 19.7a, b**).

As initiators of the neural circuitry mediating light detection, rods and cones synapse with **bipolar cell** (**BC**) interneurons in the **outer plexiform layer** of the retina, and BCs in turn synapse in the **inner plexiform layer** either directly with ganglion cells or with intermediary interneurons that connect with ganglion cells (**Figure 19.8a–c**). BCs typically form complex neural circuits with multiple photoreceptors and other interneurons like amacrine cells to generate integrated units, called **receptive fields** (**Figure 19.8c**). Within these fields are two main types of BCs: (1) **ON-center BCs** that depolarize **when light is turned on** photoreceptors in the receptive field center vs. (2) **OFF-center BCs** that depolarize **after light is turned off** in the field center. The opposing reactions to light in ON vs. OFF BCs are largely due to modulating interneuronal input as well as to **different types of receptors** expressed in BC post-synaptic membranes. ON BCs utilize **metabotropic** receptors that respond to neurotransmitter release from photoreceptors by **hyperpolarizing**, whereas OFF BCs use **ionotropic** receptors that cause BC **depolarization** when bound to neurotransmitter. Thus, the release of photoreceptor neurotransmitters **inhibits** (hyperpolarizes) ON BCs vs. **excites** (depolarizes) OFF BCs to provide redundant pathways for transmitting proper signals along the optic nerve to the brain.

BOX 19.2 REFRACTIVE SURGERY TO CORRECT VISION DEFECTS

In properly functioning eyes, light refracted by the cornea and lens is brought to a common focal point on the retina so that sharply focused images can be produced. However, if corneal and lens refractive powers do not match the eye's optical pathlength, light rays fail to focus tightly on the retina, thereby causing near- (**myopia**) or far-sightedness (**hyperopia**). Both of these conditions can typically be rectified by placing in front of the eyes corrective glasses or contact lenses with equal but opposite effects on refraction to those of the eye. (**Figure 19.5g, h**). Alternatively, the cornea can be reshaped by surgical procedures, the most common of which being **Lasik** (=laser-assisted in situ keratomileusis). During Lasik, a flap of

the anterior cornea is initially reflected back via a horseshoe-shaped laser incision (**Figure a**). The exposed remainder of the cornea is then re-sculpted via an oscillating blade or a rapidly pulsed laser beam to modify corneal curvature and thereby focus light properly on the retina without the need for corrective lenses (**Figure b**).

Another common vision defect is **astigmatism**, which involves an irregularly shaped cornea that refracts light differently in the vertical vs. horizontal planes of the eye. As with myopia and hyperopia, differential re-sculpting of the cornea to achieve a more spherical shape can also correct certain astigmatisms (**Figure b**).

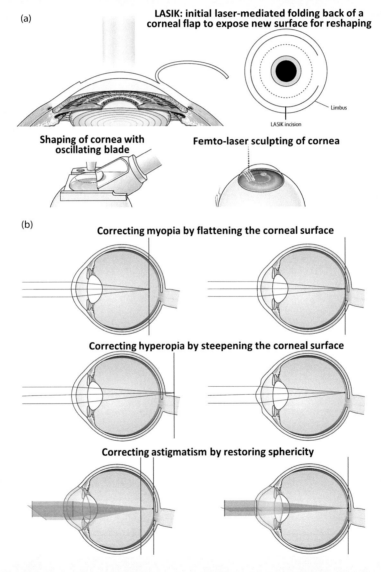

Box 19.2 (a, b) *Summary of the Lasik procedure for correcting defects in ocular refraction. (From Kim T-I et al. (2019) Refractive surgery. Lancet 393: 2085–2098, reproduced with publisher permission.)*

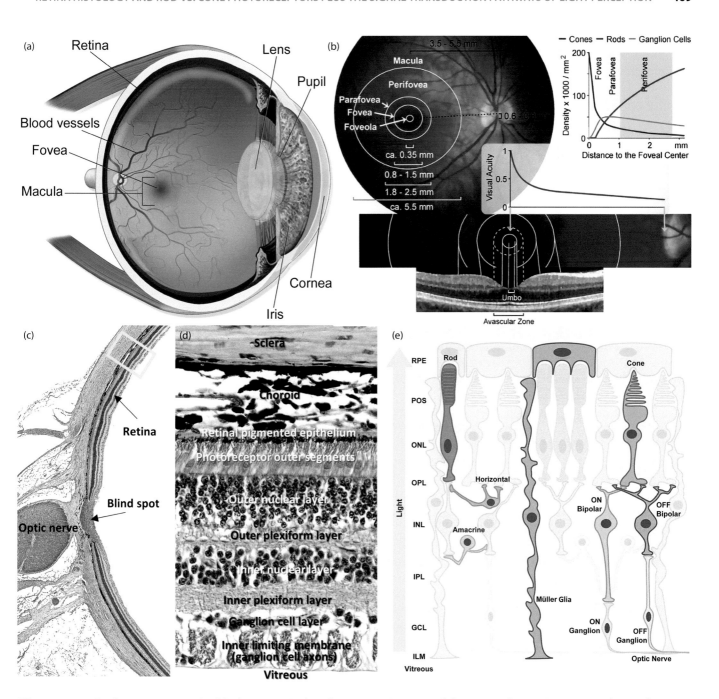

Figure 19.6 Retina anatomy. (a, b) *The retina is the photoreceptive unit of the eye and contains a central macula region that combines the thinnest retinal dimensions with the greatest density of cone photoreceptors to provide the eye's highest visual acuity. (**c–e**) The multilayered retina comprises from its inner surface to outer periphery: (1) an inner limiting membrane of mainly ganglion cell axons that exit the eye and form the optic nerve, (2) the ganglion cell layer with perikarya of ganglion cells, (3) an inner plexiform layer where interneurons synapse with ganglion cells, (4) the inner nuclear layer containing interneuron perikarya, (5) an outer plexiform layer, where photoreceptors synapse with interneurons, (6) the outer nuclear layer of photoreceptor perikarya, (7) photoreceptor outer segments where light-sensing parts of rod and cones are located, and (8) the retinal pigmented epithelium directly underlying the choroid. (c) 15X; (d) 200X. ([a]: From Blausen.com staff (2014) Medical gallery of Blausen Medical 2014. Wiki J Med 1(2):10, doi: 10.15347/ wjm/2014.01010, reproduced under CC0 and CC-SA creative commons licenses; [b]: From Bringmann, A et al. (2018) The primate fovea: structure, function and development. Prog Retin Eye Res 66: 49–84, reproduced with publisher permission; [e]: From Klapper, SD et al. (2016) Biophysical properties of optogenetic tools and their application for vision restoration approaches. Front Syst Neurosci 10, https://doi.org/10.3389/fnsys.2016.00074, reproduced under a CC BY 4.0 creative commons license.)*

CASE 19.3 THEODORE ROOSEVELT'S RETINAS

26th U.S. President
(1858–1919)

"The unforgivable crime is soft hitting. Do not hit at all if it can be avoided; but never hit softly."

(T.R.'s combative view of how to take on challenges)

The 26th President of the United States, Theodore Roosevelt, earned his reputation as a man of action by leading his troop of Rough Rider soldiers during the Spanish-American War. However, long before he had fashioned his larger-than-life persona, Roosevelt was a frail and sickly boy who suffered from debilitating asthma. Encouraged by his father to take up boxing as a teenager in order to overcome his infirmity, Roosevelt went on to become a member of the Harvard boxing team and even continued to spar after becoming President. However, his days of pugilism were abruptly ended when his retina was

detached while sparring in the White House, thereby rendering Roosevelt essentially blind in one eye. Years after that injury, Roosevelt was campaigning again for the presidency when he was shot in the chest by a would-be assassin but was able to avoid serious injury because the bullet was slowed by his glasses case and voluminous speech notes. The wounded campaigner refused treatment and instead informed his audience that "it takes more than that to kill a Bull Moose", thereby helping to popularize the moniker of Roosevelt's movement's as the "Bull Moose" party. Roosevelt finished his speech but eventually lost the election, ending the political career of the former boxer half-blinded by a detached retina.

When kept **in the dark**, photoreceptors have high intracellular **cGMP levels** that help drive continual photoreceptor **depolarization**, thereby triggering constant secretion of **glutamate** neurotransmitter at outer plexiform synapses with BCs (**Figure 19.8d**). Conversely, in the presence of light, illuminated photoreceptors undergo a complex signaling cascade that helps convert photoreceptors to a **hyperpolarized** state, thereby blocking their dark-associated glutamate secretion at BC synapses (**Figure 19.8e**). As summarized in **Figure 19.8f**, the light-induced hyperpolarization of rod cells involves: (i) light absorption in outer segment membranes converting **cis-retinal** to **trans-retinal** which causes an activating conformational change in opsin; (ii) activated opsin catalyzing GTP for GDP exchange on the trimeric **transducin** G-protein that is coupled to rhodopsin; (iii) GTP causing the α-subunit of transducin to detach from the membrane so that it can activate **phosphodiesterases** (**PDEs**); (iv) activated PDEs **hydrolyzing cGMP** into GMP; (v) reduced rod-cell cGMP levels **closing cyclic-nucleotide-gated (CNG) Na$^+$ channels** in the plasmalemma; and (vi) Na$^+$ channel closure **blocking Na$^+$ influx** to **hyperpolarize** the cell, thereby closing voltage-gated Ca^{2+} channels so that **glutamate secretion** at BC synapses is prevented. Accordingly, by blocking dark-induced glutamate release, illuminated photoreceptors **excite** (depolarize) ON BCs and **inhibit** (hyperpolarize) OFF BCs so that appropriate neural impulses transmitted by the optic nerve are perceived as light by the brain. After hyperpolarization, rod cells are returned to their dark-state by rhodopsin inactivation and the resultant hydrolysis of GTP. This reduces transducin-mediated PDE activation, thereby elevating cGMP levels so that depolarizing CNGs re-open for Na$^+$ influx and glutamate release (**Figure 19.8d**). In addition, biochemical reactions return trans-retinal to its cis configuration over several minutes and thus contribute to the period of **dark adaptation** needed to optimize vision when moving from high to low light levels.

19.6 SOME EYE DISORDERS: MACULAR DEGENERATION AND GLAUCOMA

Diseases collectively termed **macular degeneration (MD)** are characterized by an initial degradation of center-of-field vision, owing to dysfunction of the macula. The vast majority of MD cases are dry (=atrophic) forms that result from the drying out and thinning of the macula following the accumulation of debris in large deposits, called drusen. Usually, this kind of degeneration progresses more slowly than wet (=trophic) MDs that involve new blood vessel growth within the macula. Aside from dietary supplements that may slow disease progression, there are few effective medications for treating dry MD, whereas some cases of wet MD respond well to intraocular injections of antibodies against vascular endothelial growth factor (VEGF) receptors that can reduce macular vascularization.

As the leading cause of irreversible blindness, **glaucoma** is characterized by progressive degeneration of the optic nerve and its afferent neural pathway from the retina. Although several types of glaucoma can occur, the most common form, termed open-angle glaucoma, involves blockage of aqueous humor drainage through Schlemm's canal, thereby leading to high intraocular pressures. Such elevated pressures in turn damage the optic nerve head region of the retina leading into the optic nerve, which if left untreated can result in vision loss. Therapies to reduce intraocular pressures involve both medications, such as topical prostaglandins and laser surgeries to increase drainage outflow.

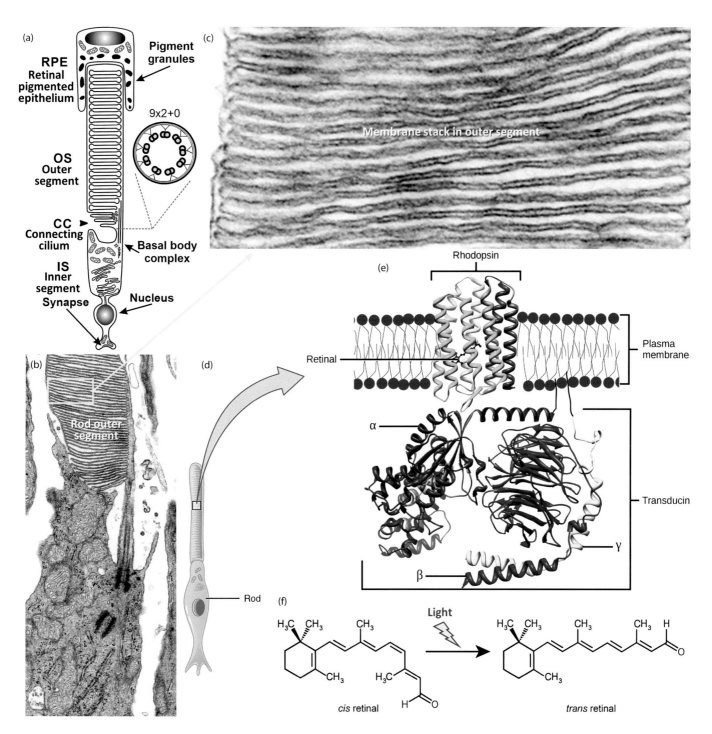

Figure 19.7 Rod microanatomy and rhodopsin structure. *(a–c) In the cylindrical outer segment of rod photoreceptors, stacked invaginations of the plasmalemma contain numerous copies of the photopigment rhodopsin. (b) 18,000X; (c) 72,000X. (d, e) Each rhodopsin comprises: (i) a large transmembrane G protein coupled receptor, called opsin, which associates on its cytoplasmic side with the G protein transducin, and (ii) retinal, which is a small vitamin A derivative that intercalates within opsin. (f) In the dark, retinal exists in an inactive cis configuration, whereas light triggers a conformational change to the active trans form of retinal. ([a–c]: From Falk, N et al. (2015) Specialized cilia in mammalian sensory systems. Cells 4:500–519, doi: 10.3390/cells4030500, reproduced under a CC BY 4.0 creative common license; images courtesy of Dr. A. Giessl; [d, e]: OpenStax College (2013) Biology 2e. OpenStax. http://cnx.org/content/ col11496/latest, reproduced under CC BY 4.0 creative commons license.)*

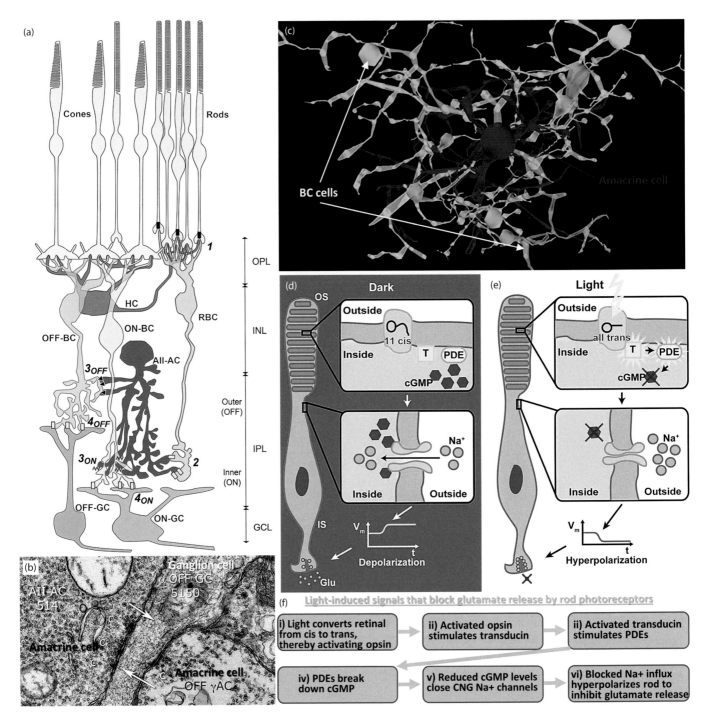

Figure 19.8 Signal transduction during light perception by rod cells. *(a) Receptive fields within the retina comprise complex neural circuits involving multiple rod and cone photoreceptors plus interneurons like bipolar cells (BCs), horizontal cells (HCs), and amacrine cells (ACs), which collectively modulate the output of ganglion cells (GCs). Two main bipolar cell types—ON- vs. OFF-BCs—depolarize to release their neurotransmitters when light is turned on vs. off in the receptor field center, respectively. (b, c) A TEM (b) and 3-D reconstruction (c) illustrate multiple synaptic connections that occur among BCs and amacrine cells of the retina. (b) 32,000X. (d) In the dark, depolarized rod cells continuously release glutamate neurotransmitter that serves as an inhibitor of downstream ganglion cell depolarization and hence prevents impulses from being sent along the optic nerve to the brain. Photoreceptor depolarization in the dark occurs due to inactive retinal in its cis form allowing cGMP to reach high levels, which in turn opens sodium channels and thereby depolarizes rod cells in the dark. (e, f) Conversely, light triggers a conformational change in retinal to its active cis state which activates transducin to stimulate phosphodiesterases (PDEs). Increased PDE activity reduces cGMP levels,*

Figure 19.8 (Continued) *closes sodium channels, and thus hyperpolarizes photoreceptors to block the release of glutamate and thereby allow ganglion cells to send action potentials to the brain. ([a]: From Tsukomoto, Y and Omi, N (2017) Classification of mouse retinal bipolar cells: type-specific connectivity with special reference to rod-driven AII amacrine pathways. Front Neuroanat 11, https://doi.org/10.3389/fnana.2017.00092, reproduced under a CC BY 4.0 creative commons license; [b, c]: From Marc, RE et al. (2014) The AII amacrine cell connectome: a dense network hub. Front Neural Circuits 10: https://doi.org/10.3389/fncir.2014.00104, reproduced under a CC BY 3.0 creative commons license; [d, e]: From Klapper, SD et al. (2016) Biophysical properties of optogenetic tools and their application for vision restoration approaches. Front Syst Neurosci 10, https://doi.org/10.3389/fnsys.2016.00074, reproduced under a CC BY 4.0 creative commons license.)*

19.7   SUMMARY—EYES

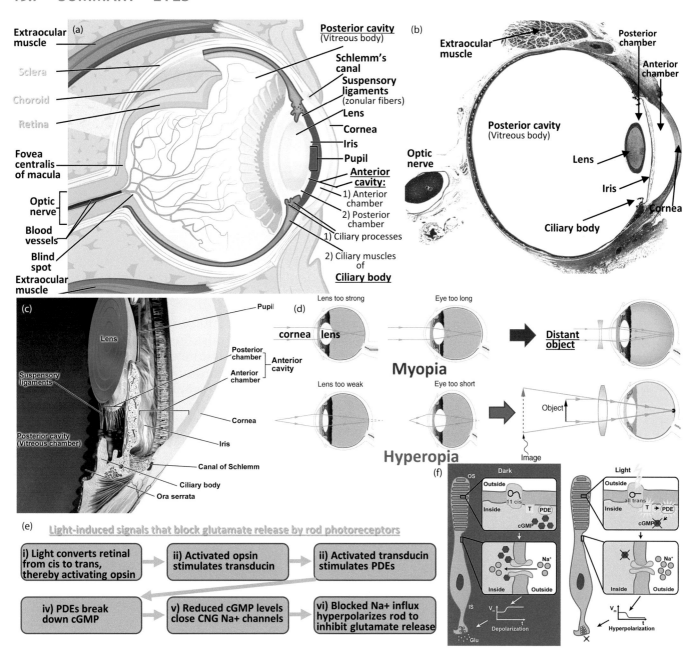

Pictorial Summary Figure Caption: (**a**): see figure 19.1f; (**b**): see figure 19.1e; (**c**): see figure 19.3a; (**d**): see figure 19.5g, h; (**e**): see figure 19.8f; (**f**): see figure 19.8d, e.

In adults, the wall of the eye comprises three concentrically arranged layers: (1) an **outer tunic**, which contains the **sclera**, **limbus**, and **cornea**, (2) a **middle tunic** (=**uvea**), which includes the **choroid**, **ciliary body**, and **iris**, and (3) an **inner** tunic consisting of the **retina** (**Figure a–c**). Medial to these tunics, the eye contains: (i) a posterior **vitreous body** of gel-like material that helps maintain the overall morphology of the eye; (ii) a **lens** which along with the **cornea** focuses light on the retina; and (iii) the **anterior** and **posterior** eye chambers that are filled with **aqueous humor** fluid for maintaining the shape of the cornea.

The eye develops from: (i) an evaginated **optic vesicle** that forms the retina; (ii) a **lens placode** invagination that generates the lens; and (iii) **head mesenchyme** that gives rise to most of uvea and vitreous body. The **sclera** (="white of the eye") is an opaque capsule that protects the eye and transitions via the **limbus** into the **cornea.** The **cornea** has dense regular collagen fibers that provide most of eye's **refractive power** (i.e., its ability to focus light). The cornea and lens normally focus light on the retina, but in **myopic** (near-sighted) vs. **hyperopic** (far-sighted) eyes, the optical pathlength is too long vs. too short for proper focusing (**Figure d**). Changing corneal shape via surgeries (e.g., **Lasik**) can correct vision defects without glasses.

The **uvea** has: (i) a **choroid** that is highly vascularized to maintain the retina; (ii) an **iris** that overlies the lens and has a central **pupil** which can be constricted or dilated to control light hitting the lens; and (iii) a **ciliary body** that makes the aqueous humor of the posterior and anterior chambers, which are normally drained via **Schlemm's canal**. The ciliary body has **ciliary processes** that: (i) attach apically to the lens via **zonular fibers**, and (ii) are underlain basally by **ciliary muscles**. Such muscles contract when viewing nearby objects during the process of **accommodation**, which rounds the lens due to reduced zonular fiber tension. The lens fails to accommodate properly during aging (**presbyopia**) and becomes opaque during **cataract** formation.

To perceive light, the multi-layered retina contains two types of **photoreceptors**— the more common **rods**, which generate low-resolution, monochromatic images under low light conditions vs. rarer **cones**, which form high-resolution, color images under bright light conditions. The **retina** also possesses: (i) **inner** vs. **outer nuclear layers** comprising perikarya of interneurons (e.g., bipolar, amacrine, and horizontal cells) vs. perikarya of rods and cones, and (ii) **inner** vs. **outer plexiform layers** where interneurons synapse with optic nerve **ganglion cells** vs. with photoreceptors.

At the beginning of the **signal transduction cascade that converts light into action potentials**, light rays hitting **rhodopsin** photopigment molecules in the outer segment of rod cells cause the **retinal** portion of rhodopsin to switch from a **cis-** to **trans** configuration. This conformational change stimulates the **opsin** component of rhodopsin, which in turn activates **transducin** to increase **phosphodiesterase** (**PDE**) activity. Elevated levels of active PDE degrade cyclic GMP thereby closing Na^+ channels to hyperpolarize the rod cell. Hyperpolarization in turn stops the continual release of **inhibitory glutamate neurotransmitter** that rods carry out in darkness and thus allows action potentials to be propagated along the optic nerve to the brain (**Figure e, f**).

The **eye** is protected by eyelids containing **Meibomian glands** that secrete the oily component of **tear films**.

Some eye disorders: **macular degeneration**: degradation of the retinal macula that causes central vision loss and can eventually lead to blindness; **glaucoma**: increased intraocular pressure that damages the optic nerve head and optic nerve, which in turn can result in vision loss.

SELF-STUDY QUESTIONS

1. What contracts during accommodation to cause rounding of the lens?
2. T/F The choroid layer is part of the sclera.
3. Which neurotransmitter is secreted by rod cells?
4. Which of the following is characterized by increased opacity of the lens?
   A. Glaucoma
   B. Accommodation
   C. Macular degeneration
   D. Cataract
   E. Retinal detachment

5. What is the G protein coupled receptor of rhodopsin?
6. Which maintains proper eye shape and helps prevents retinal detachment?
    A. Pupil
    B. Iris
    C. Cornea
    D. Vitreous body
7. What part of the eye does the optic vesicle evagination of the brain give rise to?
8. What secretes the lipid component of tears?
9. T/F cGMP closes rod sodium channels.
10. What drains aqueous humor from the anterior cavity of the eye?
11. Which of the following are interneurons?
    A. Cones
    B. Lens cells
    C. Endothelial cells
    D. Amacrine cells
12. T/F Rods are more common than cones.
13. What is the transition zone between the sclera and cornea?
14. T/F In myopia, light rays focus in front of, rather than on, the retina.

15–18: Arrange the steps of rod signal transduction outlined in 15–18 in proper order, with earliest on the left and latest on the right:
15. Transducin activation
16. Retinal cis to trans conformational change
17. Na^+ channel closing
18. cGMP decrease
19. Describe the process of accommodation.
20. Discuss how rod cells transduce light into neural impulses that exit the eye via the optic nerve.

ANSWERS

(1) ciliary muscles; (2) F; (3) glutamate; (6) D; (5) opsin; (6) D; (7) retina; (8) Meibomian glands; (9) F; (10) Schlemm's canal; (11) D; (12) T; (13) limbus; (14) T; (15–18) proper order: 16→15→18→17; (19) The answer should include zonular fiber attachments between ciliary processes and lens; relaxation of ciliary muscles for distant objects; contraction for in-close objects to reduce tension on zonular fibers; (20) The answer should include outer segment membrane stacks, cis→trans rhodopsin conformational change, transducin, PDEs, cGMP-gated Na+ channels, hyperpolarization, glutamate blockage; ON- vs. OFF- bipolar cells, amacrine cells; metabotropic glutamate receptors

Ears

PREVIEW

20.1 The **ear:** general **anatomy** and **development**

20.2 Functional histology of the **external** and **middle ear**

20.3 **Inner ear:** microanatomy of the **vestibular apparatus** used in maintaining **balance**

20.4 **Inner ear:** the **cochlea** during **hearing** and the functional histology of the **organ of Corti** as it **transduces sound**

20.5 Some disorders of ears: **vertigo** and **sensorineural hearing loss**

20.6 Summary and self-study questions

20.1 THE EAR: GENERAL ANATOMY AND DEVELOPMENT

The ear has an **outer**, **middle**, and **inner** component (**Figure 20.1a**). The **outer ear** comprises an externally positioned **auricle** (=pinna) that funnels sound via the **external auditory canal** to a **tympanic membrane** (=eardrum) bordering the middle ear. Medial to the eardrum, the **middle ear** contains three small bones (=ossicles) that transmit sound-induced vibrations to the inner ear's oval window (**Figure 20.1b**). In addition, to optimize internal pressure and mucus clearance, the middle ear cavity is connected by a **Eustachian** (=auditory) **tube** to the nasopharynx. The **inner ear** comprises two main parts: (i) the **vestibular apparatus** (**utricle**, **saccule**, and **semicircular ducts**), which helps maintain **balance**, and (ii) the **cochlea**, which mediates **hearing** by converting sound waves of varying **frequency** (=pitch) and **amplitude** (=intensity) into neural impulses for processing by the brain (**Figure 20.1c**).

Each ear begins morphogenesis during the 4th week of human gestation as an **otic placode** of thickened surface ectoderm on either side of the head (**Figure 20.1d**). In week five, nodular swellings (=**auricular hillocks of His**) arise on developing pharyngeal arches, and these eventually fuse to form an external **auricle**, which by week 20 moves to its definitive position on the side of the head. Soon after the hillocks of His develop, an invagination between the first two pharyngeal arches initially becomes filled with a meatal plug before it hollows out by 28 weeks to form the **external auditory canal** (Box 20.1 Patterns of Ear Development). Concurrently, pharyngeal arch derivatives in each **middle ear** generate cells that go on to form the three auditory ossicles (Box 20.1) as well as parts of the Eustachian tube (**Case 20.1 Juliette Gordon Low's Ears**). As the external and middle ears differentiate, the otic placode invaginates and detaches from surface ectoderm to become an inner-ear-generating **otic vesicle** (=cyst). By 8–10 weeks, the vesicle migrates internally toward the developing middle ear and differentiates into compartments that form the vestibular apparatus and cochlea (**Figure 20.1d**).

DOI: 10.1201/9780429353307-24

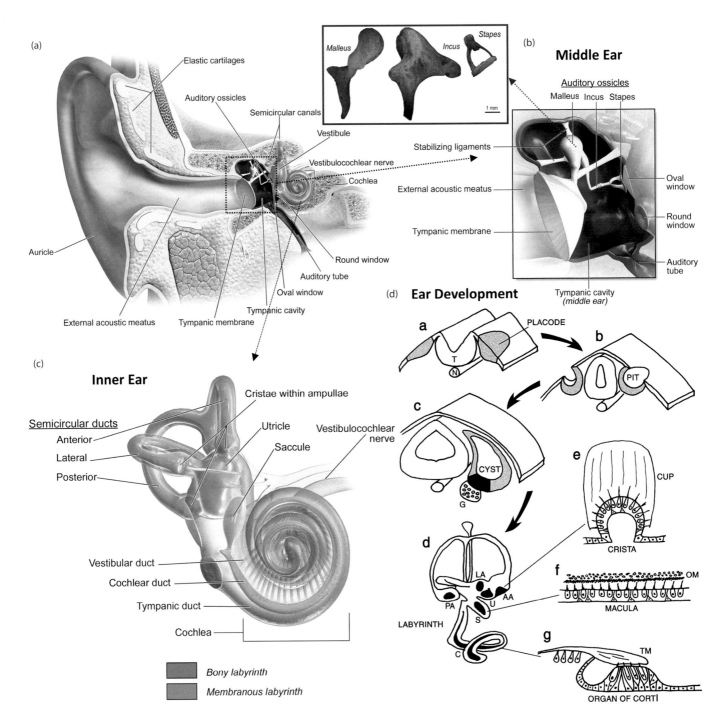

(a) Elastic cartilages
Auditory ossicles
Semicircular canals
Vestibule
Auricle
External acoustic meatus
Tympanic membrane
Tympanic cavity
Oval window
Auditory tube
Round window
Cochlea
Vestibulocochlear nerve

Malleus   *Incus*   *Stapes*   1 mm

(b) **Middle Ear**
Auditory ossicles
Malleus Incus Stapes
Stabilizing ligaments
External acoustic meatus
Tympanic membrane
Oval window
Round window
Auditory tube
Tympanic cavity *(middle ear)*

(c) **Inner Ear**
Cristae within ampullae
Semicircular ducts
Anterior
Lateral
Posterior
Utricle
Saccule
Vestibulocochlear nerve
Vestibular duct
Cochlear duct
Tympanic duct
Cochlea
Bony labyrinth
Membranous labyrinth

(d) **Ear Development**
a T N PLACODE
b PIT
c CYST G
e CUP CRISTA
d LA PA U AA S C LABYRINTH
f OM MACULA
g TM ORGAN OF CORTI

Figure 20.1 Ear anatomy and development. *(a–c) The adult ear comprises: (i) an auricle and external auditory canal of the external ear; (ii) a middle ear with three small bony ossicles (malleus, incus, and stapes) spanning from the tympanic membrane to the oval window of the inner ear; and (iii) an inner ear consisting of a vestibular apparatus (saccule, utricle, and semicircular ducts) for maintaining balance plus a coiled cochlea that converts ossicle-transmitted vibrations of its oval window into action potentials to be perceived as sounds. (d) After developing from invaginated otic placodes, otic cysts give rise to inner ear components. ([a, b]: From Blausen.com staff (2014) Medical gallery of Blausen Medical 2014. Wiki J Med 1(2): 10, doi: 10.15347/wjm/2014.01010, reproduced under CC0 and CC-SA creative commons licenses; [b, inset] SEMs from Rolvien, T et al. (2018) Early bone tissue aging in human auditory ossicles is accompanied by excessive hypermineralization, osteocyte death and micropetrosis. Sci Rep 8, doi: 10.1038/s41598-018-19803-2, reproduced under a CC BY 4.0 creative commons license; [d]: From Bryant, J et al. (2002) Sensory organ development in the inner ear: molecular and cellular mechanisms. Br Med Bull 63: 39–59, reproduced with publisher permission.)*

BOX 20.1 PATTERNS OF EAR DEVELOPMENT

Unlike the inner ear that arises from an ectodermal **otic placode** and connects to the developing neural tube via an acoustic-vestibular ganglion (**Figure a**), the **middle ear** develops from the first two **pharyngeal arches** (**PA1, PA2**) and joins the auricle of the external ear via an **external auditory meatus** (**EAM**) (=external auditory canal) (**Figure b, c**). Separated from each other by the first pharyngeal cleft (1pc)

and the first pharyngeal pouch (1pp), the first two PAs contain a mixture of mesoderm plus neural-crest derivatives from the developing brain (**Figure b**). According to the classical view of middle ear development, precursor cells from PA1 form Meckel's cartilage (MC), which ossifies to become the malleus and incus bones, whereas PA2 cells give rise to Reichert's cartilage (RC) that generates the stapes (**Figure b, c**). The three

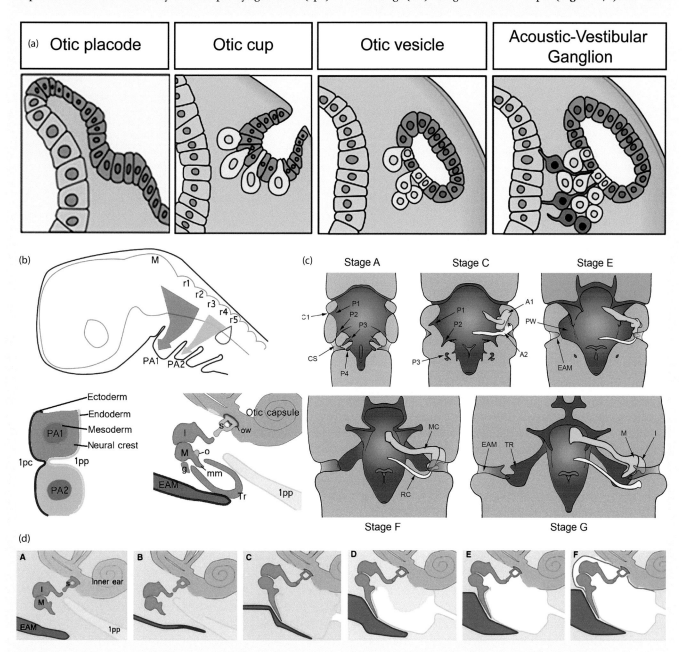

Box 20.1 (a) Diagram of early stages in vertebrate inner ear formation; **(b, c)** Lateral (b) and frontal (c) views of stages in middle ear development; **(d)** Cavitation of the middle ear. ([a]: From Margarinos, M et al. (2012) Early development of the vertebrate inner ear. Anat Rec 295: 1775–1790, reproduced with publisher permission; [b, d]: From Anthwal, N and Thompson, H (2016) The development of the mammalian middle and outer ear. J Anat 228: 217–232, reproduced by publisher permission; [c]: From Burford, CM and Mason, MJ (2016) Early development of the malleus and incus in humans. J Anat 229: 217–228, reproduced with publisher permission.)

middle ear ossicles eventually form an interconnected chain that attaches laterally to a **tympanic membrane** (eardrum) at the border between the EAM and tympanic recess (TR) (**Figure b, d**). In addition, following middle ear cavitation (**Figure d**), the ossicles span the middle ear cavity to connect medially with the oval window (ow) of the developing inner ear derived from the otic vesicle (**Figure d**).

CASE 20.1 JULIETTE GORDON LOW'S EARS

Founder of Girls Scouts of America
(1860–1927)

"The first woman I approached tried to tell me she wasn't interested. I pretended that my deafness prevented me from hearing her refusals."

(J.L. on her perseverence while seeking support to launch Girl Scouts of America)

As the founder of Girls Scouts of America, Juliette Gordon Low was able to overcome her loss of hearing to guide that organization from its initial enrollment of 18 girls toward its current membership of nearly 2 million. As a child, Low suffered numerous ear infections that led to partial hearing loss. In addition, after well-wishers threw rice at her wedding,

a kernel became so deeply lodged in her ear that she had to cut short her honeymoon to see a doctor, who ended up puncturing her eardrum and triggering an infection that left her essentially deaf. In spite her ears' inabilities to convey sounds, Low remained a dynamic force, who was inspired by an offshoot of British Boy Scouts, called Girl Guides, to create a similar program in the United States. In 1912, Low was able to assemble the first girl scout troop, and today, Low is remembered as a strong-willed role model who persevered in spite of her hearing disability.

20.2 FUNCTIONAL HISTOLOGY OF THE EXTERNAL AND MIDDLE EAR

With large ridges and intervening fossae that help direct sound waves to the external auditory canal (EAC), the **auricle** of the adult **external ear** is covered by a thin stratified squamous epithelium that closely overlies elastic cartilage or fat, thereby making the meager auricular covering susceptible to frostbite and other trauma. Proximally, the outer EAC is lined by a stratified squamous epithelium with modified apocrine glands, called cerumen glands, which help produce protective **ear wax**.

Spanning the **middle ear** cavity are the three auditory ossicles—the **malleus**, **incus**, and **stapes bones** (**Figure 20.1b**). As an interconnected chain of levers with associated ligaments and muscles, middle ear ossicles can efficiently transmit eardrum vibrations while also allowing dampening responses to protect the ear against loud noises. Such morphological innovations improved hearing acuity and provided other significant advances that helped early mammals to radiate and succeed as many of their competitors went extinct (Box 20.2 The Evolution of Mammalian Auditory Ossicles).

20.3 INNER EAR: MICROANATOMY OF THE VESTIBULAR APPARATUS USED IN MAINTAINING BALANCE

The vestibular apparatus occurs in the superior half of the inner ear and comprises five saccular organs—three orthogonally arranged **semicircular ducts** plus the **utricle** and **saccule** that both reside within a juxta-cochlear compartment, called the **vestibule** (**Figure 20.2a**). The vestibular organs, along with the scala media of the cochlea (pg. 483), form a network of membrane-enclosed sacs, collectively called the **membranous labyrinth** (**Fig. 20.1a, c**). Such sacs are in turn surrounded by a bony labyrinth within the temporal bone, which consists of three semicircular canals, the vestibule, and the scala vestibuli plus scala tympani of the cochlea (pg. 483 and **Figure 20.1a, c**). The **bony labyrinth** is filled with **perilymph** fluid that is contiguous with, and similar in composition to, cerebral spinal fluid of the central nervous system. Conversely, the membranous labyrinth contains an **endolymph** liquid that resembles intracellular fluids.

Endolymph-filled vestibular organs transmit action potentials to the brain via branches of the **vestibular nerve** (**Figure 20.1c**) and generate such input via specialized sensory cells, called **hair cells**. Such sensors possess an apical array of threadlike "hairs" comprising: (i) a single cilium (=**kinocilium**) with **microtubules**, plus (ii) several dozen flanking **stereovilli**, each of which contains a bundle of **actin** filaments (**Figure 20.2b-d**). The stereovilli closest to the kinocilium are the tallest, whereas more distantly located stereovilli are progressively shorter,

BOX 20.2 THE EVOLUTION OF MAMMALIAN AUDITORY OSSICLES

Various analyses have shown that the **malleus** and **incus** bones in mammalian middle ears evolved from the **articular** and **quadrate** bones, respectively, in jaws of basal vertebrates. The malleus and incus in turn combine with the **stapes** bone to form an innovative set of linked ossicles that are no longer a part of the jaw apparatus (**Figure a**). Alternatively, in diapsids (modern birds and reptiles), the middle ear has a single **columella** or **stapes** bone that typically remains tightly associated with the jaw (**Figure b, c**).

Via their leveraging capabilities, mammalian ossicles can transmit a wider range of sounds than are conveyed by avian and reptilian middle ears. Such enhanced auditory acuity helped improve the abilities of mammals to **detect predators** and to **hone in on prey**, particularly in the case of nocturnal mammals that sought to avoid predators during the day and thus depended less on visual cues. In addition, by separating the middle ear from the jaw, mammals, unlike reptiles, could eat while still hearing potential predators.

In addition to the direct benefits afforded by improved hearing, the evolution of mammalian ear bones also enabled jaw and dentition modifications that facilitated **mastication** and hence the ability to handle a broader array of food. Thus, it can be argued that adaptations involving middle ear bones played profound and wide-ranging roles during mammalian evolution.

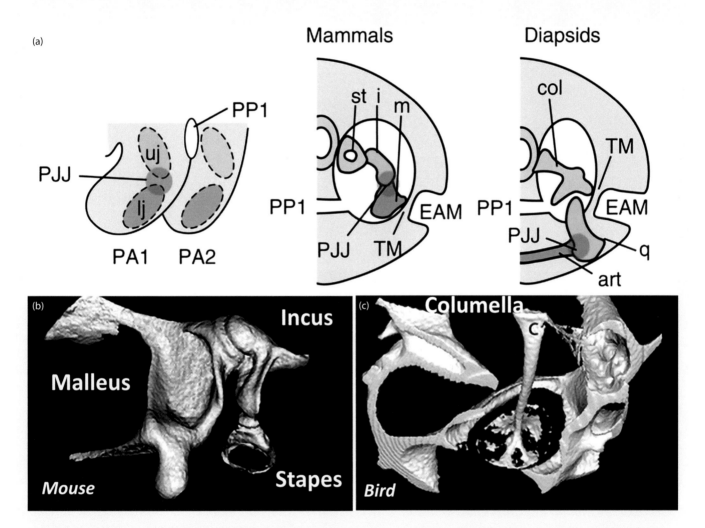

Box 20.2 *(a)* Diagram of pharyngeal arch 1 (PA1) and pharyngeal arch 2 (PA2) derivatives and how the primary jaw joint (PJJ) between the developing upper jaw (uj) and lower jaw (lj) becomes translocated from its position in diapsids (modern reptiles and birds) to where the incus (i) and malleus (m) bones differentiate in mammalian middle ears; art = articular; col = columella; EAM = external auditory meatus; PP1 = first pharyngeal pouch; q = quadrate; s = stapes; TM = tympanic membrane (from Kitazawa, T (2014). Developmental genetic bases behind independent origin of the tympanic membrane in mammals and diapsids. Nat Commun 6: doi: 10.1038/ncomms7853 reproduced under a CC BY 4.0 creative commons license); *(b, c)* Reconstructions of micro-CT scans of the developing middle ear bones of a mouse (B) vs. the columella middle ear bone of a bird (partridge) (from Anthwal, N et al (2013) Evolution of the mammalian middle ear and jaw: adaptions and novel structures. J Anat 222: 147–160, reproduced with publisher permission.)

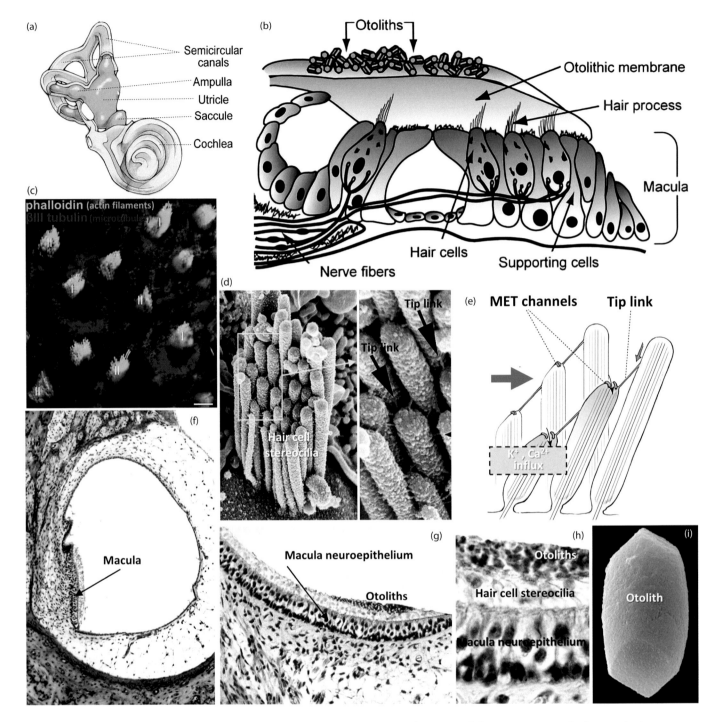

Figure 20.2 Vestibular apparatus microanatomy. (a, b) *As part of a network for maintaining balance, the inner ear's vestibular apparatus includes otolith-containing organs, called the saccule and utricle, which possess a macula with hair cells. (c) The apical processes of vestibular hair cells comprise a kinocilium with nine pairs of microtubules plus stereovilli with actin bundles. Scale bar = 5 μm. (d) Hair cell stereovilli become progressively shorter toward the side opposite to the kinocilium, and each shorter stereovillus is connected to its taller neighbor by a filamentous linkage, called a tip link (d, left) 15,000X; (d, right) 37,500X. (e) Head movements that bend stereocilia toward the kinocilium cause tip links to open mechanoelectric transduction (MET) channels, thereby allowing cation influx to depolarize the hair cell. (f–i) In an upright head, the macula of the saccule tends to be oriented perpendicular to the ground, whereas the utricle macula is positioned more parallel, allowing these two otolith-containing organs to sense different kinds of head movements. (f) 70X; (g) 140X; (h) 500X; (i) 16,000X. ([a, c, e]: From Mathur, P and Yang, J (2015) Usher syndrome: hearing loss, retinal degeneration and associated abnormalities. Biochim Biophys Acta 1852: 406–420, reproduced with publisher permission; [b]: Downloaded from https://commons.wikimedia.org/wiki/File:Otolith_organ_of_vestibular_system.jpg*

Figure 20.2 (Continued) *and reproduced as a public domain image prepared by the National Aeronautics and Space Administration of the federal government; [d]: From Taylor, RR et al. (2015) Characterizing human vestibular sensory epithelium for experimental studies: new air bundles on old tissue and implications for therapeutic interventions in ageing. Neurbiol Aging 36: 2068–2084, reproduced with publisher permission; [i] From R Kniep et al. (2018) The sense of balance in humans: structural features of otoconia and their response to linear acceleration. PLoS ONE 12: e0175769, https://doi.org/10.1371/journal.pone.0175769 and reproduced under a CC BY 4.0 creative commons license.)*

with each shorter stereovillus connecting apically to its taller neighbor by a **tip link** cable that contains cadherin and protocadherin proteins (**Figure 20.2c–e**). At the site where the tip link attaches to the shorter stereovillus are **mechanoelectrical transduction** (**MET**) channels, which can transmit cations into the stereovillar cytoplasm to trigger hair cell depolarizations. In addition to such apical features that are shared by all hair cells, two hair cell morphotypes can be distinguished based on differences in their cellular heights and neural connections. Short **type I hair cells** synapse with afferent neurons via unusual cup-like connections with abundant microtubules (**Figure 20.2c**), whereas tall **type II hair cells** connect with afferent fibers by more orthodox bouton-like synapses. Although the functional differences between these two hair cell types remain to be fully clarified, there is evidence that at least some type I cells are better suited for tracking sudden and ephemeral motions of the head.

In general, head movements that bend vestibular stereovilli **toward** the kinocilium of hair cells **open** tip-link-associated MET channels to trigger **depolarizations** via K^+ and Ca^{2+} **influx** from the endolymph (**Figure 20.2e**). Conversely, head movements that move stereovilli away from the kinocilium **close** MET channels to **hyperpolarize** hair cells. More specifically, **angular movements** of the head as it rotates around the neck tend to be perceived by hair cells in **semicircular ducts**, whereas **linear accelerations** involving either translocation or tilting of the head are sensed by hair cells in the **utricle** and **saccule**. Collectively, input from these vestibular regions is sent to the brain so that a robust 3-D mapping of head positioning can help maintain balance.

For these functions, hair cells of the saccule and utricle occur in a thickened neuroepithelial plate, called the **macula** (**Figure 20.2f–h**). Each macula is directly overlain by a gel-like **otolithic membrane** that is secreted by supportive cells of the macular neuroepithelium. Lying above the membrane are numerous small **otoliths** (=otoconia), which is why the utricle and saccule are sometimes referred to as **otolith organs** (**Figure 20.2b, g–i**). Otoliths are composed mainly of calcium carbonate crystals and serve to form a heavy slurry, thereby weighing down the otolithic membrane and its underlying hair cells when the head is stationary. However, as the head undergoes linear acceleration along a particular direction, otolith sliding causes hair cell stereovilli to bend, and depending on which way the head moves relative to the kinocilium, hair cells may either be **excited** by MET channel openings or **inhibited** by MET channel closures. Such otolith-mediated effects coupled with distinctive hair cell polarities within the saccule vs. utricle help integrate complex neural output from each otolith organ. In the case of utricle, the macula is oriented more or less parallel to the ground for monitoring **horizontal accelerations** of the head and is most sensitive to changes in head position when the head is upright (**Figure 20.3a**). Conversely, the macula of the sacuule is oriented more perpendicular to the ground for detecting **vertical accelerations** and is most sensitive to changes in head position when the head is in a prone, horizontal orientation.

As opposed to the saccule and utricle, semicircular ducts lack otoliths and form at their proximal bases dilated **ampullae**, each of which possesses a domed sensory region, called the crista ampullaris (**Figure 20.2a**). The neuroepithelium of the crista ampullaris contains hair cells with stereovilli and kinocilia that are overlain by a gel-like mass, called the **cupula** (**Figure 20.3b-d**). As the head rotates, the cupula is moved toward the ampullar side opposite to the direction of head movement, and, depending upon which way the head moves, stereovilli embedded in the cupula can bend toward kinocilia to depolarize affected hair cells (**Figure 20.3d**). Given that the semicircular ducts in each ear are oriented orthogonal to each other (**Figure 20.3d**), such configurations ensure that each head rotation will be sensed by a duct and will trigger differing effects not only within that vestibular apparatus but also in the left vs. right ear.

20.4 INNER EAR: THE COCHLEA DURING HEARING AND THE FUNCTIONAL HISTOLOGY OF THE ORGAN OF CORTI AS IT TRANSDUCES SOUND

To optimize packing within the skull, the ~35 mm-long **cochlea** of the inner ear resembles a snail shell after coiling two and a half times as it extends away from the vestibular **oval window** (**Figure 20.4a**). In an unfurled state from its proximal **base** near the oval window to its distal **apex** away from the window, the cochlea would

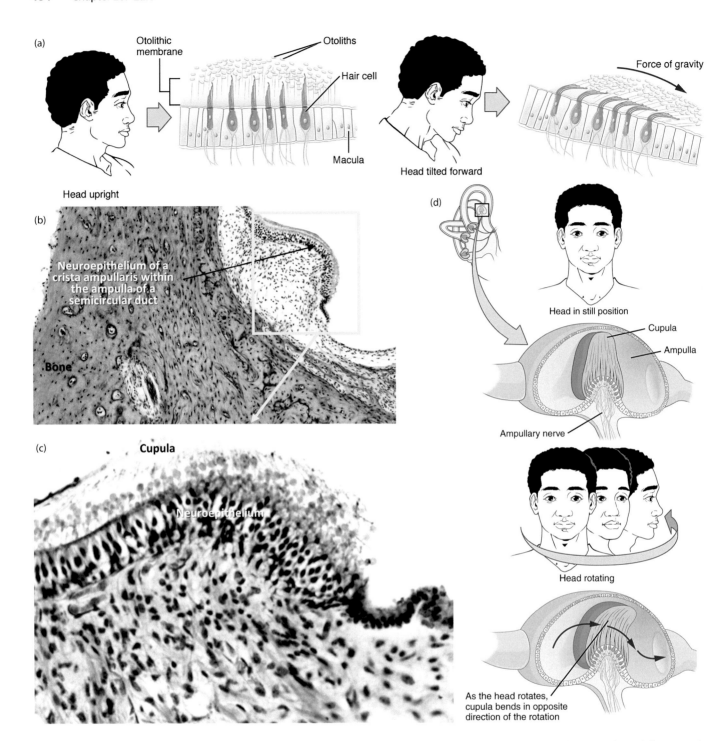

Figure 20.3 Vestibular apparatus functioning. (a) *Head movements allow otoliths overlying the maculae of the saccule and utricle to trigger hair cell depolarizations. (**b–d**) Instead of maculae, semicircular ducts have ampullae containing a crista ampullaris region, whose neuroepithelium is covered by a gel-like cupula that lacks otoliths. Head rotations can cause cupula movements and associated hair cell depolarizations. (b) 50X; (c) 250X. ([a, d]: From OpenStax College (2013) Anatomy and Physiology. OpenStax. http://cnx.org/content/col11496/latest, reproduced under a CC BY 4.0 creative commons license.)*

exhibit two fluid-filled components: (i) an inner, blinded-ended scala media (=cochlear duct) that is filled with endolymph, and (ii) the U-shaped, perilymph-filled scala vestibuli (=vestibular canal) plus a scala tympani (=tympanic canal), which folds back in the recurved apical **helicotrema** region, thereby returning toward the **round window** (**Fig. 20.4a, b**).

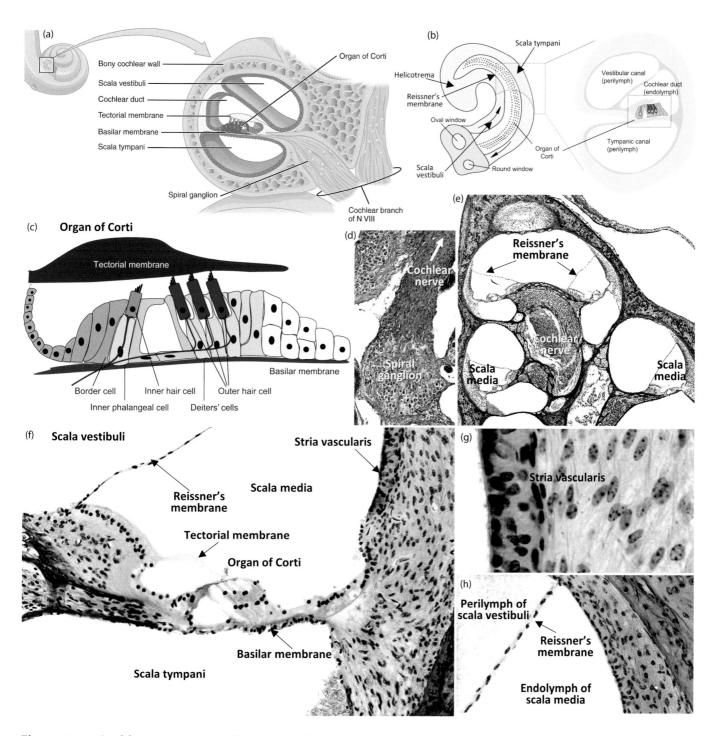

Figure 20.4 Cochlear anatomy. (a, b) *As the auditory portion of the inner ear that functions in transducing sound waves into neural impules, the cochlea is a coiled organ resembling a snail shell. In an unfurled state, the middle of each cochlea would comprise a blind-ended sac, called the scala media (=cochlear duct), which is filled with endolymph. Surrounding the unfurled scala media would be a U-shaped, perilymph-containing chamber that extends from the cochlear base, where the oval and round windows are located, before wrapping around the cochlear apex and returning to the base. The proximal half of the perilymph-containing chamber is called the scala vestibuli (=vestibular canal), which is connected via the recurved helicotrema region to the more distal half, termed the scala tympani (=tympanic canal). In a natively-coiled cochlea, the scala vestibuli lies above the scala media and is separated from it by Reissner's membrane. Conversely, the scala tympani occurs below the basilar membrane underlying the scala media. (c–f) Running along the inferior surface of the scala media, the organ of Corti contains sensory hair cells and supporting glia-like cells (e.g., Deiter's cells). Hair cells occur above the basilar membrane and are overlain by the tectorial membrane. Sound waves traveling*

*Figure 20.4 (Continued) through the perilymph can cause upward deflections of the basilar membrane, thereby rubbing inner hair cells against the tectorial membrane and depolarizing them so that action potentials are transmitted via the cochlear nerve to the brain. (d) 65X; (e) 25X; (f) 160X. (**g, h**) The stria vascularis region of the cochlear wall produces endolymph of the scala media with unique ionic composition compared to that of nearby perilymph. (g) 450X; (h) 160X. ([a]: From OpenStax College (2013) Anatomy and Physiology. OpenStax http://cnx.org/content/col11496/latest, reproduced under a CC BY 4.0 creative commons license; [b, c]: From Li, T et al. (2018) Using Drosophila to study mechanisms of hereditary hearing loss. Dis Model Mech 11: dmm031492, doi:10.1242/dmm.031492, reproduced under a CC BY 3.0 creative commons license.)*

When viewed in its natively coiled configuration, however, the cochlea exhibits three juxtaposed chambers—the **scala vestibuli**, **scala media**, and **scala tympani**. A thin sheet of cells, called **Reissner's membrane**, separates the scala media from the scala tympani, whereas a thicker **basilar membrane** forms the border between the scala media and the scala tympani (**Fig. 20.4a-h**). Collectively, the various cochlear components serve to convert sound wave energies into neural impulses for delivery via the **cochlear nerve** to the brain (**Figure 20.4d–f**). In addition, the **stria vascularis** region of blood vessels in the lateral wall of the scala media ensures proper sensory cell functioning throughout the membranous labyrinth by generating the unique **endolymph** fluid that surrounds hair cells in both the cochlea and vestibular apparatus, thereby facilitating stimulus-induced depolarizations (**Figure 20.4g, h**).

Situated at the border between the scala media and scala tympani is a sensory complex, called the **organ of Corti**, which contains both sensory **hair cells** and **supporting cells** (e.g., Deiter's, Hensen's, and pillar cells) that help maintain hair cell functionality (**Figure 20.4a, c, f**). Basally, organ-of-Corti cells are underlain by a thin **basement membrane** occurring at the top of the basilar membrane, whereas apically their surfaces are covered by a thick collagen-containing **tectorial membrane** (**Figure 20.4a, c, f**). In particular, the tectorial membrane directly overlies three rows of laterally positioned **outer hair cells** (**OHCs**) plus a single medial row of **inner hair cells** (**IHCs**), which synapse via unusual synaptic ribbons with **spiral ganglion afferent fibers** connected to the cochlear nerve (**Figure 20.5a-f**) (**Case 20.2 Helen Keller's Cochleas and Optic Nerves**). As described further below, such components generally allow sound-wave-induced upward deflections of the basilar membrane to rub IHC stereovilli against the tectorial membrane, thereby pulling on tip links to trigger action potential transmission via the cochlear nerve to the brain (**Figures 20.5g** and **20.6a**). Conversely, downward deflections close MET channels and prevent IHC depolarizations (**Figure 20.5g**).

CASE 20.2 HELEN KELLER'S COCHLEAS AND OPTIC NERVES

Inspirational Activist
(1880–1968)

"...I...had thought blindness a misfortune beyond human control, [but] found that too much of it was traceable to wrong industrial conditions, often caused by the selfishness and greed of employers."

(From a 1916 interview with H.K. published by the New York Tribune)

Helen Keller overcame her childhood loss of hearing and sight to become a trail-blazing advocate for various causes. In 1880, Keller was born in Alabama with normal hearing and sight but as a toddler was rendered deaf and blind by a fever-inducing ailment. After the illness, Keller could communicate only in rudimentary ways, but in 1887 she began working with a teacher to learn how to speak, write, and read braille, allowing Keller eventually to graduate from

college. Before dying in her sleep at age 87 of undisclosed causes, Keller became a world-renowned activist, who also garnered her fair share of criticism, because of her ardent support of controversial causes, such as eugenics and socialism. In piecing together accounts of her early life, Keller's childhood illness seems to resemble documented cases of meningitis—a disease involving CNS meningeal layers, which can also damage cochleas and optic nerves. Without medical records, it is only speculative that Keller's loss of hearing and sight was due to meningitis-mediated pathologies, but regardless of what caused her sensory impairments, Keller persevered to lead an inspirational life.

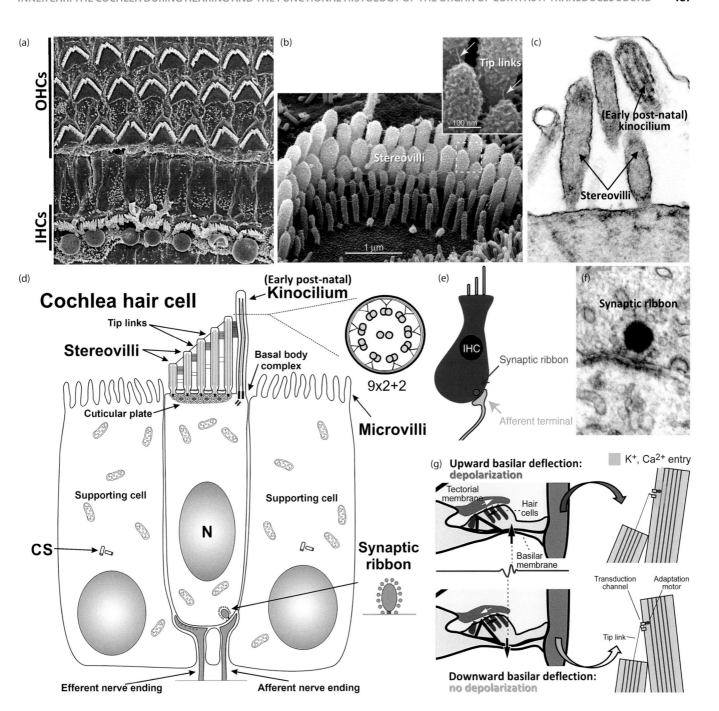

Figure 20.5 Cochlear hair cell ultrastructure. (a, b) As shown in SEMs of apical hair cell surfaces, cochlear inner and outer hair cells (IHCs and OHCs) comprise stacks of progressively longer stereovilli that are connected by tip links. (a) 2,000X. **(c, d)** Initially, cochlear hair cells also have a kinocilium, but this apical elaboration is lost after birth. (c) 9,000X. **(e–g)** Basally, cochlear hair cells can connect to afferent neurons via unusual ribbon synapses, and following upward deflections of the basilar membrane that move shorter stereovilli toward longer neighbors, tip links open MET channels thereby allowing cation-mediated depolarization of the hair cell and action potential transmission to the cochlear nerve. (f) 60,000X. ([a]: From Mathur, P and Yang, J (2015) Usher syndrome: hearing loss, retinal degeneration and associated abnormalities. Biochim Biophys Acta 1852: 406–420, reproduced with publisher permission; [b]: From Indzhykulian, AA et al. (2013) Molecular remodeling of tip links underlies mechanosensory regeneration in auditory hair cells. PLoS Biol 11(6): e1001583. https://doi.org/10.1371/journal.pbio.1001583, reproduced under a CC0 creative commons license; [c, d, f]: From Falk, N et al. (2015) Specialized cilia in mammalian sensory systems. Cells 4: 500–519; doi: 10.3390/cells, reproduced under CC BY 4.0 creative commons license; images courtesy of Dr. A. Giessl; [e]: From Asokan, MM et al. (2018) Sensory overamplification in layer 5 auditory corticofugal projection neurons following cochlear nerve synaptic damage. Nat

Figure 20.5 (Continued) *Commun 9. https://www.nature.com/articles/s41467-018-04852-y, reproduced under a CC BY 4.0 creative commons license; [g]: From LeMasurier, M and Gillespie, PG (2005) Hair-cell mechanotransduction and cochlear amplification. Neuron 48: 403–415, reproduced with publisher permission.)*

During development, each cochlear hair cell initially produces a kinocilium that organizes stereovilli into a stacked array similar to that formed by vestibular hair cells (**Figure 20.5c, d**). However, after birth, cochlear hair cells lose their kinocilium, leaving behind only stereovilli with their longest villus oriented toward the distal edge of the tectorial membrane (**Figure 20.4c**). The actual numbers and heights of OHC stereovilli vary from the cochlear base to apex. Such differences affect the stiffness of hair bundles and thus contribute to how easily MET channels will open during rubbing against the tectorial membrane. This in turn helps the organ of Corti to segregate sound frequencies via a **tuning** mechanism, in which higher frequencies are optimally detected toward the base of the cochlea, whereas progressively lower frequencies are more readily perceived toward the apex (**Figure 20.6b**). These differing sensitivities depend not only on apical-to-basal gradients of stereovillar properties, but also on variations in basilar membrane features. In particular, a relatively thick, narrow, and stiff basilar membrane occurs near the cochlear base, whereas a thinner, wider, and more pliable membrane is present toward the apex, thereby also aiding frequency discrimination (**Figure 20.6c, d**).

Along with its ability to discern differing pitches, the cochlea can also intensify low-amplitude sounds mainly via chloride-associated **prestin** proteins in OHC basolateral membranes (**Figure 20.6e**). According to one theory of sound amplification, post-depolarization changes in prestin shape involve chloride ions that can mediate rapid shortening of OHCs by several micrometers. Such shortening in turn pulls down the tectorial membrane on IHCs to amplify the effects of minor deflections that are caused by weak sound waves (**Figure 20.6e**). Conversely, in response to loud sounds, OHC shortening can be counteracted by efferent input that keeps the OHCs from depolarizing and reshaping prestin channels.

Via these features, the cochlea coordinates with the outer and middle ear in an integrated manner as broadly depicted in **Figure 20.6a** and further reviewed here: (**i**) To initiate hearing, most impinging sound waves are funneled inward from the auricle to cause tympanic membrane vibrations, which are then relayed to the cochlear oval window by middle ear ossicles. However, in addition to this air-borne pathway, a few sound waves also penetrate the skull and access the cochlea via **bone conduction**, which helps account for the finding that recorded voices often fail to match what people perceive their own voices to be. (**ii**) From both of these pathways, sounds reaching the cochlea are transmitted as waves through the perilymph of the scalae vestibuli and tympani before ultimately exiting at the round window in order to prevent retrograde reverberations that could interfere with sound detection. (**iii**) As these transmitted sound waves deflect OHCs into the overlying tectorial membrane, shorter stereovilli are bent toward longer ones, thus opening depolarization-inducing MET channels. (**iv**) This depolarization reconfigures voltage-gated prestin proteins to shorten OHCs and pull the tectorial membrane more forcefully onto IHCs as a way of amplifying the original sound-wave-induced deflections. (**v**) The particular site within the cochlea where IHCs are optimally depolarized depends on the sound frequency, with basal locations being tuned to detect high-frequency waves and apical locations more readily sensing lower frequencies. (**vi**) Once depolarized, each IHC secretes **glutamate** at its synapses with afferent fibers from the cochlear nerve. (**vii**) After binding to **ionotropic glutamate receptors** on the post-synaptic membrane of these afferents, glutamate triggers cochlear nerve action potentials. (**viii**) Such action potentials contain amplitude-specific and regionally encoded signals related to frequency so that the brain can perceive different sound intensities and pitches.

20.5 SOME DISORDERS OF EARS: VERTIGO AND SENSORINEURAL HEARING LOSS

Dysfunction of the inner ear's vestibular system can lead to balancing problems, collectively referred to as **vertigo**, in which surroundings appear to move while the affected person remains stationary. For example, in **benign paroxysmal positional vertigo** (**BPPV**), otoliths become dislodged from the utricle and pushed into a nearby semicircular duct. Such ectopic otoliths may disrupt duct functioning and trigger vertigo following sudden head movements. Exercises designed to rid semicircular ducts of otoliths can often speed recovery.

As opposed to conductive hearing loss that involves inadequate transmission of sound waves to the inner ear, **sensorineural hearing loss** (**SNHL**) results from cochlear dysfunction and/or defective neural transmission

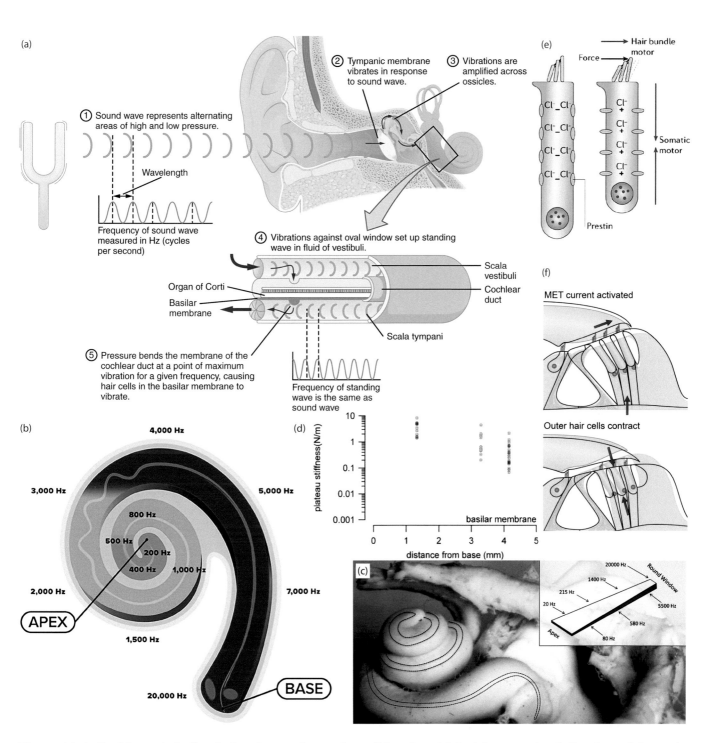

Figure 20.6 Cochlear pitch discrimination and sound amplification. (*a*) *Ossicle-transmitted vibrations of the oval window trigger perilymph waves in the cochlea to deflect the basilar membrane. Such deflections rub inner hair cells of the organ of Corti against the tectorial membrane to cause depolarizations that are sent via the cochlear nerve to the brain for sound perception. (**c–d**) To discriminate different pitches of sound, the base of the cochlea is optimized for perceiving high-frequency sounds, whereas the apex more readily reacts to low-frequency sounds. Figure (d) illustrates a model of the human cochlea with the basilar membrane outlined by dotted lines. As diagrammed in the inset, the basilar membrane is wider and thinner toward the cochlear apex, compared to near the round window at the base. Such morphological differences underlie a base-to-apex reduction in basilar membrane stiffness, which in turn contributes to location-dependent reactions to differing frequencies during pitch discrimination. (**e, f**) For sound amplification, chloride-associated prestin channels allow depolarized outer hair cells to undergo rapid shortening, thereby pulling down the tectorial membrane so that inner hair cells can be activated by relatively weak sounds. ([a]: From OpenStax College*

Figure 20.6 (Continued) *(2013) Anatomy and Physiology. OpenStax. http://cnx.org/content/col11496/latest, reproduced under a CC BY 4.0 creative commons license; [b]: From Lahav, A and Skoe, E (2014) An acoustic gap between the NICU and womb: a potential risk for compromised neuroplasticity of the auditory system in preterm infants. Front Neurosci 8, doi: 10.3389/fnins.2014.00381, reproduced under a CC BY creative commons license; [c]: From Liu, W et al. (2015) Macromolecular organization and fine structure of the human basilar membrane—relevance for cochlear implantation. Cell Tissue Res 360: 245–262, reproduced under a CC BY creative commons license; [d]: From Teudt, IC and Richter, CP (2014) Basilar membrane and tectorial membrane stiffness in the CB/CaJ mouse. JARQ 15: 675–694, reproduced with publisher permission; [e, f]: From Fettiplace, R (2017). Hair cell transduction, tuning, and synaptic transmission in the mammalian cochlea. Comp Physiol 7: 1197–1227, reproduced with publisher permission.)*

from the cochlea to the brain. As the most common type of hearing loss, SNHL can arise from various heritable mutations as well as from loud-noise-induced destruction of cochlear hair cells, which do not regenerate in mammals. Moreover, even less energetic noises are able to trigger **hidden hearing loss** in the absence of appreciable hair cell damage by causing progressive degradation of IHC synapses. Treatments of SNHL may utilize cochlear implants to enhance impulse transmissions.

20.6   SUMMARY—EARS

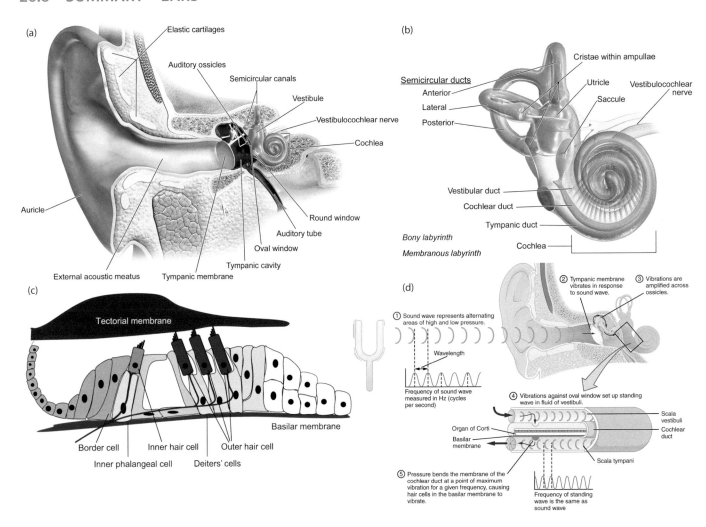

Pictorial Summary Figure Caption: (a): see figure 20.1a; (b): see figure 20.1c; (c): see figure 20.4c; (d): see figure 20.6a

The **ear** comprises **outer, middle**, and **inner** regions (**Figure a**). Sounds are funneled by the outer ear **auricle** to the **tympanic membrane** (eardrum), which vibrates middle ear auditory ossicles (**malleus**, **incus**, and **stapes**) to move the oval window of the inner ear's **cochlea** (**Figure b**). Medial to the oval window, the cochlea comprises three chambers—an upper and lower **scalae vestibuli** and **tympani**, which are filled with **perilymph** vs. a middle **scala media** that contains **endolymph**. **Stria vascularis** cells in the scala media's wall produce endolymph not only for the cochlea but also for the **saccule**, **utricle**, and **semicircular ducts** of the inner ear's **vestibular apparatus**, which functions in tracking head movements and maintaining balance. The saccule and utricle have sensory **maculae** that are covered by calcified **otoliths**, whereas semicircular ducts have **ampullae** whose neuroepithelium underlies a gel-like **cupula** that lack otoliths. Both hearing and balance utilize sensory **hair cells** with long microvilli (=**stereovilli**) that typically occur near a **kinocilium**, although cochlear kinocilia degenerate after birth. During head motions, hair cells that are moved within maculae of the **saccule** and **utricle** or within ampullae of **semicircular ducts** can pull on stereovillar **tip links** to open **mechanoelectrical transduction** (**MET**) **channels,** thereby depolarizing hair cells for neural impulses.

In order to convert sound waves into action potentials, the cochlea has three rows of **outer hair cells** (**OHCs**) and a single row of **inner hair cells** (**IHCs**) that are situated within the scala media's **organ of Corti** (**Figure c, d**). Sound waves transmitted through the scalae vestibuli and tympani can cause deflections of the organ of Corti that rub IHC hairs against the overlying **tectorial membrane**. This opens stereovillar MET channels to allow cation influx, thereby depolarizing IHCs so that action potentials can be sent via the cochlear nerve to the brain. OHCs can **amplify** weak sounds by using **prestin** proteins that rapidly shorten OHCs and thus pull the tectorial membrane closer to IHC apices for enhanced sensitivity to low-amplitude sounds. Spatially varying properties along the base to apex of the cochlea allow different **pitches** of sound to be discriminated, with the cochlear base being tuned to high-frequency sounds, and the apex being sensitized to low-frequency sounds.

Some ear disorders— **vertigos** involve imbalance due to vestibular dysfunction; **sensorineural hearing loss** is triggered by a wide variety of cochlear defects and/or abnormal neural output from the cochlea to the brain.

SELF-STUDY QUESTIONS

1. Which hair cells lose their kinocilium after birth?
2. T/F Outer hair cells can shorten during sound amplification.
3. Which membrane overlies outer hair cells of the cochlea?
4. Which of the following is NOT directly associated with the organ of Corti?
   A. Basilar membrane
   B. Inner hair cells
   C. Surrounding endolymph
   D. Otoliths
   E. Tectorial membrane
5. Which of the following would depolarize a hair cell?
   A. Bending of stereovilli away from the kinocilium
   B. Opening of MET channels
6. What part of the ear does the otic vesicle give rise to?
7. T/F The base of the cochlea is tuned to low-frequency sounds.
8. Which occurs at the recurved apex of cochlea?
   A. Helicotrema
   B. Spiral ganglion
   C. Utricle
   D. Semicircular canal
   E. Auricle

"EXTRA CREDIT" For each term on the left, and provide the BEST MATCH on the right (answers A-L can be used more than once).

9. Inferior-most cochlear chamber ___
10. Stria vascularis ____
11. Cupula ____
12. Otoliths ___
13. Hair cell shortening ____
14. Organ of Corti_____
15. External ear_____
16. Auditory ossicle____
17. Cables that open MET channels____
18. Stereovillar core___

A. Semicircular duct
B. Prestin shape changes
C. Scala tympani
D. Saccular macula
E. Scala vestibuli
F. Makes endolymph
G. Deiter cells
H. Microtubules
I. Microfilaments
J. Auricle
K. Tip links
L. Stapes

19. Describe the anatomy of the cochlea.
20. Describe how the cochlea is able to detect weak sounds and differing pitches.

ANSWERS

(1) cochlear; (2) T; (3) tectorial; (4) D; (5) B; (6) inner ear; (7) F; (8) A; (9) C; (10) F; (11) A; (12) D; (13) B; (14) G; (15) J; (16) L; (17) K; (18) I; (19) The answer should include apex vs. base of cochlea; scala vestibuli, media, tympani; oval, round windows; perilymph, endolymph, stria vascularis; organ of Corti; tectorial and basilar membranes; hair cells; spiral ganglion; cochlear nerve; (20) Consider: basilar membrane deflection, outer vs. inner hair cells, outer hair cell (OHC) prestin channels, basilar membrane differences along scala media, pitch discrimination

Figure Citations for Celebrity Medical Histories

Case 1.1: Marcello Malpighi, public domain image of painting by Carlo Cignani (1628-1719), downloaded from: http://www.gettyimages.co.uk/detail/news-photo/portrait-of-marcello-malpighi-by-carlo-cignani-17th-century-news-photo/450079329

Case 2.1: Henri de Toulouse Lautrec, public domain photograph by Paul Sescau40002, downloaded from: https://commons.wikimedia.org/wiki/File:Photolautrec.jpg

Case 2.2: Ronald Reagan, public domain photograph of official presidential portrait by unknown author; downloaded from: https://commons.wikimedia.org/wiki/File:Official_Portrait_of_President_Reagan_1981.jpg

Case 2.3: Alice Martineau, generalized thumbnail sketch; original artwork

Case 2.4: Charlie Gard, generalized thumbnail sketch; original artwork

Case 2.5: Lou Gehrig, public domain image of an uncopyrighted 1923 photograph downloaded from: https://commons.wikimedia.org/wiki/File:1923_Lou_Gehrig.png

Case 2.6: Frances Oldham Kelsey, public domain image produced by an unknown employee of the federal government of the U.S.; downloaded from: https://commons.wikimedia.org/wiki/File:Frances_Oldham_Kelsey.png

Case 3.1: Hilde Mangold, diagram of Mangold's experiments; image was created by Darizanovska and downloaded from: https://en.wikipedia.org/wiki/Dorsal_lip#/media/File:Experimental_evidence_of_the_dorsal_lip_as_the_neural_inducer_and_genetic_mechanisms..jpg and reproduced under a CC BY SA 4.0 creative commons license.

Case 3.2: Althea Gibson, public domain image produced by unknown author and downloaded from: https://commons.wikimedia.org/wiki/File:Althea_Gibson_-_El_Grafico_2033-crop.jpg

Case 3.3: Cesar Chavez, image by Joel Levine downloaded from: https://commons.wikimedia.org/wiki/File:Cesar_chavez_crop2.jpg and reproduced under a CC BY 3.0 creative commons license.

Case 3.4: Niccolo Paginini, public domain image, downloaded from: https://commons.wikimedia.org/wiki/File:Niccol%C3%B2_Paganini_ritratto_giovanile.jpg

Case 4.1: Christiaan Barnard, image by Eric Koch and released from the Dutch National Archives; downloaded from: https://commons.wikimedia.org/wiki/File:Christiaan_Barnard_(1968).jpg and reproduced under a CC BY SA 3.0 Netherlands creative commons license

Case 4.2: Elizabeth Taylor, public domain image of a non-copyrighted studio publicity photograph downloaded from: https://commons.wikimedia.org/wiki/File:Taylor,_Elizabeth_posed.jpg

Case 4.3: Harriet Tubman, public domain image, downloaded from: https://commons.wikimedia.org/wiki/File:Harriet_Tubman_1895.jpg

Case 4.4: King Tut, public domain image taken by Andreas Praefcke of a carving that may have depicted King Tut; image, downloaded from: https://commons.wikimedia.org/wiki/File:Spaziergang_im_Garten_Amarna_Berlin.jpg

Case 5.1: William Taft, public domain image of official presidential portrait by A.L. Zorn downloaded from: https://www.whitehousehistory.org/galleries/presidential-portraits

Case 6.1: Miles Davis, public domain image by William P. Gottlieb downloaded from: https://commons.wikimedia.org/wiki/File:Miles_Davis,_(Gottlieb_06851)_(cropped).jpg

Case 6.2: Yasser Arafat, image by Tibor Vegh reproduced under a CC BY 3.0 creative commons and downloaded from:

Case 6.3: Sadako Sasaki, public domain image by unknown author downloaded from: https://commons.wikimedia.org/wiki/File:Sadako_Sasaki.jpg

Case 6.4: Eleanor Roosevelt, public domain image by unknown author downloaded from: https://commons.wikimedia.org/wiki/File:Eleanor_Roosevelt_portrait_1933.jpg

Case 7.1: Frederick Douglass, public domain image by unknown author downloaded from: https://commons.wikimedia.org/wiki/File:Frederick_Douglass_(1840s).jpg

Case 7.2: Margaret Thatcher, by Rob Bogaerts released into the public domain under a CC0 1.0 creative commons license and downloaded from: https://commons.wikimedia.org/wiki/File:Margaret_Thatcher_(1983).jpg

Case 7.3: Richard Nixon, public domain image of official presidential portrait downloaded from: https://en.wikipedia.org/wiki/File:Official_Portrait_of_President_Nixon.jpg

Case 7.4: Golda Meir, public domain image by Moshe Pridan downloaded from: https://commons.wikimedia.org/wiki/File:Ahmadou_Ahidjou_with_Golda_Meir.jpg

Case 7.5: James Garfield, public domain image by unknown author downloaded from: https://commons.wikimedia.org/wiki/File:James_Abram_Garfield,_photo_portrait_seated.jpg

Case 7.6: Virginia Alexander, public domain image by D.A.B. Lindberg downloaded from: https://commons.wikimedia.org/wiki/File:Vmalexander.jpg

Case 8.1: Darius Weems, generalized thumbnail sketch; original artwork

Case 8.2: Mother Teresa, image by JohnMatthewSmith reproduced under a CC BY SA 2.5 creative commons license and downloaded from: https://commons.wikimedia.org/wiki/File:Mother_Teresa_1.jpg

Case 8.3: King Kamehameha IV, public domain image perhaps by Henry L. Chase downloaded from: https://commons.wikimedia.org/wiki/File:Kamehameha_IV_(PP-97-8-006).jpg

Case 9.1: Annette Funicello, public domain image by Walt Disney Productions downloaded from: https://commons.wikimedia.org/wiki/File:The_Mickey_Mouse_Club_Mouseketeers_Annette_Funicello_1956.jpg

Case 9.2: Abraham Lincoln, public domain official presidential portrait by G.P.A. Healy downloaded from: https://www.whitehousehistory.org/galleries/presidential-portraits

Case 9.3: Muhammad Ali, public domain image by Ira Rosenberg downloaded from: https://commons.wikimedia.org/wiki/File:Muhammad_Ali_NYWTS.jpg

Case 9.4: Marilyn Monroe, public domain image of a 20th Century Fox trailer for the movie Niagara downloaded from: https://commons.wikimedia.org/wiki/File:Marilyn_Monroe_Niagara.png

Case 10.1: Michael Jackson, public domain image by the White House photo office downloaded from: https://commons.wikimedia.org/wiki/File:Michael_Jackson_1984(2).jpg

Case 10.2: Bob Marley, public domain image by Caspiax downloaded from: https://commons.wikimedia.org/wiki/File:Bob_Marley_emancipated_from_mental_slavery_1.jpg

Case 10.3: Marie Antoinette, public domain image of a portrait possibly by J-B A Gautier-Dagoty downloaded from: https://commons.wikimedia.org/wiki/File:Marie-Antoinette,_1775_-_Mus%C3%A9e_Antoine_L%C3%A9cuyer.jpg

Case 10.4: Benjamin Harrison, public domain image of official presidential portrait by J.E. Johnson downloaded from: https://www.whitehousehistory.org/galleries/presidential-portraits

Case 10.5: Frida Kahlo, public domain image by G. Kahlo downloaded from: https://commons.wikimedia.org/wiki/File:Frida_Kahlo,_by_Guillermo_Kahlo_2.jpg

Case 11.1: Ada Lovelace, public domain image of portrait by A.E. Chalon downloaded from: https://commons.wikimedia.org/wiki/File:Ada_Lovelace_portrait.jpg

Case 11.2: Gerald Ford, public domain image of official presidential portrait by E.R. Kinstler downloaded from: https://www.whitehousehistory.org/galleries/presidential-portraits

Case 11.3: Gordie Howe, public domain image of the back of a Chex cereal box downloaded from: https://commons.wikimedia.org/wiki/File:Gordie_Howe_Chex_card_(cropped).jpg

Case 11.4: Grover Cleveland, public domain image of official presidential portrait by J.E. Johnson downloaded from: https://www.whitehousehistory.org/galleries/presidential-portraits

Case 11.5: Andrew Jackson, public domain image of official presidential portrait by R.E.W. Earl downloaded from: https://www.whitehousehistory.org/galleries/presidential-portraits

Case 11.6: Elvis Presley, public domain image, downloaded from: https://commons.wikimedia.org/wiki/File:Elvis_Presley_promoting_Jailhouse_Rock.jpg

Case 12.1: Princess Diana, image by Georges Baird reproduced by CC BY SA 3.0 creative commons license downloaded from: https://commons.wikimedia.org/wiki/File:Princess_Diana_Cannes.jpg

Case 12.3: Maurice Gibb, public domain publicity photograph by NBC Television downloaded from: https://commons.wikimedia.org/wiki/File:Bee_Gees_1977.JPG

Case 12.4: Edward Plunket, public domain image, downloaded from: https://commons.wikimedia.org/wiki/File:Portrait_of_Lord_Dunsany.jpg

Case 12.5: Corazon Aquino, public domain image by Airman G.B. Johnson downloaded from: https://commons.wikimedia.org/wiki/File:Corazon_Aquino_1986.jpg

Case 13.1: Billie Holiday, public domain image by unknown author downloaded from: https://commons.wikimedia.org/wiki/File:Billie_Holiday.jpg

Case 13.2: Mary Mallon, public domain image by unknown author downloaded from: https://commons.wikimedia.org/wiki/File:Mary_Mallon_(Typhoid_Mary).jpg

Case 13.3: Sally Ride, public domain image by NASA downloaded from: https://commons.wikimedia.org/wiki/File:Sally_Ride,_First_U.S._Woman_in_Space_-_GPN-2004-00019.jpg

Case 13.5: Sejong the Great, image by Mannaa Mohamed reproduced under a CC BY SA 4.0 creative commons license and downloaded from: https://commons.wikimedia.org/wiki/File:The_statue_of_King_Sejong_the_Great_8.jpg

Case 14.1: Amelia Earhart, public domain image by NASA downloaded from: https://commons.wikimedia.org/wiki/File:Amelia_Earhart_-_GPN-2002-000211.jpg

Case 14.2: George Washington, public domain image of official presidential portrait by G. Stuart downloaded from: https://www.whitehousehistory.org/galleries/presidential-portraits

Case 14.3: Mary Wells, public domain image by J. Kriegsmann downloaded from: https://commons.wikimedia.org/wiki/File:Mary_Wells_1965.jpg

Case 14.4: Pocahontas, public domain image by unknown artist downloaded from: https://commons.wikimedia.org/wiki/File:Pocahontas_1883.jpg

Case 14.5: Indira Gandhi, public domain image by unknown author reproduced under a CC0 1.0 creative commons license and downloaded from: https://commons.wikimedia.org/wiki/File:Indira_Gandhi_1977.jpg

Case 14.6: Catherine the Great, public domain image by M. Shibanov downloaded from: https://commons.wikimedia.org/wiki/File:Empress_Catherine_The_Great_1787_(Mikhail_Shibanov).JPG

Case 15.1: Zelda Fitzgerald, public domain image by unknown author downloaded from: https://commons.wikimedia.org/wiki/File:Zelda_Fitzgerald_circa_1919_Retouched.jpg

Case 15.2: Barbara Bush, public domain image by David Valdez as work of the U.S. federal government downloaded from: https://upload.wikimedia.org/wikipedia/commons/a/a5/Barbara_Bush_portrait.jpg

Case 15.3: Pio Pico, public domain image by unknown author downloaded from: https://commons.wikimedia.org/wiki/File:Pio_Pico.jpg

Case 15.5: Garry Shandling, image by A. Light downloaded from: https://en.wikipedia.org/wiki/Garry_Shandling#/media/File:Garry_Shandling_at_the_39th_Emmy_Awards_cropped.jpg and reproduced under a CC BY 2.0 creative commons license.

Case 15.6: John Kennedy, public domain image of official presidential portrait by A. Shikler downloaded from: https://www.whitehousehistory.org/galleries/presidential-portraits

Case 16.1: Idi Amin, public domain image by Archives New Zealand reproduced under a CC BY 2.0 creative commons license and downloaded from: https://commons.wikimedia.org/wiki/File:Idi_Amin_-Archives_New_Zealand_AAWV_23583,_KIRK1,_5(B),_R23930288.jpg

Case 16.2: Christine Jorgensen, public domain image by M. Seymour downloaded from: https://commons.wikimedia.org/wiki/File:Christine_Jorgensen_1954.jpg

Case 16.3: Franklin Roosevelt, public domain image of official presidential portrait by F.O. Salisbury downloaded from: https://www.whitehousehistory.org/galleries/presidential-portraits

Case 17.1: Adolf Hitler, public domain image from the German Federal Archive reproduced under a CC BY 3.0 creative commons license and downloaded from: https://commons.wikimedia.org/wiki/File:Adolf_Hitler_cropped_restored.jpg

Case 17.2: John Tyler, public domain image of official presidential portrait by G.P.A. Healy downloaded from: https://www.whitehousehistory.org/galleries/presidential-portraits

Case 17.3: Pablo Neruda, public domain image by A. Heinrich downloaded from: https://commons.wikimedia.org/wiki/File:Pablo_Neruda_by_Annemarie_Heinrich,_1967.jpg

Case 18.1: Coretta Scott King, public domain image, downloaded from: https://commons.wikimedia.org/wiki/File:Coretta_Scott_King_1964.jpg

Case 18.2: Eva Perón, public domain image, downloaded from: https://commons.wikimedia.org/wiki/File:Eva_Per%C3%B3n_Retrato_Oficial.jpg

Case 18.3: Mumtaz Mahal, public domain image, downloaded from: https://commons.wikimedia.org/wiki/File:Mumtaz_Mahal.jpg

Case 19.1: Aldous Huxley, public domain image, downloaded from: https://commons.wikimedia.org/wiki/File:Aldous_Huxley.JPG

Case 19.2: Marie Curie, public domain image by H. Manuel downloaded from: https://commons.wikimedia.org/wiki/File:Marie_Curie_c1920.png

Case 19.3: Theodore Roosevelt, public domain image of official presidential portrait by J.S. Sargent downloaded from: https://www.whitehousehistory.org/galleries/presidential-portraits

Case 20.1: Juliette Gordon Low, public domain image, downloaded from: https://commons.wikimedia.org/wiki/File:Juliettegordonlow-pinning.jpg

Case 20.2: Helen Keller, public domain image, downloaded from:https://commons.wikimedia.org/wiki/File:Helen_KellerA.jpg

Index

Note: Locators in *italics* represent figures and in **bold** indicate tables in the text.